LIMNOLOGY

Inland Water Ecosystems

JACOB KALFF
McGill University

Prentice Hall
Upper Saddle River, New Jersey 07458

Library of Congress Cataloging-in-Publication Data

Kalff, Jacob.
    Limnology : inland water ecosystems / Jacob Kalff.
      p. cm
    Incudes bibliographical references.
    ISBN 0-13-033775-7
    1. Limnology. I. Title.

QH96 .K24 2002
577.6—dc21   2001   2001032745

Executive Editor: *Teresa Ryu*
Production Editor: *Joanne Hakim*
Manufacturing Buyer: *Michael Bell*
Manufacturing Manager: *Trudy Pisciotti*
Interior Design: *Anne Flanagan*
Art Director: *Jayne Conte*
Cover Photo: *Courtesy of Ed Snucins, Killarney Lake, Killarney Provincial Park, Ontario.*

 © 2002 by Prentice-Hall, Inc.
Upper Saddle River, NJ 07458

Printed in the United States of America

Reprinted with corrections January, 2003.

10 9 8

ISBN 0-13-033775-7

Pearson Education LTD., *London*
Pearson Education Australia PTY, Limited, *Sydney*
Pearson Education Singapore, Pte. Ltd.
Pearson Education North Asia Ltd., *Hong Kong*
Pearson Education Canada, Ltd., *Toronto*
Pearson Educación de Mexico, S.A. de C.V.
Pearson Education—Japan, *Tokyo*
Pearson Education Malaysia, Pte. Ltd.
Pearson Education, *Upper Saddle River, New Jersey*

LIMNOLOGY

About the Author

Jacob Kalff is Professor of Biology at McGill University, Montreal and Director of its Limnology Research Center. He has worked on both fresh and saline waters and done research in the Arctic, the temperate zone, and the tropics. He has published on the structure and functioning of biotic communities (aquatic plants, bacteria, zooplankton, and fish), nutrient and contaminant cycling, and sediment attributes as they impact the biota and contaminant distribution, as well as the role of lake morphometry and drainage basin characteristics in determining the functioning of aquatic systems. The results have been published in more than 150 scientific papers, many of which reflect his interest in the application of fundamental research to environmental management. Dr. Kalff was born in the Netherlands and received his higher education at the University of Toronto (Canada) and at Indiana University (U.S.A.).

Brief Contents

Contents

LIMNOLOGY

C H A P T E R

1

Inland Waters and Their Catchments: An Introduction and Setting

1.1 Introduction

Limnology, the study of lakes, rivers, and wetlands as systems, may well be the most successful of the ecological sciences. The distinct borders of lakes in particular have always suggested the possibility of studying them as units. Limnology has, as a result, attracted a disproportionately large number of ecologists interested in the behavior of whole systems (Wetzel 1983) and has provided a disproportionate number of guiding concepts in ecology. More specifically, limnology has long been preeminent in work on energy and material flow, in the manipulation of large enclosures and whole systems, as well as in the use of experimental stream channels and ecosystem modeling. Increasingly, the lake, wetland, or river is seen and examined as a component of an integrated land–water system.

The aesthetic appeal of lakes and the apparent ease of sampling the small organisms of the open water have long drawn a wide variety of outstanding ecologists, including those interested in questions posed at the population and community level of biological organization and their experimental testing. Consequently, limnology has contributed much to ecology at those two scales of biological organization. It is, however, only partially an ecological science. Lakes, rivers, and wetlands have drawn scientists with backgrounds in chemistry, physics, and geology, and limnology has long been an exciting multidisciplinary science. In addition to being intellectually stimulating, limnology is of great practical importance in that the limited supply of fresh water must be shared by a burgeoning human population, thus becoming increasingly subject to pollution and depletion. The fairly recent development of an applied limnology, preoccupied with the remediation of polluted waters and the preservation and wise exploitation of aquatic resources, has added an applied dimension to the science—one that is exceptional in ecology. The roots of applied limnology are, however, much older and date back at least 150 years to research on the sewage pollution of waterways, depleted fish stocks, and fish culture that had not been considered part of limnology.

The substantial funding for research with a more or less direct application has led to heated debate among limnologists about the goals of science, the relative merits of fundamental versus applied science, and the importance (in management) of work carried out on whole systems over often many years versus the commonly short-term basic research on specific components under simplified conditions. The extent of debate is unprecedented in ecology, but there is no doubt that limnology has been stimulated and enriched by its applied component. In addition, funding for applied research has frequently allowed for work that contributed to the advancement of fundamental science.[1] Applied limnology has permitted many more scientists

---

[1] • "As pure science becomes harder to justify and fund we must make every effort to derive general principles from the study of applied problems. Ecologists should not be afraid of applied problems, they can tell us much about general principles." (Harris 1994)

• "Resource management efforts often constitute very interesting large-scale and long-term manipulations of communities and ecosystems that could be exploited by research-oriented ecologists." (Lodge et al. 1998)

1

to be involved with limnology than would have been the case if it had remained an overwhelmingly academic discipline. The management of polluted lakes, rivers, and wetlands and the managing of freshwater recreation (above all, for sportfishing) have become multibillion-dollar enterprises of great public interest.

While the importance and success of limnology may justify reading a textbook on the subject, it certainly does not facilitate writing one. I quickly discovered six problems that must be faced by all authors of science textbooks and that are highly relevant to the students experiencing them. I therefore urge the student reader not to skip the rest of this chapter; you will find it useful in interpreting not only this book but also other science courses.

The first issue is the problem of how much detail to present. The science of limnology is incomplete, as is evident from the large number of new findings being made every year. These findings often provide a new twist to previous interpretations and sometimes completely change existing ideas about how aquatic systems or components of the biota function. An introductory textbook can not, and should not, present the last word or last interpretation of the field; that is the job of reviews produced *by* specialists *for* specialists. However, there is also the opposite danger of presenting limnology as if the results and interpretations discussed are the final word on the subject. Taking the latter approach enables a textbook author to sound more authoritative, thereby comforting you, the student, by making the subject seem relatively simple, rendering this a tempting route for an author to take. In addition, it makes the writing, and therefore the subject, easier to comprehend. Such authoritative textbooks present what are, at best, partial truths as facts and would not sufficiently sensitize the reader that limnology, like any vital science, is a field in flux. Today's "truth" is next year's "half truth" and will likely be forgotten as irrelevant 10 or 20 years from now. It is important, even at the undergraduate level, to appreciate this flux. The recognition of the incompleteness of any science makes it possible for undergraduates to look at textbooks and scientific articles with a critical eye rather than think of them as a series of unquestionable truths. Some undergraduate readers of the present text may even become sufficiently excited and challenged by the incompleteness of the field to consider a career in limnology or aquatic management.

The author of an introductory text must remember Einstein's advice: "Everything should be as simple as possible but no simpler." If I have frequently erred on

the side of oversimplification, a common and necessary limitation of textbooks, I trust that this will be recognized as such by more experienced limnologists. I have tried to reduce the problem by providing more information than can be reasonably assigned in a one-quarter or a one-semester course by indicating which sections or portion of sections—marked with an ▲— could be skipped in short introductory courses. This will also serve to make the book useful for longer, more advanced courses. The ▲ designation allows professors of a shorter course the option of dealing with certain topics and not others, or to disregard them altogether. The many references provided serve as sources to statements made and conclusions drawn. They also serve as a possible starting point for term papers and research projects. If the reader is using the book as a text in a short introductory course s(he) may now elect to turn to Section 1.2.

▲ I recommend that all tables and figures in sections not assigned in a particular course be glanced over nevertheless. An ability to scan, to extract the nuggets of gold, is an important skill that needs to be acquired by all beginning scientists. There are simply far too many papers in limnology, or even in any of its major subspecialties, to be able to read every one. I suggest that the footnotes presented be similarly scanned. These sometimes provide additional information that is less central than the text itself. The quotations, which reflect the sometimes strong opinions of scientists, will hopefully offer a view of the opinions scientists hold. These viewpoints help reveal scientists as human beings with strong emotions and commitments rather than as totally dispassionate observers presenting the unassailable and absolute truth at all times. Not all quotations or all research quoted in the text will be referenced in order to limit their number. The initials of authors indicate that the publication is not referenced. However, the initials plus the year of publication allow most of the non-referenced papers to be readily located in databases.

The second major problem I have had to confront, one undergraduates are rarely aware of, is how to balance opinions and interpretations with facts. The conclusions scientists draw from a particular data set are greatly influenced by their backgrounds and scientific perspectives. Even if particular facts (the data) stand the test of time, their interpretation continues to change in light of new findings. The general ideas that guide individual limnologists (and all other scientists) in their research are conceptual rather than strictly based on data and are the product of their particular

Figure 1–1 Log–log scatterplot of experimental duration versus area of an individual unit of experiments in limnology. Each circle represents one experiment. For lotic studies, experimental unit area was estimated as mean stream width multiplied by the length of reach under study. For plotting, volumes (V) were converted to areas (A) by $A = V^{2/3}$. Note that a large majority of experiments are carried out over a period of less than 10 months and cover a unit area of less than 10 m^2, probably less than 1 m^2. *(After Lodge et al. 1998.)*

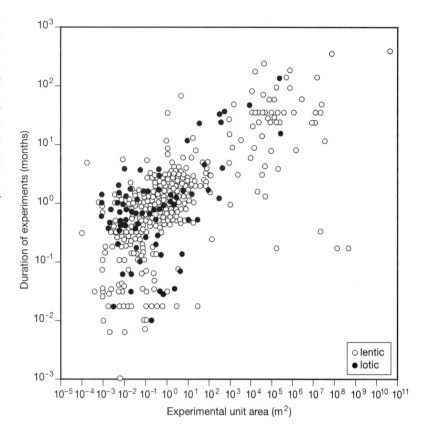

education, outlook, background, and professional experience.[2] Not surprisingly, each scientist works within a framework of ideas and beliefs that help determine not only the kind of research considered interesting but also the specific research questions to be posed. This should be recognized and is further developed in Chapter 2, outlining the history and development of limnology.

Textbooks tend to minimize the different viewpoints held to reduce possible confusion engendered by disagreements. Limnology textbooks therefore, commonly present specific examples as if they are generalities. This contributes to a desirable simplicity, but at the cost of preventing an appreciation of ongoing arguments in the literature often stemming from the tendency to draw broader conclusions than are strictly allowed by the data.[3] Thus, a problem for anyone using the literature, and particularly for students perusing books and conference proceedings that have not benefited from the critical editing done by first-rate journals, is to decide where the facts end and the opinions begin. However, the above comments should not be interpreted to mean that scientific conclusions are simply a matter of opinion. Science as an enterprise is merciless in weeding out early conclusions that are not well supported by follow-up research and, therefore, do not withstand the test of time.

Third, the vast majority of limnological (and ecological) studies are carried out over an area less than 10 m^2, over a period of less than a year (Fig. 1–1), and frequently only over a few months. This leaves the question open as to how general or applicable the results and interpretations are to the system as a whole

---

[2]"It has often been observed that different scientists may draw entirely different, sometimes dramatically opposed conclusions from the same facts. How can this be? Evidently, such divergence of interpretation is the result of a drastic difference in the ideologies of the respective scientists." (E. Mayr 1982. *The Growth of Biological Thought*)

---

[3]"In no less than 74 of a sample of 149 articles selected from ten highly regarded medical journals conclusions were drawn that were not justified by the results presented." (S. Schor and I. Karsten 1966) Strongly held views supported by weakly substantiated conclusions are similarly widespread in limnology.

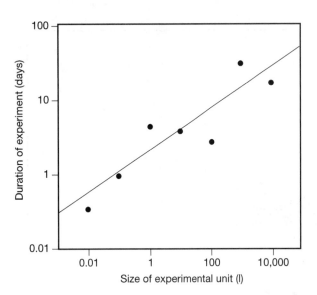

Figure 1–2 The relationship between the size and duration of experiments published in papers on marine microbial ecology between 1990 and 1995. The results are equally applicable to equivalent work in inland waters where identical techniques are used. *(Modified after Duarte et al. 1997.)*

Table 1–1 **Number of replicates required to achieve an estimate of the seasonal mean with a coefficient of variation of 20%.**

Variable	Mean	Replicates
Water		
Chlorophyll	7 mg m^{-3}	3–7
Zooplankton	10 l^{-1}	2–8
Benthos		
Lentic	300 m^{-2}	3–11
Lotic	300 m^{-2}	11–48
Phytobenthos	100 m^{-2}	6–21
Macrobenthos	32 g m^{-2} dw	8–16
Sediments		
Phosphorus	1 mg g^{-1}	2
Organic Matter	0.5 g g^{-1} dw	2

Source: Modified from Peters 1990.

or even to the same area during another season or much wetter or sunnier year, after the stocking of a new fish species, or after changes in the acidity of the precipitation. How relevant are the findings to other such aquatic systems located in somewhat different drainage basins and having a somewhat different species composition? Most of the experimental work is based on examining and manipulating a minuscule volume of the water, sediments, etc., followed by an extrapolation of the results, for example, to the whole of the plankton (Fig. 1–2). Studies are often based on too few replications to be able to draw unambiguous conclusions[4] (Table 1–1). I have tried to reduce all these problems by emphasizing, where possible, data and interpretations based on (1) interannual studies of individual ecosystems, (2) *comparative* (cross-system) analyses of aquatic ecosystems, (3) whole-system experimental manipulations, and (4) searching for patterns linking the functional properties of aquatic organisms to their size-linked metabolism.

Patterns revealed by assembling data from studies carried out over a variety of temporal and spatial scales form the basis for making predictions about how other, unstudied, inland waters or organisms could be expected to behave. If subsequent research confirms expectations, the generality of the original model or pattern receives additional support.[5] If, however, a pattern is not general enough to fit the additional data, the original model (idea) and its interpretation must be reconsidered and possibly discarded.[6] New data that do not fit existing models are invaluable in questioning the assumptions made that in turn lead to new hypotheses that should be tested. An ability to predict is the ultimate test of how well scientists understand what transpires in nature and is central to aquatic management.

Unfortunately, the number of well-documented and well-quantified patterns (regularities) is still modest, and those that have been produced are largely based on temperate-zone lakes. The number of gener-

---

[4]"When a study fails to demonstrate significant relationships, this can mean either that such relationships truly do not exist or that there was insufficient statistical power to detect those relationships that do exist." (D. J. Currie et al. 1999)

---

[5]"Hypotheses must be generated that can be tested by experiment or validated by new observations. If experimental evidence is contrary, then the hypothesis must be rejected, formulations of a secondary hypothesis to rescue a beloved primary one must be avoided." (J. Ringelberg 1980)

[6]"I have steadily endeavoured to keep my mind free so as to give up any hypothesis, however much beloved . . . as soon as facts are shown to be opposed it. I cannot remember a single first-formed hypothesis which had not, after a time, to be given up or greatly modified." (Charles Darwin, a letter.)

alities that are strong enough to have predictive power is even smaller. Most generalities available deal with time-averaged conditions (year, season) rather than the dynamic daily actuality. The generalities further pertain largely to static properties (e.g., the relative or absolute abundance of organisms or community biomass) rather than with their dynamics over time and space.

Generalities presented in the text cannot always be based on large data sets. Some are, of necessity, based on studies of only a few ecosystems in a single region. Others are based on a moderate number of species in the laboratory or field, yet others may be anchored on quantitative or qualitative studies of a single ecosystem or species over time, thereby suggesting the existence of a broader regularity that remains to be confirmed. Clearly, the predictive power of generalities varies considerably; some apply worldwide, while others may apply to only a particular climatic zone or lake type. I would be pleased to receive data from readers that will strengthen or refute generalities proposed in the text, as well as information about important patterns that were overlooked or not yet published at the time of writing. jacob.kalff@mcgill.ca.

One of the major problems in limnology, and ecology as a whole, is that it is obviously much easier and much less time consuming to articulate notions or hypotheses without any data at all, or to create hypotheses after the fact to explain the data collected, than to test hypotheses to see if they have merit.[5, 7] Therefore, the number of available ideas in the literature increases rapidly, but with their merits frequently untested. An even greater problem is that ideas (hypotheses) presented are often so loosely phrased that they are not really testable or potentially refutable. The greatest danger of all, however, is that untested notions and ideas about aquatic systems and their biota may be repeated often enough to become accepted as fact and entered into the conceptual framework used by individual scientists.[8]

The fourth problem I have had to confront, and equally an issue for students, is that limnology is a truly multidisciplinary science—one that includes physics, chemistry, geology, and engineering (hydrology), in addition to biology. Each of these disciplines examines aquatic systems from a different perspective. Ecologists dominate among the biological limnologists, yet they too differ in their perspectives. The physiologists, autecologists, and population ecologists focus on individuals or populations and their functional properties. To them the lake, river, or wetland need not be more than a convenient place or environment in which to examine these functions. System ecologists operate at the opposite end of the simplified spectrum of organization. Many of them conceive aquatic ecosystems as operational units, or black boxes, with inputs and outputs of energy and materials, but pay little or no attention to the species components or processes involved.[9] Other researchers set themselves aside from any consideration of hierarchical levels of organization and focus instead on particular processes, such as sedimentation rates in aquatic systems or the flux of nutrients.

The fifth issue to be dealt with in the writing and reading of a textbook is how comprehensive such a book should be. A textbook on inland waters should ideally treat *flowing water* (**lotic**) systems and wetlands in the same detail as *standing water* (**lentic**) systems with which they share drainage basins. Unfortunately, the attempt would make the book unwieldy since it already pays much more attention to drainage basins that nourish inland water and to environmental and management issues than is the rule in limnology texts. More importantly, the attempt would fail because much of the literature on flowing water and wetlands has lacked the same ecosystem context that has guided most of the research on lakes, and that guides this book. Work on lotic systems and wetlands has until recently been largely concerned with components of individual systems (hydrology, benthic insects, plant community structure, etc.) and biologists working there have often identified themselves more with core disciplines, such as ecology and fish biology, than limnology. Consequently, most biologists working on lotic systems or wetlands have considered themselves

---

[7]"The verification of ideas may be the most treacherous trap in science, as counter examples are overlooked, alternative hypotheses brushed aside and existing paradigms manicured. The successful advance of science and the proper use of experimentation depends on a rigorous attempt to falsify hypotheses." (P. K. Dayton and R. L. Oliver 1979)

[8]"One of the chief causes of poverty in science is usually imaginary wealth. The aim of science is not to open the door to infinite wisdom, but to set a limit to infinite error." (Galileo in B. Brecht's play, *Life of Galileo*)

[9]"The difference between limnology and the disciplines that contribute to it is one of motivation and integration rather than context. An invertebrate zoologist can measure respiration in aquatic crustaceans, a chemist can quantify the speciation of metals, and a physicist can produce new equations for fluid motion. Such information is not limnological unless it is cast in a form that sheds light on the functioning of aquatic systems." (Lewis et al. 1995)

(aquatic) ecologists rather than limnologists. In fact, to equate the study of inland waters with limnology might well be considered presumptuous to many flowing water and wetland scientists. But in the absence of a better name for the science of inland waters, I have used the term limnology by default.

Rivers, as integrated systems, are briefly discussed in Chapter 8, and flowing water systems are touched upon in all chapters, but they do not receive the same attention in terms of figures, tables, and examples as lakes. Even so, the role of lotic systems as sources of water, nutrients, organic matter, and contaminants to lakes and wetlands receives considerable attention throughout the book, particularly in Chapters 5, 8, 9, 13, 17, 18, 26, 27, 28, and 29. Wetlands and their biota receive particular attention in Chapters 8, 24, and 29.

Fortunately, there are no discrete borders between the science of lakes and the science of lotic systems or wetlands. The three types of aquatic systems form a continuum, with rivers grading into wetlands and slow-flowing rivers forming a continuum with rapidly flushed lakes, which makes information on one relevant to the others. Lotic systems and groundwater nourish the receiving lakes and wetlands; these in turn decisively impact the outflowing rivers, as well as downstream lakes and wetlands, which they provide with species, organic matter, nutrients, and contaminants. Consequently, no aquatic system can be understood in the isolation of upstream lotic and lentic systems and the drainage basins that they flow through and that characterize them. No individual scientist can be expected to be an expert on every aspect, but an appreciation of the linkages are a prerequisite for ecosystem-level research, for placing one's research in a larger context, and for collaboration with other aquatic scientists, soil scientists, hydrologists, biogeochemists, and terrestrial ecologists.

Scientists most interested in questions posed at the level (scale) of whole systems, whether they be lakes, streams, or wetlands, are often not equally interested in the specific mechanisms and processes employed by individual species, unless these act as **keystone species** (e.g., *Daphnia*, zebra mussels), species that when abundant exert a disproportionate impact on the behavior of the system as a whole. The reverse is equally true,—the findings made at particular hierarchical scales (Fig. 1–3) of biological organization (organism, population, community, ecosystem) are not easily extrapolated (other than conceptually) from one level of organization to the next, or from one particular temporal or spatial scale to other scales (Sec. 2.6).

Although virtually all biologists studying inland waters consider themselves limnologists or aquatic ecologists, they differ greatly in (1) the questions they pose, (2) the techniques they use to answer them, and (3) the interpretation of the results;[2] thus the kinds of research they find worth reading or quoting in the articles they publish also differ. Differences are typically even greater between scientists focused on lakes and those ecologists/biologists working on streams or wetlands. Naturally, all this is confusing to a newcomer to limnology and its literature.[10]

The considerable emphasis in limnology on lakes, rivers, and wetlands as systems, and the often very large number of species present, requires lumping together many species into so-called *functional groups* (e.g., herbivorous zooplankton). Similarly, the thousands of dissolved organic compounds in waterways are normally lumped as dissolved organic carbon (DOC) and the different forms of phosphorus as total phosphorus. Despite the widespread use of such aggregates in the literature, there is no doubt much of the most fascinating work in nature and the laboratory has been carried out at the scale of individual species or other components. Such studies must receive attention because they form an integral and important component of limnology.

Important and clearly-posed questions asked and answered at any one scale or level of biological organization (e.g., as biological or chemical species) provide data that are not only invaluable in their own right but also provide ideas that contribute to research at other levels of biological organization (Fig. 1–3). The emphasis in the text on inland waters as systems and the lumping together of many components into functional groups must not be interpreted as suggesting that studies at these lumped or integrated scales are the most important. Rather, they simply reflect the fact that it has been easier to draw generalizable conclusions at these broad levels than to make generalizations about the equally important components. Similarly, the attention given to comparative research is not due to the inherent importance of work at this

---

[10]"Don't be intimidated by an unfamiliar literature, whether your background is in biology, engineering, or perhaps geology, physics, or even medicine. If you insist on limiting your questions to areas in which you are fully equipped and ready to go, you'll not go far." (Vogel 1994)

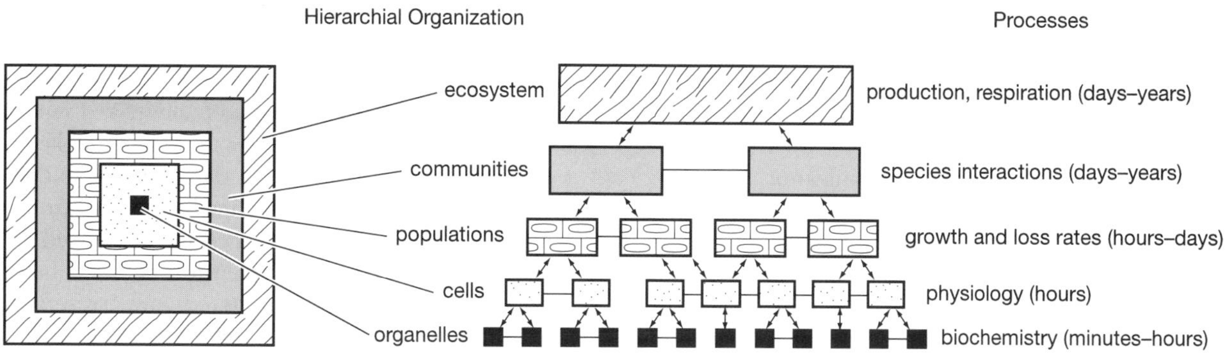

Hierarchial Organization   Processes

ecosystem — production, respiration (days–years)

communities — species interactions (days–years)

populations — growth and loss rates (hours–days)

cells — physiology (hours)

organelles — biochemistry (minutes–hours)

Figure 1–3 A diagrammatic representation of the hierarchy of process and organization in phytoplankton ecology with typical time courses of study. *(Modified from Harris 1986.)*

particular scale, but because it provides a useful framework for more detailed research on individual systems or specific components of systems.

Aquatic systems cannot be comprehended by only examining events below the shoreline. Understanding the behavior of aquatic systems requires an appreciation of the role of their drainage basins and the overlying atmosphere. The links between inland waters and their terrestrial and aerial catchments have traditionally received too little attention in limnology texts. The relatively modest amount of information provided on streams or rivers and wetlands as systems will hopefully be sufficient to whet the appetite for additional reading in texts geared specifically to them.

The sixth, and last, difficulty in attempting to write a comprehensive textbook that is useful for all climates is that limnology was long dominated by studies on a relatively small number of freshwater lakes, streams, and wetlands in the north temperate zone, which are subject to large and fairly regular seasonal changes in solar radiation, temperature, and hydrology compared to systems at lower latitudes. As a result, most of the fundamental ideas about the functioning of inland waters spring from the study of relatively small, deep lakes and small streams set in relatively undisturbed drainage basins in one climatic zone. It is not clear to what extent ideas and conclusions drawn from these studies can be extrapolated to (1) smaller water bodies (ponds) or, in the case of temperate-zone lotic systems, to larger rivers in that zone; (2) different types of lakes (e.g., shallow lakes) and streams (e.g., lowland streams in agricultural catchments) in the same climate zone and, even more so, to freshwater lakes, streams, and seasonal wetlands in the tropical and the semiarid portions of the world, some

of which are ephemeral (temporary). Nor is it clear to what extent the findings are applicable to (3) polar systems; (4) large rivers or the great lakes of the world; (5) saline lakes; or (6) rapidly flushed reservoirs. For a lighthearted view of the 'typical' lake and stream, see Section 4.3.

As inland waters lie along gradients of area, depth, climate, drainage-basin characteristics, and human disturbance, it seems essential to look for patterns among them without the constraints imposed by national borders or by continents. Data and insights obtained within one region or country can be interpreted best when seen from a wider perspective than a regional or national one. I will, therefore, make a point of presenting information and insights from polar as well as tropical regions. However, a high proportion of data are still being collected in the well-watered portion of the temperate zone and it is inevitable that limnology continues to have a temperate-zone bias.[11]

Limnology has flourished, particularly in the glaciated portions of northern Europe and North America with their many lakes and rivers and long limnological traditions. Nevertheless, tropical lakes, permanently frozen Antarctic lakes, saline lakes, basins that hold water only part of the year, heavily polluted lakes on all continents, as well as rivers and wetlands are an important part of limnology. The reader will be exposed to all of them to gain an appreciation of their variety, the many attributes they share,

---

[11]"Modern limnology is excessively concerned with the study of freshwaters in the northern temperate region. Limnology has been and is unbalanced in its interests and emphases and, as a result, many widely held limnological concepts need revision." (W. D. Williams 1988, an Australian limnologist)

and the differences ultimately imposed by climate, geology, and time.

Because large lakes and rivers often form the border between two or more countries, and spelling out country names is space consuming, I frequently use the International Standards Organization country code abbreviations [e.g., FR for France, RU for the Russian Federation, BR for Brazil, and CA for Canada (rather than California, as those from the US might think)] given in Appendix 1, with the countries always presented in alphabetical order. Occasional reference is made to marine work, either because good data from inland waters are lacking or because a comparison with marine work is appropriate. Furthermore, references to marine literature serve as a reminder of the close links between limnology and oceanography.

Human impacts on inland waters and their drainage basins are increasing rapidly and have started to affect bodies of water everywhere. The ability to predict the effects of human impacts and their mitigation is of great importance to the progress of limnology as a science and is also a matter of urgency in environmental management. The contribution of inland waters to human welfare is enormous, but its value has been captured only to a minor extent in monetary terms. A recent and courageous attempt (Costanza et al., 1997) shows—regardless of assumptions made to generate figures—the enormous value of aquatic systems. Services provided globally by lakes and rivers total about US $8,000 ha^{-1} of water surface (1 ha = 0.01 km^2 or 10,000 m^2). The single largest "service" provided is the regulation of discharge and supply of water for agriculture and industry (~US $5,000 ha^{-1}), but the role of catchments, reservoirs, and aquifers in the storage and supply of water is important as well (~US $2,000 ha^{-1}). The overall contribution of wetlands is estimated to be even greater (~US $15,000 ha^{-1}). The largest wetland contributions are in the regulation and dampening of fluctuations (flood control, storm protection, etc.) (~US $5,000 ha^{-1}), water supply (storage and retention) (~US $4,000 ha^{-1}), and waste treatment (wastewater treatment, pollution control, detoxification) (~US $4,000 ha^{-1}).

Well-trained limnologists will have to play a much larger role in the management of precious aquatic resources if their future is to be safeguarded. Aquatic management depends on an underpinning of appropriate science and there will have to be an increased emphasis on fundamental long-term research, above all at the whole-system level at which most environmental problems are largely recognized. As environ-mental degradation progresses relentlessly on a global scale, the appropriate science will require, more than ever, an ability to provide useable data, pose clear and realistic questions, analyze data appropriately, and report the results not only to peers but also to a wider public.

As the behavior of aquatic systems can't be understood in isolation from the drainage basins and atmosphere that nourish them, both greatly influenced by human activities, I pay more attention than has been the rule in limnology textbooks to drainage basins and human impacts. Indeed, the effect of humans on the environment is becoming so pervasive that it would be inappropriate not to include a specific chapter on acidification (Chapter 27), another on contaminants (Chapter 28), and a third on the man-made lakes we call reservoirs (Chapter 29). In many chapters, I touch briefly upon the implications of findings for aquatic management in order to link fundamental (pure) science with applied research and resource management.

1.2   The Setting

It is obvious to biologists that the **biological properties** of aquatic systems are to an important extent the net outcome of interactions between the species or species assemblages that make up the biota, which is reflected in the flow of organic matter and nutrients (Fig. 1–4). This long-standing recognition has meant that, historically, most biological limnologists have addressed themselves primarily to the study of competitive and predator–prey interactions. Even so, the energy and nutrients that make biological interactions possible are ultimately derived from beyond the shoreline. The light energy needed for photosynthesis and the subsequent flow of energy and nutrients from plants to animals is, together with heat energy, obtained via the atmosphere. The nutrients necessary for growth are ultimately derived from terrestrial drainage basins, with a sometimes important contribution from beyond delivered via the atmosphere (Chapter 8). Much terrestial-derived organic matter reaches aquatic systems and supplements that produced by aquatic plants. Biological properties exhibited below the water surface are, therefore, to an appreciable extent determined ultimately by **regional** and **catchment properties** (Table 1–2).

Catchment properties constrain not only biological properties directly but also indirectly through **catchment attributes** that are a major determinant of the physical/chemical **water properties.** The

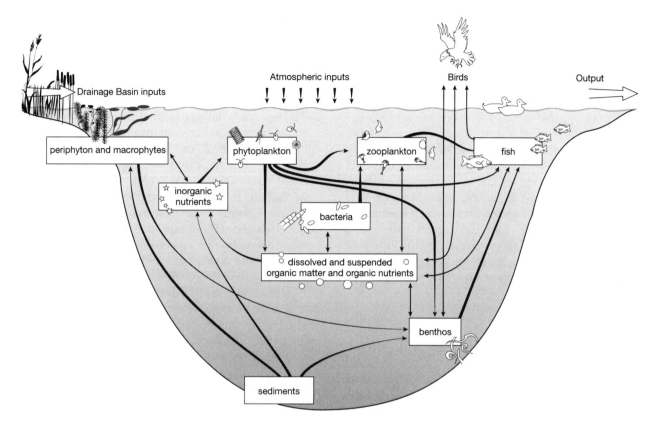

Figure 1–4 A conceptual view of the major pathways of energy and nutrient flow between the drainage basin, atmosphere, and principal biotic communities in lakes. *(Modified from Håkanson and Ahl 1976.)*

Table 1–2 **A hierarchy of attributes and properties that influence aquatic ecosystems and important human impacts on them.**

Regional properties	climate	geology	topography	
Catchment attributes	vegetation	soil	hydrology	
System attributes	morphometry	lake stratification	flushing rate, sedimentation	
Physical/chemical water properties	light/ temperature	turbidity, salinity, discharge	humic substrates	nutrients	toxins
Biological/ecological properties	biomass	productivity	trophic structure	biodiversity
Human impacts	habitat destruction, harvesting	nutrient and sediment input	toxic substances	hydrological alterations	climate change

biological/ecological properties are consequently the result not only of biological interactions below the waterline but by properties and attributes at higher levels in the hierarchy (Table 1–2), which are increasingly modified by **human impacts.**

A variety of regional and localized properties and attributes together create the 'stage' upon which the biological 'actors' play out the drama; modifying the shape and size of the original stage and changing the play as it evolves. Human activities increasingly impact both the stage and the actors themselves. Fundamental research on largely pristine systems provide an important baseline against which major human impacts can be measured. Both fundamental and applied research complement this by measuring the impact of various degrees of human activity on aquatic systems that provides the basis for environmental management.

1.3   Organization of the Text

No field can be understood without reference to its history and tradition and Chapter 2 is devoted to the development of limnology and major ideas that have guided its development. A few relatively short chapters then describe the characteristics of water, water resource distribution, and hydrology. In Chapters 6 and 7, the book turns to the origin and form (**morphometry**) of lakes and rivers, followed by a chapter on rivers and the export of materials from the land and atmosphere to the water (Chapter 8), and another that links drainage basin attributes to inland water functioning (Chapter 9).

The second, informal, section (Chapters 10–20) addresses primarily the physical and chemical properties of aquatic systems, but points out the role of the biota therein. There are chapters on light input and its distribution, lake stratification, water movements, the distribution of salts and dissolved oxygen, as well as on the cycling of important nutrients in the water, whose role in lakes is largely relevant to lotic systems and wetlands. Organisms are not only affected by their physical/chemical environment, they greatly modify their environment. The section ends with a chapter on the sedimentation of particles, sediment characteristics, and what dated sediments can reveal about the past. The stage has now been set for a more explicit consideration of the biota (Chapters 21–26) that, as pointed out, not only act on the stage but also modify it by their activities.

The third portion contains chapters devoted to the phytoplankton, the bacterioplankton, the zooplankton, the benthic plants, the animals associated with substrates, and ends with a chapter on fish and aquatic birds (Chapter 26).

The fourth, and last, informal division of the book addresses major human (**anthropogenic**) impacts on aquatic systems and consists of three chapters. It begins with Chapter 27 on the effects of acidifying precipitation on aquatic systems. The next deals with contaminants (other than nutrients), their sources, distribution, and roles in inland waters (Chapter 28). The last, Chapter 29, evaluates similarities and differences between natural and man-made lakes to integrate research on reservoirs more firmly into limnology and remind the reader that rapidly flushed reservoirs have as many lotic as lentic attributes.

Terminology will be kept to a minimum, as it is often more a hindrance than a help to newcomers to a field and discourages communication between disciplines. Older, but important references will be used when newer ones do not materially add to the information already available. This is done both to acknowledge important early work and to remind the reader that "Rome (or Cairo) was not built in a day." Particular attention has been paid to publications that attempt to quantitatively integrate new findings into the existing literature, thereby contributing to a synthesis. However, the literature is so vast that only a tiny fraction of it could be cited and the work of many excellent scientists could not be acknowledged.

A few basic terms and concepts follow immediately below so that, even before they are specifically discussed in subsequent chapters, they will be understood. The "Highlights" presented at the end of each chapter serve as reminders of important issues examined in the text. Their memorization in conjunction with Figure 1–5 will greatly aid in the appreciation of the chapters to follow:

Aphotic or **tropholytic zone:** The volume of water or the area of sediments where the photosynthetically available radiation (PAR) is <1 percent of that entering the water and where plant respiration is larger than plant photosynthesis.

Benthos: The community associated with the bottom—refers most commonly to the animal community.

Catchment, drainage basin, or (in North America) **watershed:** The area of land that drains towards an aquatic system. The term *paralimnion* has been used occasionally.

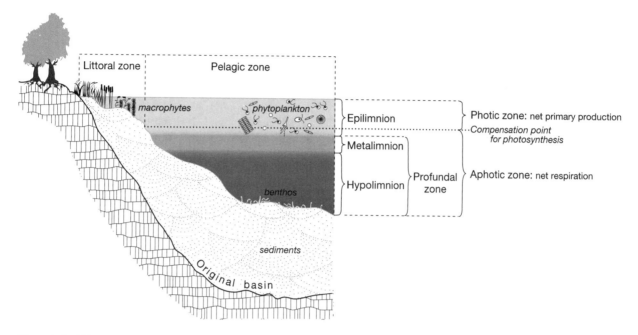

Figure 1–5   Diagrammatic cross section of a freshwater lake and its drainage basin showing the zones that become established in summer as a result of thermal stratification and a light gradient. *(Modified after Frey 1966.)*

Ecosystem: The unit of organization in which all living organisms collectively interact with the physical/chemical environment as an integrated system.

Epilimnion: The surface-mixed (turbulent) layer in those lakes that exhibit a vertical temperature stratification, with the lower boundary in contact with the metalimnion (see definition below).

Hypolimnion: The deep, cooler layer of a stratified lake lying below the metalimnion, characterized by a greatly reduced turbulence and usually insufficient light to allow algal growth.

Lentic system: Standing water systems (ponds, lakes) in which the flow is primarily imposed by wind and heat and is not primarily unidirectional.

Littoral zone: The near-shore region of lakes and lowland rivers where the sediments lie within the photic zone, and where the shallow water flora is frequently dominated physically by macrophytes.

Lotic system: Primarily unidirectional flowing water systems (streams and rivers) imposed by gravity.

Macrophytes: Community of multicellular emergent and submerged large plants dominating the shallow portions of littoral zones, lakes, slow-flowing rivers, and wetlands.

Metalimnion: The transition layer of water, between the epilimnion and hypolimnion, in which the temperature declines with increasing depth. Also known as the *thermocline*.

Mixed layer: The upper water layer recently mixed by wind or temperature-induced currents, also known as the *surface layer*. Equivalent to the epilimnion in stratified lakes.

Pelagic, lacustrine, or **limnetic zone:** The open water region beyond the littoral zone.

Photic, Euphotic or **Trophogenic zone:** The volume of water in which algal photosynthesis is, on a diurnal basis, greater than algal respiration. Operationally defined as the zone in which PAR is ≥ 1 percent of that entering the water.

Plankton: The microscopic and small macroscopic community of the open water adapted to suspension and subject to passive movements imposed by wind and currents. Composed of *phytoplankton* (plant plankton), *bacterioplankton*, and *zooplankton* (animal plankton.)

Profundal zone: The deep region (hypolimnion) of stratifying lakes, but mostly used with reference to deep-water sediments and their biota (e.g., profundal benthos).

Wetlands: Transition zones between terrestrial and aquatic systems where the soils are waterlogged for at least part of the year or covered by shallow water, and which are typically occupied by rooted aquatic vegetation (macrophytes); not all wetlands are physically connected to lakes or lotic systems. The littoral zone of lakes and rivers forms a continuum with wetlands. Wetlands can have both lotic and lentic attributes and, at times, may lack standing water entirely.

Acknowledgments

This book is dedicated to the memory of two exceptionally fine scientists and friends, Rob Peters and Frank Rigler. They kept reminding me of the importance of phrasing scientific questions as testable hypotheses rather than as somewhat vague, but interesting, notions. They also sensitized me to the artificiality of trying to distinguish between fundamental (basic) science and a more applied science. I am very grateful to my closest colleagues, the late Robert Peters, William Leggett, Joseph Rasmussen, and Neil Price, for their insights, advice, and for the scientific stimulation they have provided me over the years. I acknowledge my MSc and PhD supervisors, Drs. H. R. McCrimmon (University of Toronto and University of Guelph, Canada and D. G. Frey (Indiana University, US), for their support and confidence in me. Dr. Henri Decamps and his staff at the Centre d'Ecologie des Systemes Aquatiques (CNRS) in Toulouse, France, as well as Dr. Carlos Duarte and colleagues at the Centro de Estudios Avanzados de Blanes (CSIC) in Blanes, Spain; are thanked for their hospitality and support during sabbatical years of writing and research. I am grateful, above all, to my present and past graduate students, as well as post-doctoral fellows, for their stimulation and their repeated demonstration that ideas and notions that I held dear were, in the final analysis, either wrong or less simple than I had thought. Special thanks go to Elena Roman, Daina Brumelis, Michael Ditor, Grace Cheong, Anton Pitts, Anouk Bertner, Jason Derouin, and to various other technicians and student assistants for data collecting and graphics over the years of writing. The book would never have been written without them and the secretarial help with word processing provided during the early stages by the Department of Biology, McGill University. In addition, I acknowledge C. H. Fernando, University of Waterloo, D. Jeffries, Canada Centre For Inland Waters, K. Murphy, Hydro-Québec, V. G. Richardson, Canada Centre For Inland Waters, W. D. Williams, University of Adelaide, and M. Braun, International Commission for the protection of the Rhine, for data and/or advice. I further acknowledge, with admiration, the professionalism and dedication of the following Prentice Hall editors: Teresa Ryu, acquisitions editor, Joanne Hakim, production editor, and Jocelyn Phillips, copy editor. Most of the following people provided comments on Chapter 1. A number of people reviewed individual chapters and I thank the following for their insights and suggestions, although I alone bear responsibility for errors and omissions.

G. and I. Ahlgren, (ch. 21), *Uppsala University*

J. Bloesch, (ch. 20), *EAWAG, Switzerland*

A. S. Brooks, (chs. 8, 22, and 28) *University of Wisconsin*

W. R. DeMott, (chs. 8, 22 and 28) *Indiana University*

P. J. Dillon, (chs. 5 and 27), *Ontario Ministry of The Environment and Trent University*

J. A. Downing, (chs. 8, 9, and 18), *Iowa State University*

C. M. Duarte, (chs. 2 and 24), *Instituto Mediterraneo de Estudios Avanzados, (CSIC), Spain*

J. Fresco, (ch. 3), *McGill University*

J. M. Gasol, (ch. 22), *Institut de Ciències del Mar, (CSIC), Spain*

E. Gorham, (chs. 2 and 13), *University of Minnesota*

L. Håkanson, (ch. 7), *Uppsala University*

S. K. Hamilton, (ch. 15), *Michigan State University*

R. W. Kortmann, (chs. 16 and 19), *Ecosystem Consulting Service, Inc., Coventry, USA.*

W. M. Lewis, Jr., (ch. 11), *University of Colorado*

O. Lind (chs. 8, 22, and 28) *Baylor University*

S. C. Maberly, (ch. 14), *Centre for Ecology and Hydrology, UK*

S. MacIntyre, (ch. 12), *University of California, Santa Barbara*

J. Melack, (chs. 4 and 6), *University of California, Santa Barbara*

M. L. Ostrofsky, (chs. 8, 22, and 28) *Allegheny College*

M. L. Pace, (ch. 23), *Institute of Ecosystem Studies, USA*

R. H. Peters, (ch. 2), *McGill University*

Y. T. Prairie, (ch. 17), *University of Quebec at Montreal (UQUAM)*

J. B. Rasmussen, (chs. 26 and 28), *McGill University*

H. Regier, (ch. 26), *University of Toronto*

C. S. Reynolds, (ch. 21) *Institute of Freshwater Ecology, UK*

D. Roeder, (chs. 8, 22, and 28) *Simon's Rock College of Bard*

K. Sand-Jensen, (chs. 10 and 24), *Freshwater Biological Laboratory, Denmark*

J. P. Smol, (ch. 20), *Queen's University, Canada*

M. Straškraba, (ch. 29), *Academy of Sciences of The Czech Republic*

D. Strayer, (ch. 25), *Institute of Ecosystem Studies, USA*

2

The Development
of Limnology

2.1   Limnology and Its Roots

Limnology[1] came of age in 1901 when Francois Alphonse Forel (1841–1912) published the first textbook on the subject. In the book Forel, who was Professor of Physiology at the University of Lausanne (CH), presented the results of more than 30 years of research on Lac Léman (Lake of Geneva), drawing also on measurements made on other lakes by himself and other scientists. The textbook was the outgrowth of his 1869 book on deep-water-sediment fauna and included a synthesis of his three volume *Le Léman: Monographie Limnologique*, published between 1892 and 1904. The first two volumes described the geology, physics, and chemistry of the lake; the last dealt with the lake's biology. Forel was not just the founder of limnology and creator of its name, he was also a first rate early ecologist who recognized the relationship between organisms and their climatic, hydrological, geological, physical, and chemical environments. In other words, he saw lakes as integrated natural units or ecosystems, as had Stephen Alfred Forbes in America (1887), and as anticipated by Thoreau, who was not a scientist, while making observations on Walden Pond in the 1840s and 1850s. Forel was not, however, the first person to do what is now called limnology, and his definition of limnology as the "oceanography of lakes" acknowledges the influence of preceeding developments in marine science. Earlier limnological research of the late 18th and early 19th centuries had been primarily concerned with the physical attributes of lakes and river discharge (Table 2–1), followed by a flood of biological papers starting around the time that Forel (1869) published his first work on the bottom fauna of Lac Léman.

Starting in 1888, a number of biological stations for limnological research were, within a few years of each other, established in both Europe and North America. The first systematic comparisons of different lakes and rivers, and early measurements of phytoplankton and zooplankton population dynamics, were made before the turn of the century (Table 2–1). At about the same time, the first limnological measurements were being made in remote tropical African lakes and rivers; the development of sampling equipment and chemical techniques for water analysis was not far behind.

Prior to the First World War, nearly all limnologists were interested in descriptions of physical, chemical, and biological aspects of (usually) single lakes, thereby laying the groundwork for subsequent attempts to organize the growing mass of information into logical schemes of lake and river classification. Probably the most influential European limnologist during the first decade of the 20th century was C. Wesenberg-Lund (DK, 1867–1955). He was far ahead of his time, recognizing the importance of experimentation and manipulation to obtain clearcut answers, and he saw the need for comparative (among-system) analysis of aquatic systems involving not one but a number of lakes (C. Wesenberg-Lund 1905). In 1910 he wrote an impressive synthesis, in

---

[1]From Greek *limne* = lake, pool, or swamp, + *logos* = discourse or study.

Table 2–1   **Milestones in the development of limnology.**

Year	Development
1650	A categorization of four lake types based on the presence/absence of stream inflows (B. Varenius, DE/NL).
1674	Description of a filamentous green alga (*Spirogyra*) from a Dutch lake and an early recognition of seasonal differences in algal populations as well as descriptions of rotifers (A. van Leeuwenhoek).
1779–1796	Using a heavily insulated thermometer, H. B. de Saussure (CH) noted that deep waters in some Swiss lakes were much colder than surface waters and were also near the point of maximum density (4°C).
1787	Depth determination of some English lakes (J. Clark).
1800–1810	First systematic measures of river discharge (Rhone, CH; Göta, SE).
1819	Discovery of what is now called the metalimnion (thermocline) in Lake Geneva [Lac Léman] by H. T. de la Bèche (GB).
1826	First scientific description of an algal bloom, in Lac de Morat, by L. de Candolle (CH).
1833	Recognition of the importance of the water balance in determining lake size, salinity, and sediment retention (J. P. Jackson, GB).
1841	Diurnal oxygen cycle; linking low dissolved oxygen levels with fish kills, and high concentrations of oxygen linked with high algal concentrations (A. Morren and C. Morren, BE).
1850	Measurement of the turbidity of Lake Superior with a tin cup, which disappeared from view at 12.8 m (L. Agassiz, US).
1852	Acid rain is linked to coal burning in England (R. A. Smith), 100 years before acid rain was recognized as a widespread problem in Europe and eastern North America.
1865	Development of the Secchi disc for measuring water transparency by Commander Cialdi and Professor P. A. Secchi, priest/scientist on the *SS Immacolata Concezione*, a steam corvet of the papal navy traveling in the Adriatic Sea.
1867	Distribution and ecology of crustacean zooplankton in Danish lakes (E. Muller).
1869	Publication of a paper on the bottom fauna of Lac Léman (F. A. Forel, CH).
1871	Establishment of the U.S. Commission of Fish and Fisheries in response to the decline of commercial fish stocks in the Laurentian Great Lakes.
1877	Classification of river zones on the basis of dominant fish species present by, among others, V. D. M. Borne (DE). Beginning of detailed study of the Illinois River and the development of biological indicators of water quality (S. A. Forbes, US).
1887	Recognition of lakes as functioning units or "microcosms" (S. A. Forbes, US) and independently in 1892 by F. A. Forel (CH). The development of nets for quantitative capture of marine plankton plus the rudimentary statistics to analyze data by V. Hensen (DE). Description of the vertical migration of zooplankton by A. Weismann (DE).
1888	Development of the (still) widely used Winkler technique for dissolved oxygen determination (L. W. Winkler, DE) and used in 1895 by F. Hoppe-Seyler (DE) to explore the relationship between dissolved oxygen and the biota.
1888	Establishment in Bohemia (CZ) of the first freshwater laboratory followed in 1891 by laboratories at L. Glubokoe (RU), and Plön (DE), and at Lake St. Clair (US) in 1893.
1891	Description of what we now call thermoclines and how they are formed, by E. Richter (DE), and independently by E. A. Birge (US).
1892	Quantitative phytoplankton studies by C. Apstein in Germany following upon the largely qualitative plankton (and benthic) studies of plant and animal species distribution, which flourished during the 1870–1910 period in Europe and eastern North America. In the following year G. N. Calkins (US), reported a regular recurring spring and fall algal bloom in Massachusetts lakes. Scientists recognize the importance of competitive interactions to the distribution of some stream invertebrates (W. Voigt, DE).
1892–1904	Publication of F. A. Forel's three-volume monograph, entitled *Le Léman: Monographie Limnologie*, based upon work done on Lake of Geneva (Lac Léman), CH.

Table 2–1   *(continued)*

Year	Development
1895	Depth distribution of submerged macrophyte communities by E. Warming (DK).
1896	First use of the mark-recapture technique to estimate the size of fish stocks in the sea (S. G. J. Peterson, DK), followed by the use of fish scales to determine age of mature fish (C. Hoffbauer, DE, 1898). Experimental in-situ work on the relationship between light and the growth of diatoms (G. C. Whipple, US). C. Apstein (DE) proposed a relationship between nitrogen availability and blooms of cyanobacteria.
1897	Zooplankton population cycles in Lake Mendota (US) by E. A. Birge; a visionary recognition of the importance of small algae (nanoplankton) not caught well with the coarse plankton nets that were available by C. A. Kofoid (US), and a similar important recognition by G. C. Whipple (US) of the importance of wind-driven mixing of lakes to the development of algal blooms.
1897–1909	Inter-lake comparisons in Scotland (J. Murray and L. Pullar) followed in 1904 by a comparative analysis of Scottish and Danish lakes by C. Wesenberg-Lund (DK).
1898	Recognition of the importance of lake morphometry (lake depth) by showing that shallow lakes have greater plankton abundance than deep ones (H. Huitfeldt-Kaas, NO), a concept elaborated upon by D. Rawson (CA) between the 1930s and 1960s. First evidence, presented by A. Delebecque (FR), that the chemical characteristics of lakes are importantly determined by the geology of their catchments, reported in a book entitled *Les Lacs Français*.
1901	Publication of the first limnology text, *Handbuch der Seenkunde. Allgemeine Limnologie*, by F. A. Forel (CH), followed 25 years later by a successor text written by A. Thienemann (DE), and in 1935 by the first text written in English with the word limnology in its title (P. S. Welch, US).
1902	Seasonal cycles of planktonic bacteria (A. Pfenniger, CH) in Lake Zurich by means of plate counts, followed in 1930 by quantitatively more correct microscopical counts (S. S. Kusnetsov and G. S. Karzinkin, USSR), and in the 1970s by a superior counting procedure based on fluorescent staining of bacteria.
1903	First comprehensive studies of river plankton and associated physical/chemical conditions by C. A. Kofoid (US), following 1893 work on zooplankton (rotifers) in the Rhine River by R. Lauterborn (DE).
1904	Description of internal waves or thermocline seiches in Loch Ness (E. E. Watson, GB), following pioneering work in 1897 "On gravitational oscillations of rotating water," by Lord Kelvin (GB).
1905	Experimental demonstration that submerged aquatic angiosperms take up substances via their roots, and that herbivorous invertebrates feed primarily not on the macrophytes but on the associated algae, bacteria, and detritus (R. H. Pond, US).
1908	Establishment of the first limnological journal, *Internationale revue der gesamten Hydrobiologie und Hydrographie*, encompassing limnology and hydrology. A first systematic attempt to use a component of the biota (diatoms) as indicators of (stream) water quality. R. Kolkwitz and M. Marsson (DE).
1910	The linking of sewage inputs to algal blooms (R. Lauterborn, DE). A. Steuer (DE), while working on fish culture, reported the effect of fish predation on the zooplankton community, something not recognized in fundamental limnology until 1958.
1911	Development of the *Ekman dredge* for quantitative sampling of lake sediments and their fauna (S. Ekman, SE) and later modified to its present form by E. A. Birge (US); demonstration of the abundance of the small phytoplankton that had been largely overlooked (H. Lohmann, DE).
1912	Development of an instrument for measuring the underwater solar energy in lakes (E. A. Birge, US). Work on the longitudinal distribution of particular invertebrates in streams as a function of temperature and competitive interactions (A. Thienemann, DE). Development of "standards" for sewage effluents entering rivers (GB).
1915	Recognition of lakes as open systems with inputs and outputs, when E. A. Birge (US) and W. Schmidt (DE) independently developed lake heat budgets. Use of glass slides to study the attaching organisms in nature by E. Naumann (SE), and independently in 1916, E. Hentschel (GE).

(continued)

Table 2–1   *(continued)*

Year	Development
1920	The recognition of distinct, annually deposited sediment layers in Lake Zurich by F. Nipkow (CH), followed thirty years later by the development of paleolimnology as a subdiscipline.
1921	Development of an air-bubble system to destratify lakes and improve oxygen concentrations and water quality (W. Scott and A. L. Foley, US); seldom used until the 1970s.
1922	Establishment of the International Association for Theoretical and Applied Limnology (SIL) with 401 founding members from nearly all continents; quantification of links between the flora of lakes and the morphometry and geology of their drainage basins by W. H. Pearsall (GB).
1923	Experimental proof of the importance of phosphorus in determining algal growth in lakes and seas (W. R. G. Atkins, GB).
1925	Development of the first mathematical model to predict water quality: an equation that predicts how much the dissolved oxygen concentration in rivers will decline at given points downstream of a waste discharge (H. W. Streeter and E. B. Phelps, US).
1926	Publication of important research on the relationship between distributions of sediment-dwelling invertebrates and environmental factors by J. Lundbeck (DE).
1927	Publication of the dissolved oxygen method for determination of phytoplankton production (T. Gaarder and H. Gran, NO) and anticipated by A. Putter (1924) in Germany. Recognition of the importance of dissolved oxygen in the distribution of stream invertebrates (E. Hubault, FR), followed in 1937 by the identification of the importance of water-velocity determined substrate characteristics to stream invertebrates (G. Nietzke, DE).
1927–1929	Development of tropical limnology with the British Fisheries and Limnological survey of Lake Victoria and other Rift Valley lakes (in East Africa) by M. Graham, P. M. Jenkins and E. B. Worthington (GB) and the German-Austrian Sunda (Indonesia) expedition led by A. Thienemann (DE) and F. Ruttner (AT). Development of an accoustic method for detection of fish by K. Kimura (JP) in 1929.
1930	Use of artificial streams to study effect of nutrients and other pollutants on the biota (H. W. Streeter, US), and again in 1957 for ecological purposes (H. T. Odum and L. M. Hoskin, US). Estimation of epilimnetic primary production by the measurement of hypolimnetic oxygen consumption rates (K. M. Strøm, NO). Elaboration of the notion that the concentration of nitrogen and phosphorus (and silica in diatoms as well) determine phytoplankton species composition and abundance in lakes (W. H. Pearsall, GB).
1932	Use of inorganic N:P ratios as indicators of nitrogen versus phosphorus limitation in lakes (W. H. Pearsall, GB). His ideas were quantified in 1966 by M. Sakamoto (JP).
1934	Direct measurements of planktonic primary production and respiration rates in lakes, using changes in dissolved oxygen concentration in plankton samples placed in suspended glass bottles (G. G. Vinberg, USSR), and independently developed by J. T. Curtis (US) in 1935.
1936	Studies on control of the concentrations and cycling of elements (iron and phosphorus) in lakes by W. Einsele (DE), followed in 1941 by equally important studies on the biogeochemistry of a number of elements by C. H. Mortimer (GB).
1938	The first manipulation of a whole lake through the addition of fertilizers to study the effects on plankton and fish by C. Juday and C. L. Schloemer (US), followed 10 years later by recognition of the need for reference (control) lakes (R. R. Langford, CA), and for more controlled experiments by A. Hasler and associates (US) in the 1950s.
1939	The use of sediment traps to estimate plankton loss rates from the zone of primary production in lakes (J. Grim, DE).
1941	Addition of phosphorus to a lake and the study of its transformation and cycling (W. Einsele, DE); a mass balance (input–output) model of phosphorus in Linsley Pond (G. E. Hutchinson, US). Mass balance modeling is now used extensively in lake management, but apparently was first used for this purpose in 1957 (F. Baldinger, CH).
1942	The view of lakes as functioning units organized along trophic lines (R. L. Lindeman, US), and partially anticipated by A. Thienemann (DE), and in 1939 by V. S. Ivlev (USSR).

Table 2–1 (*continued*)

Year	Development
1947–1950	Development of radioisotope techniques to study phosphorus cycling in lakes by G. E. Hutchinson and V. T. Bowen (US), followed by similar work in lotic systems a decade later (R. C. Ball and F. F. Hooper, US). Introduction of phosphorus-based detergents and their release into waterways. Manipulation of food chains (R. C. Ball and D. W. Wayne, US). Experimental stream fertilization (A. G. Huntsmans, CA). Development of large in-situ enclosures (limnocorals or mesocosms) anchored in lakes, permitting experimental manipulations at a scale intermediate between a large flask and a whole lake (M. Stepanek and M. D. Zelinka, CZ). Characterization of fish zones in lotic systems based on river morphometry (M. Huet, BE).
1952	Development of radioisotope techniques for the measurement of phytoplankton photosynthesis (E. Steemann Nielsen, DK), bacterial sulfur metabolism (M. V. Vanov, USSR), and in 1954 for determining food consumption by zooplankton (A. G. Rodina and A. S. Troshin, USSR).
1953	Regional chemical budgets of rivers, including the effect of precipitation (P. J. Viro 1953, FI). Experimental and observational work on limiting nutrients and eutrophication in Swiss lakes (E. A. Thomas, CH).
1955	Demonstration of the effect of acid rain on the chemistry of lakes, bogs, and soils (E. Gorham 1955, CA/US), and in 1959, on fish stocks (A. Dannevig, NO).
1956	Development of a technique to estimate river metabolism by E. P. Odum (US), followed by his influential 1957 study of the dynamics of a lotic system. Demonstration of the biomagnification of contaminants in food chains by means of stable isotopes (L. A. Krumholz, US).
1958	Effects of fish feeding on plankton community structure (J. Hrbáček, CZ).
1960–1970	Wide recognition of the importance of rate measurements rather than measures of abundance and biomass alone; rapid development of electronic and analytical equipment; early measurements of trace metals in inland waters and wider recognition of the importance of bioconcentration and biomagnification of radionuclides and organochlorine pesticides in food chains; rapid growth in funding and the number of limnologists in the western world; major shift in scientific emphasis from collection of primarily observational data to experimental research; introduction of computers and development of dynamic modeling of systems; growing public concern about sewage and agricultural pollution of lakes, resulting in the development of applied limnology (lake and river management) and models linking nutrient export from drainage basins to lakewater concentration and phytoplankton biomass; demonstration of the increasing acidity of waterways in Scandinavia and Canada.
1970–2000	Much wider appreciation of the importance of wetlands and the first scientific treatise on stream ecology (N. Hynes, 1970. *The Ecology of Running Waters*). Funding shifts that encouraged more applied research on nutrient pollution, acid rain, and toxic chemicals—with an associated relative reduction in the pre-eminence of academic and fundamental science in determining research directions; funding provided by mission-oriented agencies encourages closer links between fundamental and applied limnology; major advances in microbial ecology and paleolimnology; significant progress in quantifying links between community structure and functioning and the role of fish therein, resulting in a greater reintegration of fish ecology in limnology; elucidation of the pathways, cycling, and fate of phosphorus, nitrogen, and contaminants; remote sensing of water temperature, turbidity, and plant biomass; growing interest in wetlands as systems.

English, of early limnological research. He not only did fine work on plankton and the littoral animals of Danish lakes but also worked on lotic systems. At about this time stream ecology found its roots in a German study of mountain streams by P. Steinmann (1907). Stream ecology continued to develop in the 1920s with work by A. Thienemann (DE) and a book by K. E. Carpenter (GB).

The development of a discipline can be measured roughly by the first appearance of textbooks on the subject. The first lotic textbook with a whole-system perspective, *The Ecology of Running Water* by H. B. N.

Hynes (1970), appeared 70 years after Forel's text on lakes. Its wetland counterpart was published about 15 years ago (Mitsch and Gosseling 1986, rev. 1993).

2.2   Limnology Between World War I and World War II

Following the hiatus produced by World War I, a new generation of limnologists sought to recover lost momentum. In Europe, two prominent individuals among them were August Thienemann (DE, 1882–1960) and Einar C. L. Naumann (SE, 1891–1934), who were instrumental in starting the International Association for Theoretical and Applied Limnology in 1922, an organization which still unites limnologists worldwide (Fig. 2–1). The inclusion of both fundamental and applied research in the name reflects their recognition that the two approaches have much to offer each other. This recognition was not widely shared and the more applied aspects, such as fisheries biology, wastewater biology, and hydrology drifted away from the rest of the field, as did flowing water ecology until recent concerns about the environment and interest in modeling whole systems brought them somewhat closer.

Between the two World Wars limnology developed rapidly along with the number of practitioners, and by the end of this period at least some limnologists had estimated the input and output of materials in lakes, measured primary production, and recognized that changes in the nutrient supply brought about changes in the biota and affected the geochemistry of iron. Hydrology flourished and stream ecology developed with an emphasis on fish (Hawkes 1975). Among the most prominent limnologists of the period were E. A. Birge (Fig. 2–2) and C. Juday (US); see more detailed discussion in Section 2.4.

Microbial ecology and biogeochemistry flowered in the former USSR. There was excellent Soviet work on energy flow in aquatic systems between and after the two World Wars, but it did not have the impact it deserved because of language barriers and political ideology. Early Soviet contributions to microbial limnology are discussed in Chapter 22. The interwar period also encompassed the first specifically limnological expedition to the tropics of East Africa and present Indonesia (Table 2–1). The Indonesia expedition, led by August Thienneman and Franz Ruttner (AT), did ten months of field work during 1928 and 1929, which ultimately was published in 11 volumes with nearly 10,000 pages of text and about 3,000 tables and figures (Rodhe 1974). The two tropical expeditions ended the period of almost exclusively temperate-zone activity.

Contributions of earlier developments in limnology are worth acknowledging. Reading only recent literature might give you the erroneous impression that virtually all worthwhile research was begun about five to 10 years before an author set out on their own research career. Two observations stand out from an examination of early research. The first is that contributions to limnology came from all corners of Europe and North America and, from the late 1920s on, from

Figure 2–1 August Thienemann (a) and Einar Naumann (b) in the 1920s. *[Photo (a) courtesy of the family of August Thienemann; photo (b) courtesy of Max-Plank-Institute für Limnologie.]*

(a)   (b)

Figure 2–2 Edward Birge (left) and Chancey Juday (right) on a zooplankton sampling trip on a Wisconsin Lake in 1920, using a brass trap designed by Professor Birge. *(Photo courtesy of State Historical Society of Wisconsin.)*

Japan. In Japan, where limnology was much influenced by the German school. Lake classifications that attempted to include local lake types were developed by D. Miyadi, S. Yoshimura, and others. S. Yoshimura was a major figure who also, with colleagues, did excellent work on water chemistry and lake stratification, but whose work on volcanic lakes with low pH added a new dimension to limnology.

It is evident that the development of limnology began about the same time in economically more developed parts of the world; this is apparent from the establishment of research stations in North America and western, central, and eastern Europe (Table 2–1). It is not surprising that the same or similar findings were made independently at about the same time in widely separated countries, something that is not fully evident from the small number of examples given in Table 2–1.

A new concept or promising technique (Table 2–1) is rarely accepted immediately and incorporated into limnological thinking. Rather, time lags of a decade or more were common during the 20th century before useful ideas, techniques, and approaches became widely accepted. Some ideas are produced long before science is "ready" for them. One example is a book by J. Jackson (1833) with the wonderful title *Observation on lakes, being an attempt to explain the laws of nature regarding them; the cause of their formation and gradual diminution; the different phenomena they exhibit, etc. with a view to the*

advancement of useful science. This book, based on observations in India, describes lakes as open systems with water inputs and outputs determining lake size, salinity, and sediment load. Another example is the work of R. A. Smith (1852) on acidifying precipitation and its local distribution. His research, including a remarkable book entitled *Acid rain: The beginning of chemical climatology,* 1872), was nearly forgotten for almost a century before the problem was rediscovered (Cowling 1982), and acid rain was shown to be affecting large areas of northern and central Europe, and an even larger land mass in north-east North America (Chapter 27).

▲   2.3 The Development of Ideas:[2] Europe

Lake Classification

August Thienemann (DE) and Einar Naumann (SE), who helped start The International Association for Theoretical and Applied Limnology in 1922, also be-

---

[2]"A knowledge of the history of changing ideas in the past, far from being a luxury, is essential as a means of accustoming and preparing us for the possibility of ideas changing in the present and the future, and it may make us the more ready to experiment on enlarging or revising our concepts." (K. Picken 1960)

came the intellectual leaders of a movement to provide structure for the understanding of lakes. During a 30-year period between the two World Wars and just after, they and their associates in Europe attempted to classify lakes (*Seetypen Lehre* or Lake Typology) virtually as if lakes were functioning units or superorganismal units that should be identifiable with taxonomic keys. As A. Thienemann (1925) put it, "If one investigates a lake typologically, one cannot expect the impossible from nature who, if she is ever to be scientifically comprehended to any degree at all, her peculiarities, patterns and processes, must be categorized." Their preoccupation with classification of lakes into fixed categories was understandable because most limnologists were taxonomists by training, and the need for categorizing the rapidly accumulating mass of limnological information was becoming acute. In addition, scientists were, and sometimes still are rooted in the 19th century notion of science as a purely automatic revelation of the realities of the world following the accumulation of facts. Lake classification (typology) and the characterization of rivers, based on fish or invertebrate assemblages, were assumed to lead to important advances. That science progressed most rapidly by posing and testing clearly articulated questions (hypotheses) was not widely appreciated. Nor, with the exception of visionary work by W. H. Pearsall in England, was it well recognized that aquatic systems and their biota lie along a large number of physical, chemical, and biological gradients, and that the position of any body of water into this multidimensional space depends on the climate, geology, land use of the drainage basin, system morphometry, and species present. In hindsight, it is not surprising that the lakes studied did not fit well in the simple and idealized classification schemes proposed. Initially, Thienemann based his scheme on the presence or absence of particular chironomid larvae (*Diptera, Insecta*) in the sediments, which served as indicators of environmental conditions and, especially, dissolved oxygen levels.

The problem with the classification schemes gradually became apparent to Thienemann, Naumann, and others because newly investigated lakes would inevitably differ from the relatively few on which the original classification was based. This led to the development of increasingly complex schemes, ones that might include dissolved oxygen content, other organisms, water color, lake morphometry, sediment quality, and even "maturity" of the lakes. These schemes failed when more lakes and lakes in different lake districts were examined. In other words, the schemes lacked sufficient generality to have predictive power.[3] Furthermore, the fact that classifications were largely qualitative meant that different users of any one scheme might come to different conclusions. These problems were compounded because tests for determining statistically significant differences were either nonexistent or not readily available. Equivalent attempts by E. Naumann and others to characterize lotic systems were even less successful (Hawkes 1975).

Although forgotten today, complex lake classification schemes were a milestone in limnology because they helped draw the different subdisciplines together, provided an important stimulus for among-lake studies, and showed that limnologists had developed a well-defined approach to the field. The associated notion of lakes as superorganisms or integrated units did not disappear completely and is held in a modified form today by those who examine inland waters as units characterized by inputs and outputs, and those interested in whole-system metabolism.

Thienemann

The urbane Thienemann was an outstanding ecologist and the prime force in the development of research not only on individual insect species (*autecology*) but also on functional groupings (producers, consumers, and reducers) which, as early as 1914, laid the groundwork for later research on energy flow in aquatic systems (Vollenweider and Kerekes 1980) and anticipated the ecosystem concept (Rodhe 1979). His personal research contributions to the development of limnology had largely ended by the mid-1920s, when he began working primarily on the autecology, distribution, and taxonomy of a group of aquatic insect species (*chironomids*) in lotic and lentic systems. The goal of his research on the ecology of particular lentic chironomid species was to use them as indicators to characterize aquatic systems, not simply because the species were fascinating or might be used to appreciate the ecology of all insect species.

Thienemann's influential ideas on nutrient cycling within lakes and food-web structure (A. Thienemann 1925) were acknowledged by Lindeman (1942) in his influential paper on trophic levels and energy flow in aquatic systems (Sec. 2.4).

---

[3]"No scientific theory is worth anything until it enables us to predict something that is going on. Until that is done theories are a mere game of words." (J. B. S. Haldane 1937)

Thienemann's view of lakes as systems or units was shared by most contemporary limnologists in Europe, and to some extent by E. A. Birge and C. Juday, the pre-eminent North American limnologists of the time (Sec. 2.4). Thienemann's influence in Europe was immense between the two World Wars, somewhat overshadowing other major limnologists such as Naumann, who died early, C. Wesenberg-Lund (DK), F. Ruttner (AT), K. M. Strøm (NO), W. H. Pearsall (GB), L. L. Rossolimo (USSR) or S. Yoshimura (JP). In addition to his prolific writings (460 publications), Thienemann singlehandedly edited the major journal *Archiv für Hydrobiologie* for more than 40 years. His power and influence was such that Berg (1951) felt it necessary to remind the limnological community that the groundwork of limnology had been laid well before the ascent of Thienemann. Berg's postwar article was an emotional plea against what he considered to be rewriting history and provides an interesting and impressive listing of major findings prior to the Thienemann era.

Naumann

Naumann's enduring contribution to science was largely based on intuition. He proposed (1919) a direct relationship between the phosphorus, nitrogen, and calcium supply to lakes and the amount and composition of the resulting phytoplankton. Since the nutrients could not be readily measured at that time, he believed the phytoplankton composition was an indicator of the nutrient conditions. He borrowed the terms **oligotrophic** (poorly nourished) and **eutrophic** (well nourished), and later added **mesotrophic** (medium nourished), to describe lakes of low, high, and medium nutrient content. This particular classification scheme has survived the test of time, and was accepted but modified by Thienemann (1925), who saw the division as inappropriate for lakes high in colored organic-matter (colored lakes) known as **humic** or **brown-water** lakes. Humic waters characterize many poorly-drained freshwater catchments all over the world. Thienemann called such lakes **dystrophic** (defectively nourished) lakes. In contrast to the three categories based on nutrient content, this fourth group refers strictly to water color even though most dystrophic lakes are also oligotrophic.

Naumann divided lakes on the basis of their implied nutrient content, and in the process acknowledged them to be open systems linked to their catchments or drainage basins through the supply of nutrients from the land. His ideas anticipated subsequent notions about material fluxes and lake productivity and were of particular importance to those scientists interested in both *comparative limnology* (comparing aquatic ecosystems) and environmental management. C. Wesenberg-Lund (1926) pointed out that the details in Naumann's proposal were uncertain and not backed by data. Naumann was, at the time, less influential than Thienemann, who saw lakes primarily as closed systems—simple functioning units, or "microcosms" as labelled by S. T. Forbes (1887)—whose functioning would be properly understood or characterized by their inherent structure alone. Forbes' conception of lakes, shared by contemporaries, as simple and readily comprehensible systems has drawn a disproportionate fraction of (major) ecologists to the study of lakes rather than running waters or wetlands over the last century. The concept of lakes as functioning units focused attention on events below the waterline and reduced limnologists' interest in the effect of drainage basins and the atmosphere on aquatic systems; this focus dominated limnological thinking until well into the 1960s (Vollenweider 1979), and may do so even today.

▲ 2.4   The Development of Ideas: North America

Birge and Juday

At the turn of the 20th century, North American limnologists were more influenced by two highly pragmatic scientists, Edward A. Birge (1851–1950) and Chancey Juday (1871–1944) and their Wisconsin school than by the continental European preoccupation with order and logic. Starting in the 1890s, other important centers of limnological activity could be found in Illinois (S. Forbes, C. A. Kofoid), Michigan (J. Reighard), Indiana (D. S. Jordan), Ohio, New York (J. G. Needham), and Massachusetts (G. C. Whipple) (Frey 1963, and Bocking 1990). Ontario (R. Langford) and Saskatchewan (D. Rawson) became important Canadian centers of research between about 1920 and 1960. However, none of those centers has had a legacy as large as the Wisconsin school.

When Birge joined the University of Wisconsin in 1876, he worked on zooplankton population dynamics. In 1904, he shifted his attention to the physical, chemical, and biological characteristics of hundreds of Wisconsin lakes. He was joined four years later by Juday. Their comparative studies of hundreds of lakes was encouraged and financed by the Wisconsin Natural History Survey. Their university connections and their personalities allowed them to draw upon chemists, biologists, bacteriologists, physicists, and instrument makers (Mortimer 1956). The multi- and interdisciplinary collaboration they encouraged was unheard of at the time and gave them a broad perspective on limnology. With their associates, they built many new instruments and developed a wide variety of techniques. Their experience with many lakes that differed greatly in size and morphometry and that lay in geologically varying basins made them deeply suspicious of the possibility of extrapolating from the relatively few non-humic lakes—such as those accessible to Thienemann and coworkers in northern Germany—to lakes in general. By examining many lakes, they were able to distinguish between what they called *autotrophic lakes* and *allotrophic lakes*, among others. The autotrophic type derived organic matter mainly from lake primary production and the allotrophic type from both the drainage basin and within lake processes. They also recognized differences between lakes without stream inflows or surface outflows (**seepage lakes**), which receive most of their water and nutrients directly from the atmosphere or via groundwater, and lakes with a larger catchment area able to nourish in- and outflowing streams (**drainage lakes**) that in turn provide much of the water and nutrients derived from the land. Birge and Juday were far ahead of their time in recognizing the existence of both internally **autochthonous** and externally **allochthonous** derived carbon sources underpinning lake food chains.

Birge and Juday also differed from their European colleagues in using a wide array of variables to characterize the lakes they sampled. They recognized that lakes lie along a complex variety of physical, chemical, and biological gradients (axes) that prevented the simple and simplistic European categorization from being very useful. Birge and Juday developed more quantitative comparative methods (e.g., histograms and plots of concentrations over time and depth) than their European counterparts because they had so many lakes to compare, but they lacked even the simplest techniques for ascertaining the statistical significance of any differences noted. They shared with the Europeans an interest in lakes as super-organismal systems or units (microcosms), and they too had faith that generality would somehow emerge. Their achievements rest, as described by Mortimer (1956),

"On the extent, frequency, and detail of their observation which enabled them to arrive at more penetrating and balanced interpretations than more superficial studies would have done. They were not summer-vacation limnologists; their approach was opposite of dilettante. They were by no means adverse to speculation; but first of all they assiduously collected the facts."

Birge in particular, appears to have realized that in order to see patterns among many lakes he had to bypass the possible complexities as much as possible, and seek what Mortimer (1956) called **integrating properties:** the properties "that record the end result or the sum of the combined influences of an unmanageable number of large factors." Birge and Juday used the distribution of dissolved oxygen, which is the outcome of a host of mechanisms and interacting factors, to describe the functioning of lakes. Changes in carbon dioxide and organic matter content were other integrating factors, or **surrogate properties** they subsequently used to characterize lakes seasonally and vertically. In 1928 Juday, Birge, and coworkers added phosphorus, a major plant nutrient, to the list. This element has become the principal proxy used to predict aspects of lake functioning (Chapters 8, 9, 17, 21).

It is easy now, seventy years later, to recognize major weaknesses in the approaches of Birge, Juday, and most of their contemporaries. The absence of clearly articulated hypotheses, the near absence of statistical tests, lack of recognition of the importance of replication, and the 19th century notion that the data would speak for itself resulted in too little interrogation and interpretation of their abundant data.

Hutchinson

The person who dominated limnology and ecology more than anyone else between about 1940 and 1980 was G. Evelyn Hutchinson (1903–1991). He recognized and appreciated the contributions made between the two World Wars, but being the product of a more modern era and different scientific culture, he was very much aware of the need for a theoretical underpinning of research and the posing of specific questions. Unlike his major limnological predecessors, he was capable of, and interested in, applying mathematics to biology. Hutchinson recognized the importance of both modeling and statistical analysis to integrate data more closely. Born and educated in England, with a love for music, Latin, embryology, limnology, ecology, and mathematical modeling, he was the antithesis of Birge and Juday in approach and outlook.[4] He worked in South Africa, Indian Tibet, and on lakes in the western and northeastern US, but he is best known for important studies on lake metabolism, biogeochemistry, paleolimnology, lake classification, and phytoplankton diversity, and for his monumental *Treatise on Limnology* in three volumes (1957–1976). A fourth volume (1993) based in part on notes he left behind, appeared posthumously. In the first three volumes he synthesized and reviewed major aspects of limnology. Given the rapid growth of this field and the increased specialization of its practitioners, it is unlikely that anyone will ever attempt to produce such a synthesis again. Often surrounded by good graduate students and associates who clearly benefited enormously from his original ideas, his inquisitive mind, and his wide-ranging interests Hutchinson has had an enormous influence on limnology and ecology as a whole. He shared with his eminent predecessors an interest in lakes as functioning units, but during the most influential portion of his career, he too paid little attention to the influence of events beyond the waterline.

Components and the Whole

Hutchinson differed fundamentally from his great predecessors in Europe and North America in approach. He did not consider lakes as systems to be understood as a unit. Instead, he was hopeful that it would be possible to understand the system components (life history, population dynamics, competition, feeding, and so on of species and their interaction with other species) and assemble these components with the ultimate aim "to try to build up the whole from them" (Hutchinson 1964). In this, he and his contemporaries had to assume implicitly that the *whole* (the community property of interest) is the sum of the component properties. For this to be true, interaction and feedback between populations of different species has to be

minimal. If, as the result of interactions and feedback, community properties differ appreciably from the sum of individual-species properties, the whole cannot be constructed by a summing of the components.[5]

Hutchinson's view that the structure and functioning of lakes and other systems might reflect the properties of the components (species and population attributes) was particularly stimulated by work of the mathematician Volterra on competition and predator–prey relationships (Hutchinson 1979). His simple equations were used to describe the competition between two species for limited and limiting resources that were thought to be responsible for structuring communities. The idea of competition for resources as the major structuring force in nature and the one controlling the evolution of communities—the "balance limits of nature" notion—had its roots in earlier ideas from Charles Darwin and his contemporaries on the existence of an equilibrium between resources and population size. The notion that biotic interactions are the primary determinants of the structure and functioning of communities became a central tenet of ecological thinking for more than 50 years. The idea that biological interactions below the waterline were of overriding importance in determining the behavior of lakes fitted well with the earlier idea that lakes are functional units, and downplayed the importance of external factors (drainage basin attributes and weather).

There are, however, few really major ecologists who can be easily pigeonholed as primarily interested in studying inland waters either as integrated systems (Thienemann, Birge, and Juday) or through an assembling of the components (Hutchinson). Until at least the 1950s, Hutchinson was also interested in whole-lake properties. His development of an empirical relationship between dissolved oxygen consumption in hypolimnia and lake productivity as well as his work on the "intermediary metabolism" of lakes (Hutchinson 1941), in which lakes were seen as having super-organismal attributes, attest to his broad scope as a scientist. His career as a researcher ended not long thereafter, when he turned primarily to the development of ideas about the determinants of community composition and the development of ecological

---

[4]In his autobiography, Hutchinson (1979) refers to the puritanical outlook and conservative approaches of Birge and Juday after he spent a week at one of their Wisconsin field stations. "I had learned a fabulous amount about limnological technique but had come away with two feelings of dissatisfaction. One was that it would be nice to know how to put all their mass of data into some sort of informative scheme of general significance; the other was that it would be nice to have either tea or coffee, without seeming decadent and abnormal, for breakfast. I now suspect a connection."

---

[5]"It is simply not possible to understand the properties of complex systems by looking just at the parts, because now we have to understand not only the parts themselves but also the interactions between them." (Harris 1999)

theory based on evolutionary principles articulated by Charles Darwin. While pondering and writing about components, Hutchinson was able to conceptualize elegant and logical schemes about lakes as systems of interacting species or trophic groupings, linked to their physical/chemical environment. His conceptual ideas have formed the basis for work used by many others.

Hutchinson made good use of Birge and Juday's data in the formulation of his hypotheses, and in writing Volume I of his *Treatise on Limnology* (Hutchinson 1957). His interest in the components of systems and the possibility of assembling them was stimulated by the rapidly growing amount of information on lake physics, chemistry, and biology, increasingly sophisticated instruments, and the development of mathematical models and statistical techniques that made studying components much more rewarding. His interest in species and species interactions, and in physical and chemical processes below the waterline grew out of dissatisfaction with the slow progress being made in limnology at that time by the dominant school of American limnologists.[6] That school, which practiced an observational science based on sample collection and analysis of field data, lacked interest in ecological (evolutionary) theory, mathematical modeling, and hypothesis testing. Hutchinson's approach to the study of lakes was clearly the antithesis of that practiced by Birge, Juday, and others during the first half of the 20th century in North America and, to a smaller extent, in Europe as well. It is not at all surprising that when Professor Juday was asked to comment on the suitability of Professor Hutchinson for promotion he wrote, "Professor Hutchinson has some very good ideas and, if his mathematical treatises were based on much larger amounts of observational data, his contribution to limnology would be much more valuable in our opinion." (Hutchinson 1979).

Raymond Lindeman's milestone 1942 paper on energy flow in a single lake was based on a mathematical foundation largely developed by Hutchinson (Cook 1977). That paper was a major early attempt to explain the functioning of the whole through an appreciation of major aggregated components (trophic levels). This new approach conflicted with the established view that a search for pattern among systems

was the best way to make progress. The elderly and eminent Chancey Juday was consulted to evaluate the merits of the Lindeman manuscript when it was first submitted for publication. He was unimpressed with the paucity of the data, the mathematical modeling, and attempts to extrapolate from the few components studied, among thousands possible, to gain a general picture of how lakes might be organized.

Despite the objections of Juday and P. S. Welch (the 1935 author of *Limnology*, the first North American textbook on the subject), the Lindeman paper was published; its publication signals, more than anything else, the ascendancy of scientists preoccupied with biological mechanisms, processes, and rates that allow species to prey upon and out-compete others. This paper marked the beginning of the end of dominant philosophies or paradigms[7] in Europe and North America. Those concerned with the components of lake systems, their interactions, and ultimate assembling into a whole almost totally dominated limnology (and ecology) for the next 30 to 40 years. The question of how it could be possible for so many species to co-exist in the plankton ("The paradox of the plankton," Hutchinson 1961) has probably directly or indirectly preoccupied the professional life of more aquatic ecologists than any other, and has stimulated much important autecological and population level research on individual plankton species, as well as on the competitive interactions that are thought to regulate ecosystem dynamics.

2.5   Limnology after World War II

The post-war period heralded an unprecedented rapid growth of limnology starting in the mid to late 1950s (Table 2–1). Its growth was fueled in the western world by increased funding, the opening of new universities and research centers, and the development of much better analytical (electronic) and sampling

---

[6]Thomas Kuhn's (1970) analysis of scientific change argues that new approaches are accepted only when the traditional avenues have led the researchers in a field to a dead end. The crisis comes about when they realize that the old approach no longer raises interesting questions.

[7]"Science is arranged in a number of overlapping invisible colleges each made up of scientists with similar interests and outlook on their field" (D. J. De Solla Price 1986). Those who belong to a particular college share a philosophy or paradigm (T. Kuhn 1963) and thereby share "The body of theory accepted unquestioningly by a scientific community and the entourage of beliefs and behaviour patterns that accompany the theories." (Rigler 1982). The body of theory and the necessary assumptions associated with a particular paradigm are virtually never articulated because they are self evident and acceptable to the members of a college or scientific tradition.

equipment that was the result of technical advances made during the war. Isotopes became readily available as tracers for measuring energy and nutrient fluxes, and the increasing availability of computers, starting in the 1960s, greatly facilitated statistical analyses and other types of modeling. Increased ease of travel now allowed much more research on polar, tropical, and saline lakes, making limnology somewhat less a temperate-zone enterprise. This period of unprecedented growth—and the increasing number of limnologists in Europe and North America—was fueled by environmental concerns in the 1960s and 1970s; coming to an end in most countries toward the late 1970s or the early 1980s.

Not surprisingly, during the first three decades after World War II many of the major advances in examining aquatic systems from an ecosystem perspective were made in economically strong North America rather than war-ravaged Europe or Japan. Major studies included (1) work on the metabolism and structure of rivers by H. T. Odum and associates, US, (2) linking stream functioning with catchment attributes (H. B. Hynes and students, CA, and G. E. Likens and associates, US), (3) development of models predicting the response of lakes to events on their catchments (R. A. Vollenweider, CH, CA), (4) research by W. T. Edmondson (US) on the recovery of Lake Washington following the removal of wastewater, research that greatly influenced lake management, and (5) the response of whole lakes to experimental changes in nutrient and contaminant inputs by D. W. Schindler and associates (CA). With the exception of the work by Hynes, all of the projects were large-scale, long-term empirical studies funded for work with a direct application. Even so, the studies have had a major impact in limnology and ecology, and stimulated more narrowly focused research on system components and underlying mechanisms.

In the period from 1960 to 1980, there were major changes in how limnology was carried out. Until then, projects had been conceptualized and researched almost exclusively by solitary scientists, sometimes assisted by one or two graduate students or a technician. During those two decades, and stimulated by funding made available for the International Biological Programme (Rzoska 1980), multidisciplinary research teams were established in economically strong countries. The teams were able to pose and tackle much larger questions than was possible before. Studies of whole lakes or streams and their catchments became possible. The resulting blossoming of *ecosystem ecology*

then led to much greater insights into links among functional components of aquatic systems. The large data sets provided impetus toward the development of sophisticated quantitative models that describe many components of the systems under study. There is, however, an ongoing and inconclusive debate about whether limnology, and ecology in general, is better served in the long term by making more funds available to individual scientists or teams. Whatever the merits of the team approach, most scientists continue to work alone or in small groups and most research continues to be carried out in a non-ecosystem context concerning small portions of the overall limnological pie.

Fundamental and Applied Science

The retrenchment or slowing of growth of limnological findings in the western world during the 1980s and 1990s has been painful, although world limnology remains stronger today than it has ever been. The 1960s and 1970s also saw the beginning of research funding shifts away from mainly fundamental (basic) research, where direction and questions are chosen by the scientist. Increasingly, governments push research toward more applied problems, leaving a smaller fraction of funds available for basic research. Whereas close to 100 percent of funding for limnological research in the US had been for fundamental research, this is now reduced to about 20 percent, albeit of a much larger pie (Lewis et al. 1995). The impetus for this shift has been environmental concerns, starting with the nutrient pollution of temperate lakes and rivers by sewage and agricultural fertilizers in the 1960s and continuing with concerns about acid rain, and most recently spurred by concern regarding organic contaminants, toxic trace metals, and climate change on lakes, rivers, and wetlands. These concerns have not only directed much more money towards research with specific goals and environmental management but the more applied research has contributed a great deal to fundamental limnology. Recent concerns about human impacts on the global ecosystem and the response of aquatic (and terrestrial) systems to large-scale perturbations may be shifting funding allocations back slightly toward more basic research.

The primary purpose of Table 2–1 is to show that, today's limnology is based on a long history of development. While the present era is unsurpassed in terms of equipment, ease of travel, technical backup, computing power, and development of large databases, most limnological concepts were developed prior to World War II and some prior to World War I. It is the availability of sophisticated equipment, emphasis on quantification, and the ability to mathematically model the data obtained, together with an increased blurring between fundamental and applied research, that primarily distinguishes the present science from the earlier one. As part of this quantification there have been tremendous advances in analytical techniques that allow limnologists to detect compounds and ions in much smaller concentrations than was hitherto possible. For example, the detection limits for toxins shifted from the parts per thousand (ppt or gl^{-1}) range in the 1950s, to parts per million (ppm or $mg1^{-1}$) during the 1960s. The median detection limits rose from the $mg1^{-1}$ range in the 1970s, the parts per billion (ppb or μg^{-1}) scale in the 1980s, and have reached the parts per quadrillion (ppq or pg^{-1}) range in the 1990s for contaminants such as extremely toxic dioxins (Chapter 28). New technologies that allow a radically new approach to a recognized limnological problem coupled with the increased precision of measurements continue to fuel major advances in limnology. Technological advances and new insights garnered will inevitably force a reevaluation of cherished ideas and beliefs. First in related, then in more distant areas of limnology until they cascade through the discipline as a whole.

▲ The influence of Hutchinson and his school lives on, but the underlying paradigms have been modified. The notion of competition for resources as the major structuring force in the evolution of natural communities (Sec. 2.4) is increasingly called into question—especially during the last two decades. Disturbance can keep communities in nonequilibrium states with population densities too low for resources to be a limiting factor.[8] Contemporary biologists interested in aquatic populations and communities largely support the idea that predation rather than competition for resources is the dominant nonequilibrium force in structuring communities. The idea that seasonal and short-term (nonequilibrium) changes in stream discharge have a direct effect on the substrate and biota is similarly affecting research on flowing-water (lotic) systems (Reice 1994). Even so, the many adherents of the competition paradigm and the nonequilibrium predation paradigm

agree on the importance of species interactions as the dominant force in structuring communities.

The assumption that species interactions are the prime determinants of community structure was challenged during a massive research effort on the effect of plant nutrients (phosphorus in particular), productivity, community structure, and management of northern temperate-zone lakes. The idea that the nutrient (resource) supply from drainage basins has an important effect on the community structure of oligotrophic lakes has become widely accepted today. This conflicts in some ways with the autecology/population/trophic-level-based paradigm of biological interactions between species, as articulated by G. E. Hutchinson and those inspired by him.[9]

The more applied proponents of the phosphorus paradigm (Sec. 2.6) in the 1970s and 1980s saw the lake as a kind of biological reactor, or even once again as a superorganismal unit, characterized by nutrient inputs and outputs that largely determine both the biomass and the community structure of the plankton, and indirectly, developments at higher trophic levels. Emphasis on aquatic systems as open systems—with productivity and community structure determined by *externally* derived nutrients, physical forces, and system morphometry—naturally conflicts with the views of scientists who focus on species and trophic interactions below the waterline as being the driving structural force in nature.[10]

A Synthesis

It is today widely accepted that externally derived resources primarily determine community production rates and the maximum seasonal biomass, but food-web manipulations in whole lakes, rivers, and enclosures have

---

[8]"The common condition for most communities is to be recovering from the last disturbance. Only when the return interval for the next disturbance is long, relative to the lifespan of the resident species, can something approaching equilibrium be attained. I argue that this nearly never happens." (Reice 1994).

[9]Edmondson (1990) notes that in the first approach mentioned above ". . . The appropriate unit of measurement is the individual, and one is concerned with the activities of individuals in groups. This leads to concepts of populations, population dynamics, population genetics, and the ways in which populations of different species interact and form communities. People interested in evolutionary ecology usually think this way." The second approach "treats aggregations of species into groups according to function . . . This approach leads to concepts of ecosystems, nutrient cycling, energetics and production."

[10]"There is clearly a continuum between population ecology and ecosystem ecology, but there is a tendency for research projects to be clustered towards the two ends of the scale, creating the impressions of a divergence, both conceptually and in people's assessment of what is useful." (K. H. Mann 1988)

also decisively illustrated the importance of food-web structure and species interactions in determining community structure. There is still much debate and disagreement over the relative importance of resource availability (nutrients, light), versus biotic interactions (predation, competition, disease) as determinants of community structure.[10] The purposefully provocative charge by adherents of biotic interaction paradigms that there is a chemical bias in limnology today (Persson et al., 1988), was a response to the rise of the nutrient paradigm. Most adherents of the biotic interaction paradigm accept the paramount importance of a "top-down," predator-mediated (population-based) impact on ecosystem structure and function, whereas those who stress that community/ecosystems function and structure are primarily determined by resource availability (nutrient limitation) have a "bottom-up" bias. The undoubtedly correct intermediate position is that top-down control is superimposed on the potential productivity determined by nutrient supply (Carpenter et al. 1985). Unfortunately, this recognition is of little practical help in synthesizing the various models.[11]

The present chapter has been a history of some major paradigms that have marked the growth of limnology. An awareness of the existence of differing views on how best to study aquatic systems and their modeling will help undergraduates approach textbooks and scientific papers with a greater appreciation for the conceptual underpinnings of the research.

2.6   Scales and Patterns: A Conceptual Exploration

Limnologists differ greatly as to the level of biological organization they consider most interesting and worthy of study. However, it is increasingly recognized that questions posed at different "levels" or "scales" of biological (hierarchical) organization (e.g., cell, organism, population, community, ecosystem, catchment), and further posed at different spatial, and temporal scales are best studied with the tools and approaches shown to be effective at that particular level. This

recognition implies an acceptance that there is no single approach to the study of inland waters. While such an acknowledgment should reduce unnecessary conflict between adherents of different paradigms, it does not itself facilitate the integration—except conceptually—of research carried out at different scales. The three examples below illustrate the importance of scale.

Time Scales

▲ The dynamically modeled (but not observed) relationship between phytoplankton and zooplankton (Fig. 2–3) shows that the correlation, as well as the imputed "cause-and-effect" relationship between the production of the phytoplankton prey and their presumed zooplankton predators changes with the **temporal** (time) **scale** employed. The relationship shown was negative over certain time intervals, positive over other intervals, and insignificant over yet others. Depending on the sampling interval, one could reasonably conclude that the observed negative relationship was the result of zooplankton predation on the phytoplankton. Conversely, the positive relationship noted over a different time interval may point to the positive effect of nutrients excreted by the zooplankton on the growth rate of the phytoplankton. Alternatively, it

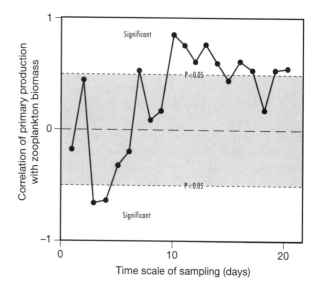

Figure 2–3 Pearson product–moment correlation coefficient of primary production with zooplankton biomass vs time (days) for a time series derived from a model. Fine dashed lines denote upper and lower statistically significant correlations ($P < 0.05$); middle line denotes zero correlation. *(After Carpenter and Kitchell, 1987.)*

---

[11]"A real synthesis is difficult, simply because the problem is one that bridges hierarchical no-man's land across which both methods and goals change radically. Population ecologists normally deal with relatively simple, definable systems and make models of high predictive value, while ecosystem ecologists [interested in dynamic modelling] deal with systems of high complexity and make models that are descriptive but highly condensed and heuristic rather than predictive." (L. R. Pomeroy 1991)

Figure 2–4 Doubling times for some marine and freshwater organisms as a function of equivalent body diameter. Phytoplankton, P; macrozooplankton, MZ; macroinvertebrates, MI; fish, F; bacteria, B. *(Modified after Sheldon et al. 1972.)*

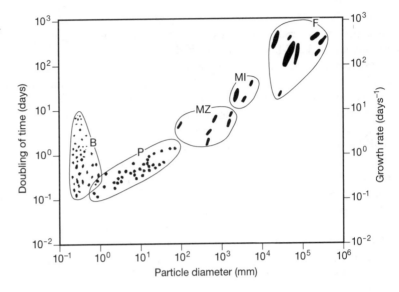

could be interpreted as proof that zooplankton did not feed much on the algae. There is ample evidence in the scientific literature for all three scenarios in nature, but it is unclear to what extent these reports are the result of the particular time scale examined. Sampling on days when the phytoplankton–zooplankton relationship was insignificant allows yet other interpretations.

The time scale selected is usually based on the time available to the investigator and tradition. Most limnological studies of plankton in single lakes appear to sample at weekly or biweekly intervals in summer. This interval usually has little or no bearing on the life span or growth rate of the organisms studied (Fig. 2–4), or the time span over which the relevant environmental factors operate. Thus the sampling interval and number of replicates (Table 1–1) raise major concerns about the validity of the conclusions.[12]

Spatial Scales

The effect of the **spatial scale** on conclusions drawn is possibly even larger. For example, climatic factors (annual irradiance, length of growing season, mean air temperature) tend to be the most important determinants of the among-lake biomass or species composition of, for example, large aquatic plants (macrophytes) on a *continental* and *global* scale (Fig. 2–5). This is because climatic factors exhibit enormous variability (large data range) at this spatial scale. Similarly, patterns of runoff and the hydrology of streams and rivers differ greatly over global and continental scales (Fig. 8–4). Climatic factors are less important at a *subcontinental* scale (e.g., eastern North America, western Europe) where other differences are observed among systems, including lake morphometry, river discharge (catchment morphometry), turbidity, or nutrient concentrations.

Over a smaller spatial scale, such as a *single lake district*, the rivers, lakes, and wetlands experience relatively little variability in climate, age, catchment soil development, mode of creation, and aspects of lake morphometry. Here other factors, such as water chemistry (geology), steepness of catchment slopes with their effect on stream velocity, and underwater slope (morphometry) of lakes vary disproportionately and typically assume the principal role in determining macrophyte distribution (Chapter 24). Within *single systems*, differences in water chemistry or transparency are usually minor, allowing yet other environmental factors to express themselves and exert the greatest influence. Important environmental factors at this scale include water depth, which may serve as a surrogate for the light received by plants growing at different depths, perpendicular to shore. The light affects the dominant

---

[12]"Attempts to infer process from pattern are fraught with the usual pitfalls associated with the interpretation of correlations. While an experimental approach goes far in side-stepping this pitfall, experiments may be difficult, expensive or impossible when the patterns and processes of interest apply to systems of large scale, such as entire lakes, catchments, or landscapes." (S. G. Fisher 1994)

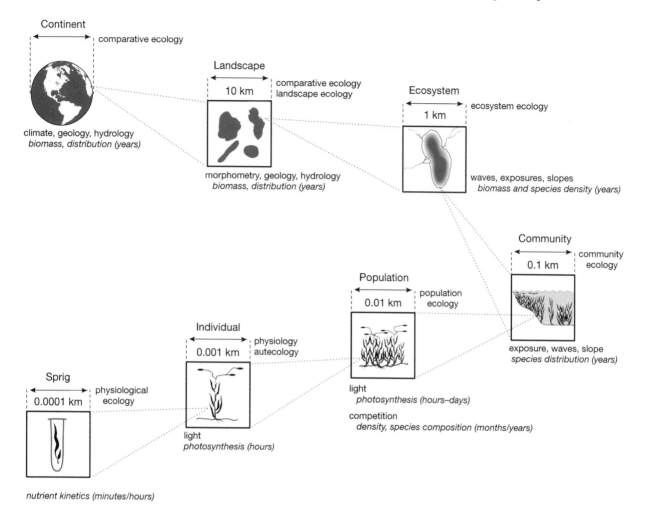

Figure 2–5 Patterns in macrophyte spatial and temporal scales, the subdisciplines that address them, and associated changes in the most important environmental determinants.

plant growth form (tall vs short species), maximum possible growth rate, and determines the depth where the largest plant biomass is to be expected (Chapter 24).

Where the research is carried out over an even smaller scale, at a *single depth* within a single system over which the light varies little, different potential determinants of, for example, the macrophyte biomass or community structure (species composition) are able to express themselves most strongly. At this scale the differences encountered in the sampled quadrants, in turbulence and sediment characteristics, the life history of different species, or competition for light between tall- and low-growing species might become

important determinants of plant abundance and distribution[13] (Chapter 24).

When the spatial scale is further reduced and the determinants of variation in the growth of individuals of a *single species* inhabiting one or few quadrants at one site at a single depth is examined, differences linked to variation in turbulence or sediment characteristics are

---

[13]"It is quite likely that aquatic ecologists have an inflated impression of the importance of biotic interactions. I say this because most small, temporal, taxonomic, and environmental scale studies overemphasize the influence of interspecific interactions by ignoring abiotic variation, by accentuating the magnitude and duration of biotic interactions, and by studying a few species in isolation." (Hinch 1991)

likely to be small. At this scale, the age and vitality of the individual plants, abundance of predators, and/or daily variation in weather best explain differences in observed growth rates.[14] Finally, in experimental flasks—where it is possible to minimize environmental variability and biological interactions—factors imposed by the investigator, such as variation in nutrient supply, light history, or respiration of the experimental twigs used, play an important and easily measured role in growth. The difficulties encountered in interpreting and integrating studies carried out over *different* spatial scales are compounded by the fact that changes in the time (temporal) scale of studies usually accompany changes in spatial scale (see Figs. 2–3 and 2–5).[15]

It is evident that changes in time and spatial scales are associated with changes in the mix of environmental variables. Important variables at one scale may or may not be important at the next scale in a nested hierarchy of scales.[16] Changes in scale are accompanied by (1) changes in the type of determinant or "cause"; (2) the type of measurements needed; (3) the techniques employed; (4) the most appropriate type of modeling; and (5) the specific subdisciplines of ecologists who find questions posed at one particular scale most attractive. Note that the importance of biological interactions becomes increasingly more visible as the spatial and temporal scales of the research decrease.[17]

---

[14] • "The problem of scale predetermines our questions, techniques, results and conclusions. Large-scale work does not necessarily lend itself to small scale understanding" (P. K. Dayton 1994)

• "Research workers should, from this time forward, try to think explicitly about whether the proposed scale of their next project is appropriate to the solution of the problem that society funds them to investigate." (K. H. Mann 1995)

[15] "We may understand process-level interactions for a leaf or a 2×2 m plot very well, but which processes are critical for a watershed-based model?" (G. E. Likens 1998)

[16] • "The proper first answer to the question `What controls algae?' . . . is, it depends on the scale." (Fisher 1994)

• "Processes operate at a number of levels and at a number of scales so it is absolutely necessary to be clear about what processes and scales are being studied and to make sure that interference from processes at other scales can be eliminated." (Harris 1994)

[17] • "The problem must dictate the methods to be employed for its solution, not the reverse." A. C. Redfield, 1958.

• "Every piece of science—experiment, lecture, paper, or treatise—should include an answer to the fundamental question: 'What is the purpose of this work?' For the proponent, clarity of purpose directs the design of the study- and the evaluation of the whole. For the granting agency, editor, and referee, clear statements of purpose are essential for an evaluation of the intentions behind a given work and the success of the finished piece in achieving those aims." (R. H. Peters et al. 1996)

Temporal Scales

There is not only a hierarchy of spatial scales along which the most important causal factors and conceptual/technical approaches change but also a hierarchy of **temporal scales** over which a process or mechanism can be examined. For example paleolimnologists, who use lake and wetland sediments to chronicle the history and evolution of biota over time (Chapter 20), are contributing a great deal to understanding the importance of long-term changes in climate or land use on sediment accumulation and the biota (Fig. 2–6). Over shorter periods of a decade or so, interannual differences in weather (sunshine, wind, runoff, ice cover) are major determinants of how much organic matter and nutrients reach a stream before being carried to a lake or wetland, how much organic matter is produced and, presumably, how much organic matter is sedimented. In studies carried out over periods of days or weeks, the variation of observed sediment accumulation may be the result of the sedimentation of a particular algal bloom, the resuspension and transport of sediments following a period of wind-induced turbulence, the seasonally changing abundance or activity of plankton and bottom-dwelling invertebrates, or as the result of particles carried in from the drainage basin after a large storm.

Range Scale

The **range scale** or **degree of resolution scale** is also important. In the same way that the changing effect of smoking on human health is most clearly evident when studies encompass a wide range in the number of cigarettes smoked among individuals, the effect of different environmental factors on macrophytes or other organisms in nature require considerable variation (often orders of magnitude) of the predictor variables examined. The resolution (range) needed for the chosen variable (nutrients, predation, toxins, etc.) is naturally much smaller under controlled conditions precisely because other potentially confounding variables are purposefully held constant, something not possible or desirable in nature.

The well known relationship between observed phosphorus concentration and algal biomass (Fig. 8–16) or the relationship between the solubility of organic toxins in water and the extent to which they accumulate in the biota (Fig. 28–9) is only evident over a wide concentration range (of phosphorus, solubility, etc.). This is because only over such a wide range is the resolution (importance) of predictor variables not

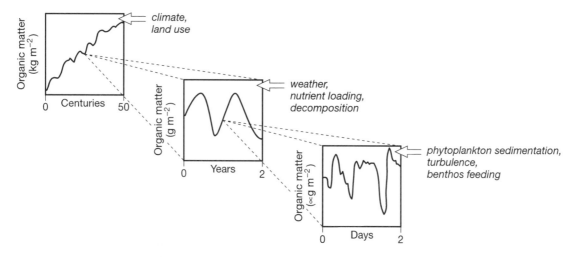

Figure 2–6   Changes in the dynamics of a lake-sediment system with changes in time scale. On a scale of centuries, the impact of climate or forest cutting is evident, while seasonal decomposition processes are apparent on a scale of years. An observation window of days reveals rapid fluctuations in sediment accumulation due to resuspension and changes in plankton sedimentation. *(Modified after Sollins et al. 1983.)*

confounded by any one of multitudinous environmental factors that also affect the algae or contaminant uptake in nature. In experimental work, the strength of manipulations has a determining effect on whether a clear response can be expected. Different conclusions about the effect of a particular environmental variable on the biota are frequently the result of an insufficient appreciation of the effect of the resolution scale used.

Conclusions Drawn

The interpretation of "cause" for any pattern observed is highly dependent on the spatial, temporal, or resolution scale examined. As pointed out by Fisher (1994), results and interpretations made at one scale may be inappropriate or wrong at another. This is still not widely recognized, and is a source of unnecessary conflict between scientists belonging to different subdisciplines or paradigms working at different scales and who prefer to model the data using different approaches.[18]

There is no "best" scale for enquiry. The importance or quality of research has nothing to do with the

particular scale at which it is carried out and everything to do with the quality and clarity of particular questions and incisiveness of the answer. Disagreements between those who favor analytical rather than empirical modeling approaches toward the behavior of ecosystems (Table 2–2) usually reflect differences of opinion about the most interesting scale. The disagreements between Juday and Hutchinson (Sec. 2.4) were largely based on differences of opinion about the most promising spatial and temporal scales.

Those limnologists primarily interested in large-scale comparisons tend to favor statistical modeling and are usually satisfied with *seasonal* or *annually averaged* responses of highly aggregated properties (e.g., phytoplankton biomass, fish yield, phosphorus release) that are examined among aquatic ecosystems in a single climatic/geologic zone. At the particular scale, differences in nutrient supply, stream hydrology, and lake or wetland morphometry assume major importance. In contrast, those scientists primarily interested in mechanisms, processes, and interactions between populations focus mostly on single systems and examine them experimentally over short-term scales (hours or days) where mechanisms and processes are least confounded by extraneous enclosure effects or environmental changes. The associated modeling usually explicitly considers time (see dynamic simulation modeling, Sec. 17.8). It is not surprising that the two groups have different goals, work

---

[18]"Some of the most vociferous disagreements among ecologists arise from differences in their choice of scale. Those who study the long-term dynamics of a particular community at a single site often reach very different conclusions from those reached by individuals conducting short-term studies of similar communities at many different sites distributed over a large area." (J. Wiens et al. 1986)

Table 2–2   **Approaches to constructing models of environmental problems.**

Labels	Definitions
Data-based/Theory-based approaches	*Data-based approach*: a simple model derived solely from data collected from the particular system of interest.
	Theory-based approach: adapting an existing model to the system of interest and incorporating as detailed a description of that system as the art of ecological theory allows.
Black-box/White-box approaches	*Black-box approach*: a model designed solely as a quantitative description of how change in input variables affects output of the system.
	White-box approach: a model dealing explicitly with the causal relations linking system inputs, internal system states, and system outputs.
Holistic/Reductionist approaches	*Holistic approach*: a model that explicitly includes properties of the system working as an integrated unit.
	Reductionist approach: a model incorporating as many details of the system as possible to fully capture its behavior.
External/Internal approaches	*External approach*: observing (describing) some integral behavior of the system of interest and then trying to understand how that behavior is brought about.
	Internal approach: assembling known behavior of individual subsystems to reveal something useful about the overall behavior of the system of interest.
Empirical/Analytical approaches	*Empirical approach*: a model whose sole function is to predict future states of the system of interest.
	Analytical approach: a dynamic model which not only predicts future states but also embodies a detailed explanation of why the system of interest behaves as it does.

Source: After Shuter and Regier 1989.

at different spatial and temporal scales, use different kinds of models (Table 2–2), and arrive at different conclusions about the most important determinants of system functioning, leading to disagreements about the feasibility of scaling up from short-term studies of components to the long-term behavior of whole systems. Human concerns about freshwater resources are almost always expressed at the whole-system scale, while conservation issues have most commonly been addressed at the population scale.

At present, most quantified generalities (or patterns)—usually with only modest predictive power in terms of the behavior of particular systems—that have emerged in limnology (and other sciences) are the result of deliberately ignoring much of the inherent complexity. Most generalities are based on linking *averaged* values of important environmental determinants (e.g., nutrient concentrations, mean depth, average river discharge) to *time-integrated* measures of some simplified biological response (e.g., annual/seasonal production among systems or functional groups of organisms). In the process, complexities associated with changes over time and space have been lost (re-scaled) and species aggregated into collective assemblages (Neill 1994). Whatever the degree of aggregation desired, the particular scale selected has a major effect on the ways in which communities and systems are perceived and modeled.

Scientists and Scales

It is apparent that many scientists believe the only way to understand aquatic systems is through assembling detailed mechanistic studies of species, but there is wide disagreement as to the likelihood that the resulting ecosystem-level models will ever have predictive power (C. J. Walters 1998). Any such predictive models would require an a priori knowledge of the relative and absolute importance of the most important mechanisms at the temporal and spatial scales at which they operate.[19] However, at the organism level, there are promising developments in the predictive modeling of groups of organisms on the basis of their form and size-related metabolism rather than species composition. Even so, in a thoughtful and provocative discussion, Steel (1997) concludes that "good predictions and explanations are not likely to be simultaneously achievable." The contentious issue of appropriate modeling approaches is developed further in Section 17.8.

This book emphasizes empirical (statistical) patterns (Table 2–2), usually based on aggregated variables, to provide a broad quantified framework into which research at less aggregated scales can be positioned. The generalities presented serve not as an endpoint but merely as one step toward research that is able to encompass more of the actual complexity of natural systems.

A better understanding of the complexity of natural systems, which requires additional research at scales ranging from species biology to ecosystem function, answers the deeply felt human need to link effects to their underlying causes and mechanisms, and provide explanations for processes and population interactions in nature. The synthesis of information on components into dynamic (analytical) models (Table 2–2), used not only for simulation (interrogation) purposes but also for prediction (extrapolation), is a long-term goal.

Before ending this section, it is worth reemphasizing that today's relative abundance of quantified and roughly predictable generalities about aquatic systems and their biota, largely based on the physical/chemical and aggregated biological attributes of inland waters and their drainage basins, is not so much a reflection of the greater importance of such generalities but rather reflects the current state of limnological development. More quantified generalities have emerged from work on aggregated (averaged) scales than at the relatively short-term and site-specific scales over

which dynamic reality is least confounded by changes in the physical/chemical environment.

Challenges

The greatest challenge for limnology is to integrate excellent research carried out at different scales and develop predictive models that incorporate much more real-world complexity. By doing so, limnology will contribute to the advancement of science and better management of aquatic ecosystems that are increasingly threatened. Another important challenge is to better communicate scientific findings with environmental managers and the public. Presenting environmental findings and conclusions explicitly in a scale context would reduce the number of apparently contradictory findings preventing managers from drawing the appropriate conclusions[19] and that conveniently permit politicians to continue allowing environmentally destructive land use and emission practices. Accepting the first challenge requires wider acknowledgement by scientists that there is no "best" scale and that research carried out at different scales generally asks different questions, uses different modeling approaches, and reaches different conclusions. The significance of research lies not in the scale at which it is carried out, but in the importance and clarity of the questions posed and the incisiveness of the answers.

Highlights

- Limnology came of age in 1901, when F. A. Forel (CH) published the first textbook on the subject.
- Limnology developed rapidly between the two World Wars, mostly in Europe, North America, and Japan; because of its emphasis on fundamental research, the more applied components of limnology (fisheries, waste-water biology, hydrology) went their own way. This was followed by their partial reintegration, during the last 30–40 years.
- Among the most prominent limnologists are A. Thienemann (DE) and E. Naumann (SE) who led a 30-year effort, largely between the two World Wars, to classify lakes into distinct categories. Their great counterparts in North America, E. A. Birge and C. Juday (US), attempted to characterize lakes and lake types along environmental gradients. The descriptive phase ended mainly because of the rise to eminence after World War II of G. E. Hutchinson

---

[19]• "Are results generated for one spatial scale misleading to management when applied to another spatial scale?" (G. E. Likens 1998)

• "Managers [and the public] often misunderstand science and expect it to deliver a truth that is non-arguable." (P. Cullen, 1990)

(US). He stimulated a great deal of research between 1945 and 1980 about, among other topics, understanding the behavior of whole ecosystems through assembling work on biological components (species) and their interactions with each other and their environment.

- The rapid development of applied limnology, starting in the 1960s and 1970s, continues to contribute much to fundamental limnology, but has naturally stimulated much healthy argument about the goals of science.

- ▲ The results obtained and conclusions drawn from any properly designed study is highly dependent on the temporal, spatial, and resolution scales selected. Both fundamental research and environmental management would benefit from a greater recognition of this fact.

3

Water: A Unique and Important Substance

3.1 Introduction

Without water, there would not only be no lakes, rivers, and wetlands but also no life on earth, because both photosynthesis and animal metabolism depend on the availability of water. Its extraordinary properties allow it to take part in and mediate the multitude of physical-demand interactions in soils, aquatic sediments, and water that allows life within it (Table 3–1). As water falls from the atmosphere to the land and enters the soil, it undergoes chemical reactions with the soil and rocks that profoundly modify its composition. When water leaves the soil for rivers, lakes, and wetlands (Chapters 8 and 13), its new composition has a major effect on the abundance of the biota, its community structure, and productivity. The evaporative loss of water (Chapter 5) not only returns water to the atmosphere but affects the concentration and composition of the remaining water and thus its suitability to the biota. In turn, the water that returns to the atmosphere has a major impact on the climate by its effect on atmospheric heat transfer, as well as through the effects of condensation and evaporation on the temperature of aquatic systems and surrounding land.

The physical characteristics of water are responsible for the existence of aquatic systems, and also determine to a very large extent, the time lags exhibited in heating and cooling that, among other factors, affect the stratification of lakes; and are responsible also for formation of ice on the surface rather than the bottom of water bodies (Chapter 11). Furthermore, water characteristics have a major impact on the sinking rate of particles (Chapter 20) and on energy expended by fish and other organisms in catching their prey.

3.2 Characteristics of Water

Water is a compound in which hydrogen and oxygen atoms are covalently bonded, with one oxygen and two hydrogen atoms sharing electrons. The bonding is such that the oxygen atoms maintain a slightly negative charge, and the hydrogen atoms a small positive charge. A positively charged hydrogen atom in one water molecule is attracted to an unshared and negatively charged oxygen atom in another water molecule, forming a hydrogen bond. The result is a weak bond between the oxygen atom of one water molecule and a hydrogen atom of a nearby water molecule. The bonding strength between molecules is only about 1/16th that of the covalent bond between hydrogen and oxygen within a molecule, but is strong enough to cause water molecules to form clusters. The latter are, however, continually broken and reestablished in a random pattern.

Liquid water is a liquid crystal rather than a true fluid. The hydrogen bonding between adjacent water molecules gives water its unusual properties. The complex hydrogen bonding that most other liquids do not have, allows water to maintain its integrity as a liquid to a much higher temperature than compounds such as H_2S or NH_3, which exist only in vapor form at room temperature. The melting point (0°C) and boiling point (100°C) at sea level are also much higher than for other elements placed near oxygen in the periodic table. As is evident from Table 3–1, water differs in nearly all properties from other liquids.

Table 3–1   **Physical and chemical properties of liquid water and their importance in aquatic systems.**

Property	Comparison with other Substances	Importance to Aquatic Systems
Density	Under standard pressure, maximum density is 3.94°C, not 0°C; expands upon freezing	Allows lake stratification, and surface freezing rather than bottom freezing
Melting and Boiling Points	Both properties unusual; very high	Allows water to exist as a liquid
Viscosity	Moderate	Influences the case of water mixing, provides resistance to the movement of organisms, and helps determine the sedimentation rate of particles
Specific Heat or Heat Capacity	Highest of any liquid other than ammonia	Moderates (buffers) temperature extremes
Heat of Vaporization	One of the highest known	Important to heat transfer in inland water and atmosphere
Surface Tension	Very high	Increases the difficulty of surface waves breaking and thereby slowing the rate of heating and cooling in lakes, allows certain insects to walk on water surfaces
Absorption of Radiation	Large in the infrared region, but moderate in the photosynthetic/visible region	Allows greater heat absorption in surface water, but reduced surface absorption at shorter wavelengths, allowing a greater penetration of photosynthetically available radiation
Solvent Properties	Dipolar nature makes it an excellent solvent for salts and polar organic molecules	Important in dissolution and transport of dissolved substances from catchments and the atmosphere to aquatic systems

Source: Modified from Sverdrup et al. 1942, and Berner and Berner 1987.

Density

The density (weight/volume) of water is about 775 times greater than air. Since the density of the biota is close to that of water, it serves as a buoyant medium. Therefore, large aquatic plants and animals do not need the strong stems or skeletons required on land. Free-floating macroscopic plants, such as duckweeds that live at the surface of ponds and wetlands, do not sink. The buoyancy provided by water also makes it easy for mobile organisms to move every which way. The effect of changes in temperature on the density of water affects organisms and nonliving particles not only directly but also indirectly.

When water changes from a liquid to a solid crystal upon freezing, the temporary, hydrogen-bonded lattice structure, with its many twisted or broken hydrogen bonds, is fixed into a more regular crystalline structure, with its molecules arranged in a cleaner fourfold (tetrahedral) pattern (Fig. 3–1) that places them further apart than in liquid water. The frequent groaning or cracking of expanding lake ice is due to expansion of the water surface upon freezing. The greater distance between water molecules means that ice also has a lower density (0.9170 g cm^{-3}) than liquid water at 0°C (0.9999 g cm^{-3}), allowing it to float. As a consequence, organisms living in the denser, warmer water below remain protected from freezing temperatures.[1] The more regular structure is destroyed upon melting and some of the hydrogen bonds are twisted or broken, allowing the resulting nonhydrogen bonded water molecules to come in closer proximity

---

[1]Turbulent streams and rivers can cool to ≤0° without ice forming on the surface, but with *anchor ice* able to form at the sediment–water interphase where turbulence is negligible. Many stream limnologists have slipped and painfully encountered anchor ice.

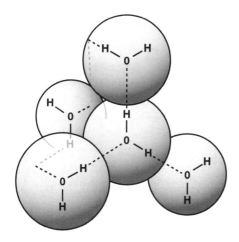

Figure 3–1   Tetrahedral arrangement of water molecules in ice. Each circle (sphere) represents a water molecule *(Redrawn from R. A. Horne 1969, in Berner and Berner 1962.)*

once again. The result is that a given volume of liquid water contains more molecules and has a higher density than the same volume of ice.

Water differs from practically all other substances in being more dense as a liquid than as a solid and hav-

ing its greatest density (0.99997g ml) at 3.94°C, not 0°C, at normal atmospheric pressure. Two opposing forces are responsible for its maximum density above 0°C. The distance between water molecules, and between the molecules of all other liquids and gases, increases with rising temperature. This process alone would cause water to be densest at 0°C. However, the second process involves water molecules' tendency to group together in progressively more icelike arrangements as the temperature approaches 0°C, this alone would cause pure water to be at its smallest density at 0°C. The effect of the two processes together results in the closest packing of molecules at 3.94°C.

The relationship between temperature (*T*, °C) and density (*D*) is nonlinear, (Fig. 3–2a) with the density readily estimated from the equation

$$D = 1 - 6.63 \times 10^{-6} (T - 4)^2 \qquad \text{EQ. 3.1}$$

Seawater behaves differently. Its high salt content and resulting high density (1.025) means that density decreases with increasing temperature, without a density maximum at 3.94°C. Seawater, with a salt content of 3.5 g l^{-1} or 35‰, has its freezing point at −1.91°C.

Density declines with increasing water temperature but as a consequence of the nonlinear relation-

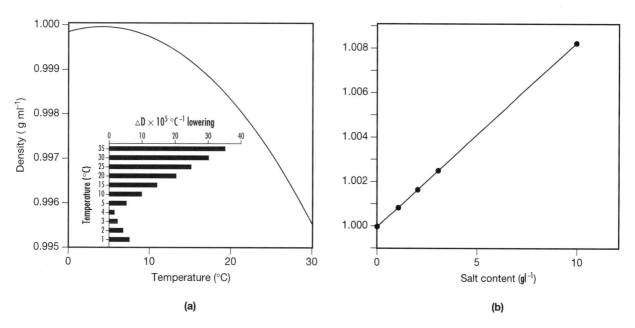

(a)   (b)

Figure 3–2   (a) Dependence of freshwater density (*D*) on temperature (*T*) [$D = 1 - 6.63 \times 10^{-6}$ $(T - 4)^2$], and the changes in density per degree of temperature difference (inset). (b) Dependence of freshwater density on salt content (*S*) in parts per thousand (g l^{-1}), based on salt composition representative of ocean waters. $D = 1 + 0.000816 S$. *[Part (a) modified after Straškraba et al. 1993, and Vallentyne 1974; part (b) based on data from Ruttner 1963.]*

ship, the density difference between pure water at, for example, 10°C and 11°C is 12 times greater than for water between 4°C and 5°C. The resulting curvilinearity in density is responsible for stratification patterns observed in lakes and for all the biological events associated with this (Chapter 11). In a relatively few lakes, the observed density differences are not attributable to temperature, but to major salinity differences between more dense saline water at depth and overlying fresher waters (Chapter 11), or to an elevated density resulting from the inflow of silt-laden and, therefore, denser river water (Chapter 29). The temperature of maximum density declines at a rate of ~0.2°C per g l^{-1} increase in salinity, lowering the freezing point of saline waters. (Fig. 3–2b)

While the temperature of maximum density for fresh water at sea level is 3.94°C, the great hydrostatic pressure in very deep lakes is able to compress the water sufficiently to lower the temperature of maximum density at a rate of about 0.1°C per 100 m of depth. Thus the bottom water of Lake Baikal (RU; Z_{max} = 1,741 m) would, in the absence of turbulence, be expected to be 2.2°C. However, the bottom water is 3.2°C and there is evidence for sufficient turbulence to permit the exchange of some deep water with warmer overlying water (Chapter 12).

The lowered density of water upon freezing results in about 9 percent expansion of its volume. This plays an important role in the mechanical breakdown of rock, known as **physical weathering**, in regions that experience temperatures below 0°C. Water contained in cracks expands upon freezing, forcing the rock apart. Periodic freezing and thawing is effective in shattering rock and exposes more rock surface to dissolution, known as **biological/chemical weathering**, following contact with water. This process is also responsible for winter potholes in paved roads of northern temperate cities subject to freezing and thawing cycles.

Viscosity

Dynamic viscosity is a measure of liquid's resistance to flow and is responsible for the frictional resistance or viscous drag that water offers to swimming fish or sinking algae. The more viscous the substance, the less "fluid" it is. An important consequence of viscosity is that the boundary layer surrounding objects, in which the velocity of water ranges from being identical at the object's surface to independence from it some distance away. Nonturbulent boundary layers reduce the gas and ion exchange rate between organism and surrounding water (Chapter 24).

The frictional resistance encountered by swimming fish or zooplankton at 10°C is about 100 times greater than for organisms living in air. The high dynamic viscosity of water is a particular problem for the smallest motile organisms who, because they have so little momentum, come to an almost immediate halt when they stop swimming. Similarly, the momentum of a food particle swept towards the gullet of a copepod zooplankton is very small, making food capture enormously complicated (Vogel 1994).

In fresh water, dynamic viscosity is largely a function of temperature and is affected little by density, with viscosity declining by a factor of 1.7 as the temperature rises from 5°C to 25°C (Table 3–2). Consequently, dissolved gases and salts diffuse more rapidly, and algal cells or other particles sink 70 percent faster out of the euphotic zone in the summer, when water is 25°C rather than 5°C. Even so, the much higher dynamic viscosity and density of fresh water results in resistance to diffusion that is 10^4 times higher than in air.

The change in dynamic viscosity with a change in temperature is not linear, but instead rises particularly rapidly as the temperature declines below 10°C (Table 3–2). The result is easier water mixing at lower temperatures than higher ones when exposed to winds of equal force. Consequently, viscosity has a significant effect on water movement (Chapter 12) and lake stratification (Chapter 11).

The ratio of dynamic viscosity to density is known as **kinematic viscosity**. This determines how readily fluids flow, steepness of the velocity gradient around an object, and how easily adjacent water layers with different temperatures become turbulent.

Specific Heat

The **specific heat** of a substance is the amount of heat needed to raise or lower the temperature of 1 g of a particular substance by 1°C. The specific heat of water is higher than nearly all other liquids, and is defined as unity. It takes 4.187 J (one calorie) to heat 1 g (or 1 cm^3) of pure water by 1°C (at 15°C). The reason for the high specific heat of water is the high heat energy required to first stretch and then break the hydrogen bonds of water molecules, heat which is then unavailable for raising the water temperature.

Table 3–2 **Decrease in Dynamic Viscosity. (Pascal second; 1 Pa · s = 1 kg m^{-1} s^{-1}) and kinematic viscosity (m^{-2} s^{-1}) of fresh water with temperature at atmospheric pressure.**

Temperature (°C)	Dynamic Viscosity (Pa · s)	Kinematic Viscosity (m^{-2} s^{-1})
0	1.792×10^{-3}	1.792×10^{-6}
5	1.519×10^{-3}	1.519×10^{-6}
10	1.308×10^{-3}	1.308×10^{-6}
15	1.140×10^{-3}	1.141×10^{-6}
20	1.005×10^{-3}	1.007×10^{-6}
25	0.894×10^{-3}	0.896×10^{-6}
30	0.801×10^{-3}	0.804×10^{-6}

Source: Hodgman et al. 1959.

Water has a very high **heat of evaporation** or **heat of vaporization** (2,454 J g^{-1} at 20°C), the heat or energy necessary to break hydrogen bonds allowing liquid water to become water vapor. As a result of its high specific heat, water temperature in tropical waters rarely exceeds 30°C, protecting the biota from overheating and preventing rapid water loss. When water vapor condenses to form liquid, the same amount of heat is released as was utilized for the vaporization. The heat of vaporization and condensation drives the water cycle. Liquid water absorbs a lot of heat at low latitudes and evaporates. When the water vapor in the atmosphere is transported to a higher altitude or latitude, it cools, condenses, and releases heat (see Chapter 5).

The **latent heat of sublimation**, the heat required to directly convert ice or snow to water vapor, is even higher (2,843 J g^{-1} or 679 cal g^{-1}) than the heat of vaporization. Although the **latent heat of fusion and melting** of water (334 J g^{-1}) is relatively low, it is high in absolute terms showing that a great deal of heat is needed for fusion of water molecules at 0°C to form ice. Much energy is lost in producing a winter ice cover and the same quantity of solar energy is required to thaw the ice in spring. The heat used to melt the ice is unavailable for heating the water, which is why polar lakes with a thick ice-cover do not warm much beyond 4°C after ice-out (Chapter 11). While the specific heat of ice is high in absolute terms, it is only about half that of water, and at 0°C ice forms and thaws much more readily and with less change in heat than is required to raise or lower the temperature of liquid water. A high specific heat allows water to store great quantities of heat and makes it relatively resistant to shortterm changes in air temperature.

The slow water-temperature changes resulting from a very high specific heat buffer inhabitants of aquatic systems against rapid temperature change and, at the same time, temper the climatic extremes experienced on land surrounding larger lakes.

Dielectric Constant

Another important physical attribute of liquid water is its exceptionally high dielectric constant, which makes water an outstanding solvent for salts. Water molecules in contact with a salt crystal orient themselves to neutralize (reduce) the attractive forces between the positively and negatively charged ions in the crystal structure of the salt. The liberated ions are then hydrated by the water molecule, preventing their recombination and recrystallization. There is hardly a substance that has not been found in solution in the earth's waters. Were it not for this solubility of essential nutrient salts in water, plants would not be able to take up nutrients, and animals would not be able to take up dissolved compounds, and their tissues would be unable to release wastes.

Water is a good solvent for inorganic and polar organic molecules because it has a high dipole moment with the positively-charged hydrogen atoms of each molecule and the oxygen atom negatively charged. This allows for hydrogen-bond formation with polar molecules (amino acids, sugars, alcohols, ammonia). Although, for example, calcium and bicarbonate ions in solution are strongly held together by electrostatic attraction, this is not the case when present as a salt and, as a result, $Ca(HCO_3^-)_2$ readily dissolves in water. It does so because the attraction to ions of oppositely charged ends of the water molecule is greater than the attraction of the oppositely charged ions to each other. Many polar organic molecules, such as sugars and organic acids, dissolve in water because they too form bonds with the two opposite poles of water molecules.

Surface Tension

Water has high surface tension, which is a measure of the strength of the water's surface film. The strong attraction between water molecules is such that the surface tension of water is exceeded only by that of mercury among common liquids. It is the high surface tension that permits surface-living insects to walk on water and light-weight debris, like pollen, to float, resulting in accumulations of wind- or current-derived particulate matter in the surface film of lakes and seas.

Absorption of Solar Radiation

Pure water scatters solar radiation in the photosynthetic and visible range at a rate 1,000 times greater than air, thereby explaining the rapid extinction of light in water columns. Pure water also absorbs solar radiation in a characteristic fashion; it effectively scatters longer wave solar radiation, including the red portion of visible or photosynthetically available radiation (PAR) and the even longer-wavelength infrared radiation (IR). Infrared alone—contributing about 50 percent of total radiation energy received at the water surface—is totally absorbed in the immediate surface layer of water (Chapter 10). Heating the surface water imparts a density difference between the upper layer and the deeper cooler water which is an important determinant of how well lakes mix (Chapter 11), and the strength (density) of currents (Chapter 12). Shorter-wavelength radiation in the green and blue portions of the PAR portion of the spectrum is much less readily scattered by pure water, thereby allowing photosynthesis and vision at much greater depths than if all wavelengths of solar energy were absorbed at an equal rate.

Shorter-wavelength radiation, in the ultraviolet (UV) portion of the energy spectrum, is readily dissipated in the surface layer of turbid or colored lakes, but this high-energy and injurious radiation is transmitted to considerable depth in highly transparent waters (Chapter 10).

Highlights

- Weak bonding between adjacent water molecules allows formation of short-lived clusters that make pure water a liquid crystal. This weak bonding gives liquid water its unusual properties.
- The high density of water, compared to air, allows it to serve as a buoyant medium for the biota, while the nonlinear relationship between temperature and density makes the stratification of freshwater lakes possible.
- The (dynamic) viscosity of water imposes a functional resistance to swimming organisms that is much higher than encountered in air and is a particular problem for the smallest organisms who lack much momentum. The viscosity also produces a boundary layer around objects that reduce the exchange rate of gases and ions between organisms and water.
- The high specific heat of water means that much heat is required to raise or lower the temperature of liquid water, thereby protecting biota from changes in temperature and, in tropical water, from overheating.
- Exceptionally high dielectric constant makes water an outstanding solvent for inorganic substances and certain organic molecules, allowing solubilization and availability for uptake and release by the biota.
- The total absorption of long-wavelength (infrared) solar radiation in the immediate surface layer of water is responsible for surface heating and establishing density differences between the upper layer and deeper, cooler water, as well as for establishing (density) currents. Shorter-wavelength solar radiation in the green and blue portions of PAR spectrum is much less readily scattered in pure water, allowing photosynthesis and vision to much greater depth than would be possible if all wavelengths of solar energy were absorbed at equal rates.

4

Water Resources, Water Pollution, and Inland Waters

4.1   Introduction

Only about 2.6 percent of the world's total volume of water consists of fresh water (see Table 4–1). Freshwater lakes, rivers, and wetlands, which are main sources for human consumption and the habitat of other organisms, contain roughly 113×10^3 km^3 or about 0.3 percent of total global freshwater reserves (Table 4–1). With only a tiny fraction of the total freshwater resources of the world readily accessible as surface water, it is not surprising that the 113×10^3 km^3 of fresh water is of great economic, social, and scientific value. The remaining 99.7 percent of fresh water is largely unavailable, locked in glaciers and ice caps (76.4%) or deep groundwater pools (22.8%) (Table 4–1).

Water Residence

The average water residence (renewal) time of the world's freshwater lakes is long, about 17 years (Table 4–1), and is greatly affected by the slow flushing of the largest lakes holding most of the water. The vast majority of freshwater lakes are small and flushed much more rapidly (Chapter 9), with the renewal rate a function of lake morphometry (Chapter 7), catchment size (Chapter 9), climate, and runoff from drainage basins (Chapter 5). Even so, the *water residence time* (WRT) of the world's lakes is long compared to the global average residence time for wetlands (~5 years) and rivers (32 days), but is short in comparison with the global WRT of groundwater, dominated by pools of deep groundwater. The average WRT for rivers

and wetlands (Table 4–1) as well, is much affected by a few large systems. Most lotic systems are small and flush more rapidly (in hours or days). The noted differences in WRT have a major impact on the biota and chemical attributes of inland waters. As most macroscopic organisms have a life cycle longer than one or two weeks (Fig. 2–4), the macroscopic biota of the open water is most diverse in lakes (Chapter 21) and much less so in streams and small, rapidly flushed wetlands, where the WRT is normally shorter than the doubling time of macroscopic organisms living in the water column. Consequently, macroscopic organisms in these streams and wetlands have to be associated with sediments or rooted plants so they are not flushed away.

It is apparent that the relatively small water volume in individual rivers and wetlands (which are the aquatic systems most closely coupled with the land) show the most seasonal variability of physical and chemical characteristics in response to terrestrial events and changes in both precipitation and air temperature. In contrast, changes in lakes, with their much larger water volume, normally occur more slowly. As a consequence, pollutants are flushed rapidly from flowing streams and wetlands (unless adsorbed to the sediments), and more slowly from lakes and large pools of deep groundwater.

Groundwater

A pollution barrier provided by soils and an impermeable subsurface layer means there has been little or no contamination, on a global scale, of the very large pool of deeper groundwater by man's industrial activities.

Table 4–1   **World water resources (km³). ND = Not Determined.**

System	Best Estimate (10^3 km³)	Range of Published Estimates[1]	Freshwater (% global reserves)	Global Water Residence Time (WRT) ($\bar{x}$)
Oceans	1,350,000	1,320–1,370	0	2,500 years
Atmosphere	13	10–15	0.036	8 days
Surface Water				
Rivers and Streams	1.7	1.02–2.12	0.005	32 days[2,3]
Wetlands	11.5		0.032	5 years[3]
Freshwater Lakes	100	30–177	0.279	17 years[3]
Saline Lakes	105	85.4–125	0	ND
Subsurface Water				
Soil Moisture	70	16.5–150	0.195	1 year[3]
Groundwater (mostly fresh)	8,200	7,000–330,000	22.8	1,400 years[3]
Glaciers and Icecaps	27,500	16,500–48,020	76.6	1,600 and 9,700 years
Biota	1.1	0.6–50	0.003	hours

[1]Large ranges reflect major differences in thickness of the subsurface groundwater zone and depth of waterbodies considered.

[2]As the global water residence time (WRT) averages ~16 days within free-flowing channels, ~32 days represents doubling the age of water in river systems as the result of reservoir construction. The WRT in *regulated river* systems (those with dams) has increased more than threefold (Vörösmarty et al. 1997).

[3]The average WRT given are, atypically long because they are dominated by a relatively few large systems. The WRT of most streams and rivers range from hours to days, most lentic systems from months to a few years, and groundwater aquifers from years to centuries (Chapter 5).

Source: From World Resources Institute (1988), Korzun (1978) and Shiklomonov (1993).

However, localized pollution of near-surface groundwater with inorganic and organic chemicals is a rapidly growing concern in industrialized regions. Groundwater contamination is extremely difficult and costly to clean up. Plumes of contaminants, carried by the groundwater may extend many kilometers (km) from the source. A contaminated aquifer may be unusable for decades or longer and during that time pollutes surface water fed by the groundwater. The long WRT of groundwater and slow release of contaminants adsorbed onto particles means that once polluted, the recovery of groundwater pools (**aquifers**) will normally be slow.

Groundwater and surface water contamination are in two categories—*point sources* and widely distributed or *nonpoint sources*. Some significant point sources are municipal landfills, industrial waste disposal sites, the smoke stacks of smelters, incinerators and powerplants, mine tailings, leaking pipelines carrying petroleum products, underground gasoline storage tanks made of steel and highly subject to corrosion, and livestock wastes, especially from industrial-sized livestock barns and feedlots. Nonpoint sources include fertilizers and pesticides used on agricultural lands or forests[1] and, at higher latitudes, road salt spread on highways in winter. (See Sec. 5.2 and Chapter 28.)

4.2   Water Resources

Distribution and Pollution

Surface water is unequally distributed over the face of earth, and quantities potentially available per capita in different countries show enormous differences (Table 4–2). The Amazon River basin and parts of south and southeast Asia are endowed with substantial seasonal rainfall, while the Middle East, large parts of Africa,

---

[1]"The fundamental policy problem in lake restoration is that those who cause non-point pollution do not benefit from reduced pollution, especially in large agricultural catchments. Conversely, the beneficiaries of non-point pollution control are not those who cause the pollution, except in some urban lakes. This mismatch between polluters and beneficiaries is the root of institutional shortcomings that prevent the success of non-point pollution control programs." (S. R. Carpenter and R. C. Lathrop 1999.)

Table 4–2   Annual internal freshwater supply, as river flow (m^3 capita^{-1}) and percent withdrawn for population, industry, and agriculture purposes in selected countries in 1990. Note that since only a fraction of the water supply is available for human use, per capita supply overestimates per capita availability in the most populated regions.

Country	Supply m^3 Capita^{-1}	Water Resources Withdrawn (%)	Withdrawn by Sector (%)		
			Population	Industry[2]	Agriculture
NEW ZEALAND	117,490	< 1	46	10	44
CANADA	109,370	1	11	80	8
NORWAY	96,150	< 1	20	72	8
BRAZIL	34,520	1	43	17	40
FINLAND	22,110	3	12	85	3
ARGENTINA	21,470	3	9	18	73
SWEDEN	21,110	2	36	55	9
AUSTRALIA	20,480	5	65	2	33
USSR (former)	15,220	8	6	29	65
INDONESIA	14,020	1	13	11	76
USA	9,940	19	12	46	42
AUSTRIA	7,510	3	19	73	8
SWITZERLAND	6,520	6	23	73	4
JAPAN	4,430	20	17	33	50
MEXICO	4,030	15	6	8	86
ITALY	3,130	30	14	27	59
FRANCE	3,030	22	16	69	15
SPAIN	2,800	41	12	26	62
CHINA	2,470	16	6	7	87
INDIA	2,170	18	3	4	93
DENMARK	2,150	11	30	27	43
UNITED KINGDOM	2,110	24	20	77	3
CZECHOSLOVAKIA (former)	1,790	6	23	68	9
SOUTH KOREA	1,450	17	11	14	75
SOUTH AFRICA	1,420	18	16	17	67
GERMANY (western)	1,300[1]	26	10	70	20
POLAND	1,290	30	16	60	24
BELGIUM	850[1]	72	11	85	4
ALGERIA	750	16	22	4	74
NETHERLANDS	680[1]	16	5	61	34
KENYA	590	7	27	11	62
HUNGARY	570[1]	5	9	55	36
ISRAEL	370	88	16	5	79
EGYPT	30[1]	97	7	5	88

[1]Countries having a low internal supply but with an important external supply from upstream countries.
[2]Includes hydroelectric use.

Source: After World Resources Institute 1992.

north/central Asia, and much of Australia are characterized by low but highly variable runoff. The differences reported among countries (Table 4–2) are somewhat deceptive because water received from upstream countries is not included. This exclusion results in a major underestimation of the water supply for a few countries like the Netherlands, which gets an additional 5,350 m³ per capita from large inflowing rivers. Egypt receives a mere 30 m³ per capita, primarily from precipitation, but gains an additional 1,075 m³ via the inflowing Nile River. Furthermore, large countries such as the United States, Russia, Brazil, China, and India have regions of both high runoff and great water shortage. In addition, Table 4–2 does not reveal the degree of seasonality in runoff in different parts of the world. Nor does the table include quality of wastewater treatment, if any, or distribution of water in relation to human needs. Without these factors, the per capita *supply* reported in Table 4–2 cannot be equated with the per capita *availability* of water. Even so, Table 4–2 provides an indication of how much, on average, the water in any country is likely to be used and abused. A high withdrawal of water resources for primarily agricultural purposes indicates an increased probability of water pollution from fertilizers, biocides, and in semi-arid regions, high salt input in rivers caused by salts flushing from irrigated soils. Conversely, a primarily industrial withdrawal (other than for generating hydroelectricity) suggests the probability of industrial pollution. A combination of low supply per capita and high domestic water withdrawal points to polluted waterways, unless offset by high-quality wastewater treatment. The continued rapid increase in fertilizer use implies an increased eutrophication and contamination of lakes and rivers particularly in the economically developing world (Table 4–3) where consumption is expected to rise by another ~70 percent between 1989 and 2020. Only a small increase is expected in the economically developed countries.

Water Demand and Supply

The per capita freshwater requirements for drinking water, hygiene, living, and food, plus industrial use, lies between 1,400 m³ and 1,800 m³ globally. This amount provides the minimum needed for households, industry, and agriculture in a moderately developed country, unless water management is highly advanced. The volume could be reduced to about 1,000 m³ by increasing the efficiency of water use in agriculture (Gleick 1993). Below 500 m³ there is severe water scarcity unless crops (which require a great deal of water to produce) are imported.

The per capita availability of water is not expected to change much in Europe and North America, where better management is expected to lower the growth in demand to two to three percent annually. However, globally, water withdrawals for human use have been increasing about four to eight percent per year, most of it in the developing world (World Resources 1992). As a result of rapid human population growth and its associated agriculture, the per capita water availability is expected to decline rapidly in low-rainfall regions of the Middle East, central Asia, the northern rim of the African continent, eastern and southern Africa (Gleick 1993), and in water-poor northern China. The problem will be exacerbated by the largescale depletion of nonrenewable groundwater pools on which about 10 percent of the world's agriculture depends. However, so much water from all sources is wasted irrigating crops in water-poor regions that less wasteful irrigation (costly) and more recycling (also costly) would free water for other uses (Table 4–2). Similarly, a major investment in wastewater treatment would allow water to be reused.

Table 4–3   **Growth in fertilizer use in kilograms per hectare (ha) of cropland. (1 ha = 0.01 km² = 10,000 m².)**

Area	1975–1977	1985–1987	% Increase
World	67	91	36
Africa	14	19	36
North/Central America	84	83	−1
South America	28	39	39
Asia	42	93	121
Europe	207	228	10
Oceania	34	34	0

Source: Modified from World Resources Institute (1990).

The per capita supply indicated in Table 4–2 greatly overestimates the per capita water supply available in most countries because much of the water discharge is remote from major population centers, and about two-thirds of all river runoff is flood runoff and generally unavailable for human use in water-poor regions. There is also a rapidly increasing quantity held back by dams (World Resources 1988). Dams always greatly affect aquatic systems (Chapter 29).

Water Limitation

Water availability and drinking water quality will become, more than ever, a critical limiting factor in economic development and human health (both as drinking water and removing and diluting human waste), therefore affecting political stability. Better reuse and management of remaining available water resources and less wasteful use of irrigation and urban water supplies can mitigate these factors. Binding treaties between upstream and downstream countries concerning the quantity and quality of water to be supplied to downstream countries will become more important in water-poor regions of the world because upstream countries are increasingly building dams on major rivers and expanding their own irrigation agriculture. Even if the amount of water available to downstream countries is sufficient in normal rainfall years, it may not be sufficient in dry years or during months when crops downstream need more water.

▲ The limnologist or stream biologist of the past was seldom affected by either the relative poverty of lentic or lotic systems in many parts of the world or the presence of polluted systems. A sufficient number of unspoiled bodies of water, traditionally favored for fundamental research, was available. A hundred years ago, water pollution was a localized problem, derived from local human and animal wastes. The much smaller amount of industrial effluent also contained few, if any, synthetic organic chemicals. In contrast, surface water now contains synthetic chemicals, toxic metals, and fertilizers produced by modern industrial and agricultural practices, as well as urban centers. Contaminants arrive from distant sources via waterways, atmospheric deposition, or directly upon release into lakes and rivers from sources within the particular catchment (Chapter 8). Many toxic organic chemicals are only slowly biodegradable and adsorb strongly to soil particles that are washed into waterways (Chapter 28). Wetlands and lakes play a crucially important role in organic chemical degradation and storage and thus in water purification (Chapter 24). For a quantification in monetary terms of the importance of aquatic systems in water storage and purification, see section 1.1.

Not only are all aquatic systems impacted, to some extent, by substances carried via the atmosphere from afar but physically unmodified streams and rivers are virtually nonexistent today in low-rainfall portions of the world. But not only there have the lotic and lentic systems they nourish become greatly modified as the result of (1) increased catchment nutrient release, resulting in eutrophication; (2) deforestation; (3) withdrawal for human and agricultural use; (4) inter-basin water transfers; (5) pollution of fresh waters with pesticides, organic waste, and/or irrigation water of unsuitably high salinity; (6) dam construction, resulting in regulation of river discharge; and (7) modification and drainage of riverine wetlands, and degradation of the original land–water interphase for agriculture and other purposes. The land–water interphase plays a crucial role in retention of materials derived from the land (Chapter 8) and as habitat used for feeding, breeding, and hiding for a large number of species, including fish and water birds (Chapter 26). Although limnologists have long acknowledged aquatic systems to be closely linked to the terrestrial and atmospheric drainage basins that feed them, they have paid all too little attention to their effects on aquatic systems. Yet, aquatic systems and their biota cannot be understood scientifically or managed effectively when viewed in isolation from their terrestrial and aerial catchments.

Concerns about environmental pollution and its effect on environmental quality and human health, together with a very rapid increase in demand for fresh water to support rapidly expanding agriculture, industrial production, and drinking water and recreation, has led to the development and blossoming of applied limnology during the past 40 years. Applied limnology is primarily concerned with multiuse management of aquatic systems and predicting how systems will respond to remedial measures. Applied limnology is also investigating the cycling and fate of pollutants and toxins, as well as with their effects on the biota and drinking water quality. Limnologists are increasingly involved in wetland reconstruction and the remeandering of channelized rivers. Despite much larger financial resources available to applied components of limnology,

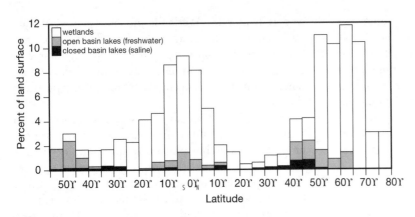

Figure 4–1 Global distribution of lakes and wetlands. Note that many small lakes are likely to have been reported as wetlands. The importance of saline waters at 40–50°N reflects the enormous Caspian lake. Also note the importance of wetlands around the equator and in the boreal forest and subarctic region of the north temperate zone. (*Data for larger lakes from Schuiling 1976, and wetlands from Crill 1996.*)

applied limnology continues to depend mostly on ideas and data produced by limnologists with less applied goals. Conversely, those with less applied goals are becoming more influenced in their choice of research and dependent on funds provided for (more or less) applied research.

▲ 4.3 Lakes, Rivers, Wetlands and Their Global Distribution

Lake and Wetland Distribution

Lakes and wetlands in particular are not evenly distributed over the earth's surface. A high proportion of the largest freshwater lakes (>100 km²) are located be-

tween 40° and 50° north and south latitude, with a secondary peak near the equator (Fig. 4–1). The Southern Hemisphere saline lakes center around 30°S, but their distribution in the Northern Hemisphere is affected by the higher latitude location of the Caspian Lake (Sea). Wetlands are equally found around the equator and in the boreal forest region of Eurasia and North America.

The 1,522 largest lakes (>100 km²) contribute importantly to the total lake surface area (and volume). However, numerically they are an insignificant fraction of the roughly nine million lakes and large ponds (>0.01 km²) worldwide, plus an even larger number of smaller waterbodies, which happen to be concentrated between 40° and 70° north latitude (Fig. 4–1, Table 4–4). In Sweden alone there are over 100,000 lakes

Table 4–4 **Classification of lakes and ponds based on surface area, with extrapolated estimates of the number of lakes worldwide in each size category and the total surface area per lake or pond size category, 1km² = 100 ha. ND = Not Determined.**

Type	Surface Area km²	Number of Lakes	Total Surface Area km²
Great lakes	>10,000	19	997,000[1]
Large lakes	10,000–100	1,504	686,000
Medium lakes	100–1	139,000	642,000
Small lakes	1–0.1	~1,110,000[2]	~288,000[2]
Large ponds	0.1–0.01	~7,200,000[2]	~190,000[2]
Other ponds	<0.01	ND	ND

[1]Caspian Sea alone contributes 374,000 km²
[2]Low accuracy, based on extrapolations.

Source: Modified after Häkanson 1977, and Meybeck 1995.

and ponds, with 96 percent of these made up of ponds and small lakes (0.1–1 km²) (Håkanson and Karlsson 1984). In the former USSR, with 2.8 million lakes and larger ponds, again the vast majority (98%) are ponds and small lakes. More water bodies are detected, and the fraction of pond-sized or smaller water bodies increases, whenever higher resolution maps are used. Satellite images generally yield the highest resolution (~ 0.001 km²). *Landsat 5* images of Atlantic Canada have shown two-thirds of the water bodies to be smaller than 0.01 km² (1 ha). If this proportion is applicable globally, it suggests the existence of some 18 million bodies of water smaller than one hectare. Lotic systems are also numerically dominated by small streams (Table 8–4).

Approximately 2.1 percent (2.8 × 10⁶ km²) of the world's land area (excluding ice caps) is covered by lakes and ponds exceeding 0.01 km² (Meybeck 1995, and see Chapter 6). The fraction of land covered ranges from < 0.1 percent for France, China, or the nonglaciated portions of the United States (Sec. 6.2) to around nine percent of the surface area of Scandinavia. Some 40–70 percent of the boreal and subarctic regions of Russia are covered by wetlands while about 34 percent of the enormous Northwest Territories of Canada (before its recent division into two territories) is covered by wetlands, lakes, and rivers.

Fresh and Saline Waters

The unequal distribution of lake water is not only evident from the spatial distribution of lakes and ponds but also from the fact that 95 percent of the total surface-water volume is contained in only 145 lakes. Data on the 20 largest lakes in terms of area, volume, and maximum depth are given in Table 4–5. About half the water volume in the world's lakes is fresh and half is saline (Table 4–1). The Caspian Lake (Sea) stands out among large lakes for its enormous area and great depth, with the result that its volume alone contributes 75 percent of the world's saline lake-water volume. As a result, saline lakes contribute only a minor fraction of surface area and volume in remaining lakes and ponds. However, saline lakes dominate in semiarid regions (Sec. 13.7).

The economic importance of saline lakes is relatively small because the water is too saline for drinking or irrigation, but their precipitated salts (mostly sodium chloride) are sometimes extracted. In addition, saline inland waters are scientifically important; not only because they contain unusual species but also because they form one end of a salinity spectrum important to resolving how community structure and functioning changes systematically along salinity gradients. They further serve as models for predicting the effects of increasing salinization in rivers draining vast areas of poorly irrigated semiarid land and projected impacts of increased evaporation during global warming.

While Lake Superior (CA, US) has by far the largest surface area among freshwater lakes (Table 4–5), the depth of the second largest lake (Baikal, RU) is so much greater that it holds a much larger water volume. Baikal contributes about 20 percent of all the surface fresh water worldwide. Lake Baikal, Lake Superior, and Lake Tanganyika (East Africa) together contain nearly half (44%) of the entire surface freshwater in the world.

Among the morphometrically most unusual lakes are the disproportionately deep Crater Lake (US) and Salsvatn (NO) (Table 4–5). Among those varying most in surface area is Tonle Sap (Grand Lake) in the Mekong River delta (KH) and Lake Chad (CD, NE, NI). Shallow Lake Chad ($\bar{z}$ = ~1.5 m) in the semiarid region of western Africa has been shrinking in recent decades into a number of isolated basins (wetlands) totaling ~2,500 km², thereby threatening the livelihood of millions of subsistence farmers, fishermen, and livestock herders. The lake had a maximum surface area of about 25,900 km² early in the 20th century after a series of wet years (Herdendorf 1982). The surface area expanded to some 300,000 km² to 400,000 km² during a much higher-rainfall (pluvial) period about 6,000 years ago. At that time the Sahara desert was well-vegetated and dotted with shallow lakes and rivers (Beadle 1981, Pachur and Kröpelin 1987).

Globally, rivers hold less than 2 percent of the water held in freshwater lakes but annually carry 2,700 km³ to the oceans (Chapter 5). Information on discharge at the mouth, and average suspended sediment load and yield for the 25 largest rivers is given in Table 4–6. Rivers, plus their bordering wetlands and linked aquifers, are home to an extensive biota (Chapter 8) while rivers themselves are the major source of water for irrigation, agriculture, industry, human consumption, and water purification (Sec. 1.1).

Table 4–5   **The 20 largest lakes of the world ranked in order by surface area, volume, and maximum depth. Country of origin is included; geologic origin, if known, is given as follows: [G] glacial, [T] tectonic, [C] coastal, [V] volcanic, [F] fluviatile, or [M] meteoric.**

Name	Surface Area (km²)	Name	Volume (km³)	Name	Maximum Depth (m)
Caspian [T] [IR, RU]	374,000	Caspian	78,200	Baikal	1,741
Superior [G+T] (CA, US)	82,100	Baikal	22,995	Tanganyika	1,471
Aral [T] (KZ, UZ)	43,000[1]	Tanganyika	17,827	Caspian	1,025
Victoria [T] (KE, TZ, UG)	62,940	Superior	12,230	Malawi	706
Huron [G] (CA, US)	59,500	Malawi	6,140	Issykkul (KG)	702
Michigan [G] (US)	57,750	Michigan	4,920	Great Slave	614
Tanganyika [T] (BI, TZ, ZR, ZM)	32,000	Huron	3,537	Matana (ID)	590
Baikal [T] (RU)	31,500	Victoria	2,518	Crater [V] (US)	589
Great Bear [G] (CA)	31,326	Great Bear	2,292	Toba [V+T] (ID)	529
Tonle Sap [F] (KH)	30,000[2,3]	Great Slave	2,088	Sarez [F] (TJ)	505
Great Slave [G] (CA)	28,568	Issykkul [T] (KG)	1,738	Tahoe [T] (US)	501
Chad [T] (CD, NE, NG, CM)	25,900[4]	Ontario	1,637	Hornindalsvatn [G] (NO)	514
Erie [G] (CA, US)	25,657	Aral	1,451	Chelan [T] (US)	489
Winnipeg [G] (CA)	24,387	Ladoga	908	Kivu [T+V] (RW, ZR)	480
Malawi (Nyasa) [T] (MW, MZ, TZ)	22,490	Titicaca [T] (BO, PE)	827	Quesnel [G] (CA)	475
Balkhash [T] (KZ)	22,000	Reindeer [G] (CA)	585	Adams [T] (CA)	457
Ontario [G] (CA, US)	19,000	Helmand (AF, IR)	510	Fagnano (AR, CL)	449
Ladoga [G+T] (RU)	18,130	Erie	483	Mjosa (NO)	449
Bangweulu (ZM)	15,100[2]	Hovsgol (MN)	480	Salsvatn (NO)	445
Maracaibo [T, C] (VE)	13,010	Winnipeg	371	Manapouri (NZ)	443

[1]Water extraction from inflowing rivers reduced area to ~ 24,200 km² and the volume by 84 percent (see Sec. 5.7) in 2000.
[2]Flood control and irrigation have reduced area to about 11,000 km².
[3]Wide fluctuations due to seasonal flooding.
[4]Drought has reduced the area to about 2,500 km².

Source: Modified from Herdendorf 1982.

▲ 4.4   A Look at "Typical" Lakes and Streams

Lakes

This section is purposely exaggerated, however, there is enough truth in it to sensitize the reader to the small lake, small stream, and temperate zone bias and perspective of research, traditionally carried out on pris-tine systems. All scientific disciplines have histories and traditions (Chapter 2) that influence what is done today, how it is done, how it is viewed, and what is discussed in textbooks. Although the Great lakes (>10,000 km²) and largest rivers of the world are impressive indeed, most of what is known and believed about the structure and functioning of aquatic systems has been learned during the last 100 years through the study of a relatively small number of north temperate

Table 4–6   **The 25 largest rivers in the world, plus some other well-known rivers; they are ranked in terms of their discharge at the mouth, drainage area, average annual suspended sediment load, and mean annual specific sediment yield from their catchments.**

River	Country	Mean Discharge (10^3 m^3 s^{-1})	Drainage Area (10^3 km^2)	Suspended Load (T $\times 10^6$ yr^{-1})	Sediment Yield (T km^2 yr^{-1})
1 Amazon	BR, CO, PE	212.5	6062	406	67
2 Congo/Zaire	AO, CG, ZR	39.7	3968	72	18
3 Yangtze	CN	21.8	1013	561	553
4 Brahmaputra	BD, CN, IN	19.8	553	813	1469
5 Ganges	BD, IN	18.7	1047	1626	1551
6 Yenisei	RU	17.4	2471	11	4
7 Mississippi[1]	US	17.3	3185	350	109
8 Orinoco	BR, CO, VE	17.0	939	97	103
9 Lena	RU	15.5	2680	80	30
10 Paraná	AR, BO	14.9	2278	91	40
11 St. Lawrence	CA	14.2	1274	4	3
12 Irrawaddy	CN, MM	13.5	362	336	927
13 Ob	RU	12.5	2448	16	6
14 Mekong	KH, LA, TH, VN	11.0	387	190	491
15 Amur	CN, RU	11.0	1822	52	28
16 Tocantins	BR	10.2	896	—	—
17 Mackenzie	CA	7.9	1784	15	8
18 Magdalena	CO	7.0	262	172	656
19 Columbia	CA	7.2	266	10	
20 Zambezi	AO, BW, MZ, NA, ZM, ZR, ZW	7.1	1280	100	78
21 Danube	AT, BG, DE, HR, HU, MD, RO, SK UA, YU	6.2	806	22	27
22 Niger	BJ, GN, ML, NE, NG, SL	6.1	1100	5	5
23 Indus	CN, IN, PK	5.6	1231	489	396
24 Yukon	CA, US	5.1	921	88	96
25 Pechora	RU	4.1	322	7	22
33 Nile	EG, ET, SD, UG	2.8	2944	124	42
38 Rhine	CH, DE, FR, NL	2.2	145	1	7
42 Rhone	CH, FR	1.7	94	32	340
45 Tigris\Euphrates	IQ, TR, SY	1.4	1048	863	823
46 Po	IT	1.4	54	17	315
47 Vistula	PL	1.1	191	2	10

[1]A map of water erosion rates in the US is available at the following Web site: http://www.nhq.nrcs.usda.gov/land/index/erosionmaps.html.

Source: After Welcome 1985.

freshwater systems. These are usually located an hour or two from a university or research institute. Saline systems, generally situated far from population centers, have received disproportionately little attention. At the university or research institution a limnologist or two does most research during the spring and summer periods. The primarily glacial lakes they and their students have examined tend to be small (0.1–1.0 km²), allowing easy exploration with a rowboat, or a boat powered by a small outboard engine. At the same time, most of the lakes selected have been sufficiently deep in relation to their surface area so they stratify stably in summer, allowing deepwater (profundal) events to be investigated. As a result, the vastly more abundant shallow ponds, large lakes, and frequently abundant flowing-water (lotic) systems nearby have traditionally received much less attention.

A combination of relatively small surface area and large maximum depth implies that stereotypical lakes tend to have steep underwater slopes. This means they normally have only a small near-shore zone suitable for rooted aquatic plants, benthic algae (Chapter 24), and associated animal biota (Chapter 25). The relatively small shallow water zone (**littoral zone**) and absence of wetlands along steep shorelines further helped draw attention to the open water (**pelagic zone**), about which much more is known than the structurally more complex littoral. This is in spite of the fact that the vast majority of inland waterbodies are both smaller and shallower (Table 4–4) than the "typical" lake and are characterized by a littoral zone much larger than the area of open water (Fig. 4–2).

Limnologists were, until the 1960s, almost exclusively interested in unpolluted and aesthetically pleasing lakes and streams set in well-vegetated, little-disturbed catchments. In the case of lakes, this led to research emphasis on relatively transparent waterbodies of moderate to low nutrient content. Such lakes are most commonly encountered in the upper, steepest, best-drained areas of drainage basins. The upper portions of drainage basins are small and often well-vegetated. Such basins either lack a significant outflowing stream to feed the receiving lake, or have one or more inflowing streams that carry relatively little water, sediment, nutrients, and (colored) organic matter from their small catchments (Chapters 8 and 9). The lakes studied have therefore been relatively transparent, and their disproportionately large depth (and volume in relation to surface area) means that they flush slowly—as compared to most other small,

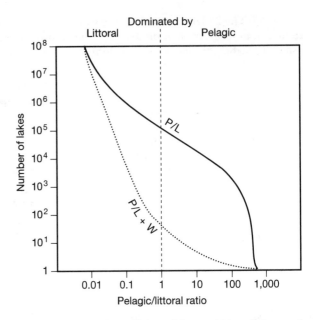

Figure 4–2   Number of lakes of the world in relation to the ratios of pelagic (P) to littoral (L) and wetland (W) regions. Water with a pelagic to littoral ratio < 1 is dominated by the littoral zone, a ratio >1 by the pelagic zone. *(From Wetzel 1990.)*

less deep lakes nearby and larger, proportionately shallower lakes located at lower elevations, which drain a much larger land area and tend to be richer in nutrients (Chapter 9).

Stereotypical north temperate lakes and streams experience strong and fairly regular seasonality in both irradiance and water temperature. Strong seasonality has, with good reason, received much attention; the resulting insights are less relevant at lower latitudes where seasonality in irradiance is reduced, but where aquatic systems experience strong and often highly variable periodicity in rainfall, runoff, river discharge, soil erosion, and wind speed.

Current Priorities.   Traditional views held by limnologists have undergone rapid changes during the last 40 years as the result of well-funded temperate zone research on shallow and largely unstratified lowland lakes and streams. These systems tend to lie in large catchments with low slope and nutrient-rich soils that are often further enriched from agricultural and/or urban activities. Yet theories based on small nutrient-poor temperate systems dominate limnology and its textbooks.

The Great Lakes of the world, and above all the Laurentian Great Lakes of North America, are now recognized as important resources that are increasingly under threat. As a result good, well-funded research on these and other inland seas (e.g., Lake Baikal, RU) has drawn the attention of lake limnologists to the importance of water mixing and turbulence (physical limnology), whose influence has been largely examined in large, wind-exposed waterbodies. The traditional focus on small lakes meant that the effect of lake size on turbulence, stratification, light climate experienced by the plankton, species richness, and foodweb structure had been insufficiently recognized. Research on the Great Lakes relies on both limnological and oceanographic approaches (Fig. 4–3) and has helped draw the two disciplines together.

The wide range of lake types studied, including saline, polar, and low-latitude lakes, provide a salutary reminder to limnologists that, while findings made on

the so-called typical small north temperate lake have provided an invaluable basis for limnology, neither the data nor the conceptual framework they have provided should be automatically assumed to be applicable to systems elsewhere (Williams 1988). Even so, the broad patterns that are gradually emerging from studies of a wide range of aquatic systems allow insights obtained from the study of "typical" lakes in lake-rich north temperate Europe and eastern North America to be seen and interpreted in a wider context.

Recognition that aquatic systems are primarily nourished by their drainage basins and the overlying atmosphere is gradually transforming limnology from a science mostly concerned with what transpires below the water surface to one that views aquatic systems as open systems closely linked to their catchments, their wetlands and inflowing rivers, as well as the atmosphere above (Chapters 8 and 9).

Streams and Rivers

The running-water counterparts of the typical lake are well-studied north temperate zone streams and small rivers draining the upper portions of little-disturbed forested catchments. Those lotic systems have provided most of the ideas and data in the literature. The preference (and bias) in lotic limnology has also been toward temperate zone systems of modest size that can be sampled with hand-held devices. Benke (1993) examined 159 published estimates of invertebrate production in lotic systems as a function of latitude, stream size, water discharge and water temperature, which serves as an indicator of the season the research was done. Further analysis of his data shows 83 percent of the research was carried out at mid latitudes (30°–50°), one percent was tropical or subtropical (<30° N or S), and three percent was carried out at higher latitudes (>60°).

Lotic limnologists too carry out most of their research in summer (67%), as indicated by stream temperatures between 10°C and 20°C at time of sampling. Winter work (<5°C) is rare (4%), as is winter work on lakes or wetlands. Small systems are preferred with 85 percent of the studies dealing with flowing waters with a summer discharge of <10 m^3 s^{-1} and only four percent based on mid-sized rivers having a discharge >100 m^3 s^{-1}.

"Typical" streams, like "typical" lakes, exhibit strong seasonality in discharge, irradiance, and temperature. Streams also receive organic matter—large amounts of it in autumn—from nearby forest or over-

Figure 4–3 Canada's principal Great Lakes research ship, CSS LIMNOS, is a general-purpose scientific vessel owned by the federal Department of Fisheries and Oceans. Based at the Canada Centre for Inland Waters in Burlington, Ontario. *(Photo courtesy of Fisheries and Oceans Canada. Reproduced with the permission of the Minister of Public Works and Government Services, 2000.)*

hanging forest canopy. The canopy allows reduced irradiance to reach the water. Lotic primary productivity is therefore modest, with a high proportion of organic matter available to aquatic microbes and invertebrates derived from the drainage basin. While stereotypical streams and small rivers[2] receive much catchment derived organic carbon, they receive disproportionately little in terms of inorganic nutrients, wastewater, and soil particles from their well-vegetated and little-disturbed catchments. Furthermore, the typical stream or small river is *perennial* (flowing all year), the result of a year-round pattern of precipitation and moderate rates of catchment evapotranspiration.

The typical stream is quite different from equivalent-sized systems in agricultural areas, from large lowland rivers lacking a canopy, and receiving much ground- and streamwater plus nutrients laterally from their drainage basins; as well as from dammed rivers and from high turbidity lotic systems in poorly vegetated regions everywhere. In addition, lotic systems and wetlands in well-watered regions differ greatly from their counterparts in arid and low latitudes, which are frequently *ephemeral* (temporary) rather than perennial and exhibit an enormous, but irregular, within- and among-year variability in river discharge.

Well-examined streams with low discharge tend to be shallow, they are readily sampled with hand-held devices developed for work on such waterbodies. The same equipment is ill-suited for work on deeper, wider

---

[2]"We know very little about large rivers ... defined as those which are large enough to intimidate research workers." (D. P. Dodge 1989.)

rivers flowing in the lower portions of large drainage basins. Larger systems are more readily sampled with equipment used on lakes and are frequently studied by lentic limnologists interested in water column processes. (see Secs. 8.2–8.4)

Highlights

- Only 2.6 percent of the world's total volume of surface water consists of fresh water. Lakes, rivers, and wetlands hold about 0.3 percent of total global freshwater reserves. Roughly half the water volume contained in the world's lakes is composed of fresh water.
- Fresh water is very unevenly distributed over the earth and quantities available in various regions in different countries show enormous differences. A high percentage of available water is used for agriculture in water-poor countries.
- The development of applied limnology during the last 40 years, with a mandate to manage aquatic resources has, at the same time, stimulated fundamental research with possible applications in environmental management.
- Aquatic systems and their biota cannot be understood scientifically or managed effectively when viewed in isolation from other aquatic systems upstream and their terrestrial and aerial catchments.
- The vast majority of lotic and lentic systems are small and shallow.
- Most limnological knowledge is based on a relatively few fresh waterbodies in the north temperate zone; the major challenge is to evaluate the extent to which results and interpretations are relevant to other types of aquatic systems, and especially to aquatic systems in other parts of the world.

5

Hydrology and Climate

5.1 Introduction

The quantities, temporal distribution, pathways, and residence time of precipitation in drainage basins determines river discharge as well as residence time of water in the catchment, receiving lakes, and wetlands. The water, together with salts and organic matter released during its passage through catchments, has a major impact on the chemical characteristics and the aquatic biota. The details of the hydrological cycle differ among climatic zones and geology. However, on average about 111,000 km^3 of moisture falls on the world's landmass every year, of which nearly two-thirds (71,000 km^3) then leaves by evaporation and transpiration of the biota without becoming river flow. The balance (40,000 km^3) flows into the world's rivers, lakes, and wetlands, replacing an equal volume of water returned to sea (Table 5–1).

5.2 Water Movement in Catchments

Runoff and Groundwater

The fraction of precipitation falling on catchments that is not lost by evaporation from land or transpired by vegetation will leave drainage basins by three routes. Most of the nonevaporated precipitation will infiltrate (percolate) and leave the soil as **subsurface runoff**, carrying dissolved materials, to streams (Fig. 5–1). Some will percolate downward through normally unsaturated substrate until it reaches either the **water table**, the level where the substrate is permanently saturated with water and joins the underlying **ground-**

water pool or **aquifer** (Fig. 5–1), or some precipitation percolates down until it reaches an impermeable layer. At this point the water flows laterally through the substrate in the direction of least resistance. When the lateral flow intersects the land surface it produces springs, streams, and fills basins to create lakes and wetlands. During periods of little or no rainfall, when the stream discharge is below its long term median value, groundwater provides most of the discharge and the flow is then known as **base (basal) flow** (Fig. 5–2). Not all groundwater is part of the present hydrological cycle and renewable. Nonrenewable deep pools of groundwater, increasingly being tapped in arid regions, are fossil waters deposited thousands of years ago when climatic conditions were more favorable.[1] Thus, the enormous Great Artesian Basin (GAB) aquifer in Australia was deposited 300,000 years ago.

---

[1] • Except where river water can be tapped, irrigation schemes in drier regions often rely on water from deep fossil aquifers that receive little or no recharge.

• The best known tracer for dating recent (young) groundwater is tritium ($^3H^+$), a radioisotope that was released as an atmospheric pulse during the peak of nuclear bomb testing in the early 1960s. It then entered groundwater as part of the precipitation. With an isotopic **half-life** (time in which half the radioactivity decays) of only 12.3 years the pulse is becoming difficult to detect. But, it is also possible to measure helium (^3He), the decay product of $^3H^+$; the combined methods allow continued dating of the water ranging from weeks to several decades. Carbon-14 decays slowly enough to trace total inorganic carbon in groundwater for up to ~30,000 years before present (BP). Several noble gas isotopes are available for dating periods ranging from days to many hundred thousands of years (Hofer et al. 1997). Radioisotopes are also used in sediment dating (Chapter 20).

Table 5–1   **Fluxes (km³ yr⁻¹) in the global water cycle.**

Process	Best Estimate	Published Range
Evaporation and Transpiration from Land	71,000	68,000–73,000
Precipitation on Land	111,000	99,000–119,000
Evaporation from the Ocean	425,000	383,000–505,000
Precipitation on the Ocean	385,000	320,000–458,000
Runoff from Land to Oceans	40,000	34,000– 47,000
Rivers	27,000	27,000– 45,000
Direct Groundwater Runoff	12,000	0– 12,000
Glacier Runoff (water and ice)	2,500	1,700– 4,500
Net Moisture Transfer from Marine to Terrestrial Atmosphere	40,000	

Source: Modified from World Resources Institute 1988.

Low slope catchments with deep sandy soils and fractured bedrock has a much greater capacity to store groundwater than upland basins characterized by steep slopes and shallow soils overlying impermeable bedrock. Consequently, upland streams on thin substrates are dominated by surface and subsurface runoff, whereas lowland streams and rivers on thick and porous substrates tend to carry more groundwater, with the discharge of some rivers dominated by groundwater. Forest vegetation cover promotes infil- tration, with tree roots providing infiltration channels and a layer of litter capable of intercepting surface flow. Vegetation cover also stabilizes soil and reduces erosion, of particular importance in semiarid regions (Sec. 9.7 and Chapter 29). But forest vegetation also produces high transpiration that reduces runoff. Cut- ting forests lowers both infiltration and transpiration rates, allowing rivers to rise rapidly after heavy rains, but also returning to baseflow more rapidly than be- fore (Fig. 5–2). Rapid rise and fall characterizes

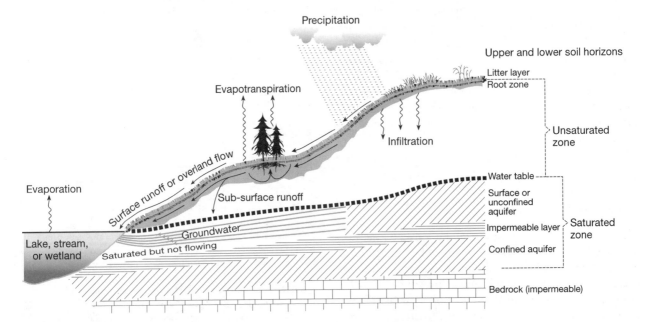

Figure 5–1   Schematic view of water movement in and on catchments, including the flow of groundwater into lakes, streams, or wetlands. Agriculture and other human activities contribute plant nutrients, organic matter, pesticides and other contaminants to runoff and groundwater. *(Greatly modified after Gibbs 1987.)*

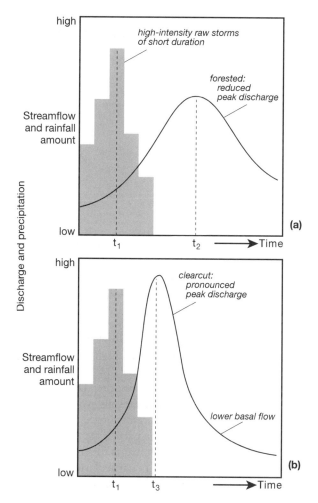

Figure 5–2 The effects of vegetation clearance on discharge. For equal rainfall, there is greater infiltration in the naturally vegetated catchment (a) than in the cleared catchment (b). The result is a larger but shorter flood, followed by a reduced base flow in the cleared catchment. However, high evapotranspiration by low latitude forests (in semi-arid regions) will reduce the annual net discharge as well as the base flow below that of clear-cut catchments during the season of lowest run-off. *(Modified from Whitlow 1983, in Allanson et al. 1990.)*

streams in all regions where shallow or impermeable soils dominate. The time lag between the beginning of rainfall and an increase in stream discharge, as measured using a weir with a **hydrograph** (water-level recorder), is stream specific and also depends on whether the soil moisture content was high or not before the rain event and, at higher latitudes, whether or not the soil was frozen (Figs. 5–3 and 5–4).

Figure 5–3 A calibrated weir used to determine stream discharge and housing for the water level recorder used to measure changes in discharge over time. *(Photo courtesy of J.D. Cunningham, Visuals Unlimited.)*

The residence time of groundwater in soil has a major impact on the quantities of plant nutrients exported. Short groundwater residence times reduce contact between the water and substrate, reducing possibilities for solubilization of chemical elements (biological/chemical weathering) and lowering export to rivers, lakes, and wetlands. Solubilization is mainly brought about by carbonic acid and other acids released during microbial and invertebrate metabolism as well as by plant roots (Chapter 14).

While precipitation and stream input to lakes and wetlands can be measured accurately, this is not the case for groundwater entering directly from the surrounding land without becoming part of stream flow. Groundwater is normally estimated by the difference in a **mass–balance equation:**

$$
\begin{aligned}
\text{groundwater input} = &\ \text{system outflow} + \Delta \text{ system storage} \\
&+ \text{precipitation on surface} \\
&- \text{evaporation} \qquad\qquad \text{EQ. 5.1} \\
&- \text{system inflow}
\end{aligned}
$$

None of the terms on the right side of the equation is measured without error, the groundwater estimates obtained by difference are then somewhat problematic. Yet the results of comparisons between *carefully* obtained mass–balance estimates of groundwater show good agreement with studies in which the groundwater component is measured directly. Groundwater

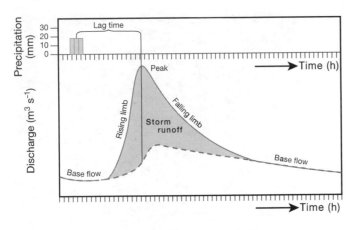

Figure 5–4 Diagrammatic stream hydrograph resulting from single storm event. (*After Beaumont 1975.*)

input to lakes range from negligible in basins underlain by hard rock with few fractures permitting transport of water, to about 50 percent in some **drainage lakes**, lakes with an inflowing and outflowing stream, highly porous basins, with deep soils or sands (Staubitz and Zarriello 1989, Shaw and Prepas 1990). The groundwater contribution may be higher in **seepage lakes**, lakes without stream inflow or outflow, located on a porous basin where the only other source of water is precipitation on the lake surface. Where the substrate is thin, groundwater inputs are much less important and virtually all of it must enter lakes near shore (Fig. 5–1).

While it is possible to measure groundwater discharge accurately, the same is not the case for nutrient and contaminant loads carried because some of the substances transported in **anoxic** (no dissolved oxygen) or **hypoxic** (low dissolved oxygen) groundwater are precipitated in lake or stream sediments when groundwater comes in contact with the typically better-oxygenated surface sediments or overlying water. Consequently, direct groundwater inputs of nutrients and contaminants to lakes are rarely taken into account in nutrient loading models.

Apart from groundwater and subsurface flows, there may be **surface runoff** (**overland flow**) to streams and lakes, especially during periods of high or prolonged precipitation when the rate of water supply to the soil is greater than the rate that can be absorbed (Fig. 5–1), or during snowmelt when underlying soils are still frozen. Surface runoff can be an important source of turbidity, carrying organic and inorganic particulate matter, plus adsorbed nutrients and contaminants to streams, lakes, and wetlands. Surface runoff is particularly important in semiarid regions at lower latitudes, areas characterized by short intense rainfalls and poor vegetation cover (Fig. 5–5). The receiving rivers there carry enormous sediment loads (Table 4–6), depositing them in down-

Figure 5–5 High sediment concentrations in a Kenyan river traversing a semiarid region during the rainy season. The reddish rust color of the water, produced by the highly weathered and easily erodable soils, rich in oxidized iron and high in aluminum (Chapter 13), that characterizes large areas of the tropics. (*Photo courtesy of L. Rue, Jr., Photo Researchers, Inc.*)

stream reservoirs or coastal estuaries. However, it is subsurface runoff that, during typically less extreme high rainfall events characterizing the temperate zone, is responsible for most water carried in summer by streams on moderately permeable substrates.

Path and Timing of Discharge

▲ It is important to recognize that the specific path water takes is of great importance in determining the soil contact time and thereby the chemical composition and export of materials to inland waters. In most regions subject to significant soil freezing, the maximum runoff and peak nutrient loads enter waterways in spring (Fig. 8–4), when light conditions are favorable, temperatures are rising, seasonal stratification of lakes is beginning, (Chapters 11 and 22), and phytoplankton and bacterial growth is on the upswing (Chapters 21 and 22). In contrast, at a lower latitude where soils and lakes are subject to little or no freezing, water and nutrient loading in aquatic systems is often maximal during winter (Chapter 11). But since winter irradiance and water temperatures are at the annual low, the efficiency by which nutrient loads can be utilized for plant growth is consequently low as well (e.g. Stockner and Short-reed 1985). Water and nutrient export from catchments at low latitudes is maximal during the rainy season, however, terrestrial evapotranspiration and stream evaporation rates are high and precipitation falling directly on lake surfaces becomes disproportionately important as a source of water and nutrients (Sec. 5.6).

Rivers, which normally carry most of the water, nutrients, and contaminants to lakes and wetlands, are far more than simple conduits of catchment-derived substances. Nor are rivers simply flowing lakes (Table 8–3). Indeed, river discharge regimes, catchment attributes, including the geology and the linked porosity of the substrates, and the biota have a major impact on the quantity, timing, and form (e.g., dissolved vs particulate, oxidized vs reduced) in which substances enter the receiving lakes and wetlands (Chapter 8). (Cushing et al. 1995, and Allan 1995).

5.3 Humans and the Hydrologic Cycle

The freshwater stocks that sustain life on the continents make up only a tiny fraction of the global water stock (Table 4–1). These life-sustaining stocks are highly unequally distributed over the world, yielding a vast surplus over human needs in, for example, the Amazon basin and the rivers draining towards the Arctic Ocean, and great deficits in semiarid and arid parts of the world. Humans have significantly modified the regional global water cycle by (1) constructing reservoirs on rivers that retain runoff and allow much evaporation from their surfaces, thereby changing the hydrologic cycle (Chapter 29); (2) transfers of river water from one river to another (**interbasin transfer**) that, among other effects, increases the salinity and endangers the survival of downstream saline lakes and wetlands receiving a reduced water input and (3) tapping river and groundwater for consumption, industrial use, and irrigation of crops (Table 5–2). The preoccupation of climate modelers is predicting if, and by how much, human-produced global warming will affect the hydro-

Table 5–2 **Used water (km^3) returned to waterways by region in the 1980s and projected for the year 2000.**

Region	Domestic and Municipal		Industry		Irrigation		Projected Total Increase (%)
	1980s	2000	1980s	2000	1980s	2000	
Europe	38	48	174	173	15	20	6
Asia	35	100	88	262	320	350	61
Africa	3	12	5	25	35	50	102
North America	46	68	265	310	115	130	20
South America	10	20	24	82	15	20	149
Australia and Oceania	3	4	1	3	3	5	71
former USSR	18	30	105	122	80	90	19
Total	153	282	662	976	583	665	38

Source: Modified after World Resources Institute 1990.

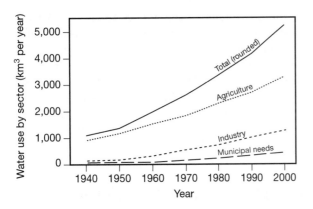

Figure 5–6 Estimated water use by demand sector from 1940 to the year 2000. Note that agriculture increasingly consumes the lion's share. *(After UNEP 1991.)*

logic cycle. If projected modeling scenarios turn out to be correct, the effects will be enormous (Sec. 11.12).

On average some 40 percent of the water used by humans worldwide is returned, often in polluted form, to local streams and rivers, but the balance (60%) is lost as a result of evaporation and removal in the crops harvested. Human benefits received from the water extracted are real, with 18 percent of the world's irrigated cropland producing about one-third the world's food supply. Irrigated cropland continues to grow in area after having already tripled between 1950 and 1985 (World Resources 1988). The large scale withdrawal of water for irrigation and industry (Fig. 5–6), together with interbasin transfers, has led to local

changes in the annual and seasonal discharge of rivers as well as water quality. Changes are also created by the construction of reservoirs in areas of low runoff to manage available water better (Chapter 29).

Hydrology and Land Use

Seasonal variations in discharge are further modified by cutting forests and draining wetlands, which increase the rate of runoff and river discharge, and change its timing (Figs. 5–2 and 5–4). Clearcutting of forests also increases soil erosion and raises the export of dissolved nutrients, sometimes for years (Fig. 5–7). Moreover, in the first few years after forest cutting in semiarid regions both annual and dry season discharge increases substantially because of reduced evapotranspiration. The outcome is more precious water sent to downstream users and disputes about the desirability of reforestation (Scott and Smith 1997). Extensive deforestation in high rainfall regions of tropical and subtropical Asia has allowed (1) more rapid runoff resulting in increased flooding downstream; (2) reduced discharge during the dry season; and (3) increased soil erosion (Fig. 5–7, Chapter 9). The increased sediment loads are reducing (4) river habitat quality and diminishing fish catches (Dudgeon 1992); (5) the volume of downstream lakes and reservoirs; and (6) sediment loads negatively affect plankton and fish through increased watercolumn turbidity, and the benthos by providing sedimenting particles very low in organic matter (Chapter 29). Poor land-use practices are not

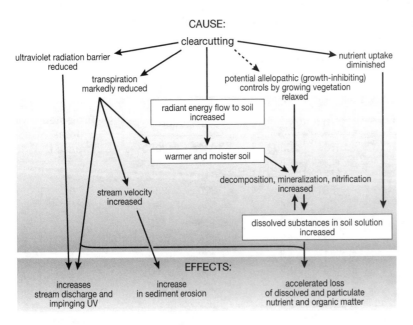

Figure 5–7 Hydrological and biogeochemical responses of forest ecosystems during the first years after clearcutting. *(Modified after Likens 1985.)*

Figure 5–8   Forested buffer strips along the Aroostook River, Maine (US). Note the narrowness and incomplete nature of the buffer strips. *(Photo courtesy of R. Perron, Visuals Unlimited.)*

restricted to low latitudes. A frequent absence of well-vegetated buffer strips between waterways and plowed fields, often highly fertilized and used for crop growing in the temperate zone, leads to soil erosion, the loss of habitat for the biota in the degraded land-water interphase, and the export to waterways of nutrients and pesticides used (Fig. 5–8).

Humans are affecting the quality of lake and river water in ways quite apart from our increasing local and global impact on the hydrological cycle. Rapid growth in agricultural and industrial activities, particularly in the economically less developed regions of the world, plus an enormous increase there in the amount of wastewater produced by a rapidly growing human population and its livestock, are placing tremendous demands on the water resources available (Table 4–1 and Fig. 5–6). Increasing demand for often limited supplies of fresh water (Table 5–2) plus the increasing local and global pollution of water returned to waterways—with not only human and livestock wastes, but also pesticides, fertilizers and industrial chemicals—is drawing limnology, once largely an academic science, into research linked to water pollution and multipurpose management of lakes, rivers, and wetlands.

▲   ## 5.4   Global Patterns in Precipitation and Runoff

Amounts of precipitation, evaporation, runoff (Fig. 5–9) and irradiance (Fig. 5–10) differ greatly with latitude. Continuous heating near the equator (Fig. 5–11

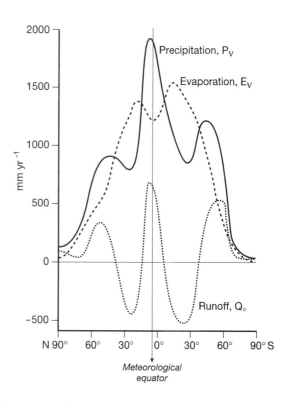

Figure 5–9   World pattern of average annual mean evaporation (Ev), precipitation (Pv), and runoff (Q$_o$) in mm yr^{-1} as a function of latitude. Location of the meterological equator is indicated by thin vertical line. (Modified after Straškraba 1980.)

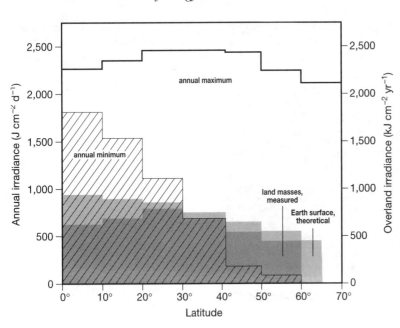

Figure 5–10 Annual irradiance measured over land masses (kJ cm^{-2} yr^{-1}) and the annual maximum and annual minimum irradiance (J cm^{-2}d^{-1}) as a function of latitude. *(Modified after Lewis 1996.)*

and Sec. 10.1) causes air there to rise. The rising air cools with increasing distance from the warm earth; cooling air is able to hold less moisture and the resulting seasonal rainfall is generally high in equatorial regions. This air is replaced by air flowing to the equator from the north and south. The now drier, cooler air is transported by air currents to about 15° to 35° north and south of the equator, where it descends (Fig. 5–11) and is again heated upon approaching the warm earth. However, here the dry air generally absorbs rather than releases moisture, resulting in little precipitation at these latitudes. The combination of low precipitation and strong solar heating results in a high percentage of already low amounts of precipitation to be lost by surface evaporation (Fig. 5–9), typically yielding runoff of only one to eight percent of what fell, so there are very few locally nourished waterbodies and wetlands (Fig. 4–1). The low rainfall regions are further characterized by enormous seasonal and interannual variability in precipitation and runoff that lacks pronounced seasonality. On an annual basis, the negative hydrological budget is associated with extreme stochastic variability in precipitation and runoff. The effect is much larger interannual variability in size, flushing, and biological and chemical characteristics of the receiving aquatic systems than ever encountered by temperate-zone limnologists working in much better watered regions.

The water shortage in low latitude regions of low runoff has stimulated construction of many reservoirs on rivers which usually carry water from distant areas with higher rainfall (Chapter 29). The reservoirs are naturally subject to major evaporative water losses. In the case of the lake Nasser/Nubia Reservoir on the Nile River (EG, SD), the average evaporative loss may appear moderate (13%) when expressed as a percentage of the enormous Nile River inflow. However, in absolute terms the loss is huge (~11 km^{3}yr^{-1}), with the water unavailable for use in electricity generation or irrigation downstream (Sec. 5.6).

A portion of air reaching latitude 15° to 35° returns toward the equator, while the remainder of the reheated and rising air mass is transported by air currents to higher and cooler latitudes, picking up moisture evaporated from the oceans (Fig. 5–11). Upon arrival in the temperate zone (35°–60°), the air is subject to only moderate surface heating by the land and associated moderate evaporation and evapotranspiration. These masses of air are further subject to rapid cooling upon contact with air arriving from polar regions, resulting in frequent supersaturation and considerable precipitation. As a result, the runoff (Fig. 5–9) is relatively high and sufficient to allow rivers to flow year-round, filling the enormous number of glacier-produced basins of temperate zones, and maintaining permanent wetlands (Fig. 4–1). Typically, windspeeds are higher over temperate zone continental lakes in summer than over their equatorial counterparts (Fig. 5–11). Some of the moderately warmed, moist temperate zone air is—following transport,

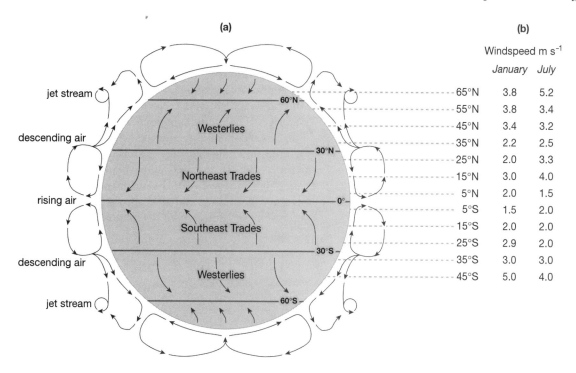

Figure 5–11   (a) Idealized global atmospheric circulation patterns showing prevailing surface winds and movement of air masses and their moisture plus dissolved substances between the equator and the poles. (b) Average January and July overland windspeeds (m s⁻¹) at different latitudes. *(Modified after MacArthur and Connell 1966, and Lauscher 1951.)*

cooling and condensation—responsible for precipitation in polar regions. The air rises again at about 60° north and south latitudes and cold (dense) air holding relatively little moisture is responsible for modest precipitation at high latitudes (Berner and Berner 1987).

Climatic Patterns

The large, relatively regular seasonal variability in the temperate zone of irradiance and temperature, as well as variation in the distribution of precipitation *seasonally* is periodically made less regular by changes in zones of high and low pressure over the North Atlantic ocean (the **North Atlantic Oscillation, NAO**) and the equivalent system operating between eastern Asia and South America (the **Southern Oscillation, SO**) that have pronounced effects on heat flow, precipitation, nutrient cycling, and thereby, on the biota worldwide. Precipitation and streamflow in regions as far apart as New Zealand and eastern North America are marked by periodic Southern oscillation (SO)–**El Niño (EN)** or **ENSO** events. El Niño, occuring every three to seven

years, affects the length of time the ice cover stays on the Great Lakes of North America, with strong El Niño years shortening the ice-cover period and raising total phosphorus concentration, as measured in Lake Huron (Nicholls 1998). NAO events have been linked to periodic changes in precipitation, air and water temperatures, period of lake stratification, changes in timing of the phytoplankton spring bloom in a Swedish and German lake, and zooplankton–phytoplankton interactions (Sec. 23.7) in western Europe (see Gerten and Adrian 2000), as well as the survival of young perch and the period of ice cover in lakes ranging from western Europe to central Asia (Lake Baikal) and as far as northern China. (Livingstone 1999)

In addition to interannual variability there are longer-term patterns, on the scale of decades, in air circulation linked to the NAO that yielded warmer than normal winter temperatures over much of Europe between ~1900 and ~1930, followed by a period of lower temperatures between the early 1940s and early 1970s. A sharp reversal in the NAO index since 1975 has yielded several record warm winter temperatures since 1980 that have been linked to wetter-than-normal winter conditions over northern Europe, and drier conditions over southern Europe and the Mediterranean (Hurrell 1995). The decades-long wet and dry cycles over the Volga River basin and the Caspian sea have also been linked to NAO events (Sec. 5.6, Livingstone 1999).

Pressure systems over the Atlantic and Pacific oceans affect aquatic systems globally.

Contaminent Transport

Movement of air masses from low to high latitudes and laterally across oceans (Fig. 5–12), carry not only moisture, gases, and dust but also volatile organic contaminants and trace metals (Chapter 28). This movement includes high altitude transport of chemicals from industrialized and agricultural regions of the temperate zone and low latitude to distant boreal forest and polar regions, between Africa and Central America, and between northern Asia and North America. The significant increase of toxic mercury and of certain semivolatile organochlorines in high latitude and high altitude waterbodies (Fig. 5–13), many of which are used as pesticides at lower latitudes, is thought to be the result of reduced vapor pressure in cold regions that allows their deposition, retention, and subsequent accumulation in the biota (Chapter 28).

Figure 5–12 July wind flow across North America. Note that the upwind continents include South America, Asia, and Europe. High altitude winds, the Jet Stream (not shown), are positioned above the Pacific air wedges. *(From Bryson and Hare 1974, quoted in Commission for Environmental Cooperation 1997.)*

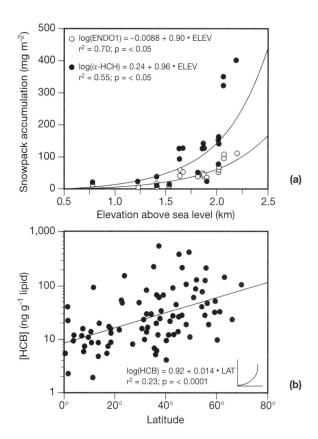

Figure 5–13 Accumulation of (a) Endosulfan-1 and α-Hexachlorocyclohexane (mg m^{-2}) in 1994–1995 snowpack samples from different altitudes in the Rocky Mountains of Alberta and British Columbia, CA and (b) hydrochlorobenzenes in tree bark samples (ng g^{-1} lipids) from different latitudes. *(Part a modified after Blais et al. 1998, and part b after Simonich and Hites, 1995.)*

5.5 Runoff and the Presence of Waterbodies

River Discharge

The average amount of precipitation that becomes runoff in different climatic zones reflects major latitudinal differences in evaporation (Fig. 5–9). When the average *specific runoff* from an area (m^3 km^{-2} yr^{-1}), obtainable from hydrological atlases, is multiplied by the area of the catchment of interest, a first useful guide to the longterm average annual river discharge is obtained. Longterm records of average annual precipitation and runoff naturally do not consider the substantial interannual or interdecade variation in runoff (Fig. 5–14) that affects not only the nutrient supply and algal biomass, the organic matter supply to aquatic system, and the sediment export from drainage basins to rivers but also the magnitude of river discharge and river size (Fig. 5–15 and Chapter 8). Discharge and evaporation determine water residence time of aquatic systems (Chapter 9). River discharge also largely determines the size of wetlands (Table 6–3 and Sec. 24.2), and the surface area and volume of saline lakes (Sec. 5.6).

▲ Average precipitation in the north temperate zone as a whole has increased during the 30-year period (1950–1980) for which data are available, accompanied by a decrease in the tropics and subtropics (Fig. 5–16). Thus, changes observed during this interval in inland waters of the north temperate zone that are attributed to human impacts, must have been exacerbated by the effects of generally higher runoff in most areas and resulting increased nutrient loading. Similarly, declines in river discharge, wetland size, lake levels, and salt content of low latitude inland waters attributed to human activity (Secs. 5.7 and 5.8), were accentuated by precipitation declines during the same interval.

Precipitation and Waterbodies

The number and size of rivers, lakes, and wetlands in any climatic zone is partly determined by the volume of runoff per unit area, known as the **specific runoff** (m^3 km^{-2} sec^{-1}), and by the number of basins suitable to hold the water. A combination of an enormous number of basins produced by the last and relatively recent glaciation (~7,000–12,000 years BP) in the present boreal forest zone and adjacent deciduous forest zone (now largely converted to agriculture), together with a relatively high runoff and modest evaporation (Fig. 5–9) are responsible for the abundance of lakes and wetlands between about 40° and 60° N and S (Fig. 4–1 and Chapter 6). Conversely, runoff from ancient and well-eroded landscapes at lower latitudes encounters few suitable basins deep enough to hold water year round. As a result, water at lower latitudes is mostly held in ephemeral lakes and ponds and long and winding rivers containing riverine lakes and associated wetlands in their floodplains.

The principal exception to river, rather than lake, dominated landscapes in the lowland tropics is the Rift Valley of East Africa. Lakes Malawi, Tanganyika, and Victoria (Fig. 5–17 and Fig. 6–1) lie in basins that have not had sufficient time to be filled with sediment from their large, but generally low runoff drainage basins and the lakes are deep and transparent. The vast majority of subtropical and tropical lakes are located in well-eroded drainage basins with low relief

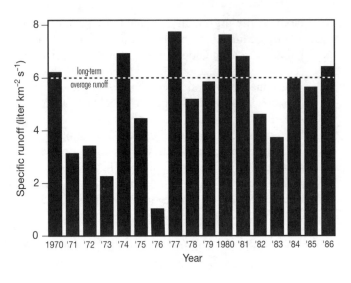

Figure 5–14   Annual mean water export coefficient (or mean specific runoff, l km^{-2} s^{-1}) from the catchment of a north temperate river between Lake Vallentunasjön and Lake Norrviken (SE). Dotted line indicates the longterm average runoff. The nine-fold interannual variation is modest compared to that observed in semi-arid regions at low latitude. *(After Ahlgren 1988).*

(Chapter 6). They are shallow and highly subject to wind, temperature or river-flow induced currents that promote mixing (Chapter 12).

5.6   Water Inputs and Outputs

Precipitation and Evapotranspiration

Aquatic systems in the low precipitation, low latitude, semiarid or seasonally arid zone are subject to high evaporative losses. Evaporative losses from land and water surfaces are sufficiently high to yield a periodically reduced or zero discharge of local streams and rivers, as well as large variation in the size of normally shallow lakes and seasonal wetlands. In the semiarid zone zero discharge from lakes and wetlands is common. The river water entering these systems is typically lost through evaporation, the salts in the inflowing water are retained. Lakes with no stream or subsurface seepage outflow are known as **closed basin** or **endorheic lakes**, and are more or less saline. In **endorheic regions**, river water does not reach the ocean via outflowing rivers. Lakes located in higher runoff catchments, or those characterized by lower surface evaporation, have a stream and/or a ground-

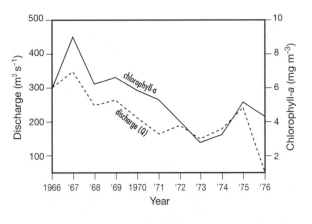

Figure 5–15   Mean spring river discharge (Q$_{Jan–May}$, m^3 s^{-1}) and mean summer (May–Oct.) algal biomass (chlorophyll-*a* at 1m depth, mg m^{-3}) in central Lake Mälaren, SE (LA = 1140 km^2 $\bar{z}$ = 11.9 m; $\bar{x}$ WRT = 2.2 yr). Note the importance of river discharge in determining the summer nutrient supply, which is normally the principle determinant of the phytoplankton biomass (Chl-*a*). *(Modified after Wiederholm, 1978.)*

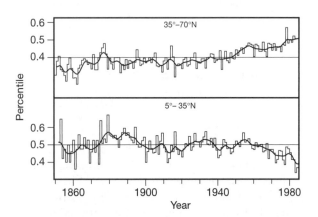

Figure 5–16   Precipitation indices for zones 35° to 70°N and 5° to 35°N. Note the variation over time and the large changes observed since 1960, with changes more pronounced in Africa, the US and the former USSR than in Europe, southeastern Asia, or the equatorial region (0°–5°N). *(After Bradley et al. 1987.)*

water (seepage) outflow and are known as **open basin** or **exorheic lakes**. Here wetlands tend to be inundated year round.

The paucity of natural lakes (and wetlands too), most of which have closed basins, between ~15° to 35° N and S, is more evident from their distribution in the Southern than in the Northern Hemisphere (Fig. 4–1). In the Northern Hemisphere the relative distribution of nonmarine saline waters is somewhat distorted by the enormous surface areas of the Caspian Sea and, until recently, the Aral Lake (Table 4–1 and

Sec. 5.8). The principal saline lake regions of the world are shown in Fig. 5–17.

Precipitation, Evaporation, and Discharge

High precipitation and relatively low evaporation characterize the exorheic portions of Europe, resulting in the largest lakes receiving an average of 84 percent of their water through stream plus groundwater inflows while losing only ~12 percent through surface evaporation (Fig. 5–18) Higher summer temperatures

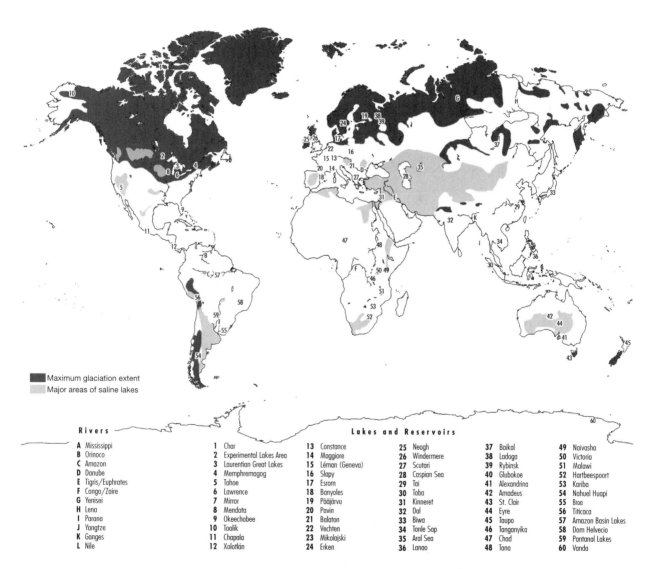

Figure 5–17 The maximum extent of the last glaciation (dark grey area), and the principal areas where saline lakes occur (light grey area), and some large or limnologically well known lakes, river and reservoirs. Most but not all lakes in the saline lake districts are saline. Saline lakes [S] also exist outside the major endorheic regions. *(After Snead 1980, and Williams 1995.)*

over large areas of continental North America allow for greater catchment and lake surface evaporation than in western Europe and, proportionally greater rainfall directly on lake surfaces. The large freshwater lakes of tropical South America and Africa lose a great deal of water through surface evaporation, and as a result, have outflow volumes that are only a small fraction of the inputs. The difference between these regions is that large South American lakes and wetlands tend to lie in a seasonally high rainfall zone and receive most of their water through inflowing rivers, whereas their much lower rainfall African counterparts are subject to high catchment evapotranspiration and are more dependent on seasonal rainfall directly onto their surfaces than on river inflow. On average, only ~12 percent of water input in large African and South American freshwater lakes leaves via outflow as the result of enormous evaporative losses. In contrast, on average ~88 percent of large lake input leaves via outflow and only 12 percent is lost by evaporation in cooler European regions. (Fig. 5–18).

As plant nutrients, organic carbon, and contaminants are supplied to lakes by inflowing rivers and groundwater, and direct precipitation/deposition on their surfaces, it is evident that large European and northeastern North American lakes must receive a much higher fraction of nutrients from their catchments than their much less well-studied African counterparts. The latter are more dependent on plant nutrients and contaminants received via rainfall, supplemented by dustfall. The intermediate South American large lakes are expected to have a more equal nutrient loading from river discharge and rain and dustfall. High surface evaporation, which characterizes tropical and subtropical catchments and water surfaces, concentrates solutes and presumably affects nutrient cycling in inland waters to a much larger extent than in the temperate zone (Chapters 14, 17, 18, 19).

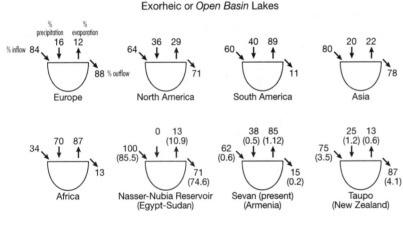

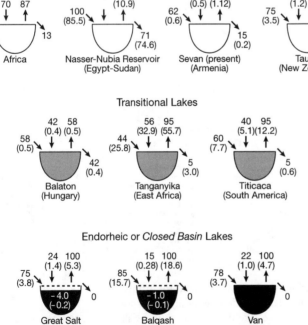

Figure 5–18 The long-term average water inputs and outputs, expressed as a percentage of the total for large lakes and reservoirs (>1,000 km^2) on different continents, plus data for individual lakes that also show the absolute values (km^3 yr^{-1}) in brackets. Evaporation determined by difference in a mass–balance. The transitional lakes presented are currently open, but would close under only slightly drier (lower precipitation) conditions. Great Salt Lake and Balqash decreased in volume during the intervals investigated. (*From data in Voskresensky 1978, and Vikulina et al. 1978.*)

The average input–output (mass–balance) patterns presented in Fig. 5–18 cannot reveal whether any particular lake or wetland receives its water and inorganic nutrients primarily via land or atmosphere. The relative importance of land versus atmosphere for individual lakes and wetlands is determined in part by how readily the catchment rock and soils are weathered (Sec. 8.3 and Chapter 13), the hydrology, soil development and land use, and the size of the catchment area relative to that of the lake—the **CA:LA ratio**. More about the importance of this ratio in Chapters 8 and 9 and in later chapters.

Closed Basin Lakes

The long-term relative distribution of water input and output for the three closed basins presented broadly reflect the pattern for all such lakes (Fig. 5–18). They are characterized by surface evaporation that, over a longer period, is larger than or equal to the sum of river—plus groundwater—and precipitation inputs. During a series of particularly dry years or centuries, evaporation greatly exceeds water input and lake levels fall, the sequence is reversed during years of above-average rainfall. The resulting increased lake surface area enhances evaporation (a surface phenomenon), which at some point will offset water input and thereby terminate the increase in surface area and volume. Conversely, a reduced surface area will retard evaporative losses during drought periods something long appreciated by hydrological engineers. For example, to reduce evaporative losses from Lake Sevan (Fig. 5–18), and thereby increase lake discharge for electricity generation and irrigation, Sovjet engineers lowered the lake level by 20 m. This reduced the lake surface area by 13 percent, decreased evaporation 4.5 percent and increased discharge 29 percent, after subtracting from the increased discharge the amount resulting from higher rainfall on the catchment after its lowering and not linked to manipulation. The economic gain was considered, but the enormous negative environmental and aesthetic consequences were not.

▲ Shallow, closed lakes such as Lake Chilwa (LA 750 km^2, lake range ~0–2500 km^2 plus 390 km^2 of associated wetlands) with a mean depth of about 2.5 m during a series of wet years, is located south of Lake Malawi (Fig. 6–1). It dried up seven times during the 20th century with disastrous results for the local fishery, which contributes about half of the Malawi catch during highwater years (Beadle 1981). Lake Eyre

(0–8200 km^2) and Lake Torrens (0–6000 km^2) in Australia are more extreme; they only hold water during a few exceptionally wet years each century and are dry salt pans the rest of the time. Another unusual lake is Lake Chad (west Africa). This once enormous lake lacks a surface outflow, but remains open and fresh as the result of seepage outflow that carries water and salts from the lake.[2] After subsurface evaporation, the salts are precipitated in soils well away from the lake (Beadle 1981).

Transitional Lakes and Wetlands

In addition to saline closed-basin lakes and low salinity (fresh water) open-basin lakes, there are **transitional lakes** and **wetlands** that oscillate between being open and closed as the result of changes in climatic conditions. Lakes Tanganyika (east Africa), Titicaca (South America), and Balaton (HU) lie in different climatic regions and differ in relative distribution of water inputs and outputs (Fig. 5–18). However, all of them have a long-term lake discharge slightly larger than the sum total of the water inputs minus evaporation, and a lengthy water replacement time (WRT). Consequently this type of lake becomes increasingly saline during dry periods (lasting from decades to centuries), but becomes increasingly fresh during wet cycles. Their inflowing rivers also show long-term changes in discharge and salinity. Some lakes, such as Turkana, Nakuru, and Magadi in the Rift Valley of east Africa (Fig. 6–1) are closed and saline today (Chapter 13), but were fresh and much larger during the last major pluvial or high rainfall period (~8000 BP). Conversely, Lake Tanganyika and Lake Kivu (Fig. 6–1) are open basin lakes under the present climatic regime, but Kivu has experienced much lower water levels and been closed several times during the past 11,000 years (Hecky 1978). Lake Valencia (VE) (Bradbury et al. 1981) is another example of a lake that was open and fresh 200 to 300 years ago, but is now closed and exhibits rising salinity.

It is not only the somewhat arbitrary decisions as to whether to characterize a particular lake as fresh or saline (Sec. 13.8), but it is above all the existence of transitional lakes that makes it evident that studying

---

[2]While all endorheic lakes lacking seepage outflow have elevated salinity and nearly all saline lakes are closed (endorheic), there are some exhorheic saline lakes, fed by streams draining highly soluble beds of salt.

saline lakes is an integral part of limnology. Limnology is, therefore, properly defined as the study of inland waters as systems rather than as the study of fresh water, or the study of freshwater systems. Among lakes, transitional and saline lakes are most susceptible to changes in lake size and salinity as the result of climatic change and human activity. River diversions, for irrigation or hydroelectric purposes have a particularly large long-term effect on transitional and saline lakes. The impact of anthropogenic and climatic effects is most graphically evident from research on the Aral Sea and Caspian Sea, two saline lakes in the southern portion of the former USSR.

▲ 5.7   The Aral Sea

This moderately saline lake, located about 500 km east of the Caspian Sea (KZ, UZ) (Fig. 5–17), has suffered greatly from water diversions. In 1960 it was ranked the world's fourth largest lake by area (Table 4–5). By 1987 its area had decreased 40 percent to 41,000 km², its volume reduced by 66 percent, and its water level by 13m because of 70 percent reduction in river inflow brought about by withdrawal of irrigation water. This decline has continued with the estimated surface area down 64 percent and volume down 84 percent in 2000 (Figs. 5–19 and 5–20).

Available data show that the two principal inflowing rivers did not even reach the lake every year during the dry 1980s. From 1960 to 1987 the irrigated land area in the lower portion of Aral Lake's catchment more than doubled to 6.4×10^4 km².

The effects of the massive reduction of freshwater inflow on the lake and its environment have been disastrous. The increase in lake salinity from about 10,000 mg l⁻¹ (10‰) to 34,000 mg l⁻¹ (34‰) in 1994 (Golubev 1996) has led to major changes in the biota, loss of nearly all fish species, and decline of the annual commercial fish catch from 48,000 tonnes to zero today. Major reduction in phytoplankton biomass occurred by the mid 1970s when average salinity first exceeded 11,000–14,000 mgl⁻¹ (Aladin et al. 1993). Phytoplankton as well as zooplankton and benthic communities lost their freshwater and brackish species. The first components of the zoobenthos to disappear were low-salinity (**oligohaline**) species such as insect larvae and oligochaete worms. Some were replaced by higher salinity (**euryhaline**) species, some of which were introduced by people. The lowering lake level and drying delta wetlands of inflowing rivers has led to massive decline in emergent macrophytes and their associated fauna (including waterfowl). However, submerged macrophytes continue to grow to considerable depth in this now more highly transparent oligotrophic lake, with benthic plants

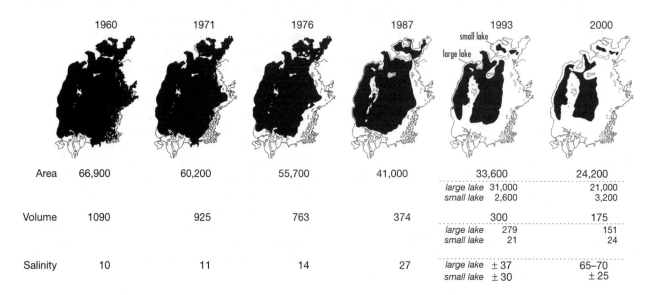

	1960	1971	1976	1987	1993	2000
Area	66,900	60,200	55,700	41,000	33,600	24,200
					large lake 31,000	21,000
					small lake 2,600	3,200
Volume	1090	925	763	374	300	175
					large lake 279	151
					small lake 21	24
Salinity	10	11	14	27	large lake ± 37	65–70
					small lake ± 30	± 25

Figure 5–19   The changing surface area (km²) volume (km³) and salinity (‰) of the Aral Lake, from 1960 to 1993 and a prediction for the year 2000. (*Modified from Micklin 1988, and Micklin, personal communication.*)

Figure 5–20 The Aral Sea in the year 2000. Note the band of recently precipitated salts along the edge. The presently separated small lake(s) on the left in the photo are partly osbcured by clouds. *(Photo courtesy of NASA/Johnson Space Center.)*

contributing more than 90 percent of the plant biomass (Aladin et al. 1993).

Gradually increasing salinity and declining wetland area first affected the spawning of fishes of fresh water origin, but was exacerbated by the construction of dams on inflowing rivers that prevented upstream spawning and by stocking alien fish species. Commercially-fished species virtually ceased to reproduce by the mid 1970s when salinity rose above 14,000 mg l^{-1}. Aral Sea lost its commercial fisheries in the early 1980s when salinity had risen above 18,000 mg l^{-1}.

Observations in the Caspian Sea (Sec. 5.8) make it clear that not all changes are attributable to the deleterious effects of increasing salinity. Rising salinity also reflects declines in the input of nutrients and organic matter carried by progressively smaller river inflows. Futhermore, the introduction of new species of fish and invertebrates, beginning in 1927, was followed by changes in foodweb structure. Extensive use of pesticides in the catchment may also have contributed to the changes observed. (Aladin et al. 1993).

Much of the ~43,000 km² of previous lake bottom, now exposed, is covered with enormous quantities of precipitated salt (NaCl 56%, $MgSO_4$ 26%, $CaSO_4$ 15%) deposited as the result of surface evaporation during the shrinking (Fig. 5–20). About 40–150 × 10⁶ tonnes of these salts are, as the result of annual wind erosion, blowing over a land area of 150,000–200,000 km², with salt storms of such magnitude they are visible on satellite images. The blowing salt plus lowering (in part of the catchment) the groundwater level (3–8 m) through irrigation and evaporation, caused rapid

desertification and reduction in the carrying capacity of the land for livestock and wildlife. Excessive irrigation and leaking irrigation channels in the delta region of the two principal inflowing rivers created new lakes and wetlands in the desert (Golubev 1996).

Poor-quality drinking water from wells containing high levels of pesticides and fertilizers, plus salt-laden air appears to have contributed to sharply increased human mortality rates since the 1960s. Throat cancer in one region has increased 7–10 times (Mandych 1995). As a final result of the much smaller lake surface area, the ameliorating influence of the lake on the regional climate (Sec. 11.12) has been reduced, resulting in a dangerously shortened growing season for cotton (hotter summers, colder winters).

In the Soviet era, plans were proposed for a huge 2,500-km-long canal, with associated dams and pumping stations to transfer part of the flow of two western Siberian rivers to the Aral Sea (Micklin 1988). However, the enormous cost of this endeavor led to its cancellation. Although demands for fresh water are projected to continue growing, the question of whether the lake will continue to shrink at estimated rates will depend on climatic conditions, such as precipitation and runoff. Newspaper reports claim the lake surface area was 41,000 km² in 1998, identical to 1987 (Fig. 5–19), but a little larger than in 1993; reflecting an apparent increase in precipitation and catchment runoff in recent years (Kindler 1998). If the continued increase in the demand for water is coupled with a return to the low levels of runoff recorded in the 1980s, and if no significant attempt is made to improve irrigation efficiency, the present two lake basins will probably separate into a larger number of separate saline lakes totaling a mere 9,000 km^2 by 2015.

▲ 5.8   The Caspian Sea

The Caspian Sea is the world's largest lake, with a surface area the size of Great Britain (Table 4–5, Fig. 5–17). It is also the most voluminous inland waterbody on earth. In contrast with Aral Sea, its water levels have not been affected very much by human activity. The Volga River supplies about 75 percent of the inflow to the lake. Water level changes are largely determined by longer cycles of precipitation and runoff that are influenced by long-term patterns in atmospheric circulation over the North Atlantic Ocean imposed by the North Atlantic oscillation (NAO).

After falling sharply between the late 1920s and 1977, lake levels rose approximately 2.5 m by 1994, reflooding a 32,000 km^2 area. Nevertheless, water levels still remain 1.5 m lower than during the period of higher precipitation experienced between 1840 and 1880 (Golubev 1996). Human influence on the biology of the Caspian Sea has been important because of changes made in the discharge regime of inflowing rivers (Rozengurt and Hedgpeth 1989). The Sovjet government built eleven large hydroelectric irrigation reservoirs plus several hundred smaller reservoirs on the inflowing Volga River and its tributaries over many years. The principal human effect on the lake has been a shift in timing of the maximum river discharge, rather than the modest (~12%) long-term reduction in annual discharge noted in 1977.

Before the reservoirs were built, maximum runoff was in the spring, this was reduced by nearly 40% when maximum runoff shifted to summer. The effect of this shift on the fisheries was dramatic; Fish species breeding in fresh and brackish water lost most of their spring breeding habitat in the formerly low salinity deltas of inflowing rivers, until mitigated by the recent period of increased discharge. The springtime upstream spawning migration of fish species, such as salmon and the six species of Caspian sturgeon which spawn in fresh water was disrupted by reduced spring inflows, interference of spawning migrations by dams, and poorly documented large fish kills. Until recently, the most severe effects on fish stocks were observed during the driest years, when spring discharges were particularly small. Unregulated fishing and poaching

since the collapse of the USSR has led to the collapse of sturgeon and salmon stocks.

Reduced river discharge during the decades of low runoff caused the north Caspian Sea's salinity to increase from 8,000 to 11,000 mg l^{-1} (8–11‰) which may have been responsible for a major reduction in the benthic invertebrate biomass. Larger invertebrates are the principal food resource for economically important sturgeon and salmon.

The cascade of reservoirs on the Volga serve, like reservoirs everywhere, serve as sediment and nutrient traps (Table 9–4 and Chapter 29). The reservoirs increased the water retention time of river water 8–10 times, totaling almost 180 days, allowing thick layers of inorganic sediments to accumulate (~60% of the total), and retention of about 50 percent of the previously delivered phosphorus. The input of soluble silica, required primarily by diatoms (Chapter 21), declined greatly because much of the river-transported dissolved silica is taken up by diatoms in the reservoirs and largely retained after sedimentation. It is believed that reduced phosphorus loading during the flood season was responsible for, or contributed to, the 50 percent reduction observed in primary production of the north Caspian during the low-water decades. The silica decline may have been responsible for the observed shift in phytoplankton species composition from traditional diatom dominance in spring and summer biomass dominated by blue-green algae, to one dominated by the large filamentous green alga *Spirogyra*. However, data are too fragmentary to allow more than speculation about causes of these changes (Kosarev and Yablonskaya 1994).

Human activity was blamed for most of the changes observed in the Caspian Sea during the decades of reduced river discharge. Unfortunately, the hydrological and human effects are confounded because the period of greatest human impact, which started in the 1930s, coincided with a period of reduced precipitation in the Volga River drainage basin. The post-1930 construction on the exposed lake bed or close to what was the shoreline—of now obsolete petrochemical, chemical, and metallurgical plants which leak contaminated water into the lake, plus leaking oil wells that are progressively inundated—are major ecological problems. Another problem is a near-shore nuclear power station that uses the lakewater for cooling, and whose liquid and solid nuclear waste storage sites pose a major potential threat to the lake if water levels continue to rise

(Dumont 1995). Unrelated to human activity, the recent large inflow of low conductivity (low density) fresh water, which lies on top of the much higher salinity (high density) water is reducing water exchange between surface and deep water. If the observed decline in the presently modest (<3 mg l^{-1}) deep water dissolved oxygen concentrations (Kosarev and Yablonskaya 1994) continue as expected, the deep waters are predicted to become anoxic by about 2006, with large consequences for the biota, increased nutrient recycling from deep water to the epilimnion, and the foodweb structure (Chapters 11, 12 and 15). Lakes that do not fully mix are known as *meromictic* and are discussed in Section 11.10.

Highlights

- Precipitation that is not evaporated or transpired in catchments will leave as subsurface runoff and groundwater, carrying primarily dissolved substances, and as surface runoff carrying particulate matter.

- The specific path water takes on land is of great importance in determining the soil contact time and thereby the chemical composition and export of dissolved materials to inland waters.

- Humans have significantly modified the regional and global water cycle through construction of reservoirs on rivers. These allow additional evaporation and increase water residence time in the system. Other major modifications include the interbasin transfers of river water, tapping river and groundwater for household and industrial use and, most significantly, through water withdrawal for irrigation. Water is often returned polluted.

- The number and size of rivers, lakes, and wetlands in any climatic zone is determined in part by the annual volume of runoff per unit area, the specific runoff (m^{3}km^{-2}yr^{-1}), and by the number of lake and wetland basins suitable for holding water.

- Lakes and wetlands located where water loss through evaporation is greater than or equal to outflow are saline to some degree. Lakes without stream or seepage outflows are known as closed basin or endorheic lakes, while those with an outflow are known as open basin or exhorheic lakes. With few exceptions, open basin lakes are fresh. Transition lakes oscillate between open and closed. They become gradually saline during dry periods (decades–centuries) and increasingly fresh during wet cycles.

- ▲ Since 1960, water diversion for agriculture from inflowing rivers has reduced the surface area of the Aral Sea—once the fourth largest lake in the world—by an estimated two-thirds and 84 percent in volume. The salinity has more than tripled, the fisheries have collapsed, the biota has changed dramatically, and human conditions have deteriorated greatly. The Aral Sea disaster stands as an extreme example and warning that (1) a protection of aquatic resources is essential if natural resources and human welfare are to be safeguarded, and (2) aquatic resources cannot be managed in isolation from their drainage basins.

- ▲ The Caspian Sea is an example of how the construction of reservoirs on rivers, with associated changes in timing of the seasonal discharge patterns of rivers, has major impacts on the biota.

6

Origin and Age of Lakes

6.1 Introduction

A small number of very deep and very ancient lakes, such as Baikal, Tanganyika, Malawi, Caspian, and Biwa, were formed somewhere between two and twenty million years ago (Meybeck 1995), and a slightly larger number formed >100,000 years ago (e.g., Tahoe), but the vast majority of lakes were created between about 15,000 and 6,000 years before present (BP), following the retreat of continental glaciers from higher latitudes of the present temperate zone (Fig. 5–17). Most coastal lakes are also recent, because present sea level stabilized only about 6,000 years ago.

Lakes continue to form today for various reasons, such as filling basins left by the retreat of glaciers from mountain valleys; Coastal lakes that emerge following the continuing rebound of land in response to removal of the weight of continental glaciers. In the semitropics and tropics waterbodies are created by depressions filling with water during the rainy season or following a series of dry years. Lowland rivers continue to create lakes by cutting off river sections during their meanderings across their floodplains (Sec. 6.5). Lowland rivers also refill or enlarge abundant floodplain lakes when they overflow their banks during the flood season. Both humans and beavers create lakes and ponds (reservoirs) by damming rivers and streams.

Lake Filling

All lakes receive sediment from their catchments (Sec. 9.7), which, together with particulate matter produced as the result of within-lake synthesis of organic matter, will ultimately fill the basins (Chapter 20) unless earth movements interfere. The filling of basins is particularly slow in lakes on small, well-vegetated catchments (Fig. 9–9). Ancient Lake Tanganyika with more than 6,000 m of sediment (Scholtz and Rosendahl 1988) has slowly reduced its basin depth by three-quarters since its formation, but at an average rate of only 0.3 mm yr^{-1} (Chapter 20). In sharp contrast, reservoirs built on large, poorly vegetated catchments may yield three or even ≥ 40 cm yr^{-1}, filling most of their basins within less than a century (Sec. 9.7).

Excellent work in northern England shows sharply increased (4×) rates of sediment accumulation in lakes starting about 1,000 years ago, the result of deforestation and cultivation of their catchments allowing substantially increased soil loss to waterways (Pennington 1981, and Fig. 9–9). Consequently, a substantial number of shallow lakes have filled in and disappeared during historical times in Europe, the Middle East and in southern and eastern Asia.

Sedimentation rates vary as a function of the productivity of lakes and particularly in response to the amounts of inorganic and organic particulate matter carried from their catchments (Chapters 9 and 29). However, rates between 0.1 and 1.0–2.0 mm yr^{-1} would encompass a high percentage of lakes in vegetated catchments at higher latitudes (Chapter 20), providing some idea of realized filling rates for natural lakes in well-watered regions.

Lake Types

In an elegant chapter, Hutchinson (1957) has discussed all major lake types, their origin, morphometric characteristics, and their distribution over the

Table 6–1   **Geologic origin and an order of magnitude estimate of the abundance of the major lake types by origin (> 0.01 km² or 1 ha) and total lake surface area based on regional averages.**

Origin	Number of Lakes[1]	Percent of Total Lakes	Total Lake Area (km²)	Percent Total Area
Glacial	3,875,000	74	1,247,000	50
Tectonic	249,000	5	893,000[2]	35
Coastal	41,000	<1	60,000	2
Riverine	531,000	10	218,000	9
Volcanic	1,000	<<1	3,000	<<1
Miscellaneous	567,000	10	88,000	4
Total	5,264,000	~100	2,509,000	100

[1]Approximate only. The number of large ponds (0.01–0.1km²) that dominate numerically are particularly uncertain because of the widespread availability of only coarse-scale maps. Extrapolations based on the size distributions of lakes in the well documented larger size categories suggest a more plausible total of about 8.4×10^6 lakes >0.01 km² of which ~85% are in the 0.01–0.1 km² size category, with a total lake area of $~2.8 \times 10^6$ km². There are in addition millions of smaller ponds (Table 4.4).
[2]The Caspian Sea (374,000 km²) accounts for 42% of the global tectonic lake area.

Source: After Meybeck 1995.

earth. He recognized 11 major types and 76 subtypes of lakes.[1] Six of these types have been identified as the major causative factor in the formation of present lake basins (Table 6–1), and these plus manmade lakes (reservoirs) will be discussed in this chapter. Timms (1992) provides a good, recent discussion of lake types and lake formation and the names of some well known lakes and their origins are given in Table 6–2 and Fig. 5–17. Note that the relatively poorly studied large ponds (0.01–0.1 km²) numerically dominate (~85%) the total (Table 6–1).

6.2   Glacial Lakes

Glacial lakes contribute about three-quarters of the standing (lentic) waterbodies greater than 0.01 km² worldwide (Table 6–1). There are uncounted millions of smaller waterbodies that are remains of glacial movement over the landscape at latitudes higher than

[1]Additional major lake types listed by Hutchinson include landslide lakes; aeolian lakes, which owe their origin to wind action; lakes formed by organisms, for example, cattle hollows; and meteorite impact lakes. Probably the most unusual lake type of all is the relatively recently discovered freshwater ponds that form during the summer meet on the McMurdo ice shelf. This glacier shelf floats attached to land on the Antarctic ocean (Vincent 1987), and there is a similar process in the arctic.

~40°N in Europe and North America, and ~60°N in Asia. For the names of some well known lakes see Table 6–2.

At it's maximum, 20,000 years BP, the last glaciation—called the "Wisconsin" in North America and "Weichsel" or "Varsovian" in central Europe, plus their equivalent in the Southern Hemisphere—covered about one-quarter of the world's land area (Fig. 5–17), this area is now reduced to about 10 percent. The impact of glaciation and its significance in lake formation is evident mostly in the Northern Hemisphere because there is relatively little land in the south temperate zone, other than Antarctica. Only a small area of South America (>36°S), Tasmania, the southern tip of mainland Australia, and New Zealand's South Island were located in areas of continental glaciation.

Types of Lakes

Glacial activity produced four types of lakes. The development of a lake is caused by (1) an ice barrier; (2) glacial erosion; (3) glacial deposition; and (4) a mixture of glacial activity and nonglacial processes.

Lakes formed by ice barriers (type 1) are small lakes associated primarily with the Greenland icecap and Antarctica. Such lakes, known as **proglacial lakes**, were common when the continental glaciers retreated

Table 6–2 **Geological origin of some large or limnologically well known lakes in different parts of the world. A secondary origin, if known, is included in brackets.**

Glacial (G): Constance (AT-CH-DE), Ontario (CA-US), Como (IT), Loch Neagh (UK), Linsley Pond (Connecticut US), Great Bear (CA), Mirror (US), Memphremagog (CA-US), Windermere (GB), Léman (CH-FR), Erken (SE), Loch Leven (GB), Peters (Alaska, US), Stechlin (DE), Mikolajski (PL), Ladoga [T] and Glubokoe (RU), Ruwenzori lakes (UG), Gangabal (IN), Nahuel Huapi (AR), Wakatipu (NZ), St. Clair (AU), Heywood (Signy, AQ)

Tectonic (T): Victoria (KE-TZ-UG), Balaton (HU), Okeechobee (US), Thingvallavatn [G, V] (IS), Abaya (ET), Kivu [V] (RW-ZR), Malawi (MW-MZ-TZ), Tanganyika (BI-TZ-ZR-ZM), Tahoe (US), Paijanne [G] (FI), Urmia (IR), Baikal (RU), Issyk-Kul (KG), Ziling (Tibet, CN), Dead Sea (IL-JO), Eyre (AU), Nicaragua [V] (NI), Valencia (VE), Great Salt (US)

Coastal (C): Sibaya (ZA), Alexandria (AU), Eskimo North (CA), Selawik (US), Aby (CI), Kurisches (LT), Luang (TH), Chilka (IN), Terminos (MX), Patos (BR), Chirique (PA), Ablation (AQ)

Fluvial (F): Dong Ling (CN), Poyang (CN), Hung Ze (CN), Tonle Sap (KH), Mai Ndombe (ZR), Ponchartrain (US), Manzala (EG), Edku (EG)

Volcanic (V): Crater (US), Pavin (FR), Zuni (US), Viti (IS), Kauhako (Hawai, US), Atitlan (GT), Transimeno (IT) Pulvermaar (DE), Bunyoni (UG), Bishofter (ET), Toba (ID), Lanao [T], (PH), Haruna-Ko (JP), Nyos (CM), Taupo [T], (NZ), Barrini (AU)

Solution: de la Girotte (FR), Scutari [T] (AL-YU), Banyoles (ES), Tyrrell district (AU), Nelson district (NZ), Gorki district (RU), Bottomless (New Mexico, US), Central Florida and Kentucky (US)

---

and blocked natural outflows.[2] Lake Agassiz in the Prairie region of Canada is the largest proglacial lake known, once covering an area of 350,000 km². The lake, with a maximum depth of 200 m, lasted in various forms for a few thousand years (from 12,000 to 8,000 yrs BP) until the ice retreated far enough northward, causing it to suddenly drain. One of its remnants is Lake Winnipeg (24,500 km²).

Advancing continental glaciers scraped the land, pushing rock and earth ahead or transporting it within their mass. In the process, they not only reworked but scoured the landscape (type 2). When the last glacial age ended the glaciers modified the landscape again as they retreated and left behind layers of stones and fine particles of variable thickness on top of the underlying bedrock, referred to as **glacial drift** (**drift**). When it is evident that the drift was derived from sediment transported by glacial ice, with little sorting by water, it is known as **till**. Drift (till) is typically composed of a mixture of minerals and rock types that reflects the glacial transport of rock types from the source regions. Today's soils developed from the physical and biological/chemical breakdown (weathering) of the drift layer and the subsequent incorporation of organic matter into it.

Drift covered large blocks of ice that had broken off from the glaciers during the last glacial retreat. When these blocks melted they left behind an enormous number of what are called **kettle** or **pothole lakes** (type 3). Kettle lakes are relatively deep and have small drainage basins. This lake type, together with shallow basins formed by irregularities in the drift, dominates on the large glaciated plains of North America and Eurasia. They are frequently located in well-developed, nutrient-rich soils and tend to be productive.

Most abundant of all are the enormous number of **ice-scour lakes** (type 2) produced on the vast precambrian shield region of North America and Europe, which roughly coincides with boreal forest region on the two continents. In this region, underlain by extremely hard rock and little overlying drift, the glaciers scraped and gouged the jointed and fractured bedrock, creating a vast number of shallow basins now occupied by lakes, ponds, and wetlands (Meybeck

---

[2]A lake that fits no category is Lake Vostok, a vast body of fresh water (14,000 km², z_{max} = 500 m, estimated WRT ~50,000 years), similar in size to Lake Ontario, but located four km below the Antarctic ice sheet, warmed by heat flow from the earth below. It is at least 2 million years old and may even predate Antartica's glaciation, which occurred ~40 mya. As ice samples taken from above the lake contain bacteria, algae, and fungi, it is plausible that the enormous lid of ice is releasing living organisms into the water as it slowly melts; organisms that are 500,000 years older than when they were deposited on the Antarctic ice cap! (A. Kapitsa et al. 1996.)

1995). Located in drainage basins characterized by low average air temperatures and poorly developed soils, the lakes tend to be nutrient poor. The low relief further implies poor (slow) drainage, encouraging development of peat bogs (wetlands) that release highly-colored dissolved organic matter into the outflowing streams and seepage water, yielding brown water (humic) rivers, lakes, and wetlands (Chapter 8).

Glaciers scoured deep basins where advancing icesheets moved through preexisting valleys with relatively soft or highly fractured rock. When subsequent climatic cooling allowed a temporary readvance, the glaciers bulldozed dams. They deposited sufficient material during their final retreat to block (dam) valleys with what are known as **terminal** or **end moraines**. Drift deposited at the edge of major valleys, known as **lateral moraines**, closed tributary valleys, thereby allowing additional lakes to be created (type 2, Fig. 6–1). Glacially widened, deepened, and dammed valleys have produced large, beautiful lakes and lake districts in, for example, New Zealand, Tasmania, southern Argentina/Chile, New England, northwestern England, southcentral Sweden, upstate New York, southeastern Québec, areas of British Columbia, and northwestern Russia. Today, these lakes play an increasingly important role in recreation and water supply to urban areas.

▲ Among basins formed directly or indirectly by glacial erosion are lakes formed by mountain glaciers at the upper end of mountain valleys. The excavation there of a cirque (bowl), is attributed to the rotational action of the glacier ice, together with freezing and thawing of a rocky concavity just below the headwall. The resulting small, round lakes are located in a kind of amphitheatre and are known as **cirque lakes**. They are familiar features in presently or formerly glaciated mountains, among others in the European Alps, Norway, Himalayas, New Zealand, the Andes of South America, the Rocky Mountains of North America, and Tasmania.

The Laurentian Great Lakes of North America (Superior, Michigan, Huron, Erie, Ontario) were also formed by glacial erosion and the subsequent deposition of end moraines in deepened and widened valleys. However, their present surface area is the result of an approximately 200 m gradual raising of the land (**isostatic uplift**, a tectonic process that continues today) after the weight of the overlying ice disappeared (type 4). The uplift became sufficient to allow the establishment of the present outflows and lake surface areas about 5,000 years ago (Hough 1958, Fulton and Andrews 1987).

The wet coastal lowlands (wetlands) of arctic Eurasia and North America are covered in summer with innumerable polygonal ponds. They are formed within ice-wedge polygons that grow in the underlying frozen ground (**permafrost**) after summer meltwater seeps through cracks in the overlying soil (Hobbie 1984). Eventually these shallow ($\bar{x} < 0.5 - 1$ m) **polygon ponds** form a polygonal network of low ridges ranging from about 10 to 50 m in diameter (Fig. 6–2). When the ponds coalesce as more of the underlying permafrost melts, they form larger ponds or shallow **thaw lakes**.

Figure 6–1   Glacial Lake Baldegg (CH) separated from downstream Lake Hallwill by an end moraine (15,000 years BP). The hills parallel to the lakes are lateral moraines. (*Photo courtesy of Stadelmann P., Butscher E. and Buergi H. R. 1997 Massnahmen zur Sanierung des Baldeggersee, GWA Journal des Schweizerischen Vereins des Gas—und Wasserfaches (SWGW), Zuerich, 77, 1-18 (1997).*)

Figure 6–2 Polygon shaped ponds in the arctic coastal plain near Barrow, Alaska (US). *(Photo courtesy of J. Kalff.)*

6.3   Tectonic Lakes

The forces that bring about warping of the earth's surface, resulting in mountain formation or lowering of an area, are called **tectonic forces**. These forces have been most active in areas of low rainfall (central and southwest Asia, east Africa, Andes) and consequently there are few tectonically produced basins that are presently holding water. Most of the larger tectonic lakes are old. An interesting exception are a number of Icelandic lakes that were formed 2,000–10,000 years ago. These lakes continue to deepen and widen because they are located in the spreading zone between two continental plates.

Although freshwater tectonic lakes are few in number compared to glacial lakes (Table 6–1), they hold the record for size and volume. Lakes Baikal (18%) and Tanganyika (14%) alone hold about 32 percent of all the fresh water contained in the world's lakes (Table 4–5).

Lake Baikal (RU, Fig. 5–17) and the east African Great Lakes (Fig. 6–3) lie in flat-bottomed valleys or fault-troughs, known as **grabens**, produced by tectonic activity (Fig. 6–4). The Great Lakes are located in the two arms of the Rift Valley, about 800 km long and 75 km wide (Fig. 6–3). The valleys were produced during the Miocene–Pliocene period when land along the eastern and western edges of the two sets of faults was raised about 1,000–2,000 m. The uplift of the western edge blocked the originally westward-flowing rivers, creating Lake Victoria (Fig. 6–3).

Tectonic forces are similarly responsible for the creation of the largest saline lake basins and many saline lake districts. The Caspian Sea (Fig. 5–17), containing about 75 percent of the world's saline lakewater, and nearby Aral Sea were submarine depressions that became basins following lifting of the seabed during mountain formation in the Miocene (12–20 million years BP). The enormous basins now lack an outlet to the sea and contain salts overwhelmingly derived from their drainage basins.[3] Aral and the Caspian Seas are discussed in Secs. 5.7 and 5.8. Lake Titicaca (PE–BO), the largest (8,030 km²) high-altitude (~3800 m) lake in the world (Table 4–5), arrived where it is today from rapid uplifting of a low-altitude valley during the formation of the Andes.

▲ A much more modest upraising of land in the area of original outflows to the sea produced very large, but shallow ephemeral saline lakes in Southern Australia (Lake Eyre and Lake Torrens). These two lake basins fill partially or totally during high rainfall years once every century or so; Lake Eyre can briefly extend over an area of nearly 100,000 km². However, the largest number of tectonically produced saline lakes are found in the endorheic plateau region of central Asia, the southeastern portion of the former USSR, Tibet (CN), Mongolia and adjacent southwestern China. In addition, there are significant saline lake districts in

---

[3]The Caspian Sea was a freshwater lake during the early stages (~17,000 yr BP) of the last deglaciation receiving meltwater from the North Russian Ice Cap. The lake drained to the present Black Sea, which was a freshwater lake for much longer. The Aral Sea was fresh water until at least 1559, and drained to the Caspian Sea (Meybeck 1995).

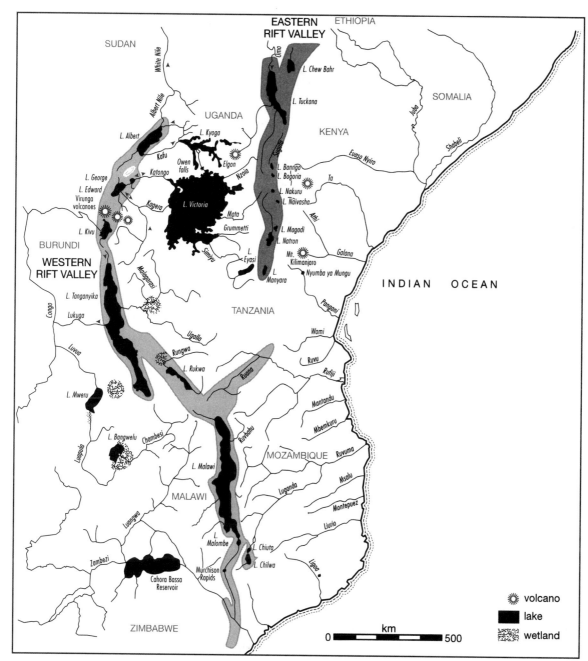

Figure 6–3   Eastern tropical Africa showing courses of the two Great Rift Valleys and their drainage systems. *(Modified after Beadle 1981.)*

the Prairie region of Canada, the southwestern United States, eastern Spain, southern Australia and the Andes region of South America (Fig. 5–17). Like virtually all other saline lakes, these lakes are characterized by an annual inflow of fresh water that is, on average, smaller or equal to the evaporation from the lake surface (Chapter 5).

Lake Biwa in Japan (Fig. 5–17), about 2.5 million years old, has one of the best known lake histories among ancient freshwater lakes. Its history has been deciphered through examination of a sediment core, approximately 1,400 m long, extracted by means of a drilling tower placed at a 68 m deep site. The core is providing a wealth of information on the history of

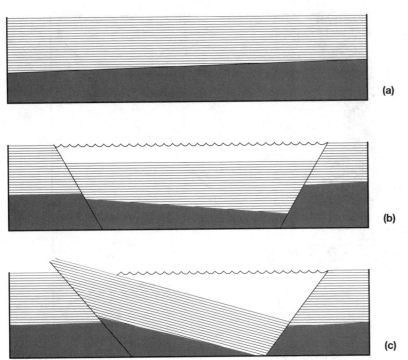

Figure 6–4   Formation of graben lakes. (a) the rocks before faulting; (b) lake occupying a symmetrical graben; (c) lake occupying a graben with a tilted fault block. (*After Cole, 1983.*)

the lake and past climates of the region. Analysis has shown that the lake reached its present depth about 700,000 years ago. (See Horie 1984). There has long been talk of deep-coring other ancient lakes, such as Lake Tanganyika in east Africa and Lake Baikal, whose kilometers-thick sediments contain a record of their history and regional climates over millions of years.

▲ 6.4   Coastal Lakes

Coastal lakes are recent because sea levels only stabilized 6,000 years ago, after glacial melting ended. They are usually formed when a spit or bar builds up between headlands of marine or very large freshwater bays such as the ones in the Laurentian Great Lakes. Such **embayment lakes** come about when an along-shore current carrying sediment encounters a quiescent bay or indentation and deposits part of its load. If the resulting bar is not destroyed by storms it grows until a coastal lake forms behind it. Some coastal lakes become fresh through flushing by inflowing streams or groundwater, but most remain brackish from partial inundation of the bar during storms or through the entry of saltwater via the outflow channel at times when freshwater outflow is small. If the lake is too deep to be readily wind mixed,

the less dense inflowing fresh water will lie on top of a denser, heavier brackish or salt water layer below (Chapter 11). Among other coastal lake types are **drowned-valley lakes** in which the basins were created before sea level rose. Lakes are then created when sand-sized particles and waves create barriers enclosing the basins (Timms 1992).

6.5   Riverine Lakes

There is a wide variety of **riverine** or **fluvial lakes**, and this lake type dominates at low latitudes. Best known and most common are lakes formed in floodplains or river deltas. These are **oxbow lakes**, known as **billabongs** in Australia and **baors** in parts of India and Bangladesh. They develop on wide floodplains where rivers are allowed to meander across the floodplain. Sediments are deposited at the slow-flowing inside bend and eroded from the quick-flowing outside bend (Fig. 6–5). When either the deposition becomes sufficient for the water to forge a new route or erosion is sufficient for the river to break through a narrow isthmus between two succeeding curves, a new section of river channel is created. The old cutoff section may remain as a crescent-shaped shallow lake fed by groundwater or seasonal flooding.

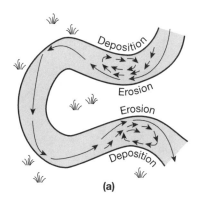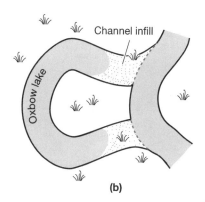

Figure 6–5   Conceptual view of a meandering river section on its wetland floodplain, showing (a) two areas of deposition and erosion leading to (b) creation of both a new river channel following the neck cutoff and an oxbow lake. *(Modified after Timms 1992, and Allan 1995.)*

The second most common type of fluvial lakes are known as **blocked-valley lakes**, formed where rivers run through a relatively narrow floodplain confined by bedrock. Lakes are formed in lateral tributary valleys when the main river deposits sediment (*aggrades*) faster than inflowing tributary streams, with the result that tributaries become blocked. These lakes are typically small and shallow (Timms 1992).

Very large fluvial lakes are a third type of lake formed where rivers flow through large **internal deltas**, low-slope depressions away from coasts. During the rainy season, when inflowing rivers deliver water more rapidly than it can be evacuated, the lakes flood far beyond their normal boundaries. Probably the single most impressive lake located in an internal delta is Tonle Sap (Great Lake) in the Mekong River delta (KH) (Fig. 5–17). It has a surface area of ~2,500 km² during the dry season, but expands to ~11,000 km² during the rainy season.

A fourth type of riverine lake is found in very large internal and/or coastal deltas of rivers such as the Niger River in west Africa or the Amazon and Magdalena Rivers of South America (Fig. 6–6). During the flooding period, rising water carrying dissolved and particulate matter, spills through gaps in the river banks and floods large portions of the low-lying deltas behind. A large number of shallow lakes emerge when water levels begin declining rapidly at the end of the rainy season. In Brazil these lakes on the floodplain of non-humic rivers are known as **várzea lakes**. Lakes that partially or totally dry up during the dry season also experience decomposition of the frequently abundant macrophyte vegetation produced during the rising phase, plus terrestrial vegetation produced during the low water phase. As a result, the lakes become *hypoxic* (low dissolved oxygen) or *anoxic* (zero dissolved oxygen) (Chapter 15, Junk and Weber 1996). Rivers draining large and well-vegetated wetlands may periodically contain little

or no dissolved oxygen, resulting in fish kills (Chapter 15) followed by changes in the foodweb structure.

A fifth type of riverine lake is formed in river valleys such as the Paraná River of South America or the Danube River in Europe before channelization. Lakes form during seasonal flooding of the wide strip of low-

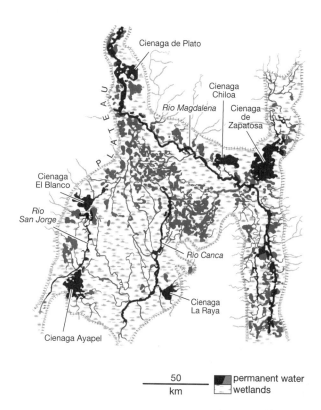

Figure 6–6   The internal delta of the Magdalena river at its confluence with the Canca and San Jorgé rivers in Colombia during the low water period, showing the numerous permanent várzea floodplain lakes, known locally as cienagas, and local rivers. They and associated wetlands are flooded during the season of high runoff. *(From Welcomme 1985.)*

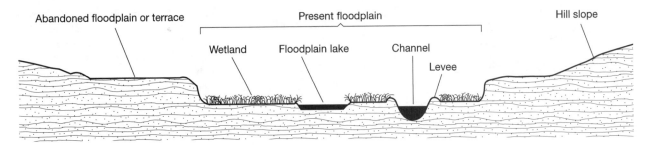

Figure 6–7 Diagrammatic cross-section of a valley showing present river channel, levees, which are sites of greatest sediment deposition at the point where the river overflows its banks and the turbulence declines, and a floodplain with a floodplain lake and a seasonally inundated wetland. A terrace representing a previous floodplain established during an earlier period of higher discharge is also shown. *(Modified after Allan, 1995.)*

lying land between the river at low discharge and the edge of the floodplain. The strip, 20–30 km wide and about 1100 km long in the case of the Paraná River, is dotted with várzea lakes and ponds during the period of declining water level that were filled during the high water season. When water levels decline 2–3 m below those of peak flow, the same area is characterized by many dry basins, isolated lakes fed by local streams, and larger lakes still connected to the main river.

▲ When a delta's principal rivers spill over their immediate banks during flood periods, the heavier particles transported are immediately deposited because of reduced flow outside the main channel. These deposited sediments create crests or **levées** above the floodplain surface. Shallow lakes or **backswamps** form between these levées and the edge of the floodplain (Fig. 6–7). Backswamps are typically covered with floating and rooted aquatic macrophytes (Chapter 24).

Várzea lakes and backswamps tend to be very productive because of their shallowness, annual fertilization with river-carried nutrients, and the release of nutrients from the flooded soils that are anoxic during the one-to-six-month inundation period. (See Chapters 17, 18, and 26.)

Tropical riverine lakes, and to a lesser extent their higher latitude counterparts (e.g. MacKenzie River delta, CA), exhibit remarkable seasonal changes and among-lake variation in salinity (Chapter 13), total phosphorus (TP), total nitrogen (TN), dissolved silica concentrations, inorganic turbidity, macrophyte development, and the planktonic biota. For example Lake Batata (BR), connected year round to an Amazonian river, undergoes a distinct annual phytoplankton cycle linked to flushing, water depth, and water column stability (Huszar and Reynolds 1997).

Some várzea type lakes are filled with river water derived from a large, distant catchment at the end of high water season, while others are largely filled with water from local rivers. Groundwater and local inflowing streams become increasingly important in determining the chemical and biological composition of riverine lakes as principal river discharge declines. In the Amazon region, those floodplain lakes having relatively modest catchment area to lake area ratio (CA: LA<~20) are particularly variable in chemical composition at the end of the dry season (Chapter 9). At that point, they contain varying mixtures of groundwater, local stream water, and principal river water (Forsberg et al. 1988). Their nutrient levels are further modified by within-lake decomposition processes and release of sediment nutrients to the overlying water column (internal loading, Chapter 17), as well as local rain and dust (Forsberg et al. 1988). Conversely, lakes with a large CA:LA ratio receive sufficient inflow from their disproportionally large drainage basins even during the low-water phase, to have water chemistry most closely resembling that of the principal inflowing river. The water chemistry of all várzean lakes and rivers is furthermore modified by evaporation during the dry season. In addition, as water levels decline, inorganic turbidity of shallow lakes typically rises because of progressively greater turbulence induced by winds.

Manmade changes to major floodplain lake–wetland systems have made them a nontemperate zone phenomenon at mid-latitude, hardly appreciated by temperate zone limnologists. Floodplain systems that are untouched or modified little are located mostly in Africa and South America. In contrast, the enormous south Asian floodplains are increasingly subject to flood control measures (Table 6–3) and growing

Table 6–3   **Area of major floodplain lakes and wetlands (km². ND = not determined.**

Continent	River	Flood Plain	Area at Flood	Area at Low Water
Africa	Niger	Internal (central) delta (ML)	20,000[1]	3,900[1]
	Okavango	Internal delta (BW)	17,000	3,100
	Lualaba	Internal delta depression "Kamulondo" (ZR)	11,800	6,600
	Nile	Internal delta "Sudd" (SD)	31,800	16,300
Asia	Tigris and Euphrates	Coastal delta (IQ)	20,000[2]	5,000
	Ganges and Brahmaputra	Coastal delta (BD)	93,000[2]	14,000
	Irawaddy	Coastal delta (MM)	31,000[2]	ND
	Mekong	Internal delta (KH)	53,000[2]	4,000
	Oga and Komering	Coastal delta (ID)	500,000[2]	ND
Europe	Danube	Fringing floodplain (RO)	5,000[3]	ND
North America	Peace and Athabasca	Internal delta (CA)	2,600[4]	ND
South America	Magdalena	Internal delta (CO)	20,000	3,300
	Atrato	Coastal delta (CO)	5,300	ND
	Catatumbo	Coastal delta (VE)	5,000	ND
	Amazon	Internal (central) delta (BR)	50,000	ND
	Amazon	Coastal delta (BR)	25,000	ND
	Orinoco	Internal delta (VE)	70,000	ND
	Orinoco	Coastal delta (CO, VE)	20,000	ND
	Paraguay	Internal delta "Gran Pantanal" (BO, BR, PY)	~90,000	11,000
	Paraná	Fringing floodplain (AR, BR, PY)	20,000	ND

[1]Reduced in recent decades of low precipitation and runoff in its subSaharan drainage basin.
[2]Increasingly subject to flood control and rice farming.
[3]Much larger floodplain before damming, flood control and land reclamation.
[4]Reduced to a series of mudflats following the damming of the Peace River.

Source: Modified largely after Welcomme 1985.

crops.[4] Long European (the Danube) and North American (the Mississippi) rivers originally characterized by lateral floodplains have been so modified by channelization and dam building that little evidence of their original wetlands remains. Massive diversion of water from dry land (semiarid) rivers for irrigation has greatly reduced the abundance of floodplain lakes and their wetlands in Australia, Africa, and semiarid portions of Asia and North America. (Chapters 8, 24 and 29.)

[4]Deepwater rice is grown today in many tropical river basins during the months of inundation even where water level control through diking is not possible, as in the deeper portions (0.5–3.5 m) of flooded floodplains. The practice is most developed in south and east Asia, covering ~11 × 10⁶ ha. As a result, there have been great changes in the natural macrophytic vegetation and associated fauna and flora. (Whitton et al. 1988.)

▲ 6.6   Volcanic Lakes

Volcanic lakes include lakes located in craters—formed after an eruption—and those resulting from flowing lava damming river valleys. Eruption (explosion) crater lakes are of three kinds. Those formed after direct ejection through a volcanic cone of underlying material. A second type is produced following underground explosions brought about by hot lava (**magma**) coming in contact with groundwater or by degassing of the magma. These are small in area (~2 km diameter), have small drainage basins, and are often deep. There are hundreds of such small volcanic lakes, known as **cone lakes** in Africa, with the greatest number (89) along the Uganda–Zaire border (Melack 1978) in the Virunga volcano region (Fig. 6–3); there are a respectable number in Japan, Cameroon, and Central America (Fig. 11–17), but they are also present on all continents (see Table 6.2). Some lakes lie at the bottom of classically conical craters, but other craters never had a high rim. In others the rim has been eroded so their origin is not immediately evident, as in the Eiffel region of Germany (Fig. 6–8) and Auvergne region of France, and in central Italy. The third kind of craters are much larger and result from collapse of the earth's surface overlying an area where magma has been ejected. These basins are called **calderas** and, under the right climatic conditions, permit establishment of **caldera lakes**. Crater lake (Oregon, US) was formed about 6,600 years ago, creating a circular lake with an area of 54 km^2 and a maximum depth of 589 m (Fig. 8–15, Table 4–5). A much larger, but shallower caldera lake, is Lake Taupo (NZ), with an area of 610 km^2, but a Z_{max} of only 159 m. Its basin has been further modified by tectonic forces. Lake Kivu (RW, ZR) lies in a sunken valley or **graben** that was dammed by lava flow, creating a lake basin of 2220 km^2 with a Z_{max} of 480 m (Fig. 6–4).

▲ 6.7   Solution or Karst Lakes

Solution or **karst lakes** are typically small, lying in basins of highly soluble rock—mostly limestone, $CaCO_3$ and $MgCO_3$. Limestone is solubilized upon contact with slightly acidic waters, created when carbon dioxide released during respiration in overlying soils combines with water from precipitation yielding carbonic acid (Chapter 14). Solution lakes tend to be circular and are characterized by very steep underwater slopes and a high relative depth ratio (maximum depth/mean diameter, Sec. 7.4). Karst lakes and associated caves are common in the Balkan region of Europe, but also occur in the European Alps, Spain, Russia, in areas of Kentucky, Indiana, Tennessee, and Florida of the United States, as well as in Tasmania in Australia. Lake Saskkol (YU, AL) is the only karst lake included among the Great Lakes of the world (>500 km^2 Herdendorf 1982) because it was subsequently enlarged by tectonic faulting.

▲ 6.8   Manmade Lakes or Reservoirs

The number of reservoirs with dam walls over 15 m high has increased from about 5,000 in 1950 to 36,562 in 1986 (World Resources 1992); 64 percent were in

Figure 6–8 Two volcanic lakes, Schalkenmehrener and Tolenmaar, in the Eifel region of Germany. *(Photo courtesy Landesmedienzentrum Rheinland-Pfalz. ©LMZ RP/Bruno Fischer.)*

Table 6–4   **The maximum surface area and volume at full capacity of selected reservoirs.**

Name	Maximum Surface Area (km^2)	Maximum Volume (km^3)
Volta (GH)	8,480	148
Kuibyshev (RU)	6,450	58
Bratsk (RU)	5,500	169
Nasser (EG, SD)	5,120	157
Rybinskev (RU)	4,550	25
Kariba (ZM, ZW)	4,450	160
Caniapiscau (CA)	4,285	72
Sanmensia (CN)	3,500	65
Cabora Bassa (MB)	2,700	66
Bennett (CA)	1,660	108
Ilha Solteira (BR)	1,230	21
Three Gorges[1] (CN)	1,084	39
Keban (IT)	750	30
Silto Grande (UR)	–	20
Lake Mead (US)	637	38
Cerros Colorados (AR)	620	44
Hendrik Verwoerd (ZA)	372	6
Randsfjord (NO)	136	6

[1]under construction

Source: Largely after Kopylov et al. 1978.

Asia, 21 percent in the Americas, and 12 percent in Europe. Reservoirs under construction in 1986 or later brings the total of such reservoirs to 40,000, an eightfold increase over 1950. If we define reservoir size by volume, there are more than 60,000 large reservoirs (>0.1 km^3), most dating from after World War II. The largest in terms of surface area, is the Volta reservoir (8,480 km^2), whereas the Bratsk Reservoir on the Angara River draining Lake Baikal, has the largest volume (Table 6–4).

If the largest reservoirs are compared with well-known natural lakes, the surface area of Brask Reservoir would place it in thirty-first position. It has larger volume than Lake Tahoe (US) and close to twice the volume of Lac Léman (CH–FR).

While the number of large reservoirs is impressive and reflects man's ability to modify the earth's surface, the water contained within them is still less than 10 percent of the volume contained in natural freshwater lakes, but it is rising rapidly. During the mid1970s, the equivalent of eight percent of the global annual runoff was held within reservoirs, with this fraction rising to ~14 percent a decade later. Reservoirs impose drastic and generally negative effects on the riverine biota and downstream lakes (Sec. 5.8 and Chapter 29).

In addition to large reservoirs there are innumerable smaller manmade lakes referred to as **impoundments**, but are called **dams** in southern Africa and Australia. Smaller reservoirs were created for water supply, flood control, fish farming, irrigation, or recreation and are largely located in more arid portions of North America, Australia, southern Africa, and Europe, as well as in southern and eastern Asia. In China, there are 87,000 mostly small impoundments with a combined storage capacity (volume) nearly twice that of the country's natural freshwater lakes (Jin 1994). There are another 86,000 impoundments in the United States and a similar number in northeastern Brazil. Reservoirs of all sizes, and their modified outflowing rivers and associated wetlands, have received disproportionately little attention from limnologists (Chapter 29).

Highlights

- The vast majority of lakes (74%) were formed ~15,000–6,000 BP as the result of glacial action, primarily at higher latitudes in the Northern Hemisphere (>40–60°N). A small number of very deep, ancient lakes formed 2–20 mya, and again ~100,000 years ago, as the result of tectonic action.
- The dominant lake type at low latitudes is riverine (fluvial) lakes within lowland rivers or located on their floodplains.
- Tectonic lakes, plus tectonically modified lakes (Laurentian Great Lakes), are few in number, but include those with the largest surface area and greatest volume (Caspian, Baikal, Tanganyika).
- ▲ More than 60,000 large reservoirs (>0.1 km^3) and a vast number of smaller reservoirs (impoundments) have been constructed on river systems, primarily during the last 50 years, with large reservoirs alone holding >14 percent of global annual runoff.

7

Lake and Catchment Morphometry

7.1 Introduction

Lake outlines range from near perfect circles, observed in volcanic or meteoric craters, to perimeters with very high length-to-width ratios, such as those found in narrow glacial valleys or tectonic faults (Chapter 6). Regardless of how lakes are formed, their surface shape, surface area, underwater form, depth, and the irregularity of their shoreline have a major impact on turbulence (Chapter 12), lake stratification (Chapter 11), sedimentation and resuspension (Chapter 20), and the extent of littoral-zone wetlands (Chapter 24) that determine lake functioning—a fact partially recognized a century ago in Germany by H. Huitfeldt-Kaas (1898). Catchment morphometry is commonly the principal determinant of lake morphometry, with the size and steepness of drainage basins determining average runoff and rate of runoff in any one climatic zone. It similarly determines the size and steepness of rivers and possibility of wetland development (Chapter 9). Catchment size, together with geology and climate, has a major impact on the nutrient and organic matter exported from land to aquatic systems (Chapter 8).

7.2 The Bathymetric Map

The **bathymetric map** is the standard way of recording the morphometry of lakes. Where topographic maps are not available, a lake's outline is normally created from aerial photographs that show the shoreline clearly. Next, the scale of the photograph is ascertained (e.g., 2 cm = 100 m). Bathymetric or **contour maps** (Fig. 7–1) are created from depth soundings made along transects, followed by connecting the points of a particular depth with a contour line (see Håkanson 1981). Soundings are conveniently made with small, relatively inexpensive echosounders of the type used by commercial and recreational fishermen. The echosounder is a great advance over the traditional **sounding line** or **lead line**, a marked line with a weight attached. Such lines require a completely stationary boat to avoid large errors, a major problem on all lakes but those protected from the wind. Winds can also force survey boats off course, yielding inaccuracies in contour lines produced on the basis of echosoundings. This problem can be almost eliminated by using a handheld *global positioning system* (GPS) device which permits accurate positioning based on signals received from a stationary satellite array.

Bathymetric maps allow morphometric parameters to be calculated. The *lake or wetland surface area* (A) and the surface area of each depth contour can be determined with a *planimeter* or computer scanner. A simpler method is to trace the area encompassing each depth contour onto a semitransparent piece of paper. The area is then cut out and weighed. A square piece of paper representing a known area is also weighed (Fig. 7–1). The area of a contour interval or the lake is obtained from the ratio of the two weights.

Lake volume (V) is most simply determined by constructing a graph of the area enclosed by all con-

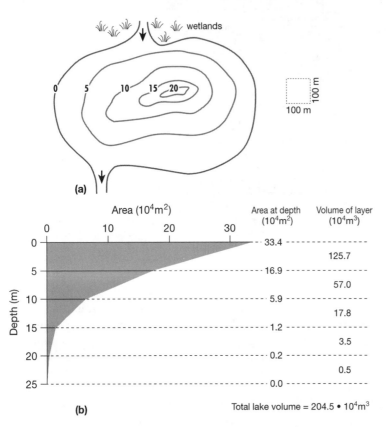

Figure 7–1 Method for determining the volume of a lake. The area enclosed by each contour line on a morphometric map (a) is determined, these areas can then be plotted to yield the depth–area hypsographic curve (b). The shaded region on the graph represents lake volume, calculated using the equation: $V_{(x_1 x_2)} = ((A_{x_1} + A_{x_2})/2) \cdot (X_2 - X_1)$ for each depth interval and adding the resultant volumes.

tours against depth (Fig. 7–1). The area encompassed by this graph corresponds to basin volume. The **mean depth** in meters ($\bar{z}$) is

$$\bar{z} = \frac{V(m^3)}{A(m^2)}. \qquad \text{EQ. 7.1}$$

7.3   Lake Surface Area

The maximum length or surface area of lakes governs the distance over which the wind can blow, which subsequently determines the wave height of both surface waves and internal waves (Chapter 12). Both lake surface area and maximum depth together help determine whether a particular lake can be expected to stratify (Chapter 11). In lakes that stratify, lake surface area will largely determine the thickness of the epilimnion (Secs. 11.7 and 11.8), which influences the light climate experienced by planktonic organisms (Fig. 10–12). The number of plant, fish, and invertebrate species rises with lake surface area (Chapters 24 and 26), as does the length of foodchains.

The distance over which the wind can blow and bring about turbulence is referred to as the **fetch**. Fetch can be calculated in a variety of ways (Håkanson 1981). Some use the maximum length (L') of lakes, others take length and width (W) into account [$(L' + W)/2$], and yet others use lake surface area.[1] Whatever the definition, fetch is not only a useful indicator of the depth of the thermocline (Chapter 11) but also of the depth to which particles of varying size and density can be resuspended into the water column and transported to deep areas of low turbulence (Chapter 12). Håkanson and Jansson (1983) developed the **dynamic sediment ratio** (DSR) as the square root of the area (A) divided by the mean depth ($\bar{z}$)

$$\text{DSR} = \frac{\sqrt{A}}{\bar{z}}, \qquad \text{EQ. 7.2}$$

[1]Lake surface area (A or $A^{0.5}$) has operational advantages in measuring average fetch because it is less arbitrary than trying to estimate average length or width of lakes with convoluted shorelines, or having to decide whether the total length is also the "effective" length in lakes with one or more islands.

used to estimate the area of lake bottom subject to re-suspension of sediments and organisms, and transport of fine (low-density) sediments followed by deposition in nonturbulent deep water. Sediment sorted by wind and waves has a major influence on particle size distribution and other characteristics of sediments (Chapter 20), diffusion of dissolved oxygen into sediments and nutrients out of the sediments into the overlying water, food resources available to the benthos, and suitability (density) of sediments for physically supporting invertebrates of different sizes (Chapter 25).

7.4   Lake Depth

Mean depth is probably the single most useful morphometric feature available. It serves as a surrogate (proxy) for most morphometric attributes and a host of biological processes, but no correlate can provide unambiguous information on underlying causes (Sec. 2.6). Among lakes within any climatic/geologic zone, mean depth ($\bar{z}$) is correlated with the maximum depth (z_{max}) (Table 7–1), probability of stratification (Chapter 11), water flushing rate (Fig. 7–2) and therefore with nutrient loading (Fig. 9–3), and the fraction of particles sedimented rather than flushed (Chapter 17). Relatively deep lakes typically have a disproportionately small fraction of lake surface area with a well-illuminated littoral zone and tend to be unproductive. Conversely, transparent shallow lakes and wetlands have a disproportionately large surface area occupied by algae growing on sediment and large plants rooted

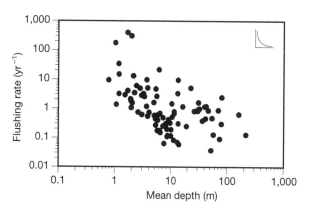

Figure 7–2 Flushing rate (FR) versus mean depth for 88 Canadian glacial lakes. The considerable scatter is probably a function of variation in catchment size of lakes of a particular mean depth and variation in the relationship between the mean depth of lakes and their volume. WRT = 1/FR. *(Modified after Chow-Fraser 1991.)*

in them. Water nutrient levels reflect nutrient loading and tend to decline with an increasing mean depth in both unpolluted freshwater lakes (Fig. 7–3) and saline lakes (Chow-Fraser 1991). The algal biomass or its surrogate, chlorophyll-*a* (underpinning much of the energy flow in foodwebs), declines with increasing depth. As a consequence, fish catches also decline with increasing lake depth (Fig. 7–3).

Although there are three lakes in the world (Table 4–5) with a **maximum depth** (z_{max}) greater than 1,000 m, and another nine with a z_{max} greater than 500 m, most well-studied lakes have a maximum

Table 7–1   **The relationships of mean depth ($\bar{z}$) to maximum depth (z_{max}) for largely glaciated lakes in different geographical regions.**

Region	Regression Equation $\bar{z}=$	Maximum Lake Depth Included (m)	Number of Lakes	Authors
World	$0.47\,z_{max}$	614	107	J. Neumann (1959)
Scotland glacial drift basins[1]	$0.75\,z_{max}^{0.80}$	80	137	E. Gorham (1958)
World	$0.46\,z_{max}$	1,741	71	E. Gorham (1964)
N. Europe Baltic lowland	$0.65\,z_{max}^{0.82}$	83	85	D. Ventz (1973)
C. Europe Alps	$0.62\,z_{max}^{0.95}$	370	39	D. Ventz (1973)
Japan	$0.42\,z_{max}^{1.03}$	425	231	S. Horie (1962)
New Zealand	$0.59\,z_{max}^{0.94}$	444	39	J. Irwin (1972)
Canada	$0.45\,z_{max}^{0.94}$	115	60	Chow-Fraser (1991)

[1]Sec. 7.4.

Source: Modified from Straškraba 1980 and Chow-Fraser 1991.

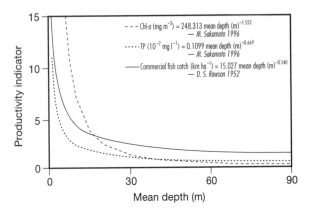

Figure 7–3 Empirical relationships between indicators of lake productivity and mean depth. *(Modified after Straškraba 1980.)*

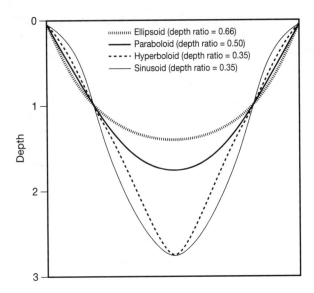

Figure 7–4 Vertical cross sections through four forms of lake basin with identical surface areas, mean depths, maximum depths, and volumes, but different shapes. Vertical scale is in multiples of mean depth. The depth ratio is $\bar{Z}/Z_{max}$. *(After Carpenter 1983.)*

depth of less than 50 m. The vast majority of lentic waterbodies worldwide are ponds (Table 4–4) with a maximum depth of a few meters or less.

Neumann (1959) concluded that **depth ratio** ($\bar{z}/z_{max}$) provides a useful approximation of lake form. The average glacial lake (in his data set) is most closely approximated by an elliptical sinusoid ($\bar{z}/z_{max}$ = 0.35), but with lake form ranging from elliptical cones ($\bar{z}/z_{max}$ = 0.33 – 0.35) to ellipsoids ($\bar{z}/z_{max}$ = 0.66 – 0.67). An ellipsoid form characterizes shallow lakes with flat bottoms (Fig. 7–4). Ratios below 0.35 are generally encountered in large glacial lakes having one or more small areas of relatively great depth in basins otherwise characterized by a more typical conal or elliptical sinusoid form. The mean depth of lakes in glaciated regions is fairly close to half the maximum depth (Table 7–1).

▲ An alternative measure of lake form is provided by **volume development** (D_v). This is the ratio of lake volume to a cone of basal area equal to the surface area, and its height equal to the maximum depth. Rather than making actual calculations, D_v is readily computed as three times the $\bar{z}/z_{max}$ ratio. Lakes with D_v = 1 are perfectly cone-shaped. A lake with a flat floor and steep sides has value of $> \sim 2$. Laboratory beakers and petri dishes both have D_v = 3, showing that D_v is poorly related to lake depth (Timms 1992).

Differences in lake form are not only linked to the extent that the deepest portions have been filled with sediments but also to the geological formation in which a lake lies. Shallow Scottish lake basins (*ice-scour lakes*) scraped out of hard rock by glaciers have a high

depth ratio and are characterized by flat bottoms (Table 7–1). Using published data and assuming the average surface form resembles an ellipse, the maximum length (L) in kilometers of glacial lakes examined is statistically related to both z_{max} and $\bar{z}$ (Straškraba 1980).

$$z_{max} = 26.6 \sqrt{L} \qquad \text{EQ. 7.3}$$

$$\bar{z} = 12.1 \sqrt{L}. \qquad \text{EQ. 7.4}$$

A more informative alternative to the depth ratio ($\bar{z}/z_{max}$) for estimating lake form is the **relative hypsographic curve**, also known as the **depth–area** or **depth–volume curve**. The curves represent percentage plots of the relative area or cumulative volume versus the relative depth (Figs. 7–5 and 7–6). An extremely convex lake of the f(–3) type is disproportionately shallow, with only about 20 percent of the area found at depths greater than 10 percent of the maximum depth. Such a lake has one or more deep holes that occupy a small percentage of the area, and are characterized by a very high ratio in the $\bar{z}/z_{max}$ scheme. This type of lake is highly subject to sediment resuspension and is expected to have a wide littoral zone. Conversely, an extremely concave lake

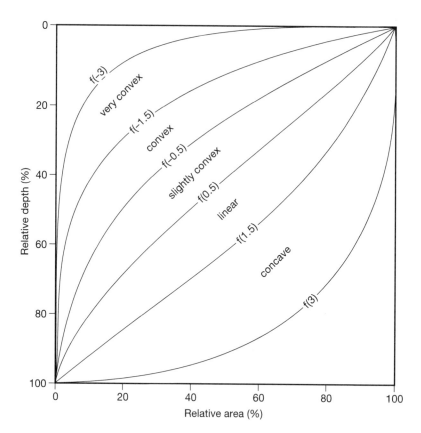

shape, f(3), reflects a trough-like basin, with steep walls and a flat bottom (Fig. 7–6). Volcanic Lake Gnotuk (AU) (Fig. 7–7) is characterized by exceptionally steep underwater slopes, and a very small fraction of its volume is subject to wind-induced turbulence. Lakes such as Windermere (GB) and Vanern (SE) are convex with a modest area of relatively great depth (Fig. 7–7) and a high proportion of the volume above a relatively shallow depth (Fig. 7–8, p. 92). Note that the hypsographic terminology represents the curves rather than the lake forms given by Neumann (1959) and shown in Fig. 7–4. Håkanson (1981) reports that only 13 percent of lakes in glaciated Sweden are characterized by either very convex or very concave curves, with the single largest group (38%) slightly convex and the rest equally divided between linear and concave.

Relative hypsographic curves are useful in comparing lake forms, but cannot be used to calculate actual areas or volumes of specific lakes above or below a particular depth unless the lake area and/or volume is known. For that, **absolute hypsographic curves** are needed. They represent the actual depth in meters plotted against the cumulative area or cumulative volume (Figures 7–1 and 7–9 on p. 92). The availability of absolute hypsographic curves makes it possible, for example, to obtain a measure of the consumption rate of the hypolimnion as a whole by scaling up measurements of hypolimnetic dissolved oxygen consumption rates (mg O_2 l^{-1} d^{-1}). Similarly, hypsographic curves allow phytoplankton primary production measurements, per unit volume (l) or area (m^2), to be readily expressed on a whole-lake basis for interlake comparisons of the relationship between whole lake primary production (kg yr^{-1}) and whole lake fish yields (tonnes yr^{-1}), for example. Finally, measurements of seasonal changes in hypolimnetic concentration of, say, dissolved oxygen, total phosphorus, or total iron (μg l^{-1}) can be readily scaled up to yield changes in their mass in whole epilimnia or hypolimnia with the aid of an hypsographic curve. Absolute hypsographic curves should always be referenced, or if unavailable, presented in publications. They provide important information on area and volume distribution; information that is much more cumbersome to obtain from the more commonly presented bathymetric maps.

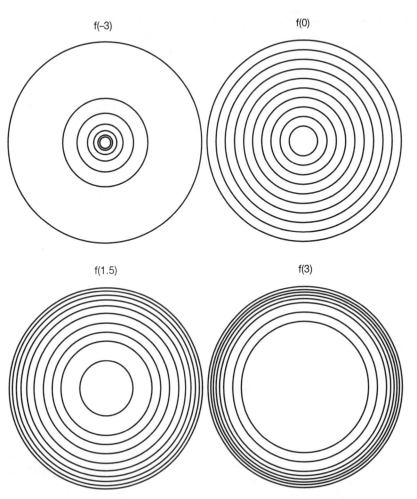

f(-3)   f(0)   f(1.5)   f(3)

Figure 7–6 Schematic bathymetric interpretation of four of the lake forms presented in Fig. 7–5. *(After Håkanson 1981.)*

The last lake form parameter to be discussed is **relative depth ratio** (z_r), defined as the ratio of maximum depth (z_{max}) to mean diameter (represented by the square root of the lake area in km² $\sqrt{A_0}$)

$$z_r(\%) = \frac{z_{max}\sqrt{\pi}}{20\sqrt{A_0}}. \qquad \text{EQ. 7.5}$$

Large, shallow lakes have low relative depth ratios (<2%), whereas small, deep waterbodies have high ratios. High z_r lakes have a small surface area available for wind induced turbulence, and such lakes tend to be stably stratified. Inland lakes at mid and low latitudes with exceptionally high z_r (~4–5%) become so strongly stratified that they do not completely mix (circulate) on an annual basis (Sec. 11.10) and have permanently anoxic hypolimnia. Medium-high and high z_r lakes usually lie in small, steep drainage basins that give protection from wind, but their small catchments export little organic matter and plant nutrients.

Lakes with a high relative depth ratio are typically nutrient poor and highly transparent unless disturbed by human activity (Sec. 9.5). This high transparency is usually sufficient to allow development of a phytoplankton and sometimes a photosynthetic bacterial biomass or production maximum in the metalimnia, and occasionally in the upper portion of the hypolimnia (Figs. 21–26 and 22–14). Dissolved oxygen liberated during phytoplankton photosynthesis in metalimnia may even be sufficient to allow development of a dissolved oxygen maximum (Sec. 15.4, and Eberly 1964).

7.5 Lake Shape

A simple way of expressing the influence of lake shape on the size of the littoral zone is provided by the ratio of volume (V) or surface area (A) to the shore length (L'). A series of lakes with the same volume or surface

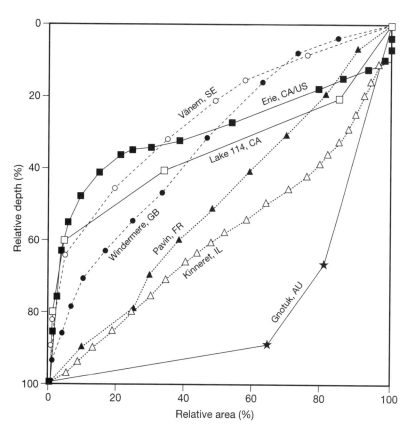

Figure 7–7 Relative hypsographic curves for six lakes of relative lake area versus relative lake depth.

area, but increasing shore length, will be progressively more dominated by their littoral zones. How irregular the shoreline is determines the number of wind-protected bays and associated wetlands suitable for feeding and breeding of waterfowl and fish (Fig. 7–10), as well as for macrophyte colonization.

The degree of irregularity is most frequently estimated from the **shoreline development factor** (*D*), which is defined as:

$$D = \frac{L'}{\sqrt{2A\pi}}, \qquad \text{EQ. 7.6}$$

where L' = shore length and A = the circumference of a circle with an area equal to that of the lake. *D* reflects the extent to which the measured shore length is greater than the length of the circumference of a circle of an area equal to that of the lake. A perfectly circular basin is $D = 1$ and many volcanic crater lakes have a value only slightly higher. In practice, values between 1.4 and <3 range from subcircular to elliptical, and most lakes have values between 1.5 and 2.5. Reservoirs created by flooding of a valley that include a number of tributary valleys will have many branching arms and an indented (*dendritic*) shoreline with $D > 3.5$. Lake

Mälaren, SE, one of the most irregular natural lake basins in the world, has a value as high as 10.0 (Håkanson 1981). Strongly dendritic reservoirs may have values approaching 20 (Fig. 29–3). A pronounced elongation is more important than high irregularity (sinuosity) in producing exceptionally high values. Lakes and reservoirs with high shoreline development have an extensive littoral zone and contain many bays that are likely to differ appreciably from the pelagic zone in temperature, flushing rate, water chemistry, and biota.

▲ 7.6   Underwater and Catchment Slopes

The *underwater slope* (in percent or degrees) along a transect is of considerable importance because it affects (1) the steepness and extent of the littoral zone, (2) sediment stability and structure, (3) whether sediment accumulation is possible at all (Chapter 20), and (4) the angle by which waves and currents impact the lake bottom. The slope, therefore, has a major effect on (5) how well aquatic plants with roots can establish

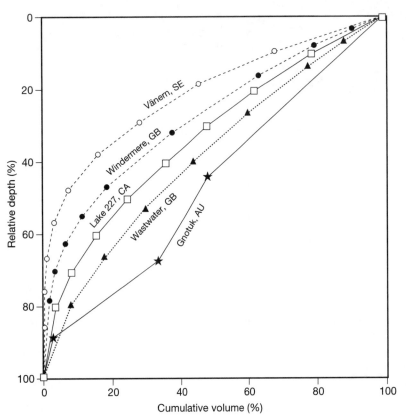

Figure 7–8 Relative hypsographic curves for four lakes of cumulative lake volume versus relative depth.

and maintain themselves in littoral zones (Chapter 24), (6) abundance and distribution of aquatic invertebrates in sediments (Chapter 25), (7) whether the littoral zone is suitable for waterfowl (Fig. 7–10), and (8) suitability as a habitat for fish feeding, hiding, and breeding (Chapter 26). The underwater slope(s) in percent is

$$S = \frac{L}{h} \times 100, \qquad \text{EQ. 7.7}$$

where L = distance between two points and h = difference in water depth. The average *whole lake slope* can be approximated from the equation (Timms 1992):

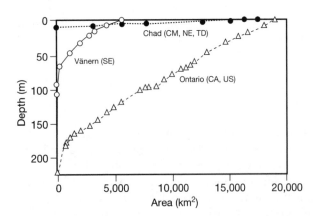

Figure 7–9 Absolute hypsographic curves of lake area versus depth for three lakes.

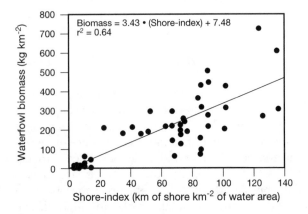

Figure 7–10 Relation between waterfowl (ducks and geese) biomass and shore-index. *(Based on data from Patterson 1976, and Nilsson 1978. After V. Connolly and J. Kalff, unpublished data.)*

$$S = \frac{100 \, z_{max}}{\sqrt{A/\pi}}.$$ EQ. 7.8

The average *channel slope* of lotic systems, affecting water velocity and turbulence, is the average drop per unit horizontal distance. Maximum velocity determines whether particles of a particular size can be transported downstream, thereby determining the substrate structure and its suitability for particular types of organisms (Figs. 20–3 and 20–4). The average *catchment slope* or *relief* (discussed in Chapter 8) affects the number of streams per unit area (Fig. 8–12), possibilities for both soil development and type of soil development, resulting vegetation, and export of particles and their adsorbed nutrients, toxins, and organic and inorganic particles to aquatic systems. Indices of basin relief include **maximum basin relief** (H_m), in meters, which is the difference between the highest and lowest point in a catchment. The **average basin relief** ($H_{\bar{x}}$), in meters is

$$H_{\bar{x}} = \frac{H_m}{\sqrt{CA}}$$ EQ. 7.9

where CA = catchment area (km²). More common is the **relief ratio** (R_h)

$$R_h = \frac{H_m}{L},$$ EQ. 7.10

where L = horizontal distance (km) along the longest dimension of a basin parallel to the principal drainage line (Gregory and Walling 1973).

Highlights

- Attributes of lake morphometry include surface area and shape, depth, and underwater slope. These work together directly and indirectly to exert a major impact on lake functioning.
- In any climatic/geologic zone, catchment morphological attributes are the single most important group of predictors of lake, river, and wetland morphometry; another is the export of plant nutrients and organic matter to aquatic systems.
- Lake surface area governs, among others, the impact of wind-induced turbulence affecting the possibility of lake stratification, thickness of the epilimnion, underwater light climate, and size distribution of sediments.
- Mean depth, through its correlation with most morphometric attributes, is probably the best predictor in environmental research and management of a host of biological processes.
- The relative depth ratio, which combines a measure of depth (z_{max}) and lake surface area, is a good indicator of whether mid and low latitude inland lakes will stratify stably and, if so, whether they will stratify so strongly that they do not totally mix at least annually. Those that do not annually mix usually lie in small, steep drainage basins that provide wind protection.
- The shoreline development factor provides a measure of shoreline irregularity and represents the extent to which the measured shore length is greater than the circumference of a circle of an area equal to that of the lake or reservoir. Systems with high shoreline development have an extensive littoral zone and many bays suitable for macrophyte colonization and as waterfowl and fish habitat.

8

Rivers and the Export of Materials from Drainage Basins and the Atmosphere

8.1 Introduction

It has long been known that nutrients are transported from drainage basins to their receiving waters.[1] The effect of the geology of catchments on the chemistry of lakes and rivers was first noted by A. Forel 100 years ago, acknowledged in research by I. Naumann (Chapter 2) early in the 20th century, and incorporated in limnology by W. H. Pearsall (GB) a few decades later. The next major step, a conceptualization and early quantification of links between critical plant nutrients derived from drainage basins, lake morphometry, and lake productivity, was articulated by Rawson (1939). Some decades later, his ideas were further developed and much better quantified by R. Vollenweider (1968) in response to public concern about increasing algal turbidity of lakes draining rich agricultural and urban areas (Sec. 17.6). All of the scientists named above were concerned with lakes and paid scant attention to lotic systems, other than recognizing them as conveyors of nutrients from catchments.

Vollenweider's work, and subsequently most other research concerned with the enrichment of waters with plant nutrients (**eutrophication**), focused on phosphorus and nitrogen. These two elements were already known or suspected to be the nutrients that normally limit production rates and biomass in plant communities (Table 8–1). In other words, among the elements required by aquatic plants, the supply of nitrogen and/or phosphorus to aquatic systems is normally disproportionately low in relation to the demand (need) of the plants. In oligotrophic temperate-zone fresh waters at least, phosphorus normally has the lowest supply:demand ratio and it, rather than nitrogen, is considered the primary limiting nutrient for phytoplankton (Table 8–1). This conclusion is supported by a high N:P supply ratio from undisturbed and well-watered temperate zone soils, which is so much higher than the demand ratio of benthic algal protoplasm in streams and phytoplankton protoplasm in lakes (Table 8–2) that it is likely that the phytoplankton will be phosphorus limited, even if a considerable fraction of the phosphorus supplied is not in an immediately available (PO_4) form. However, the N:P supply ratio in eutrophic systems everywhere is usually sufficiently low to allow a primary nitrogen limitation (Chapter 21). Algae growing under sufficiently poor light conditions will not be nutrient limited at all.

The experimental addition of sufficient quantities of phosphorus and nitrogen to whole lakes and large enclosures anchored in lakes, known as **limnocorrals** or **mesocosms** ($> \sim 1 \ m^3$), has repeatedly demonstrated that with rare exceptions no other elements need to be added to raise the algal biomass sufficiently to make phytoplankton light rather than nutrient limited in the mixed layer of lakes, and to greatly elevate the benthic algal biomass in streams (Stockner and Shortreed 1976). Other elements are required for growth (see Table 8–1), and their relative abundance may affect species composition, but their supply rates to fresh waters are sufficiently large in relation to demand that it does not normally limit the algal *community* growth rate and biomass.

---

[1]"The Land flouds [floods] doe carry away the fatnesse [fertility] from the arable land, and all high grounds, in huge quantities into the sea." (G. Plattes 1639)

Table 8–1 **Concentrations of essential elements for growth (selenium omitted) of freshwater plants (representing demand) and in mean world river water (representing supply) plus the plant:water ratio (demand:supply).**[1]

Element	Symbol	Demand Plants (%)	Supply Water (%)	Demand:Supply Plants: Water (approx)
Oxygen	O	80.5	89	1
Hydrogen	H	9.7	11	1
Carbon[2]	C	6.5	0.0012	5,000
Silicon[3]	Si	1.3	0.00065	2,000
Nitrogen[2]	N	0.7	0.000023	30,000
Calcium	Ca	0.4	0.0015	<1,000
Potassium	K	0.3	0.00023	1,300
Phosphorus[2]	P	0.08	0.000001	80,000
Magnesium	Mg	0.07	0.0004	<1,000
Sulfur	S	0.06	0.0004	<1,000
Chlorine	Cl	0.06	0.0008	<1,000
Sodium	Na	0.04	0.0006	<1,000
Iron[4]	Fe	0.02	0.00007	<1,000
Boron	B	0.001	0.00001	<1,000
Manganese	Mn	0.0007	0.0000015	<1,000
Zinc	Zn	0.0003	0.000001	<1,000
Copper	Cu	0.0001	0.000001	<1,000
Molybdenum	Mo	0.00005	0.0000003	<1,000
Cobalt[4]	Co	0.000002	0.000000005	<1,000

[1]Concentrations listed for freshwater plants were derived from approximately 20 scientific papers dealing with chemical composition of algae and rooted aquatic plants. Averages were weighted for annual production ratios of about 7 units algae to 1 rooted plant, with diatoms comprising 30% of total algal production. Concentrations of the same elements in water (except for phosphorus) were derived from estimates of the composition of mean world river water (Livingstone 1963). The estimate for phosphorus supply is an intuitive estimate for mean world river water.

[2]Concentrations of carbon, nitrogen, and phosphorus in water are given for inorganic forms only.

[3]Silicon is essential for growth of diatoms, some flagellates, and perhaps some higher plants (*Equisetum*). It plays no essential role in the growth of other plants.

[4]The listed value of the demand:supply ratio is deceptively low, in particular for iron and cobalt, since in water much of these elements are not in a form readily available to plants and their true supply values may be, at times, sufficiently low to be growth limiting (Hecky and Kilham 1988).

Source: After Vallentyne, 1974.

This chapter deals with the supply of P and N, as well as organic carbon to lotic systems and lakes. The role of phosphorus and nitrogen in plant production will be discussed in Chapters 17, 18, 21, and 24, whereas the role of these nutrients and organic carbon in microbial metabolism will be examined in Chapters 16 and 22.

8.2 Flowing Water Systems

Flowing water or **lotic systems** are not simply conduits for water, plant nutrients, and organic matter obtained from land and conveyed downstream to nourish lakes and wetlands. Nor are they flowing lakes (Table 8–3). Rather, they are complete ecosystems intimately coupled with their drainage basins. Lotic systems reflect the climate, as well as geomorphology and land use of their drainage basins. Catchment and stream characteristics (width, depth, slope, velocity, discharge, discharge variation, substrate, and immediate shoreline vegetation) help structure stream communities and determine their productivity (Fig. 8–1a, p. 98). The stream biota, in turn, has a major impact on the processing of autochthonously and allochthonously produced organic matter and, consequently, on the release rate, timing of release, and form of organic

Table 8–2 **Selected average N:P supply ratios by atoms (moles) of potential nutrient sources for fresh water systems plus the average demand ratios of organism protoplasm (Chapters 21, 23 and 24). Divide by 2.21 to convert to mass ratios.**

Source	N:P
Runoff from unfertilized fields	547
Export from medium fertility soils	166
Export from forested areas	157
Export from rural areas and croplands	135
Export from fertile soils	74
Groundwater, pristine	63
Precipitation	54
Tropical forest	52
Export from agricultural catchments	44
River water (Mississippi)	27
Macrophytes/crops	24
Algae, not P deficient	22
Fertilizer, average	17
Phytoplankton (balanced growth; Redfield ratio)	16
Macrozooplankton excreta	15
Feedlot runoff	14
Sewage	14
Sediments, mesotrophic lake	12
Macrozooplankton excreta	11
Pasture land and urban runoff	10
Bacterioplankton protoplasm	10
Septic tank effluent and sewage	6
Sediments, eutrophic lake	6
Gull feces	2
Sedimentary rock	2
Earth's crust	<0.2

Source: Modified from Downing and McCauley 1992, Downing et al. 1999, Fagerbakke et al. 1996, and Duarte 1992.

matter and plant nutrients exported to the receiving lakes or wetlands.

The growing recognition of lotic systems as integrated units is much more recent than for lake studies, which have been examined as systems since the 1920s. Until the last few decades, stream biologists saw lotic systems and rivers primarily as a habitat for organisms of interest rather than a system of interacting physical, chemical, and biological processes (Sec. 1.1).

Stream biologists, the majority of whom study aquatic insects or fish, have characterized lotic systems on the basis of dominant fish communities or insect assemblages. A less demanding physical classification, based on linear geomorphic characteristics of lotic systems, known as **stream order**, is most often used today (Fig. 8–2, p. 99). Small, branched, upper tributaries of a permanently flowing river system are classified **first order**, streams receiving two or more first-order streams are **second order**, and so on (Fig. 8–3, p. 99).

The addition of a number of second-order tributary streams to a third-order stream will increase the discharge and catchment area, but not its order. Stream order was originally designed to characterize the distribution of stream channels in the landscape of homogeneous physiographic and climatic regions. Within such a region, stream order is correlated with catchment area, basin relief, stream slope, stream length, and stream discharge, in the process allowing stream order to characterize some stream attributes important to the biota. But stream order designations were not developed to provide information about the climatic (temperature), geological, or hydrological conditions *among* regions that have a major impact on abundance, distribution, and growth rates of aquatic organisms (Fig 8–4, p. 100).

The upper ephemeral portions of stream-order 1 systems in the well-watered north temperate zone are sometimes designated as stream-order 0, but that designation is not informative when applied to a stream-order 5 river in a semiarid region whose streambed is dry for part of the year. Stream order also cannot be applied to lotic systems emerging from springs, wetlands, or lakes (e.g., the St. Lawrence River which drains Lake Ontario). Nor is stream order an index of organic matter and nutrient export from drainage basins, and thus is also not a suitable surrogate for the impact of catchments on the biota *among* lotic systems differing in climate, geology, hydrology, and land use. The final difficulty with this classification scheme is that stream order changes with map scale.

Hughes and Omernik (1981) recommend using the mean annual discharge per unit area (water discharge coefficient, or specific discharge, $m^3 \ km^{-2} \ yr^{-1}$) as the most informative drainage basin characteristic, allowing a comparison *among* lotic systems and their biota *within* and *among* climatic and geologic regions of the temperate zone. Specific discharge is followed, in descending order of importance, by mean annual discharge, drainage basin area, and mean annual discharge range. Lotic and lentic waters are dominated by small systems (Tables 8–4 and 4–4).

Table 8–3 **A comparison of selected attributes of primarily larger north temperature rivers and lakes.**

Attribute	Rivers	Lakes
Dominant water movement	Horizontal, downstream	Vertical, circulatory
Force initiating movement	Gravity downstream	Wind induced
Distribution of substrate	Determined by water currents; gravity driven	Determined by wind-induced currents
Water level fluctuations	Large, flooding	Minor (major in tropics)
Flooding effects on biota	Traumatic, reset event	Minor (major in tropics)
Water residence time	Short (days–weeks)	Long (months–years)
Shape	Long, linear	Short, oval
Catchment:area	High	Low
Groundwater:surface water drainage ratio (summer)	High	Commonly moderate to low
Channel or basin shape	Changes spatially	Stable
Mean depth	Shallow	Deep
Thermal stratification	Rarely	Common
Vertical chemical profile	Homogeneous	Stratified
Dissolved oxygen (summer)	High	Variable
Nutrient concentrations	High, increase downstream	Low, temporally variable
Turbidity	High	Low (higher in tropics)
Dissolved organic carbon	High, allochthonous	Variable, autochthonous
Nutrient retention	Low	High

Source: Modified after Ryder and Pesendorfer 1989.

River Continuum Concept

The development some twenty years ago of the **river continuum concept** (RCC) (Vannote et al. 1980) by biologists interested in stream insects, and based on the pioneering work of Hynes (1975), was a major step in developing a conceptual framework for viewing lotic systems as *longitudinally integrated* systems intimately linked to their drainage basins and with downstream functioning closely linked to processes occurring further upstream. The RCC links stream size (or its surrogate, stream order), organic matter (energy) inputs, and the processing of organic matter and structuring of communities by invertebrates within the channel from headwater to mouth.

The RCC is based on a synthesis of ideas largely derived from research on small (low order), largely forested streams in the north temperate zone (Fig. 8–5, p. 101). It portrays lotic systems as a longitudinally physically changing "stage" upon which the organisms act out the "play," with particulate organic carbon as the energy source for microbes (fungi and

bacteria) (Bärlocher 1992) and macroinvertebrates, with the microbes serving as an important source of food for the better-investigated invertebrates. The abundance and biomass of invertebrates changes seasonally in response to (1) patterns of growth, and emergence in the case of flying insects; (2) food availability; (3) predators; (4) **drift**, in which invertebrates leave the substrate and are carried to sites downstream or lost from the system,[2] and (5) major floods resulting in a large, but temporary and involuntary, reduction in species richness, density, and invertebrate biomass (Quinn and Hickey 1990).

▲ Allochthonous organic matter was considered to be the main energy source of stereotypical, undisturbed, low-order streams in the north temperate zone that are shaded by their forests. The allochtho-

---

[2]"Drift rates [in an Austrian mountain brook] of animals and LPOM [large particulate organic matter, > 200 µm] ranged from 3200700 individuals and 148.9 kg dry weight per hour at flood peak to 17440 individuals and 0.2 kg dry weight per hour at base flow." (K. Tockner and J. A. Waringer 1997)

(a)

Figure 8–1 (a) Upper Lehigh River, Pennsylvania (US). Note the undisturbed land-water transition zone and the partially forested seasonal wetland *(Photo by Michael Gadomski/Courtesy of Animals Animals/Earth Scenes.)*. (b) Planned residential community created by dredging and filling a former wetland on the Neuse River, North Carolina (US). Note the complete absence of a transition zone and lack of connectivity between land and water. *(Photo courtesy of Animals/Earth Scenes.)*

(b)

nous organic matter from the catchment and immediately bordering (and overhanging) forest was thought to overwhelm any autochthonous primary production, yielding a photosynthesis/community respiration ratio (P:R) smaller than one.[3] The wider medium-order

---

[3]Oligotrophic headwater lakes in forested catchments also receive sufficient allochthonous organic matter and other reduced compounds from their inflowing streams to yield a pelagic P:R < 1. During the growing season phytoplankton production is low as the result of a modest supply of inorganic nutrients released from drainage basins (Chapter 21). Where catchments release much colored organic matter, photosynthesis will also be impeded by poor light climate (Chapter 10).

(~3–5) systems typically lack a closed forest canopy, allowing more autochthonous primary production by attached algae (*periphyton*; Sec. 24.9), mosses, liverworts growing on immobile boulders or bedrock, and rooted angiosperm macrophytes (Sec. 24.1).

The P:R of medium-order streams is stereotypically larger than unity according to the RCC concept (Fig. 8–3), but in reality mostly well below one (Table 8–5). The RCC concept visualizes more turbid higher-order systems (> 5) as characterized by poor light climate, resulting in reduced primary production and a P:R < 1. In reality, many slowly flushed, nutrient-rich lowland rivers are autotrophic (P:R > 1)

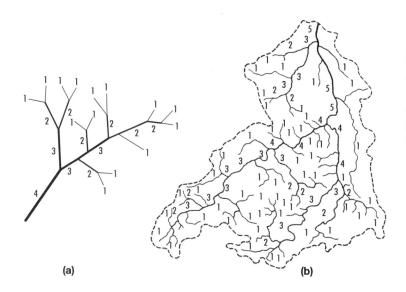

Figure 8–2 A drainage network illustrating stream-order classification developed by A. N. Strahler (1957) for (a) a fourth-order drainage basin in a well-watered area and (b) its application to the Logone River, at Moundou (Chad). Note that, depending on topography, a stream of any one order in a single large drainage basin may drain subbasins of different size and have a different discharge. *(After Welcomme 1985.)*

(Descy and Gosselain 1994). Biotically, such systems closely resemble rapidly-flushed and typically shallow lowland lakes of similar nutrient richness. But where temperate zone lowland rivers have a P:R < 1 (Table 8–5), it is usually the result of increased respiration attributable to downstream lateral input of wastewater or agricultural runoff, enhanced respiration at higher downstream temperatures (Webster et al. 1995), and substantial respiration below the immediate sediment surface (not normally measured) (See Sec. 8.3). Even so, the RCC provides a useful, if idealized, picture of (north temperate) streams in forested and undisturbed

drainage basins, as well as a useful framework for discussing research needs in limnology.

Nutrient Spiralling and Other Riverine Concepts

A second organizing concept for research on flowing water, one that has received much less attention than the RCC, is the **nutrient spiralling concept** (NSC). It is based on the uptake and release of plant nutrients (mostly P, N, and organic carbon) traveling down-

Figure 8–3 The upper portion of a third order German stream, named the Steina. *(Photo, A. Gadtgens, © Prof. Dr. J. Schwoerbel, Universitat Konstanz Fachbereich Biologie, Limnologisches Institut.)*

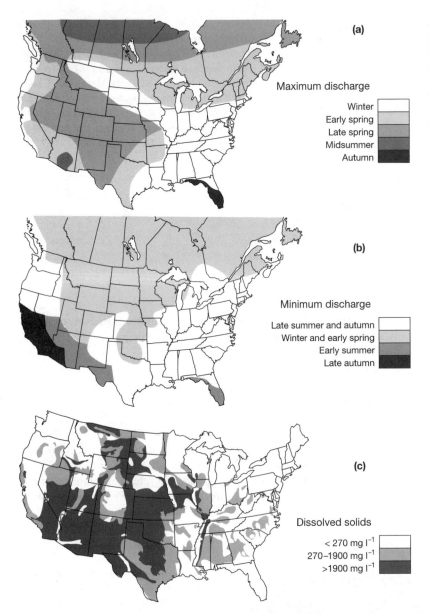

(a)

Maximum discharge

Winter
Early spring
Late spring
Midsummer
Autumn

(b)

Minimum discharge

Late summer and autumn
Winter and early spring
Early summer
Late autumn

(c)

Dissolved solids

< 270 mg l^{-1}
270–1900 mg l^{-1}
>1900 mg l^{-1}

Figure 8–4 The normal timing of (a) maximum and (b) minimum discharge in North American streams and rivers, and (c) their annual average suspended solid load. Note (1) the inorganic turbidity and the spatially changing discharge regimes of streams when viewed on a continental scale. These two variables are linked to the spatial variability of solar radiation, water temperature, and nutrient export from the land; (2) variables such as climate, geology, hydrology (discharge), ionic composition (dissolved solids), and content of suspended inorganic particles. All have an influence on the biota, with the most important determinants changing temporally and spatially. *[(a and b) after Langbein and Wells 1955, in Whitton 1975, (c) from Thornton 1984.]*

stream (Newbold et al. 1983). The nutrient or carbon **uptake length** (m) is defined as the average distance traveled by a substance once released by the sediment biota until it is once again accumulated. The uptake length is used to compare nutrient or carbon retention within and among streams under different environmental conditions. NSC describes downstream cycling of nutrients and/or organic carbon between the dissolved (transport) phase, and the biota-controlled utilization (retention) phases. Sequential uptake and release allows downstream movement to be visualized as spirals. **Spiral length** is a measure of system nutrient retention and

is influenced by the supply:demand ratio for nutrients and organic carbon by microbes and plants, with microbes responsible for 80 percent or more of the phosphorus uptake in some small American streams (Elwood et al. 1981). Spiral length shortens as the amount of organic matter available increases, which serves as an energy source for microbes (Chapter 22) and invertebrates (Allan 1995). It also declines as temperature increases, but increases along with stream discharge as the possibility of particle sedimentation declines. The shorter the spiral, the longer a particular nutrient will be retained by a stream or river.

Table 8–4   **The average number and length of US stream and river channels as a function of the stream-order scheme divised by A. N. Strahler (1957). According to Benke (1993) the logarithm of discharge (m³ s⁻¹) roughly corresponds to stream order in well-watered north temperate systems, with ~0.001 m³ s⁻¹ representing a first-order stream, 0.01 m³ s⁻¹ a second order stream, etc.**

Order	Number	Length (x̄ km)	Total Length (km)	Total Length (%)
1	1,570,000	1.6	2,528,000	48.3
2	350,000	3.7	1,304,000	24.9
3	80,000	8.5	676,000	12.9
4	18,000	19	354,000	6.8
5	4,200	45	187,000	3.6
6	950	103	98,000	1.9
7	200	237	48,000	0.9
8	41	544	23,000	0.4
9	8	1251	10,000	0.2
10	1	2898	3,000	0.1

Source: From Leopold et al. 1992.

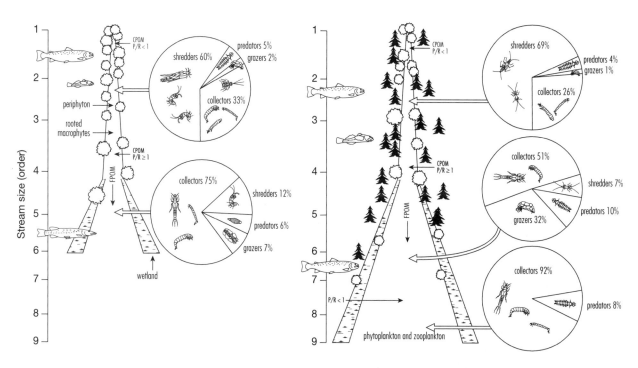

Figure 8–5   A conceptual representation of ecosystem properties in the river continuum concept (RCC) for both a north temperate zone deciduous-forest lotic system and a deciduous-coniferous system with a characteristic distribution of fish and the major functional feeding groups of aquatic insects. Collectors are dependent on filtered detrital particles; shredders are ingestors of vascular plant material; grazers (scrapers) scrape food from the substratum; predators are ingestors of other organisms (Chapter 25). *(Modified after Peterson et al. 1995, in Cushing C. E. et al. 1995.)*

Table 8–5 **Gross primary production (GPP24) and community respiration (CR24) in selected lotic systems on one or more dates during summer, obtained from diurnal oxygen changes in the open water. Note, the majority of measurements in the literature are obtained with paired transparent and opaque sediment chambers placed on the sediment surface, which do not sufficiently capture the subsurface (hyporheic) respiration. (See Fig. 15–2 for a graphical representation of the GPP:CR ratio.)**

Stream/River	Stream Order	GPP24 ($g\ O_2\ m^{-2}\ d^{-1}$)	CR24 ($g\ O_2\ m^{-2}\ d^{-1}$)	GPP:CR
Walker Branch (US, Tennessee)	1	0.14	1.45	0.1
Hugh White Creek (US, North Carolina)	2	0.07	3.41	0.02
Little Sandy Creek (US, New York)	3–4	0.9–10.2	5.0–14.0	0.2–0.7
Black Creek (US, Georgia)	4	0.5–1.5	3.0–6.0	<0.2–0.3
Sycamore Creek (US, Arizona desert)	4	9.4–10.3	10.1–10.9	0.9–1.0
Deep Creek (US, Idaho)	ND[1]	8.5–11.9	6.9–10.8	1.2–1.1
Baker River (US, New Hampshire)	4	0.4	1.9	0.2
Neuse River (US, North Carolina)				
at Smithfield	ND	1.8	2.4	0.75
at Kingston	ND	1.3	5.6	0.23
Necker River (CH)	6	$\bar{x} = 2.5$	$\bar{x} = 3.5$	0.71

[1] ND = No data

Source: See Mulholland et al. 1997, and Uehlinger and Naegeli 1998.

Small streams remove dissolved nutrients more effectively than large rivers, possibly because small streams have a large surface area of sediment in relation to the volume of overlying water that allows efficient benthic uptake (removal) and, therefore, short spirals (Rutherford et al. 1987, and Butturini and Sabater 1998). Not surprisingly, the demand for inorganic nutrients becomes a more important determinant of spiral length than discharge among systems varying modestly in discharge and temperature (Marti and Sabater 1996), attesting to the importance of scale in determining cause and effect (Sec. 2.6).

The RCC and NSC, based primarily on streams and small rivers, spawned three other riverine models that address how large systems function (Thorp 1994): the **serial discontinuity concept** which incorporates the effect of large dams and reservoirs on the RCC, the **flood pulse concept** involving the impact of transport by large (tropical) rivers of nutrients and sediments onto their downstream floodplains and the subsequent processing by a biota adapted to seasonal variation in hydrology, and the **riverine productivity model** which stresses the importance of local autochthonous production and organic inputs from the land–water transition zone of tributaries.[4]

Organic Carbon and Inorganic Nutrient Limitation

▲ Stream biologists working on forested streams that receive much allochthonous organic matter and relatively low irradiance have considered organic matter from catchments to be the primary energy source for the biota. However, it is evident from work on sun-exposed streams that autochthonous production can be of great importance (Kjeldsen et al. 1998). Recent work, employing stable carbon isotopes as tracers, is showing that in shaded streams production by algae growing on substrates (*periphyton*; Chapter 24) is more important as a source of organic matter than had been surmised (France 1995). Rapid algal growth followed

---

[4]"Models are to be used but not to be believed." (H. Theil 1971)

by equally rapid removal by grazers will leave a small and barely visible algal biomass (Huryn 1996, and Kjeldsen et al. 1998).[5]

The rapid response of periphytic algae to experimental fertilization in a number of pristine lotic systems has shown low-order systems to be primarily nutrient rather than light limited, and to have lotic foodwebs that can readily shift from heterotrophic to autotrophic mode upon nutrient addition (Peterson et al. 1985). The systems further exhibit increased organic matter decomposition upon addition of inorganic nutrients, and then have enhanced fish production (Johnston et al. 1990). But lotic research is still dominated by work in the water-rich north temperate zone, characterized by relatively predictable biological responses to fairly regular hydrological seasonality, (Fig. 8–4) and there is little appreciation of impacts on the biota of exceptionally large inter-year variation in river discharge and intra-year variability in seasonal timing of peak and minimum water flows (Fig. 8–6) occurring over much of the globe (Puckridge et al. 1998).

8.3   Rivers and Their Ecotones

The transition zones between adjacent ecological systems are known as **ecotones**. These encompass sharp gradients in environmental factors, dominating processes, and species composition. The transition zone between an aquatic system and the adjacent land, is known as the **riparian zone**. Riparian zones have major, but poorly quantified, effects on inland waters. The transition zone between lentic or lotic systems and adjacent wetlands form a second important ecotone. Deltas formed at the mouth of rivers where they enter lakes or reservoirs are a third type of interphase between lotic and lentic ecosystems.

The fourth important ecotone type is the transition zone in the vertical plane between the stream-channel water and the groundwater, known as the **hyporheic zone** or **hyporheos** (Fig. 8–7). This zone, an important habitat for the biota, contains a variable mixture of surface water and groundwater characterized by strong gradients in flushing, dissolved oxygen

concentration, and chemistry. The lower boundary has been defined operationally as the depth to which 10 percent or more channel water (< 90% groundwater) is present (Triska et al. 1993). The hyporheic zone not only extends vertically below the stream bottom but, depending on hydrological conditions and the definition used, may even extend beyond the stream or river margins into the groundwater zone of the **floodplain** when substrate permeability (porosity) allows the passage of (oxygenated) water, organic matter, and organisms (Fig. 8–7).

The difficult to sample hyporheic and floodplain zones are home to communities of aquatic microbes and invertebrates, including some *endemic* macroinvertebrates (species found nowhere else). Strayer et al. (1997) reported an average of nearly 600 species per three-liter pump sample taken in the hyporheic zone. The species were numerically dominated by copepods, nematodes, and microannelid worms, but included typically larger insects and molluscs.[6] Hyporheic organisms, including microbes, apparently play a major (but little assessed) role in retention and transformation of organic matter and the cycling of nutrients and gases between land, water, and the atmosphere. Work on a sixth order prealpine Swiss river allowed the conclusion that the gravel-dominated hyporheic zone and its microbes contributed far more (76–96%) of whole system respiration than the typically better-investigated surficial sediments of the stream bed (Naegeli and Uehlinger 1997).

Larval aquatic insects (stoneflies, Fig. 25–3) have been collected in groundwater monitoring wells up to two kilometers from a lowland river bordered by a thick substrate of high porosity, showing how extensive the floodplain ecotone can be (Stanford and Ward 1993). Stoneflies use the *aquifer* (groundwater) to return to the river, there emerging from the water as adults for breeding and egg deposition. The fraction of macroinvertebrate production associated with the immediate channel bottom, rather than with the hyporheic zone or floodplain, is determined by substrate characteristics and is further affected by the degree of connection (**connectivity**) of compartments which in turn are affected by discharge through the hyporheic zone and the water residence time there. North tem-

---

[5]Incorrect conclusions drawn about community energy and nutrient flow on the basis of the (relatively easy to measure) biomass (*standing stock*) rather than on growth and loss rates (more difficult to measure and interpret) have long held back the development of ecology.

[6]"One of the most sensational zoological discoveries of the late 19th and early 20th century was that groundwaters contain a diverse fauna of endemic, morphologically distinctive invertebrates. Thousands of specialized groundwater animals, "stygobionts," including crustaceans, worms, mollusks, and other invertebrates are now known . . ." (D. L. Strayer et al. 1995)

Figure 8–6 The main North basin of the Sahaka wetland complex in Madagascar, during (a) the dry season, and (b) the wet season, characerized by flooding and much floating (and submerged) vegetation. *(Photos courtesy of R. Safford, Royal Holloway Institute for Environmental Research.)*

(a)

(b)

perate zone first-order streams on a solid substrate typically lack a hyporheic and floodplain zone, implying that all invertebrate production is associated with the channel surface. In contrast, at least half of the system production of invertebrates ($g\ m^{-2}\ yr^{-1}$) in two Virginia streams is associated with hyporheic and floodplain ecotones (Smock et al. 1992). Even though these two ecotones tend to be less productive per square meter than the channel surface, the ecotone area is disproportionately large. From a whole-system perspective, 67–95 percent of the production was estimated to be associated with the hyporheos and floodplain, leaving only 5–32 percent for the immediate channel surface.

The scale at which studies are carried out has a major bearing on the conclusions drawn (Sec. 2.6).

▲ Animal densities in surficial sediments tend to be greatest in oxic (> 1 mg O_2 l^{-1}) high organic matter sediments (Table 8–6). Detailed work on a permanently flowing (*perannual*) New Zealand pasture stream indicated that macroinvertebrate production in the stream itself was insufficient to support measured trout production, unless the hyporheic invertebrates and terrestrial invertebrates falling into the water were also considered (Huryn 1996). Terrestrial arthropods can contribute up to 90 percent of the summer food of salmonid fishes (Hunt 1975), whereas adult aquatic in-

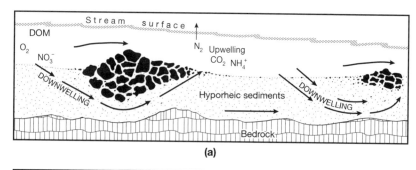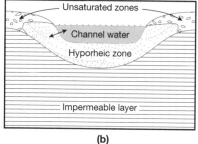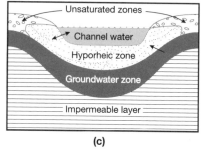

Figure 8–7 Hyporheic zone. (a) Longitudinal section of streambed showing surface to hyporheic zone hydrologic and chemical exchange. (b) Cross section of streambed showing the potential for channel to hyporheic zone exchange without a groundwater presence, and (c) hyporheic zone to groundwater hydrologic exchange. *(Part a modified after Boulton 1993, parts b and c modified after White 1993.)*

sects play a role in transferring energy and nutrients from aquatic to riparian systems (Fig. 23–20 and Jackson and Resh 1989). It is apparent that land and water are closely coupled.

Ecotones are far easier to conceptualize than define, a difficulty shared with ecosystems in general. Ecotones change with space and time scales, they are sometimes viewed as ecosystems in their own right (Bretschko 1995) and tend to be rich in species. Land–water and hyporheic ecotones appear to play an important role in cycling, transforming, and spiralling downstream of nutrients and organic matter. When streambeds in semiarid regions become dry, the hyporheic zone and any adjacent groundwater zone are the only habitats available to aquatic organisms in *ephemeral* (temporary) systems.

Fringing (riparian) wetlands bordering lowland rivers, floodplains, and wetlands are ecotones of great importance as feeding, breeding, and nursing areas for fish and wildlife (Siole 1984, and Chapter 26). They are also of inordinate importance in regulating runoff and purifying wastewater (Sec. 1.1), as well as sources of organisms to rivers.

The most spectacular wetlands are the enormous inland and coastal deltas of major tropical river systems (Table 6–3). Water levels in the principal inland delta of the Amazon River rise to 15 m above base flow during the annual 4–10-month flood period and inundate an area of about 50,000 km². Seasonal flooding by "flood–drought rivers" fills basins and enlarges or fuses existing lakes (Secs. 6.5, 24.2). Plant nutrients and organic matter are carried from both main and

Table 8–6 **Abundance, species richness, and species diversity of aquatic insects found in five habitats, characterized by their substrate, in a small, first-order deciduous forest stream (CA). Values are annual averages.**

Habitat	Abundance (m^{-2})	Number of Species	Diversity[1]
Sand	920	61	1.96
Gravel	1,300	82	2.31
Cobbles and pebbles	2,130	76	2.02
Leaves	3,480	92	2.40
Detritus[2]	5,680	66	1.73

[1]Diversity here = $(S-1)/\log_e N$
[2]Finely divided leaf material in pools and along stream margins.

Source: From MacKay and Kalff 1969.

local rivers into wetlands, and much organic matter is transported into rivers from the flooded floodplain. Inflowing nutrients that are in high demand are largely retained by wetlands everywhere, fertilizing the riparian zones and the wetlands beyond (See Sec. 6.5 and Chapter 24, Naiman and Décamps 1990, Breen et al. 1988).

▲ 8.4 Rivers, Their Banks, and Human Activity

Human activity (forestry, agriculture, road construction, fires) greatly affects lotic systems. Overhanging trees and shrubs in undisturbed catchments reduce the irradiance received, with shoreline vegetation having a deciding influence on maximum water temperature and the suitability of higher latitude streams for cold-water fish species (Barton et al. 1985). The roots of trees, shrubs, and native grasses stabilize river banks, while overhanging branches are a source of fresh organic matter (leaves and arthropods) to the biota. Additional organic matter enters from catchments and is carried by streams, wind, overland flow, subsurface seepage, and groundwater (Sec. 5.5). Fallen trees and organic debris produce dams that help create deeper scour pools between riffle areas, thereby increasing habitat diversity for microbes, plants, invertebrates, and fish (Fig. 8–8), and also helping to regulate export of organic matter from forest streams (Bilby 1981).

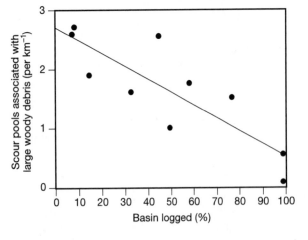

Figure 8–8 Frequency of pools associated with large woody debris in 10 Oregon coastal streams with different logging histories. Scour pools are an important habitat for fish and certain types of invertebrates. *(After Bisson et al. 1992.)*

Human Activity

Forest cutting and forest fires open the canopy, reduce input of allochthonous organic matter over the medium term, raise water temperatures, and allow for increased autochthonous primary production (Fig. 8–9) by algae attached to substrates (periphyton; Sec. 24.9). The invertebrate and fish abundance/biomass increase (Bisson et al. 1992) unless the more favorable photosynthetic conditions are offset by soil erosion (turbidity) where slopes are steep; and in lowland crop growing areas, where frequent plowing to the edges of streams results in increased turbidity and loss of pool-forming woody debris.

Most rivers and larger streams in economically advanced countries not only receive large quantities of agricultural and urban wastewater runoff that greatly affects stream metabolism but the streams and rivers have also been channelized by straightening bends to facilitate water transport and/or drainage. The resulting increased stream gradient (slope) enhances discharge rates, and reduces variation in river depth. Increased discharge rates then permit increased channel and bank erosion during high water phases, followed by lowering of the groundwater table. Draining adjacent wetlands for agricultural or urban purposes and diking principal river channels to prevent flooding of the original floodplain greatly reduces the size of land–water ecotones and contact of the river with its banks (Fig. 8–1b). The role of ecotones as a buffer and habitat for the biota is similarly reduced. The previous wetlands not only served to store floodwater and thus reduce peak water discharges[7] and catastrophic floods but further acted as traps for sediment, organic matter, and for both dissolved and particulate nutrients carried with the floodwater, allowing their utilization in the wetland ecotones (Pinay et al. 1990). Moreover, many rivers have been dammed, affecting not only the migration of fish and invertebrates but also water discharge, timing, and flow variability in the system below the dams (Sec. 5.8 and Chapter 29). The resulting cessation of annual floods eventually leads to the reduction or disappearance of downstream flood-

---

[7]Relatively low rates of evapotranspiration in the north temperate zone mean that the role of riverine wetlands in water management is one in which floodplains are reduced but little water is lost. In contrast, semiarid zone floodpeaks allow much evapotranspiration by plants, wetted soils, and water surfaces, resulting in a much lower outflow than inflow. For example, only about 54 percent of the on average $7.11 \times 1{,}010$ m^3 entering the inland central delta of the Niger River (west Africa) (Section 6.5) is returned to the river (Welcomme 1985).

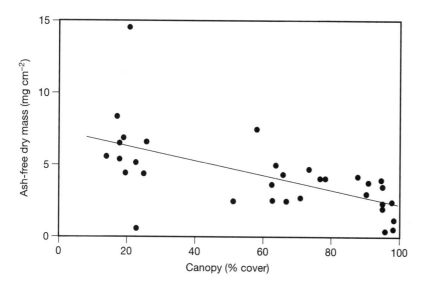

Figure 8–9 Relationship between the riparian canopy cover and periphyton biomass on tiles placed on platforms in a coastal California (US) stream. Ash-free dry weight refers to the organic matter weight minus the minerals retained in the ash following burning (oxidation). Note that the effect of the canopy is maximized and clearly evident because the variation imposed by substrate variability has been removed (controlled) by using standardized tiles positioned on platforms. (*After Feminella et al. 1989.*)

plains. It has become more evident in economically advanced countries that the benefits gained from diking rivers and draining the adjacent wetlands for agriculture and housing may not be cost effective. The damage wrought by extensive lowland flooding during rare years of exceptionally high runoff upstream, with former wetlands unavailable for temporary water storage, can greatly exceed return from agricultural and other human activities.

Forests are frequently cut right to river or stream edges, maximizing the amount of land available for agriculture or tree harvesting. This has negative effects on stream habitat quality and on the suitability of the streams for silt-sensitive invertebrate and fish species (Figs. 5–7 and 8–10). Fortunately, the importance of an undisturbed buffer zone between rivers and their catchments is now recognized. Well-vegetated buffer zones reduce streambank erosion, trap soil particles and adsorbed nutrients on the land (Table 8–7), and reduce irradiance (heat) reaching the water. The width of the strips (often 10–40 m), beyond a certain minimum, appears less important than

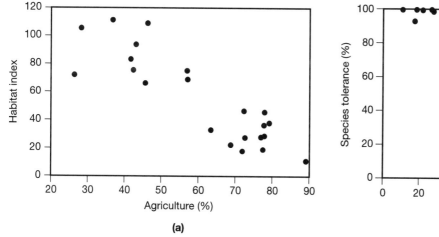

(a)

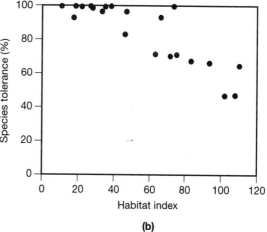

(b)

Figure 8–10 Influence of agricultural land use at 23 stream sites in southeastern Michigan. (a) Instream habitat quality declined with increasing percentage of agricultural land upstream. The habitat quality index is a composite of 10 variables that include habitat heterogeneity and evidence of degradation. (b) The percentage of fish species able to cope with silty and degraded habitat conditions varies directly with instream habitat quality. (*Modified after Roth 1994, in Allan, 1995.*)

Table 8–7 **Percent change in sediment and nitrate loads over a 15-year period under different forestry practices in hilly Oregon, US. Buffer strips retain sediment and stimulate denitrification and the loss of nitrogen as N_2 gas to the atmosphere, in the process reducing stream nitrogen levels.**

Forestry Practice	Suspended Sediments	Nitrate
Clear cut	205	214
Clear cut with buffer strips along streams	54	0
Controls	0.1	0

Source: Modified from Moring 1975, in Karr and Schlosser 1978.

their quality and continuity (Fig. 5–8 and Pinay et al. 1990). The riparian zone (including wetlands) plays an additional role in promoting denitrification (Sec. 18.4, and Hedin et al. 1998), thereby reducing or preventing much of the groundwater NO_3 in agricultural drainage basins from entering lotic systems. Moreover, the wetlands of the riparian zone serve as an important site for denitrification of wastewater nitrate released directly into streams (Table 8–8).

Recent attempts at in-river "restoration" by re-meandering and reestablishing riffle and pool habitats, as well as by recreating waters-edge irregularities, show promise as a management tool and provide excellent possibilities for fundamental research. Increased variation in water depth and flow rates resulting from restoration efforts yielded greater retention of organic matter in the ecotone of a stretch of the Melk River (AT), allowing the number of fish species to double within three years (Jungwirth et al. 1995). Fish abundance and biomass more than doubled as well, reflecting an improved invertebrate and fish food-resource base, plus probably the provision of new spawning grounds, even though there had been no ecotone restoration beyond the immediate river edge (Fig. 8–11).

The re-meandering of streams reduces the discharge rate, thereby increasing groundwater levels and extending riparian zones with their characteristically anaerobic soils, encouraging denitrification of NO_3–N and release of N_2 and N_2O to the atmosphere from the recreated wetlands. Increased flooding allows increased sedimentation and retention of particulate phosphorus and particulate organic matter, fertilizing the recreated wetlands and improving riverwater quality (Iversen et al. 1995, in Skotte-Møller et al. 1995). However, stream restoration and the reconstruction of wetlands is expensive and may be politically difficult. Conservation is a superior alternative where streams and wetlands have not yet been damaged or drained.

8.5 Drainage-Basin Export of Nitrogen and Phosphorus

High retention of phosphorus and nitrogen by undisturbed and well-vegetated catchments means quantities exported to rivers and lakes are relatively small.

Table 8–8 **Nitrogen retention, including denitrification, computed by mass–balance from the difference between terrestrial input and stream water output.**

Location	Input (kg ha^{-1} yr^{-1})	Nitrogen Retention (%)
Little River (US)	119 TN	89
Chowan River Swamp (US)	58 TN	64
Louge River (FR)	–	100
Rhode River (US)	83 TN	89
Beaverdam Creek (US)	35 NO_3N	85
Stevns River, wet meadow (DK)	59 NO_3N	97
Rabis Brook, wet meadow (DK)	711 NO_3	56
Gjern River fen (DK)	884 NO_3	99
Glumsø reed swamp receiving wastewater (DK)	5046 NO_3	54
Stor River, restored wet meadow receiving wastewater (DK)	979 NO_3	48

Source: Modified after Triska et al. 1993, and Iversen et al. 1995.

Figure 8–11 Fish abundance and biomass of the total fish stock in the Melk River, AT, prior to restructuring (1987) and after (1988 and 1990). *(Modified after Jungwrith et al. 1995.)*

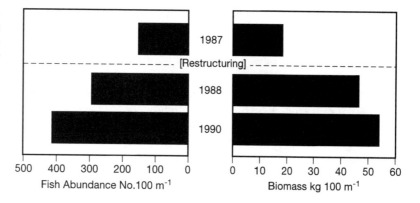

Forested streams receiving large amounts of allochthonous organic matter, but little N and P from their drainage basins will be characterized by high C:P and C:N nutrient ratio discharge. Receiving lakes and wetlands will receive little N and P and therefore be oligotrophic, with only a modest phytoplankton biomass and community primary production (Chapter 21). Low phytoplankton biomass and low production rates result in both high water transparency in low humic waters, and low community productivity of zooplankton (Chapter 23), benthic animals (Chapter 25), and fish (Chapter 26). The nutrient export from land to aquatic systems is particularly low at high latitudes, which are characterized by poorly developed soils that are frozen as well for an appreciable part of the year (see Table 8–9).

Conversely, aquatic systems located in agricultural drainage basins with rich soils receiving fertilizer applications (and those basins draining urban areas) receive very high inorganic nutrient loadings (inputs) (Table 8–9). Tile drainage systems, installed below the surface to facilitate drainage of croplands, export exceptionally high levels of nitrate and phosphorus to receiving waterways. Extreme values are recorded in the Netherlands with its heavy fertilization of cropland and extreme livestock densities. Typical export coefficients there are in the order of 30,000 kg N and 1,000 kg P km^{-2} yr^{-1} (D. T. Van der Molen et al. 1998). Since most crop-growing areas were covered with either forest or densely vegetated grassland before conversion to agriculture, it is apparent that the water was once much more oligotrophic than today.

Nutrient Export Rates

Nutrient export coefficients (NECs) represent the annual loss of nutrients per unit area of drainage basin ($g\ m^{-2}\ yr^{-1}$, or $kg\ km^2\ yr^{-1}$). NECs vary with the water export coefficient ($m^3\ km^{-2}\ yr^{-1}$), climate, and regional geology. Export coefficients are more precisely re-

Table 8–9 **Typical total nitrogen and total phosphorus export coefficients (kg km^{-2} yr^{-1}) and the average N:P ratio as mass for selected catchment types. 100 ha = 1 km^2.**

Land Use	TN	Range	TP	Range	N:P	Climate
Tundra	27	–	1.4	–	19	arctic
Boreal forest	97	–	4.1	–	24	subarctic
Forests	300	130–500	10	5–12	38	temperate
Agriculture	660	500–1000[1]	18	5–50[1]	13	temperate
Forest/Savanna	126	–	7.7	–	16	semiarid tropical
Forest	998	–	46	–	22	moist, tropical
Urban	700	500–880	100	30–1660	7	temperate

[1]Higher nitrogen coefficients are characteristic of rich agricultural soils (Fig. 8–13b), and for sorbed phosphorus from rich soils subject to soil erosion.

Source: Schindler et al. 1974, Jansson 1979, Lesack et al. 1984, Lewis 1986.

ferred to as **specific export coefficients**, with *specific* referring to export per unit area. Nutrient export coefficients vary with land use (Table 8–9). But as each land-use category encompasses a variety of different soil types, local geologies, and morphometry (size and steepness) of the drainage basins, there is much within land use variation. Moreover, individual catchments further exhibit considerable variation in seasonal and interannual precipitation (mm yr^{-1}), and thus in water export (runoff) and the amounts of nitrogen, phosphorus (Table 8–9), and other materials exported per unit area and per catchment over time.

▲ Some of the variation in NECs reported must be attributable to inaccuracies in measuring the total amount of a nutrient leaving the examined drainage basins (kg yr^{-1}). Accurate determinations of nutrient export depend on accurate measurements of both river discharge (m^3 d^{-1}) and nutrient concentration (g m^{-3}) at different discharges. Poor estimates of one or both variables yields export estimates that can be in error by more than 100 percent (Scheider et al. 1979). NEC inaccuracies are usually largely attributable to uncertainties in discharge determinations because it tends to vary much more over time than nutrient concentrations [concentration × discharge = export (kg yr^{-1})]. Unrecorded storms cause important errors in nutrient or DOC budgets because half or more of the annual export may occur during the days or weeks of highest discharge. The export (kg yr^{-1}) divided by the catchment area (km^2) yields the specific export coefficients reported (Table 8–9).

Reliable discharge data require a quantified relationship between discharge (m^3 d^{-1}) and water height measured with a staff gauge. Rivers for which the relationship is known and water height is recorded daily or continuously are known as **gauged rivers**. Most large rivers are gauged, but very few of the streams and small rivers studied by limnologists are. Where a daily or continuous discharge record is available, the principal uncertainty in determining the nutrient export shifts to the variation in nutrient concentration over time (Stevens and Stewart 1981).

Published nutrient export coefficients give a very useful first estimate of the predicted export from catchments and urban centers for which no direct measurements are available. However, reported NECs (Table 8–9) are normally averaged values and, if not, are based on studies of individual systems whose relevance to the catchment of concern is unknown. Furthermore, interannual differences in NECs determined for single catchments can be large due to differences in average and peak runoff. Interannual differences in runoff were responsible for the two- to threefold change in the measured nutrient export attributable to livestock during a three-year study of several Iowa drainage basins (Jones et al. 1976). Interannual differences in runoff were similarly responsible for large differences observed in the nutrient export from six Swedish catchments (Forsberg and Ryding 1979). Even so, in the absence of direct measurements, appropriate NECs are very useful.

The quantity of nutrients exported rises with increasing catchment area because the export (kg yr^{-1}) is the product of the NEC (kg km^{-2} yr^{-1}) and catchment area (km^2). Catchment area and **catchment relief** (slope) are also related, with average slope greater in small than in large drainage basins when examined over a large spatial scale (Fig. 8–12a). Furthermore, larger catchments have reduced **stream** or **drainage density** (total length of streams/catchment area) (Fig. 8–12b). This means that a particle has to travel further to reach a stream in a large drainage basin before it can be exported. The combination of a reduced drainage density and low catchment slope, which increases the possibility of retaining particles on land, appears to be responsible for the inverse relationship in small catchments between catchment size and the NEC for elements that are partially or largely adsorbed to particles (Fig. 8–13a). Most persistent organic contaminants and toxic trace metals adsorb strongly to particles and their catchment export is typically small in well-vegetated regions. Export coefficients are greatest for small, relatively steep catchments. Where most adsorbing contaminants and nutrients are retained on land the relative importance of direct atmospheric deposition of nutrients and contaminants on lake and wetland surfaces is relatively large (Chapters 18 and 28).

Dissolved and Particulate Nutrients

Phosphorus exported from agricultural drainage basins is overwhelmingly adsorbed to particles. For example, an average 16 percent (range 2–56%) of P exported from 116 agricultural catchments was not adsorbed to particles (Prairie and Kalff 1986). In contrast, most of the total nitrogen (TN) in well-drained (aerobic) agricultural basins is in the soluble NO_3 form and readily exported to streams (Keeney and DeLuca 1993), with the basin export coefficient a function of the fraction of a catchment in cropland (Fig. 8–13b). Ammonium (NH_4^+), as a fraction of the

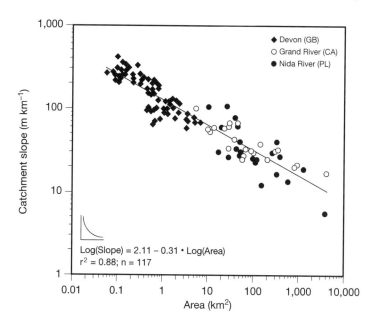

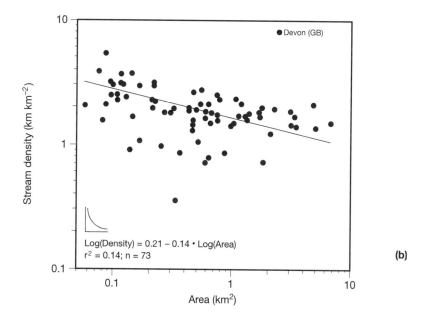

Figure 8–12 Relationship between (a) catchment area and catchment slope, and (b) catchment area and stream (or drainage) density in north temperate zone drainage basins. Note the rapid decline in drainage basin slope with increasing area and the smaller decline in the stream density as the catchment size increases, implying a greater particle retention on larger catchments. *(Modified after Gregory and Walling, 1973, and Mahon, 1984.)*

total nitrogen export, tends to be small because NH_4^+ ions adsorb to soil particles, but NH_4^+ greatly dominates in poorly drained anoxic substrates where NO_3 is lacking (Forsberg and Ryding 1979).

As most of the TN in well-drained (well-aerated) agricultural basins is exported in the NO_3 form, its export coefficient does not normally decline with increasing drainage basin size. The amount of TN exported (kg yr^{-1}) is, therefore, linked to the amount of runoff, surficial geology (soils), and catchment area (Duarte and Kalff 1989) as:

$$\log \text{TN export} = 2.6 \ln \text{runoff} + 2.5 \text{ Geo} - 0.18 (\text{Geo} \cdot \ln \text{area}) - 15.1$$

EQ. 8.1

$$R^2 = 0.56 \qquad \text{SE est.} = 2.68 \qquad n = 46$$

where TN export = kg yr^{-1}; runoff = 1 m^{-2} yr^{-1}; Geo = 1 on igneous substrate; Geo = 2 on other non-carbona-

Figure 8–13 (a) Relationship between the total phosphorus export coefficient and catchment area under different land use regimes, based on a world data set dominated by temperate zone drainage basins, but excluding areas of intense agriculture. (b) Relationship between the nitrogen export coefficient and cropland as a percent of the total drainage basin area (%C) in the Chesapeake Bay region of the northeastern USA. (*Part a from Prairie and Kalff 1986, part b from Jordan et al. 1997.*)

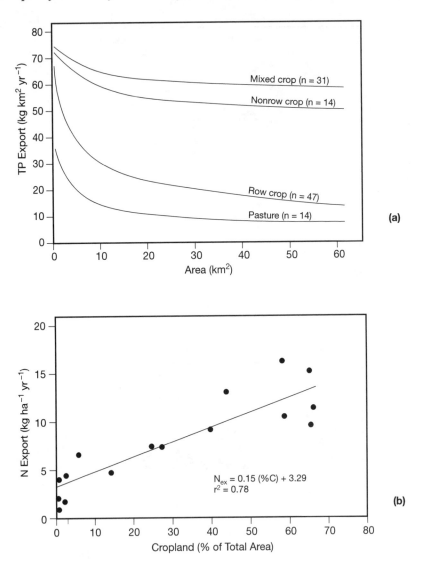

ceous substrates; Geo = 3 on calcareous substrates and glacial drift; catchment area = km². Note that the catchment area in Eq. 8.1 is involved through an interaction with geology rather than directly, as in the case of the TP export. The modest fraction of the variance explained (R^2) indicates that these surrogate variables are not very good predictors of export at a continental scale, probably because agricultural and other human activities are not explicitly considered.

The contribution of livestock to nutrient export is extremely variable, depending on stocking densities, stream bank trampling, catchment slope, overgrazing, soil type, runoff, and manure application to agricultural fields. Manure application (feedlot runoff N:P ~14 as atoms) to agricultural fields are calculated on the basis of the N requirements of crops which are characterized by a N:P demand ratio of ~24 by atoms (Table 8–2). Consequently, manure (N:P ~14) applications have to be high. This allows for loss to waterways of phosphorus sorbed to eroding soil particles and of nitrogen as NO_3^- following oxidation of the manure NH_4^+ to soluble NO_3 and volatile NH_3. The problem is most severe in seasons when uptake by crops is small or lacking. Modern regulation prohibits manure spreading on frozen ground because it washes almost totally into waterways during the spring melt. In England, the three environmental determinants used in Eq. 8.1 vary only moderately and the highly variable stocking density is the single most important determinant of nutrient export from pastures. The experience there suggests that one hectare (0.01 km^2) of typically low-slope pasture land there can efficiently absorb the waste of 2.5 milking cows, 25 pigs, or 250 laying hens, with each group excreting about 400 kg N yr^{-1} (Cooke and Williams 1973).

It is not surprising that water pollution can be a major problem where livestock is kept at exceptionally high densities in barns or feedlots (Jones et al. 1976) or where, as in the Netherlands, high densities require importing much feed (organic matter and nutrients therein), with the wastes overwhelming the absorptive capacity of soils (Table 8–9). When abundant, migrating or resident geese, swans, and ducks excrete sufficient nutrients obtained from fields and waterways elsewhere to significantly enhance the local nutrient loading of lakes and wetlands (Manny et al. 1994), but the waterfowl contribution is typically small compared to livestock.

Sewage Wastewater

Sewage wastewater is by far the most important source of nutrients to rivers and lakes near large urban centers. Humans release ~12 g N and ~1.5 g P per person per day, plus additional phosphorus contained in detergents and food particles. Where populations are large and wastewater treatment is nonexistent or poor, nearly all of the nutrients released by urban humans, their animals, and garden fertilizers enter waterways via sewers. However, in economically advanced countries using good sewage treatment, the nutrient contribution of agriculture dominates.

In countries with advanced wastewater treatment, chemicals [usually $Al(SO_4)_3$, $FeCl_3$ or $Ca(OH)_2$] are commonly added during **primary sewage treatment** to flocculate and precipitate more nutrients, organic matter, and adsorbed contaminants than would be the case without them (Fig. 8–14). The precipitated material is collected before the treated wastewater is released. More advanced (and more expensive) treatment systems (**secondary treatment**) allow microbially-induced flocculation and sedimentation of the organic matter, the nutrients contained therein, and the adsorbed contaminants (Fig. 8–14). The sedimented solids must be buried in landfills that do not leak, or they have to be incinerated, with the attendant release of metals and organic matter to the atmosphere.

Wastewater lagoons and wetlands serving as nutrient traps for wastewater from villages and livestock farms can be a less expensive alternative to sewage treatment plants (Sec. 24.2). They are most effective

at lower latitudes where light and temperature are sufficiently high to allow a year-round growing season and much denitrification, where sediments have a high adsorbing capacity, and quantities of wastewater are not so high as to overwhelm the retention capacity of the systems. Typically shallow wetlands allow ready trapping of particles by the sediments and nutrient uptake by the aquatic vegetation and sediment microbes. (Sec. 24.2.)

8.6 Atmospheric Deposition of Nutrients

Not all the nutrients that nourish inland waters are derived from their **terrestrial catchments**. Some enter as difficult-to-measure micron-sized aerosol particles (*dryfall*) and gases, complementing the quantities measured in precipitation (*wetfall*) from the overlying atmosphere or **aerial catchment** (Chapter 27). Aquatic systems are therefore not only products of their terrestrial drainage basins but, to varying degrees, the product of events far away. The role of the aerial catchment is probably best understood with reference to highly acidic wetfall and dryfall that may be produced locally as well as hundreds, if not thousands, of kilometers from the system upon which it falls, whose fate it helps determine (Chapter 27). Even so, most aquatic systems with a substantial catchment that are remote from urban, industrial, or agricultural regions are overwhelmingly dependent for nutrients on the more easily defined terrestrial catchment.

Terrestrial catchments serve as receptors of aerially-derived materials from elsewhere, some of which is subsequently reexported to receiving waterways. Schindler et al. (1976) argue that all of the catchment phosphorus carried to remote Rawson Lake (CA) by its streams must have been derived from the atmosphere because no phosphate containing rock has been found in the drainage basin. There is long-range transport in the upper atmosphere of gases and the dust containing nutrients from the Saharan desert to the Amazon basin of South America, and from central Asia to the Hawaiian Islands and North America; there is also transport of pollutants from the industrialized temperate zone to polar regions (Fig. 5–13 and Chapter 28).

Nitrogen and phosphorus entering aquatic systems via the atmosphere (kg km^{-2} yr^{-1}) rise from low quantities in polar regions to a broad maximum in agricultural regions of the temperate zone, characterized by

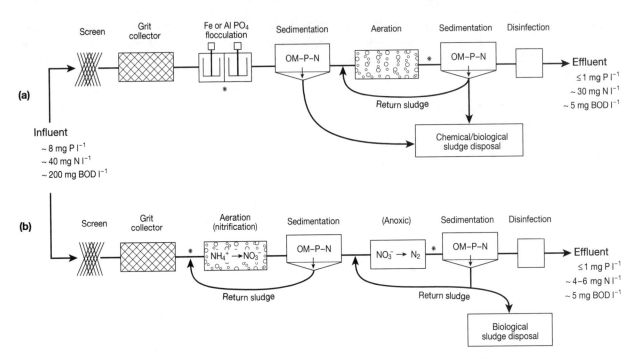

Figure 8–14 Two types of a variety of waste water treatment plant configurations involving biological/chemical treatment. A screen removes larger objects from the influent. Biological treatment involves an aeration tank to allow a rapid aerobic growth of bacteria, and oxidation of organic matter. The aeration tank promotes flocculation and subsequent sedimentation of particulate organic matter—microbial aggregates (flocs) plus associated nutrients and contaminants as sludge. Sludge is recycled to the aeration tank as a bacterial innoculant. This procedure is followed by sludge disposal, including incineration or drying, followed by transportation to landfills. Increased nitrogen removal involves a nitrification–denitrification step. For chemical treatment, precipitating agents, typically Fe or Al salts, are added at one of the various points (*) to convert soluble phosphorus to a particulate form. This step is followed by sedimentation of the precipitate and adsorbed organic matter, and sludge disposal. Finally, disinfection of the effluent can be achieved by chlorination or UV radiation before discharge. The organic matter removal decreases the biological oxygen demand (BOD) in the effluent and the probability of anoxia in the receiving waterways. *(Modified after Harremoës 1985.)*

nutrient-rich windblown soils that are often further enriched by inorganic fertilizer additions (Table 8–10). Dust from plowed agricultural soils raises atmospheric phosphorus levels (Table 8–10, Brezonik 1996), with these levels much higher during dry warm periods than when soils are wet or covered with snow. Large amounts of nitrogen (as NH_3 and N_2O) are released into the atmosphere where livestock densities are high and cropland receives manure or fertilizer applications. The nitrogen emitted for transport and deposition elsewhere is supplemented by nitrogen oxides produced in internal combustion engines and by industry (Chapter 27). The atmospheric oxidation of

NH_4^+ to NO_3^- yields nitrate deposition rates in western Europe about three times greater than levels in the 1890s, and these rates doubled between the late 1950s and early 1970s (Rodhe and Rood 1986). Lake concentrations continue to rise because of aerial loading in regions far from agricultural and industrial activity. For example, nitrate concentrations in remote Lake Superior (CA, US), located more than 600 km downwind of the major agricultural and industrial region of the United States midwest, have risen nearly sixfold since 1906 at a rate of 2 percent per year, with deposition rates continuing to double every 34 years (Fig. 18–1).

Table 8–10 Mean atmospheric bulk deposition of nitrogen and phosphorus (kg km^{-2} yr^{-1}) and the mean N:P ratio (as mass) as a function of climate and increasing distance from the North Pole. Bulk deposition may underestimate total atmospheric deposition of nitrogen by 30–40% by not capturing aerosol particles and gases quantitatively. 100 ha = 1 km^2.

Location	TN	TP	N:P	Authors
Char Lake Northwest Territories, CA	90	5	18	Schindler et al., 1974
Kuokkel, northern SE	115	6	19	Jansson, 1979
Kevo, northern FL	140	6	23	Happala, 1977, in Persson and Broberg, 1985
Experimental Lakes, northwest Ontario, CA	480–730	24–53	19	Schindler et al., 1976
central Alberta, CA	424	20	21	Shaw et al., 1989
Lammi, south FI	524	20	26	Happala, 1977, in Persson and Broberg, 1985
England, GB	870–1900	20–80	28	Allen et al., 1968
northwest Czech Republic (1979–1994)	1000–1400	15–24	62	Kopáček et al., 1997
north central Iowa, US	1200	260[1]	5	Downing, J., pers. comm.
south central Ontario, CA	1130	22	51	Molot and Dillon, 1993
north central Florida, US	660	45	15	Brezonik et al., 1969, and Riekerk, 1983, in Boring et al., 1988
Valencia, VE	750	170[1]	4	Lewis and Weibezahn, 1983
Rio Negro, BR	560	20	28	Ungemach, 1970, in Likens et al., 1977
Naivasha, KE	480	67[1]	7	Kalff, J., unpubl.
Australia, temperate	ND	39	ND	P. M. Hallam, pers. comm. in Likens et al., 1977
South Africa	1063	47	23	Archibald and Muller, 1987
New Zealand	370	38	10	Rutherford et al., 1987

[1]Includes windblown soil dust

Catchment Area:Lake Area Ratio

▲ The relative importance of atmospheric nutrient loading directly onto water surfaces is determined by the **drainage ratio** (catchment area:lake area, CA:LA), but is modified by geology and land use. Atmospheric loading predominates only in systems with either a very small CA:LA or where atmospheric nutrient loading is exceptionally high. A small CA:LA is most commonly encountered in the upper subcatchment portions of large catchment systems that, in landscape ecology parlance, are known as positioned *high in the landscape* (Kratz et al. 1997). Lotic systems have a very small surface area to catchment ratio and are almost invariably nourished entirely by their catchments.

However, topographically determined catchment areas only provide a rough measure of subsurface drainage patterns. The link is weakest in ancient and highly eroded landscapes of low relief where topographic features may poorly reflect subsurface and groundwater drainage patterns.

An extreme example of the importance of atmospheric loading is Crater Lake (US) with a CA:LA ratio of 0.2 (Table 9–3 and Fig. 8–15). It's disproportionately small terrestrial drainage basin means it receives almost all its externally derived nutrients from the atmosphere. The gradual decrease in its transparency (Sec. 10.8), resulting from a larger algae biomass, appears to be the result of increases in atmospheric nutrient loading from distant agriculture sources during

Figure 8–15   Aerial view of Crater Lake (U.S.), showing its disproportionately small drainage basin (CA:LA = 0.2). *(National Park Service photo.)*

the last 50 years, and possibly from recent tourism (Larson 1984). Computations based on terrestrial and atmospheric export coefficients (see Tables 8–9 and 8–10) suggest that Lake Superior in North America (CA:LA 1.6), and Lake Victoria in east Africa (CA:LA 1.8; Table 9–3), also receive several times more nitrogen and phosphorus via the atmosphere than inflowing rivers or groundwater. High catchment evaporation means that the fraction of the Lake Victoria catchment able to provide water and nutrients to receiving waterways is smaller than suggested by the nominal size of the drainage basin. Lakes with semiarid catchments like Lake Victoria are more dependent on the atmosphere than the nominal CA:LA ratio suggests.

The proportion of nitrogen and phosphorus obtained from land rises in step with the CA:LA ratio and usually quickly overwhelms aerial contributions in most lakes (Sec. 8.7). Work on two hard-rock (igneous) catchments, far from major aerial sources and direct human impacts, in Sweden (Broberg and Persson 1984) and Canada (Schindler et al. 1976) suggests that a CA:LA ratio of about six is where the atmospheric phosphorus contribution is greater than or equal to the terrestrial contribution. Conversely, remote lakes lying in well-vegetated catchments on fertile soils located upwind from agricultural regions have a disproportionately low atmospheric contribution. For example, in central Alberta (CA), the atmospheric supply of phosphorus and nitrogen will only

exceed the terrestrial supply when the CA:LA is well below six (Shaw et al. 1989).

Many of the limnologically best-known lakes have a moderately high CA:LA ratio (> 10, Table 9–3) and are located in areas characterized by well-watered soils, rich in nutrients. For these lakes, the importance of the atmosphere as a source of nutrients has to be small. It is likely the nutrients supplied by the atmosphere are in a more available form, and the contribution of these nutrients is probably higher during growing season than suggested by the CA:LA ratio.

8.7   Nutrient Export, Catchment Size, Lake Morphometry, and the Biota: A Conceptualization

The CA:LA ratio serves as a useful surrogate for the effect of catchment size on nutrient loading, a subject further developed in Chapter 9. The **nutrient export** from a drainage basin (kg yr^{-1}) and the resulting **nutrient loading** of the receiving waterway—expressed per m^2 of waterbody surface per year (g m^{-2} yr^{-1})—is reflected in stream and lakewater concentrations, and ultimately in the phytoplankton biomass (chl-*a*). Nutrient loading is much more difficult and time consuming to determine than lakewater concentration. Thus concentration is commonly used to estimate the phytoplankton biomass (Fig. 8–16) in lakes where the main constraint on primary production is shortage of

Figure 8–16 The relationship between total phosphorus (TP) and algal biomass as chlorophyll-*a* (chl-*a*) for 133 north temperate lakes. The solid line represents the LOWESS (Locally Weighted Regression) fitted trend line. The dashed line represents the linear regression line given by the equation Log(Chl-*a*) = –0.39 + 0.874 Log(TP) (r^2 = 0.69). The LOWESS line represents the best fit of the relationship between TP and Chl-*a* for every concentration of TP. Note the relationship is curvilinear when very high concentrations of TP are included. Plots such as this one form the basis for lake eutrophication management. Most such plots exclude: (1) humic water, and waters with a high inorganic turbidity where light rather than a nutrient is the limiting factor, and (2) water with particularly low N:P ratios (< ~10 by mass or < ~22 by atoms), that suggest a probable nitrogen limitation and, if true, which would yield less Chl-*a* per unit of phosphorus (Table 8–2). (*Modified from Prairie et al. 1989.*)

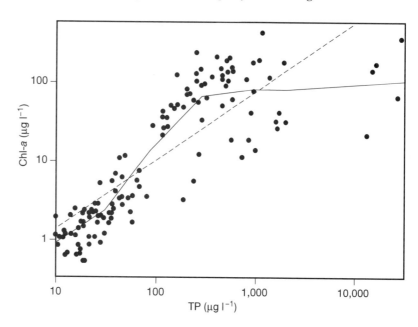

nutrients rather than light limitation (Chapters 10 and 21).

The computations (simulations) in Table 8–11 point to the impact of catchment size, land use, and atmosphere on nutrient loading and thereby on the biota. This table shows that (1) catchment phosphorus export to receiving lakes (and streams) rises with catchment size, (2) terrestrial exports are several times larger from agricultural drainage basins than forested basins of the same size, and would be even larger in areas with rich soil and high-intensity agriculture, (3) estimated atmospheric contributions of phosphorus dominate in forested drainage basins with the smallest CA:LA ratio, but is negligible in most agricultural catchments, (4) predicted algal biomass and fish yields rise systematically with increased catchment size and (5) predicted algal biomass (chl-*a*) and fish yields are expected to be substantially greater in agricultural rather than forested drainage basins, unless the waters are unsuitably turbid or periodically anoxic.

These simulations do not offer specific predictions, which depend on the particular empirical equations selected and other assumptions made, rather, they illustrate that drainage basin attributes, and

sometimes the atmosphere, have an important impact on inland water nutrient levels, the productivity of aquatic systems, and thereby directly and indirectly on the community structure of inland waters.

8.8   Organic Carbon Export from Drainage Basins

Organic carbon produced on land is released by soil leaching to waterways and in forested catchments by particulate organic matter (insects, leaves) falling directly into waterways. Roughly one percent of terrestrial primary production is exported to rivers and lakes on a global scale (Table 8–12). However, the fraction exported is much larger from anoxic soils or wetlands where the organic matter decompostion by microbes is incomplete and a disproportionately large fraction is exported in dissolved form. Wetlands serve not only as partial traps for both allochthonously and autochthonously produced **total organic matter** (TOM) or **total organic carbon** (TOC) but also serve as major sources to receiving waterways. The *dissolved*

Table 8–11 Simulated annual phosphorus export (kg yr^{-1}), phosphorus loading (g m^{-2} yr), predicted euphotic zone annual average phosphorus concentration (μg l^{-1}), and predicted biological attributes for hypothetical midlatitude lakes with a lake area of 1 km^2 and a fixed depth, but a systematically changing catchment area covered by forests and crops, respectively. Atmospheric loading is assumed to be 30 kg km^{-2} yr^{-1} (see Table 8–10). Catchment export and loading include atmospheric loading onto the land.

Type	Catchment area (km^2)				
	5	10	20	40	80
Forested					
Catchment export (kg yr^{-1})	40	79	157	312	617
Catchment loading (g m^{-2} yr^{-1})	0.04	0.08	0.16	0.31	0.62
Total loading (g m^{-2} yr^{-1})	0.07	0.11	0.19	0.34	0.65
Atmospheric loading (%)	43	28	16	9	5
P concentration (μg l^{-1})	4.8	5.6	7.1	10.0	15.8
chl-*a* (μg l^{-1})	0.7	0.9	1.2	2.1	4.0
Fish yield (kg ha^{-1} yr^{-1})	0.36	0.41	0.52	0.75	1.19
Agriculture					
Catchment export (kg yr^{-1})	228	390	666	1,139	1,946
Catchment loading (g m^{-2} yr^{-1})	0.23	0.39	0.67	1.14	1.95
Total loading (g m^{-2} yr^{-1})	0.26	0.42	0.70	1.17	1.98
Atmospheric loading (%)	12	7	4	3	2
P concentration (μg l^{-1})	8.4	11.5	16.7	25.7	41.1
chl-*a* (μg l^{-1})	1.6	2.5	4.3	8.1	15.9
Fish yield (kg ha^{-1} yr^{-1})	0.6	0.9	1.3	2.0	3.2

Source: Based on empirical equations from Prairie and Kalff 1986, Ryding 1980, Dillon and Rigler 1974, and Hanson and Leggett 1982.

organic matter (DOM) exported, or DOC when expressed as carbon,[8] is highly correlated with the percent area covered by peat bogs (Fig. 8–17) and loading with the water residence time of lakes (Fig. 8–18). In steeper forested catchments, lacking wetlands, DOC exported is largely derived from leaf litter and organic matter in the soil.

The highly organic and largely anoxic sediments of poorly drained (low slope) permanent wetlands allow anoxic microbial processes to dominate carbon decomposition, with much of the carbon released in gas form. The world's wetlands are major sources of the greenhouse gases methane (CH_4) and carbon dioxide (CO_2) for the atmosphere (Table 16–1, Eq. 6). Decomposition end products not so lost are retained or flushed into receiving waterways in the form of large-molecular-weight organic acids, plus a wide variety of lower-molecular-weight molecules (Chapter 22).

The TOC exported to waterways is divided operationally into **particulate organic carbon** (POC) and **dissolved organic carbon** (DOC) after filtration, usually through 0.2 or 0.45 μm-membrane filters. The POC is sometimes further subdivided by means of 1 mm-mesh screens into **fine** (FPOC) and **coarse** POC (CPOC). The fraction of TOC exported as POC rather than as DOC varies widely. In lowland fluvial systems and wetlands where most TOC is in DOC form, the POC:DOC ratio is close to 0.1 (Schlesinger and Melack 1981); a ratio now assumed to be widely applicable (Wetzel 1983). However, upland or mountain streams subject to considerable erosion from their much steeper catchments are dominated by particles and thus have a much higher POC:DOC ratio. Furthermore, the ratio in any stream changes seasonally and with precipitation events, with the highest POC:DOC ratio during periods of maximum discharge and the lowest POC:DOC ratio during long dry spells.

Reliable POC export rates from land are not easy to obtain. A high proportion of particles is transported

---

[8]Carbon contributes typically 45–50 percent of plant dry weight and results are readily interconvertible.

Table 8–12 **Typical dissolved and particulate organic carbon concentrations, organic carbon export coefficients from drainage basins, and terrestrial primary production.**

Vegetation zone	Median DOC (range in brackets) (mg l^{-1})	TOC Mean Export Coefficient (kg C km^{-2} yr^{-1})	Mean Terrestrial Net Primary Production (kg C km^{-2} yr^{-1})
Tundra alpine	2 (1–5)	6,000	65,000
Boreal Forest	10 (8–25)	2,500	360,000
Cool temperate	3 (2–8)	4,000	225,000–585,000
Wet tropical	6 (2–30)	6,500	315,000–90,000
Semiarid	3 (2–10)	300	32,000
Temperate wetlands	25 (5–60)[1]	~20,000	-

[1]Extreme values include a south Swedish stream with a DOC concentration of 91 mg l^{-1} and color of 1,200 mg Pt l^{-1} (Petersen et al. 1995).

Source: Modified after Meybeck 1982, Schlesinger and Melack 1981, and Thurman 1985.

during short periods of high discharge that may not be represented in sampling programs. For example, the annual export of POC in three gauged forest streams was underestimated by 30–70 percent on the basis of discrete (discontinuous) water samples, compared to a continuous sampling of discharge and POC concentrations (Cuffney and Wallace 1988). Discrete measurements are unable to sample shortlived, but very important storm events and do not sample the *bedload* at all, which is composed of larger particles trans-

Figure 8–17 Mean dissolved organic carbon (DOC, ○) and total phosphorus (TP, ●) specific export coefficients from 20 Ontario (CA) catchments from 1980–1992 as a function of the fraction of the catchments covered by peatlands. In better drained areas where peatlands are lacking, the DOC and TP measured are derived from the leaf litter and upper soil profile. Note (1) percent peatland varies greatly (25 fold) here, whereas other potentially important variables such as runoff, climate, vegetation type, and geology vary only modestly. However, runoff becomes an important predictor of DOC export where the mean long-term runoff varies greatly and where peatlands are few (Fig. 8–18); (2) the exceptionally high C:P ratio in the receiving streams. Phosphorus is largely retained by wetlands and terrestrial systems rather than exported in streamwater (See Table 9–2). *(Data from Dillon and Molot 1997.)*

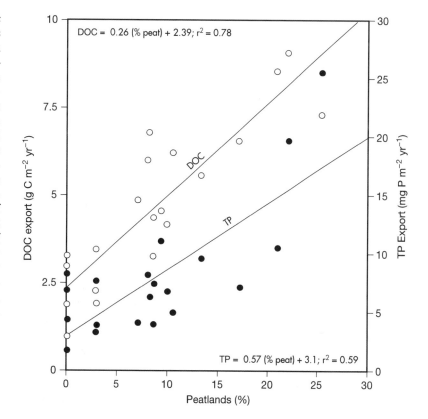

DOC = 0.26 (% peat) + 2.39; r^2 = 0.78

TP = 0.57 (% peat) + 3.1; r^2 = 0.59

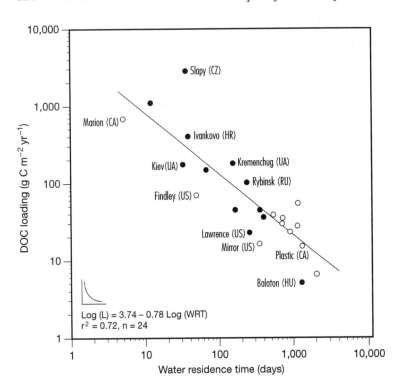

Figure 8–18 Relationship between mean water residence time and specific DOC loading for a variety of temperate zone lakes and reservoirs in primarily non-forested (●) and forested (○) landscapes. *(Data from Groeger and Kimmel 1984, and Dillon and Molot 1997.)*

ported in association with the stream bed rather than with the water column. Even so, allochthonously derived CPOC is widely believed to be largely degraded on the land and to sediment rapidly (Chapter 20), playing a minor role in the planktonic carbon cycle of lakes. Nor is much known about the fate of the sometimes abundant ultrafine allochthonous POC (<50 μm) in lakes. But the allochthonously derived CPOC is recognized of great importance in lotic systems where it is an important substrate for bacteria and fungi and serves, together with the attached microbes, as a source of food for invertebrates (Chapter 25). Following its sedimentation in lakes, the CPOC and FPOC may be the principal source of organic carbon for sediment microbes of oligotrophic lakes (Chapter 22).

Catchment-derived DOC is highly colored in well-watered, well-vegetated regions. Water color there is enhanced by oxidized iron (Fe^{3+} adsorbed to the colored DOM (Sec. 19.4). Therefore, water color serves as an easy-to-measure surrogate for DOC concentration (Cuthbert and del Georgio 1992). The color (DOC) of temperate zone lotic systems is a function of the morphometry of their catchments, while the color of oligotrophic lakes is roughly predictable from catchment and lake morphometry in the well-

watered north temperate zone (Rasmussen et al. 1989):

$$\log \text{water color} = 1.599 + 0.344 \log \text{CA:LA}$$
$$- 0.568 \log \bar{z} - 0.178 \log \text{lake area}$$
$$- 0.368 \text{ catchment slope}$$

EQ. 8.2

$$R^2 = 0.58 \qquad \text{SE est.} = 0.34 \qquad n = 287$$

Equation 8.2 shows that rivers and lakes with disproportionately large drainage basins, that are also relatively low sloped (Fig. 8–12), contain more (colored) DOC, but that DOC concentration in receiving lakes depends on the extent that the DOC supplied is diluted in the receiving lake volume, decomposed within the lake or flushed. Lake area (m^2) and mean depth ($\bar{z}$, m) together serve as surrogates for lake volume in the model. The amount of (colored) DOC supplied increases as catchment slope (%) declines; presumably because the organic soil layer that supplies most of the lake DOC tends to increase in thickness with decreasing slope, and because more TOC is converted to (colored) DOC on land before being exported.

Where waters contain much colorless DOC, color is naturally a poor surrogate for DOC. The colorless portion of DOC normally increases as total DOC declines in well-watered and well-vegetated regions of northern Europe and northeastern North America. Conversely, in slowly flushed, and in closed saline lakes in the prairie region of North America, the colorless fraction increases with increasing DOC (Curtis and Adams 1995). This is probably the result of long water retention (slow flushing), allowing evaporative concentration of DOC, particle formation and sedimentation, respiration, and photobleaching (Chapter 10) of the entering organic matter.

Not all allochthonous TOC (and inorganic plant nutrients) is received via inflowing streams or rivers. Some is received via groundwater, DOC deposition in precipitation, and as POC in the atmosphere deposited on the water surface (~ 0.1 g m^{-2} yr^{-1}) (Sec. 22.6).

Highlights

- Lotic (flowing-water) systems are complete ecosystems intimately coupled to their drainage basins. They are more than conduits of organic matter and nutrients from catchments to receiving lakes and wetlands.
- A widely used and easily determined surrogate for stream size is stream order, a linear geomorphic characteristic of lotic systems.
- The river continuum concept is a useful framework for viewing lotic systems as longitudinally integrated systems, linking stream size (order), energy input, and the processing of organic matter by the biota from headwater to mouth.
- The land–water transition zone (ecotone, or riparian zone) is characterized by sharp gradients in environmental factors, species composition, and dominating species. The riparian zone is important in cycling and transforming nutrients and organic matter. Well-vegetated riparian zones are of great importance, reducing streambank erosion, trapping soil particles and absorbed nutrients on land and, in agriculture, stimulating denitrification ($NO_3 \rightarrow N_2$). All of these help maintain water quality and the traditional biota, including fish.

- The hyporheic zone is the transition zone in the vertical plane between the stream channel and subsurface sediments. It is an important habitat for the biota and contains a mixture of surface water and groundwater. This zone may extend laterally beyond the river margin into the floodplain.
- ▲ Rivers and their associated wetlands receive, with increasingly fewer exceptions, large quantities of wastewater and agricultural runoff that impact the biota. Superimposed is widespread channelization (straightening bends, resulting in more rapid discharge), channel deepening, bank erosion, as well as separation of rivers from their floodplains through dike construction. Rivers are often dammed, reducing maximum discharge and its timing, the size of downstream wetlands, and riverine biota. Promising attempts at "river restoration," including reconnection with former (drained) wetlands, are underway.
- The export of nutrients from catchments (kg yr^{-1}) increases with increasing catchment size, slope, and runoff. It is further affected by surficial geology (soil structure) and land use. Nutrient exports per unit area (km^2) are low from well-vegetated catchments and exceptionally high from fields with seasonal crops, high livestock densities, and in urban areas.
- Additional nutrients (and contaminants) enter waterways via the atmosphere. Atmospheric nutrient loading (kg km^{-2} yr^{-1}) is low at high latitudes and rises to typically very high downwind from rich agricultural and urban areas. Lakes with small catchments, in relation to the surface area of water, are particularly dependent on atmospheric inputs.
- The export of organic matter from land to the receiving water increases with catchment area and is particularly high from wetlands. Dissolved organic matter (DOM) from well-watered, vegetated regions is typically colored.

9

Aquatic Systems
and their Catchments

9.1 Catchment Size

Inland water functioning is tightly linked to the characteristics of their drainage basins and the **ecoregions** in which they lie.[1] The average slope (relief) of drainage basins decreases as the area of catchments increases (Fig. 8–12a) resulting in large basins within a region having the lowest slopes, whereas small catchments tend to be relatively steep. Similarly, drainage density (the total stream length per unit area) normally declines as catchment size increases (Fig. 8–12b). Therefore, particulate matter with adsorbed plant nutrients or contaminants leaves land at a higher rate per unit area, in a small catchment (Sec. 8.4), even though the total amount of water and particulate matter exported (tonnes yr^{-1}) will be smaller. At the same time, more rapid runoff in smaller, steeper drainage basins means reduced contact time between water and the substrate, permitting less biological/chemical breakdown of substrate (weathering). Soils are expected to be thinner because the rapid runoff erodes and transports particles toward less steep portions of the catchment for at least partial storage there. The net result is typically lower nutrient export

(kg yr^{-1}) (Sec. 8.5) and lower dissolved salt content (Chapter 13) in lotic systems draining small, rather than large, catchments within a particular climatically and geologically homogeneous region. Low-slope drainage basins may also have areas of impeded drainage, forming wetlands that release much dissolved and colored organic matter in higher rainfall zones (Fig. 8–18).

▲ 9.2 Catchment Form

The topography (geomorphology) of the landscape helps determine stream morphology as well as the surface shape and size of lakes, their underwater form and depth (Sec 7.4), and the abundance of wetlands (if any). The system attributes affect a variety of physical and chemical characteristics of water that, together with the system morphology, have a deciding effect on the productivity and the types of organisms present.

Catchment topography (slope) is a function of catchment size. Within any climatic/geologic region the steepness and total length of river channels typically decline as drainage basin size increases. Catchment area and lake area are also highly correlated (with the correlation coefficient r = 0.82), as shown in a study of 215 North American lakes (Table 9–1). Other correlations include those between lake depth and catchment area, depth and nutrient concentrations, phosphorus and nitrogen, and nutrient concentrations and algal biomass (chl-a). However, lake area is not only a function of catchment area but also of catchment geology because lakes on dense igneous

---

[1] • Ecoregions describe ecological zonation across the landscape, with an ecoregion commonly based on similarity of climate, geology, land surface form, soil formation, primary production, terrestrial vegetation, and energy flow. (Landers et al. 1998)

• "The ecosystem, considered by many to be the fundamental ecological unit, is not definable in a way to allow nutrient processes to be assessed. The watershed is a much more realistic ecological unit." (D.W. Schindler et al. 1976)

Table 9–1   **Correlation matrix between morphometric, chemical, and biological attributes of 215 northeastern North American lakes. All variables were log transformed prior to analysis and only significant correlations ($p < 0.01$) are shown. Note the extensive covariation.**

	Lake Area	Catchment Area	Mean Depth	Conductivity	Total Phosphorus	chl-*a*	Total Nitrogen
Lake Area	*	0.82	0.59	-	-	-	-
Catchment Area		*	0.46	-	-	-	-
Mean Depth			*	-	−0.49	−0.59	−0.66
Conductivity				*	0.79	0.71	0.73
Total Phosphorus					*	0.82	0.84
Chl-*a*						*	0.77
Total Nitrogen							*

Source: From Duarte and Kalff 1989.

bedrock (Chapter 13) have, on average, smaller surface areas for a particular catchment size than those on other noncarbonaceous sedimentary substrates. The latter have a smaller CA:LA than those on more easily eroded carbonaceous substrates (Duarte and Kalff 1989). The high correlation between total phosphorus and the total nitrogen ($r = 0.84$) makes it difficult to distinguish their effects on the biota without experiments under controlled conditions.[2]

9.3   Catchment Soils and Vegetation

Vegetation and soils determine the fate of gases and particles entering drainage basins from the atmosphere. They also determine what happens to elements after they are liberated following the breakdown (weathering) of soil and other substrate particles.

The first, possibly best, investigated catchments are those of the Hubbard Brook drainage basin in Massachusetts (US). There, the inputs, outputs, and storage of elements was reported for an undisturbed forest ecosystem over an 11-year period (Table 9–2).

---

[2]Intercorrelations (covariation) make it impossible to discern what is "cause" and what is "effect." For that, specifically designed experiments, with controls, are required. A particular effect observed in nature may be the outcome of a combination of causes that change over time and space. This problem is recognized in other sciences, for example, areas of physics and astronomy where controls are similarly lacking. In a paper entitled *On the Notion of Cause* the physicist/philosopher Bertrand Russell (1913) writes: "It is not in any sameness of causes and effects that the constancy of scientific laws consists, but in sameness of relationships."

Soil and substrate weathering was demonstrated to be the major source of elements, with the exception of sulfur and nitrogen that, at this site not far from urban areas, were primarily derived from the atmosphere. The elements in highest demand by terrestrial biota (N, P, K) in relation to supply are those most effectively retained on land rather than lost to outflowing streams. In contrast, elements in low demand by terrestrial biota relative to their supply rate are readily lost to streams (Table 9–2).

While it is widely known that most plant nutrients are made available through the weathering of soil and rock particles, the importance of the atmosphere (Sec. 8.6) as a source of plant nutrients is less well recognized. Parent rock usually contains virtually no nitrogen (Table 13–2); in most drainage basins, it is mostly obtained through microbial nitrogen fixation in soils, except when downwind from agricultural, industrial, or other sources that emit fixed nitrogen into the atmosphere (Chapter 18).

Details of the retention and export of elements differ between different types of drainage basins, and nutrient retention on land is much lower in rich agricultural areas where supply is large in relation to demand by the biota. Even so, catchments are the principal determinant of chemical characteristics of rivers and most lakes and wetlands located in substantial catchments (Chapters 8 and 13). Soil chemical attributes in agricultural areas are further modified by fertilizer additions. Downwind from major industrial areas emitting acid precursors (SO_2, NO_x, Chapter 27) plus trace metal and organic contaminants, the land and waterways reflect this contamination (Chapter 28).

Table 9–2 **Sources and sinks of elements in an undisturbed forest ecosystem at Hubbard Brook, US. Note the disproportionately high fraction of incoming phosphorus, distantly followed by nitrogen, retained in the catchment and incorporated in the vegetation or forest floor. * = not measured.**

Sources and Sinks (%)	Ca	Mg	Na	K	N	S	P
Sources							
Wet deposition	9	15	22	11	31	65	1
Gas or aerosol input	*	*	*	*	69	31	*
Soil and rock weathering	91	85	78	89	*	4	99
Sinks							
Stream water output	59	79	98	28	19	90	<1
Net incorporation in vegetation	35	17	2	68	43	6	82
Net incorporation in forest floor	6	5	<1	4	37	4	18

Source: Modified after Likens et al. 1977.

9.4 Water Residence Time

Lake and Wetlands

Not only the morphology of lakes and wetlands but also their **hydraulic water residence time** (τ_w, yrs), which is the average time required to refill a basin with new water if it were to be emptied, is in any climatic zone determined by basin and catchment morphology (Chapter 7). Large catchments receive more total precipitation and the resulting larger runoff ($m^3 \, yr^{-1}$), plus any groundwater input, is able to replace basin volume more quickly than an identical lake or wetland located in a smaller drainage basin nearby. Naturally, two headwater lakes lying in identically-sized drainage basins—but differing in geology and therefore, area, depth, and volume—would have different water residence times even if the climate and runoff were identical. In practice, the hydraulic water residence time, the **retention time**, the **flushing time**, or simply the **water residence time** (τ_w, yrs) is most easily obtained from the basin volume (V, m^3) and the average annual water outflow (Q, $m^3 \, yr^{-1}$)

$$\tau_w = \frac{V}{Q} \qquad \text{EQ. 9.1}$$

Using outflow to estimate water residence time eliminates having to determine surface evaporation in addition to various inflows.[3] The flushing time of lakes is

often referred to as the **hydraulic water residence time** rather than water residence time (WRT) to acknowledge that stream water entering an epilimnion will not be mixed with the whole lake's volume in lakes with vertical temperature or chemical stratification (Secs. 11.2 and 11.10). Consequently, an inflowing water molecule will leave by the outflow more quickly than the average water molecule in the system. This is of no concern when lakes are unstratified, but is important in those lakes or reservoirs that receive most of their inflow during a lengthy period of stratification. An interesting example of this is limnologically well-known Lake Constance (AT, CH, DE), which receives most of its water in summer from melting glaciers when it is stratified. The lake's epilimnion has an average WRT of about 0.13 yr (1.5 months) whereas the average whole-lake WRT is about four years.

Lentic Systems

▲ The calculated WRT is fortunately not much less than the observed water residence time in most temperate zone freshwater lakes that stratify because most of the river inflow occurs during nonstratified periods (winter, spring, fall). Even so, the calculated flushing time is a poor measure of the average time spent by a water molecule in those lakes that destratify only in some years (e.g., Lake Maggiore, IT), or those that don't do so completely (e.g., Lake Tanganyika, BI, TZ, ZM, ZR) (Table 9–3). Overall WRT is also somewhat deceptive in large lakes composed of major basins that are not uniformly mixed and are flushed at different rates. A good example is the enormous Lake Baikal in Siberia (RU), with an overall average water residence

---

[3]There is little difference between outflow and inflow based definitions of WRT in rapidly flushed lakes or wetlands where surface evaporation is modest. But WRT is much longer when calculated on the basis of outflow in slowly flushed lakes and particularly so for wetlands in regions of high surface evaporation.

Table 9–3   The longterm average water residence time (WRT) or flushing time (years) for selected lakes and reservoirs [R] (ranges in brackets), mean depth ($\bar{z}$, m) and the associated ratio of catchment area (minus lake area) to lake area (CA:LA), as well as the average summer or growing season Secchi disc depth ($\bar{Z}_{SD}$, m). The WRT presented is most commonly based on inflow, precipitation on surface and lake volume data for long WRT lakes and on outflow measurements and volume in short WRT systems.

Lake	Time (yrs)	$\bar{z}$	CA:LA	z_{SD}
Tanganyika (BI, TZ, ZM, ZR)	6,000	557	7.1[1]	14
Malawi [Nyasa] (MW, MZ, TZ)	1,225	290	3.0[1]	20
Titicaca (BO, PE)	1,183	107	3.7	7
Tahoe (US)	700	313	1.6	28
Baikal (RU)	327	730	12.1	11
New Québec Crater (CA)	330	145	0.4	33
Caspian (AZ, IR, KZ, RU)	210[1]	182	8.3	12
Superior (CA, US)	191	148	1.6	9
Crater (US)	150	329	0.2	29
Bracciano (IT)	137	89	2.4	6
Great Bear (CA)	131	72	5.0	30
Victoria (KE, TZ, UG)	123	40	2.8	4
Michigan (US)	99	84	1.6	7
Llanguihue (CL)	74	182	1.8	17
Vättern (SE)	58	39	3.3	12
Shikotsu (JP)	51	266	2.8	20
Stechlin (DE)	40	23	2.9	7
di Garda (IT)	27	136	6.4	12
Waldo (US)	30	39	2.1	32
Huron (CA)	22	61	2.1[4]	8
Ladoka (RU)	13	47	15.0	2
Esrom (DK)	18	12	11.8	2
Taupo (NZ)	12	91	5.4	17
Léman [Geneva] (CH, FR)	11 (9–15)	153	13.7	6
Cayuga (US)	9	54	11.8	2
Kinneret (IL, JO)	7	24	16.0	3
Erken (SE)	7	9	5.7	4
Ontario (CA, US)	6	86	3.4[4]	3
Mjösa [R] (NO)	6	153	45.2	4
Biwa (JP)	5	41	5.6	7
Balaton (HU)	5	3	8.7	0.5[3]
Mendota (US)	4	12	16.7	3
Maggiore (CH, IT)	4	177	30.0	5
Constance [Bodensee] (AU, CH, DE)	4	100	20.4	4
Ihawashiro (JP)	3.8	37	6.8	8
Lake 227 (CA)	3.0 (2.0–4.0)	4	8.8	2[5]
Paijanne (FI)	2.9	17	23.7	3
Kariba [R] (ZM, ZW)	2.8	30	87.8	5
Erie (CA, US)	2.6	18	2.3[4]	5

(continued)

Table 9–3   (*continued*)

Lake	Time (yrs)	$\bar{z}$	CA:LA	z_{SD}
Washington (US)	2.4 (1.7–6.1)	33	17.0	8
Mälaren (SE)	2.6	12	19.8	4
Zurich (CH)	2.0 (1.5–3.0)	44	26.0	5
Vechten (NL)	2.0	6	0.4	3
Memphremagog (CA, US)	1.8	16	17.6	4
Neagh (IE)	1.2 (1.0–2.0)	9	11.4	1
Norvikken (SE)	1.3 (0.6–5.0)	5	34.1	1
Mirror (US)	1.1	6	6.9	6
Windermere (GB)	0.8	21	14.6	4
Hartbeespoort [R] (ZA)	0.7 (0.5–1.1)	8	207.0	0.8
Lawrence (US)	0.6	6	7.1	5
Mikolajskie (PL)	0.4	11	389.0	2
Manasbal (IN)	0.4	4	10.1	3
Leven (GB)	0.4	4	10.9	0.6
George (UG)	0.3	2	39.0	0.4
Lundzer Untersee (AT)	0.3	20	37.0	10
Tjeukemeer [R] (NL)	0.2	2	142.3	0.3[3]
Broa [R] (BR)	0.1	3	41.2	1
Slapy [R] (CZ)	0.1 (0.05–0.26)	21	93.8	1
Myvatn [S Basin] (IS)	0.07	2	34.4	0.5[3]
Cameron (CA)	0.06	7	24.0	4
Talquin [R] (US)	0.125	4	113.2	1
Marion (CA)	0.015	2	976.0	ND
Frances [R] (US)	0.007 (.001–.04)	1	710.0	1

[1]WRT based on surface evaporation and the assumption, incorrect, of constant lake volume.

[2]Catchment evapotranspiration results in a smaller effective catchment area yielding runoff than the nominal ones listed. The particular lakes are permanently stratified (meromictic) and WRT for the upper mixed layer only is much shorter than indicated.

[3]Partially attributable to sediment resuspension.

[4]The immediate catchment only, excluding catchments of upstream lakes. When these are included the CA:LA ratios are: Huron = 6.4, Erie = 17.8, Ontario = 27.5.

[5]Prior to fertilization.

time of 327 years (Table 9–3), but the northern, central, and southern basins have an average WRT of 505, 221, and 92 years, respectively (Verbolov et al. 1989). Another example includes rapidly-flushed reservoirs where the water in the main stem is flushed much faster than the tributary arms (Chapter 29). Finally, in saline lakes or wetlands without an outlet, the designation of hydrological water residence time is problematic because the system serves as a sink for water entering, with water molecules (but not their associated ions) leaving by evaporation only (Fig. 5–12, and Sec. 13.7). Exceptions like these have led some limnologists to prefer the term **theoretical water residence time.**

While the average flushing time within any climatically and geologically homogeneous region can be estimated from catchment area and lake morphometry, the model produced will be a poor predictor for lakes elsewhere. Thus, small Newfoundland lakes (CA), lying in an area of high rainfall and low evapotranspiration, have flushing times ranging from 0.02 to 1.26 years (Knoechel and Campbell 1988), whereas morphologically and geologically very similar lakes at about the same latitude in northwestern Ontario (CA) have a much longer average flushing time of 2.6 to 3.8 years (Brunskill and Schindler 1971). Longer WRT in continental Ontario is primarily attributable to precipitation about half as large as in ocean-affected Newfoundland, plus higher evapotranspiration. Similar-sized lakes further west in the much more arid southern portion of the Prairie Provinces may have WRTs that exceed 50 or 100 years, or may lack an outflow altogether. (Mitchell and Prepas 1990).

Lotic Systems

The WRT of lotic systems represents the time the water has been in the system or its "age". It may only be a few hours in small streams, but averages about seven days (ranging 3–19 days) for some midsized (> 5th order) Ontario rivers (Basu and Pick 1996). The WRT is about 18 days for a set of larger US rivers (Søballe and Kimmel 1987). Catchment area (*CA*) is the principal determinant of river WRT at the scale of a single climatic zone. But over larger spatial scales, the effect of precipitation and evapotranspiration differences on discharge (*Q*) and WRT of lotic systems has to be taken into account. At a continental scale WRT is roughly predictable (Leopold et al. 1992, Søballe and Kimmel 1987) with the formula

$$\text{WRT} = \frac{0.08\text{CA}_d^{0.6}}{Q^{0.1}} \qquad \text{EQ. 9.2}$$

where WRT is the residence time at sampling site (d), CA = catchment area (km^{-2}) upstream from the sampling site, and Q = river discharge ($\text{m}^{-3}\text{s}^{-1}$).

9.5   Nutrient Concentrations, Trophic State, and WRT

The WRT or its inverse, the flushing rate, is one of the most important predictor variables in limnology. It is coupled to **water loading** on the water surface (q_s, m^3 km^{-2} yr^{-1}) (Chapter 17), which represents the number of

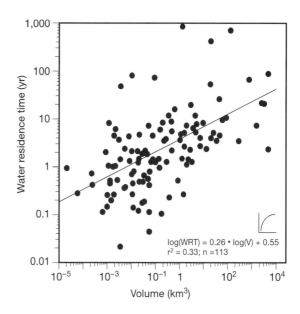

Figure 9–1   Water residence times and volumes of temperate zone freshwater lakes. Note that much more than volume is needed to predict WRT. (*Data from Lerman and Hull 1987.*)

meters of water that is, conceptually, stacked on a lake or wetland surface annually. This water contains nutrients, organic matter, and contaminants, making the WRT an ideal surrogate for (1) their supply to aquatic systems, (2) the time allowed for particles to sediment in lakes rather than flush out, (3) an indicator of the productivity of aquatic systems and, finally, (4) a rough measure of how much time is available for planktonic communities to develop a community structure before being flushed from lotic and rapidly-flushed lentic systems.[4] WRT is, however, a poor measure of the time available for plankton communities to develop in lakes with a thin (small volume) epilimnion in which even a moderate inflow will flush the epilimnetic volume more rapidly than suggested by the whole-system WRT.

The WRT among lakes increases with lake volume (Fig. 9–1) or its surrogate, mean depth, and declines in any climatic/geologic region as the size (water-catching ability) of catchments increases. The large variation in WRT at any one volume is probably in part attributable to unaccounted for differences in

---

[4]"It has long been recognized that lentic-lotic distinctions are arbitrary (Odum 1959, Margalef 1960), and our results suggest that a more useful strategy is to view aquatic systems as occupying positions along a continuum ordered by water residence time." (Søballe and Kimmel 1987)

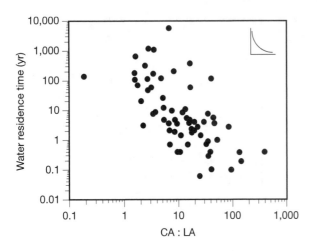

Figure 9–2   The relationship between the theoretical water residence time and the catchment to lake area (CA:LA) ratio of 66 world lakes. The data scatter must be attributable in part to variation in (1) precipitation, (2) catchment evapotranspiration and resulting runoff, (3) catchment area and (4) lake area to volume ratio, which affects evaporation, a surface phenomenon. *(Derived from Data Book of World Lake Environment. International Lake Environment Committee. Otsu (JP) 1988.)*

runoff from drainage basins located in different regions varying considerably in runoff, and to variation in lake morphometry (surface area to volume ratio). Lakes with a disproportionately long WRT are consequently characterized by a catchment area (CA) that is relatively small, on average, in relation to the lake surface area (LA) (Fig. 9–2) or to its correlate, lake volume. Lake volume is a more powerful predictor of WRT, but it requires bathymetric maps whereas its proxy, surface area (LA), is readily obtainable from topographic maps. The CA:LA ratio is known as the *drainage ratio* (DR). A small DR implies a relatively small land area able to collect precipitation and release nutrients. The result is relatively small stream water discharge and water loading onto the receiving lakes and wetlands (m^3 km^{-2} yr^{-1}), relatively long WRT (Fig. 9–2), plus relatively modest catchment exports of inorganic nutrients to lentic systems (Fig. 9–3).

WRT and Trophic State

Unpolluted lakes with a long WRT have a low algal biomass, which together with an associated small supply rate of dissolved (and usually colored) organic matter, makes them highly transparent, permitting a thick water column and deep littoral zone in which photosynthesis is possible (Chapter 10). Long WRT lakes receive much of their water and ion input from the atmosphere, making them prone to acidification where precipitation is highly acidic (Chapter 27). In contrast, a large CA:LA (or volume) ratio is typically associated with relatively rapid lake flushing and higher nutrient and major ion inputs, allowing elevated rates of phytoplankton production and planktonic respiration (Fig. 9–4). More rapidly flushed lakes, located in well-watered regions, also receive much dissolved organic matter (allochthonous DOM) from their large and well-vegetated catchments (Fig. 8–18). Their counterparts in semiarid regions, charac-

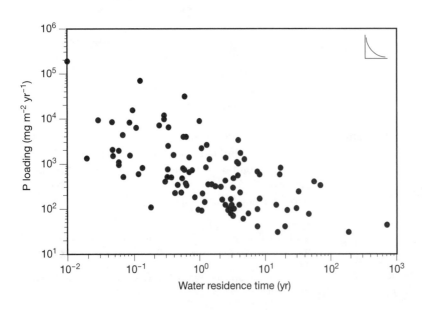

Figure 9–3   The relationship between theoretical water residence time and phosphorus loading for 66 temperate zone lakes lying in well-watered drainage basins. The typically low nutrient (and sediment) loading in lakes with a long water residence time is reflected in a low algal biomass and a high transparency, unless modified by human activity or a high water color. Note, short WRT lakes tend to be nutrient rich and turbid (high algal biomass; Table 9–3) as a result of a high nutrient export from the land carried by nutrient rich lotic systems and ground water. *(Data from Turner et al. 1983, and Nürnberg 1984.)*

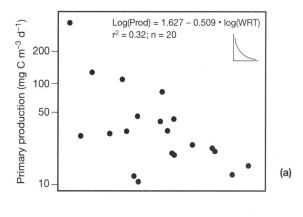

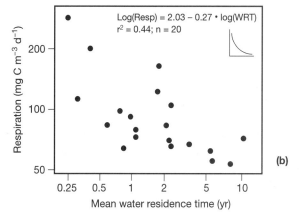

Figure 9–4 The relationship between mean water residence time and (a) mean phytoplankton primary production and (b) mean summer epilimnetic plankton respiration in 20 southern Quebec (CA) lakes. The wide data scatter, reflected in a low r² indicates that over the relatively small spatial scale (southeast Quebec) and small range scale (<2 order of magnitude) of production and respiration examined, variables other than WRT (e.g. catchment attributes, size of littoral zones) play an important role in primary production and planktonic respiration that is not captured here. *(After del Giorgio and Peters, 1994.)*

terized by poorly-vegetated catchments, carry silt washed from the catchments during rainy periods. Lotic systems are the most rapidly flushed and most intimately connected with their catchments of all aquatic systems. They are characterized by a minuscule water surface area or volume in relation to catchment area.

Long lake WRT is typically associated not only with relatively low water and nutrient loading from its catchment but also with more time for within-system oxidation by microbes of the allochthonous DOM (Chapter 22) and sedimentation of particles not re-

spired in the watercolumn. Consequently, the system serves as a sink for incoming particles as well as for those produced in situ, little of the incoming material is lost through outflow. A long WRT also allows more time for photobleaching of colored DOC, and the photooxidation of DOC in general, by ultraviolet radiation (Chapter 10) to gaseous end products and colorless organic molecules of smaller size. Unpolluted long WRT lakes therefore are particularly transparent (Table 9–3). The primary producers in highly transparent lakes are highly dependent on atmospheric deposition of nutrients directly on the surface (Sec. 8.6) and on efficient recycling (reuse) of the plant nutrients received.

▲ The most transparent lakes are remote lakes lying in small hard-rock drainage basins. The two most transparent lakes in the world may be New Québec Crater Lake (CA) and Waldo Lake (US) (Table 9–3). The New Québec Crater Lake, a meteor-impact crater located in the subarctic, has a tiny drainage ratio (0.4). In addition, its catchment is frozen for about three-quarters of the year, further reducing nutrient export from its drainage basin. This, combined with its great distance from agricultural or industrial regions, resulting in very low atmospheric input (Table 8–10) may make New Québec Crater Lake the clearest, most oligotrophic lake in the world today, with *average* transparency of 33 m during the growing season (Table 9–3). Its rival—with a transparency of ~32 m— is Waldo Lake, a tectonically-formed lake in the Rocky Mountains of Oregon lying in a small, largely barren igneous basin (DR = 2.1, Larson 2000). Crater Lake (Fig. 8–15) was formed about 6,500 yrs BP in a collapsed caldera crater, and was once even more transparent (max. 38 m), but has seen a 25–30 per cent decline in its average transparency over recent decades (Larson et al. 1987). This decline is probably due to an increase in the atmospheric input of phosphorus attached to dust particles and ionic forms of nitrogen derived from distant agricultural areas. Wastewater from tourism may be adding additional nutrients. The great average transparency (now ~28 m) of Lake Tahoe (US), characterized by a small drainage ratio (DR = 2.6, Table 9–3), has been reduced as well. The rapidly increasing human population in its small catchment coupled with an increase in atmospheric loading has allowed an average five percent annual increase in primary production in recent decades. The resulting increase in the number of light-absorbing (algal) particles, and their release of light-absorbing DOM is responsible for the reduction in transparency (Fig. 9–5).

Among lakes with a large drainage ratio and moderately short WRT (< 1–2 yrs), short WRT (< 0.5–1 yr), or very short WRT (< 0.5 yr) are many of the limnologically best-known lakes (Table 9–3). Other very short WRT lakes include most riverine lakes and reservoirs (WRT = days to months, Chapter 29), as well as those lakes positioned at the bottom of large, well-watered drainage basins (CA>>LA). The US National Surface Water Survey reports a median drainage ratio of 15 in the recently glaciated well-watered portions of the American northeast and midwest compared to a ratio of 44 in the ancient and, as a result, much more eroded landscape of the southeast. The lowest average ratio (six) is encountered in Florida, which has a particularly high proportion of its surface covered by open water or wetlands.

WRT and the Biota

Waterbodies with a WRT < 2 yrs experience the greatest interannual (see Table 9–3) and seasonal variation in flushing rate caused by considerable variation in runoff from their relatively large catchments to generally shallow (small volume) lakes (Fig. 7–2), which are characterized by a large surface area to volume ratio and high evaporation (Fig. 9–6). The planktonic community composition (structure) becomes affected when the system flushing rate (1/WRT) is high enough to exceed the growth rate of some, but not all, planktonic species (see Fig. 2–4). The effect of flushing on the composition of temperate-zone phytoplankton communities is greatly affected by temperature and is usually restricted to lakes with WRT of < 60–100 days during the growing season (Søballe and Kimmel 1987, Sec. 29.4). Well-studied lakes with very short average WRTs are Windermere (GB), south basin $\bar{x}$ = 100 d, Blelham Tarn $\bar{x}$ = 42d), the epilimnion of Lake Constance (AT, CH, DE, $\bar{x}$ = 47d), Estwaite water (GB, $\bar{x}$ = 90d), and the entire volume of the Überlingen basin of Lake Constance ($\bar{x}$ = 80d). Shallow lakes everywhere in areas of high runoff are rapidly flushed (Fig. 7–2). Eight of 16 very short WRT Danish lakes have a mean theoretical flushing time of less than 29 days (Windolf et al. 1996).

Macrozooplankton (Chapter 23) typically have a much longer life cycle than the phytoplankton, and their community structure WRT is impacted at WRT greater than those at which the phytoplankton community structure or biomass is affected (Fig. 2–4 and Sec. 29.4). The effect of flushing on herbivorous macrozooplankton has an additional *indirect* effect on their phytoplankton prey over a much longer time interval (weeks) than the *direct* effect of flushing has on the phytoplankton community biomass and structure.

Lakes and reservoirs begin to resemble rivers when WRT becomes reduced to days or weeks. Summertime flushing rates in shallow unstratified lakes with a favorable light climate for photosynthesis and WRT of < ~5–10 days approach or exceed the growth rate of

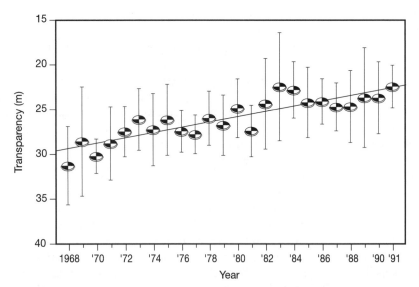

Figure 9–5 The 24-year record of declining Secchi disc transparency in Lake Tahoe, California (US). Each point represents the average of approximately 35 weekly or biweekly Secchi depth measurements at an index station. Note (1) a cyclical interannual variation in transparency that is linked to variation in climatic conditions and the linked upwelling of deepwater nutrients, and (2) considerable intra-annual variation in seasonal transparency linked to ENSO events (Sec. 5.4). The bars represent one standard deviation. *(After Goldman 1993.)*

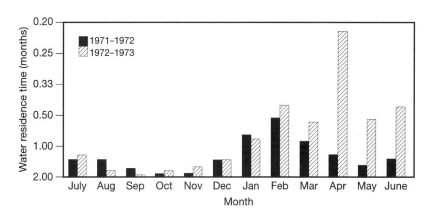

Figure 9–6 A two-year comparison of monthly water residence time for Lake Talquin (Illinois, US), a reservoir (LA = 39 km^2, $\bar{z}$ = 4.1m). The differences reflect variation in the amount and distribution of precipitation, evapotranspiration, and resulting runoff. *(Modified after Turner et al. 1983.)*

even the most rapidly growing small phytoplankton cells (Chapter 21) and protozoan species, resulting in wash-out from the water column. Shallow rivers with an even thinner water column and even more favorable light climate for growth experience a phytoplankton biomass decline in summer only when WRTs declined below three days (Basu and Pick 1995). The effect of **wash-out** on aquatic systems entails not only an absolute decline in species diversity and biomass during periods of rapid flushing but also a reduction in diversity relative to less rapidly-flushed systems nearby of similar size and trophic status. But the (epilimnetic) WRT is rarely considered by plankton ecologists when interpreting their findings.

The WRT of lotic systems and wetlands in well-watered regions is typically short compared to nearby lakes (Table 4–1, Søballe and Kimmel 1987, Basu and Pick 1996). Most streams have WRTs shorter than the ~3-day minimum summer doubling time of planktonic algae and protozoa, and WRTs much shorter than the doubling time of most macrozooplankton. In fact, it has long been recognized that only long or slow-flowing rivers—with WRTs of several weeks or more—are able to develop a diverse plankton community (Søballe and Kimmel 1987). Yet, small rivers, characterized by WRTs of a few days, contain a considerable phytoplankton and zooplankton biomass, with the algal biomass linked to water nutrient levels (Basu and Pick 1996). The extent to which the plankton biomass represents detached benthic algae and/or growth of planktonic organisms in embayments with longer WRT or reflects organisms washed in from the land–water interphase (wetlands) must be system specific. However, the importance of bays, wetlands, and the hyporheic zone (with a much longer WRT than the main channel) as sources of organisms, is con-

firmed by the presence of some macrozooplankton in rapidly-flushed lotic systems. (Fig. 29.7)

9.6 ▲ Retention of Dissolved and Particulate Materials by Lakes and Reservoirs

Inland waters with a long water residence time (τ_w) have, by definition, a slow water renewal rate. Changes in nutrient input resulting from changes in agricultural practices, human density, or improved wastewater treatment will be gradually reflected in the nutrient content of the water. The time needed for a new equilibrium concentration to become established can be approximated with the following empirical equations:

$$t(95\%) = 3\,\tau_w\,(1 - R) \qquad \text{EQ. 9.3}$$

where t (95%) = time (yr) for reaching 95 percent of the concentration equilibrium; τ_w = WRT (yr); R = annual retention coefficient, or the fraction of a specified incoming substance that does not leave via the outflow (OECD 1982).

The time required for 90 percent of the concentration equilibrium to be reached will be

$$t(90\%) = 2.3\,\tau_w\,(1 - R) \qquad \text{EQ. 9.4}$$

The value of R will be close to zero for substances such as NaCl that are highly soluble and for which the biota has little or no need. R may also be near zero for elements such as magnesium, which are required by the biota, but supply is usually so great in relation to demand that the concentration is little affected by organismal uptake (Table 8–1). Such substances are termed **conservative**. For example, a lake with WRT = 1 yr polluted by highly soluble road salt would be predicted to take about three years before t (95%) is reached following a change in winter salting, whereas it would be predicted to take 60 years before equilibrium would be approached in a lake with WRT = 20 yrs. In contrast, **nonconservative** substances, such as nutrients in high demand by organisms in relation to supply or toxins adsorbed to sedimenting particles, will be largely removed from the water by sedimentation and retained by the system. Their retention will be manifested by an R between > 0 and 1.0 (Fig. 9–7). Nonconservative substances will therefore reach a new equilibrium much more quickly.

A wastewater diversion scheme or the installation of a sewage treatment system is expected to yield a new phosphorus equilibrium t(95%) in about 2 years in a lake with aerobic sediments (Sec. 17.3), WRT = 3 yrs, and phosphorus retention coefficient, R = 0.8. Conversely, a change in winter road salting (NaCl, R = ~0) in the same drainage basin would predict a new equilibrium concentration after about nine years because the NaCl in the system needs time to flush whereas a nutrient in high demand is subject to uptake by organisms and relatively rapid removal by sedimentation (Sec. 20.4).

The nonlinear relationship observed between WRT or water loading and phosphorus retention (Fig. 9–7) is interpreted as the outcome of increasingly effective lake retention of nonconservative elements over time. In nutrient-poor lakes, living organisms hold much of the limited pool of critical nutrients in their protoplasm, as WRT increases the planktonic organisms and their nutrients are increasingly retained through sedimentation rather than flushing from the system.

The ability to predict the supply of plant nutrients and their loss to sedimentation and outflow is of great importance in eutrophication management of lakes, and in predicting nutrient levels in the outflowing rivers. Taxpayers need to be assured that a recovery can be expected after a reduction in external nutrient loading to lakes (Sec. 17.6). Not only the retention of inorganic nutrients but also of externally supplied (allochthonous) dissolved organic matter (DOC) is reduced during periods when low volume lakes and wetlands are rapidly flushed (Fig. 9–8). The fraction retained in sediments or lost to the atmosphere following respiration is, at any one water residence time, lower for DOC from forested and wetland catchments—which release disproportionately little nitrogen and phosphorus—than for agricultural catchments rich in inorganic nutrients. The importance of low C:P and C:N ratios for the microbial decomposition (respiration) of organic matter will be discussed in Sections 22.9 and 24.8. However, there is evidence from forest lakes that more of the carbon retained (not lost via the outflow) is respired, rather than

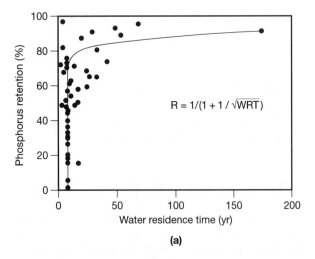
(a)

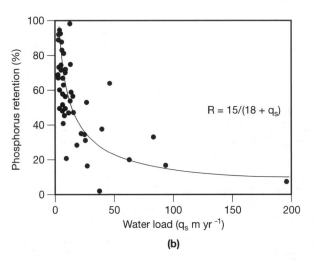
(b)

Figure 9–7   Observed phosphorus retention (R) as a function of (a) water residence time and (b) the annual water load on the lake surface (q_s, m yr^{-1}) for 54 largely oligotrophic to mesotrophic lakes with oxic hypolimnia. Retention is roughly predicted by the equation $R = 1/(1 + 1/\sqrt{WRT})$ developed by D. P. Larsen and H. T. Mercier (1976) and $R = 15/(18 + q_s)$ as developed by Nürnberg (1984). Phosphorus retention is appreciably smaller for the same WRT in lakes than in reservoirs with a deep (hypolimnetic) outflow (M. Straškraba 1996). *(Modified after Nürnberg 1984.)*

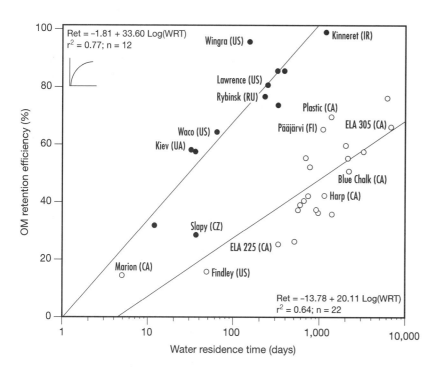

Figure 9–8 Relationship between mean water residence time and organic matter (OM) retention efficiency [(input–output)/input, as %] for a variety of lakes and reservoirs in non-forested (●) and largely forested (○) landscapes. Retention is here defined as material not lost via the outflow, and includes organic matter sedimented as well as the gaseous fraction subsequently lost to the atmosphere following decomposition (respiration). Selected lakes are identified. The exceptionally high organic matter loading of Slapy reservoir (see Fig. 8–18) is reflected in a relatively low retention efficiency, but a high absolute retention. *(Data from Groeger and Kimmel 1984, Dillon and Molot 1997, and Curtis and Schindler 1997.)*

stored in sediments in higher alkalinity (nutrient richer) lakes (Mollot and Dillon 1996).

Not only phosphorus and other nutrients but also many toxic metals and persistent organic contaminants readily adsorb onto organic and inorganic particles (Chapter 28) and are then piggybacked onto the sediments. Inorganic particles derived from catchments have R values that approach one in all except rapidly-flushed (< ~1 yr) lakes (Borg et al. 1989). But even systems with WRTs of weeks to months trap a significant portion of incoming sediments and their adsorbed nutrients and contaminants. Consequently, those lakes and rivers receiving water from upstream lakes, reservoirs, or large wetlands receive reduced sediment, organic matter (Fig. 8–18) and nutrient loads (Sec. 8.5). The trapped materials are then unavailable to outflowing rivers, downstream lakes, or oceans (see Table 9–4).

▲ 9.7   Sediment Loading to Aquatic Systems

Lotic systems draining well-watered, well-vegetated north temperate zone catchments, further characterized by smaller seasonality in precipitation than is

Table 9–4   **The influence of reservoirs on sediment discharge (10^6 t yr^{-1}) and adsorbed nutrients of major rivers to oceans.**

River	Before	After	Dam
Mississippi	500	210	6 larger dams on the Missouri
Rio Grande	20	0.8	4 large dams
Colorado	125–150	0.1	Several dams, including Hoover Dam
Zambezi	50–75	20	Kariba, Cabora Bassa
Nile	125	3	Aswan High Dam
Niger	40–65	5	Kanji Dam
Indus	250	50	Tarbela, Mangla, and other irrigation schemes

Source: Adapted from Milliman and Meade 1983, in Meybeck et al. 1990.

common at lower latitudes, receive among the lowest sediment loads in the world. Average annual sediment yields (t km^{-2} yr^{-1}) are low (Table 4–6) and inorganic sediment concentrations in the water column average between ~20 and 200 mg l^{-1} (Milliman and Meade 1983). Typically low sediment concentrations in receiving undisturbed streams, rivers, and lakes has meant little attention has traditionally been accorded to the effects of inorganic sediments on the biota. Nevertheless, high sediment yields characterize areas plowed for crop cultivation, particularly on slopes, and where cultivation occurs right to the water's edge when there are no buffer strips (Fig. 5–8). High sediment concentrations in waterways can also be caused by mines. Temporarily high concentrations result from road construction, forest clear cutting, and urban development. Rivers that drain mountain glaciers may carry high sediment loads, known as **glacial flour**, giving them a gray-white appearance.

The conversion of land from forest to agriculture increases catchment sediment yields dramatically. Dated lake-sediment cores show yields from temperate zone drainage basins to have increased three to tenfold following conversion (Fig. 9–9). Although

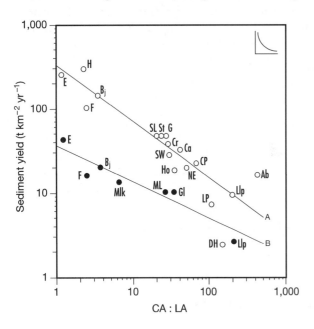

Figure 9–9 Relationships between sediment yield per unit area, also known as the export coefficient, and the catchment to lake area ratio for 20 temperate zone lakes presently under cultivation (line A) and when forested (line B), with prehistoric forested rate for presently cultivated catchments obtained from dated sediment cores. *(From Dearing and Foster 1993.)*

lakes with a large catchment to lake area ratio (CA:LA) receive more sediments than those in a small drainage basin (t yr^{-1}), sediment yields from land are lower per unit area (t km^{-2} yr^{-1}), as the result of a greater sediment trapping by large, low-slope basins. Small, generally much steeper, catchments deliver more sediment with adsorbed nutrients (Fig. 8–12) and contaminants per unit area to waterways, but lower annual loads. High sediment loads and associated nutrient exports from intensively farmed, typically lowsloped basins have a deleterious effect on the native biota of receiving rivers, (Fig. 8–10) and cause massive algal blooms in the shallow lakes. Additional results are the loss of submerged macrophytes and undesirable changes in the zooplankton and fish communities (Chapter 24). Applied research on the effect of buffer strips and other land use practices on nutrient and sediment export from drainage basins in intensively farmed areas of western Europe and North America provide wonderful opportunities for experimental research on manipulated and reference systems.

On a global basis, the highest sediment yields from drainage basins and the highest concentrations in receiving rivers are found in southern and eastern Asia and South America (Table 4–6), regions where the rivers drain an ancient and highly weathered tropical or subtropical landscape lacking dense vegetation cover following deforestation. Average sediment concentrations of 1,000–10,000 mg l^{-1} are typical, and orders of magnitude higher than for large north temperate rivers. Some Chinese rivers carry annual average sediment concentrations as high as 50,000 mg l^{-1} (Milliman and Meade 1983). High sediment concentrations have highly negative effects on the biota of the rivers and the reservoirs built upon them (Nontji 1994, Chapter 29). Such reservoirs, built primarily for generating electricity, experience a rapid reduction in volume and thus, utility.

Floodplain lakes and reservoirs in the well-watered upland areas of south and southeast Asia are particularly prone to siltation after deforestation and resulting erosion from the denuded catchment slopes. An extreme example is the Wonogere reservoir in central Java (ID), designed for a life of 100 years, but will likely have an effective life of only 18 years because of the annual accumulation of 6.6×10^6 m^3 of topsoil from its drainage basin (Nontji 1994). Among the highest sedimentation rates, as measured by changes in maximum depth, are decreases of 41 cm yr^{-1} (22 m in 54 yrs) in Lake Tondano, and 51 cm yr^{-1} (from 28.6 m to 1.5 m) in Lake Lumboto between 1934 and 1987

(Nontji 1994). The transparency of such reservoirs is typically ~20 cm and sedimentation rates are orders of magnitude larger than in lakes and reservoirs in the glaciated portion of the temperate zone. Although these examples are in Indonesia, there is no reason to believe that environmental management there is appreciably worse than elsewhere in southeast Asia.

Highlights

- The functioning of inland waters is tightly linked to the characteristics of their drainage basins.
- The average slope (relief) of drainage basins declines as the catchment area increases, as does the length of streams per unit area within a climatic/geological region. Therefore, particulate matter with adsorbed plant nutrients or contaminants leaves the land at a lower rate (t km^{-2} yr^{-1}) from large than small catchments even though the total export (t yr^{-1}) is greater from large than small drainage basins.
- Wetlands are restricted to low-slope drainage basins or the low-sloped portions of large basins where drainage is impeded.
- The elements with the highest demand/supply ratio in relation to the terrestrial biota (N, P, K) are most effectively retained on land in undisturbed basins.
- The water residence time (WRT) or water renewal time is a useful indicator of the "age" of water in a system, nutrient input, time available for nutrient cy-

cling and respiration of organic matter in the water column, sedimentation of particles, probability that nutrients in high demand will be flushed, and whether particular planktonic species can reproduce quickly enough to offset flushing.

- A small catchment to lake area (CA:LA) or catchment area to volume ratio implies a particularly small area able to collect precipitation and released nutrients. Stream discharge is relatively small as is the resulting water and catchment organic matter and nutrient input to lakes or wetlands. Receiving lakes therefore have a low algal biomass and a low color. The lake and wetlands will be proportionally more dependent on atmospheric nutrient inputs directly onto the water surface than more productive lakes draining similar, but larger, drainage basins in the same climatic/geological zone.
- WRT in lotic systems and wetlands is typically short compared to nearby lakes, it ranges from hours to days for most streams while ranging from days for shallow riverine lakes and some reservoirs, to years, and even centuries for some freshwater lakes.
- Lotic systems draining well-vegetated north temperate zone catchments receive among the lowest sediment loads (t yr^{-1}) in the world. Deforestation and conversion to agriculture (crops) greatly raises sediment and nutrient yields regionally with negative effects on the native biota.
- ▲ Globally, the highest sediment yields are from subtropical and tropical catchments draining ancient and highly weathered soils lacking dense vegetation cover.

10

Light

10.1   Introduction

Solar radiation not only drives the photosynthetic process and permits vision, it also determines the different amounts of solar energy received annually at different latitudes, (Fig. 5–10), producing a heating gradient over the earth (Fig. 10–1). This gradient is ultimately responsible for both latitudinal difference in precipitation (Fig. 5–9) and the winds that bring about currents and turbulence in lakes (Fig. 5–11). It is wind energy, together with local heating, that determines if lakes will stratify and at what depth.

Wavelengths transferring appreciable energy from the sun to the earth extend from about 100 to 3,000 nm, but virtually all of the energy is about 300 to 2,000 nm (Fig. 10–2). About three percent of the energy arriving at the earth's atmosphere is in the ultraviolet region (100–400 nm), with the remainder about equally split between visible, or photosynthetically available radiation (PAR, ~400–700 nm), and infrared radiation (~700–3000 nm).

Solar energy can be considered either a continuous flow of electromagnetic waves or as discrete packets of energy. Irradiance is normally expressed in nanometers (nm) when described by wavelength, whereas photochemical reactions, including photosynthesis and vision depend on the number of quanta absorbed. A quantum of light is called a **photon**. Although the behavior of solar radiation in the atmosphere and water is best explained by assuming it to travel as waves, photochemical reactions including photosynthesis and vision can only be explained by viewing the irradiance as discrete packets of energy (**quanta**). It is, therefore, important to know how the two are related.

▲ The energy (E) content of photons varies inversely with the wavelength according to the formula:

$$E = h\nu, \qquad \text{EQ. 10.1}$$

where h is Planck's universal constant (6.625×10^{-34} J) and ν is the frequency. As a result of this relationship, the energy content of individual photons rises with increasing frequency (ν) and, because frequency is inversely proportional to wavelength, the energy content of photons declines with increasing wavelength. Thus, one mole of photons of red light at 670 nm has an energy content (quanta) of 176×10^4 joules, whereas the same quantity at 470 nm has an energy content which is 43 percent higher at 251×10^4 J.

The wavelength visible to humans (380–770 nm) and the **photosynthetically available radiation** (PAR, ~380–710 nm) lie in that portion of the electromagnetic spectrum (blue to red) where photons have the correct amount of energy to affect the outer valence bonds of the atoms of the pigment molecules, thereby allowing vision and photosynthesis. Two of the grand rules of light-photosynthesis research are that eight photons of PAR are required for the release of one molecule of dissolved oxygen (DO) in photosynthesis and the maximum efficiency of photon use (**quantum efficiency**) in photosynthesis is ~0.09–0.10 mol carbon fixed per mol photons absorbed, where 1 mol = 6.022×10^{23} (Avogadro's number). At wavelengths greater than ~700 nm, the lower-energy photons of infrared energy are perceived as heat. In the ultraviolet range (< ~400 nm) the flux of very high energy photons may cause structural damage to DNA and proteins. Very high energy photons also play a role in photooxidation and photobleaching dissolved

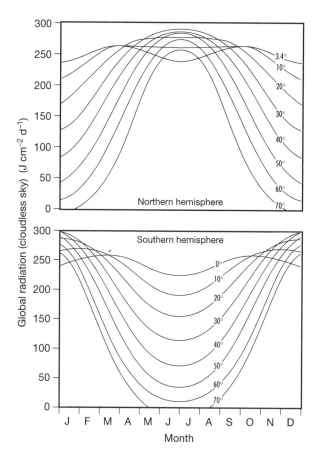

Figure 10–1 Annual variations in radiation reaching the ground on totally clear days. Note the increasing variation with increasing latitude, producing a heating gradient. *(Modified after Straškraba 1980.)*

organic matter (Sec. 10.5). The widely used term "light" refers, strictly speaking, to radiation visible to humans.

It is not only because irradiance is measured both as waves (or frequencies) and photons that conversion factors are needed (Table 10–1) but also because the internationally acceptable units in which energy flux is expressed have fairly recently been changed from calories (cal) to joules (J). The energy flux of incoming irradiance has also changed units from cal cm^{-2} time^{-1} to watts (W) m^{-2}. The number of photons absorbed, the **photon flux density** (PFD),[1] is the unit used by those interested in photochemical reactions. Energy flux and the PFD are closely related, but, as photons

energy changes with wavelength, the conversion of moles to energy is operationally made at the midpoint region of 400–700 nm (~550 nm, see Table 10–1). The mole is conventionally described as mol m^{-2} s^{-1}, but is sometimes referred to as Einsteins (1 mol = 1 E) in physiological plant ecology.

▲ 10.2   Detectors

Virtually all solar energy reaching the earth and its water surface is found between 300 nm in the ultraviolet and 3,000 nm in the infrared regions of the spectrum. This is the fraction of incoming energy that is recorded by weather bureaus (mW cm^{-2} or cal cm^{-2} time^{-1}). Between 45 and 50 percent of the incoming solar energy (300–3000 nm) that is recorded is in the photosynthetically available range (Fig. 10–2); the exact value is a function of weather conditions. Since availability is so close to 50 percent, the data reported by weather bureaus are normally divided in half to obtain a measure of photosynthetically available radiation (PAR), and the fraction of energy visible to the human eye that reaches the water surface. Prior to the 1980s, measurements of visible irradiance and PAR were usually made with **photometers**. The sensor was typically a selenium barrier cell that converted incident solar energy to an electrical signal that could be read with a meter or transmitted to a recorder or integrator. Today, nearly all measurements are made with **quantum sensors** (Fig. 10–3) that are sometimes operated as photometers. Most sensors are *flat plate* (2π), also known as **cosine collectors**, in which the response is proportional to the cosine of the angle from the vertical. Under water, these measure downwelling light when facing up. Some collectors are spherical **scalar collectors** and their almost 4π sensors are able to collect light from all directions, a better representation of the light climate perceived by planktonic organisms. Cosine collectors are the most appropriate sensors for algae growing on surfaces or for plant leaves.[2]

---

[1]"In our survey a unit was rarely paired with its correct physical quantity. Thus, of eight authors who measured light in µEm^{-2} s^{-1}, five called the quantity light intensity, one said it was irradiance, one said it was photon flux, and only one correctly said it was a photon flux density." (L. D. Incoll et al. 1977)

[2]Underwater photometers with cosine collectors are often incorrectly calibrated, with the absolute values obtained frequently differing by 25–30% among instruments. Scalar sensors are more difficult to calibrate because they receive energy fluxes from all directions that yield much higher energy or photon fluxes—up to 70% higher where particle-induced light scattering is high—than those obtained with more common cosine collectors (D. H. Jewson et al. 1984).

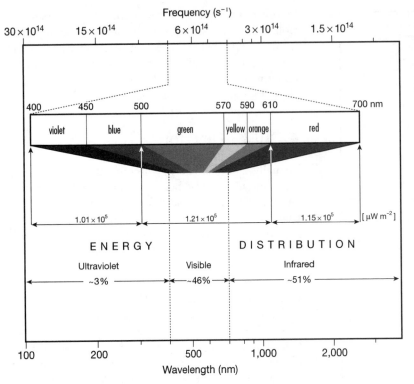

Figure 10–2   Wavelengths of solar radiation, average daily energy distribution at sea level in the temperate zone and the percent distribution of energy at sea level. *(Modified after Reifsnyder and Lull 1965.)*

The early data on underwater irradiance were obtained in the US by Birge and Juday between 1928 and 1931, and in Germany by F. Sauberer. Birge and Juday measured light transmission as a function of depth in 50 Wisconsin (US) lakes, using a submersible array of thermocouples connected to instruments on board ship or onshore. They measured the spectral composition (350–750 nm) in 43 of the lakes using ten different colored filters mounted on a rotating wheel over the detector (Birge and Juday 1932). Underwater measurements were greatly simplified by the development of submersible photocells at about this time. However, a major problem with photocells was they were not equally sensitive to different wavelengths of PAR impinging on them. In other words, they did not have a flat or near flat spectral response to energy of different

Table 10–1   **Some conversions relative to the properties of light. Photosynthetic units, based on the sensitivity of the human eye, are no longer used in aquatic science, but 1,000 lux in daylight (PAR) = ~19.5 μmol m^{-2} s^{-1}.**

Convert	To	Multiply by	Reciprocal
calories	joules	4.19	0.24
calories cm^{-2} min^{-1} (PAR)	μmol m^{-2} s^{-1}	3.17×10^3	3.15×10^{-4}
calories cm^{-2} min^{-1} (PAR)	watts m^{-2} (PAR)	6.98×10^2	1.43×10^{-3}
joules m^{-2} d^{-1}	watts m^{-2}	2.31×10^4	4.33×10^{-5}
μmoles (μEinsteins)	photons (quanta) m^{-2} s^{-1}	6.02×10^{17}	1.66×10^{-18}
watts	ergs s^{-1}	1×10^7	1×10^{-7}
watt m^{-2} (300–3,000nm)	μmol photons m^{-2} s^{-1} ($\bar{x}$ PAR at 550 nm)	2.11	0.47
watt m^{-2} (PAR)	μmol photons m^{-2} s^{-1} ($\bar{x}$ PAR at 550 nm)	4.67	0.21

Source: After W. G. Biggs, in Genser 1986, and Pennycuick 1988.

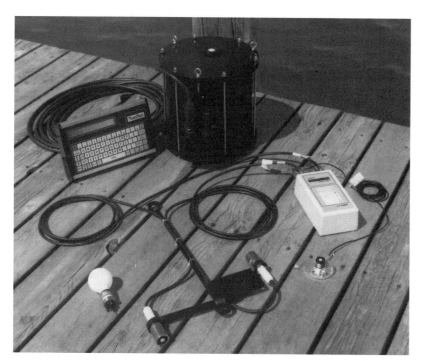

Figure 10–3   Light measuring equipment. In the foreground, a lowering frame holding two cosine (2π) underwater quantum sensors to measure the downwelling and the less frequently measured upwelling radiation. Electrical cables connect to a data logger (shown on right) or to a quantum radiometer (not shown) for instantaneous measurements. An above surface quantum sensor is attached to the data logger. To the left of the lowering frame, an almost spherical (4π) scalar sensor. At the rear, an underwater spectroradiometer plus a portable terminal. *(Photo courtesy of LI-COR, Inc., Environmental Division.)*

wavelengths in the PAR region. When such cells were lowered through the water column, the electrical signals recorded were affected not only by the rapidly decreasing amount of energy received with depth but also by their differential sensitivity to the underwater spectral changes encountered with changes in depth (Sec. 10.4). The problem has been resolved by a modification of the now more sensitive sensors to make them equally sensitive to incoming solar energy (or quanta) over the whole photosynthetic range. The more recent development of **spectroradiometers** (Fig. 10–3), with cosine or scalar collectors, are able to scan and resolve an underwater light field at roughly 8 nm intervals over the photosynthetically available spectrum (PAR) (Sec. 10.4), and some models are able to record measurements from 280–800 nm, thereby including the UV-A and UV-B ranges.

10.3   Light Above and Below the Water Surface

Although about 50 percent of the incident solar radiation reaching the water is in the photosynthetic range, the absolute amount arriving at the surface of a particular waterbody each day or year is greatly affected by the latitude (Fig. 10–1). The contrast in irradiance received by aquatic systems at different latitudes is at-

tributable to differences in minimum daily irradiance during the hemispheric winter; the differences are small during the hemispheric summer (Fig. 5–10). Locally and regionally, the amount of cloud cover, surrounding topography, and atmospheric conditions affect the irradiance received. Comparison of weather bureau readings, obtained on a totally sunny and totally cloudy day, make this clear. Full cloud cover composed of cirrus or cumulus clouds reduces the incident solar radiation by about 35 percent and 25 percent respectively, whereas the equivalent stratus and nimbus clouds or fog reduces radiation 85 percent and 75 percent respectively over that obtained under cloudless conditions (Straškraba 1980). Lakes and streams in mountain valleys or small lakes and streams surrounded by tall trees may be sufficiently shaded to experience greatly reduced incident radiation over that recorded in an open location nearby.

The fraction of PAR entering ice-free lakes is determined by the sun's angle, waves, and by the ice and snow cover at higher latitudes in winter. The greater the sun's elevation, the smaller the reflection from the water's surface. However, the fraction reflected (400–700 nm) is minor (about 6%) except during the few hours before sunset or after sunrise under both sunny and cloudy conditions. The effect of waves alone is even smaller (~2%), with the result that an average total of eight percent of the midday incoming energy is lost through reflection plus wave action during the passage of PAR (~400–700 nm) from air to 5 cm under the water surface (Schanz 1983).

The Effect of Snow and Ice

▲ The principal impediment to PAR entering higher-latitude lakes in winter is the result of reflection and the high absorption of solar energy by snow and ice. **Albedo**, the ratio of reflected light to total incident light, is highest at about 0.7–0.9 for freshly fallen snow, meaning that 70–90 percent of the solar energy is reflected (Ragotzkie 1978). Deposition of atmospheric dust on the surface and melting of ice cover reduces the albedo.

The incoming PAR able to pass through snow cover must next pass through the ice layer to allow vision or algal photosynthesis in the water below. However, light absorption (extinction) and reflection is low in clear ice with few air bubbles (similar to window glass) Fig. 10–4. This type of ice is known as **black ice** to distinguish it from **white ice**, which is formed in early winter following the deposition of snow on a thin ice cover. The weight of the snow is sufficient to bend the ice and lowers the snow layer below the hydrostatic water level. It becomes saturated with water seeping through cracks in the ice, produced by the weight of the wet snow or contraction as the ice sheet expands and contracts with changing air temperature, and a layer of milky-white ice is formed when it refreezes. The snow cover and, secondarily, the layer of white ice are responsible for the extremely rapid extinction of incoming PAR. Thus, the percent transmission of PAR through 10 cm of compacted dry snow is only about 2–4 percent (higher for wet snow and lower for fluffy dry snow), whereas it is about 44 percent for the same thickness of white ice and 98 percent for black ice that is virtually free of air bubbles (Fig. 10–4).

The albedo plus the light absorbed by the snow and ice usually reduces the flux of PAR able to enter the water column to a negligible level. The duration of near darkness under the ice may range from a few days in a midlatitude country at sea level, such as the Netherlands or southern New England, to periods of one to six months (longer at higher altitudes) in the northern portion of the midwest region of the United States, southern Canada, central Scandinavia, and the northwestern portion of the former USSR and Antarctica.

A lengthy period of snow and ice cover, in conjunction with a small amount of daily irradiance reach-

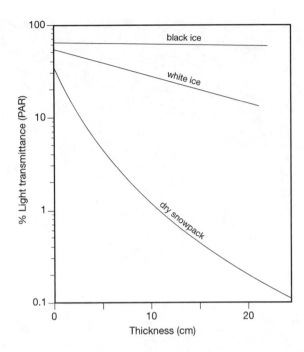

Figure 10–4 Percent light transmittance as a function of substrate thickness determined for a polar lake. 100% represents photosynthetically active radiation at the frozen surface, before reflectance. Note the exceptionally low absorbance of light in polar black ice, indicating a virtual absence of air bubbles. *(Modified from Welch et al. 1987.)*

ing the surface of lakes and wetlands at higher latitudes (Fig. 10–1), permits only negligible rates of photosynthesis and associated dissolved oxygen (DO) production under the ice. But winter rates of respiration are not lowered to the same extent causing DO levels to decline; drastically so in eutrophic lakes and in shallow, organically-rich wetlands covered for months by snow and ice cover. In such systems, or in ice-covered rivers draining nearby wetlands, the DO may be totally consumed through respiration by microbes and other organisms, resulting in death (*winter kill*) for fish and invertebrates (Chapter 15). The loss of fish then has a major impact on the community structure during the next ice-free period, one that may extend for years.

10.4   Absorption, Transmission, and Scattering of Light in Water

PAR entering the water is partly **attenuated (absorbed)** by organic particles and partly by colored DOM, as well as by very small silt- and clay-sized inorganic particles derived from drainage basins. Very small inorganic particles dominate in the highly weathered soils of low rainfall regions where rivers carry high sediment loads from surface runoff (Fig. 5–5 and Sec. 5.2) during rainfall periods (Sec. 9.7). Some photons are scattered rather than absorbed upon striking particulate matter. There is additional scattering produced by inorganic solutes and water molecules.

In the most highly transparent lakes and in the open ocean, both characterized by very few particles and almost no DOM, the red end of the PAR spectrum (600–700 nm) is largely attenuated by water molecules. At the opposite end of the PAR spectrum (Fig. 10–2), blue light and shorter-wavelength ultraviolet radiation (UV < 400 nm) are particularly quickly attenuated by colored DOM. Water molecules and solutes dominate absorption only in exceptionally transparent lakes containing very small amounts of DOM and inorganic particles.

While blue light is transmitted best in highly transparent waters, it is subject at the same time to high scattering because scattering in pure water increases inversely with the fourth power of the wavelength $(1/\lambda)^4$, making blue and UV portions of the spectrum most prone to scattering. Although light scattering by water molecules occurs in all directions, the part that is scattered up toward the observer is responsible for the blue color of these highly transparent lakes.

▲ Even a very modest amount of colored dissolved and particulate organic matter will rapidly attenuate photons in the blue portion of the visible or photosynthetic spectrum, thereby allowing green light to be transmitted best. The latter then becomes most prone to scattering upward, giving the clear water a greenish appearance.

In humic or turbid water, shorter wavelengths are rapidly absorbed with the result that the small amount of downwelling light becomes quickly dominated by long wavelength PAR (orange and red > ~600 nm). The ubiquitous, often highly colored DOM has an important effect on the light climate of virtually all lakes.[3] DOM, composed 45–50 percent of carbon, is alternatively expressed as carbon and then referred to as dissolved organic carbon (DOC). Colored DOM or DOC is frequently referred to as **color, gilvin**, or **gelbstoff** (German: yellow substance).

Earlier measurements of light quality usually employed three broadband color filters (blue, green, red) placed over underwater sensors to describe selective light attenuation over different portions of the PAR spectrum. The generalities derived from these studies were elaborated upon in much finer detail during the last two decades when spectroradiometers (sometimes equipped with submersible quartz-fiber probes) came into wide use.

Spectral Attenuation

Spectroradiometric measurements made in a series of Tasmanian (AU) lakes provide an example of changes in attenuation with depth in and among lakes. Even measurements made in the clearest of the lakes (Perry, with an average transparency (Z_{SD} = 15 m) show it contains enough color to shift the wavelengths of minimum attenuation (maximum transmission) away from the blue toward the green portion of the spectrum (Fig. 10–5a). It is only in ultra-oligotrophic lakes such as Crater Lake (US), high mountain lakes located in small and barren catchments, and the oligotrophic portions of the world's oceans is minimum attenuation found in the blue portion of the PAR spectrum. Note the rapid PAR attenuation in deeply colored Lake Chisholm (Fig. 10–5c). The few photons able to penetrate to any depth in this lake (or any inland waters containing much silt or clay) are in the red portion of the spectrum. At a distance below the surface in all lakes the light will be almost exclusively of one color (monochromatic), the result of selective absorption by particles and colored DOM.

Particles not only absorb photons but also add to light attenuation (extinction) as the result of light scattering. Scattering increases the path length of the down and upwelling photons as they are bounced around, thereby increasing the possibility of absorption. While the light absorption capacity of particles is a function of their abundance or mass, it is also affected by size distribution, with light attenuation larger when the mass is composed of many small particles rather than fewer large particles (Fig. 10–6). This finding was also noted in Kirk's (1994) theoretical evaluation of light interception by particles, which suggested that a chlorophyll-*a* concentration (a mea-

---

[3] All systems contain some colorless DOM, but the fraction is highest in slowly flushed freshwater lakes and shallow endorheic (saline) lakes (Sec. 8.8). Measurement in Canadian saline lakes and wetlands show very high concentrations of colorless DOM and low light attenuation, which is attributed to photobleaching of the more highly colored DOM arriving from drainage basins. (Arts et al. 2000)

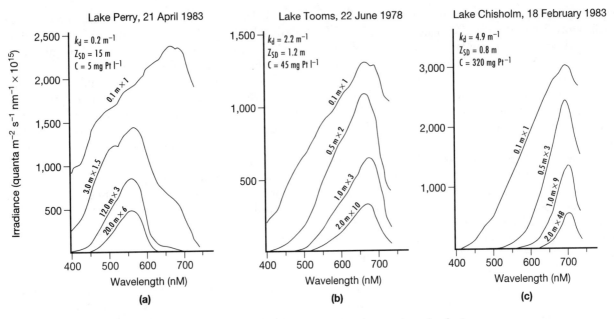

Figure 10–5 Spectral distributions of downwelling PAR (400–740 nm) at various depths in three Tasmanian (AU) lakes that vary in vertical light extinction coefficient (k_d), Secchi disc transparency (Z_{SD}), and water color (C). Note the difference in scale. *(After Bowling et al. 1986.)*

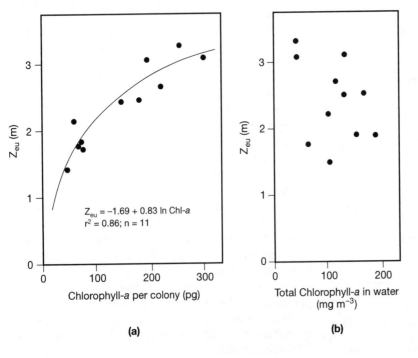

$$Z_{eu} = -1.69 + 0.83 \ln \text{Chl-}a$$
$$r^2 = 0.86; \; n = 11$$

Figure 10–6 Relationships in Hartbeespoort reservoir (ZA) between (a) eutrophic zone depth (z_{eu}) and mean *Microcystis aeruginosa* colony size (chlorophyll-*a* content per colony); and (b) z_{eu} and chlorophyll-*a* content in the water for the same points as in part (a), indicating that it is particle (colony) size rather than algal biomass (as chl-*a*) that determines the z_{eu} in systems where turbidity is primarily produced by algae. *(After Robarts and Zohary 1984.)*

sure of algal biomass) of 3 mg l^{-1} results in a 29 percent deeper euphotic zone for colonies with a diameter of 58 μm than for small single-celled algae with a diameter of only 8 μm.

The absorption of photons by algae is largely a function of the amount of photosynthetic pigments. However, the fraction of cell volume packed with pigment declines as cell size increases (Agustí 1991), and there is less light attenuation in lakes dominated by large algae species rather than the same biomass of small species. Under conditions of ample nutrients (eutrophic lakes) large algal cells or colonies are able to develop a greater maximum biomass than when the community is composed of small-sized units.

10.5 Ultraviolet Radiation and Its Effects

The development of arctic and antarctic ozone holes attributed to anthropogenically-induced ozone levels has led to concerns about the reduction in the earth's shielding from damaging **ultraviolet radiation** (UV). Sensors, not available when E. Steemann Nielsen (DK) and I. Findenegg (AU) explored the effect of UV radiation on algae in the 1950s and 1960s, have greatly facilitated research on UV effects on the biota and organic matter in general.

UV radiation is subdivided between UV-C (40–280 nm), which is so strongly absorbed by the atmosphere that negligible quantities reach the earth; UV-B (280–320 nm), middle ultraviolet radiation which is extremely injurous to organisms by damaging DNA and disrupting many photosynthetic processes as well as pigment stability; and the near-ultraviolet radiation UV-A (320–400 nm), shown to cause minor photodamage and reduce algal, bacterial, and protozoan growth rates in laboratory experiments. UV-A contains less energy per photon and is less injurious than UV-B, but a much greater fraction of energy lies in the UV-A than UV-B portion of the UV spectrum. As a result, the negative effect of UV-A on living organisms may be as large or larger than the UV-B effect. UV-A is also better transmitted (less attenuated) than UV-B in transparent water columns (Sommaruga et al. 1997).

PAR quantum sensors and most spectroradiometers do not measure the UV portion of the spectrum because it requires special sensors. Extensive scattering plus a very high photon attenuation by even small

concentrations of colored organic substances means that a few milligrams per liter of colored DOC provides an effective shield against UV radiation for aquatic organisms, and virtually all of the UV-B is absorbed in the top 20–50 cm (Fig. 10–7). At the same time, small changes in DOC over the < 2–3 mg l^{-1} range has a major effect on UV-B attenuation. Williamson et al. (1996), concluded that projected changes in climate—affecting runoff, river discharge, and lake water residence time, as well as acid deposition which affects DOC solubility and its concentration (Ch. 27)—is probably much more important than stratospheric ozone depletion in determining future UV radiation in inland waters. Negative UV effects on the biota, unless offset by the development of protective pigments or ability to escape by hiding in deeper waters or shallow-water sediments, are greatest at high latitudes in highly transparent (mountain) lakes far from regions of high aerosol production. It is also an important factor in shallow lowland systems (ponds, rivers, and wetlands) in which UV-B attenuation is modest in the thin water layer.

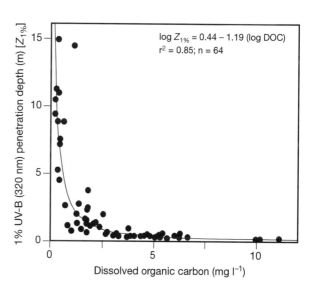

Figure 10–7 Relationship between 1% UV-B (320 nm) attenuation depth and dissolved organic carbon (DOC), based on a survey of 64 data points taken from 59 glacial lakes in North and South America (two data points are cut off from the plot). Values were calculated using the equation $z_{1\%} = -\ln(0.01)/k_d$. Note the rapid increase in UV-B penetration as DOC declines below ~2 mg l^{-1}. (*Data from Morris et al. 1995.*)

The negative effect of UV-B and most UV-A radiation on phytoplankton photosynthesis was detected in the top one-third (6 m) of the euphotic zone of transparent Lake Michigan (US, mean spring Z_{SD} = 5–10 m), reducing the primary production in the water column by an estimated 13 percent below the (unrealistic) absence of *any* UV-B and most UV-A radiation[4] (Gala and Giesy 1991). Although the negative effects of high levels of UV radiation on the biota are undeniable, after modeling their data (Gala and Giesy 1991) they concluded that a projected further decline in stratopheric ozone levels over Lake Michigan would *not* add significantly to present UV-B levels.

Plant and animal species vary widely in their sensitivity to UV rays, and their effect on the structure of natural communities is not easy to interpret. For example, experiments in stream mesocosms have noted the greatest effects of UV radiation on animal predators of algae, resulting in larger numbers of phytoplankton and benthic algae (Bothwell et al. 1994). Diatoms appear to be more sensitive than blue-green algae among phytoplankton examined experimentally (Gala and Giesy 1991). This suggests that UV flux should be added to the present, already extensive, list of environmental factors directly affecting natural community structure. There are also indirect UV effects that may be difficult to separate from direct effects. For example, UV radiation increases the photobleaching of DOC that reduces light extinction and thus increases the thickness of both the epilimnion and the euphotic zone (Sec. 10.7). Photodegradation of DOC affects the size structure of organic molecules and their availability to microbes (Sec. 22.5). Photodegradation of DOC also yields reactive oxygen species (O_2^-, H_2O_2, OH^{-1}) that serve as powerful oxidants and are toxic to the biota. Furthermore, photodegradation of methylmercury reduces mercury toxicity and enhances the loss of mercury to the atmosphere (Sec. 28.9, Sellers et al. 1996. For a recent review see de Mora et al. 2000).

---

[4]The significance of the experiment depends on the relationship between short-term rates of carbon uptake during photosynthesis and daily or weekly rates of algal growth and is still not well documented. Furthermore, primary-production experiments are almost without exception carried out in glass bottles. These transmit little UV-B radiation (~5%), but do transmit 50–60% of incident UV-A. UV-B shielding results in a modest, but nevertheless unrecognized overestimation of photosynthetic rates measured in water near the surface.

10.6   Light Attenuation

The vertical attenuation of downwelling PAR is most simply expressed as a percent reduction through a water layer of specified depth

$$\frac{100(I_o - I_z)}{I_o},$$ EQ. 10.2

where I_o = the light intensity over a specified wavelength interval just below the surface and I_z = the light intensity at depth z. Vertical attenuation of PAR with depth is exponential, but much more rapid in highly colored lakes. Thus, a 50 percent reduction in PAR is reached at about 2 m in Lake Perry, whereas the same percent reduction was attained in the top 10 cm of Lake Chisholm. Yet, even transparent Lake Perry is considerably more colored, with greater vertical attenuation, than Crater Lake (US)—one of the world's clearest lakes, in which the 50 percent PAR attenuation is only reached at about 10 m (Fig. 10–8).

▲ Rather than expressing the amount of underwater PAR as a percentage of surface PAR, it is more useful to compute the downwelling radiation by means of the **vertical extinction** (or **vertical attenuation**) **coefficient** (designated as k or k_d), the slope of the line formed when the natural logarithm of the energy flux is plotted against depth. The exponential curve is described by the equation

$$I_z = I_o e^{-kz},$$ EQ. 10.3

where I_z is the energy flux or photon flux density at depth z (m), I_o the intensity a few cm below the surface, e is the base of the natural logarithm value of 2.303, and k_d equals the vertical extinction coefficient (ln units m^{-1}). When Equation 10.2 is converted to the more convenient natural or logarithmic$_{10}$ form, expanded and rearranged it becomes

$$k_d = \frac{\ln I_o - \ln I_z}{z},$$ EQ. 10.4a

or

$$k_d = \frac{2.303 (\log_{10} I_o - \log_{10} I_z)}{z}.$$ EQ. 10.4b

The mean vertical extinction coefficient (PAR) is very useful in determining the rate of light absorption whereas k_d, determined at different wavelengths with a spectroradiometer, generates a detailed characterization of optical attributes. The coefficient expresses the

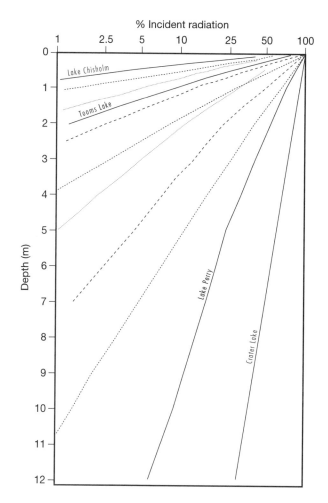

Figure 10–8 Attenuation profiles of downwelling PAR (400–700 nm) for selected waters. (**A**) Lake Chisholm, (**B**) Tooms Lake, and (**C**) Lake Perry, AU; and (**D**) Crater Lake, US, whose attenuation rate further declines in even deeper waters. *(Modified from Bowling et al. 1986, and Smith et al. 1973.)*

transparent lake such as Perry contains many more particles, much more dissolved organic matter, and has a much higher k_d than ultra-oligotrophic Crater Lake (Oregon, US). Crater Lake's mean k_d (PAR) of about 0.05 m^{-1} and an even lower 0.03 m^{-1} in the amictic antarctic Lake Vanda (lying in a catchment with almost no organic matter, Vincent et al. 1998) resemble those recorded in oligotrophic portions of the world's oceans.[5,6]

Highly eutrophic, but shallow lakes in the western part of the Netherlands typically have summer k_ds of 4 to 8 m^{-1}, with roughly half of the attenuation attributable to phytoplankton and the rest divided between resuspended bottom sediments and dissolved organic matter (Blom et al. 1994). Dissolved organic matter dominates attenuation in brown-water lakes with k_ds up to 13 m^{-1} (Chambers and Prepas 1988). Silt-laden semiarid reservoirs commonly have k_ds of greater than one or two. It seems the record is a k_d of 57 m^{-1}, measured in a South African reservoir with exceptionally high levels of inorganic particles (Allanson et al. 1990).

The mean attenuation coefficient (k_d, PAR) is a composite measure of attenuation by water, suspended particles, chl-*a*, and colored dissolved plus colloidal organic matter (DOM). Apart from characterizing the optical characteristics of waterbodies, mean k_d is useful for predicting light flux (I_z) at any depth (Eq. 10.3).

10.7 Light Attenuation and Photosynthesis

Biological limnologists and oceanographers are particularly interested in the portion of the water column in which phytoplankton photosynthesis is greater than phytoplankton respiration. This zone is known as the **euphotic**, **photic**, or **trophogenic zone**. The bottom

normally straight line formed when light intensities at different depths are plotted on semilogarithmic paper. As an example of such a computation, the energy transmitted to 10 m below the surface was about 14 percent of the flux recorded at the surface of highly oligotrophic and transparent Lake Perry (Fig. 10–8). Using Equation 10.4 and taking the logarithm of both $I_o = 100\%$ and $I_z = 14\%$ yields an average vertical extinction coefficient of 0.20 m^{-1} over that 10 m. In highly eutrophic and colored Lake Tooms the extinction rate is large, as reflected in a coefficient of 2.2 m^{-1} (Figs. 10–5 and 10–8). However, even a highly

---

[5]Very high scattering in pure water and rapid absorption of UV radiation by colored DOM and particles implies a much higher k_d (UV) than k_d (PAR). Measurements made in Lake Michigan show this to be the case with an average k_d (UV-A) of 0.43 and k_d (UV-B) of 0.72, about two and three times larger than the k_d (PAR) of 0.23 (Gala and Giesy 1991). These absolute extinction rates are still much higher than in ultra-transparent Lake Vanda, AQ, with a k_d (UV-A, 380 nm) of 0.02 and k_d (UV-B, 305 nm) of 0.08, and sufficient UV penetration through the 3.5-m-thick ice cover to allow photoinhibition in the upper water column. (Vincent et al. 1998)

[6]Natural water k_ds can be compared with those reported for ice and snow by J. T. Scott (1964) in Ragotzkie (1978), here presented per m^{-1}: black ice = 1.5; white ice = 6.7; fresh snow = 24–34.

of this zone, the **compensation depth**, is where phytoplankton photosynthesis just balances respiration on a daily basis. Below this level is found the **aphotic or tropholytic zone**. As a rule of thumb, the compensation depth of the phytoplankton community is normally considered to be the depth where one percent of the surface or immediate subsurface incident radiation remains. The true (physiological) compensation depth is frequently measured at ~0.1%, but the amount of primary production below one percent is usually so small that no appreciable error is incurred by omitting it in calculations of integrated production (mg m^{-2} d^{-1}) in the euphotic zone. Note the large variation in depth of the one percent level among lakes in Figure 10–8.

▲ Experimentally determined compensation intensities (I_c) typically vary from 1–10 µm mol^{-2} s^{-1} (2.4 – 24 × 10^{-2} Wm^{-2}, PAR) in a 12 hr light/dark cycle, but lower compensation intensities are possible in nature when algae or mosses are given sufficient time to adapt, or when they photosynthesize at low temperatures in the laboratory (Table 10–2).[7] Compensation intensities are much higher for macrophytes and mats of benthic algae (Chapter 24). Macrophyte leaves are several cell layers thick and absorb large amounts of light. Overlapping leaves help rapidly attenuate downwelling irradiance. In mats of benthic algae the one percent light level is typically reached within the top 1–4 mm (Sand-Jensen 1989).

The thickness of the euphotic zone (z_{eu}) is easily computed by setting $I_o = 100\%$ and $I_c = 1\%$ in the equation

$$z_{eu} = \frac{\ln 100}{k_d} = \frac{4.6}{k_d}. \qquad \text{EQ. 10.5}$$

Using highly-colored (humic) Lake Chisholm (AU) and exceptionally transparent Crater Lake (US) as examples, (Fig. 10–8) this equation yields euphotic zone depths (1%) of 0.8 m and 90 m, respectively. However, the true (physiological) compensation depth in Crater Lake appears to lie well below the computed 90 m. Water column chl-*a* maxima in summer were noted

between 120 and 140 m (Larson et al. 1987) and, presuming growth there, point to a physiological compensation depth at the ~0.1% level. One population of moss and a hepatica species were found living at 127 m.

The thickness of the euphotic zone (z_{eu}) is often less than the thickness of the mixed layer (z_{mix}, Sec. 10.9). Exceptions to the rule are not uncommon and include Crater Lake (US) in which the euphotic zone extends into the meta- or hypolimnion (Fig. 10–8). Other such exceptions include small wind-protected lakes that are highly transparent as the result of their location in small, steep catchments that release little organic matter.

When a phytoplankton community is not confined to a particular depth, (as in the metalimnion), but the organisms are instead assumed to circulate through an epilimnion or mixed layer of depth z—spending equal time at each depth—the mean irradiance (I_e) or the **effective light climate** is given by the equation:

$$I_e = \frac{(I_o - I_z)}{\ln (I_o/I_z)}, \qquad \text{EQ. 10.6}$$

where I_o and I_z are the PAR received at the surface and bottom, respectively, of the epilimnion or mixed layer.

▲ 10.8 Light Attenuation and Lake Stratification

Turbid or colored lakes, characterized by high vertical extinction coefficients, absorb and convert to heat not only the infrared (> ~700 nm) and ultraviolet (< ~400 nm) portion of radiation but also the PAR fraction (~400–700 nm) within a short distance of the water surface (Fig. 10–8). In contrast, work on transparent Wisconsin (US) lakes by Birge and Juday (1929), and subsequent work elsewhere, showed that only about half (40–65%) of the incident irradiance (300–3000 nm) was converted to heat in the top 10 cm. The PAR transmitted into deeper water is gradually converted to heat (absorbed) (Hutchinson 1957).

The depth at which PAR is absorbed has important limnological implications beyond those pertaining to photosynthesis and vision. All nontransparent lakes convert virtually all irradiance into heat near the surface. Therefore, such lakes stratify earlier in spring and have colder hypolimnia than transparent lakes of similar size and wind exposure in the same climatic zone. A longer period of stratification increases the

[7] Even though laboratory studies are not strictly part of limnology, and receive less attention here than research in nature, such studies are of great importance in determining not only the limits of what can be expected in nature but also investigating mechanisms and processes under controlled experimental conditions. Laboratory research complements both experimental work on enclosures in nature and observational studies in nature.

Table 10–2 **The order of magnitude of some threshold photon flux densities (PFD, $\mu mol\ m^{-2}\ s^{-1}$) for photosynthesis (PAR), growth, vision, and activity. The order of magnitude conversion between PFD and the units in which the data were originally presented are based on Table 10–1. Full moon = $\sim 1 \times 10^{-3}\ \mu mol\ m^{-2}\ s^{-1}$, allowing fish to feed visually in surface waters at night.**

Organism	Conditions	Flux	Author
Phytoplankton growth, at light saturation	temperate phytoplankton, mostly summer	~50–120	see Harris 1980
Lotic periphyton photosynthesis, at light saturation	laboratory, natural, and artificial rock substrate	100–400	Boston and Hill 1991
macrophytes, long-term survival	nature, north temperate	45–90	Sand-Jensen and Borum 1991
macrophyte growth, light saturation	laboratory, low temperature	~50	Sand-Jensen and Madsen 1991
Phytoplankton biomass, onset spring increase	polar lake, 0.1° (on a 1-hour daylight basis)	~15	Kalff et al. 1972
Algal growth onset	laboratory	0.1–1	see Middelboe and Markager 1997
Photosynthetic sulfur bacteria, growth	meromictic lakes	~0.3	Pfennig 1989
Chaoborus (Insecta) 4th instar, maximum vertical migration, leading edge evening[1]	temperate lake	$\sim 5 \times 10^{-1} - 1 \times 10^{-2}$	see Haney et al. 1990
Rainbow Smelt, maximum aggregation (schooling)	temperate lake	$\sim 4 \times 10^{-1} - 4 \times 10^{-3}$	see Appenzeller and Legget 1995
Daphnia hyalina (Crustacea) onset maximum vertical migration, evening[1]	temperate lake	$\sim 3 - 15 \times 10^{-3}$	Ringelberg et al. 1991
Young Pike, onset of reduced feeding on zooplankton	experimental	$\sim 2 \times 10^{-4}$	see Blaxter 1970
Perch, onset of feeding on *D. hyalina*	temperate lake	$\sim 1.5 \times 10^{-2} - 1.5 \times 10^{-4}$	Ringelberg et al. 1991
Mysis relicta (Crustacea) light avoidance threshold	temperate lake	$\sim 4 \times 10^{-6} - 7 \times 10^{-7}$	Gal et al. 1999

[1]Migration is better correlated with relative change in flux rather than absolute flux

possibility of hypolimnetic anoxia. Turbid or colored lakes will also have a much shallower epilimnion because, with all the solar energy absorbed in the surface layer, such lakes quickly establish a large temperature (density) difference between the warmed surface layer and much colder, denser water below. (See Sections 11.7 and 11.8.) The thickness of the epilimnion affects, among others, the time needed for nonmotile organisms or other particles to traverse the water mass before sinking out of the mixed layer. Thin epilimnia increase the probability of particulate organic matter and associated plant nutrients being lost from the euphotic zone (Sec 20.4), reducing nutrient reycling into the mixed layer.

10.9   The Secchi Disc and Its Utility

The Secchi disc was the first instrument used for providing a measure of the water transparency. The still widely used disc is named after Professor P. A. Secchi who, together with commander Cialdi, was among the first to use it (Table 2–1). The white or black-and-white disc, which is normally 20 cm in diameter,[8] is attached to a marked line and lowered; the depth where it just disappears or reappears provides the measured transparency. Recorded transparencies range from a few cm in the hypereutrophic saline lakes of east Africa (Kalff 1983) and sewage lagoons everywhere, to a growing-season average of about 32–33 m in New Quebec Crater Lake (CA), Waldo Lake (US), and (until recently) Crater Lake (US) and Lake Tahoe (US) (Table 9–3). Annual *maxima* in the last two lakes once extended to ~40 m. These ultratransparent lakes lack a terrestrial catchment of sufficient size to supply appreciable plant nutrients and colored organic matter (Sec. 9.5).

The trophic state of lakes can be characterized by their Secchi disc transparency (Z_{SD}) when the turbidity is largely attributable to phytoplankton rather than inorganic particles or color (Table 10–3). Conversely, the low transparencies ($\leq$ 1.5 m) recorded in many South African reservoirs and their inflowing rivers, or other semiarid regions dominated by shallow, wind-swept lakes and reservoirs, are usually due to high loading of inorganic particles during the rainy season and wind-induced particle resuspension (Walmsley et al. 1980).

The Secchi disc is of great utility for both research and aquatic management because it provides a broad characterization of the transparency of lakes and deeper rivers. The simplicity of its use and its durability make it possible to involve interested laypersons in seasonal and long-term monitering of many more lakes and deeper rivers than would be obtained otherwise, providing abundant data that are very useful in aquatic management. Moreover,

Table 10–3 **Secchi disc transparency and the trophic classification of nonhumic lakes and lowland rivers low in inorganic suspended matter.**

Trophic Class	Transparency (m)	
	mean	min.
Ultra-oligotrophic	$\geq$ 12	$\geq$ 6
Oligotrophic	$\geq$ 6	$\geq$ 3
Mesotrophic	6–3	3–1.5
Eutrophic	3–1.5	1.5–0.7
Hypertrophic	$\leq$ 1.5	$\leq$ 0.7

Source: OECD 1982.

involving private citizens and organizations in the collection of data encourages interest in the functioning of aquatic systems, and through them the wider public is educated. There are, however, limits to every technique. The depth of disappearance of the Secchi disc (Z_{SD}) is often incorrectly thought to bear a fixed relationship with light extinction. The Z_{SD} is as a result commonly, but inappropriately, used to estimate the light extinction coefficient (k_d) and the thickness of the euphotic zone in clear systems, as well as humic and highly turbid systems. The next section explains why there is no fixed relationship between Z_{SD} and light extinction and why the common assumption of a fixed relationship results in major errors in estimating the light climate in aquatic systems.

▲   10.10   Limitations of the Secchi Disc

In the absence of underwater photocell or spectrophotometer measurements, the vertical light extinction coefficient (k_d) is often estimated from the depth at which the Secchi disc disappears from view (Z_{SD}). The conversion factor

$$k_d = \frac{1.7}{Z_{SD}} \qquad \text{EQ. 10.7}$$

was developed by H. H. Poole and W. R. G. Atkins (1929) for relatively transparent marine waters, but was shown to be equally applicable to natural waters low in color, and some highly turbid waters with a Z_{SD} as low as 9 cm (Idso and Gilbert 1974).

The widely used Equation 10.7 (Wetzel 1983) proclaims a systematic relationship between k_d and Z_{SD}.

---

[8]Larger discs (75–100 cm) are used on the great lakes and oceans because a 20 cm disc becomes a mere speck when lowered and is difficult to follow when viewed from the deck of a high vessel. However, in small boats, the size of the disc has little effect on the recorded depth of disappearance in all but the clearest lakes (Hutchinson 1957).

More recent work by Koenings and Edmundson (1991) makes it clear that there is no fixed conversion factor, but rather a continuum of conversion factors, ranging between 0.5 and 3.8. The median value of 1.9 for transparent lakes low in color (< 10 mg Pt l^{-1}) and turbidity (< 5 NTU) is indeed close to the long-used conversion factor of 1.7 (Eq. 10.7) that was apparently developed on lakes of this type. But a median conversion factor for systems in which Z_{SD} is primarily determined by color or turbidity are very different at 3.0 and 1.3, respectively (Table 10–4). The commonly assumed fixed conversion factor of ~1.7 then greatly underestimates light extinction in humic waters and overestimates it in turbid ones. There are further ramifications. Estimates of the thickness of the euphotic zone (z_{eu}), derived from a Secchi disc-estimated k_d rather than a k_d obtained directly from light meter measurements, will produce large errors when subsequently used to estimate the thickness of the euphotic zone of turbid and colored waters (Eq. 10.5 and Table 10–4). Unfortunately, Secchi disc derived estimates of z_{eu} based on a single conversion factor are common in the literature and help confound reported relationships between z_{eu} and phytoplankton or macrophyte photosynthesis, which are based on a combination of direct and Secchi-based determinations (Sec. 10.11).

Work on Lake Zurich (CH)—a lake low in watercolor, but varying seasonally in turbidity (algal biomass)—provides a good demonstration of the changing and nonlinear relationship between Z_{SD}, k_d, and z_{eu} (Fig. 10–9); as well as between Z_{SD} and the fraction of incoming irradiance (PAR) remaining at Z_{SD} (Fig. 10–10). The fraction of surface irradiance (I_o) remaining at the depth of disappearance of the Secchi disc (I_{SD}) is often—but also incorrectly—assumed to be around 10 percent (Wetzel 1983), a value that is indeed appropriate for the nonhumic lakes for which the generality was developed (Table 10–5). Assuming

Lake Zurich to be a good example of such a nonhumic lake, the 10 percent value appears to be based on research on primarily mesotrophic lakes of low color with Z_{SD} = ~4 m (Fig. 10–10). The earlier suggestion by R. A. Vollenweider that roughly 15 percent of the surface light remains at Z_{SD} suggests that he worked on slightly more turbid European lakes with summer Z_{SD} = ~3 m (Fig. 10–10). If the research forming the basis for these incorrect generalities had been carried out on much more turbid lakes with Z_{SD} = 2 m, the Lake Zurich data suggests that the fraction of I_o remaining at Z_{SD} would have been accepted as close to 30 percent (see also Table 10–5).

The Secchi disc is a valuable tool for broadly characterizing the transparency of individual systems seasonably and over a series of years, and for making broad comparisons among groups of lakes and deeper rivers. The median conversion factors presented for clear, turbid, and colored lakes (Table 10–4) may even be useful for exploratory research and comparisons of the nutrient status of groups of lakes that differ primarily in an algae-produced turbidity. But fixed conversion factors should not be used to estimate light extinction and the thickness of the euphotic zone. Every technique has its limitations as well as its uses.

▲ 10.11 Light and Primary Production

It is evident to limnologists that the phytoplankton biomass of large wind-exposed deep lakes, such as the Laurentian or African great lakes, does not respond as strongly to a particular nutrient increase as small lakes. The smaller response of transparent inland seas is partially the result of their very thick epilimnia (Sec. 11.7), which provide much room for light scattering and light absorption by the modest quantities of colored

Table 10–4 **Comparison of $k_d \times Z_{SD}$, z_{eu}:Z_{SD}, I_{SD}:I_o, and R$_f$, the ratio of upwelling to downwelling PAR (%), median values from humic (> 10 Pt l^{-1}, < ~5 natural turbidity units, NTU), clear (< 10 Pt l^{-1}, < ~5 NTU) and turbid waterbodies (< 10 Pt l^{-1}, > 5 NTU) located throughout the world. The ratio of turbidity to color was determined for Alaskan lakes.**

Parameter	Median Conversion Factor		
	Humic	Clear	Turbid
$k_d \times Z_{SD}$	3.0	1.9	1.3
z_{eu}:Z_{SD}	1.3	2.4	3.3
I_{SD}:I_o (%)	3.2	10.4	22.6
R$_f$ (%)	1.4	4.6	33.7
Turbidity (× 10): color	0.4	1.4	16.2

Source: After Koenings and Edmundson 1991.

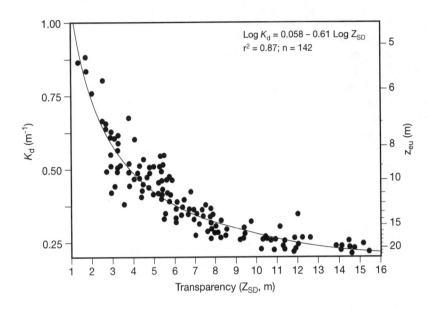

Figure 10–9 Relationship between the transparency (Z_{SD}, m) measured with an underwater viewer, the vertical extinction coeficient (k_d) and the thickness of the euphotic zone (z_{eu} = 1% I_o), computed based on the equation z_{eu} = 4.6/k_d, for nonhumic, but mesotrophic, Lake Zurich (CH) during the period 1979–1981. *(Modifed after Schanz 1982.)*

DOC present. The same scattering and absorption possibilities are reduced in the thinner epilimnia of wind-protected small lakes of equal transparency. The PAR available for photosynthesis within the water column, the effective light climate, (Eq. 10.6) therefore improves with decreasing lake size (Fig. 10–11).

Light extinction among lakes and large slow-flowing rivers increases with increasing algal biomass (Fig. 10–8), inorganic turbidity, and watercolor; (Fig. 10–5) resulting in a thinner euphotic zone (z_{eu}), declining ratio of euphotic depth (z_{eu}) to mixing depth (z_{mix}), and an increasing fraction of the mixed layer located below the euphotic zone (Fig. 10–12). The phytoplankton are moved (Chapter 12) through the mixed layer by currents and consequently spend a larger fraction of their time in the aphotic zone in low z_{eu}:z_{mix}

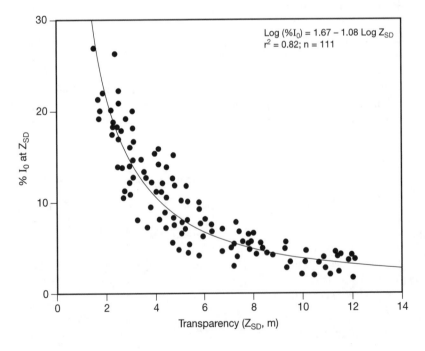

Figure 10–10 The relationship between the transparency (Z_{SD}, m) and the fraction of irradiance (PAR) remaining at the Secchi disc depth (% I_o at Z_{SD}) in nonhumic, mesotrophic Lake Zurich (CH) during 1979–1981. *(Modified after Schanz 1982.)*

Table 10–5   **The average growing-season euphotic zone depth (z$_{cu}$) to Secchi depth (Z$_{SD}$) ratio, and percent irradiance remaining at the Secchi depth for selected lakes. See also Fig. 10–10.**

Lake	z$_{eu}$:z$_{SD}$	% Irradiance at Z$_{SD}$
Torneträsk (SE)	1.3	2.5
Zurich (CH)	1.6	8.0
Constance (AT, CH, DE)	1.7	7.0
Léman (FR, CH)	1.9	10.0
di Garda (IT)	2.1	20.0
Mariut (EG)	2.2	10.0
Nine Japanese lakes	2.4	15.0
Finstertaler (AT)	2.5	16.7
Chad (CD, CM, NE, NG)	2.7	17.0
Kariba (MZ, ZW)	3.6	22.7
Tarfala (IT)	3.8	26.0
Neusiedler/Fertö (AT, HU)	4.6	35.7

Source: Modified after Dokulil 1979.

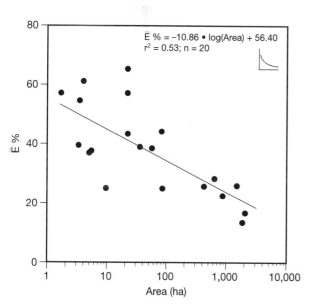

Figure 10–11   Mean irradiance in the mixed layer (*effective light climate*, Sec. 10.6) as a percent of incident irradiation (E%) and area (ha) for 20 Minnesota (US) lakes. Note (1) the decline in the effective light climate with increasing lake area (fetch) and, presumably, an increase in the thickness of the mixed layer (epilimnion), allowing more light absorption and scattering than in thin mixed layers; and (2) the data scatter is attributable to variation in color and/or turbidity among lakes of similar size. (*Modified after Sterner 1990.*)

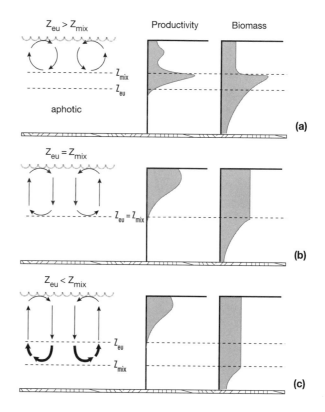

Figure 10–12   Diagrammatic presentation of the influence of euphotic zone depth (z$_{eu}$) and mixed-layer depth (z$_{mix}$) on the vertical distribution of phytoplankton productivity and biomass. (*Modified from Thornton et al, 1990.*)

ratio lakes, where respiration is not offset by photosynthesis. The $z_{eu}:z_{mix}$ ratio, with its effect on the light climate experienced by phytoplankton, exhibits large short-term fluctuations at scales of minutes to days in single lakes (Fig. 10–13). Large changes in $z_{eu}:z_{mix}$ (the light climate) are the result of both temporal changes in incoming irradiance plus changes in water column stability resulting from differences in heat input and wind strength (Chapter 12). Changes in light climate are reflected in variability of photosynthetic rates among the individual algal species and, ultimately, in species composition (Sephton and Harris 1984, Steinberg and Hartmann 1987).

The biggest difficulty in determining the **critical mixing depth**, the depth of mixing at which the respiration (carbon ion) over a 24 hour period of a depth-integrated phytoplankton community equals the photosynthetic rate (carbon gain), is determining the phytoplankton respiration. The respiration rate measured in water samples includes the respiration of microbes, protozoans, and other zooplankton. However, by assuming that the phytoplankton community respiration is 10 percent of the photosynthetic rate at the depth of maximum photosynthesis, Cloern (1987) estimated that the net phytoplankton of turbid estuaries could be expected to be light limited when the $z_{eu}:z_{mix}$ ratio declines below 0.2–0.5. Talling (1971) and Grobbelaar (1985) had earlier reported similar critical $z_{eu}:z_{mix}$ ratios of 0.20 (1:5) in a few English lakes and 0.18 (1:6) in a subtropical South African reservoir. These fresh water ratios imply that net photosynthesis declines to zero where the aphotic zone is between five and six times thicker than the euphotic zone. Even so, the critical mixing depth value is expected to vary among lakes with algal type and physiological condition, wind exposure (turbulence), transparency, and climate. Nevertheless, it is evident that even high primary production rates in the euphotic zone may be insufficient to offset respiration of the phytoplankton community in the mixed layer if the z_{mix} is disproportionately large. Therefore, lakes with a deep mixed layer (e.g., the Laurentian or African great lakes and Sec. 11.7) or those with a particularly low transparency (e.g., humic or turbid lakes) respond less to nutrient additions (eutrophication) than do shallow unstratified lakes or transparent lakes with a high $z_{eu}:z_{mix}$ ratio (Reynolds 1984a). The phytoplankton in deeply mixed, but also colored lakes—such as Loch Ness in Scotland, GB—experience a sufficiently unfavorable light climate to be primarily light, rather than nutrient limited even when stratified (Jones et al. 1996).

The onset of temperature stratification brings about a sudden reduction in the z_{mix} and thereby increases the $z_{eu}:z_{mix}$ ratio and thus the effective light climate (I_e, Eq. 10.6) experienced by the phytoplankton community. Phytoplankton growth initiation has been noted in both lakes and the sea when the I_e exceeds ~7 mol photons m^{-2} d^{-1} of PAR (Carignan and Planas 1994).

Autumnal algal blooms are uncommon in deeper stratified lakes at higher latitudes because of rapid autumnal decline in received daily irradiance and thus in the thickness of the euphotic zone, resulting in a rapid decline in the $z_{eu}:z_{mix}$ ratio and the I_e. An autumnal increase in the z_{mix} during destratification further decreases the $z_{eu}:z_{mix}$ ratio and the probability of autumn blooms in deep lakes at higher latitudes.

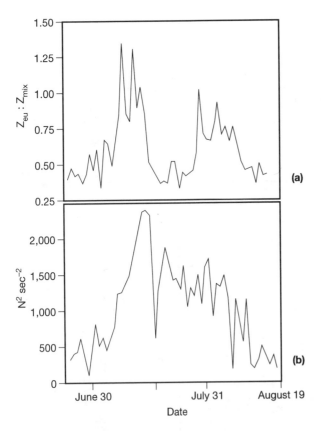

Figure 10–13 Daily changes (a) in the $z_{eu}:z_{mix}$ ratio and (b) in N^2 sec^{-2}, a measure of water column stability (Sec. 12.4), in Hamilton Harbor (Lake Ontario, CA) during the summer of 1979. Note periods of low stability and resulting deepening of the mixed layer (Z_{mix}) are reflected in an unfavorable photosynthesis $z_{eu}:z_{mix}$ ratio. (*After Sephton and Harris 1984.*)

▲ 10.12   Underwater Vision

The instantaneous photon flux densities needed for migrating zooplankton to respond to daylight or for fish to perceive their prey are one or more orders of magnitude lower than the flux required for phytoplankton to offset their respiration at low temperature (Table 10–2). Laboratory experiments with planktivorous fish show that their ability to perceive (larger) freshwater zooplankton prey, known as the **reaction distance**, falls off rapidly below about 0.6 μmol m^{-2} s^{-1} (0.147 W m^{-2}) (Vinyard and O'Brien 1976, Confer et al. 1978). The critical flux is equivalent to only about 0.04 percent of PAR entering a lake around noon in midsummer, but with the critical flux higher for smaller prey.

The critical depth (z_p) to which freshwater fish and other visual predators can effectively search for macrozooplankton can be estimated in the same way as the z_{eu} (one percent light level) is determined (Eq. 10.5) assuming that laboratory determined reaction distances are also relevant to other species in nature:

$$z_p = \frac{7.8}{k}.$$ EQ. 10.8

Kitchell and Kitchell (1980) hypothesized that if fish predation is indeed constrained by light (and DO), such predation should no longer affect the diurnal zooplankton distribution at a depth where the light flux is lower than the required reaction distance of the fish. They confirmed this theory for *Daphnia pulex* in Riley Lake (Wisconsin, US); a high proportion of small *Daphnia*, not subject to predation, was present above the critical depth whereas the largest number of big individuals was found in deep water, apparently beyond the reach of the fish.

Highlights

- About three percent of the energy arriving at the earth's atmosphere is composed of ultraviolet radiation (UV, 100–400 nm), with the remainder about equally split between visible, or photosynthetically available radiation (PAR, ~400–700 nm) and infrared radiation (IR, ~700–3000 nm).

- Photons in the blue and green portions of the PAR spectrum are rapidly lost (attenuated) in colored or turbid waters while photons in the red portion of the spectrum are best transmitted there (least attenuated). The reverse is true in highly transparent waters further characterized by a relatively low UV extinction coefficient.

- High-energy UV radiation photons have a negative effect on the biota, unless offset by protective pigments or an ability to hide. Plants and animals vary widely in their sensitivity to UV radiation and as indirect effects on the biota include photobleaching and photooxidation of organic molecules, its effects are difficult to interpret.

- Turbid and colored lakes, characterized by a high vertical extinction coefficient, convert to heat not only the IR but also the UV and PAR close to the surface causing them to stratify earlier in spring than transparent lakes.

- The depth where the Secchi disc disappears from view provides a useful measure of transparency and is widely used in aquatic management to characterize the trophic status of waters where algae, rather than color or inorganic particles, are primarily responsible for the observed transparency.

- ▲ In turbid or colored waters where the thickness of the euphotic zone (z_{eu}) is much less than the thickness of the mixed layer (z_{mix}), phytoplankton spend considerable time, while circulating through the mixed layer, in the aphotic zone with negative consequences for water column-integrated production. Where $z_{eu} \ll z_{mix}$, phytoplankton primary production is primarily light, rather than nutrient, limited.

- ▲ The photon flux densities needed for vertically migrating zooplankton to respond to daylight, or fish to perceive their prey are one or more orders of magnitude smaller than the flux needed for algal growth.

11

Temperature Cycles, Lake Stratification, and Heat Budgets

11.1   Introduction

The temperature and stratification regime of lakes has interested limnologists since its beginning. Early temperature measurements showed not only strong seasonal variation but also a seasonal pattern of stratification that allowed lake types to be distinguished and classified. This interest in temperature reflects much more than just an interest in classification and order, it also shows recognition of the importance of temperature and season in determining the structure of biotic communities and productivity of aquatic systems.

Stratification separates water into three zones in freshwater lakes (where permitted by climatic conditions and depth). The deep water zone, the **hypolimnion**, is largely nonturbulent and separated from contact with the atmosphere. It is dominated by respiratory processes that utilize organic matter derived from the surface layer, the **epilimnion** (Fig. 1–5). The epilimnion is in contact with the atmosphere and is frequently turbulent. This is also the zone where primary production normally dominates respiration, providing for much of the energy needs of the heterotrophic animals and microbes throughout the system. The transition zone between the surface and deep layers is the **metalimnion**. It is characterized by a temperature gradient and is commonly referred to as the **thermocline**. Formation of the metalimnion and the restrictions it imposes on nutrient and gas circulation, as well as in preventing the phytoplankton from spending considerable time in the aphotic zone, makes

stratification the single most important physical event for the biota (Sec. 10.11).

An unstratified lake becomes stratified when wind-induced currents are unable to mix the solar energy received at the lake surface throughout the water mass, thereby preventing the system from maintaining a uniform water temperature. A lake stratifies when surface warming increases the temperature difference and resulting density difference (Fig. 3-2) and Chapter 12) to the point where resistance to mixing becomes greater than the mixing power of wind-imparted turbulence (Chapter 12). At that point the epilimnion, hypolimnion, and metalimnion become established. During the temperate zone autumn or the cool season in the tropics, this process is reversed and surface cooling reduces the density difference until resistance to mixing becomes smaller than the mixing energy imparted by currents and waves. At that point the whole lake volume mixes once again. The mixed portion of the water column is also known as a **mixed layer** or **surface layer**, which during stratification is restricted to the epilimnion or a portion thereof.

A successive series of lake classification schemes were based on the frequency with which lakes experience periods of stratification followed by periods of mixing, known as **overturn** or **circulation periods**. The only other widely accepted classification scheme was developed by E. Naumann (1919) and described lakes according to supply or concentration of plant nutrients. These two classification schemes have probably persisted because they are based on two of the most fundamental characteristics of lakes—nutrition and seasonal variation—however, they are simplifications. In reality, lakes lie along gradients of nutrient concentration, morphometry, and duration of stratification rather than being readily catagorized into a few distinct groupings.

11.2 Types of Stratification and Mixing

Francois Forel in his 1892 monograph on L. Léman (Geneva) proposed a classification scheme based on water temperature that recognized three types of lakes: (1) temperate lakes, which show a period of summer—and inverse winter—stratification, separated by two periods of mixing at the temperature of maximum density when the water temperature is the same everywhere; (2) tropical lakes, characterized by one period each of stratification and mixing, with the temperature never declining below 3.94°C; and (3) polar lakes, exhibiting an inverse temperature stratification except for a period of summer mixing and having a water temperature never higher than 3.94°C. His scheme was quickly found to be unsatisfactory, because it would classify many temperate lakes as tropical and create many polar lakes in nonpolar regions. Forel's scheme was changed and elaborated by G. C. Whipple (1898) and S. Yoshimura (1936), among others, and was greatly revised by G. E. Hutchinson and H. Loffler (1956). It was most recently modified by Lewis (1983), whose scheme is presented below.

A. Amictic Lakes

Amictic lakes (Fig. 11–1) are permanently covered by ice. However, with the exception of a very few proglacial polar lakes (Sec. 6.2) at the edge of the permanent Greenland and Antarctic ice caps, all other amictic lakes develop at least a summer band of open water or *moat* along the shore even though the center of antarctic lakes remains covered by up to 3.5 to 4.5 meters of ice. The moat is produced by heat flow from the warmer snowless shores, by warm subsurface and stream runoff of dilute meltwater from glaciers, and by the absorption of solar energy by shallow-water sediments when lakes are free from snow and light energy is able to penetrate the ice cover, heating the water and shallow sediments below (e.g., Welch and Bergmann 1985).

Light penetration in the open water (pelagic) zone of lakes generates heat below the ice, and creates density currents (Chapter 12) that allow some mixing even though the lakes remain mostly or totally covered with ice. In addition, inflowing streams impart not only heat but also turbulence and density currents during the summer. Consequently, the term amictic (not mixed) is a misnomer even though it is true that

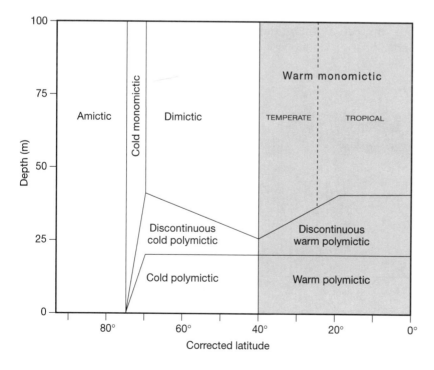

Figure 11–1 Estimated distribution of eight lake types in relation to latitude (corrected for elevation) and water depth. Each 200 m increment in elevation is equivalent to about 1°C between 40° and 50°N and S, and 0.6°C from 0° to 20°N and S. Note that some continental lakes are dimictic in the temperate zone, but may be monomictic in regions dominated by high winds and the tempering influence of nearby oceans. *(Modified after Lewis 1983.)*

the influence of wind in mixing is very much smaller in amictic lakes than in lakes that become totally ice free. Reduced mixing restricts the degree of vertical mixing of the water column, increases the probability of incomplete mixing, and increases the chance for anoxic conditions and linked biogeochemical processes in both profundal water and sediments. The best known amictic lakes are Vanda and Bonny, AQ. Others, located at slightly lower latitudes in both hemispheres are amictic during cloudy summers, but become ice free most years (e.g., Char Lake, CA), at which time they are characterized as cold monomictic.

B. Cold Monomictic Lakes

These high latitude lakes (Fig. 11–1) are covered with ice most of the year, but become ice-free during the summer. Temperatures then do not rise much above 4°C, with wind-induced turbulence sufficient to allow the lakes to remain unstratified (Fig. 11–2). These lakes are characterized by a single (mono) period of mixing (mixis). Examples include Char and Meretta Lakes in Canada (Schindler et al. 1974), Schrader Lake in Alaska (Hobbie 1961), and Latnjajaure Lake in Swedish Lapland (Nauwerck 1978).

C. Cold Polymictic Lakes

The polymictic (multiple mixing) lakes are either relatively shallow or very wind-exposed and ice-covered part of the year, but ice-free during the summer. Shallower lakes (< ~20 m) weakly stratify on sunny days, but turn over at night, whereas deeper ones stratify for periods of days to weeks during the summer. Examples include the vast number of shallow polar ponds (Kalff 1967), subarctic and subantarctic lakes (Light et al. 1981), as well as large numbers of shallow and/or wind-exposed lowland and mountain lakes and ponds in north temperate zones, plus a more modest number on South Island (NZ) and the southern portion of South America. The deeper lakes that stratify for weeks form a transition group between the polymictic and dimictic categories (see E. below) designated by Lewis (1983) as **discontinuous polymictic**.

D. Warm Polymictic Lakes

These shallow or relatively shallow lakes are distinguished from their cold polymictic counterparts by their lack of an ice cover at any time. The deeper ones are highly wind exposed and/or located in particularly windy regions that, with more wind protection, would be warm monomictic (e.g., NZ, see Lewis 1983). The period of stratification ranges from daily to a few days in shallow tropical lakes with a small heat storage capacity, subject to regular periods of relatively high daytime winds and/or nighttime cooling. Detailed work on a Brazilian floodplain lake showed that daily destratification ended and a seasonal thermocline developed when the water depth exceeded 3 m (MacIntyre and Melack 1988). Lakes that become stratified for periods of days to weeks are designated as

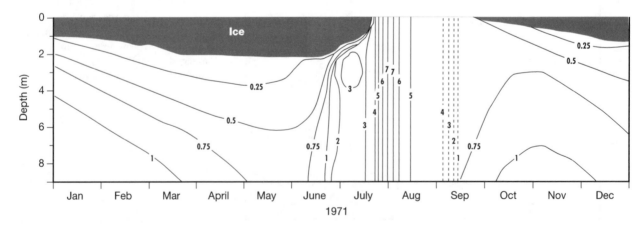

Figure 11–2   Temperature isopleths for Meretta Lake (CA), 1971. *(After Schindler et al. 1974.)*

Table 11–1 **Regression equations predicting the number of ice-free days (IFD) from mean air temperature (TEMP), and/or mean depth ($\bar{z}$) are based on 59 North American low altitude lakes from latitude 41°N to 75°N.**

Equation	R^2
ln IFD = 0.0624 (TEMP) + 5.168	0.88
ln IFD = 0.153 (ln $\bar{z}$) + 5.056	0.13
ln IFD = 0.06 (TEMP) + 0.073 (ln $\bar{z}$) + 5.005	0.90
ln IFD = 0.0615 (TEMP) + 0.0318 (ln fetch) + 5.129	0.89

Source: From Shuter et al. 1983.

discontinuous warm polymictic. This category grades into the warm monomictic type (see F. below).

E. Dimictic Lakes

This lake type has an ice cover during part of the year and is stably stratified another portion of the year, with two periods each year (di yearly) of mixing at the transitions. The number of days, on average, dimictic lakes are expected to be ice-free (IFD) is predictable from the mean annual air temperature (Table 11–1). It is evident that the annual mean air temperature has to be below 8-10°C for an ice cover to develop (Fig. 11–3). At higher air temperatures lakes lack an ice cover and then have only one mixing period (au-

tumn to spring) and are classified as *warm monomictic* (see F. below).

It is evident from Table 11–1 that on a *continental scale* air temperature is the best determinant for predicting if ice cover should be expected on lakes at low altitudes. Morphometry (fetch and mean depth) plays a lesser role in determining the stratification regime at a continental scale. This does not mean morphometry is unimportant, but only that mean air temperature alone encompasses nearly all the variation explained statistically over the very large temperature range encountered at the spatial scale of a continent. However, within a *single climatic zone* over which differences in air temperature are relatively modest, major differences in lake volume (or its surrogates, mean depth and maximum depth) exhibit the largest variation and become the single best predictor of the stratification regime. But volume variation is too small to be an important predictor of the among-lake stratification regime for lakes of similar volume within a climatic zone. Among such lakes other predictor variables emerge and dominate, such as nutrient or DOM loading which affect where in the water column photosynthetically available radiation (PAR, ~400–700nm; Chapter 10) is absorbed and converted to heat. At the even smaller spatial scale of *individual lakes*, interyear differences in weather (temperature, runoff), or abundance of zooplankton filtering particulate matter (seston) in free water, or size distribution of phytoplankton will become the dominant factor influencing

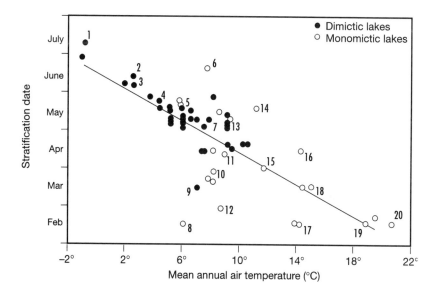

Figure 11–3 Relationship between mean annual air temperature and date of onset of spring stratification for 70 North temperate zone lakes. Some points are hidden. Identified lakes are: 1) Baikal, RU; 2) Paijanne, FL; 3) St-Jean, CA; 4) Shagawa, US; 5) Vanern, SE; 6) Michigan, US; 7) Stechlin, DE; 8) Ikeda, JP; 9) Lunzer Untersee, AT; 10) Lomond, GB; 11) Windermere, GB; 12) Zurich, CH; 13) Mendota, US; 14) Washington, US; 15) Maggiore and Lugano, IT; 16) Shinji, JP; 17) Sagami and Okutama Reservoirs, JP; 18) Biwa, JP; 19) Kinneret, IL; 20) Phewa, NP. Date = 162 − 6.05 Temp. r^2 = 0.60; p < 0.01; n = 70. (*After Demers and Kalff 1993.*)

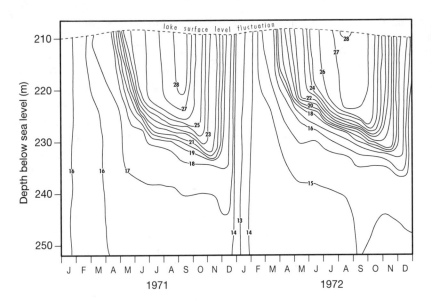

Figure 11–4 Thermal regime of Lake Kinneret (IL, JO) in 1971 and 1972. The upper layer represents the actual water level. The lake is located well below sea level. (*After Serruya and Leventer 1984.*)

the stratification regime. The importance of spatial scale in determining the depth of the thermocline or mixed-layer depth is discussed in Section 11.7.

F. Warm Monomictic Lakes

Deeper temperate and tropical lakes that lack ice cover, but have an extensive single stable stratification period during the warmest part of the year while mixing the rest of the time, are known as warm monomictic (Fig. 11–4). This lake type is widespread in southern Europe, the windy British Isles, and those parts of the temperate zone where the mean air temperature remains > ~10°C (Fig. 11–3). It also encompasses subtropical and tropical lakes deep enough to allow stable stratification for a significant period. But as maximum and minimum summer water temperatures decline with increasing latitude and altitude, temperature differences between the epilimnia and hypolimnia become reduced (Fig. 11–5), the stratification period shortens, and polymictic conditions develop.

Table 11–2 **Selected models predicting the planar thermocline depth (z_t) and the mixing depth (z_{mix}) to the top of the thermocline, in meters. ML = maximum length (km); MEL = maximum effective length (km); MEW = maximum effective width (km); F = fetch; A = area (km^2)**

Region	Model	r^2	n
Worldwide	$z_t = \ln\left(\dfrac{A^{0.5}}{0.043}\right)^{2.35}$	0.66	150
North America	$z_t = 0.298 \cdot \ln(MEL) + 1.82$	0.66	73
Cameroon	$z_t = 9.94\,(F)^{0.300}$	0.83	52
Argentina	$z_t = 23.68 + 1.60\,(A)^{0.5}$	0.81	26
Japan	$z_{mix} = 6.22\,(A)^{0.152}$	0.53	36
New Zealand	$z_{mix} = 7.00\,(MEL)^{0.42}$	0.79	33
Poland & Canada	$z_{mix} = 4.6\left(\dfrac{MEL + MEW}{2}\right)^{0.41}$	0.85	88

Source: Modified from Hanna 1990, Kling 1988, and Baigun and Marinone 1995. ($z_t = z_{mix} - 2.4m$ in data from Hanna 1990.)

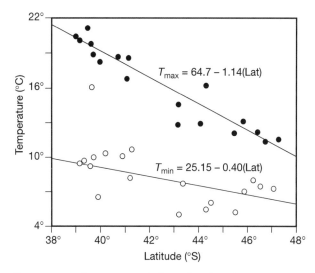

Figure 11–5 Variation with latitude of surface (● = T_{max}) and deep water (○ = T_{min}) temperatures of deep (z_{max} > 90 m) Chilean lakes. Note the difference between epilimnetic and hypolimnetic temperatures and how the stability and thus the duration of stratification decreases with increasing latitude. *(After Geller 1992.)*

▲ ## 11.3 Morphometry and Stratification

The distribution pattern of lakes presented in Fig. 11–1 is based on latitude (climate) and depth. Lakes at higher altitudes are incorporated by increasing, for classification purposes only, their latitude by 0.6–1.0°C for each 200 m of altitude (adjusted latitude, Fig. 11–1). The depth component of Figure 11–1 allows separation between shallow polymictic lakes on the one hand and deeper dimictic and warm monomictic lakes on the other. The latter two types have sufficient depth (volume) to absorb the heat needed (Sec. 11.11) for stable stratification (Table 11–2). Conversely, shallow polymictic lakes lack the required heat storage capacity (volume), to maintain stratification for 24 hours. Deeper ones may be stratified for periods of days to weeks and can be categorized as either discontinuous warm or cold polymictic. To maintain the desired simplicity, the generalized classification scheme (Fig. 11–1) does not consider exposure to winds (lake area or catchment relief) and their role in stratification. However, it is apparent that lake surface area (km²) or the distance wind can blow unimpeded over the water (**fetch**, km) (Chapter 12) has a significant bearing on whether a lake at a particular latitude will stably stratify.

Small, wind-protected temperate lakes stratify when the maximum depth exceeds a mere 3 m, with the minimum depth required rising to about 20 m for lakes of around 20 km² (2,000 ha) (Fig. 11–6). Larger, deeper lakes at higher latitudes, with sufficient volume to have a large heat absorbing capacity, may not be able to absorb sufficient heat (Sec. 11.12) after ice-out to stably stratify and will thus be considered cold monomictic, whereas slightly smaller lakes nearby are dimictic. Local climate also has an effect on stratification. Lakes in particularly windy regions of the tem-

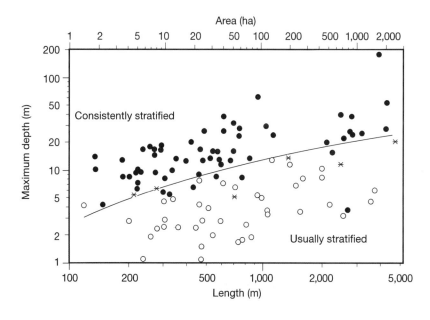

Figure 11–6 The relationship between lake size (ha) and the maximum depth required for Ontario (CA), Polish, and Minnesota (US) lakes at low altitudes to stratify. Closed and open circles represent stratified and usually unstratified lakes, respectively. Asterisks represent Minnesota lakes that stratify in some years. The stratification boundary was determined by eye. *(After Gorham and Boyce 1989.)*

perate zone (e.g., Scotland, Argentina, New Zealand, and South Island) are normally warm monomictic, but they would be dimictic in less windy regions of the same climatic zone. A mean annual air temperature of ~10°C separates most monomictic lakes from their dimictic counterparts (Fig. 11–3).

There are lakes that fit one category in some years and another during particularly warm or cold years. Some large lakes in the subarctic have multiple basins composed of a deep cold monomictic basin adjacent to a shallow polymictic basin, with shallower and more readily heated arms that are dimictic (Vincent and Hobbie 2000). Differential heating of the basins results in sharp horizontal temperature gradients called **thermal bars** (Sec. 11.4) that hinder the exchange of water and materials between basins. As mentioned earlier, arctic Char Lake is cold monomictic during sunny summers but amictic during cloudy ones. Dutch and Maryland (US) lakes deep enough to stably stratify are warm monomictic in most years, but dimictic during years with cold winters. Nevertheless, most lakes fit one category or another, and their mixing designation indicates the probable presence or absence of a hypolimnion and the potential development of an anoxic hypolimnion.

11.4   Seasonal Temperature Cycles and Stratification

When dimictic lakes become ice-free, the water temperature is close to the temperature of maximum density (3.94°C) and the lake is of uniform temperature (**isothermal** or **homothermal**). During the following days and weeks the temperature rises rapidly because the incoming heat is no longer needed to melt the ice. More than half the incoming solar energy is typically absorbed in the upper meter of water (Sec. 10.8) and the temperature profile should show an exponential decline with depth except for the mixing power of the wind. It provides the kinetic energy to create the turbulence needed (Chapter 12) to mix the incoming heat throughout the water column and keep the lake isothermal. Cool spring nights permit surface cooling and sinking of the newly cooled, denser surface water, creating *convection* or *density currents* (Chapter 12) that contribute to mixing and continuation of the mixing period (overturn), especially in shallow lakes. However, as the water temperature continues to rise, the difference in density between the daytime warmed surface water and deeper water becomes progressively

more difficult to overcome by wind- and density-induced turbulence (Fig. 3–2). Mixing of the whole water column usually ends following several sunny days with little wind, increasing the temperature difference and resistance to mixing sufficiently to prevent continued total mixing.

Stratification

The date by which dimictic and temperate zone warm monomictic lakes can expect to stably stratify is a function of air temperature (Fig. 11–3). A better prediction is possible by also taking into account the surface area to mean depth ratio, making this ratio a surrogate for the impact of wind on lakes and heat absorption capacity (Demers and Kalff 1993). Once a lake becomes stably stratified, the depth of the hypolimnion will change relatively little until destratification commences. However, even stably stratified large lakes show some thickening of the epilimnion as the result of summer storms eroding the surface of their metalimnia. This effect is greatest in less stably stratified lakes, in which the temperature (density) differences between the epilimnia and hypolimnia are small. For example, in weakly stratified Lake Erken (SE) the hypolimnion temperature rises several degrees during the stratification period and varies considerably between years (Fig. 11–7). Irregular step-like

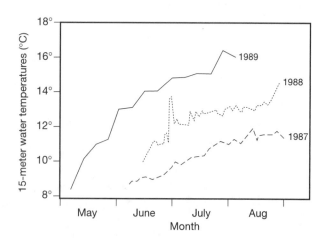

Figure 11–7   Seasonal variations in the hypolimnetic temperature at 15 m in Lake Erken, SE (LA = 22.9 km², z̄ = 9 m). The large increases in hypolimnetic temperature during stratification are the result of the intrusion (entrainment) of warmed epilimnetic water into the hypolimnion of the weakly stratified lake (compare with Fig. 11–8) following storms. (*After Pierson et al. 1992.*)

rises in the temperature profiles point to large mixing events, while more gradual rises appear to be the outcome of a large number of smaller events.

After temperature stratification becomes established, the epilimnion or surface layer continues to be subject to mixing by wind-induced turbulence. The turbulence is enhanced by convection currents established during cooler nights as well as during cloudy and cool days. The metalimnion, the transition zone separating the epilimnion from the hypolimnion, exhibits not only a temperature gradient (thermocline) but also, in strongly stratified lakes, a large density gradient and associated large resistance to mixing (Fig. 11–8). The **planar thermocline** depth, defined as the depth where the temperature gradient is maximal (Hutchinson 1957), has been argued to be easier to identify than the upper surface of the thermocline as used by Birge and others.[1]

The depth of the thermocline within lakes changes seasonally, particularly rapidly during storms, when nutrient inputs into the epilimnion from the hypolimnion can be appreciable (Fig. 11–9) as the result of thermocline erosion. The average depth of the thermocline increases with increasing exposure to winds (wind stress) at the among-lake scale, but the rate of increase in thermocline depth declines as the fetch increases. However, empirical regression models developed in one region to predict among-lake thermocline depths will not be applicable where water temperature and wind speed differ substantially. Nor should measurements on small and medium-sized lakes be extrapolated to much larger (or smaller) lakes not included in the development of the original model. Nor can empirical models developed in one country or region with a particular temperature and wind regime be assumed to make equally good predictions elsewhere. Thus, the empirical stratification models in use (Table 11–2), based on smaller lakes overestimate the thermocline depth of large temperate lakes that have a maximum length of greater than about 25 km. The Laurentian Great Lakes (CA, US)

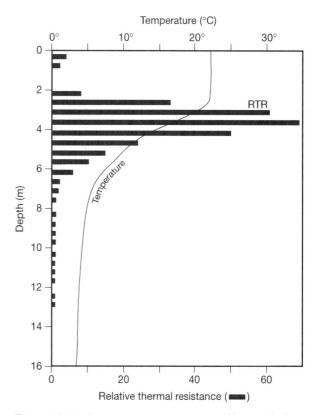

Figure 11–8 A summer temperature profile (single line) and relative thermal resistance (RTR) to mixing (bars) for Little Round Lake, Ontario (CA). The RTR is defined by the density difference at adjacent depths compared to the density difference between 4° and 5°C. Note the largest RTRs are found at the depth of maximum temperature change. *(After Vallentyne 1957.)*

have midsummer thermocline depths of 20 m to 25 m, whereas extrapolation of models developed for smaller lakes suggest that thermocline depths should be expected at 35 m to 40 m. Even so, the regression models based on small- and medium-sized lakes were useful in identifying very large lakes as "outliers," thereby raising questions as to why they might be so.

Fall Overturn

Temperate zone lakes start losing heat in late summer when the daily mean air temperature declines below that of the epilimnion. During the onset of destratification (overturn) the density difference between the cooled epilimnion and the upper metalimnion declines below the kinetic energy imparted by the wind, resulting in thermocline erosion (see below). Epi-

---

[1] In the late-19th century, E. A. Birge defined the thermocline as the depth interval over which the temperature changes more than 1° per meter. This definition is inappropriate for those large temperate lakes in which the gradient is so long and gradual that the > 1° m⁻¹ limit is never attained. It is equally inappropriate for tropical lakes that may have a temperature difference between the epilimnion and hypolimnion so small that a > 1° m⁻¹ is never reached even though the lakes are stably stratified. In them, the thermocline has sometimes been defined as the depth interval over which the temperature changes more than 0.1 or 0.2° m⁻¹.

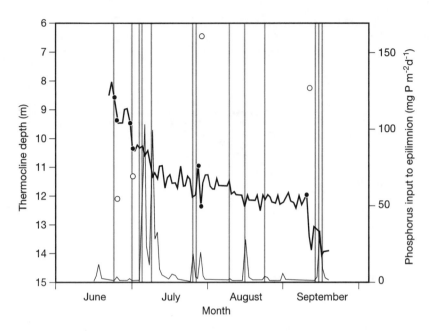

Figure 11–9 Daily thermocline depth (thick line) and days of thermocline migrations ≥ 1 m d^{-1} (solid circles), phosphorus inputs from the catchment (thin line) and by entrainment from the hypolimnion (open circles, calculated for days in 1993 on which the epilimnion deepened by more than a meter) in Lake Mendota, US (LA = 39.9 km^2, $\bar{z}$ = 12.7 m). Days classified as storms are shown as vertical gray bars. Note the progressive thermocline erosion, with large changes (≥ 1 m) normally the result of storms. The resulting nutrient (phosphorus) entrainments were typically much larger than external summer loadings by streams. *(Modified after Soranno et al. 1997.)*

limnia rapidly thicken as the epilimnetic temperatures decline because they progressively incorporate more of the metalimnion with the same temperature as the cooled epilimnion (Fig. 11–10).

The incorporation of metalimnetic water in the mixed layer is brought about largely by wind-induced currents, but is supplemented by density currents, the latter exerting their greatest relative effect in wind-protected lakes and low wind areas. As the epilimnetic temperatures decline, the density difference between the epilimnion and hypolimnion and mixing resistance also declines, until it becomes small enough to be overcome by a storm of sufficient force. At that point, the fall overturn is complete and the lake has once again become isothermal, remaining so until the onset of the next stratification period in warm monomictic lakes, or until ice cover becomes established in dimictic lakes (Fig. 11–10).

The approximate date by which low-latitude north temperate lakes can be expected to be isothermal in autumn is partly determined by the temperature of the summer hypolimnion, the major determinant of density difference between the epilimnia and hypolimnia. The larger the temperature difference, the later the

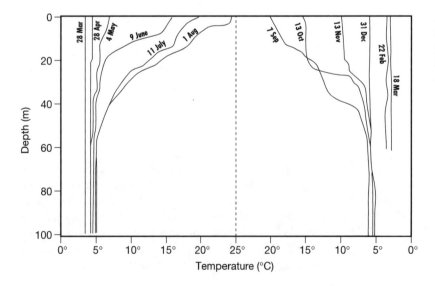

Figure 11–10 Seasonal temperature profiles showing the warming and cooling patterns of Cayuga Lake (New York, US) in 1951–1952. *(Modified after Henson et al. 1961.)*

overturn. The second important variable is mean depth, which serves as a surrogate for the volume of water that needs to be cooled and the inertia of the water mass to mixing (Nürnberg 1988):

$$\log (\text{overturn date}) = 2.62 - 0.116 \cdot \log (\text{hypotemp})$$
$$+ 0.042 \cdot \log (\text{mean depth})$$
$$- 0.0002 \cdot (\text{adjusted latitude})$$
$$R^2 = 0.67; \ p < 0.0001; \ n = 89.$$

EQ. 11.1

Clearly, with only 67 percent of the variance explained by Eq. 11.1, there are other factors that help determine the turnover date. The most important among these are probably interyear variation in late summer and autumn air temperatures, and wind strength.

Winter Conditions

▲ There is no reverse thermocline when freshwater lakes are cooled below 3.94°C, because resistance to mixing is easily overcome by even light winds (Fig. 3–2). The water, therefore, continues to cool to below 3.94°C in wind-exposed lakes prior to freezing over,[2] with water temperature at the time of freezing predictable from the fetch (Fig. 11–11). Small lakes freeze over much earlier and have a higher under-ice water temperature than larger lakes more exposed to wind. Frozen-over freshwater lakes exhibit a temperature gradient, but not a temperature stratification, ranging from water of 0°C at the ice–water interface to the maximum temperature in the water column when the ice cover was established.

The normally small amount of solar energy able to pass through snow and ice (Chapter 10) on sunny days permits development of small density (convection) currents immediately below the ice. More important are density currents produced by heat flow from lake sediments warmed the previous summer in the adjacent catchment or from inflowing groundwater and streams. The resulting density currents allow slow mixing even in ice-covered lakes.

The relatively stable winter arrangement is disturbed during the spring melt when the water imme-

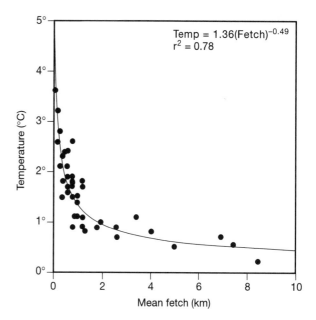

Figure 11–11 Relationship between water temperature at freeze-up and mean fetch for lakes in Wisconsin, US. *(Data from J.T. Scott 1964, in Ragotzkie 1978.)*

diately under the ice is warmed following increased penetration of solar energy into it (Chapter 10), accompanied by warmed water from surface ice-melt draining through increasingly porous ice. Convection (density) currents increasingly mix the water column as melting continues. Mixing is further enhanced by inflowing streams carrying warmer, less dense water which mixes with the < 4°C water directly below the ice and increases its density. The resulting turbulence, combined with an increased amount of photosynthetically available radiation (PAR) able to enter the water column as the snow layer disappears, may even allow the development of a spring algal growth period before ice-out. Algal **blooms** (maxima) under the ice are possible in those lakes where the mixing depth (z_{mix}) does not greatly exceed the thickness of the euphotic zone (z_{eu}, Sec. 10.11).

Thermal Bar

Once dimictic lakes become ice-free, the period of spring overturn or circulation begins. More rapid warming of the shallow waters along the perimeter, supplemented by the inflow of warmer river water, leads to the development of a *thermal bar* that is most clearly evident in large or multibasin lakes with basins of different depth (volume) (Fig. 11–12). Where the

[2]The temperature at which water freezes declines with increasing salinity, and saline lakes freeze later than freshwater lakes in the same area. Probably the most extreme manifestation of a freezing point depression is Deep Lake, AQ which, with a salinity about 10 times that of sea water, remains ice-free most winters with a minimum winter water temperature of about −17°C. (Ferris and Burton 1988)

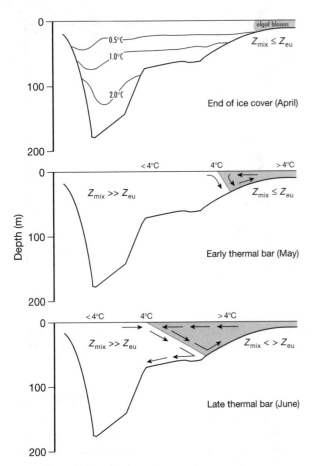

Figure 11–12 The development of a spring algal bloom and current directions (arrows) at the thermal bar in Lake Ladoga, RU (LA = 18,130 km², $\bar{z}$ = 50 m). *(Modified after Petrova 1986.)*

warmed (> 4°C), therefore lighter, nearshore water meets the colder (< 4°C) water offshore, mixing produces 4°C water which, being most dense, sinks. The consequences are four-fold: (1) separation of the two water masses and retention of inflowing nutrients and pollutants along the shores until the bar dissipates; (2) development of a spring algal bloom in the shallow, well-illuminated, warmed, and stably stratified nearshore water column enriched by nutrients entering from the land, occuring well before the bloom period gets under way in the pelagic zone. (Fig. 11–12); (3) development of a complicated frontal mixing pattern at the bar; and (4) gradual offshore movement of the bar as spring heating proceeds. Movement commences once the difference between the temperature of the nearshore and offshore water is sufficiently

large to allow lighter (warmed) onshore water to move offshore by flowing over the still cold open water. Spring algal blooms that developed within the thermal bar then serve as a major innoculum for the open water.

The bar may exist for more than a month in large dimictic lakes, is limited to a few days in small dimictic lakes, but remains all summer between dimictic arms and the primary cold monomictic basins of subarctic lakes (Sec. 11.3). However, the bar leaves an effect in large dimictic lakes even after it disappears. Moving offshore, the bar water becomes subject to the earth's gravity force (Coriolis effect, Sec. 5.4) and is forced to the right (in Northern Hemisphere lakes), producing a counterclockwise-moving band of warm water at the edge. The temperature-driven band, which in Lake Ontario persists into the summer, contributes to prolongation of nutrient and contaminant retention in nearshore regions (Boyce et al. 1991).

For most midlatitude lakes the onset of stratification is determined by the solar irradiance received (or its proxy mean annual air temperature), water volume that needs to be heated, degree of protection from wind and, for any particular year, weather.

The dimictic pattern presented above is sufficient to illustrate why cold monomictic lakes do not stratify, why warm monomictic lakes are stably stratified for only one period of the year, and why resistance to mixing is insufficient to allow stable stratification in polymictic lakes.

▲ 11.5   Stability of Stratification

The computation of thermal stability (*S*) requires a determination of the amount of work needed to fully destratify a lake instantaneously, without having to consider the addition or removal of heat required over time. It represents the susceptibility to wind mixing of the entire water column and thereby the potential for introducing hypolimnetic water and nutrients into the epilimnion (Chapter 12). The Schmidt stability index (see Wetzel and Likens 2000) is calculated from daily temperature profiles:

$$S = \frac{1}{A_0} \int_{z_0}^{z_m} (p_z - p_m)(z - z_g) A_z dz, \qquad \text{EQ. 11.2}$$

where S = thermal stability (kJ cm⁻²), A_0 = surface area (m²), A_z = area at depth z (m²), p_m = mean density at complete mixing (g cm³), p_z = density at depth

z (g m^{-1}), z_g = depth of center of gravity (mean density) at complete mixing (m). The summation is taken over all depths (z) at a specified interval (d_z).

Stability of stratification (resistance to mixing) is more closely linked to the maximum or mean depth, surrogates for volume, than to the surface area of lakes (Kling 1988). The energy needed to overcome inertia of the water mass to mixing rises sharply with increasing depth, but beyond a mean depth of about 40 m adds little to the stability (Viner 1984). Shallow lakes of a particular surface area are much less stably stratified than their deep counterparts, and lakes with small seasonal variation in water temperature tend to be less stably stratified then those showing much vertical change in water temperature (Fig. 11–7).

A first and simple estimate of whether temperate lakes will exhibit summer stratification is based on the thickness of the epilimnion (z_e) and the maximum lake depth (z_{max}) (after Patalas 1984, and Davies-Colley 1988)

$$\frac{z_e}{z_{max}} < 0.5 \quad \text{3-layer lake (epi-, meta-, hypolimnion) Stable seasonal stratificaiton.}$$

$$0.5 > \frac{z_e}{z_{max}} < 1.0 \quad \text{2-layer lake (epi-, metalimnion). May be turned over by strong winds.}$$

$$1.0 > \frac{z_e}{z_{max}} < 2.0 \quad \text{Mixed lake. Intermittent stratification during calm periods.}$$

$$\frac{z_e}{z_{max}} > 2.0 \quad \text{Turbulent lake. No stratification.}$$

11.6   Stability of Temperate vs Tropical Lakes

The density (temperature) difference between epilimnia and hypolimnia is usually much smaller for tropical lakes than for their stably-stratified warm monomictic and dimictic counterparts (Fig. 3–2). The maximum density difference calculated for 52 tropical African lakes averaged 0.65 g l^{-1} (Kling 1988), which is only one-third to one-half the value noted in north temperate lakes. For Lake Victoria (east Africa, Figs. 5–17 and 6–3), with as little as a 1 C° (25–24°C) difference between the epilimnion and hypolimnion, the density difference is a mere 0.25 g l^{-1}. Consequently, only modest seasonal changes in temperature are needed to mix tropical lakes. Wind-induced evaporative cooling drives much of the seasonal water temperature variation in tropical Africa, but is further abetted during the hemispheric winter by deposition of cooler rain and, often, reduced irradiance. Even so, the stability of all lakes is greatest during the period when heat content and density difference between the epilimnia and hypolimnia is maximal (Fig. 11–13). Mixing in deep tropical lakes during the cool season or hemispheric winter is often enhanced by stronger winds plus greater turbulence produced by increased river discharge. River discharge is the principal destratification determinant in riverine lakes and many reservoirs.

Considering an idealized lake at different latitudes, Lewis (1987) concluded that the stability of stratifica-

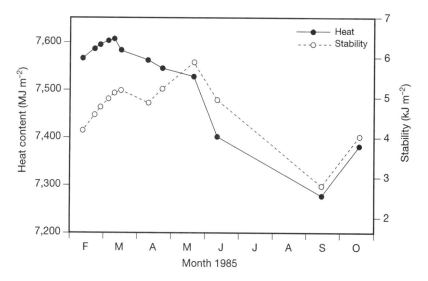

Figure 11–13 Water column stability as calculated with Eq. 11.3 and changes in heat content of tropical crater lake Barombi Mbo (CM) (LA = 4.15 km^2, z_{max} = 110 m). Note that the lowest heat content and stability occur during the hemispheric summer period of low solar input (high cloud cover) and greater precipitation. *(Modified after Kling 1987.)*

tion should be lowest between 0° and 20° N and S, then rise sharply to a broad maximum between ~25° and 40° N and S. At those latitudes, lakes have both high absolute temperatures and large vertical temperature gradients. Stability declines towards the poles because surface water temperature declines, reducing the density difference between epilimnia and hypolimnia (see Fig. 11–5), yielding weakly stratified lakes such as Lake Erken, SE (Fig. 11–7).

▲ 11.7   Thermocline Depth

The depth of the thermocline, or the mixed-layer depth, is determined primarily by exposure to winds (fetch) when lakes of very different surface areas, but similar water temperature, are compared (Table 11–2). A modifying factor is regional wind strength, with lakes in oceanic and windy New Zealand, northern Great Britain, and southern South America having a deeper thermocline than their more continental counterparts (Kling 1988). Thus, the average depth of the thermocline in a series of deep, wind-exposed Argentinean lakes averaged 38 m, and increased with increasing latitude (Baigún and Marinone 1995). Not surprisingly, exposure becomes a less important pre-

dictor of thermocline depth when lakes of similar surface area within a climatic zone or lake district are compared. Over the smaller spatial scale other attributes, such as drainage basin characteristics with their effect on lake water color, inorganic turbidity, nutrient export, and lake morphometry acquire an importance they did not have before (Sec. 11.2 and Fig. 11–14).

More solar energy is trapped in the surface water of colored or turbid lakes, and the resulting increased temperature difference between surface and deep water yields a shallower thermocline than in transparent lakes of roughly similar fetch (Jones 1992). Conversely, the introduction of filter feeding zebra mussels (Sec. 25.6) in a small US reservoir increased the transparency (z_{SD} = +0.8 m) and the thermocline depth by nearly 1 m in two years (Yu and Culver 2000). At a yet finer interval scale (Sec. 2.6), over which both lake morphometry and color/turbidity vary relatively little, the plankton community structure influences light penetration and thus depth of the thermocline. An algal community composed of a relatively few large cells or colonies absorbs much less irradiance (has a lower light attenuation) than where the same biomass is distributed over many smaller phytoplankton units (Fig. 10–6, and Mazumder et al. 1990).

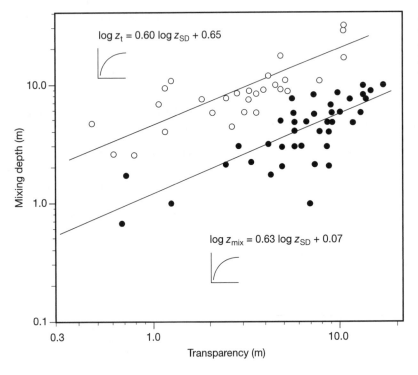

Figure 11–14 Plot of mixing depth (z_{mix}) vs transparency depth (z_{SD}) for small African crater lakes and Japanese lakes. Mixing depths are represented by the top of the thermocline (z_{mix}) in Japanese lakes and by the planar thermocline (z_t) in the tropical lakes. The different intercepts are, at least in part, attributable to z_t being located below z_{mix} (see Table 11–2). Note the important effect of water transparency on mixing depth in the absence of much variation in surface area. *(Data from Kling 1988, Melack 1978, and Yoshimura in Hutchinson 1957.)*

(figure axis labels and equations:)

$\log z_t = 0.60 \log z_{SD} + 0.65$

$\log z_{mix} = 0.63 \log z_{SD} + 0.07$

Mixing depth (m) — 10.0, 1.0, 0.1

Transparency (m) — 0.3, 1.0, 10.0

Tropical vs. Temperate Lakes

The typically deeper tropical thermoclines cannot be attributed to higher wind speeds which, away from coasts, are if anything, lower than in the temperate zone (Fig. 5–11), but rather to a mixture of wind-induced and convection currents. Whatever the reason, the much lower stability of tropical lakes allows a more ready erosion of their thermoclines and thickening of the mixed layer. In tropical lakes of moderate volume, the nighttime heat loss may be sufficient to allow much diel (24-hour) variation in thermocline depth. In Lake Calado, a Brazilian varzéa lake (LA = 2.8 km^2, z_{max} = 1–12 m), the thermocline depth in January and February ranged from 1.5–6 m at sunrise, became progressively shallower as the day progressed, ranging from 0.5–1.5 m at noon; followed by nocturnal cooling. Nighttime heat losses followed by production of convection currents was the principal mechanism responsible for mixed-layer thickening (MacIntyre and Melack 1988).

Large tropical lakes near the equator should be less affected by the Coriolis effect than their temperature counterparts (Sec. 5.4). Wind-produced surface currents in the tropics are less deflected from the direction of the wind, and thus more effective in mixing the upper layer than in the temperate zone. Lewis (1987) estimates that for a given wind speed the mixing current will be twice as great between N and S latitude 0° and 10° as it would above latitude 20°. The combined effects of enhanced lower stability and reduced Coriolis effect on mixing in large tropical lakes is dramatically evident when temperate zone and tropical inland seas are compared. Midsummer thermocline depth in the Laurentian Great Lakes (US) is found at 15–25 m, but is between ~70 m and 80 m in the tropical great lakes Tanganyika and Malawi (Kling 1988).

It is evident that the annual cycle of stratification and mixing in the temperate zone is ultimately controlled by a large seasonal variation in radiant energy reaching the water surface, sufficient to produce large temperature and density differences between surface and deep waters and strong water column stability in summer. The cycle in tropical lowland lakes deep enough to stably stratify is linked to minimum water temperatures (heat content) during the hemispheric winter on either side of the equator, establishing an annual mixing season. The mixing period for tropical monomictic lakes is typically much shorter than for their equivalent temperate counterparts. Furthermore, since the density difference between epilimnetic and hypolimnetic water is much smaller in tropical lakes (Sec. 11.6), they are not only less stably stratified and more prone to diel erosion but also interseasonal thermocline erosion and mixed layer thickening that is the outcome of fairly regular weather variations and associated heat losses. The next period of heat gain will reestablish a new, thin, mixed layer marked by a secondary thermocline. The periodic thickening of the mixed layer affects nutrient cycling and the plankton dynamics of deep tropical lakes through a periodic return of nutrients to the mixed layer and euphotic zone, resulting in greater variation in phytoplankton biomass and production than in more stably stratified temperate systems (Lewis 1996).

High quality measurements made on hot windless days in low latitude lakes and tropical reservoirs show not only the existence of what has been termed the **parent thermocline** at the bottom of the mixed layer but also the daily development of a **diurnal** or **breeze thermocline** in the surface one or two meters (Fig. 12–2). Recent work has led Lewis (1996) to abandon wind-based terminology for an interpretation of mixing depth based on changes in net heat flux that is based on mass–balance modeling (input–output). In it the principal role of wind is viewed through its effect on evaporative processes (heat loss) rather than turbulence. Yet, wind-based terminology is useful for identifying multiple thermoclines and reminding us that phytoplankton will, during the day, be trapped either above or below the diel thermocline and exposed to $z_{eu}{:}z_{max}$ conditions quite different from those in which the water column mixes down to the parent thermocline on a daily basis (Sec. 10.11).

▲ 11.8   Thermocline Shape

The shape of thermoclines is a function of exposure of particular lakes to wind and recent wind conditions, as well as density differences between the epilimnia and hypolimnia, and morphometry of basins, with morphometry affecting current patterns and turbulence. Small north temperate lakes (traditionally best studied) have a modest exposure (fetch) and, as a result, are able to stratify fairly early in spring before the water warms greatly. Consequently, they tend to exhibit a steep temperature gradient between the warmed surface and cold deep waters (Figs. 11–8 and 11–15), allowing for stable stratification. The steep temperature gradient in small, often wind-protected, lakes has

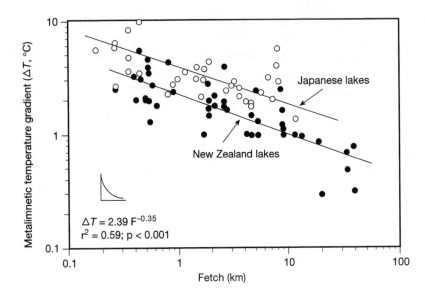

Figure 11–15 Comparison of the relationship between the maximum temperature gradient in the metalimnion and mean fetch for New Zealand and Japanese lakes. Note that Japanese lakes typically have a steeper thermocline than the lakes of the same fetch in windy New Zealand. Large lakes are more wind-exposed and have a more extensive thermocline (see Fig. 11–16). Clearly, more than fetch alone determines the gradient. Additional determinants would include (1) solar radiation, (2) the principle lake axis relative to prevailing winds, (3) shoreline morphometry and islands, (4) wind protection by hills or mountains, and (5) water color and turbidity (see Fig. 11–14). *(After Viner 1987.)*

yielded the text-book type of thermocline (Fig. 11–8) also evident in Lock Awe, Scotland, GB (Fig. 11–16).

A longer mixing period (turnover time) in larger, more wind-exposed lakes allows for higher hypolimnetic temperatures upon stratification and a modest temperature difference between and epilimnia and hypolimnia. Lakes may exhibit an **extensive thermocline** (Loch Lomond) or a **poorly defined thermocline** (Loch Ness) (Fig. 11–16).

11.9   Meromictic Lakes

Whereas all the lake types discussed so far (Sec. 11.2) mix to the bottom each year and are known as **holomictic** (complete mixing) lakes, there are others that do not mix completely and are designated as **meromictic** (partial mixing). Nearly all meromictic lakes have sufficiently elevated deep-water salinity to produce a density difference between deep and surface

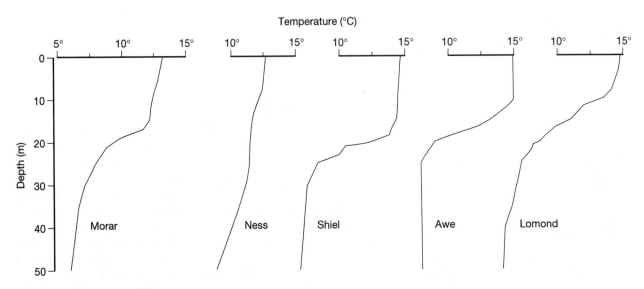

Figure 11–16 July temperature profiles and thermocline shapes in five deep Scottish lochs (lakes). *(After George and Jones 1987.)*

waters too large to be overcome by wind mixing or density currents. But not all meromictic lakes have a pronounced salinity gradient. Lake Baikal (RU) is so deep (z_{max} = 1,741 m; Table 4–5) that, even in the absence of a salinity gradient, the kinetic energy imparted by wind is insufficient to overcome the inertia of its enormous water mass.

Lake Malawi, the southernmost of the large Great Rift Valley lakes of east Africa, and its neighbor Lake Tanganyika (Figs. 5–17 and 6–1), are meromictic and anoxic below about 250 m. The salinity difference between the upper and lower water masses is small in Lake Malawi (~20 μS cm^{-1}). However, the modest density difference, together with a density difference imposed by about a 3°C temperature difference between the two water masses, combined with the inertia provided by their enormous water volume is sufficient to allow these two lakes to be meromictic (Pilskaln and Johnson 1991).

The overlying layer of fresh water in most meromictic lakes is thick enough to allow both temporary (seasonal) and stable (long-term) temperature stratification, making it possible to apply the Lewis classification scheme (Fig. 11–1). Thus, Lake Malawi is a warm monomictic as well as a meromictic lake and Fayetteville Green Lake (New York, US) is a dimictic meromictic lake (Brunskill and Ludlam 1969).

▲ Unique terminology has developed for describing the different density zones of meromictic lakes. The upper fresh water layer, subject to mixing and possibly to temperature stratification, is known as the **mixomolimnion**, whereas the deep and normally higher density layer that is not mixed (and anoxic) is known as the **monimolimnion**. The two are separated by a layer of rapid salinity change known variously as the **chemocline**, **halocline**, or **pycnocline**.

Inland meromictic lakes are few in relation to holomictic ones. Some twenty-five years ago, Walker and Likens (1975) were able to report only 44 such lakes (coastal brackish waters excluded) in North America, 33 in western Europe and a small number elsewhere. Many have been added since then, particularly among the small crater lakes of Africa (e.g., Melack 1978) and in Antarctica (e.g., Burke and Burton 1988). Yet, meromictic lakes have received more attention than their abundance merits since I. Findenegg (1935) started their systematic study in Austria. The disproportionate attention accorded to them is not the result of their unusual stratification but rather because the anoxic monimolimnetic sediments and

ionic gradients in the chemocline are interesting from a geochemical point of view. In addition, a fascinating community of photosynthetic bacteria and associated protozoa develops in the anoxic portion of the oxic–anoxic interphase in the chemocline of those lakes where the mixomolimnion is sufficiently transparent to allow photosynthesis at the interface (Secs. 7.4 and 22.10).

Anoxic sediments are undisturbed because of the absence of burrowing (**bioturbation**) by aerobic (requiring oxygen) macroscopic animals and by a lack of wind-induced turbulence. Undisturbed sediment cores have been extracted yielding a nearly perfect record of changes over time in the plankton—and even the fish—community, based on organismal remains identified in the sediments (see Fig. 14–6 and Sec. 20.5). Sediments are readily dated by counting annually deposited laminations (layers), known as **varves** (Fig. 14–6). Cores provide an important record of the history of lakes and their biota; pollen from terrestrial plants preserved in dated sediments reveal changes in the vegetation of the area, while patterns in the deposition of manmade pollutants provide a picture of changes in emissions and the long-range transport of anthropogenic substances (Sec. 28.5).

▲ 11.10   Development of Meromixis

A large majority of the inland lakes that are meromictic have a high **relative depth** (z_r), which represents the maximum depth to mean lake diameter ratio (Eq. 7.5). Nearly three-quarters of the noncoastal meromictic lakes in North America have a z_r larger than 0.05 (5%), whereas holomictic lakes have low relative depth ratios of around 0.001 (0.1%). Those meromictic lakes with a lower z_r must lie in exceptionally wind-protected basins. One such lake, Lake Mahoney (CA), has a z_r of 3.6 per cent and is protected not only by local hills and cliffs but also by having its major axis at a right angle to the prevailing winds. A morphologically similar lake parallel to valley winds is holomictic (Northcote and Hall 1983). Meromictic lakes, such as Mahoney lake, lying in small catchments that export little nutrients or sediments are highly transparent and characterized by a layer of photosynthetic bacteria in the chemocline (Secs. 7.4, 22.10 and Fig. 22–15). Recent explosion crater lakes typically have a very high relative depth, are highly wind protected and strongly meromictic even without a supply of ions and/or gases via underwater vents (Fig. 11–17).

Highly wind-protected holomictic lakes may become meromictic because of increased density brought about by the decomposition of sedimented organic matter in their hypolimnia and sediment layer. The production of particles in the upper water, followed by settling and, at least partial, dissolution in hypolimnia and sediments, result not only in an increase of water density at the bottom but also in decreased density in the mixed layer and resulting increased water column stability. The respired organic matter liberates ions, and the release of CO_2 dissolves sediment $CaCO_3$ or sedimenting calcite ($CaCO_3$) particles (Eq. 14.4) (if present). The $Ca(HCO_3)_2$ that is released contributes to increased salinity (Chapter 13). Only very small density differences are required to allow meromixis to become established in lakes with a high z_r. For example, the artificial destratification of Hemlock Lake (Michigan, US) by aeration (Fast and Tyler 1981) was followed within a year by the reestablishment of meromictic conditions. Clearly, both physical and biogeochemical processes affect the mixing regime of lakes.

Types of Meromixis and Duration

During the development of meromixis the increasing density difference between the miximolimnion and the monimolimnion requires increasingly larger storms or density currents to be overcome. Meromixis

developed in this fashion is, depending on the scientist's background, referred to as **endogenic** (produced within), **biogenic** (produced biologically) or **morphogenic** (resulting from lake and catchment morphometry) **meromixis**. Morphogenic meromixic is a particularly appropriate designation for tropical lakes Tanganyika and Melawi, whose meromixis is only partly determined by salinity and temperature differences, but is principally the result of wind energy being insufficient to fully mix these very deep and large waterbodies (Talling 1969). Even so, isotope studies have shown that mixing in Lake Melawi is sufficient to incorporate up to 20 per cent annually of its monimolimnetic volume into the mixomolimnion, in the process returning nutrients to the epilimnion and preventing the establishment of a true chemocline. Similarly, about 12 per cent of holomictic Lake Baikal's deep (but oxic and cold) water is renewed each year by means of convection currents (Weiss et al. 1991).

Meromixis is not always of long duration. Small (LA < ~0.03 km², maximum length 0.6–1.2 km) wind-protected humic lakes in the southern boreal forest of Eurasia and North America often experience rapid spring heating under windless conditions immediately after winter ice cover melts. Stratification then becomes established without total mixing of the water column, and in some cases it is maintained during autumnal cooling (Salonen et al. 1984, and Molot et al. 1992). The resulting meromixis is sustained for one or more years until an unusual windy summer destroys it.

Figure 11–17 Lake Mfou (CM) (LA = 0.08 km², z_{max} = 58 m, z_r = 0.18), a meromictic explosion crater lake with a 200 m crater wall. *(Photo courtesy of G. Kling.)*

Yet another meromictic pattern is seen in very deep ($\bar{z}$ = 177 m, z_{max} = 370 m) Lake Maggiore (CH, IT), a lake protected from winter winds by the Alps. In most winters it mixes to only 100–150 m, but it completely turns over during cold and windy winters.

Long-term meromixis, lasting centuries, can develop under appropriate conditions. However, even lakes that have been meromictic for many years or even centuries have been known to turn over as the result of severe storms and the input of more saline water (or a reduction in fresh water input) into the miximolimnion, which reduces density differences. The release of gas accumulated into the monimolimnion can also produce turbulent mixing, occasionally with dramatic effects. Hydrogen sulfide or other reduced compounds (e.g., CH_4, NH_4^+, Fe^{2+}) that have accumulated in the monimolimnia, consume most or all dissolved oxygen in the mixed layer after overturn and produce major fish kills, but a fresh supply of deep water nutrients encourages the development of algal blooms. The sudden massive release of CO_2 from the monimolimnion of the west African volcanic crater lake Nyos (CM) killed about 1,700 people and 3,000 cattle by asphyxiation in 1986 (Kling et al. 1987) and is the most dramatic recorded overturn of a meromictic lake. Meromixis resulting from the introduction of gases or higher-salinity groundwater resulting from chemical processes in the earth's crust is known as **crenogenic meromixis**.

Humans and Meromixic

When conditions are borderline for the establishment of meromixis, man's influence can be decisive. After examining a dated core taken from Längsee, AT, Frey (1955) concluded that the lake had become meromictic about 2,000 years before present (BP). During the transition period between holomixis and meromixis the core contained bands of clay and pollen from agricultural plants and weeds. Frey surmised that cutting the beech forest for agricultural purposes led to increased erosion of fine clay particles, whose slow settling from the hypolimnion increased the water density enough to induce meromixis.[3]

Many meromictic lakes are the result of not endogenic or crenogenic but **ectogenic meromixis**

(outside-produced). Examples include the periodic invasion of sea water in coastal lakes and lagoons, with dense water sinking to the bottom and covered by low-salinity fresh water entering from the drainage basin as runoff and direct rainfall. In inland regions, saline water can be the result of lake evaporation, solution of bedrock, or in volcanically active areas, deep vents allowing higher salinity groundwater to enter.

The recent development of meromixis in Little Round Lake (Ontario, CA) is attributed to winter road salting, followed in spring by the addition of a low salinity surface layer derived from melting snow in its catchment (Smol et al. 1983). Ectogenic meromictic lakes can have a much smaller relative depth than their endogenic counterparts and still remain meromictic because the density difference between the freshwater and saline portions alone are normally large enough to impose the necessary resistance to mixing (Fig. 3–2).

Man's interference with local hydrology can bring about meromixis or its destruction. A recent example is Mono Lake (California, US; LA = 160 km², recent z_{max} = 45 m), where water diversions from inflowing streams lowered the lake level 14 m and nearly doubled its salinity during the last century. The lake, characterized by a historically modest catchment runoff, had been a holomictic (monomictic) saline lake, but briefly meromictic during years of exceptionally high runoff. For laudable aesthetic and environmental reasons, water diversions to Los Angeles were halted in 1994 to allow the lake level to rise a certain amount before more limited diversions were to be allowed again. The consequence of this plus record levels of runoff the following spring, was the unintended creation of a meromictic lake. A modeling exercise, based on new regulations and historical runoff patterns, predicts it will last for decades (Jellison et al. 1998). The now permanently anoxic hypolimnion (monimolimnion) is unsuitable to an aerobic biota. Strong chemical stratification greatly reduces turbulent mixing across the chemocline (Chapter 12). This, together with the characteristic development of a microbial community in the chemocline (Chapter 22), is expected to greatly reduce the previous seasonal nutrient recycling from deep water to the euphotic zone during turnover and lower lake productivity (Jellison et al. 1998). A second example of **anthropogenic meromixis** is Big Soda Lake (Nevada, US), a highly saline, but holomictic lake until 1905. Agricultural irrigation led to increased runoff and a layer of fresh water overlying the originally saline water and the development of meromictic conditions (Kimmel et al.

---

[3] Recent multidisciplinary work on a new 8.8-m-long core taken from Längsee suggests that Frey's interpretation was wrong because meromixis appears to have become established > 15,000 BP. (Löffler 1997)

1978). An example of a meromictic lake becoming holomictic is the Dead Sea (IL, JO), a hypersaline lake which fairly recently became holomictic after approximately 300 years of meromixis (Stiller and Chung 1984). Diversion of fresh water from the inflowing Jordan River starting in the 1960s, together with continuing surface evaporation led not only to a dramatic fall in water level but also to a gradual decrease in the density difference between the miximolimnion and monimolimnion, with the lake turning over for the first recorded time in 1979.

Some of the most fascinating lakes are Antarctica's extremely saline and ectogenic meromictic lakes. Organic Lake is a shallow (LA = ~0.05 km^2, z_{max} = 7.5 m), wind-exposed lake with a highly unfavorable relative depth (z_r) which nevertheless remains meromictic during its three-month ice-free period, the result of the enormous density (~1.15 kg l^{-1}) of its monimolimnion. With a salinity about six times that of sea water, the density difference between it and the overlying low-salinity snowmelt water is so great that mixing is prevented (Franzmann et al. 1987, and Burke and Burton 1988). For more on meromixis see Hutchinson 1957 and Walker and Likens 1975.

▲ 11.11   Heat Budgets

Annual heat budgets provide quantification of the energetics of heating and cooling determining the resistance of lakes to mixing. Such analyses show that temperate zone monomictic lakes absorb roughly the same amount of heat during the spring and summer as is released to the atmosphere during fall and winter. Large seasonal changes in heat gained and lost explain the dampening influence of large deep lakes on local climate in the temperature zone. Their large heat-storage capacity allows them to freeze late and to have a moderating effect on the temperature of the surrounding land in both spring and fall. This is in contrast to their small, shallow counterparts nearby which rapidly lose their small pool of accumulated heat in autumn and freeze over early (Fig. 11–11). These lakes also heat up rapidly in spring and, if deep enough, stratify early.

The amount of heat that lakes can take up and the maximum temperature they reach is partly determined by climate, surface area available to lose heat, and lake volume available to store heat. At night, shallow tropical lakes normally lose all the heat gained during the day and exhibit a daily stratification–destratification pattern. Others, polymictic lakes, periodically exhibit a sufficient net heat loss to allow destratification. Yet others remain seasonally stratified, but lose enough heat at night to exhibit diel variation in mixed-layer thickness (Sec. 11.7). However, all tropical lakes lose nearly as much heat on a daily basis through evaporation (latent heat of vaporization = 2,454 J g^{-1} at 20°C) and back radiation from the surface as they gain from inputs. Evaporative cooling power prevents lakes from heating to levels unsuitable to the biota. The constraints imposed on the amount of heat that tropical lakes can gain and hold (Fig. 11–13), together with moderate seasonal changes in heat input and low absolute stability—and the large effect of small temperature changes on that stability—are ultimately responsible for the major seasonal and diurnal changes in stability and stratification observed in holomictic lakes at low latitudes (Sec. 11.7). Heat-budget calculations for lakes with seasonal ice cover must consider the latent heat of fusion in the formation and melting of ice (latent heat of fusion and thawing = 334 J g^{-1}) (Welch et al. 1987).

Birgean Heat Budgets

The most commonly reported analyses in limnological literature are **annual Birgean heat budgets** (ABHA), named in recognition of the 1914/1915 pioneering work by E. A. Birge. ABHBs are not really mass–balance budgets at all, they give the maximum heat-storage capacity of lakes, representing the amount of heat required to raise water temperatures from their annual minima to their annual maxima. The heat required to melt ice is not usually considered in ABHB computations and is obviously not an issue in warm monomictic or polymictic lakes. It is a minor issue in dimictic systems with a brief and thin ice cover. It is important, however, in lakes with a thick ice cover, whose spring heat income is used to melt the ice, lowering the heat available for raising the water temperature from 4°C to its summer maximum. For example, each 1-m-thick layer of ice on a polar lake requires about 279 MJ m^{-2} to convert the ice to water, an amount larger than the net annual heat accumulation by lowland tropical lakes (Table 11–3) experiencing very large evaporative heat losses (latent heat of evaporation of water = 2,454 joules g^{-1} at 20°C, but varies with temperature). The long period of ice cover on, for example Chandler Lake, Alaska (Table 11–3), means that only about 20 percent of the heat accumulates during the ice-free period, resulting in a low ABHB. But low water temperatures during the ice-free period also mean low evaporative losses. Thus, high latitude lakes accumulate heat efficiently in comparison with

Table 11–3 **Average annual Birgean heat budgets, and ranges, where available, that do not include the heat required to melt any ice-cover.** $\bar{z}$ = mean depth (m); 4.1868 J = 1 cal.

Lake	ABHB MJ m^{-2} yr^{-1}	$\bar{z}$ (m)
Low Latitude		
Nkuruba (UG)	45[1]	16
Aranguadi (ET)	130	18
Valencia (VE)	126–251	19
Lanao (PH)	188–304	60
Victoria (KE, TZ, UG)	377–461	40
Titicaca (BO, PE)	510	135
Kariba [Basin 1] (MZ, ZW)	628	13
Mid Latitude		
Wingra (US)	176	2
Nimetön (FI) [humic]	228–333[2]	11
Bullenmerri (AU)	805	39
Washington (US)	502–1181	33
Fure (DK)	716	12
Mendota (US)	934–1089	12
Como (IT)	1,340	153
Vattern (SE)	1,340	39
Kinneret (IL, JO)	1,403	24
Ness (GB)	1,557	133
Michigan (US)	2,194	77
Baikal (RU)	2,742	730
High Latitude		
Chandler (US)	241[2]	14
Deep Lake (AQ)	1,023	20

[1]Turbid and highly wind-protected small volcanic crater lake with heat exchange restricted to a thin surface layer; mean temperature 0–2 m = 23.3°C ± 0.7°C.

[2] > 50% and 79%, respectively, of the annual heat income is spent melting the ice on Nimetön and Chandler, thereby greatly reducing the ABHB.

those at lower latitudes. An extreme example of this is Deep Lake, AQ (Table 11–3), which is so hypersaline that it usually remains ice-free during the severe Antarctic winters. In the process, its minimum temperature reaches about –17°C, guaranteeing negligible evaporative heat loss to the atmosphere during the heat accumulation period and a ABHB similar to many midlatitude lakes of similar mean depth (volume).

The highest annual Birgean heat budgets are observed in deep, relatively transparent large north temperate lakes such as monomictic Lake Michigan (US) or dimictic Lake Baikal (RU) Table 11–3. A combination of reasons for this include (1) a very large volume of cold water to be heated; (2) the end of winter water temperatures close to 0°C that allows heat accumulation before being offset by major evaporative losses; (3) a relatively high transparency, allowing spring heat to be distributed over a thick water layer, thereby delaying the onset of stratification (Sec. 10.8) and minimizing evaporative losses following heating of the epilimnia; and (4) a large fetch that guarantees a thick epilimnion that is able to accumulate heat before the temperature rises enough to allow substantial evaporative losses.

Low ABHBs are evident in lowland tropical lakes (e.g., Lanao and Valencia) exhibiting little annual variation in water temperature (Table 11–3). These lakes experience enormous evaporative losses that exert a very large cooling power, offsetting all or most of the gains. The heat-holding capacity of shallow (low volume) lakes that mix daily is particularly low. Lowest of all is the heat-holding capacity of turbid and wind-protected deeper lakes with only a thin (small volume) mixed layer subject to heating and located at low latitudes (e.g., Nkuruba or Aranguadi, Table 11–3). As large evaporative losses allow only a modest net heat gain in large, wind-exposed tropical lakes, these experience not only a relatively long period of mixing but also disproportionately deep mixing and a thick epilimnion upon stratification.

Dimictic lakes in the sunny interior of continents typically receive a surge of heat following melting of the ice cover, allowing rapid warming, particularly if the mean depth (volume) is moderate. Near-surface heating of somewhat colored and/or turbid lakes of modest or moderate volume helps establish an early temperature gradient that prohibits deep mixing. As in turbid and wind-protected tropical lakes, the resulting shallow mixed layers (volumes) severely constrain the amount of subsequent heat gain (e.g., Nimetön). That lake volume is important is evident from the difference between two lakes in close proximity: Wingra ($\bar{z}$ = 2 m) and Mendota ($\bar{z}$ = 12 m) (Table 11–3). Lake volume (V) and its surrogate mean depth ($\bar{z}$) are, not surprisingly, the best determinants of temperate zone annual Birgean heat budgets where climatic variation is modest, but where mean depth and volume vary greatly (Gorham 1964):

$$\text{ABHB} = 18.8\log_{10}\sqrt[3]{v} - 29.7$$
$$r^2 = 0.67; \quad SE = 6.4; \quad n = 71;$$

EQ. 11.3

$$\text{ABHB} = 18.4\log_{10}\bar{z} - 2.52$$
$$r^2 = 0.59; \quad SE = 7.2; \quad n = 71.$$

EQ. 11.4

11.12   Climatic Change and Aquatic Systems

A number of climatic models predict increased summer temperatures and decreases in soil moisture over much of the north temperate zone during the present century. If the predictions are correct, the lakes will (1) be a little warmer; (2) have a reduced period of ice cover; (3) have earlier onset of the summer stratification period; (4) have a longer period of stratification; and (5) experience reduced runoff; resulting in (6) longer water residence time; (7) lower water levels; (8) reduction in the size of upstream, adjacent, or downstream wetlands; and (9) reductions in hydroelectric-power generation by dammed rivers and water availability for drinking water and irrigation reservoirs. The higher evapotranspiration rates on catchments would (10) increase the concentration of ions in inflowing streams, with inlake and wetland concentra-

tions further enhanced by increased evaporation and slower flushing. The reduced runoff would also (11) lower organic matter and mineral nutrient supply rates (Sec. 8.5), thereby reducing the phytoplankton biomass and changing the species composition; (Chapter 21) as a result (12) reducing the abundance of planktonic organisms at higher tropic levels.

▲ An excellent and unusual 20-year record of climatic, hydrologic, and ecological records, made on Lake 239 at the Experimental Lakes Area (ELA) of northwest Ontario (CA), during a period when the average annual air and water temperatures increased about 2°C (Schindler et al. 1990, 1996) may provide an indication of the type of limnological changes that are expected to emerge. The observed temperature increase is similar to the anticipated increase over the next several decades by some, but not all, climatic models (World Resources Institute 1990). Changes observed in the lake are confounded because of accompanying changes on the catchment—forest fires—that are impossible to distinguish from other changes that have been observed. Forest fires should indeed become more common in boreal forest regions if warming reduces the soil moisture content, but fires would naturally not have the same effect on lakes in nonforested regions. The value of the Lake 239[4] study lies not so much in its being a harbinger of what might be expected elsewhere but rather in providing a reminder that changes in only one environmental factor (temperature) have wide ramifications that are easy to appreciate after the fact and would have been impossible to predict a priori.

The increase in air temperature (Fig. 11–18a) and average volume corrected lake temperature of Lake 239 (Fig. 11–18b) as well as a reduction in precipitation led to an average 20-day increase in the length of the ice-free season (Fig. 11–18d). Lower precipitation

---

[4]The advantages gained by studying the response of whole lakes, rather than of a limited number of large (and expensive) containers (> 1 m^3) anchored in lakes, and known as *mesocosms* or *limnocorrals*, to perturbation are great and undeniable. Yet, the lack of replication plus the lack of true controls makes work on one or a few whole lakes or limnocorrals unsuited for predicting the magnitude of responses elsewhere. Conversely, the easy replication possible by studying perturbations in smaller containers (*microcosms*) suffers the major handicap that these are not miniature lakes and, furthermore, that the vessels are highly subject to container effects that make their extrapolation to nature even more difficult than from limnocorrals. Both whole lake and smaller volume approaches, carried out over different spatial and temporal scales, have their constituency of scientific adherents and all approaches have possibilities and limitations.

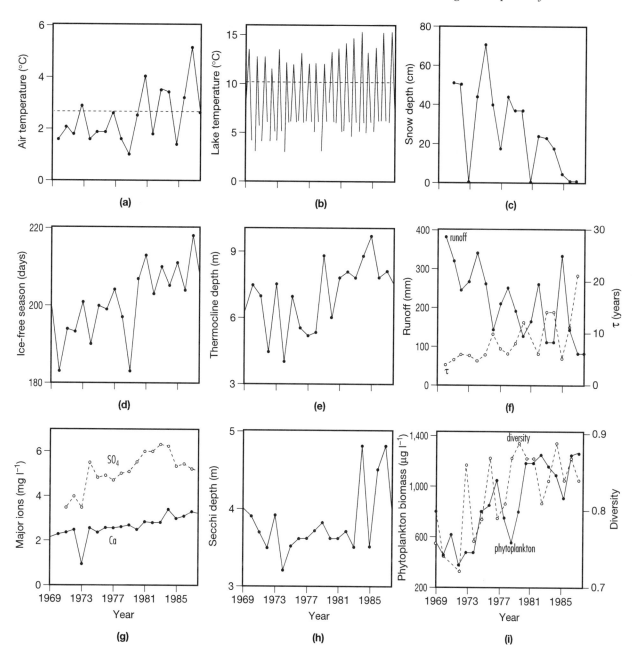

Figure 11–18   Records of physical and biological variables for Lake 239 at the Experimental Lakes Area in northwestern Ontario from 1969 through 1988. (a) Mean annual temperature at Rawson Lake meteorological station in the watershed. (b) Volume-corrected average lake temperature during the ice-free period. Each of the spikes is a plot of monthly water temperatures in the ice-free season for a single year. The horizontal dashed line is the mean temperature for ice-free seasons for the entire period. (c) Snow depth in the watershed at the end of March. (d) Duration of the ice-free period. (e) Average thermocline depth in the ice-free season. (f) Annual runoff from the watershed (●) and water renewal time for the lake (τ, ○). (g) Mean annual volume-weighted concentrations of Ca and sulfate. (h) Average secchi disc readings during the ice-free season. (i) Average phytoplankton biomass in the epilimnion during the ice-free season (●) and Simpson's index of diversity (○). *(Modified after Schindler et al. 1990.)*

led to thinner snow cover on the lake (Fig 11–18c) and earlier spring ice melt.

Burning much of the catchment, linked to reduced precipitation, decreased the wind protection afforded by the surrounding forest and also raised the wind speed over the surface which led, not surprisingly, to an increase in the depth of the thermocline (Fig. 11–18e). The 25 percent reduction in precipitation and a major increase in catchment evaporation further yielded much smaller runoff (Fig. 11–18f), allowing once-permanent streams to become intermittent in summer. The number of days small (first order) streams were dry during summer increased from less than 10 days to about 40 days during the 20-year drought (Schindler et al. 1996), but the effect of higher temperatures on stream functioning and reduced runoff on upstream wetlands were not investigated. Reduced runoff greatly increased water residence time (reduced the flushing rate) (Fig. 11–18f). Reduced runoff and longer water residence time concentrated chemical solutes (Fig. 11–18g) and raised the total dissolved nitrogen (TDN) concentration, but not the phosphorus level. The lake became clearer (Fig. 11–18h) as the result of smaller input of dissolved organic matter (DOM) from the terrestrial (including the reduced wetland) catchment, a longer period available for microbial decomposition in the water (Chapter 22) and photooxidation of DOC (Chapter 10), as well as the sedimentation and loss (Chapter 20) of previously dissolved organic matter following coagulation (flocculation). Phosphorus and silica concentrations in lakes and streams at ELA declined, presumably because longer water residence time allowed more time for biological removal processes (Schindler et al. 1996). Greater water clarity increased the thermocline depth (Fig. 11–18e) and, in humic Lake 239, doubled the depth to which the one percent level of UV-B radiation penetrated from ~0.3 m to ~0.6 m. Phytoplankton biomass and diversity increased (Fig. 11–18i), probably from increases in some nutrients, raised temperature, and increased water clarity.

Climate Change, Animals, and Humans

The ELA study did not explore effects on organisms other than algae, but suggests that observed increases in temperature would be expected to eliminate higher temperature-intolerant species from some temperate zone lakes. Cold-water fish such as salmonids and lake whitefish, as well as their lotic counterparts, would suffer while epilimnetic species such as perch and bass would be favored (Regier et al. 1990, and Rahel et al. 1996). Changes would come about not only as the result of a northward shift of the southern latitudinal boundary of cold-water species but also because greater water clarity and depth of light transmission would increase the depth of the thermocline (Fig. 11–18e), in the process reducing or removing the hypolimnetic summer refugium for cold-water fish in shallow lakes. Even where hypolimnia remained, smaller volumes together with a longer period of stratification, would increase the probability and frequency of hypolimnetic anoxia and fish kills (Chapter 15).

The predicted warming could lead to a northward-range extension of relatively large numbers of species found south of the boreal forest region. An evaluation of the relationship between temperature and species richness of North American crayfish and amphipods (Fig. 11–19) suggests that a mean annual temperature rise of 3°C would, more than double the number of these crustacean species over time. Since the present distribution pattern was established over thousands of years and, as crossings between drainage basins are slow (particularly so if more streams become intermittent), invasions would lag temperature changes unless they were introduced by humans.

The predicted decline in catchment-derived DOM in streams and lakes and the greater time for inlake removal processes would allow a substantial increase in the penetration of UV radiation (Fig. 10–7). In fact, the effect of the projected climate change on reducing export of colored DOC, and the resulting increases in UV penetration, will probably be much larger than the effect on the biota of the projected increase in UV radiation resulting from ozone depletion (Chapter 10).

Reduced runoff and river discharge will not only decrease the water supply but also the dilution of wastewater entering rivers, which together with higher water temperatures, reduced discharge (turbulence), and a longer water residence time would allow more algal growth, greater bacterial activity, and a higher resulting *biological* or *biochemical oxygen demand* (BOD, Chapter 17) as well as lower dissolved oxygen levels. More lotic systems would become intermittent, groundwater flows would decline, and wetlands would decrease in number and area.

The greatest impact of global warming on aquatic systems is expected in semiarid regions (mean precipitation 200–500 mm yr^{-1}) where enhanced evapotran-

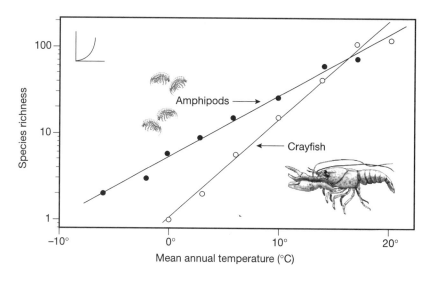

Figure 11–19 Relationship between mean annual air temperature and species richness of North American crayfish and amphipods. *(Modified after France 1991.)*

spiration would have an even more dramatic effect on river discharge, wetland size, and the ability of the water supply to support increasing human populations. Reduced inflow of fresh water and continuing high lake surface evaporation would lead to further reduction in flushing rate and elevated salinity (Chapter 13).

Saline lakes and both freshwater and saline wetlands of mid and low latitudes are already threatened by increased diversion of inflowing waters for irrigation and other human activities. Global warming would reduce their size, increase salinity, or dry them up altogether. Salinity increases lead to changes in species composition (Fig. 13–9) and salinity-linked reductions in algal biomass can be expected, particularly at the low to moderate end of the salinity gradient (~0.3–10 g l^{-1}) (Williams 1993). For more on the projected effects of climate change on aquatic systems see amongst others, McKnight et al. (1996) and Schindler (1997).

Highlights

- Lakes are divided, on the basis of their mixing regime, into those in which the whole water mass mixes at least once per year (holomictic lakes) and those that do not fully mix annually (meromictic lakes).
- Holomictic lakes, in turn, are divided on the basis of frequency of mixing into at least six categories that intergrade. Among them, temperate lakes with an ice cover that are deep enough to stably stratify have two mixing periods (spring and autumn) and are desig-

nated as dimictic. Their counterparts at lower latitudes, where an ice cover is lacking, are characterized by a single mixing period during the hemispheric winter and are known as monomictic. Polymictic lakes mix more than twice a year.
- Periods of mixing, followed by periods of relative stability, have a major effect on the cycling and availability of mineral nutrients to phytoplankton, light climate, temperature variation in the water column, and the possibility of deep-water anoxia; these affect primary production, foodweb structure, and whole-system metabolism.
- Wind and density currents create turbulence and mixing. Stratification of holomictic lakes commences when temperature-imposed density differences become too large to be overcome by turbulence. Destratification begins when the density difference between epilimnetic waters and the top of the thermocline become sufficiently small to be overcome by turbulence.
- Tropical lakes are, on a seasonal basis, less strongly stratified than their temperate counterparts. Their stability of stratification is also much more sensitive to modest seasonal and short-term changes in temperature.
- ▲ The depth of the thermocline, or thickness of the mixed layer, is in a particular climate zone primarily a function of the exposure of lakes to winds (lake length, lake area) among systems ranging widely in surface area or length.
- Nearly all meromictic lakes have deep water salinity sufficiently elevated to produce a density difference between surface and deep water that cannot be overcome by wind mixing and density currents.
- ▲ Heat income, heat loss, wind speed, and morphometry of lakes and their catchments determine

whether lakes will stratify, and if they stratify, also determine the depth of the thermocline and stability of stratification.

- Climate change models predict that over the next century temperate zone lakes will be warmer, have a reduced period of ice cover, an earlier onset of stratification, and a longer period of stratification. They are expected to receive less water and nutrients from their catchment basins resulting in reduced lake productivity and changes in the biota. Reduced river discharge as well as wetland and saline lake dessication will be other manifestations of global warming, with negative effects on both the human population and management and preservation of aquatic systems and their biota.

12

Water Movements

12.1 Introduction

Waterbodies are always in motion; this characteristic is of the greatest significance for the distribution of heat, gases, nutrients and organisms, both in the water and sediments.

There are two kinds of turbulent motions in lakes: **waves**, exhibiting periodicity, but little or no forward flow; and **currents**, which lack periodicity, but exhibit unidirectional flow. The relative importance of waves and currents in producing the turbulence necessary for effective distribution of chemicals and organisms in lakes depends on lake morphometry (depth and surface area). In all lakes except those riverine lakes and certain reservoirs where river flow imparts most of the kinetic energy, the two principal physical processes bringing about horizontal and vertical mixing are the wind at the air–water interface and heat loss (cooling) at the surface, which produces density currents. Within rivers, motion is ultimately the result of currents produced by gravity (slope).

Wind produces not only the typical waves and currents observed in the surface water of lakes but also induces much larger internal waves or *seiches* that oscillate along the thermocline and produce currents in the hypolimnia. In addition to internal seiches, there also exist very long surface waves or surface seiches with a periodicity of hours or days. Surface seiches have a periodicity too long to be evident when studying the effects of typical surface waves that reach the shore every few seconds, properly called *surface gravity waves*. Water movement in lakes is very complex, not only do wind speed and direction change but

the effect of wind on the waves and currents that are produced is greatly modified by the outline of the lake surface area, underwater morphometry, and lake depth. Islands and bays further change the current pattern while depth influences the importance of seiches and many other types of internal waves. The principal hindrance to using water movements for modeling or predicting the biological behavior of lakes lies not in an inability to study and measure individual wave and current types over a period of hours or days, rather it lies in assessing and predicting the integrated outcome of all waves and currents over time (Fig. 12–1) on the chemistry and biology of lakes (Imberger 1985, and Boyce 1974).[1]

12.2 Laminar vs Turbulent Flow

At very slow flows, the viscosity of water is high enough to allow layers of water to slide over one another in a smooth, orderly (**laminar**) manner. This is because the viscous forces that bind the fluid outweigh the forces of acceleration. An analogy would be smoke rising from a cigarette in a draft-free room that first exhibits the same orderly flow as it rises. With increasing water flow in a lake, or increasing aerial turbulence

---

[1]Boyce (1974) discusses the accomplishments and problems encountered by physical limnologists in predicting water movements over the different temporal and spatial scales. "Although the processes are acted out beneath the keels of our ships and as we swim on our backs in the summer sun, we are not unlike astronomers explaining the whorls of a distant galaxy."

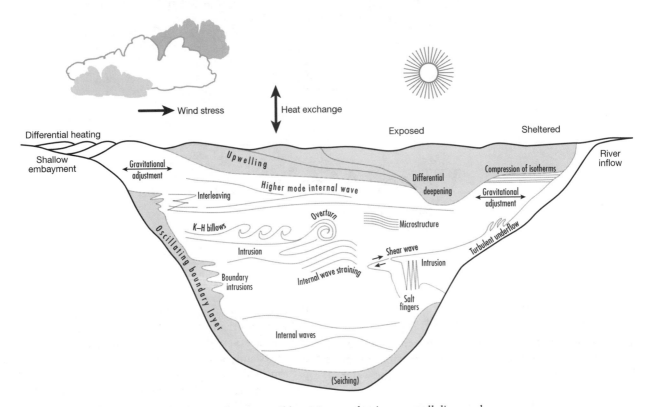

Figure 12–1 Diagrammatic illustration of some possible mixing mechanisms, not all discussed in the text, operating in a stratified lake subject to sudden wind stress. (*After Spigel and Imberger 1987.*)

near the room's ceiling, the coherent structure breaks down and a new turbulent motion of swirling and chaotic eddies is superimposed (Smith 1975). The process is propagated by eddies breaking into successfully smaller eddies until the residual energy is overcome and viscous coherence is reestablished. The interaction between the viscous properties of the liquid and the changing flow velocities impinged upon it can be defined by the dimensionless **Reynolds number** R_e of flow as

$$R_e = \frac{UL}{v}, \qquad \text{EQ. 12.1}$$

where U = mean current velocity (cm s^{-1}), L = depth or thickness of layer (cm), and v = kinematic viscosity, a property of water molecules (cm s^{-1}) (Chapter 3).

It is evident that flow is definitely laminar when R_e is < 500, and is turbulent at > 2,000, when measured in pipes. In lakes, the residual surface velocity is sufficient to guarantee some turbulence even in the absence of wind. It is the turbulent—rather than laminar—flow that is responsible for mixing heat,

gases, nutrients, and particles in the water and surface sediments of lotic and lentic systems.

Recent high-precision measurements of vertical temperature microstructure indicate that deep waters of stratified lakes are mostly characterized by laminar flow, but interrupted by periods of turbulence at the thermocline, in isolated sites in the hypolimnion, as well as over the lake bottom (Fig. 12–2). As mentioned, the wind stirring the water's surface and surface cooling of lakes produces the required **turbulent kinetic energy** (TKE). TKE is needed to break surface gravity waves where these touch the lake bottom near shore. Winds generate surface currents in the same direction as the water, but since the water has greater viscosity it moves more slowly than the wind, setting up a force in the vertical direction. This force, known as **shear**, produces a velocity gradient able to transport water by the imposed current system. Shear determines not only currents and associated turbulences but also the thickness of the *diffusive boundary layer*, characterized by laminar flow, over solid surfaces in lakes and streams (Sec. 24. 7).

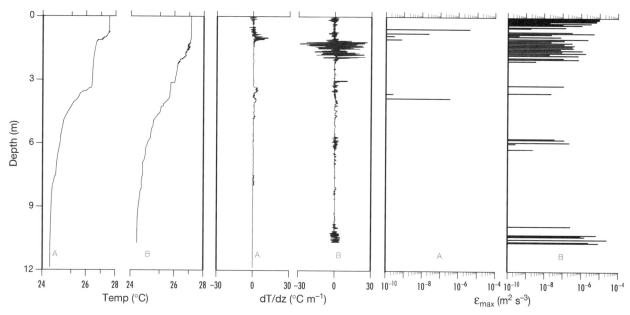

Figure 12–2   Temperature profiles (°C) and temperature gradient profiles (°C m^{-1}) in the subtropical Harding reservoir (AU) at (A) 1127 and (B) 1317 hours and a measure of energy dissipation ε_{max} (m^2 s^{-3})] on a day of extremely low Lake number (L$_N$ = 2.2). Modified from Imberger and Patterson, 1990. Compare the temperature profiles with those of Figure 11–16. Note (1) the multiple thermoclines, (2) the near absence of turbulent eddies during the calm morning, and (3) turbulence at different depths in the water column as indicated by temperature changes and energy dissipation.

In the open water of lakes, shear disrupts the laminar flow of the surface currents because of the inability of denser water to move as rapidly as the flow of the wind. Work on shearing stresses in lotic systems (laboratory tanks) has shown that only a low current velocity is needed to change flow from laminar to turbulent.

12.3   Surface Gravity Waves

Surface gravity waves are the typical waves seen on the surface of lakes and oceans and are the most immediately obvious manifestation of kinetic energy imparted by the wind. The waves have characteristic lengths, heights, periods, and frequencies (Fig. 12–3a). Wave period refers to the time needed for the passage of two crests or troughs to pass a fixed point. The wave height in lakes ranges from occasional ripples on calm days to waves with a wavelength of 5–10 m and associated wave height of one meter or more during periods of sustained strong winds over large lakes. Wavelength and period are largely determined by the

maximum fetch (Fig. 12–4), the longest distance (km) over which the wind can impart its kinetic energy to a lake. Alternatively, the surface area (km^2) of lakes can serve as a convenient surrogate for fetch. The wavelength in the open water of lakes is approximately 20 times wave height (H)

$$L = \sim 20\,H. \qquad \text{EQ. 12.2}$$

The maximum wave height (H_{max}, m) encountered can be estimated from the maximum fetch (F, km) by the empirical relationship (Wetzel 1983):

$$H_{max} = 0.332\,F^{0.5}. \qquad \text{EQ. 12.3}$$

Surface gravity waves bring about the oscillation of water particles primarily in the shape of an ellipse having only minor horizontal motion (Fig. 12–3a). The energy imparted by the wind is largely dissipated through the creation of a vertical series of ellipses that rapidly decline in size with increasing depth. The circular motion is interrupted when the waves "touch bottom" which happens where the water is shallower than about half the wavelength. Movement is then

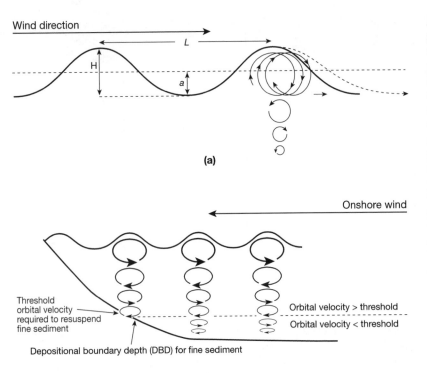

Wind direction

(a)

Onshore wind

Threshold
orbital velocity
required to resuspend
fine sediment

Orbital velocity > threshold

Orbital velocity < threshold

Depositional boundary depth (DBD) for fine sediment

(b)

Figure 12–3   (a) Diagram of the motion of water parcels and the various linear definitions of a rhythmic wave. The water parcels oscillate elliptically, but exhibit only minor forward motion. The more important vertical movement decreases exponentially with depth. L = wavelength, H = wave height, a = wave amplitude ($\frac{1}{2}$h). Motion of the water parcels is not to scale. (b) Diagram of the orbital velocity of the elliptical water parcels and their ability to resuspend and transport fine sediments along shallow slopes. Coarse sediments are found above the orbital velocity threshold where slopes are steep and transport to less turbulent water possible. *(Part (b) after Rasmussen and Rowan 1997.)*

translated in a forward direction until the waves break near the shoreline at a point where the lake depth is about $\frac{4}{3}$ the wave height (Smith 1979). Whitecaps seen in lakes of all sizes when wind speed exceeds 4–5 m s^{-1} are the result of extreme turbulence (frothing) at the air–water interface, not wave size (Smith 1979).

▲ In contrast to the major horizontal and vertical currents that will be discussed in Secs. 12.7 and 12.8, water movement associated with surface gravity waves contributes only modestly to the overall turbulence of the mixed layer of smaller lakes (Fig. 12–4). However, vertically imparted energy is sufficient to play an im-

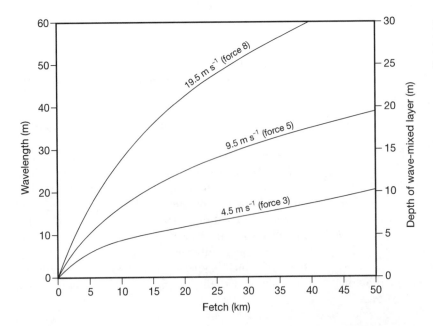

Figure 12–4   Surface gravity waves. Relation between depth of the wave-mixed layer (wavelength/2), fetch, and wind speed. *(After Smith and Sinclair 1972.)*

portant role in the resuspension of fine sediment particles in larger lakes during the largest annual storms (Fig. 12–3b) from depths considerably greater than those where waves break at the shoreline (Rowan et al. 1992a). Above the *threshold orbital velocity*, fine particles are resuspended and then transported by **advection** (horizontal movement) to deeper, less turbulent waters. Fine particles are deposited permanently beyond the depth where the energy remaining is sufficient to allow their continued resuspension, known as the **depositional boundary depth** (DBD). See Fig. 12–3b. Advection determines the depth separating zones of coarse and fine sediments (Chapter 20) and the resulting distribution, abundance, and type of organisms living on and in the substrate (Chapters 22, 24, and 25). Advection also determines how easily gases and other dissolved substances diffuse into and out of sediments with different porosity.

▲ 12.4 Turbulent Flow and Measures of Stability

The *mixed layer* or *surface layer* is defined as the water layer recently mixed by wind or convection currents. Whether turbulence is likely in the mixed layer and temperature stratification can be maintained depends on the relative strength of turbulent mixing and the thermally induced density gradients of the thermocline, whose stabilizing buoyancies produce resistance to mixing. If the turbulent energy is insufficient to overcome the resistance, any previously induced turbulence quickly subsides.

The magnitude of the work required to overcome density gradients relative to the energy available is denoted by the **gradient Richardson number** (R_i)

$$R_i = \frac{g(dp/dz)}{\bar{p}(du/dz)^2}$$

$$= \frac{\text{measure of water buoyancy}}{\text{measure of turbulence}},$$

EQ. 12.4

and where

$$R_i\left(\frac{du}{dz}\right)^2 = N^2;$$

EQ. 12.5

g = acceleration due to gravity (cm s^{-2}), $\bar{p}$ = mean density of water (g cm^{-3}) through a water column of a given depth, z = depth (cm), u = horizontal current velocity (cm s^{-1}), and N^2 = the square of the **Brunt-Väisälä buoyancy frequency** (s^{-2}) which is another measure of water column stability or intensity of stratification that is used frequently. The Richardson num-

ber defines the stirring force to stabilizing buoyancy force ratio. Mixing and friction are small in lakes with well-stratified shear layers with an R_i of > 0.25, where layers of waters with different densities slide past each other without turbulence. As a result, any turbulence in the epilimnion at that time will not extend into the hypolimnion. Conversely, at R_i < 0.25 there is sufficient kinetic energy available to overcome buoyancy forces. The flow then becomes unstable and disturbances at the density interface increase and ultimately break into large vortices. A time series of R_i indicates the frequency of mixing events at particular depths and times as R_i drops below 0.25. A third index of mixing potential is the static *Schmidt stability index* (Eq. 11.2).

The dimensionless **Wedderburn number** (W) and, in single lakes, the closely correlated **lake number** (L_N) are two additional and increasingly used indices of mixing potential which combine the effects of stability, wind forcing, and mixed-layer aspect ratio into a single parameter. They are used to determine the degree of tilting of a thermocline or chemocline and the associated upwelling that produces turbulence and mixing.

The Wedderburn number is obtained by multiplying the dimensionless R_i by the ratio of the mixed layer depth (H) to basin length at the thermocline (L) (Imberger and Patterson 1990). The Wedderburn number combines the effect of stability, wind exposure, and the mixed layer aspect ratio in a single parameter (W):

$$W = R_i\frac{H}{L} = \frac{g'H^2}{\mu_*^2 L}$$

$$= \frac{\text{depth-based buoyancy}}{\text{length-based wind mixing}};$$

EQ. 12.6

where R_i = the **bulk Richardson number**, which is the ratio of stability to kinetic energy for the mixed layer as a whole, H = thickness of mixed layer (m), g' = reduced gravitational acceleration due to the density jump across the base of the thermocline, L = length of lake at thermocline (m), and μ_*^2 = shear friction velocity (m s^{-1}) (Spigel and Imberger 1980).

The necessary wind shear velocity (μ_*^2) can be estimated from wind data:

$$\mu_*^2 = \left(\frac{P_a}{P_o}\right)C_D U_w^2,$$

EQ. 12.7

where P_a/P_o = 1.23 × 10^{-3}, the density of air (1.23 kg m^{-3}) divided by the density of water (1,000 kg m^{-3}), C_D = ~1.3 × 10^{-3} and ~1.0 × 10^{-3}, the drag (shear) coefficients of wind over water for large and small lakes, respectively, and U_w = wind speed (m s^{-1}).

Based on experimental work in rectangular laboratory tanks, Spigel and Imberger (1980) demonstrated that at high Wedderburn numbers (>> 1– >100) the wind energy is small compared to the strength of stratification. Thermocline tilting is therefore small, the amplitude of internal waves produced is small and there is little mixing (Sec. 12.11). But as W and L_N (utilizing the Schmidt stability index) decline to between one and five during a sustained period of strong wind, the mixing becomes stronger (MacIntyre et al. 1999). The upper and lower portions of the thermocline tilt downward at the downwind end of the lake and upward at the upwind end where metalimnetic and hypolimnetic water wells up and is *entrained* (incorporated) into the mixed layer. There is rapid deepening of the mixed layer at the downwind thermocline interface, and it is fastest in weakly stratified (Fig. 11–7) lakes and lakes with shallow mixed layers. The resulting upwellings lead to mixed layer nutrient increases and enhanced primary productivity.

Wedderburn numbers computed for southern African reservoirs under varying wind regimes yielded values for W during stratification that were typically small, indicating that stratification was weak, but W ranged up to 231 (Table 12–1), pointing to a wide variation in water-column stability, seasonal thermocline development, and associated differences in phytoplankton community structure (Allanson et al. 1990). An example of the importance of water-column stability in helping to determine phytoplankton species composition is the decline in the contribution of filamentous cyanobacteria to the phytoplankton biomass in a tropical Australian lake when L_N declined from ~ 50–100 to < 10 (Fig. 12–5). Another example (Fig. 12–6) illustrates the importance of water-column stability on the abundance of both large filamentous algae and large zooplankton (*Daphnia sp*). Water-column stability in Lake Windermere (GB) was associated with high numbers of large inedible algae and few small edible algae that are favored under less stable conditions. Since small algae provide most of the food for zooplankton, predators thrive in years of low stability linked to the North Atlantic Oscillation (NOA; Sec. 5.4).

While the calculation of Wedderburn or lake numbers is useful for determining the reaction of the thermocline to individual wind events the evaluated mechanisms cannot explain or predict the depth of the mixed layer at any particular time because the mixed layer depth (H) is the integrated outcome of turbulence induced by many individual heating and cooling events and wind episodes (Davies-Colley 1988). Even so, the *average* thickness of the summer mixed layer and the thermocline are predictable from exposure to wind (fetch) and turbidity (Table 11–2 and Fig. 11–14).

Thermoclines of stably stratified lakes examined over a vertical distance of a meter or so are characterized by high gradient Richardson numbers, indicating lack of turbulence within the thermocline, with the thermocline acting as a barrier to the rapid exchange of heat, gases, and nutrients between the hypolimnia and epilimnia. However, when thermoclines are examined in much more detail over short depth intervals they are revealed as systems of thin isothermal layers separating sheets of water that greatly differ in temperature (see Fig. 12–2). Limited evidence shows these individual sheets or layers within thermoclines (< 10 cm thick) to be zones of relatively high shear velocity and occasional turbulence (Kullenberg et al. 1974).

Table 12–1 **Measured Wedderburn numbers on individual but unspecified dates for several southern African reservoirs and lakes under varying wind regimes. H = depth of mixed layer (m) and L = length of reservoir (km). Note the relatively low stability of most systems.**

Reservoir/Lake	Country	Epi/hypo (°C)	Wind Speed (m s^{-1})	H/L (10^3 m)	W
Kariba	ZW, MZ	28.5/22	1.4	20/102	21.1
Kariba	ZW, MZ	28.5/22	11	20/102	0.3
Midmar	ZA	23/19	3.5	5/3	3.7
Sibaya	ZA	27.5/27	5.5	18/8	0.5
Hartbeespoort	ZA	22.5/20.5	2.5	6/2	8.0
Le Roux	ZA	20/14	2.2	20/2.4	231
McIlwaine	ZW	20.8/16.5	3.5	6/15.7	1.0

Source: Allanson et al. 1990.

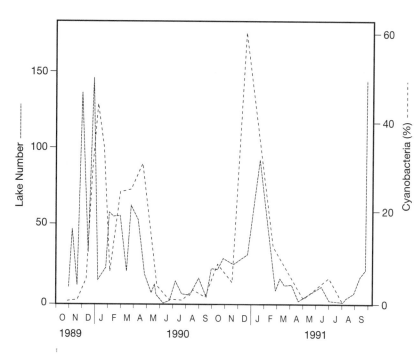

Figure 12–5 Lake Julius North East Australia. Within-year variation in Lake Number (—) in relation to the % contribution of large cyanobacteria to the phytoplankton biomass at 0.5 m depth (---). (*After Boland and Griffiths 1995.*)

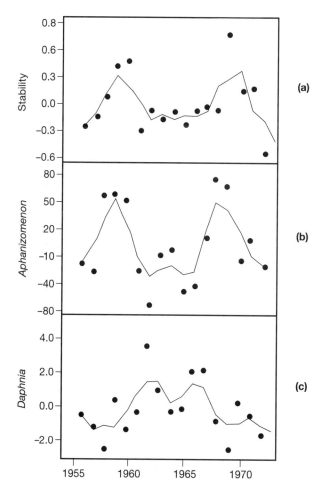

Figure 12–6 Year-to-year changes in (a) the intensity of thermal stratification; (b) the abundance of colonies of the large filamentous algae, *Aphanizomenon*; and (c) the number of *Daphnia hyalina* var. *lacustris* in Esthwaite Water (GB). The *Aphanizomenon* and *Daphnia* time-series have been corrected for change in trophic status with a simple linear model, with the data presented as residuals from the mean of zero. Note that years of high water column stability are associated with a high abundance of inedible *Aphanizomenon* colonies whereas years of low stability show a greater abundance of edible (nanoplankton) algae, coupled with a high *Daphnia* abundance. Note, quite different conclusions would have been drawn if the time scale had been limited to one or a few summers (Sec. 2.6). (*After George et al. 1990.*)

The thin sheets are fractured when seiche waves in the thermocline of larger lakes become steep and unstable or break. The resulting instability is referred to as **Kelvin-Helmholtz instability** (Sec. 12.11). Wind-generated internal waves then spread kinetic energy far from the point of production, and the resulting instability erodes the top of the thermocline. Associated thickening of the interface layers within the thermocline, known as billowing, allows much higher diffusion rates of substances across thermoclines than would be possible if the thermocline sheets could truly be characterized as zones of strictly laminar flow (non-turbulence) (Sec. 12.11). The billowing that causes mixing in the thermocline and increases exchange with the overlying water operates on a time scale of tens to hundreds of seconds; the same timespan over which nutrient uptake and photoadaptation of algae occur (Sec. 12.13).

▲ 12.5 Coefficient of Vertical Eddy Diffusion

The behavior of a turbulent water mass is conveniently expressed quantitatively by the **coefficient of vertical eddy diffusion** (K_v). This coefficient is a measure of the intensity of mixing that provides not only an indication of the vertical flux of gases and nutrients between the hypolimnion and epilimnion but also of momentum and heat (Hutchinson 1957, Spigel and Imberger 1987). Until fairly recently, it was assumed that conductivity of heat and mass were the same and that the easily obtained measure of vertical temperature change (heat) was also appropriate to estimate the more difficult to measure diffusion of gases and nutrients (mass). But when Quay et al. (1980) injected tritiated water (H^3) into the hypolimnia of two lakes, they confirmed that the lakewide vertical K_v across the thermocline for H^3 was far slower than for heat, showing the thermocline to be a greater barrier to the transfer of mass than heat.

When the K_v coefficient is multiplied by the vertical concentration gradient of a particular gas or nutrient between the epilimnia and hypolimnia, it provides a measure of the exchange rate between the two zones. Measurements have shown K_v to be highly variable over time and space and to range from as low as 10^{-6} m^2 s^{-1} (approaching molecular diffusion)[3] during periods of very strong stratification to values as high as 10^{-4} m^2 s^{-1} in epilimnia during periods of weak stratification (see Imberger and Patterson 1990). K_v can be one to four orders of magnitude greater near the shore, where the thermocline region intersects the lake bottom, than offshore—with major consequences for sediment suspension and nutrient recycling (MacIntyre et al. 1999). The critical point that characterizes the onset of turbulent conditions is reached when K_v exceeds 10^{-6}–10^{-7} m^2 s^{-1}. Conversely, when turbulent shear is reduced, viscosity becomes dominant and vertical density differences develop.

Longer Term Diffusivities

Physical limnologists are interested in instantaneously measured diffusivities and diffusivities over time within and between hypolimnia and epilimnia, as well as with the specific mechanisms and processes that bring them about. Biological limnologists (and oceanographers) are interested in measured diffusivities over time for linking physiological adaptations of organisms over the same time scale, and/or time integrated (averaged) fluxes of heat and mass over time of days or weeks; scales that involve both molecular and turbulent diffusivities and encompass the doubling rate of many planktonic organisms. To avoid confusing the short-term diffusivities (K_v) with longer term time-averaged pseudo-diffusivities—of greatest interest to those making among-system comparisons (Sec. 2.6)—these are indicated by the symbol E_v (Spigel and Imberger 1987). E_v within and across a thermocline is predictable from the lake surface area (Fig. 12–7).

The small E_v in the thermocline of wind-protected lakes indicates a physically stable environment for periods of weeks to months, or even centuries in the case of certain meromictic lakes (Sec. 11.9). Stability allows a phytoplankton and/or photosynthetic bacterial community to become established in or below the metalimnion of those lakes that are also transparent enough to permit the required photosynthesis (Sec. 22.10). Plant nutrients required for growth may be derived from turbulence-enhanced diffusion through the thermocline in the open water region of large lakes. But recent work on a large, now meromictic lake shows inshore turbulence where thermocline motions cause turbulence at the lake's edge (**boundary mixing**) that is great enough to inject sufficient nutrients into the thermocline region to nourish the deep water (metalimnetic) chlorophyll maximum throughout Mono lake (California, US). The K_v values at this depth were

---

[3]Molecular diffusion coefficients of dissolved substances in water are $\leq 10^{-7}$ m^2 s^{-1}.

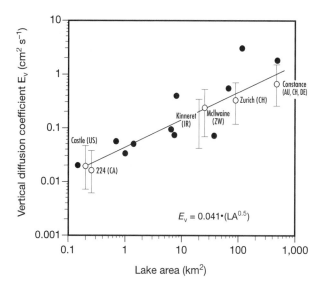

Figure 12–7 Mean values of average vertical diffusion coefficients during stratification in the upper hypolimnia of various lakes as a function of lake surface area. Bars indicate the observed range of variation. Solid circles represent older data summarized by C. H. Mortimer (1942). *(After Maiss et al. 1994.)*

two to four times greater in the inshore zone of thermocline boundary mixing than offshore at the same depth (MacIntyre et al. 1999). Similarly, Ostrovsky et al. (1996) attributed increases in chlorophyll concentrations and fish abundance in the nearshore waters of Lake Kinneret (IL, JO) to a combination of upwelling plus mixing at the top of the thermocline.

Using the diffusion rate of a naturally occurring radioisotope released from sediments (Radon 222)[4], Imboden and Emerson (1978) determined a K_v of $5–15 \times 10^{-6}$ m^2 s^{-1} in the hypolimnion[5] and $1–4 \times 10^{-6}$ m^2 s^{-1} in the thermocline of Lake Greifen (LA = 8.6 km^2, $\bar{z}$ = 18 m, z_{max} = 32 m), a eutrophic Swiss lake with a high hypolimnetic phosphorus concentration. They then used thermocline K_v values and the phosphorus concentration gradient between the anoxic hypolimnion and the oxic epilimnion to calculate the P

flux rate to the epilimnion by means of Fick's equation:

$$M = K_V A \frac{d_x}{d_z}, \qquad \text{EQ.12.8}$$

where M = metalimnetic flux (kg d^{-1} or mg m^{-2} d^{-1}), K_v = vertical diffusion coefficient (cm^2 s^{-1}), A = lake area at the thermocline z (m^2), and d_x/d_z = phosphorus concentration gradient (mg m^{-4}). This computation showed that, when river discharge from the drainage basin was relatively low in summer, more phosphorus was being returned daily to the epilimnion from the hypolimnion through *internal loading* than was obtained from the drainage basin by *external loading*. Similarly, the upward flux of P from the anoxic hypolimnion of a eutrophic Wisconsin (US) lake (Mendota) exceeded, by a factor of 10, all other fluxes to the epilimnion during stratification over a dry summer, but equaled the external loading during a wet one (Soranno et al. 1997). Internal and external phosphorus loading will be discussed in Chapter 17.

▲ 12.6 Coefficient of Horizontal Eddy Diffusion

In addition to the vertical diffusivities, **horizontal diffusivities** (K_H) are sometimes computed as well. These can be used to estimate the horizontal, rather than vertical, exchange of heat and mass. It is not surprising that, without the dampening effect of the thermocline, epilimnetic K_H are vastly greater than vertical ones (K_v) across hypolimnia, which are also lower than the hypolimnetic diffusivities. This was demonstrated by Schaller et al. (1997), for example, who measured K_H values of 0.02–0.03 m^2 s^{-1} in the open water of the epilimnion of Lake Baldegger (CH), with the coefficients four to five orders of magnitude larger than the vertical ones ($K_v < 10^{-6}$ m^2 s^{-1}). Much earlier work in Lake Ontario, (CA, US) (Murthy 1976) had shown hypolimnetic K_H to be one to two orders of magnitude lower than those in the epilimnion.

It is generally assumed that horizontal diffusivities are sufficiently large to produce a horizontal uniformity in chemical and plankton composition, thereby allowing investigators to sample the pelagic zone of most natural lakes at a single station only. Unfortunately, horizontal diffusivities and currents vary considerably over time and space in response to both

---

[4]Radon 222 is a natural radio tracer with a short half-life of 3.3 days, making it suitable for monitoring mixing events lasting a week or less. For more on it, see Section 20.6.

[5]E_z values similar to those noted in *hypolimnia* have been recorded in limnocorrals, (large plastic enclosures for experimental manipulations) that are floated in the *epilimnia* of lakes. Consequently, the turbulence in such epilimnetic enclosures is much smaller than in the adjacent epilimnia themselves (Bloesch et al. 1988).

storm-induced stirring events and localized differences in heating and cooling, allowing the development of rarely examined patchiness of the plankton and concentration gradients of all kinds between the littoral zone and open water (see Figs. 22–7 and 23–2).

The horizontal diffusivities measured are the product of wind- or heat-induced movements, apart from those imposed by river currents that dominate rapidly flushed reservoirs and riverine lakes. Since wind shear is rarely uniform over lakes, neither are currents and associated turbulence. Small lakes are often very affected by localized protection from the wind, whereas diffusivities in large lakes vary in response to spatially differing wind, heating, and evaporation, thereby contributing to the complexity of water movements (Fig. 12–1).

12.7   Horizontal Currents

The primary action of wind is to apply a force (shear) to the lake surface and thereby transfer the energy needed to set the surface in motion. Surface-floating drift bottles or depth-specific drogues suspended from a small surface float, as well as anchored current meters are used to measure the direction and rate of flow (Fig. 12–8). A number of marine and freshwater investigators have determined the **wind factor** which describes surface current speed as a fraction of wind speed over time. In Lake Windermere (GB), surface current speeds range between 1.5 and 3.5 percent of the mean wind speed (Fig. 12–9) with the percentage declining to a relatively constant one percent at wind speeds above about 6 m s^{-1}. Mean wind speed in the same lake is a good predictor of mean surface current velocity (George 1981).

$$0.5 \text{ m velocity (cm s}^{-1}) = 0.76 + 0.013 \text{ wind speed (m s}^{-1})$$

$$r^2 = 0.94 \quad n = 13. \qquad \text{EQ.12.9}$$

With increasing depth there is increasing deflection of horizontal currents—to the right in the Northern Hemisphere, and to the left in the Southern Hemisphere—brought about by the *Coriolis effect* (Fig. 5–11), which is the outcome of trying to move in a straight line on a rotating planet. The rotational effect is greatest at the poles and zero at the equator, so the effect of a particular wind speed on the direction of currents produced is expected to be greater in large equatorial lakes without a rotational effect than in their temperate zone counterparts (Sec. 11.6). The

consequence of this progressive rotation is that at some depth, termed the **depth of frictional resistance**, the flow direction is exactly opposite to the wind direction. At that point the current speed in unstratified lakes is approximately $\frac{1}{23}$ that at the surface. The resulting flow—resembling a spiral staircase—extends from the lake surface to the depth of frictional resistance and is called an **Ekman spiral**. The deflected current is called the **Ekman drift**; both are named after discoverer V. W. Ekman (1905). The minimum size lake where drift will be evident is larger than ~1 km^2 and deeper than ~4 m (George 1981). However, Ekman drift is important only in larger lakes (> ~5 km fetch) when winds blow from one direction for an extended period.

In Lake Windermere (GB) the Ekman spiral is clearly evident (Fig. 12–10) but, as usual, it is imperfect because the pattern is modified by the lake bottom, deep counter currents under isothermal conditions, and the presence of a thermocline. A reverse flow is evident just above the thermocline and a third reversal is seen in the metalimnion. When lakes are isothermal (homothermal) there is normally not only a major near surface flow roughly along the wind axis but also a transverse flow (Fig. 12–11).

Wind-induced surface currents—with typical speeds of ~2–15 cm s^{-1} in mid-sized lakes, but up to 30 cm s^{-1} in larger lakes—show an exponential decline with depth in isothermal lakes (Figure 12–10). When the surface current is known, the current speed at any depth (U) can be estimated from the equation

$$U = U_s e^{-kz}, \qquad \text{EQ.12.10}$$

where U_s = surface current, e = 2.718, z = depth, and k = the current "extinction coefficient" (analogous to the vertical light extinction coefficient Eq. 10.3). Determination of the current extinction coefficient requires one flow measurement at the surface and one at the depth below which the flow is negligible (Smith 1979). These measurements are usually made with anchored current meters that record current speed and direction, or with near-surface or subsurface floats (drogues).

Wind-Induced Circulation

Circulation patterns in elongated lakes are of two basic types according to George (1981). In elongated shallow lakes the mid-lake current flowing in the di-

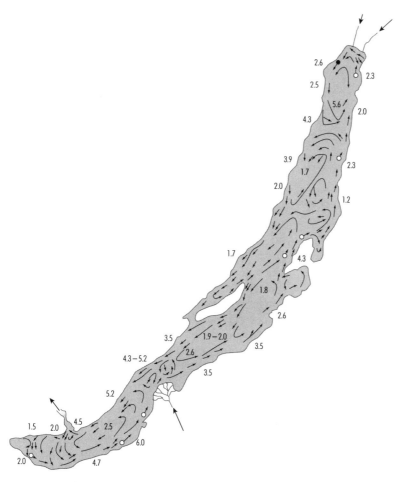

Figure 12–8 Typical current patterns and speeds in near-surface waters during summer and fall in Lake Baikal, RU (LA = 31,500 km², z_{max} = 1,741 m). Numbers recorded along the edge represent average velocities (km day⁻¹) near shore. Lower velocities characterize the open water. *(After Verbolov et al. 1989.)*

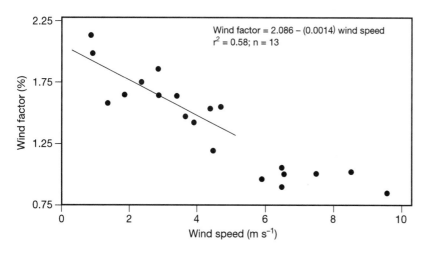

Figure 12–9 The relationship between wind speed and the wind factor, the current speed to wind speed ratio, as determined in Lake Windermere (GB) for wind speeds of up to a critical value of ~500 cm s⁻¹. *(After George 1981.)*

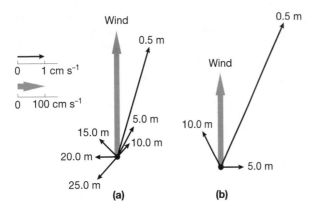

Figure 12–10 Vertical distribution of currents for monom-ictic Lake Windermere (GB) (south basin, LA = 6.7 km^2, $\bar{z}$ = 18 m) on (a) January 29, 1974 (unstratified) and (b) July 27, 1973 (thermocline at 13 m). *(After George 1981.)*

rection of the wind is balanced by return currents along the shores, whereas deeper, unstratified lakes of the same outline have return currents flowing below the surface currents. Another type of circulation occurs in stratified lakes. In Lake Windermere (GB), circulation has been conceptualized as a distorted "conveyer belt" running along the wind axis, with the counter current just above the thermocline flowing 90° to the left of the wind direction (Fig. 12–12). The resulting helical circulation means a longer residence time for pollutants or nutrients entering from catchments than would be the case if the flow direction

were closer to the longitudinal axis of the lake. Even the time-averaged near-surface current patterns made in enormous Lake Baikal (RU) show considerable complexity and, in addition, exhibit much larger current speeds in the shallow coastal regions (Fig. 12–8).

Measurements of currents in the Laurentian Great Lakes (CA, US)—proportionally much wider—do not show a distorted conveyer-belt pattern like Lake Windermere, indicating the importance of lake morphometry in creating lake-specific current patterns. However the pattern creates a strong tendency for entering effluents to remain concentrated in the nearshore regions during stable meteorological conditions (Boyce 1974). The pattern here too results in a longer residence time of wastewater and agricultural runoff than if the flow had been along the longitudinal axes of the lakes. Elevated nearshore currents point to a faster distribution of incoming nutrients and toxins along the edges, as in Lake Baikal (Fig. 12–8), than in the pelagic zone.

Density Currents and Lake Management

Not all horizontal currents are wind induced. **Density (convection) currents** also play an important role in water circulation, with velocities of the same magnitude as those imposed by wind stress (Table 12–2). Within hours, differential heating and cooling is sufficient to transport nearshore water to and from the center of small lakes, as well as between the arms and

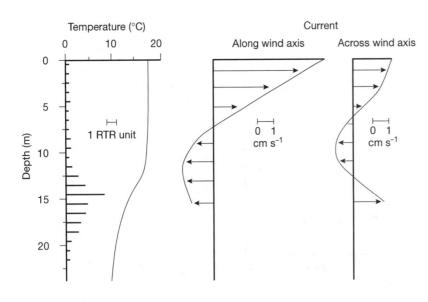

Figure 12–11 Deduced current profile in Lake Windermere, GB (south basin) during a period of steady wind and stable stratification on August 23, 1994. The temperature profile present is also presented as the relative thermal resistance (RTR) to mixing (see also Fig. 11–8). *(After George 1981).*

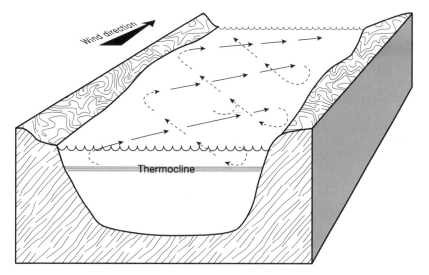

Figure 12–12 A conceptual representation of the circulation in stratified Lake Windermere, GB (south basin). The estimated pattern is one resembling a distorted conveyor belt with surface currents deflected to the right of the wind due to the Coriolis effect and countercurrents just above the thermocline moving at about 90° to the left of the wind direction. *(After George 1981.)*

main body of dendritically-shaped reservoirs (Imberger and Patterson 1990) (see Fig. 29–3). During the day, solar heating warms shallow bays and the arms of lakes and reservoirs more rapidly than the remainder of the lake. The warmed, lighter water (and materials therein) then flows as an overflow current to the pelagic zone, to be replaced by cooler epilimnetic water. Conversely, the small water volume of the littoral zone cools more rapidly on calm nights. The cooled water and materials contained therein then flows downslope until it encounters water of the same density, after which it produces a horizontal **intrusion**

current from shore (see Fig. 12–1); the littoral water and nutrients so lost are replaced by pelagic water (Monismith et al. 1990, James and Barko 1991). The thermally driven flow, the **thermal siphon**, greatly increases the flux of nutrients, contaminants, and organisms between inshore and offshore regions.

Shallow, upwind portions of lakes sheltered from wind stress experience reduced evaporative cooling during sunny days, creating density currents—also known as **buoyancy driven flows**—between upwind and more deeply mixed downwind portions of a lake. Horizontal density currents are of great interest in

Table 12–2 **Expected range of current speeds associated with different types of motions in moderate sized lakes such as Lake Windermere, GB. Higher speeds are encountered during exceptional storms and in larger lakes.**

Type of Motion	Source of Estimate	Current Speeds (cm s^{-1})
Surface seiche movements	Calculated from water level records.	0.1–2.0
Internal seiche movements	Estimated from published observations, but up to 30 cm s^{-1} in great lakes.	0.5–3.0
Inshore–offshore density currents	Calculated from hydrographic temperature distribution data and dye release.	1.0–15.0
Direct wind stress on water surface	Estimated from measured current speed and wind speed.	2.0–15.0

Source: George 1981, and Nepf and Oldham 1997.

aquatic management. For example, a river flowing through a contaminated river–wetland system near Boston (Massachusetts, US) carries arsenic to the receiving lake at a lower temperature than the lake water. During periods of base flow in summer (low discharge, primarily lower temperature groundwater), little of the arsenic enters the epilimnion because the cooler water plunges to the thermocline. However, during episodic summer storms resulting in high discharge (runoff), nutrient, and contaminant loads, the inflow temperature is closer to that of the warm air and surface-soil than the groundwater, causing surface and subsurface runoff to be warmer and less dense than the lake. The arsenic then enters at the lake surface, impacting the surface water quickly. A second, but unmeasured, contaminant flux pathway noted in the same system is between the wetland and lake—because the wetland is shallow and the water more turbid, it heats and cools more rapidly than the lake, creating an exchange flow between the two (Nepf and Oldham 1997).

Wetlands serve as traps for inflowing dissolved plant nutrients in demand and particulate matter. This characteristic is increasingly utilized in aquatic management through the construction or reconstruction of wetlands for wastewater treatment (Sec. 24.2). However, constructed wetlands may raise summer water temperatures above those of the inflowing river, known as *thermal mediation*, affecting the intrusion dynamics in the receiving lake. This factor was not considered in the design of a wetland intended to reduce the eutrophication of Lake McCarrons (Minnesota US). Although the wetland reduced the nutrient inflow, it did not improve the lake's water quality in summer because the warmed water carried the remaining nutrients directly into the epilimnion rather than allowing them to plunge into the thermocline as they did before (Andradóttir and Nepf 2000).

▲ 12.8 Long-term Surface Current Patterns

Emery and Csanady (1973) noted that the *long-term* summertime surface current pattern in large Northern Hemisphere lakes and nearly enclosed seas was counterclockwise in 29 of 30 waterbodies (see Fig. 12–8); the one exception is almost certainly erroneous. Their interpretation of the counterclockwise current patterns observed[6] is linked to the Coriolis effect, with the movement of sun-warmed surface waters is indeed to the right of the prevailing wind direction in the Northern Hemisphere (see Fig. 12–12). As a result, water on the righthand shore exhibits a smaller temperature difference between the surface water and overlying air than on the lefthand shore. This smaller temperature difference requires smaller wind stress to push the surface water than is needed at the left shore, where the upwelling of colder water further accentuates the temperature differences. The resulting long-term current pattern is, consequently, counterclockwise in the Northern Hemisphere, and clockwise in the Southern Hemisphere (Emery and Csanady 1973). The long-term counterclockwise circulation observed in the Northern Hemisphere has a significant effect on the dispersal, distribution, and water residence time of nutrients and toxins entering large lakes. A particularly clear example of this is the primarily counterclockwise movement of the insecticide Mirex along the shore of Lake Ontario (CA, US), which enters the lake with the Niagara River outflow (Pickett and Dossett 1979, and Thomas 1983).

▲ 12.9 Langmuir Currents

The existence of **Langmuir currents** was surmised by I. Langmuir (1938), during an Atlantic crossing. He noticed parallel, elongated lines of seaweed and foam where currents appeared to be converging. He subsequently examined lines of foam and their associated particles, known as **streaks** or **windrows**, in Lake George (New York, US) and noted that the streaks were oriented in roughly the same direction as the wind (Fig. 12–13). It is now clear that the slicks are formed where helical current spirals converge at the surface before starting their plunge below the surface (Fig. 12–14). The downwelling current traps natural oils and foaming agents, plant remains, and zooplankton at or near the surface (George and Edwards 1973). The area between the slicks is characterized by more diffuse, slower upwelling.

---

[6]Whereas careful observations (data) made with good equipment in nature or the laboratory are usually not in dispute, the interpretations made are often hotly disputed. Interpretations are based on observations (data) but go beyond them.

Figure 12–13 Langmuir streaks on a New York/Vermont bay of Lake Champlain (CA, US). (*Photo courtesy of J. Kalff.*)

The limnological importance of Langmuir currents lies primarily in the current spirals—with their downwelling and upwelling components—which play an important role in the diffusion of heat and mass, as well as in the vertical transport of plankton in the mixed layer. Langmuir currents are evident during windy periods on all except very small, wind-protected lakes. How the currents become established is still not completely understood, but it is evident that they are the result of interactions between horizontal surface currents (*drift*) and surface gravity waves (Leibovich 1983).

Measurements in lakes and oceans (Assaf et al. 1971) show the onset of streaks occurring at wind speeds of about 3–4 m s^{-1}, when streaks with 3–6 m spacings are common. At somewhat higher, but still moderate, wind speeds (5–15 m s^{-1}) a second major set of streaks emerges. The distance between the largest set of streaks at the highest wind speeds is about equal to the depth of the mixed layer (Harris

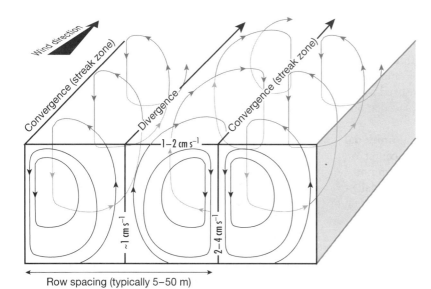

Figure 12–14 Schematic representation of Langmuir currents, helical currents with both large vertical and horizontal components. Row spacing and current speed are a function of wind speed.

and Lott 1973, and Boyce 1974). Langmuir-type circulation may reach depths of 100 m or more in Lake Ontario (CA, US) during strong winds under homothermal conditions (Assaf et al. 1971). During stratification, Langmuir currents contribute to the erosion of thermoclines. Down welling currents—typically 2–4 cm s^{-1} (Fig. 12–14)—are also more than sufficient to prevent any aggregation of either phytoplankton or zooplankton at some optimal light depth in the upper portion of epilimnia. Langmuir currents contribute to the high vertical eddy diffusion coefficients within mixed layers, and the resulting rapid mixing of dissolved substances in lakes (Boyce 1974).[7]

Langmuir and other mixing currents transport phytoplankton between the well and poorly illuminated portions of the mixed layer. In the process the organisms experience a rapidly changing light climate for photosynthesis (Sec. 10.11) that is quite different from that experienced in laboratory incubators, or when they are placed in glass bottles and suspended at fixed depths during in-situ primary production determinations. Vertical circulation times of phytoplankton in a mixed layer of about 10 m probably range from one half to a few hours, assuming reasonable values for wind speed and density gradients.

12.10   Standing Surface Waves

Standing surface waves, referred to as **surface seiches**, are free oscillations in which the water at one end of a lake rises in response to a steady wind from one direction, followed by its flowing back in the opposite direction when the wind lets up. The water that is piled up downwind flows back, but overshoots the equilibrium position in the opposite direction. The result is a standing wave which rocks back and forth. Surface seiches are most conveniently measured with a water level recording device (*hydrograph*) placed at one end of the lake near the shore.

The simplest type of surface seiche is **uninodal** (Fig. 12–15). But strong rainfall on a lake center or, more commonly, interactions between a uninodal seiche and lake morphometry can set up **binodal** or **multinodal** seiches, with the water rocking between

two or more nodes. Seiche-induced water movements are further complicated with the addition of cross-lake seiches. The rocking motions decline over time until reestablished by a subsequent wind event.

▲ Uninodal surface seiches, first reported in 1895 by F. Forel in Lake Geneva (CH, FR), exhibit maximum *vertical* transport at the two antinodes and none at the node, whereas *horizontal* transport is greatest at the nodes, with velocity a direct function of seiche amplitude. Surface seiches having a typical vertical amplitude of a few centimeters in moderate-sized lakes are of little interest in biological and chemical limnology because their energy levels are low as the result of frictional shear at the air–water interface and shores. The period of a uninodal surface seiche is lake specific and varies from minutes in small lakes to about four hours in inland seas such as Lakes Baikal (RU) and Huron (CA, US) (Hutchinson 1957). In those lakes where the lake's basin does not diverge greatly from the rectangular box for which the formula applies, the period can be approximated by

$$T = \frac{2L}{\sqrt{g\,\bar{z}}}, \qquad \text{EQ. 12.11}$$

where T = seiche period (sec), L = length of basin (m), $\bar{z}$ = mean depth of the presumed rectangular basin (m), and g = acceleration due to gravity (9.81 m s^2). More useful than the period of the measured surface seiche is the possibility of using it to calculate the mean depth of elongated basins after rearranging the seiche formula to

$$\bar{z} = 4L^2/gT^2. \qquad \text{EQ. 12.12}$$

Thus, only a hydrograph is necessary to measure the seiche period to obtain an estimate from the shore of the mean depth ($\bar{z}$), without the need for a cumbersome depth-sounding program. Stewart (1964) found excellent agreement between the two types of determinations for 10 lakes.

12.11   Internal or Thermocline Seiches

Internal seiches are of much greater importance than surface seiches because they have a much greater amplitude and effect. Where they become steep and unstable (Fig. 12–16), or are disrupted by contact with underwater mounds or lake boundaries, they bring about significant currents and turbulence in the hypolimnia (Davies-Colley 1988). Considerable vertical transport of water and nutrients from the hypolimnia

---

[7]Physical limnologists and oceanographers see Langmuir currents as large vertical eddies, whereas plankton ecologists are more interested in the organized motions and velocities affecting the distribution of organisms.

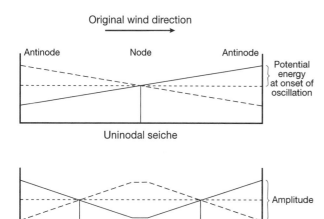

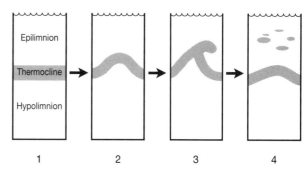

Figure 12–15 Diagram of uninodal and binodal seiches in a rectangular lake. The movement of a uninodal seiche is akin to that of a seesaw, and the binodal seiche resembles a trampoline. The potential energy relative to the still water is converted to kinetic energy when the stress (wind) is removed. *(In part after Smith 1979.)*

Figure 12–16 Schematic representation of internal wave formation and breakage of internal (thermocline) waves resulting from shear instability. (1) Condition prior to wave formation (2) An internal wave forms, (3) breaks and incorporates different water layers at the crest, and finally (4) produces water of intermediate density in the epilimnion. *(After Thorpe 1971.)*

to the epilimnia (Fig. 12–17), plus an important leakage of hypolimnetic water and nutrients into the epilimnia around the lake perimeter, is possible where the vertical amplitude of the seiche is substantial (Ostrovsky et al. 1996). Advective upward and downward transport, known as **entrainment**, not only carries water and nutrients into the euphotic zone but also brings dissolved oxygen and heat into hypolimnia. It is further responsible for the development of hypolimnetic and intrusion currents (Fig. 12–1) that carry entrained water and nutrients from lake edges toward the interior. Entrainment of nutrients into the euphotic zone has a major effect on phytoplankton production (Sec. 12.5), as does the injection of dissolved nutrients derived from nearshore upwelling into the metalimnion of lakes that are transparent enough to allow a metalimnetic chlorophyll maximum.

Internal seiches are created when the wind stress forcing epilimnetic water toward the downwind end of a lake produces thickening of the mixed layer there, causing the thermocline to become tilted (Fig. 12–17). Lessening of the wind establishes the characteristic oscillation of the thermocline described in Section 12.4. A second seiche type which dominates in some lakes, but is rarely reported, is one in which the thermocline region is periodically "squeezed" from one end of the lake to the other by the upper and lower layers (Gloor et al. 1994).

The speed of internal waves, dependent on basin length and the density gradient, may be as high as 30 cm s^{-1} in the Laurentian Great Lakes (CA, US) (Table 12–2). The velocity produces shear across the thermocline and the resulting turbulence is responsible for the exchange of water, heat, and nutrients between hypolimnia and epilimnia (Fig. 12–17 and Sec. 12.4). In a very large lake such as Lake Ontario, the combined long fetch and great depth permit an ~15m oscillation above and below the 12°C isotherm that characterizes the metalimnion (Fig. 12–18). However, the amplitude of oscillation is not simply a function of lake area but also of scale. Among lakes of similar size, it is primarily a function of the stability of stratification. The greater the density gradient (i.e., the higher the stability), the smaller the amplitude. Therefore, internal waves are larger in weakly stratified than stably stratified lakes. This is shown by recent measurements in a small, weakly stratified Alaskan lake (Toolik, US, LA = 1.5 km^2 and z_{max} = 22 m), which revealed a wide variety of internal waves. Internal waves in the 3 m thick thermocline had amplitudes of 1.5 m, while those in the hypolimnion typically ranged from 3 m to as high as 6 m (S. MacIntyre, pers. comm.). The widely held notion that internal waves and associated turbulence are of little significance in small lakes is probably a result of the traditional focus of physical limnologists on large lakes and oceans, and of much research on stably stratified small lakes (Sec. 4.4).

▲ An approximately 1000-fold density difference between air (D = ~0.001 g cm^{-3}) and surface water (D = ~1.000 g cm^{-3}), and a relatively minuscule difference

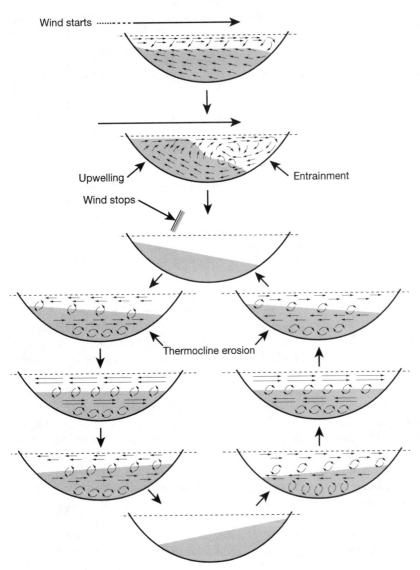

Wind starts ·······

Upwelling

Entrainment

Wind stops

Thermocline erosion

Figure 12–17 Schematic representation (exaggerated) of the role of wind on establishing oscillation (rocking back and forth) of the thermocline (gray), the direction of epilimnetic and hypolimnetic currents (arrows), turbulence at the thermocline and lake bottom, and the entrainment of nutrient-carrying hypolimnetic water into the epilimnion. (*Modified after Beadle 1981, and Mortimer 1974.*)

(0.0016) between epilimnetic water at, say, 20°C and hypolimnetic water at 10°C, is ultimately responsible for much larger internal seiches than surface seiches. Theoretically, depression of the metalimnion by the amplitude of the surface seiche is

metalimnion depression (m) =   EQ. 12.13

$$\frac{\text{surface amplitude} \times \text{metalimnion density}}{\text{metalimnion density} - \text{epilimnion density}}.$$

Using the temperature values above, it is easy to calculate that 623 times more energy is required to produce a surface wave of the same amplitude as an internal wave. Thus, a surface seiche only 1 cm above and below the mean could depress the thermocline by about 6 m (in theory, at least). However, thermocline depression in nature and the resulting observed wave height is a complex product of wind speed, lake length, underwater morphometry and water column stability; in reality, it is considerably smaller than predicted by Eq. 12.13.

The first internal seiche mode can be readily estimated, by modifying Eq. 12.14, for surface seiches in a two-layer system (Hutchinson 1957); again without considering the Coriolis effect and the fact that lakes diverge from rectangular basins:

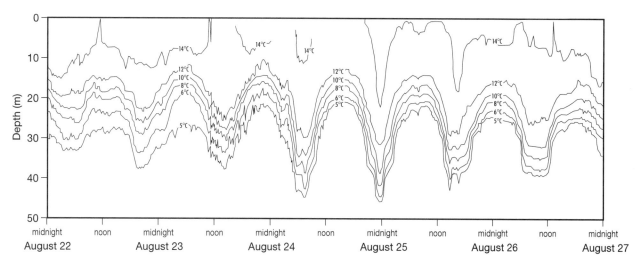

Figure 12–18   Oscillation of the Lake Ontario, (CA,US) thermocline produced by internal waves and measured with an array of thermistors moored 10 km from the north shore from August 22–27, 1972. The plot shows isotherm depths as a function of time. *(After Boyce 1974.)*

$$T = \frac{2L}{\sqrt{\left[g\,(D_h - D_e)/\left(\dfrac{D_h}{\bar{z}_h} + \dfrac{D_e}{\bar{z}_e} \right) \right]}}$$

$$= \frac{\text{distance traversed}}{\text{velocity of internal wave}}, \qquad \text{EQ. 12.14}$$

where T = period of first internal seiche (hours), L = length of a rectangular basin (m), g = gravitational acceleration (m s^{-2}), D_h = density of hypolimnetic water (g cm^{-3}), D_e = density of epilimnetic water (g cm^{-3}), $\bar{z}_h$ = mean depth of hypolimnion (m), and $\bar{z}_e$ = mean depth of epilimnion (m).

Internal seiches are not only much larger but also more slowly dampened than those at the surface. Thus, thermocline seiches in Baldeggersee, CH (LA = 5.2 km^2, $\bar{z}$ = 34 m, z_{max} = 66 m, Fig. 6–1) normally persist for more than a week (Imboden et al. 1983). A new wind event destroys or modifies a previous seiche in this and other lakes, with the result that a particular seiche is usually composed of remnants from earlier wind impacts plus the response to the most recent event. Smaller amplitude binodal, trinodal, and cross-basin waves created by the interaction of basin morphometry and wind can be measured. They contribute to the development of instabilities and turbulence at the thermocline, both in the pelagic zone and where seiches "touch bottom" at the lake edge.

The impact of the earth's rotation (the Coriolis effect) and lake morphometry on internal seiches and currents is appreciable in larger lakes (fetch > 2–3 km) at higher latitudes. The Coriolis effect on internal seiches allows the development or modification of a variety of waves. Among the best known are **Kelvin waves,** basin-scale internal gravity waves that rotate at the lake boundary in a counterclockwise direction (Northern Hemisphere) about a stationary central node in the lake. The waves produce large oscillations near the shore that create fast currents known as **coastal jets** running parallel to the shoreline. Kelvin waves in the Laurentian Great Lakes (CA,US) can tilt the nearshore thermocline by several meters, allowing hypolimnetic water and nutrients to leak into the well-illuminated and typically productive coastal zones (Fig. 12–19). When Kelvin waves become unstable and break, following contact with the lake bottom, they produce other types of waves that enhance nearshore turbulence (MacIntyre et al. 1999). Kelvin waves reverse their direction at intervals, allowing pollutants and organisms that had been concentrated in the coastal zone to be quickly transported offshore (Boyce 1977). Another type of basin-scale internal wave is a cross-basin seiche known as a **Poincaré wave.** Like Kelvin waves, these develop after the onset of thermocline seiches in large lakes, but their effect is most apparent in the pelagic zone where they create pressure on the thermocline at some locations, while releasing pressure in adjacent ones. This produces a mosaic of alternating water mounds and depressions that appear to be important mechanisms for producing currents in the hypolimnia (Mortimer 1974). When Kelvin and Poincaré waves interact with a sloping bottom, the internal wave spectrum is changed. An

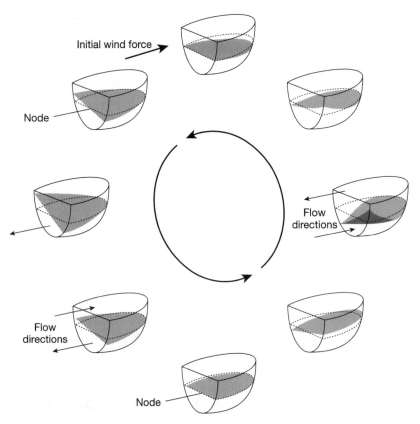

Figure 12–19 One cycle (exaggerated) of a Kelvin wave as reflected in the nearshore tilting of the thermocline, resulting in large nearshore currents. The shaded area represents the oscillating lake surface, and the equilibrium position is indicated by a dashed line. *(Modified after Mortimer 1974.)*

increase in energy and shear may be the result of wave deflection and reflection producing instabilities and mixing, in the process creating a turbulent boundary layer over the lake bottom (Saggio and Imberger 1998, and Sec. 12.12). Again, measurements in Mono Lake, California (US), allowed the conclusion that nearshore boundary mixing at the base of the thermocline, following high winds, produced horizontal currents. These currents distributed sufficient ammonium to the phytoplankton peak in the metalimnion to sustain primary productivity of the chlorophyll maximum throughout the lake (MacIntyre et al. 1999).

12.12   Internal Seiches, Hypolimnetic Currents, and Sediment Resuspension

Different internal waves do not themselves cause turbulence and currents, but they erode thermoclines, create currents in the hypolimnia, and resuspend and transport sediments and their associated materials when they become unstable and their energy is converted to turbulent energy. The conversion of wave energy to turbulent energy can be triggered by wave–wave or wave–sediment interactions and by wind shear instability (see Fig. 12–16).

Horizontal currents and associated turbulence in the hypolimnia are created when internal seiche waves—and Kelvin and Poincaré waves in large lakes—move past the roughness at the boundaries of lakes, and when internal waves in the pelagic zone become large enough for the slope of the leading edge to become unstable and break (see Fig. 12–16). The resulting turbulent billows enhance diffusion across the thermocline. This is called **Kelvin-Helmholtz mixing** (or *K–H instability*), extends beyond the metalimnia in a vertical direction, deepening the mixed layer and generating currents (turbulence) in the hypolimnia. In larger lakes, fast currents (known as **bores**) are created that race back and forth within thermocline layers of only slightly different densities, creating turbulence (Imberger 1985) and adding to the K–H instability. As the turbulence created by internal waves decays, new internal waves are produced.

Hypolimnetic Currents

▲ There are few direct measurements of currents in the hypolimnia because the minimum current speed that can be resolved with standard meters in small and mid-sized lakes is close to, or below, observed hypolimnetic current speeds. Direct measurements were obtained just above the bottom of one such lake (monomictic Baldeggersee, CH (LA = 5.2 km^2, z_{max} = 66 m) by means of an anchored, floating sphere whose deflection from the vertical was photographed every three minutes by a camera positioned above it. The deflections showed current speeds close to the bottom averaging around 1.5 cm s^{-1} in both summer and winter, but current speeds of > 2.5 cm s^{-1} were not uncommon in summer (Imboden et al. 1983). Summertime currents, which showed frequent reversals in direction, are occasionally high enough to allow bursts of intensive horizontal and vertical mixing of deep hypolimnetic water. Deep water currents of similar magnitude, also induced by thermocline seiches, have been recorded in Lake Windermere, GB (see Table 12–2).

Sediment Resuspension

The sediment resuspension observed in the profundal zone of a number of well-stratified lakes is typically attributed to even faster seichelike currents that are only produced occasionally, with measured speeds of up to 7 cm s^{-1} (Gloor et al. 1994). Not surprisingly, the possibility of resuspension is greatest when the thermocline is located not far above the lake bottom (Pierson and Weyhenmeyer 1994). The resuspension of fine (2–8 μm dia.) deep-water sediments creates a **benthic boundary layer** (BBL) or **nepheloid layer** (BNL), not only in large lakes and oceans (Eadie et al. 1984) but even in medium-sized lakes (Bloesch and Uehlinger 1986). The nepheloid layer in Lake Alpnech, CH (LA = 4.8 km^2, z_{max} = 34 m) is commonly 2–7 m thick and has a largely organic particle concentration that is two to four times greater than in the water above (Gloor et al. 1994). In the much larger lakes Michigan and Superior (CA,US), the BBL extends 5–30 m above the sediment and, in Lake Michigan, may even extend to the top of the hypolimnion.[8]

---

[8] Resuspension of sedimented organic matter can be large—75 g C m^{-2} yr^{-1}, at a depth of 100 m in Lake Michigan. The Lake Michigan resuspension, measured with deep-water sediment traps, represents about ~ 10 times larger upward flux than downward flux of photosynthetically produced particles in the euphotic zone (Eadie et al. 1984).

When the fine, resuspended sediments settle during periods of reduced seiche activity, they produce—together with sedimenting organic particles derived from the euphotic zone—a ~5-mm-thick "fluffy" layer at the sediment surface, known as the **sediment boundary layer** (SBL).

While nepheloid layers can be generated by (1) thermocline motions, other possible mechanisms include (2) sediment resuspension produced by plunging inflowing rivers, sometimes having a high sediment load; (3) deep water currents produced during isothermal conditions; and (4) the activity of benthic organisms. Whatever the mechanism, resuspended sediments captured in deep-water sediment traps produce an overestimation of the *net* sedimentation rates (Chapter 20). (Imboden and Wüest 1995).

12.13   Turbulent Mixing and the Biota

Important progress has been made, but current patterns, internal waves, and the associated turbulence are very complex and cannot yet be comprehensively modeled and predicted. Currents and waves differ not only spatially and as a function of basin morphometry but also temporally in response to changes in heat input and wind stress, as well as to lagged responses to past events. Lakes of complex morphometry and exposure are particularly subject to spatially variable heating, cooling, and wind mixing that produce horizontally changing density gradients and currents. The biota quickly responds physiologically to changes in heat, diffusion of nutrients and gases, and changes in irradiance; which, with lags, are reflected in growth rate changes. If the change is sustained long enough, it will be reflected—with a greater lag time—in changes in population size and community structure.

Time Scales and Biotic Responses

▲ The observed time scales (Fig. 12–20) of these physical processes (see Fig. 12–1) and their impact on the biota range from periods of minutes for changes in surface irradiance to affect photosynthetic rates of algae, to periods of a half-hour to several hours necessary to mix the euphotic zone and change the maximum photosynthetic rate (P_{max}, mgC mg chl-a^{-1} t^{-1}) (Cullen and Lewis 1988). The P_{max} is a good predictor of the daily depth-integrated phytoplankton production (mgC m^{-2} d^{-1}). Periods of hours to weeks or months are required to bring about major changes in

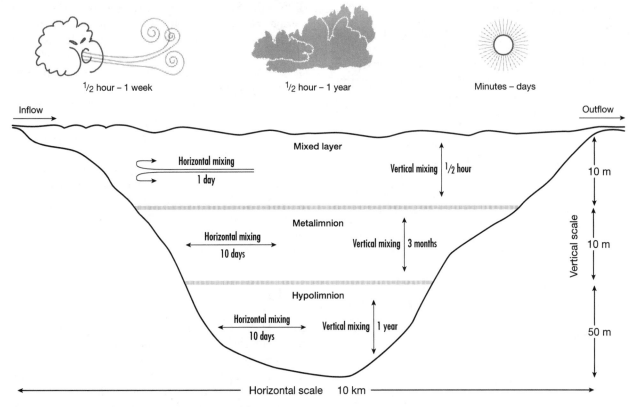

Figure 12–20   Estimates of mixing times in a medium-sized holomictic lake with a thick hypolimnion. *(Modified after Imberger 1985.)*

heating, cooling, stratification (Chapter 11), flushing, nutrient inputs (Chapter 8), and spates (high water) in lotic systems; matching the time changes needed to affect longer-term algal community growth rates (Chapter 21) which bring about changes in algal and invertebrate species composition (Chapters 21 and 23).

Physical processes that operate over time scales of months to years are responsible for changes in seasonal and interannual temperatures, stratification, dissolved oxygen concentrations in the hypolimnia, and in water and nutrient loading. They are of the same time scale as the generation times of macroinvertebrates (Chapter 25), development of submerged macrophyte beds (Chapter 24), as well as changes in invertebrate and fish species composition and the success or failure of particular fish cohorts (year-classes of the population) (Chapter 26). Physical processes operating over time scales of decades or centuries—reflected in changes in climate, land use, and pollution—bring about changes in community composition

and productivity that are best evaluated by paleolimnological techniques (Chapter 20). The principal determinants (predictors) of community structure and function, and the appropriate modeling approaches, obviously must change with the time scale under examination (Sec. 2.6). It is evident that progress in predicting population and community behavior is closely linked to advances in physical limnology.

Highlights

- There are two kinds of turbulent motion in lakes: waves, exhibiting a periodicity but little forward motion; and currents, which lack periodicity. Currents create waves, and waves break to produce currents.
- Turbulent flow, rather than laminar flow, is responsible for mixing heat, gases, nutrients, and particles in both water and surface sediments.

- Other than where gravity-imposed river flow dominates, it is wind-induced stirring of the water's surface and surface cooling—producing convection currents—that create the needed turbulent kinetic energy in lakes.
- Highly visible surface gravity waves bring about oscillations of water particles, primarily in the shape of an ellipse having little horizontal motion. Wave energy is dissipated through the creation of ellipses in a vertical direction that rapidly decline in size with depth.
- The relative strength of turbulent mixing and the thermally or chemically produced density gradients of the thermocline or chemocline, whose stabilizing buoyancy produces resistance to mixing, both determine the likelihood of turbulence in portions of the mixed layer and whether temperature stratification can be maintained.
- ▲ The coefficients of vertical (K_v) and horizontal (K_H) eddy diffusion are widely used measures of the intensity of mixing. Mixing rates across well-defined thermoclines in the center of lakes are orders of magnitude lower than horizontal mixing in mixed layers. Multiplying K_v by the vertical concentration gradient of a particular gas or dissolved nutrient provides a measure of the exchange rate between the epilimnia and hypolimnia.

- The primary action of wind is to apply a force (shear) to the water's surface and thereby transfer energy to set the surface in motion; surface current speed is a function of the sustained average wind speed.
- The Coriolis effect causes increasing deflection of currents with increasing depth. The effect of this progressive rotation—unless prevented by a thermocline or the bottom—is that at some depth flow direction is exactly opposite to wind direction.
- When internal seiches become steep and unstable or "touch bottom," they bring about significant currents and turbulence in the hypolimnia and allow vertical transport of water (and substances therein) from the hypolimnia into the epilimnia. They also allow the transport of dissolved oxygen and heat into hypolimnia.
- The wide variety of currents and waves that have been observed vary not only spatially and temporally as the result of local winds and differential heating but also in response to the underwater morphometry and, with a time lag, to changes in weather.
- ▲ The time scale of physical processes and their impact on the biota range from minutes to centuries. Progress in the prediction of population and community behavior is closely linked to advances in physical limnology.

13

Salinity and Major Ion Composition of Lakes and Rivers

13.1 Introduction

Runoff water from catchments is dominated by eight major ions (Ca, Mg, Na, and K, and the anions HCO_3, CO_3, SO_4, and Cl), each characterized by having a concentration greater than 1 mg l^{-1}. Their total concentration, expressed in mg l^{-1} or milliequivalence per liter (meq l^{-1}) yields the **salinity** of inland water. The remaining elements are either present in nonionic form (like silica) or are present in fresh water in such low quantities (e.g., Fe, N, P, Co, Mo, etc.) that they occur at concentrations of < 1 mg l^{-1} or trace levels ($\leq$ 1 μg l^{-1} or pg l^{-1}) and have a negligible effect on the salinity measured. However, NO_3N concentrations in rich agricultural areas and downstream from urban areas are sometimes high enough to consider nitrogen a major ion.

13.2 Salinity and Its Origins

The observed salinity and component ions in fresh water are usually determined by reactions that take place in the soils and rocks of the catchments. The ease of breakdown (weathering) and the resulting salinity of the water decreases (Table 13–1) as rock stability or hardness increases from soft sedimentary carbonate rock (e.g., limestone or dolomite) to somewhat harder sedimentary rocks (e.g., siltstone and sandstone), to hard igneous rock (e.g., basalt and granite),

and ultimately to very hard crystalline rock (e.g., chert and quartzite) (Table 13–2). Weathering in nonarid regions is not only the result of solubilization (chemical weathering) but also of biological weathering processes in soils and on vegetated rock surfaces. The combined processes far outstrip the physical disintegration (physical weathering) of rock. In biological-chemical weathering, the principal agents are hydrogen ions produced when gaseous CO_2 from the atmosphere, or produced by root respiration and in organic matter decomposition (primarily by microbes) combines with water to form carbonic acid (H_2CO_3) (Chapter 14). In addition, organic acids that are released to the soil by bacteria, fungi, and plant roots contribute to weathering.

When the substrate is *calcareous*, containing considerable quantities of carbonate rock, it is dissolved as

$$CaCO_3 + H_2O + CO_2 \rightarrow Ca^{2+} + 2HCO_3^-,$$
(limestone)

EQ. 13.1a

$$CaMg(CO_3)_2 + 2H_2O + 2CO_2 \rightarrow Ca^{2+} + Mg^2 + 4HCO_3^-.$$
(dolomite)

EQ. 13.1b

Not only is the resulting salinity dominated by the released Ca, Mg, and HCO_3 ions but the whole rock goes into solution. In areas dominated by hard igneous rock such as granite and gneiss, which are composed of silicates containing sodium, calcium, or aluminum, little goes into solution during initial weathering. In-

Table 13–1 **Geographic variability of dissolved major elements (mg l^{-1}) in pristine fresh waters draining the most common rock types after correcting for sea salts (μeq l^{-1}, in brackets). Electrical conductivity is in μS cm^{-1}, and SiO$_2$ is in μmol l^{-1} (in brackets). ND = not determined.**

	Conductivity	pH	Σ cations	Ca^{2+}	Mg^{2+}	Na$^+$	K$^+$	Cl$^-$	SO$_4^{2-}$	HCO$_3^-$	SiO$_2$
Granite	35	6.6	3.5	0.8	0.4	2.0	0.3	0	1.5	7.8	9.0
			(166)	(39)	(31)	(88)	(8)		(31)	(128)	(150)
Gneiss	35	6.6	4.1	1.2	0.7	1.8	0.4	0	2.7	8.3	7.8
			(207)	(60)	(57)	(80)	(10)		(56)	(136)	(130)
Volcanic Rock	50	7.2	8.0	3.1	2.0	2.4	0.5	0	0.5	25.9	12.0
			(435)	(154)	(161)	(105)	(14)		(10)	(425)	(200)
Sandstone	60	6.8	4.6	1.8	0.8	1.2	0.8	0	4.6	7.6	9.0
			(223)	(88)	(63)	(51)	(21)		(95)	(125)	(150)
Shale	ND	ND	14.2	8.1	2.9	2.4	0.8	0.7	6.9	35.4	9.0
			(770)	(404)	(240)	(105)	(20)	(20)	(143)	(580)	(150)
Carbonate Rock	400	7.9	60.4	51.3	7.8	0.8	0.5	0	4.1	194.9	6.0
			(3,247)	(2,560)	(640)	(34)	(13)		(85)	(3,195)	(100)

Source: Meybeck et al. 1989.

stead, most is retained as new, largely insoluble minerals. Ultimately, a considerable amount of silica and aluminum is released when the degraded (weathered) minerals (e.g., sodium feldspar) are hydrolyzed to a secondary mineral (kaolinite) and then to its elemental components in what is essentially a titration process.

$$2NaAlSi_3O_8 + 2H_2CO_3 + 9H_2O \rightarrow$$
(sodium feldspar)

$$Al_2Si_2O_5(OH)_4 + 2Na^+ + 2HCO_3^- + H_4SiO_4 \quad \text{EQ. 13.2}$$
(kaolinite)

The secondary (clay) minerals that are produced control the structure and chemical properties of well-drained higher latitude soils.[1] Their net negative charge is of great importance in determining the charge and quantities of major cations and aluminum sorbed onto them which then determines resistance to the acidification of soils and waterways by acidifying precipitation (Chapter 27). Chemical composition and salinity is determined by (1) the local geology; (2) cli-

mate (temperature, precipitation, evaporation); (3) the biota; (4) the length of time soil formation has been taking place; (5) distance from the sea, with its effect on ion deposition; (6) anthropogenically-produced atmospheric inputs of acids; (7) trace metals and organic contaminants, and (8) fertilizers and, at higher latitudes, road salt.

Igneous rock is particularly low in carbonates (Table 13–2) and most of the HCO$_3$ and CO$_3$ that is present in inland waters on igneous catchments is ultimately derived from atmospheric CO$_2$ following its combination with water and the formation of H$_2$CO$_3$ (Sec. 14.2). The H$^+$ of carbonic acid is the major force in weathering, not only in limestone and dolomite (Eqs. 13.1a and b) which yield an HCO$_3$/CO$_3$ dominance in the composition of the world's large rivers at their mouths but of all rock minerals.

Together, weathered rock, atmospheric precipitation, and temperature allow the development of characteristic soil types and plant communities in a particular climatic zone over time (Hem 1985). The vegetation and soil types have a major influence on the cycling and release of elements to inland waters (Schlesinger 1997).

The greatest quantities of ion-rich fresh water are produced by abundant precipitation on the thick, and as yet little weathered, soils of the geologically re-

---

[1] Ancient, highly weathered soils of tropical regions have largely lost their clays and are dominated not by partially weathered clay minerals but by oxides and hydroxides of aluminum and iron that have a positive charge under acidic conditions and a negative charge under alkaline conditions.

Table 13–2   **Average composition, in $\mu g\ g^{-1}$, of some important components of igneous rocks and the major sedimentary rock types.**

Element	Igneous[1]	Sedimentary[2]		
		Resistates[3]	Hydrolysates[4]	Precipitates[5]
Si	285,000	359,000	260,000	34
Al	79,500	32,100	80,100	8,970
Fe	42,200	18,600	38,800	8,190
Ca	36,200	22,400	22,500	272,000
Na	28,100	3,870	4,850	393
K	25,700	13,200	24,900	2,390
Mg	17,600	8,100	16,400	45,300
Mn	937	392	575	842
P	1,100	539	733	281
C (as CO_3)	320	13,800	15,300	113,500
S	410	945	1,850	4,550
Cl	305	15	170	305
Cu	97	15	45	4
Ni	94	3	29	13
Zn	80	16	130	16
N[6]	46	580	600	150
Pb	16	14	80	16
As	2	1	9	2
Hg	0.33	0.06	0.27	0.05
Cd	0.19	0.02	0.18	0.05

[1]Rocks formed upon the solidification of hot molten material or magma which, when slowly cooled, yields very dense crystalline rock of relatively insoluble minerals, or where rapidly cooled, produces porous rock of relatively soluble minerals.

[2]Rocks formed from the destruction and cementing through heat and pressure (weathering) of preexisting igneous or sedimentary rock.

[3]Rocks composed principally of sedimented residual minerals not chemically altered by the weathering of the parent rock (sandstones).

[4]Rocks composed principally of relatively insoluble minerals subsequently sedimented (shales).

[5]Rock produced by the chemical precipitation of mineral matter from aqueous solution (carbonates).

[6]Microbial nitrogen fixation in soils results in much higher nitrogen concentrations than is suggested by composition of the parent rock.

Source: From Hem 1985, and from Meybeck 1982.

cently (~10,000 yrs) deglaciated areas of the temperate zone (Fig. 5–17). Ion-rich fresh waters are normally characteristic of sedimentary catchments (Tables 13–1 and 13–2), whereas waters draining igneous catchments contain few ions. Low-ion waters are common in drainage basins dominated by very hard, and therefore highly insoluble, silica-rich rocks (Table 13–2) and sands, as well as in bogs where a thick coating of organic matter prevents contact with the inorganic substrate, and ancient, highly weathered soils in well-watered regions at lower latitudes. The soft water that is produced is susceptible to acidification (Chapter 27). The high salinity of saline lakes and wetlands is the result of evapotranspiration of ions derived from either sedimentary or igneous rock (Sec. 13.16) in basins without an outlet to the sea.

13.3 Total Salinity and Its Determination

Total salinity is the sum (mg l^{-1} or meq l^{-1}) of ionic compounds dissolved in the water.[2] Salinity can be obtained by determining the concentration of each of the individual major ions and adding them, a time-consuming procedure subject to analytical errors. A second, quick and simple indicator (surrogate) of total salinity is obtained by measuring the electrolyte content of waters. This is done by measuring the ease with which an electrical current passes between two electrodes immersed in water.[3] The greater the concentration of salts, acids, and bases in natural waters, the greater the electrical conductivity. Conductivity is normally expressed in microSiemens (μS) cm^{-1} at 25°C, milliSiemens (mS) in saline waters (1,000 μS cm^{-1} = 1mS cm^{-1}). (In North America, the equivalent was micromho cm^{-1} at 25°C, 1 μS = 1 μmho.) Measurement at a standard temperature is important because the conductivity of, for example, NaCl rises about two percent per degree. This easy to measure surrogate has become the measure of choice for salinity. A third measure is the **total dissolved solid** (TDS) content of waters which is obtained by filtering a water sample, evaporating the filtrate (< 100°C), and measuring the dry weight of the major ions plus silica remaining. Most of the weight is contributed by salts present in ionic form making the TDS another surrogate of salinity (Bierhuizen and Prepas 1985). TDS is now used primarily by geomorphologists interested in the amount of chemical erosion or weathering of different landscapes (Fig. 13–1). For well-watered north temperate regions, the salinity of streams and rivers is a function of runoff, geology of the drainage basin, and catchment size (km^2) (Duarte and Kalff 1989):

$$\ln C = 3.6 - 0.24 \ln \text{runoff} + 0.1 \ln W_s + 1.0 \text{ geo},$$

$$R^2 = 0.76 \text{ S.E. linear estimate} = 0.63 \qquad \text{EQ. 13.3}$$

where C = conductivity (μS cm^{-1} at 25°C); runoff = 1 m^{-2} yr^{-1}; W_s = catchment area (km^2); geo = geological grouping, with largely igneous catchments assigned a value of 1, largely sedimentary ones a value of 2, and largely calcareous and/or unconsolidated catchments a value of 3.

Equation 13.3 shows that, on average, conductivity (salinity) declines as runoff increases, and larger catchments yield higher conductivities for a given geology. This may be because small catchments are more frequent in upland or mountainous areas which commonly have a more igneous exposure, whereas larger ones are more likely to lie in lowland areas of more easily weathered sedimentary or glacial deposits. An alternative or complementary explanation is that larger catchments are often less steep (Fig. 8–12) and there is proportionately more time for evapotranspiration and contact of the runoff with rock and soils. A third plausible explanation is that there are often climatic differences between the generally small upland and larger lowland catchments which have an affect on soil formation, soil erosion, and salinity. A single outcome that is observed in nature may be the result of a changing combination of different mechanisms and processes operating at different spatial and temporal scales, preventing the identification of a single "cause" or mechanism for the results obtained. All three of the postulated mechanisms probably play a role and in concert determine the salinity (conductivity) observed, with their relative importance varying.

▲ In the temperate zone, the influence of surface geology on the salinity of exhorheic (open-basin) lakes, (Chapter 5) is most strikingly evident along the edge of the very ancient (> 570 million years), enormous Precambrian Shield of North America and its counterpart in northern Scandinavia and Russia. The Shield is underlain by highly insoluble igneous rock (primarily granite) and metamorphic rock of igneous origin (e.g., gneiss, granitic gneiss). Bedrock is usually found at or near the surface, and only in low lying areas are there thin glacial and other deposits overlying it. The bedrock and resulting soils are of low solu-

---

[2]When analyses are expressed as "milliequivalent per liter," unit concentrations of all ions are chemically equivalent and comparable. For elements in nonionic form (e.g., Si) with a charge of zero, an equivalent weight cannot be computed and the concentration is usually given in moles (or millimoles) per liter or mg l^{-1}. For conversions between units see Appendix 2.

[3]The salinity of fresh waters can be estimated as S = ~0.75 C where C = conductivity (μS cm⁻1) and S = total salinity based on a summation of the individual major ions (mg l^{-1}) (Golterman et al. 1988), and where salinity is expressed in meq l^{-1}, S = ~0.01 C (Golterman 1975), except in strongly acidic waters with an unusually high equivalent conductivity of H⁺. For saline waters, Williams (1986) and other scientists have developed intercalibrations between conductivity and salinity with S (mg l^{-1}) = ~0.6–0.7 C over a 5,500–100,000 μS cm⁻1 conductivity range. The relationship is nonlinear at higher salinities.

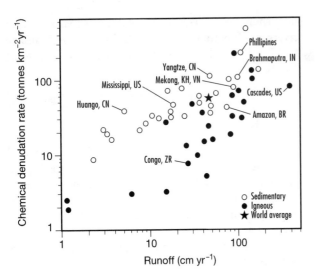

Figure 13–1 Influence of rock composition on the total dissolved solid (TDS) export coefficient (tonnes km^{-2} yr^{-1}) vs runoff per unit area (specific runoff, cm yr^{-1}) for some major world rivers and some small river basins. Note the higher denudation rates from river basins composed of highly erodable sedimentary rock, and the lower rates from hard igneous rock basins, except where the rock is highly weathered. *(Redrawn from M. Meybeck 1980, in Berner and Berner 1987.)*

bility and thus produce a very dilute (*softwater*) runoff[4] compared to adjacent areas with a substrate containing more carbonates (Table 13–2) where the water has a much higher salinity (*hardwater*).

The age of the soils in a region also determines salinity. Thus, soils formed after the retreat of the last glaciation from central Iowa (US) yield an average stream and lake conductivity 1.6 times greater than the mean value obtained for nearby lakes just to the south, located on much older and more weathered soils that were not directly impacted by that glaciation period (Jones and Bachmann 1978). For the location of this border in other areas, see Figure 5–17.

___
[4] The slightly acidic (pH = ~6) runoff from those portions of the Precambrian Shield that are not subject to acidifying precipitation is frequently high in dissolved organic matter because the relief on the Precambrian Shield is typically low (< 100 m) and the drainage is therefore often impeded, with the result that the organic soils overlying the peat release much acidic organic matter to receiving rivers, lakes, and wetlands (Chapter 9).

13.4   Major Ion Composition

Although salinity, conductivity, and total dissolved solids (TDS) reflect the total ion content, none of these measures can reveal anything about the specific ions that together produce different salinities observed in the large rivers that drain large portions of geologically and climatically varying continents (Table 13–3). Geologically different catchments differ in both the release rate of elements to streams and lakes and the relative proportions that are released (Table 13–1). Yet, for the major ions at least, the relative proportions do not greatly differ in well-watered portions of the temperate, arctic, and subtropical zones (Table 13–3). Inland lakes and rivers in these zones are generally dominated by the cation calcium and the anion bicarbonate. The most common ion sequences, as equivalent weights, is:

$$Ca > Mg > Na > K,$$
$$HCO_3 > SO_4 > Cl.$$ EQ. 13.4

There are, however, sufficiently important exceptions to this generalization[5] to make it inappropriate to call it a *standard composition* (Rodhe 1949):

1. In coastal areas and oceanic islands such as New Zealand, increased Na$^+$ and Cl$^-$ ions in particular (from sea spray and ocean-derived rain) typically result in Na being more dominant than Mg.

2. Small upland streams and lakes in small, well-watered igneous basins have an ionic composition more closely reflecting the chemical characteristics of local precipitation than a sequence based on rock weathering.

3. Over large areas of eastern North America and much of northern and central Europe, the ion sequence of streams and their receiving small lakes located on igneous drainage basins (away from coasts) that are seriously impacted by acidifying precipitation, the cation sequence is changed to

___
[5] There are exceptions to all generalizations, and scientists differ in how large and frequent exceptions to a rule can be before a generalization becomes uninteresting or (to them) even dangerous to the progress of science. Most empirically determined generalities are based on field observations in single lake districts or one climatic zone, with the applicability of the models unknown elsewhere. It is natural that the authors of any apparent generality have more faith in its broad applicability than others.

Table 13–3   **Average major ion concentration (mg l^{-1}) and composition (meq, percentage of cations and anions, in brackets) of large river systems reaching the oceans in selected climatic zones. Note (1) the elevated salinity and increased importance of Na and Cl ions in semiarid regions; and (2) the low salinity of tropical rivers draining ancient well-weathered catchments.**

Climatic Zone	Ca^{2+}	Mg^{2+}	Na$^+$	K$^+$	HCO$_3^-$	SO$_4^{2-}$	Cl$^-$	Salinity	SiO$_2$
Tundra and Taiga e.g. north central Canada, Siberia	17.7 (55.4)	4.7 (24.2)	7.0 (19.1)	0.8 (1.3)	62.3 (65.7)	13.6 (18.2)	8.9 (16.1)	115	3.7
Temperate e.g., east China, Europe, east North America	24.8 (66.4)	4.4 (19.4)	5.2 (12.1)	1.5 (2.1)	79.0 (69.8)	18.2 (20.4)	6.4 (9.7)	140	6.0
Semiarid e.g., northwest India, Australia, southwest United States, east Africa	38.3 (32.0)	13.5 (18.6)	64.5 (47.0)	5.5 (2.4)	153.0 (46.1)	54.9 (21.0)	63.4 (32.9)	393	15.4
Humid tropics (a) e.g., central Africa, east India, Southeast Asia	8.2 (42.9)	3.5 (30.2)	4.9 (22.3)	1.7 (4.6)	45.4 (78.7)	4.0 (8.8)	4.2 (12.5)	72	11.4
(b) e.g., Amazon basin, Brazil	3.2 (47.6)	1.0 (24.5)	1.8 (23.3)	0.6 (4.6)	11.2 (54.7)	3.1 (19.2)	3.1 (26.1)	24	11.4

Source: From Meybeck 1979.

$$H^+ > \Sigma Al\ ions > Ca^{2+} > Mg^{2+}. \qquad EQ.\ 13.5$$

4. On a more local scale, urban wastewater discharges and, at higher latitudes, winter road salting with NaCl affects the major ion composition. Heavy application of inorganic fertilizers also substantially raises the absolute salinity, adding SO$_4$, Cl, and NO$_3$ among the anions; decreasing the relative importance of HCO$_3$ in the process (Procházková and Blažka 1989).

5. There are seasonal, interyear, or even decade or longer changes in salinity and its composition resulting from changes in precipitation, temperature (evapotranspiration), and runoff that not only affect the salinity but also the ion composition. Interyear variation and climatic cycles frequently confound the identification of a clear response of drainage basins to changes in land use (Sec. 13.5) and acidifying precipitation (Chapter 27).

Despite these exceptions, large lowland rivers in well-watered areas of the world, draining large portions of continents and integrating local compositions, usually yield a major ion composition that is not very different from the composition observed in pristine inland catchments at higher latitudes.

▲ Gibbs (1970, 1992) summarized many of the generalities about the major ion composition by noting that most intermediate salinity fresh waters are primarily the product of rock weathering and are dominated by Ca and HCO$_3$ and therefore also have a low Na to (Na + Ca) ratio (Fig. 13–2). Secondly, precipitation effects dominate in hard-rock regions of high precipitation and runoff. The resulting low salinity waters (TDS < 10 mg l^{-1}) are often dominated by NaCl carried from nearby or distant oceans (Fig. 13–2). Lastly, in regions of high evaporation and low runoff, evaporation is primarily responsible for the high salinity observed, with highly soluble Na$^+$ and Cl$^-$ ions dominating in desert rivers and semiarid zone lakes (Fig. 13–2 and Table 13–3). For an outstanding article on the effect of atmospheric processes and geology on the chemical characteristics of inland waters along a climatic gradient see Gorham et al. 1983.

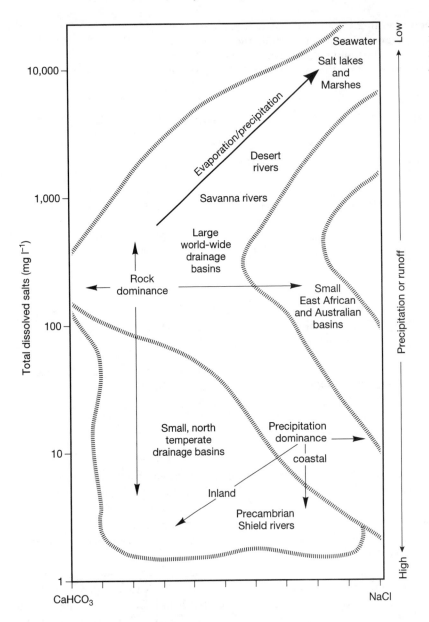

Figure 13–2 Diagrammatic representation of mechanisms controlling the chemistry of world surface waters. *(Modified after Gibbs 1992).*

Like all schemes (models), the one proposed by Gibbs is a simplification and works best at a global scale. It has little or no predictive power at small spatial scales over which the precipitation: runoff ratio and temperature vary little, allowing geological differences to dominate. For example, a rock component is apparent in Northern Ireland (Ulster)—where the climate, precipitation, and evaporation vary little, but geology varies a great deal. At the spatial scale of Northern Ireland the ascending limb of Fig. 13–2 reflects not a decline in runoff but instead a coastal–inland geological gradient, whereas the descending limb represents an altitude gradient (Gibson et al. 1995, and Gibbs 1992). Conclusions invariably change with the scale of investigations (Sec. 2.6).

The modification of runoff by sea spray is of particular concern to acid-rain researchers working in coastal areas. They must distinguish between the SO_4 in precipitation and waterways that originated from sea spray versus that derived from the combustion of fossil fuels to measure the impact of the acidifying precipitation (Chapter 27). They do this by assuming that all the Cl in precipitation in inland waters is derived from sea salt, and that the ionic ratios of the other

Table 13–4 **Rodhe's regionally-based standard composition of unpolluted north temperate fresh waters moderately affected by sea spray (Rodhe 1949) compared to world averages produced for large rivers at their mouth by D. A. Livingstone (1963); all in equivalent percentages, before (1) and after correction (2) for sea salts. Note the effect of the correction on composition.**

Component	1		2	
	Rodhe	Livingstone	Rodhe	Livingstone
Ca + Mg	84.2	76.5	88.6	87.9
Na + K	15.8	23.5	11.4	12.1
Cl	9.5	15.6	–	–
HCO_3	74.3	67.9	83.0	82.0
SO_4	16.2	16.5	17.0	18.0

Source: After Henriksen 1980.

major ions in the precipitation in coastal areas is the same as in sea water. It is then possible to subtract the marine-derived ions from the observed total, leaving the nonmarine components (Tables 13–1 and 13–4, and Fig. 13–3), thereby making it possible to quantify the acidifying chemicals produced by humans. In acid-rain research the nonmarine-derived major elements are now routinely indicated by an asterisk (e.g., Ca*).

13.5   Human Activity, Climate, and Ion Composition

The effect of human activity on major ion composition and salinity is large in the Rhine River, which drains the industrial heartland of western Europe (Fig. 13–4). Sulfate levels doubled between 1875 and the early 1970s. The extraordinary fifteenfold rise in Na^+ and eighteenfold rise in Cl^- was attributable to both salt mining in France, and industrial and sewage water release along the course of the river. More recently, declines (seen in NH_4^+ and NO_3^-) are coupled with greatly improved industrial practices and wastewater treatment. However, declines in Na^+, Cl^-, and SO_4 have been modest and concentrations remain well above those recorded in 1875 and 1928. Among the ions presented, only Mg^{2+} and HCO_3^- show little or no change and apparently continue to reflect weathering processes in the drainage basin rather than human activity. Even so, their relative contribution to salinity has declined considerably. Changes imposed by humans are not so clearly evident in the lower Great Lakes of North America.

The increased ion content of Lake Ontario (CA, US) (Fig. 13–5) seen until the 1980s was attributed to

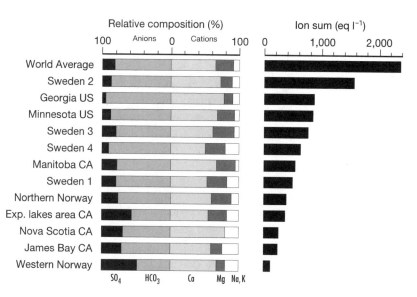

Figure 13–3 Relative ionic composition of fresh waters of moderate to low salinity (relative to the world average) lakes in Scandinavia and North America that are located in areas relatively unaffected by acid precipitation at the time of sampling, and corrected for sea spray and marine precipitation. *(After Henriksen 1980.)*

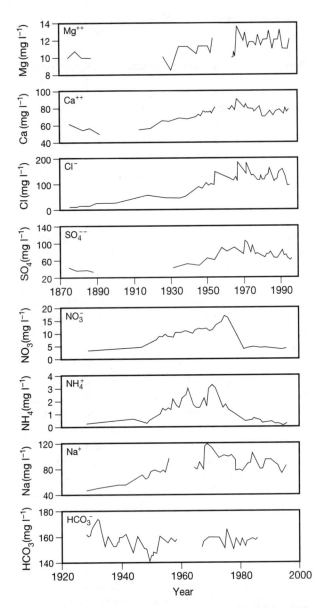

Figure 13–4 Historical record of the changes of important ionic components in the Rhine River near the German-Dutch border. Note the different scales. *(After Van der Weijden and Middelburg 1989, and data courtesy of M. Braun of the International Commission for the Protection of the Rhine, Koblenz.)*

a combination of industrial, agricultural, and construction activities, as well as to inputs of wastewater from its immediate drainage basin and upstream lakes (Weiler 1981).[6] Yet, the differential declines that commenced around 1980 during a period of high discharge provide a useful reminder that positive correlations over a certain period need not reflect cause and effect. Changes in land use and acidifying precipitation, with an effect on soil weathering and release (Kilham 1982), plus changes in runoff and evaporation and a lagged response resulting from a long water residence in upstream lakes, confounds a simple interpretation.

▲ The difficulty of evaluating the effect of human activity on the major ion composition of nonacidified lakes is more clearly evident from work on Ekoln Bay of Lake Mälaren (SE) over different time scales (Table 13–5). The salinity of the bay rose 45 percent and the concentrations of Cl doubled and SO_4 tripled between 1934 and 1974. (The two elements increased by about 92 and 76 percent respectively in Lake Ontario over the same period.) When examined over this temporal scale, these rises suggest several possibilities: increased weathering resulting from changed agricultural practices; an increase in wastewater- or fertilizer-derived Cl and SO_4; or increased deposition of anthropogenically-derived H_2SO_4, with the H^+ converting HCO_3 to CO_2 and H_2O, thereby lowering the HCO_3 content of affected inland waters (Eq. 14.10). The increase in salinity from 217 mg l^{-1} to 314 mg l^{-1} over the same period (Table 13.5) could reasonably be interpreted as indicating increased soil weathering, possibly the result of major increases in H^+ derived from the acidified atmosphere. Such an increase in H^+ loading could also explain the greater release of positively charged Ca and Mg ions to waterways, following their release from clay particles (Sec. 27.6). The increased SO_4 levels observed in the bay could also be attributable to SO_4 deposited as H_2SO_4 and SO_2.

The 1984 Ekoln Bay data provide a salutary reminder of the importance of time scale on interpretations. In 1984 the calcium concentration and salinity, but not the Na, K, Cl and SO_4 concentrations, virtu-

---

[6]The catchment export worldwide of Na, Cl, and SO_4 to the oceans has increased 30 percent over background levels (Meybeck 1979). This increase is small compared to the twofold increase in dissolved N and P in major rivers worldwide, and locally in the western world by a factor of 10–50. (Meybeck 1982 and Chapter 18)

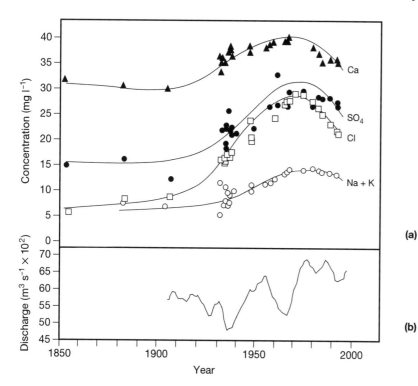

Figure 13–5   (a) Dissolved mineral quality (mg l[-1]) of Lake Ontario (CA, US). Lines fitted by eye. (b) Six-year moving average discharge of the Niagara River into Lake Ontario (m[3] s[-1] × 10[2]), exhibiting a roughly 30 year cycle. (*Part a: Data from Weiler and Chawla 1969, and V. G. Richardson at Ecosystem Health Division, Environ. Can., Burlington; and part b: Co-ordinating Committee on Great Lakes Basic Hydraulics and Hydraulic Data 1900–1998*)

ally returned to 1934 levels while HCO_3 declined even further. The changes observed in 1984 appear to be the complex outcome of increased runoff (T. Ahl, pers. comm.) plus human activity. It is evident that interannual and longer-term cycles in climate and river discharge not only affect the major ion composition and

salinity (T. Ahl 1980, and pers. comm.) but also make it difficult or impossible to identify a rapid widespread response to changes in H^+ deposition following a reduction or increase in industrial emission (Chapter 27).

Seasons and years of high river discharge result in lowering of the concentration of major ions in Ekoln

Table 13–5   **The major ion composition over time of Ekoln Bay of Lake Mälaren, SE.**

	equivalent %				mg l[-1]			
	1934	**1964**	**1974**	**1984**	**1934**	**1964**	**1974**	**1984**
Ca	64.3	65.1	64.0	64.6	37.1	47.3	57.3	41.9
Mg	20.8	15.9	16.5	15.4	7.3	7.0	9.0	6.1
Na	12.5	16.0	17.1	17.5	8.3	13.4	17.6	13.1
K	2.4	3.0	2.4	2.4	2.7	4.1	4.3	3.1
HCO$_3$	71.7	64.1	43.5	57.9	126	140	117	108
SO$_4$	19.0	23.4	41.9	26.3	26.3	40.2	88.7	38.8
Cl	9.3	12.5	14.6	15.8	9.5	15.8	20.3	17.2
Salinity					217.2	267.8	314.2	228.2

Source: Modified after Ahl 1980, and unpublished data.

Bay—that, when summed, yield the salinity.[7] The effect, as examined in Scandinavia (Ahl 1980) or Lake Ontario, is not identical for individual ions, with the contribution of Ca and HCO_3 increasing but most of the other major ions declining with increasing discharge in Ekoln Bay. Assessment of the impact of human-induced changes requires the frequent collection of chemical and hydrological data over a long series of wet and dry years. Unfortunately, there are few such long-term data sets available, making it difficult to draw unambiguous conclusions about human impact, other than where the impacts are so large—as in the Rhine River (see Fig. 13–4) or Slapy Reservoir, CZ (Procházková and Blažka 1989)—that they greatly overwhelm the normal long-term seasonal and interannual variation. To compensate for the usual absence of long-term data, comparisons are often drawn between similar systems—one affected and one as pristine as possible. Yet, how similar the compared systems were before human impact (e.g., acidifying precipitation) has been unclear. Fortunately, advances in paleolimnology (Chapter 20) and analyses of contaminants in dated sediment cores (Chapter 20) have made it possible to compare present lake and wetland conditions with chemical and biological background conditions prior to the onset of human-induced changes.

13.6   Saline Waters and Their Distribution

Saline lake and wetland districts are generally restricted to the semiarid (mean annual precipitation 20–50 cm) or arid (2.5–20 cm) parts of the world, and are largely encompassed within the endorheic zone (Fig. 5–17). More specifically, saline lakes and wetlands are found where the net evaporation equals or exceeds precipitation over the longer term (Fig. 13–6). Lakes, wetlands, and inflowing rivers are subject to both a seasonal and long-term cycle in precipitation, water level changes (Sec. 5.6), salinity (Fig. 20–12), and the selective precipitation of ions, readily evident from a band of precipitated salts along the shore-line

---

[7]Note that even substantial lowering of the absolute concentration of individual elements (mg l^{-1}) as the result of dilution by increased river discharge (m^3 yr^{-1}) nevertheless results in increased catchment export (g m^{-2} yr^{-1}) because of typically larger seasonal variation in discharge than in the concentration of individual elements (export = concentration × discharge) (Chapter 8).

following a dry period (Figs. 13–7 and 5–20). A well known example is shallow Lake Nakuru (KE), famous for its sometimes enormous population of flamingos. Its measured conductivity has varied between 9,444 and 183,315 μS cm^{-1}, but the lake has dried up altogether after a series of particularly dry years (Vareschi 1982).

Saline lakes and wetlands, as well as the local rivers, are increasingly under threat because of water diversion from the rivers for irrigation of crops in the semiarid zone. These widespread diversions are leading to the disappearance of shallow lakes and wetlands, and a reduction in size of larger ones (Sec. 5.7)—shrinkage that would only be exacerbated by a predicted global warming. Poor agricultural practices are not only leading to widespread salinization of poorly irrigated (overwatered) soils but also of the rivers receiving the drainage waters. The negative consequences of this for the conservation of saline aquatic systems, local rivers, the aquatic biota, and local wildlife are becoming disastrous as water withdrawals continue to grow globally (Lemly et al. 2000).

▲   13.7   Ionic Composition of Inland Saline Lakes and Wetlands

The chemical composition of saline lake waters (Table 13–6) is very different from both the relatively low salinity river water flowing into them and the geochemistry of the catchment soils. The ionic composition of both saline lakes and wetlands, as well as high-evaporation rivers, is the outcome of differential precipitation of the elements (Eugster and Hardie 1978). Precipitation comes about when the solubility product of particular ion pairs (salts) is exceeded during evaporation. Yet, the water of most closed lakes is not saturated with respect to one or more of the common salts present. This suggests that previously sedimented salts are only partially resolubilized when the lake basins are refilled with lower salinity water during wet periods. Another interpretation is that sedimented salts become permanently buried under windblown sediments at times when most or all of the lake bottom becomes exposed to the air. Dried salts on a wind-exposed lake bed are also blown away by the wind in a process known as *deflation*, providing a third means by which sedimented salts can be permanently removed from contact with subsequently inflowing waters. The

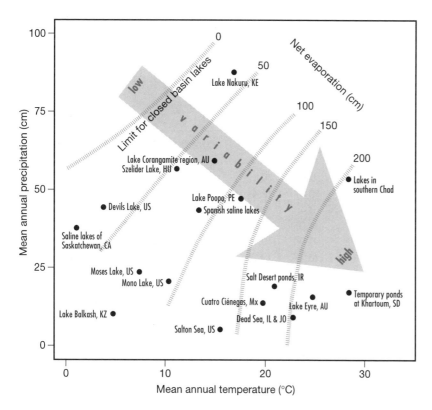

Figure 13–6   The distribution of closed-basin lakes as a function of mean annual precipitation, temperature, and net evaporation. Increased temperature and evaporation, together with a small and highly variable precipitation in arid and semi-arid regions increase the seasonal and annual variation in lake size and salinity of the typically shallow lakes. However, longer-term variation in precipitation has allowed the occasional dessication of, for example, shallow Lake Nakuru (KE), and the rare filling of Lake Eyre (AU). *(Data from Cole 1968, and Hammer 1986.)*

Figure 13–7   Laguna Lejia, a Chilean saline lake at 4,300 m in the Andes, showing salt precipitated during a lowering of the water level. *(Photo by F. Gohier/Courtesy of Photo Researchers, Inc.)*

effect of deflation is dramatically evident around the Aral Sea (KS, UZ), with detrimental effects on crops and human health (Fig. 5–20).

The ion composition of saline lakes and wetlands is predictable from the original ion ratios of the runoff and the degree of evaporation. The original ion composition of the runoff to lakes and wetlands resembles the composition of river water in higher runoff regions of the same geology. Calcium carbonate (calcite) is typically precipitated first, following its concentration during evaporation. The result is a relative enrichment of Na, Mg, Cl, and SO_4 in the remaining water. Next, $MgCO_3$ precipitates as crystals of dolomite [$CaMg (CO_3)_2$] in waters where the concentration of Ca and Mg remains relatively high. If there is further evaporation, and if some dissolved Ca remains, the solubility product of calcium sulfate will normally be exceeded next and gypsum ($CaSO_4 \cdot 2H_2O$) precipitates. Other minerals also precipitate and contribute to the permanently deposited salts (Table 13–6 and Fig. 13–8). Large sediment deposits are known as *evaporites* and some are mined commercially (Williams 1998). The water remaining after all

Table 13–6 **The range of individual measurements of salinity (g l⁻¹) and major ions (g l⁻¹ and equivalent % in parentheses) in some inland saline lakes. Note the wide variation in lake salinity and composition at the time of the sampling. Present-day values are likely to be different because of changes in hydrology, temperature, and evaporation.**

Lake/Sea	Salinity (g l⁻¹)	Na	K	Ca	Mg	Cl	SO₄	HCO₃ + CO₃	
Redberry (CA)	18	1.9 (29.2)	0.2 (1.7)	0.1 (1.8)	2.3 (67.4)	0.2 (2.2)	12.5 (93.1)	0.6 (3.2)	0.1 (1.5)
Van (TR)	23	8.1 (94.8)	0.4 (2.8)	0.0 (0.1)	0.1 (2.4)	5.9 (44.6)	2.4 (13.6)	2.4 (10.7)	3.5 (31.1)
Seawater	35	10.8 (77.0)	0.4 (2.0)	0.4 (4.0)	1.3 (18.0)	19.4 (90.0)	2.7 (9.3)	0.1 (0.4)	0.0 (0.0)
Bogoria (KE)	36	14.4 (98.6)	0.3 (1.2)	0.0 (0.2)	0.0 (0.0)	3.4 (14.1)	0.2 (0.6)	17.7 (85.3)	
Gallocanta (ES)	40	7.9 (53.1)	0.2 (1.0)	0.3 (2.6)	3.4 (43.3)	18.1 (71.3)	9.7 (28.3)	0.0 (0.0)	0.1 (0.3)
Soda (US)	82	20.0 (71.4)	1.5 (3.2)	0.6 (2.6)	3.4 (22.7)	4.1 (9.5)	50.4 (86.2)	1.5 (2.0)	0.8 (2.3)
Mono (US)	89	29.5 (96.9)	1.5 (2.9)	0.0 (0.0)	0.0 (0.2)	17.6 (32.6)	10.3 (14.1)	11.2 (12.0)	18.9 (41.3)
Eyre (AU)	116	45.8 (96.4)	0.0 (0.0)	0.9 (2.3)	0.3 (1.3)	68.0 (95.9)	2.9 (4.1)	0.0 (0.0)	0.0 (0.0)
Dead Sea (IR, JO)	295	38.5 (29.7)	6.5 (2.9)	16.4 (14.5)	36.1 (52.8)	196.9 (98.7)	20.6 (0.2)	0.2 (0.1)	0.0 (0.0)
Great Salt Lake (US)	332	105.4 (80.6)	6.7 (3.0)	0.3 (0.3)	11.1 (16.0)	181.0 (89.8)	27.0 (9.9)	0.5 (0.2)	0.3 (0.2)
Don Juan (AQ)	339	11.5 (7.9)	0.2 (0.7)	114.0 (90.4)	1.2 (1.6)	212.0 (100.0)	0.0 (0.0)	0.0 (0.0)	0.0 (0.0)

Source: From Hammer 1986.

precipitation events tends to be enriched in Cl relative to the other anions, unless the original concentration of CO₃ was disproportionately high. At the same time, sodium is enriched relative to the other remaining cations (Table 13–6).

Among inland saline waters, more are of NaCl-dominated composition than of any other composition. This is the result of the high relative solubility of NaCl and the original composition of the runoff. Sodium chloride-dominated water is particularly common in Australia, South America, and Antarctica. In regions such as east Africa—where the inflowing fresh waters are disproportionately rich in HCO₃ but relatively poor in Ca and Mg—lakes are dominated by NaCO₃ and are known as *soda lakes* (Table 13–6). However, most inland saline lakes worldwide have Na as the dominant cation. Even so, saline lakes are typi-cally characterized by both a dominant and a subdom-inant anion. Saline lakes are therefore more properly characterized as, for example, NaCO₃Cl (Bogoria), NaClCO₃ (Van), or NaMgCl (Dead Sea) lakes (Table 13–6, and Eugster and Hardie 1978).

▲ 13.8 The Salinity Spectrum and the Biota

Salinity is of interest not only to chemical limnologists and geochemists but also to biological limnologists because it has a major influence on the distribution and abundance of organisms.

Lakes lie along a salinity gradient. The lowest recorded salinity is about 1.5 µS cm⁻¹ in Rocky Mountain Lake (US), which lies in a tiny, barren igneous

Figure 13–8   A 6 cm diameter thick coating of precipitated salt on an instrument mooring line in the Dead Sea (Israel, Jordan) following a six week exposure. *(Photo courtesy of A. Hecht.)*

catchment (CA:LA = 1.2)—far from sources of ion-carrying dust or marine-influenced precipitation (Eilers et al. 1990). Waters only slightly more saline have been recorded in nearby Waldo Lake (US) (3 μS cm⁻¹) and in some Amazon River basin streams and lakes (Reiss 1973, in Day and Davies 1986). Only slightly higher values are found in the Congo River basin of central Africa. All the lakes and rivers in these two major river basins lie on extremely old (Precambrian), highly weathered (leached) tropical catchments. Lowest saline lakes closely resemble distilled water[8] and their composition reflects local atmospheric precipitation (Gorham et al. 1983). Lakes and streams averaging 12 μS cm⁻¹ are found over a large area of northeastern Quebec, CA (Lachance et al. 1985), and in adjacent Labrador (CA) lake conductivities as low as 5 μS cm⁻¹ are encountered. Here too, local precipitation rather than the influence of the drainage basin determines salinity and the ion composition (see Fig. 13–2).

Saline lakes in areas of Africa and Australia, with conductivities of up to about 270,000 μS cm⁻¹ (Livingstone and Melack 1984, Golterman and Kouwe 1980) lie at the other extreme of the conductivity/salinity spectrum. Conductivities as high as 790,000 μS cm⁻¹

[8]Ordinary good-quality single distilled water (or deionized water) has a conductivity of at least 1.0 μS cm⁻¹, with most of the observed conductivity resulting from the production of HCO_3^- and H^+ when atmospheric CO_2 dissolves into the water (Chapter 14).

have been recorded in Antarctica (Hammer 1986) where the salt content is determined by evaporative processes, but is also supplemented by the expulsion (*freezing out*) of salts into the water during ice formation. The more "typical" inland fresh waters familiar to north temperate zone limnologists, with conductivities of about 20–600 μS cm⁻¹, lie between these exceptionally dilute and highly saline lakes. This typical range encompasses water chemically dominated by the atmosphere at the low end, and affected by evaporation at the high end. The majority of well-studied inland waters have an intermediate conductivity and salinity determined primarily by weathering in their relatively well-watered drainage basins (see Fig. 13–2).

The salinity variation encountered by limnologists—ranging from virtually distilled water to extremely concentrated solutions of salts (*brine*)—is enormous (nearly six orders of magnitude) compared to the small variation encountered by oceanographers. The open oceans have a fairly narrow salinity range of 33,000–35,000 mg l⁻¹ (33–35‰), equivalent to a conductivity of about 50,000 μS cm⁻¹.

A number of schemes have been devised to distinguish between freshwater and saline systems. Among them, Williams (1967) defines saline lakes as those with a salinity (TDS) of ≥ 3,000 mg l⁻¹ (~5,500 μS cm⁻¹ or 5.5 mS cm⁻¹) at 25°C and freshwater lakes and wetlands as having a salinity of < 500 mg l⁻¹ (~670 μS cm⁻¹), with subsaline waters between these values. A salty taste becomes evident at ~3,000 mg l⁻¹. Hammer (1986) divides saline lakes into 3 categories: *hyposaline*, 3,000–20,000 mg l⁻¹ (~5,500–30,000 μS cm⁻¹), *mesosaline* 20,000–50,000 mg l⁻¹ (> ~30,000–70,000 μS cm⁻¹), and *hypersaline* > ~50,000 mg l⁻¹ (> ~70,000 μS cm⁻¹). The transition between *open basin* and the chemically most dilute *closed basin lakes* in Africa lies at a salinity of about 600 mg l⁻¹ (~1,000 μS cm⁻¹) (Talling and Talling 1965).

The Biota

The first subtle shift from freshwater to saline species in both the phytoplankton composition and algal biomass produced (per unit of total phosphorus) was noted when salinity rose to about 1,000 mg l⁻¹ (~1,600 μS cm⁻¹) (Bierhuizen and Prepas 1985). This transition was initiated at a lower salinity than in the brackish Baltic Sea of northern Europe, where a shift in phytoplankton composition was first seen between 2,000 mg l⁻¹ and 3,000 mg l⁻¹ (Niemi 1982). The difference in critical salinity observed in the two studies is probably due to differences in the criteria used to measure change, as well as in the techniques used. A very careful investigation will report earlier shifts than a more casual one, or one based on the investigation of insensitive species. However, somewhere between 1,000 and 2,000 mg l^{-1} salinity begins to manifest itself ecologically, and the latter value appears to be appropriate, in terms of the phytoplankton, for distinguishing fresh water from saline or brackish water for all but the most detailed studies. Cladoceran zooplankton are present at 1,800 mg l^{-1} (1.8‰) in one eutrophic brackish lake, and absent from another at a salinity of 3,000–4,000 mg l^{-1} (Moss 1994). However, work in Scandinavia indicates a wider cladoceran salinity range, suggesting that shifts in abundance and disappearance from brackish waters also involve predation (Jeppesen et al. 1994).

The freshwater macrophyte composition also shifts when salinity rises above ~2,000 mg l^{-1}, and beyond 4,000 mg l^{-1}, few fresh water forms remain (Hart 1991). This applies to invertebrates in Australia as well, where the number of mollusk species and their abundance declines above ~1,000 mg l^{-1}, and freshwater species were absent above ~2,200 mg l^{-1} (Hart 1991); and is interpreted to be the result of osmotic and ionic regulation difficulties, problems that are not suffered by species more tolerant of higher salinity. Beadle (1981) notes a completely freshwater mollusk fauna for Lake Turkana (KE), which has a present-day salinity of about 25,000 mg l^{-1}. Its existence illustrates the importance not only of a physiological ability of freshwater species to cope with elevated salinity but also of the time available for adaptation and the absence of both potential invading replacements and predators.[9]

The relationship between average salinity and species richness of plants and animals is particularly evident below a salinity of ~10,000 mg l^{-1} (~10‰) in hyposaline waters and changes little above ~50,000 mg l^{-1} (50‰) (Fig. 13–9, and Hammer 1986). Intriguingly, salinity appears to play a minor role in the structuring of communities in mesosaline and hypersaline waters, systems that are considered to be typically saline waters. However, organisms there have to cope not only with high salinity but also with low dissolved

---

[9]Changes in species composition and relative abundances at all taxonomic levels normally occur long before their physiological limits are exceeded, attesting to the importance of the relative competitive ability rather than the absolute (physiological) ability of species to grow and reproduce.

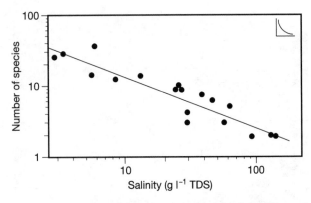

Figure 13–9 The relationship between the logarithms of the littoral macrozoobenthic species richness and salinity, as total dissolved solids, of 18 saline lakes in Alberta and Saskatchewan, CA. Note the rapid decline in species richness between 3,000 and 10,000 mg l^{-1}, and the slow decline above ~10,000 mg l^{-1}. *(After Hammer et al. 1990.)*

oxygen solubility and its further lowering at high temperatures (Table 15–3), plus an increased possibility of periods of total anoxia and dessication.

Highlights

- Runoff from catchments is dominated by eight major ions (the cations Ca, Mg, Na, and K, and the anions HCO_3, CO_3, SO_4, and Cl) that together yield the salinity of inland waters.

- As rock stability (hardness) increases from soft sedimentary rocks to very hard crystalline rock, the ease of breakdown (weathering) decreases along with the salinity (ion content) of the water released from drainage basins.

- Most intermediate-salinity fresh waters are the product of rock weathering, dominated by Ca and HCO_3; whereas the principle ions carried in precipitation (Na and Cl) dominate in hard-rock regions of high precipitation that are not impacted by highly acidified precipitation. Low and variable precipitation combined with relatively high evaporation is responsible for the salinity of waters in semiarid and arid regions.

- Human activity (wastewater, irrigation, water diversion), and short- and long-term climatic variation has a large effect on the major ionic composition, salinity, and thereby on the biota of affected inland waters.

- ▲ The chemical composition of saline waters is the outcome of both original ion ratios in the runoff and differential precipitations of elements when the solubility product of particular ion pairs (salts) is exceeded during evaporation. Among inland saline waters, NaCl dominates worldwide.

- ▲ The salinity variation of inland waters is enormous, ranging nearly six orders of magnitude, from water indistinguishable from distilled water to a briny solution of salts.

- ▲ The first subtle species shift from a freshwater to saline biota occurs in the plankton between a salinity of ~1,000–2,000 mg l⁻¹. Species richness in plants and animals declines with increasing salinity. However, the largest decline occurs under conditions of relatively low salinity (< ~10,000 mg l⁻¹) and little change in species richness occurs above ~50,000 mg l⁻¹ in what are considered to be more typically saline waters.

14

Inorganic Carbon and pH

14.1 Introduction

The concentration of **dissolved inorganic carbon** (DIC) is of tremendous importance in aquatic systems because it (1) buffers fresh waters against rapid changes in pH; (2) determines the amount of inorganic carbon available for photosynthesis; (3) provides the great bonding capacity of bicarbonate (HCO_3^-) and carbonate (CO_3^{2-}) ions for cations; (4) makes the ionic carbon concentration an important component of the anion concentration; and (5) removes inorganic carbon and adsorbed materials from the water column upon precipitation as $CaCO_3$ aggregates (Chapter 19).

14.2 Carbon Dioxide in Water

With a global average content of atmospheric carbon dioxide of 0.037 percent or 370 ppm by volume,[1] about 1.10 mg l^{-1} can dissolve in pure water at 0°C. Solubility is temperature dependent and declines to 0.65 mg l^{-1} at 15°C and 0.48 mg l^{-1} at 25°C (Hutchinson 1957). It was long thought that exchanges of CO_2 between the water and atmosphere kept aquatic systems from being highly super- or subsaturated. It is now evident that most lakes worldwide are supersaturated with CO_2, with a partial pressure that is on average three times that of the atmosphere (Cole et al.

1994). It is also clear that inorganic carbon-depletion episodes occur in productive waters (Maberly 1996), limiting rates of phytoplankton and submerged macrophyte photosynthesis.

The following reaction governs the flux and concentration of inorganic carbon:

$$CO_2 \text{ (air)} \leftrightarrows CO_2 \text{ (H}_2\text{O)}. \qquad \text{EQ. 14.1}$$

The dissolved CO_2 reacts with water to yield carbonic acid:

$$CO_2 + H_2O \leftrightarrows H_2CO_3. \qquad \text{EQ. 14.2}$$

H_2CO_3 is a weak acid and it dissociates, yielding HCO_3^-, but the dissociation declines with decreasing ion concentration (ionic strength) and pH.

$$H_2CO_3 \leftrightarrows H^+ + HCO_3^-. \qquad \text{EQ. 14.3}$$

An apparent dissociation or equilibrium constant (K_1) for the first ionization steps of carbonic acid describes the reaction equilibrium for Equation 14.3; the square brackets indicate the concentration of each ion in mol l^{-1}.

$$K_1 = \frac{[HCO_3^-][H^+]}{[H_2CO_3]} = 4.41 \times 10^{-7} \text{ at } 25°C.$$

Where the substrate is rich in precipitated or rock carbonates and the microbially produced CO_2 levels are high, the H^+ produced (Eq. 14.3) dissolves carbonates:

$$CaCO_3 + H_2O + CO_2 \leftrightarrows Ca^{2+} + 2HCO_3^- \qquad \text{EQ. 14.4}$$

---

[1]Due to human activity, this is about 87 percent higher than the 270 ppm of preindustrial times.

assuming that Ca^{2+} is the only cation for simplicity. In calcareous drainage basins, about half of the HCO_3^- released to waterways is derived from weathering (dissolution) of the substrate, whereas in systems with substrates that are very low in carbonates (e.g., igneous rock) or basins lined with peat (peat bogs), almost all of the HCO_3^- produced is the result of respiratory CO_2 production (Eq. 14.3).

The HCO_3^- that is produced will further dissociate,

$$HCO_3^- \leftrightarrows H^+ + CO_3^{2-}, \qquad \text{EQ. 14.5}$$

and yield the relatively insoluble carbonate ions which—at high HCO_3^- and high pH ($> \sim 8.5$)—is readily precipitated after the withdrawal of CO_2.

$$Ca^{2+} + CO_3^{2-} \leftrightarrows CaCO_3 \text{ (solid).} \qquad \text{EQ. 14.6}$$

Dissociation can also be described by a second equilibrium constant (K_2),

$$K_2 = \frac{[H^+][CO_3^2]}{[HCO_3^-]} = 4.7 \times 10^{-11} \text{ at } 25°C.$$

The HCO_3^- and CO_3^{2-} hydrolyze to produce OH^-:

$$HCO_3^- + H_2O \leftrightarrows H_2CO_3 + OH^-, \qquad \text{EQ. 14.7}$$

$$CO_3^{2-} + H_2O \leftrightarrows HCO_3^- + OH^-. \qquad \text{EQ. 14.8}$$

▲ If the equilibria between the different forms are disturbed—as when CO_2 is added to the system in respiration, removed in photosynthesis, or vented to the air—the reaction will shift one way or the other in an attempt to reestablish equilibrium. When CO_2 is added to the system it increases the pool of H^+ following the production of H_2CO_3 (Eqs. 14.2 and 14.3). At the same time, any CO_3^{2-} that is present consumes H^+ and yields HCO_3^- (Eq. 14.4). In the absence of CO_3^{2-}, the OH^- (Eq. 14.7) neutralizes (buffers) the added H^+:

$$OH^- + H^+ \leftrightarrows H_2O. \qquad \text{EQ. 14.9}$$

The overall result of this buffering is that there is only a slight decline in pH when H^+ is added to well-buffered waters. Conversely, when CO_2 is removed during photosynthesis or lost to the air by diffusion after a rise in water temperature and reduction in solubility, the reactions shift to the left (Eqs. 14.2 and 14.3). The H^+ concentration declines, but the CO_3^{2-} (Eq. 14.5) and OH^- concentrations (Eq. 14.8) only rise slightly, preventing a rapid rise in pH. Under extreme carbon depletion, the HCO_3^- declines in part by conversion to CO_3^{2-}. The inorganic carbon system is graphically summarized below.

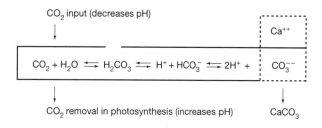

$$\text{EQ. 14.10}$$

The proportion of different ionic forms of DIC is controlled by the pH (Fig. 14–1). Most freshwater systems have a pH between six and eight, and it is evident that inorganic buffering against rapid pH changes in such waters is almost exclusively provided by the HCO_3^- component which dominates in this range. Only at a very high pH (9–10.5)—recorded in highly productive freshwater sewage lagoons and rivers, in dense productive macrophyte beds, and in highly buffered and often nutrient-rich saline lakes—is more than a minor fraction of the buffering provided by CO_3^{2-}.

In low-pH lakes (< 5.7), a high proportion of DIC is present as free CO_2 (including H_2CO_3) (Fig. 14–1). Any remaining inorganic carbon buffering is provided by HCO_3^-. Below a pH of about 4.5–5, the amount of HCO_3^- available to provide buffering is negligible (Fig. 14–1) and the small amount of buffering then observed in fresh waters is provided by organic acids and aluminium (Chapter 27). For a more complete discussion and interpretation of the equations that control the inorganic carbon concentrations and the pH of lakes, see Hem (1985), and Stumm and Morgan, 1996.

14.3 pH and Its Range in Aquatic Systems

The alkalinity or acidity of waters is defined in pH units. The **pH**, standing for *puissance d'hydrogène* or "strength of hydrogen", ranges from < 1 to 14 on a logarithmic scale. Although it is possible to express the effective concentration (activity) of the H^+ in mg l^{-1} or mol l^{-1},[2] the concentrations are very low for all but

---

[2]For the appropriate conversion factors, see Appendix 2.

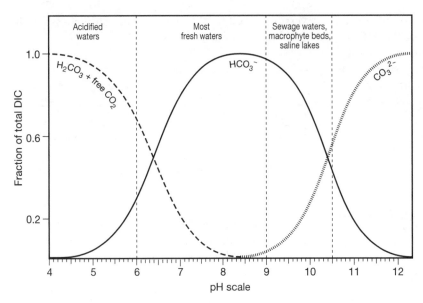

Figure 14–1 Distribution of the various forms of inorganic carbon in rivers and lakes with changes in pH. Note that at pH 6–8, bicarbonate (HCO_3^-) is the most abundant form. CO_2 and H_2CO_3 dominate at low pH, while most saline systems have a high pH and are characterized by a mixture of bicarbonate (HCO_3^-) and carbonate (CO_3^{2-}) ions. *(Modified after C. Schmitt 1955, in Golterman 1978.)*

extremely acidic (pH < 3) volcanic waters or coal mine drainage. The activity of H^+ is most conveniently expressed in logarithmic units with pH standing for the negative base-10 log of H^+ activity (mol l^{-1}).

▲ The activity quotient (K_w) is the sum of the activities of H^+ and OH^-. By convention, the ion product of water is taken to be unity. The ion activity quotient for water at 25°C is $10^{-14.00}$, with the two-place log of $K_w =$ −14.00.

$$K_W = \frac{[H^+][OH^-]}{[H_2O]} = 1.00 \times 10^{-14} \qquad \text{EQ. 14.11}$$

At neutrality, the activity of $[H^+]$ and $[OH^-]$ are equal and pH = 7.00, indicating the presence of (only) 1×10^{-7} mol l^{-1} of H^+. Remember, using a logarithmic scale for pH means that one unit change in pH represents a tenfold change in H^+ concentration.

14.4 Alkalinity of Inland Waters

The term **alkalinity** refers to the acid neutralizing capacity contributed by the sum of weak acid anions. In lake acidification research, the term *negative alkalinity* refers to the acidity of a system.

The alkalinity of fresh water is a measure of the total quantity of base present and is determined by titrating a water sample with strong acid, which neutralizes all hydroxyl, carbonate, and bicarbonate ions present (Eq. 14.10, right to left). Therefore, alkalinity can be defined as the equivalent concentration of

titratable base present. Other terms used to describe alkalinity are *alkaline reserve*, *total alkalinity*, and (most informative) *acid neutralizing capacity* (ANC). Higher ANC means a smaller change in pH in response to addition of a particular acid or base, and thus the water is more buffered.

▲ The ANC in calcareous lakes is obtained following a titration directly from the original pH to pH 4.5, at which point all HCO_3^- has been converted to H_2CO_3. As the specific buffers are not identified in the titration, it is best to describe the alkalinity in terms of the equivalent concentration of titratable base (meq or μeq l^{-1}) and not as mg l^{-1} $CaCO_3$. The latter designation, while in common use, incorrectly suggests that the buffering is provided only by CO_3^{2-} and also assumes that calcium was the only cation present. The pH of both fresh and saline inland waters is predictable from the ANC (Table 14–1).

When the HCO_3^- concentration is low, most of the buffering capacity is provided by organic acids and aluminium. A titration to an endpoint of pH 4.5 then overestimates the alkalinity, requiring a different procedure for waters with a pH of 5–6. However, the vast majority of best-studied fresh waters have a higher pH (6–8.5) reflecting HCO_3^- concentrations that typically range between 5 mg l^{-1} and 180 mg l^{-1} of HCO_3^- + CO_3^{2-}, representing a conductivity range of ~15–500 μS cm^{-1}.

A pH of between 9.5 and 10.5 is common in the extremely high alkalinity (> 100,000 μeq l^{-1} or > −6,000 mg l^{-1} HCO_3^- + CO_3^{2-}) saline lakes of east

Table 14–1   **Empirical relationships between pH and acid neutralizing capacity (ANC) as μeq l^{-1}.**

Location	Equation	n	r^2
Adirondacks, US	pH = 10.8 + 0.86log ANC	94	0.84
Florida, US	pH = 9.61 + 0.79log ANC	9	0.86
Rocky Mountains, US	pH = 11.25 + 0.86log ANC	29	0.83
Labrador, CA	pH = 9.75 + 0.79log ANC	167	0.73
Ontario, CA	pH = 10.83 + 1.04log ANC	378	0.85
Norway	pH = 9.57 + 0.77log ANC	834	0.86
Saline Lakes Worldwide	pH = 7.30 + 1.12log ANC	58	0.81

Source: From Wright 1983, and Hammer 1986.

Africa (Talling and Talling 1965, and Wood et al. 1984). The alkalinity, and the buffering provided, is so great it prevents large diurnal oscillations of pH even in highly eutrophic systems. Thus, diurnal pH variation in one such hypertrophic lake (Aranguadi, ET) changes only 0.4 pH units, between 10.2 and 10.6 (Wood et al. 1984). Conversely, large pH changes of one unit or more are encountered under a combination of low to moderate alkalinity and high algal (> ~30–50 μg Chl-*a* l^{-1}) or submerged macrophyte biomass. The resulting high rates of primary production allow a large daytime CO_2 and HCO_3^- withdrawal (depletion), resulting in a large rise in pH to values of > 10 (Maberly 1996). This particular combination of conditions is most commonly encountered in nutrient-rich waters such as sewage lagoons, fertilized fish ponds, or other hypertrophic shallow waterbodies, and in dense submerged macrophyte beds in littoral zones or wetlands. However, oligotrophic lakes, characterized by a low plant biomass that cannot appreciably affect the CO_2 pool and the buffering that it provides, exhibit little seasonal or diurnal oscillation in pH.

14.5   pH, Extreme Environmental Conditions, and Species Richness

Chemically or physically extreme aquatic ecosystems have a reduced species richness, but this reduction can almost never be attributed to a single environmental factor. The low species richness observed in acidified waters is not simply due to low pH. It is often due to low pH plus high, toxic levels of aluminium and possibly other metals; high temperatures in hot springs; high sulfur levels in streams receiving coal or other mining effluents; or to exceptionally high levels of trace metals (Chapter 27). Conversely, the high pH of eutrophic saline lakes covaries with (1) extremely high nutrient levels; (2) high levels of salt; (3) high, potentially toxic levels of trace metals; (4) regular or occasional periods of nighttime anoxia as the result of very high levels of respiration that is not offset by oxygen production by photosynthesis; and (5) low dissolved oxygen solubility requiring only modest levels of nighttime respiration to produce anoxic conditions in saline waters (Sec. 13.8 and Chapter 15). With two or more factors affecting species richness and population size at any one time, it is very difficult to attribute the abundance that is observed in nature to a particular "cause" (mechanism)—also known as a *stressor* (Chapter 15)—even though it can be identified under controlled conditions in the laboratory. Although scientists are highly interested in simple explanations, it is almost always possible to provide a plausible alternative for events seen in nature.[3]

Species richness tends to decline not only at low pH but also at either extreme from a broad optimum (pH 6–8.5) (Fig. 14–1). The saline, highly eutrophic lakes of east Africa (pH of 9.5–10.5) are frequently dominated by only a single species of filamentous blue-green algae, *Oscillatoria platensis*. Even in years when a near unialgal bloom of *Oscillatoria* is lacking, the number of phytoplankton species remains low, with species abundance only about half that recorded in nearby

---

[3]Recognition of the existence of multiple "causes" nearly a century ago allowed the scientist/philosopher Bertrand Russell to conclude that the power of science lies not in the identification of cause but in the power of the predictions made (the outcome). Even so, "cause and effect" plays a central role in exploring how nature works and in formulating research questions.

freshwater lakes (Kalff and Watson 1986); zooplankton and fish diversity are also reduced at very high pH. In the highly saline and productive Lake Nakuru (KE), with a pH of ~10.5, the lesser flamingo (*Phoeniconaias minor*) is the major predator of *Oscillatoria* (Fig. 14–2); together with one copepod, one rotifer species, and one introduced cichlid fish, these few species account for virtually all the community biomass and consumer energy flow (Vareschi and Jacobs 1985).

At the opposite end of the pH spectrum lie extremely acidic volcanic lakes, certain hot springs, and streams draining coal mine tailings rich in sulfur. Satake and Saijo (1974), in Mori et al. (1984) measured a pH of 1.8–2.0 in the Japanese crater lake Katanuma, which receives reduced (and toxic) hydrogen sulfide (H_2S) via underwater gas vents. Bacterially mediated oxidation of the reduced sulfur to sulfuric acid (Chapters 19 and 27) yielded an SO_4^{2-} concentration of about 1,000 mg l^{-1} in the oxic portion of the water column. Three species of algae and a large population of a benthic insect (*Chironomus dorsalis*) were noted and the major microbial decomposers were fungi rather than bacteria (Mori et al. 1984). Lake Rotokawa (NZ), with a pH of 2.1, is turbid with suspended sulfur particles and, in summer, has blooms of the algae, *Euglena* with a chlorophyll-*a* concentration of up to 500 μg l^{-1} and a high bacterial biomass (Vincent and Forsyth 1987). High abundances typically reflect both a high resource base and a paucity or lack of competitors and predators.

Figure 14–2 Lesser Flamingos (*Phoeniconaias minor*) filter feeding on filamentous cyanobacteria in Lake Nakuru, KE. (*Photo courtesy of J. McDonald, Visuals Unlimited.*)

The most acidic of all volcanic lakes seems to be Lake Yugama (JP), which experienced a pH of 0.6 in 1950 and contained 3,800 mg l^{-1} of sulfuric acid and 4,700 mg l^{-1} of hydrochloric acid (Mori et al. 1984). More recently, Green and Kramadibrata (1988) measured a pH of 2.5 in an Indonesian crater lake containing a comparatively modest 560 mg SO_4 l^{-1} and 540 mg Cl$^-$ l^{-1}, with the acidity again contributed by H_2SO_4 and HCl. The low pH allowed extremely high total trace-metal solubility and dissolved aluminum concentration of 20,000 μg l^{-1}, a concentration several orders of magnitude higher than the toxic levels observed in poorly buffered temperate zone fresh waters receiving high levels of acidifying precipitation (Sec. 27.8). The acidity of the Indonesian lake was sufficient to corrode the metal fittings of the survey boat below the water line. The highly acidic Indonesian lake was characterized by a unialgal bloom as in highly alkaline Lake Nakuru. The bloom was large enough to reduce the transparency to only 1.5 m. The H$^+$ concentration (mol l^{-1}) of such acidic lakes is about 10,000 times greater than the concentration measured in freshwater systems with a pH between 6.0 and 7.5.

14.6 Carbonates: Precipitation and Solubilization

The vast majority of fresh waters, characterized by moderate buffering and a pH between 6 and 9, have a very small fraction of DIC in the CO_3^{2-} form (Fig. 14–1). The $CaCO_3$ that is formed has a very low solubility product—0.87×10^{-8} mol l^{-1} at 25°C, declining further with increasing temperature—allowing precipitation of $CaCO_3$ from high alkalinity waters at pH > ~8.5. The dissolved CO_2 concentration is minimal (Fig. 14–1) and even a modest reduction of CO_2 during photosynthesis disturbs the CO_2–H_2CO_3–HCO_3–CO_3^{2-} equilibrium (Eq. 14.10) enough for the solubility product of calcite to be sufficiently exceeded to allow $CaCO_3$ precipitation:

$$Ca(HCO_3)_2 \rightleftharpoons \underset{\text{(calcite)}}{CaCO_3} \downarrow + \ H_2O \ + \ CO_2 \qquad \text{(EQ. 14.12)}$$

Calcite ($CaCO_3$) is by far the dominant crystal formed in high alkalinity waters dominated by a low (< 2) molar Mg:Ca ratio. At somewhat higher molar ratios, $CaMgCO_3$ crystals are formed, and at even higher ratios (> 12) $CaCO_3$ crystals of different structures (aragonite) are created. Those crystals that are not redissolved in the sediments are subject to further

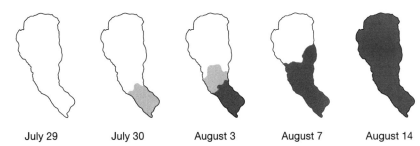

Figure 14–3   Seasonal development of a 1978 whiting ($CaCO_3$ precipitation) in Pyramid Lake, Nevada, US. The shading ranges from white for areas of low reflectance (albedo) to black for areas of highest albedo produced by $CaCO_3$ crystals. *(Modified from Galat and Jacobsen 1985.)*

chemical transformation, yielding a variety of different minerals (Kelts and Hsü 1978).

Groundwater is normally supersaturated with dissolved CO_2 from soil respiration. The excess is lost to the atmosphere where calcareous springs reach the surface. The result is precipitation of $CaCO_3$ (Eq. 14.12), evident from whitish crystalline encrustations on sediments or stones in high alkalinity streams. Similarly, the upper surface of submerged macrophyte leaves in moderate to high alkalinity lakes and wetlands are often encrusted with a $CaCO_3$ precipitate usually attributed to an uptake of CO_2 in photosynthesis and efflux of OH^- at the upper leaf surface, resulting in supersaturation, precipitation, and deposition of calcite (Eq. 14.12). The same phenomenon has been observed on the upper surface of plastic leaves, rendering this explanation insufficient. An alternative or complementary explanation for calcite deposition is that it is the result of a combination of elevated temperature in macrophyte beds and reduced $CaCO_3$ solubility (see below) at the upper leaf surface of rapidly growing apical leaves (Dale and Gillespie 1977, in Sand-Jensen 1989).[4]

The summertime precipitation of calcite crystals from the upper epilimnia of calcareous freshwater lakes and carbonate-rich saline lakes is a spectacular manifestation of calcite precipitation (Fig. 14–3). The phenomenon is referred to as **whiting** or seasonal clouding of lakes, referring to the milky-white color imparted by the crystals (Fig. 14–3). The many crystals dramatically lower the transparency.

▲ For a whiting to develop, the water must not only be supersaturated but must also contain an abundance of tiny algae or bacteria (picoplankton, < 2 μm), or other small particles serving as nuclei around which the $CaCO_3$ can precipitate (Kelts and Hsü 1978, and Thompson et al. 1997). However, high levels of Mg^{2+},

soluble inorganic phosphorus (SRP), and dissolved organic compounds reduce or retard crystal development and precipitation by serving as surface inhibitors of nucleation. These same inhibitors apparently serve to reduce the dissolution rate of crystals when these sink into the deeper subsaturated waters of the hypolimnia and sediments. Consequently, supersaturation is not synonymous with precipitation.

There is now abundant evidence that the development of whitings is greatly affected by the plankton community structure which, in turn, is linked to the trophic status of lakes. Intense fish predation on herbivorous macrozooplankton allows a sufficiently elevated phytoplankton biomass in eutrophic lakes, generating the required high level of photosynthesis (CO_2 removal) for supersaturation and calcite precipitation (Andersson et al. 1978). Conversely, the removal of zooplanktivorous fish during successful biomanipulation experiments to increase water clarity permits a much larger biomass of filter-feeding macrozooplankton to develop (e.g., *Daphnia* spp; Sec. 23.7). Their abundance reduces the algal biomass and the possibility of a whiting because of a lowered water-column-integrated primary production (Hanson et al. 1990).

In well-studied Lake Zurich (CH), the precipitation of calcite is initiated during onset of the spring diatom bloom, when high photosynthetic rates lower the dissolved CO_2 and raise the pH at a time of rising water temperatures resulting in reduced CO_2 solubility (Fig. 14–4). Crystals created usually have a diameter of between 3 and 15 μm, with their size varying somewhat between lakes and becoming smaller after the onset of precipitation (Fig. 14–5).[5]

Although whitings are almost always described in terms of events below the waterline, the ultimate causes are a function of drainage basin geology and soils, and runoff. Thus, whitings and the associated re-

---

[4]Interpretations are drawn from data (observations) and do not totally depend on them, but also on the scientific background of investigators which influences how particular findings are interpreted.

[5]The extent to which waters are saturated with calcite is usually computed rather than directly measured, with the solubility of calcite indicated by the IAP: K_c ratio (Fig. 14–4). (Kelts and Hsü 1978)

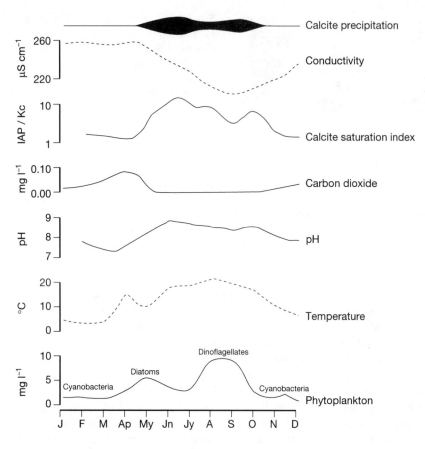

Figure 14–4 Seasonal pattern of calcite (CaCO₃) precipitation and related environmental variables in the epilimnion (5 m) of Lake Zurich, CH in 1974. *(Modified from Kelts and Hsü 1978.)*

duction in transparency in a monomictic Swiss lake were attributed to the role precipitation (runoff) plays in carrying nutrients and bicarbonate to the lake, allowing both increased primary production and a larger pool of bicarbonate for calcite production (Lot-

ter and Birks 1997). Conclusions change with the scale of investigation (Sec. 2.6).

Calcite crystals play a role in nutrient and organic matter cycling because they adsorb dissolved organic matter and phosphorus (Otsuki and Wetzel 1972,

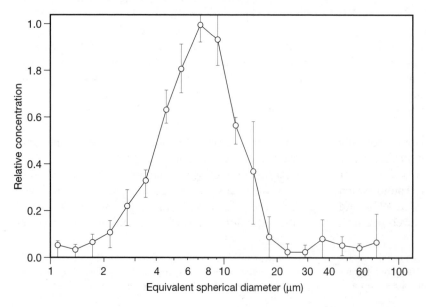

Figure 14–5 Relative concentrations of calcite particles in different size classes (μm ± SD) in Lake Michigan, US. *(After Vanderploeg et al. 1987.)*

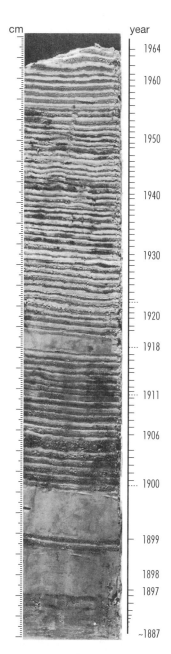

Figure 14–6   Short gravity core taken from the central portion of Lake Zurich at a water depth of 134 m. Light layers are calcite rich; dark layers are diatom and organic rich. The large sediment accumulation between 1897 and 1900 represents fine-grained redeposited shoreline sediments. *(Photo by H. Simola/After Kelts and Hsü 1978.)*

Jäger and Röhrs 1990) and produce detritus aggregates that serve as vehicles for the removal of sorbed nutrients and organic matter from euphotic zones during particle sedimentation (Chapter 20), followed by a partial resolubilization in the sediments and a return to the overlying water. The rain of crystals facilitates the sedimentation of other particles—including algae—and their incorporated nutrients (see Fig. 14–6, Koschel et al. 1983, and Hodell and Schelske 1998). Whitings, therefore, play a role in the biogeochemistry and cycling of a variety of elements and, as they occur on a seasonal basis in meromictic and poorly mixed lakes, they readily allow the dating of sediments (Chapter 20).

Highlights

- Dissolved inorganic carbon (DIC) concentrations are of great importance as a buffer against rapid changes in pH, as a source of inorganic carbon necessary for photosynthesis, and in removing inorganic carbon and absorbed materials from the water column following precipitation as $CaCO_3$.
- The proportion of different ionic forms of DIC (H_2CO_3, HCO_3^-, CO_3^{2-}) is largely controlled by, and in calcareous waters predictable from, the pH.
- Acid neutralizing capacity (ANC) is the principal among-system determinant and predictor of pH.
- ANC in inland waters ranges from extremely high alkalinity (well-buffered) saline lakes to streams draining coal mine tailings or highly acidic volcanic lakes that lack buffering and have highly negative alkalinity (high acidity).
- Species richness tends to decline at extreme pH on both sides of a broad optimum (pH 6–8.5) that characterizes the vast majority of unpolluted inland waters. The few species present can be highly abundant.
- When the solubility product of $CaCO_3$ is considerably exceeded—following a large-scale removal of CO_2 during photosynthesis or an appreciable loss to the atmosphere—calcite crystals precipitate.
- ▲ Whitings, precipitation events in lakes lower transparency and affect nutrient cycling through the removal of adsorbed organic matter, sorbed plant nutrients, and enhanced sedimentation of algae and other particles.

C H A P T E R

15

Dissolved Oxygen

15.1 Introduction

The dissolved oxygen (DO) levels in aquatic systems probably reveal more about their metabolism than any other single measurement. Concentrations reflect the momentary balance between oxygen supply from the atmosphere and photosynthesis on one hand, and the metabolic processes that consume oxygen on the other. Low DO levels not only affect the distribution and growth of fish (Chapter 26) and invertebrates (Chapter 25) but also have a major influence on the solubility of phosphorus and other inorganic nutrients (Chapters 17, 18, and 19) through the influence of DO on redox potentials (Chapter 16), including toxic trace metal solubility (Chapter 28). In addition, there is a growing interest in using diel DO changes as a measure of whole-ecosystem metabolism (Table 8–5), making it possible to see if a waterbody is net heterotrophic, with a photosynthesis:respiration (P:R) ratio smaller than one and fueled in part by allochthonous organic matter or conversely, with a P:R > 1 dominated by autochthonous primary production.

As Hutchinson (1957) put it, "a skilled limnologist can probably learn more about the nature of a lake [or river] from a series of oxygen determinations than from any other kind of chemical data. If these oxygen determinations are accompanied by observations on Secchi disc transparency, water color, and some morphometric data, a very great deal is known about the lake [or river]." This statement remains as valid today as it was more than 40 years ago. What has changed during the interval is the ease and precision by which DO concentrations can be measured. The standard

technique for determining DO levels in fresh water is the oxidation-reduction titration technique developed by L. W. Winkler (1888) over a hundred years ago. This technique has been slightly modified since then to reduce the effects of interfering substances.[1] The development of rugged oxygen meters with easily maintainable and sensitive sensors means that vertical profiles can now be obtained in a fraction of the time. The sensors are routinely calibrated against the almost constant oxygen concentration of the atmosphere, but critical calibrations still use the Winkler technique for standardization. For very precise measurement, the Winkler technique—using automated titrators with endpoint detectors—remains unsurpassed.

15.2 Solubility of Oxygen in Water

The solubility of DO in fresh water is primarily determined by water temperature. About 14.62 mg l^{-1} can be held in distilled water at 0°C declining to 11.29 mg l^{-1} at 10°C and declines to 8.26 mg l^{-1} at 25°C when in equilibrium with water-saturated air at standard pressure (Table 15–1). For example, this means that the

---

[1]The Winkler method involves two steps: DO is first reduced upon oxidation of a Mn^{2+} solution added to the water sample. Iodine is added next to reduce the oxidized Mn^{4+}. The oxidized iodine is then back-titrated with a thiosulphate solution and starch or pH meter is used for endpoint detection. The very precise automated titrators now available use a redox electrode to determine the endpoint. The technique suffers from chemical interferences in saline waters requiring the use of other, less precise, techniques.

Table 15–1 **Dissolved oxygen concentrations in pure water in equilibrium with water-saturated air at sea level (1 atm = 760 mm Hg = 101.324 Pa).**

Temp. (°C)	DO (mg l^{-1})	Temp. (°C)	DO (mg l^{-1})	Temp. (°C)	DO (mg l^{-1})
0	14.621	14	10.306	28	7.827
1	14.216	15	10.084	29	7.691
2	13.829	16	9.870	30	7.558
3	13.460	17	9.665	31	7.430
4	13.107	18	9.467	32	7.305
5	12.770	19	9.276	33	7.183
6	12.447	20	9.092	34	7.065
7	12.138	21	8.914	35	6.949
8	11.843	22	8.743	36	6.837
9	11.559	23	8.578	37	6.727
10	11.288	24	8.418	38	6.620
11	11.027	25	8.263	39	6.515
12	10.777	26	8.113	40	6.412
13	10.537	27	7.968		

Source: After Benson and Krause 1980.

warm hypolimnia of tropical lakes at 25°C at the onset of stratification holds one-third less DO per liter than similar temperate lakes at 10°C. Combined with higher community respiration rates in the much warmer tropical hypolimnia, this means that hypolimnetic waters of all but highly oligotrophic deep lakes at low latitudes (and low altitudes) become anoxic, regardless of trophic status (Thornton 1987). Low latitude wetlands and slow-flowing rivers are also more prone to deoxygenation than in the cooler temperate zone.

The solubility of any gas in a volume of liquid is proportional to the pressure that the gas exerts (Henry's law), causing the solubility of DO to decline with decreasing barometric pressure. This occurs with increasing altitude (Table 15–2) and appreciably increases with depth in deep lakes due to hydrostatic pressure.

Another factor influencing the solubility of oxygen in water is salinity. Sea water and saline lakes with a salinity of about 3.5‰ (35,000 mg l^{-1}), hold about 20 percent less DO at saturation than distilled water at the same temperature (Table 15–3). The oxygen-holding capacity is even lower in highly saline lakes. For example, in the highly saline Dead Sea (IL, JO) with a salinity of 345 g l^{-1}, only 1.45 mg l^{-1} can be dissolved at 25°C in oxygen-saturated water (Nishri and Ben-Yaakov 1990) compared to the 8.26 mg l^{-1} expected in turbulent fresh water at sea level. An impor-

tant implication of the decreased oxygen solubility in saline waters (or warm fresh waters) is that only modest levels of hypolimnetic respiration are required for them to become low in DO (*hypoxic*, operationally defined as < 2–3 mg O$_2$ l^{-1}) or turn *anoxic*. Global warming would lower gas solubility and increase respiration rates, leading to an increase in hypolimnetic and wetland anoxia, as well as in anoxic lowland rivers follow-

Table 15–2 **Partial pressure correction factors (P) and dissolved oxygen solubility factors (S) at different altitudes.**

Altitude (m)	Pressure (mm Hg)	P	S
0	760	1.000	1.00
100	750	0.988	1.01
200	741	0.976	1.02
300	732	0.965	1.04
400	723	0.953	1.05
500	714	0.942	1.06
1,000	671	0.887	1.13
2,000	594	0.785	1.28
2,500	560	0.735	1.36
3,000	526	0.692	1.44

Source: After Mortimer 1981.

Table 15–3   **Dissolved oxygen concentrations as a function of NaCl concentration.**

NaCl concentration (g l^{-1})	Temperature (°C)						
	0	5	10	15	20	25	30
0	14.60	12.79	11.34	10.13	9.10	8.22	7.49
2	14.40	12.63	11.2	10.01	9.00	8.13	7.42
5	14.12	12.39	11.00	9.84	8.85	8.00	7.30
10	13.65	11.99	10.66	9.55	8.60	7.79	7.12
15	13.20	11.61	10.34	9.27	8.36	7.58	6.93
25	12.33	10.88	9.72	8.74	7.90	7.18	6.58
50	10.37	9.22	8.29	7.50	6.82	6.24	5.75
100	7.24	6.53	5.95	5.46	5.03	4.65	4.33
150	4.96	4.54	4.20	3.90	3.64	3.40	3.21
200	3.34	3.10	2.91	2.74	2.58	2.45	2.33
260	2.03	1.92	1.83	1.75	1.67	1.61	1.55

Source: From Sherwood et al. 1992.

ing the point-source input of large quantities of organic matter from urban centers or wetlands.

The amount of DO that can be held by water in equilibrium with the atmosphere at a particular temperature, pressure, and salinity is known as the **saturation** or **equilibrium concentration**, the concentration observed in nature is expressed as a percentage of this value. Thus, inland fresh waters at a temperature of 20°C are expected to hold 9.09 mg l^{-1} DO at sea level (Table 15–1). If a particular system is shown to hold 7.02 mg l^{-1}, the water is at 77 percent saturation and is thus **subsaturated**. Conversely, if the observed concentration is 10.31 mg l^{-1}, the water would be at 113 percent saturation and is **supersaturated**.

15.3   Sources and Sinks of Oxygen

The rate DO is added to a waterbody by diffusion across the water surface is determined by the degree of subsaturation. However, even in highly subsaturated waters, measurements show molecular diffusion across a calm water surface to be of little importance. The major mechanisms for rapid DO gain (or loss) are wind-induced turbulence in lakes and gravity-induced turbulence in rivers, which thereby maintain a strong diffusion gradient. Reduced turbulence in littoral zones, and wetlands covered by floating leaves and submerged macrophytes whose stems extend to or above the water surface (Fig. 15–1), reduces diffusion,

as does reduced turbulence in metalimnia. As a result, the vertical eddy diffusion coefficient (Sec. 12.5) is low and the DO transport rate to the hypolimnia from the epilimnia is small, except when there is turbulence at the thermocline (Sec. 12.11).

The fraction of lake volume in which daytime photosynthesis is normally larger than the associated community respiration, is known as the **euphotic** or **photic** zone, whereas decomposition predominates in the **aphotic** zone. Photosynthesis by phytoplankton, benthic algae, and submerged macrophytes add DO to the euphotic zone according to the simplified formula

Figure 15–1   False water fern (*Salvinia molesta*) covering wind-protected littoral zone of Lake Naivasha, in Kenya. (*Photo courtesy of J. Kalff.*)

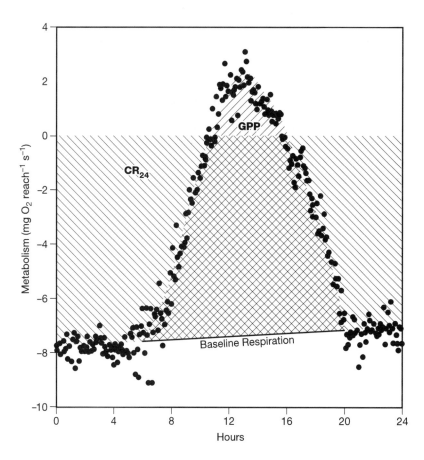

Figure 15–2   Diurnal dissolved oxygen (DO) evolution and consumption over a 62 m reach in Walker Branch, a Tennessee (US) forest stream, on 10 April, 1992. Gross primary production (GPP) represents the triangular area encompassed by the black circles dots and the baseline respiration. Community respiration (CR_{24}) represents the average night respiration scaled over 24 hours. Dissolved oxygen evolution exceeds system respiration only between ~10 and 17 hours, yielding a diel GPP: CR_{24} ratio of 0.44. Note that the ratio, like all other findings, is scale-dependent with the ratio > 1 around noon in spring and summer, and << 1 in midwinter. (*Corrected after Marzolf et al. 1998.*)

$$6CO_2 + 6H_2O \ b \ C_6H_{12}O_6 + 6O_2. \quad \text{EQ. 15.1}$$

Conversely, plant, animal, and bacterial respiration (community respiration) consumes DO, with the reaction acting in the opposite direction (Eq. 15.1).

Well-vegetated wetlands, the epilimnia of eutrophic lakes, small streams, and slow-flowing eutrophic rivers exhibit pronounced diurnal changes in DO concentration, and the whole-system 24-hour respiration rate in most streams (Table 8–6) is only partially offset by DO evolution during the day by photosynthesis. The stream shown in Fig. 15–2 only exhibits a net DO evolution between late morning and midafternoon, showing that the P:R ratio is highly sensitive to the time scale selected: negative over 24 hours, but positive midday. Similarly, DO concentrations and P:R ratios of mesotrophic and eutrophic monomictic lakes in the north temperate zone plus the rivers and wetlands exhibit not only a diel but also a seasonal pattern. The upper layer typically holds as much DO during the midwinter period of low irradiance and photosynthesis as the temperature and respiration allow (≤ 100 percent saturated). Elevated

temperatures reduce the absolute solubility in summer and the tropics, lowering DO levels, but appreciable algal photosynthesis during the stratification period causes the epilimnia of mesotrophic and eutrophic lakes, plant-rich rivers, and some wetlands to become supersaturated during the day (Fig. 15–3)—greatly so in highly eutrophic lakes (Fig. 15–4).[2] Similarly, high rates of photosynthesis in dense beds of submerged macrophytes in the littoral zone and wetlands may produce supersaturation on sunny days, but the beds become subsaturated as the result of nighttime respiration.

The fraction of organic matter produced in the epilimnia that sinks into the hypolimnia progressively lowers hypolimnetic DO concentrations as the result

---

[2]Not all supersaturation is the result of photosynthesis, it can have a physical basis. Slow-flowing and sun-exposed temperate zone streams become temporarily supersaturated when they warm rapidly in the morning, with DO loss to the atmosphere less rapid than the temperature rise. Ultra-oligotrophic polar lakes become supersaturated as the result of expulsion or "freezing out" of DO (and salts) during ice formation.

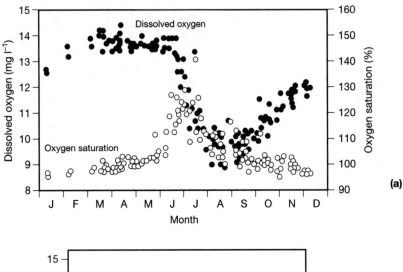

(a)

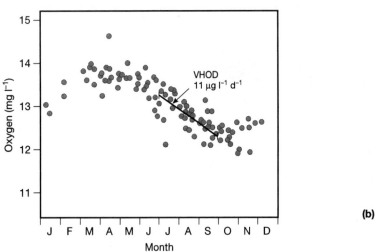

(b)

Figure 15–3   (a) Dissolved oxygen concentration (black circles) and oxygen saturation (white circles) in off-shore (≥ 100 m depth) near-surface (≤ 10 m) waters of Lake Ontario (CA, US). Mean values obtained from 124 cruises between 1966 and 1981. (b) Dissolved oxygen concentration and the volumetric oxygen depletion rate (VHOD) in the profundal zone at < 4°C. *(After Dobson 1985.)*

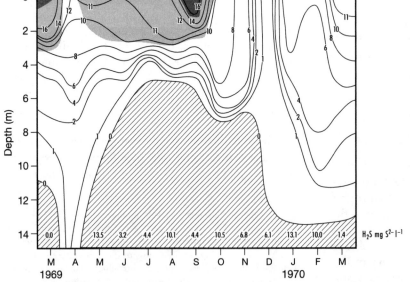

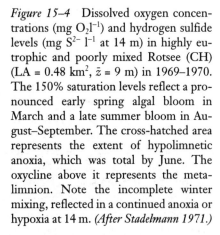

Figure 15–4   Dissolved oxygen concentrations (mg $O_2 l^{-1}$) and hydrogen sulfide levels (mg $S^{2-} l^{-1}$ at 14 m) in highly eutrophic and poorly mixed Rotsee (CH) (LA = 0.48 km^2, $\bar{z}$ = 9 m) in 1969–1970. The 150% saturation levels reflect a pronounced early spring algal bloom in March and a late summer bloom in August–September. The cross-hatched area represents the extent of hypolimnetic anoxia, which was total by June. The oxycline above it represents the metalimnion. Note the incomplete winter mixing, reflected in a continued anoxia or hypoxia at 14 m. *(After Stadelmann 1971.)*

of decomposition. It does this most effectively in deep hypolimnia where more time is available for decomposition in the oxic water column before the remaining particles are incorporated into the sediments. The hypolimnia becomes anoxic where DO consumption exceeds the mass of DO accumulated in the water during the overturn period. Where there is also an excess of reduced ionic sulfur, toxic hydrogen sulfide sulfide (H_2S) is evident (Fig. 15–4, Chapter 19).

The rate at which hypolimnetic oxygen is consumed in a given volume, (Fig. 15–3), known as the **volumetric hypolimnetic oxygen depletion rate** (VHOD), provides a useful indication of nutrient loading and primary production rates in the overlying epilimnion of nonhumic lakes (Sec. 15.5).

15.4   Photosynthesis, Respiration, and DOC

Wind-exposed freshwater lakes typically exhibit a nearly constant oxygen concentration throughout the water column during periods of complete mixing. The concentration closely approximates DO solubility at a particular temperature and atmospheric pressure (Tables 15–1 and 15–2). Following stratification, epilimnetic concentrations remain close to saturation in the open water of oligotrophic clear-water lakes containing too few phytoplankton to allow appreciable daytime supersaturation, although supersaturation is possible in shallow lakes and ponds dominated by submerged macrophytes and their epiphytes (Chapter 24). High photosynthetic rates during the day and high respiration rates at night in these water bodies yield large diel DO changes (Wylie and Jones 1987). Highly eutrophic fresh waters and saline lakes with a very high algal biomass become highly supersaturated on calm days (Fig. 15–5); DO concentrations as high as 20 mg l^{-1} develop in the top 20 cm of tropical Lake Nakuru (KE) and the supersaturation is so large there is extensive gas bubble formation on calm days (Melack and Kilham 1974).[3] Exceptionally high night-

---

[3]At equilibrium, well-oxygenated hypolimnetic waters can hold much more DO than at the surface because of pressure at depth. Below reservoir dams, the sudden release of DO-rich hypolimnetic waters under pressure yields a sudden supersaturation at atmospheric pressure and the release of bubbles of DO and other gases in the water. The equivalent production of gas bubbles in the blood vessels of fish passing through the subsurface outlet of such dams causes considerable mortality and is known as *gasbubble disease*.

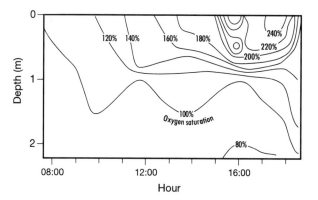

Figure 15–5   Diurnal changes in oxygen in shallow, highly eutrophic (hypereutrophic) Lake George (LA = 250 km², $\bar{z}$ = 2.4 m), UG (see Table 9–3). Percent saturation isopleths are presented at 20% intervals. *(Modified after Ganf and Horne 1975.)*

time community respiration rates in hypereutrophic ponds, shallow lakes, and wetlands is sufficient to develop anoxic conditions by early morning, resulting in fish kills (Barica 1975). But even where wetland hypoxic conditions do not result in fish kills, the low DO condition resulting from high respiration rates causes fish to avoid such habitats (Miranda and Hodges 2000).

With the onset of stratification, one of two vertical patterns develop as the result of separation of the deep water (hypolimnion) from the upper layer (epilimnion), which remains in contact with the atmosphere. The first pattern, seen in oligotrophic clear-water lakes at higher latitudes, occurs where the small amount of decomposable organic matter—produced in the euphotic zone and derived from drainage basins—is insufficient to appreciably reduce DO concentrations following sedimentation into the hypolimnia. The seasonal decline of DO is therefore small and the saturation of the hypolimnia remains close to 100 percent throughout the stratification period. The DO solubility is higher in the colder hypolimnia than in the warmed epilimnion and a characteristic concentration curve becomes established in which DO concentrations are higher in the hypolimnion (e.g., Atter Lake, AT, Fig. 15–6), with the vertical concentration gradient known as an **orthograde oxygen profile**.

The second pattern is characteristic of stratified eutrophic lakes and humic lakes where the hypolimnia receives considerable organic matter input from the epilimnia. The resulting elevated hypolimnetic respi-

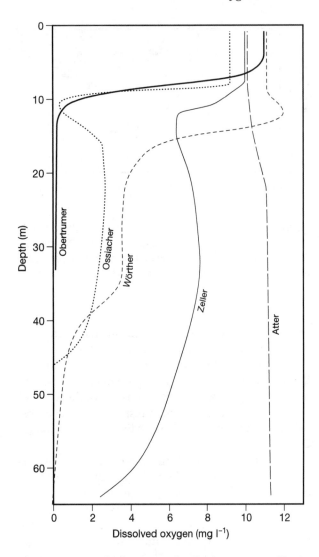

Figure 15–6 End-of-summer dissolved oxygen profiles in five Austrian lakes. *(Modified after Findenegg 1967.)*

DO captured during the last overturn and made available for decomposition) and water temperature.

The orthograde and clinograde DO patterns seen in lakes are sometimes modified by the presence of either a metalimnetic DO maximum or minimum. Lakes with a maximum DO are characterized by a **positive heterograde profile** (e.g., Wörther Lake, AT, Fig. 15–6). Observed maxima are usually the result of elevated algal photosynthesis in the metalimnia (Sec. 22.10), but can also come about because offshore density currents (in small lakes) (Sec. 12.6) carry supersaturated littoral zone water away from macrophyte beds (Dubay and Simmons 1979) located at roughly the same depth as the metalimnetic maxima. A metalimnetic DO minimum, described by a **negative heterograde profile** (e.g., Ossiacher Lake, AT, Fig. 15–6) has been linked with the respiration of high numbers of macrozooplankton (Cornett and Rigler 1987), but could also result from subsaturated water carried by density currents from low DO macrophyte beds. At other times, high rates of metalimnetic respiration and the resulting low DO levels are probably attributable to heterotrophic bacteria oxidizing organic matter sedimented from the overlying water or transported by currents from littoral zones, first postulated 90 years ago by F. A. Birge and C. Juday (1911).

Humic lakes have a clinograde curve despite typically low rates of phytoplankton primary production. High inputs of externally-derived (allochthonous) organic matter, combined with disproportionately low inputs of plant nutrients, only allow low within-lake (autochthonous) production. Disproportionately high rates of community respiration (CR_{24}), primarily by microbes, yield primary production (GPP) to (CR) ratios < 1. High rates also yield subsaturated epilimnia and anoxic hypolimnia (Fig. 29–11). Lotic systems and wetlands may be subsaturated, hypoxic, or even anoxic for the same reason (Molot et al. 1992).

Biological Oxygen Demand and Deoxygenation

▲ Slow-flowing lotic systems receiving large quantities of untreated sewage wastewater or livestock manure—applied to fields but washed into streams—may exhibit sufficiently high rates of system respiration to yield a **biological/biochemical oxygen demand** (BOD) high enough to produce hypoxic or anoxic water columns. The amount of readily decomposable organic matter has long been assessed in water pollu-

ration rates progressively lower DO concentrations from saturation levels established during overturn, typically lowering them far below levels recorded in the epilimnia. The result is a **clinograde oxygen profile**, with hypolimnetic dissolved oxygen levels typically declining with depth (e.g., Wörther Lake, AT, Fig. 15–6), sometimes to zero.

The extent to which hypolimnetic DO levels become lowered from the saturation concentration is not only a function of the amount of decomposable organic matter entering the hypolimnia but also by the length of time since the onset of stratification, the volume of the hypolimnia (which determines the mass of

tion work by means of the BOD assay, using oxic water samples usually incubated for 5 days in the dark at 20°C. The measured oxygen consumption rates serve as a surrogate for the organic matter consumed (primarily by heterotrophic microbes, Chapter 22). The long established and widely used BOD technique was originally developed to predict the level of organic wastes that could be accommodated over time and space by rivers below sewage outfalls while preventing the rivers from turning anoxic (Fig. 15–7). The modeling yielded the first dynamic-process models in aquatic science some 80 years ago. BOD determinations are also used occasionally to assess the rate of heterotrophic metabolism in lakes. Organically polluted lotic systems exhibit an **oxygen sag curve** (Fig. 15–7) where the BOD is the result of a point-source input. The first sewage treatment plants were established to remove organic matter in order to reduce the possibility of water-borne diseases such as cholera, eliminate foul smells associated from anoxic respiration, and improve the appearance of the affected rivers. River deoxygenations are greatest at high water temperatures when the holding capacity of DO is lowest (Table 15–1) and microbial respiration rate is high.

Not all river deoxygenations are the result of human activity. Pristine tropical rivers—and their associated well-vegetated wetlands inundated during periods of high seasonal river drainage—may experience anoxic or hypoxic conditions and fish kills (Hamilton et al. 1997) when hypoxic water carrying abundant organic matter, plus reduced compounds such as iron (Fe^{2+}), methane (CH_4), and ammonium (NH_4^+), enter receiving rivers and lakes. The microbial and chemical oxidation of the organic matter and other reduced compounds in the receiving waters greatly lowers DO levels (Chapter 16). One mole of DO is consumed for every four moles of Fe^{2+} that is oxidized; one mole DO per 0.6 moles of CH_4 oxidized; and one mole DO per 0.3 moles of NH_4N oxidized. DO depletion in the wetlands themselves is the result of microbial and animal respiration of the organic matter produced by macrophytes and algae and of terrestrial vegetation produced during the preceeding low-water period when large portions of tropical wetlands are dry (Chapter 24). Well-vegetated South American tropical wetlands are therefore typically hypoxic (< 2 mg O_2 l^{-1}) during the day and even more so at night (Hamilton et al. 1995). The input of low DO water plus decomposable organic matter and reduced inorganic compounds into receiving rivers may be large enough to result in anoxia and fish kills. High loadings (inputs) of respired wetland CO_2 further increase the respiratory stress on fish and sensitive invertebrates in the receiving waters (Calheiros and Hamilton 1998).

▲ 15.5 Dissolved Oxygen Consumption and Lake Productivity

Midsummer vertical oxygen distribution profiles provide an indication of the productivity of clear-water (nondystrophic) lakes (Fig. 15–6). When hypolimnetic profiles over time are examined, they provide useful information on the seasonal pattern of epilimnetic primary production in individual lakes (Fig. 15–3). The same vertical profiles over a series of years can reveal long-term changes in nutrient loading and primary production (Figs. 15–8 and 15–9). But A. Thienemann recognized as early as the 1920s the impact of mor-

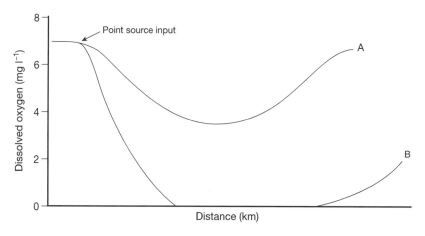

Figure 15–7 Schematic oxygen sag curves observed in two rivers receiving a modest (A) and a large (B) input of wastewater from an urban, industrial, or agricultural source. Curve A represents conditions where the dilution of the point source is sufficient to prevent bacterially generated anoxia. Curve B represents the development of anoxic conditions followed by a gradual exhaustion of the organic supply and a gradual re-aeration. The re-aeration is slowed by the consumption of DO by microbes catalyzing the oxidation of large quantities of NH_4^+ to NO_3^- (see Chapter 16).

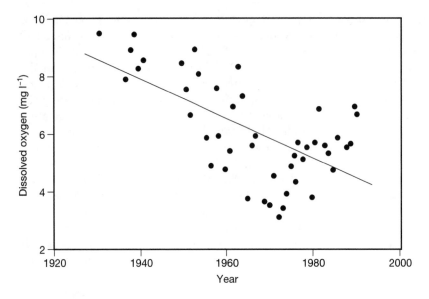

Figure 15–8 Change in the minimum dissolved oxygen concentration in the profundal zone (60–80 m) of the central part of the north basin of Lake Biwa, JP (LA = 674 km² $\bar{z}$ = 41 m) just before the onset of the autumn circulation period. Reasons for the considerable variation between adjacent years include (1) differences in the date of onset of stratification, with late years allowing less time for DO consumption; (2) between-year hypolimnetic temperature differences, established at the time of stratification, affecting hypolimnetic respiration rates; and (3) inter-annual differences in nutrient loading, planktonic primary production, and organic matter input from the drainage basin. It is unclear whether the last 25 years or so have seen a reverse of the progressive increase in seasonal deoxygenation of the hypolimnion. *(Modified from Mori et al. 1984, and unpublished data from Shiga Prefecture,*

phometry, with a thick hypolimnion allowing greater dilution of the oxidizable organic matter sedimenting into it than the same amount of organic matter sedimenting into a thin hypolimnion. He argued that it would be meaningless to use the average volumetric hypolimnic oxygen depletion rate (VHOD, g m⁻³ d⁻¹) to compare DO consumption rates of hypolimnetic water *among* lakes (Fig. 15–10) to estimate epilimnetic primary production (Fig. 15–3). This conclusion led

K. M. Strøm (1931) and G. E. Hutchinson (1938) to propose a DO depletion rate for a given area of hypolimnetic surface, the **areal hypolimnetic oxygen depletion rate** (AHOD, g O₂ m⁻² d⁻¹) as a superior alternative. The AHOD was thought to be independent of the dilution effect produced by the hypolimnia of differing volumes (Fig. 15–10).

AHOD is obtained from the VHOD for individual layers of the hypolimnion, summing them, and ex-

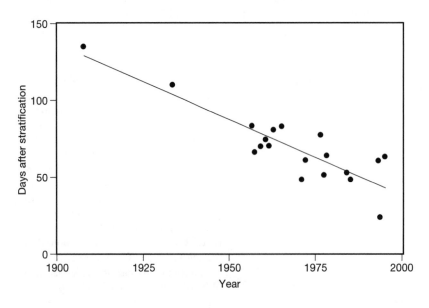

Figure 15–9 The increasingly early onset of anoxic conditions (< 0.2 mg O₂ l⁻¹) over the profundal sediments of Lake Esrom, a Danish kettle lake (LA = 17.3 km², z_max = 22 m, CA:LA = 3.4), after the onset of stratification. *(After Sand-Jensen and Pedersen 1997.)*

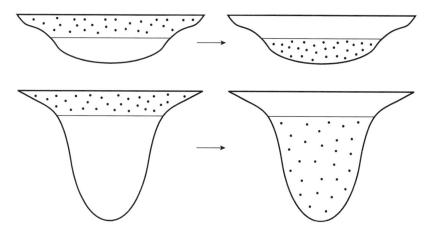

Figure 15–10 The proposed effect of lake morphometry on the hypolimnetic consumption of DO in lakes with identical epilimnetic volume and primary production, as well as identical upper surface of their hypolimnia but differing in the hypolimnetic volume to receive the sedimented organic matter. *(Modified after A. Thienemann 1928, in Jónasson 1993.)*

pressing the results in areal units (g O_2 m² d⁻¹). G. E. Hutchinson (1938) reported a good correlation between AHOD and productivity of four lakes. His finding was widely accepted as sufficient proof—although based on limited data—for the utility of the AHOD as an indicator (index) of primary production, at a time when direct measurements of primary production were difficult to obtain (Sec. 2.4). There was renewed interest in the AHOD during the eutrophication era (~1970–1985), when the effect of DO consumption/concentration on phosphorus release from anoxic sediments was of wide concern (Chapter 17). It then became apparent that the AHOD is not really a good predictor of epilimnetic production or the related phosphorus loading (Fig. 15–11).

At that time, Cornett and Rigler (1980) demonstrated that the AHOD alone accounted for less than half of the variability in primary production of 65 north temperate zone lakes. Important refinements were made, ultimately allowing 95 percent of the variability in 25 lakes to be accounted for (Charlton 1980), proving the importance of also considering water temperature—which, as discussed above, affects the rate of microbial respiration and hypolimnetic thickness. Sedimenting organic particles spend more time and are more fully oxidized (consume more DO) in deep than in shallow hypolimnia of equal volume because of longer contact with oxygenated hypolimnetic water during sedimentation (Cornett and Rigler 1987). In shallow hypolimnia, the particles quickly become part of the mostly anoxic sediments where they serve as a substrate for microbes (Sec. 22.11). Because anoxic respiration is less efficient, it ultimately consumes less DO than if these particles had spent more time in a deeper, better oxygenated water column (Chapter 16).

The "rain" of organic particles allows for microbial growth and respiration in the sediment (Chapter 22).

The newly recognized importance of hypolimnetic temperature, thickness of the hypolimnia, and rate of flushing (particle removal via the outflow) made it possible to estimate epilimnetic production and phosphorus release from anoxic sediments better than be-

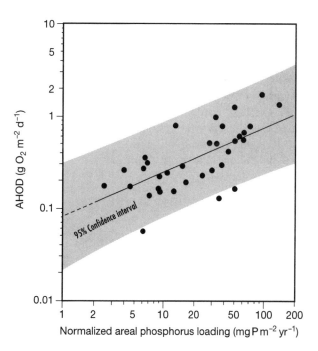

Figure 15–11 The relationship between phosphorus loading and hypolimnetic oxygen consumption rates (AHOD) in 37 US lakes. The phosphorus loading is normalized for water loading (q_s, m yr⁻¹) and water residence time (WRT, yr). *(After Lee and Jones 1984.)*

fore (Chapter 17). The epilimnetic productivity can now be determined directly, more simply and quickly and the AHOD is rarely used for this particular purpose. However, VHOD measurements are often used in aquatic management to predict whether, or when, particular hypolimnia are expected to become anoxic or hypoxic (Fig. 15–3). As pointed out above, hypoxia has negative effects on invertebrates and fish, whereas a lengthy anoxia results in the disappearance of most aerobic organisms and massive changes in food-web structure (Sec. 15.7). Furthermore, there is commonly a large-scale release of sediment phosphorus in the hypolimnia of eutrophic lakes, followed by diffusion and entrainment into the epilimnia, large algal blooms, an increase in particle sedimentation, and deoxygenation (Chapters 17 and 21).

▲ 15.6 Oxygen Depletion in Ice-covered Waters

Eutrophic lakes covered with ice and snow, wetlands, and northern rivers receiving large quantities of organic matter from their ice and snow covered wetlands, exhibit substantial DO losses during winter. A *winterkill* of fish and macroinvertebrates is common where the organic matter supply that is available for decomposition is large relative to the mass of DO available, and the waters are covered for months by snow and ice.[4] Poor light conditions permit little photosynthesis but a continuing respiration. Babin and Prepas (1985) combined existing Canadian data and found that the *winter oxygen depletion rate* under the ice (WODR, mg O_2 m^{-2} d^{-1}) is a function of the summer total phosphorus concentration (TP$_{SU}$, mg m^{-2}), and mean lake depth ($\bar{z}$, m)

$$\text{WODR} = -0.101 + 0.0247\,\text{TP}_{SU} + 0.0123\,\bar{z},$$
$$R^2 = 0.64 \quad P < 0.001 \quad n = 33 \qquad \text{EQ. 15.2}$$

with TP$_{SU}$ serving as a surrogate for the summer algal production it allows. This partially determines the pool of organic matter available for decomposition

under the ice. Larger mean depth is a proxy for larger water volume, making a larger mass of DO available for the aerobic decomposition of organic matter (Sec. 15.5 and Fig. 15–11). Although temperature has an important effect on respiration (DO consumption) the small under-ice temperature differences preclude the possibility of a measurable effect. See section 2.6 regarding the importance of the range (interval) scale in limnology.

15.7 Dissolved Oxygen and the Biota

Eutrophication brings about increased hypolimnetic DO consumption in lakes, progressive lowering of hypolimnetic DO concentrations (Figs. 15–8 and 15–9), pronounced diel DO oscillations in rivers and wetlands, and at higher latitudes, under-ice anoxia. Where hypolimnetic consumption rates are sufficiently high, most or all DO is consumed before the next overturn (Fig. 15–4). Anoxic or extended hypoxic conditions have a dramatic negative effect on aerobic organisms such as macroinvertebrates and fish (Fig. 15–12).

Unfortunately, the water quality standards necessary to prevent mortality differ among species and for the various lifestages. Fish eggs and young-of-the year, characterized by particularly high metabolic rates, are more sensitive than adults. Among invertebrates, eggs

---

[4]"Due to the oxidation of humic and other materials carried from marshes of the Vasiugan basin, and to the ice-cover which cuts off atmospheric oxygen from the water, total deoxygenation [of the Ob River, RU] occurs over an enormous area for a period of 3–4 months and fishes which cannot escape die. Such conditions, which occur annually and are relatively prolonged in duration, give fishes no opportunity to adapt physiologically." (V. M. Brown 1975, in Whitton (1975) reporting on Russian research by N. A. Mosevich)

Figure 15–12 Fish kill on the Seine River in France. *(Photographie De Sazo-Rapho/Photo courtesy of Photo Researchers, Inc.)*

tend to be less sensitive than larvae. In addition, there are long-term negative (*chronic*) effects on growth and reproduction affecting community structure when DO concentrations are well above lethal levels.

The limited data that are available suggests that maintaining healthy fish populations requires concentrations of at least 2–5 mg O_2 l^{-1} for moderately tolerant epilimnetic species and 5–9 mg O_2 l^{-1} are necessary for cold-water species (salmonids) and warm-water species with high oxygen demands in waters that do not also contain stressful levels of free CO_2 and pollutants (Alabaster and Lloyd 1982). However, most adult fish and invertebrates can cope for some time with concentrations considerably lower than the optimum—especially at higher latitudes where metabolism is reduced during periods of low water temperature (Sec. 26.6)—but even modest reductions below the optimum have long-term negative effects. For example, the food conversion efficiency and resulting growth of juvenile large-mouth bass (*Micropterus salmoides*), a warm-water species, declines dramatically when DO concentrations fall below 3–4 mg l^{-1}, and swimming speeds are reduced at concentrations below ~5 mg l^{-1} (M. L. Dahlberg et al., in Miranda and Hodges 2000). Furthermore, sustained hypoxia (< 1.5–2.0 mg O_2 l^{-1}) is lethal for most prey organisms not specially adapted,[5] although invertebrate resting eggs and resting stages exhibiting reduced respiration rates may survive lengthy periods of anoxia in the sediments (Chapters 23 and 25).

Many macroinvertebrates and some fish evolved in waters experiencing reccuring periods of hypoxia/anoxia and have developed the means to survive periods of weeks to months with little or no DO. They can enter a state of dormancy during which glycogen stores—laid down during earlier, favorable feeding periods—are consumed anaerobically. However, the conversion process is inefficient, and the incompletely utilized end products (lactate, ethanol, etc.) are released into the environment. But the ability to enter a dormant phase is no defense against sudden deoxy-

genations, such as overturns or a major entrainment that mixes large quantities of reduced, sometimes toxic, compounds from the anoxic hypolimnia into the epilimnia. Such events produce virtually anoxic surface waters under low wind conditions. Lakes and reservoirs at low altitudes may remain anoxic and hypoxic for days to weeks (Robarts et al. 1982), and wetlands probably remain so much longer.

Apart from the direct effects, there are also important indirect effects of low and zero DO on biotic communities. For example, anoxic conditions make the hypolimnia and sediments unavailable to macrozooplankton as a daytime refuge from sight-feeding predators. Anoxic and hypoxic sediments cause not only major mortality of benthic invertebrates but also the death of fish eggs layed on sediments. Their death may result in a partial or total loss of a cohort (year-class), impinging on the abundance of both their prey and predators. The outcome is a change in the planktonic and benthic food-web structure, with the effects manifesting themselves over periods of months to decades of which investigators are rarely aware and don't consider in their interpretations. The development of sediment anoxia also enhances the sediment release of phosphorus, iron, toxic hydrogen sulfide, and others (Chapters 16, 17, and 19) that affect the biota.

Proximal causes—anoxia and hypoxia as discussed here—produce a series of direct and indirect effects, sometimes referred to as **stressors**, influencing aquatic systems as much, or more, than the original stressor. Combinations of stressors and their impact changes over time and between systems making the impact of an identified direct stressor unpredictible (see also Sec. 11.12). This is the major reason for the difficulties encountered when developing badly needed dynamic models (Sec. 17.8) of ecosystem functioning and of the abundance of particular biological species or chemicals. These models are usually based on otherwise excellent studies of single stressors under controlled conditions. In other words, the validity of conclusions is highly dependent on the relevance of the scale at which the research was carried out to the scale at which anwers are sought (Sec. 2.6).

---

[5] "Extensive mortality of fish and zooplankton occurred in Lake Valencia, Venezuela, between November 29th and December 6th, 1977. The mortality was preceded by an extended calm period during which the lower part of the water column became anoxic and accumulated large amounts of reduced substances. Sudden increases in wind strength brought about complete mixing of the water column leading to greatly reduced oxygen levels and high concentrations of hydrogen sulfide, ammonia, and nitrite. Large numbers of fish . . . and about 90 percent of the total zooplankton population were killed during the sudden mixing." (A.O. Infante et al. 1979)

Highlights

- Dissolved oxygen (DO) concentrations represent the momentary balance between the DO supply from the atmosphere and in-situ photosynthesis (P) on the one hand and all the respiratory process (R) that consume DO on the other.

- The solubility of DO declines with increasing temperature and salinity and the possibility of anoxia or hypoxia is much higher in hypolimnia, wetlands, and lowland rivers at low latitudes than at high latitudes. Anoxia and hypoxia probability is highest in the hypolimnia of low latitude saline lakes.

- Rates of DO depletion in hypolimnia serve as indicators of the supply of organic matter produced in epilimnia (autochthonous production) plus that received from drainage basins and the littoral zone (allochthonous production).

- The community structure of aerobic organisms (macroinvertebrates and fish) is greatly affected by hypoxic (low DO) conditions and is decimated under anoxic conditions. The best-adapted species are able to survive periods of anoxia by means of either anaerobic respiration of stored glycogen reserves and/or by entering resting stages characterized by reduced respiration rates.

16

Oxidation–Reduction Potential

16.1 Introduction

Organisms and their abiotic environment are closely linked. Life in aquatic systems is possible because of the availability of solar energy, a fraction of which is used by aquatic plants for reducing CO_2 and oxidizing water during aerobic photosynthesis, with the reduced products stored in the form of organic matter, and supplemented by organic matter exported from drainage basins. Reduced products are oxidized during the energy yielding respiration of organisms, thereby tending to restore the oxidation-reduction equilibrium. The decomposition process recycles carbon and nutrients, and provides the reducing power to drive the oxygen, nitrogen, iron, manganese, sulfur and, indirectly, phosphorus cycles, plus a number of other cycles discussed in this and other chapters. The **oxidation–reduction potential** or **redox potential** is a composite measure of the overall intensity of the oxidizing or reducing conditions within a system and reflects the degree of balance between oxidizing and reducing processes.

Redox reactions involve inorganic plant nutrients (Fig. 16–1). On the basis of average compositions of marine phytoplankton, known as the *Redfield ratio* (Redfield et al. 1963), a more complete equation is:

$$106CO_2 + 16HNO_3 + H_3PO_4 + 122H_2O + 18H^+ + \text{trace}$$
$$\text{elements} + \text{solar energy} \underset{R}{\overset{P}{\rightleftarrows}} (CH_2O)_{106}16NH_3H_3PO_4 + 133O_2$$

EQ 16.1

Rather than presenting the protoplasmic ratio as in Eq. 16.1, it is normally presented simply as the atomic (molar) ratio of the principal elements 106C:16N:1P, and more completely as 106C:16N:1P:~0.01 (Fe, Zn, Mn):~0.001 (Cu, Mo, Co, etc.). It is the diurnally and seasonally shifting balance between photosynthesis and respiration that determines the measured redox potential of inland waters and surface sediments. Photosynthesis dominates respiration in the euphotic zone, whereas respiration dominates in the aphotic portions of the water column and sediments, but respiration processes dominate everywhere at night. The redox condition has a major impact on the distribution of the biota—from microbes to fish.

16.2 Redox Reactions and Nutrient Cycling

Any study of biological transformations in aquatic systems requires consideration of redox potentials since they have a major effect on the concentration and/or form of organic carbon (Chapter 8), dissolved oxygen (Chapter 15), nitrogen (Chapter 18), sulfur (Chapters 19 and 27), iron and manganese (Chapter 19), and to a varying extent, trace metals (Chapters 19 and 28). The oxidation state of redox elements supplied by terrestrial and aerial catchments determines whether they will be found dissolved in water, or will be insoluble and removed by sedimentation. This determines whether or not particular nutrients are available to aquatic plants for uptake and growth which fuel pro-

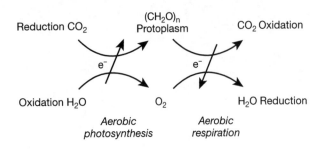

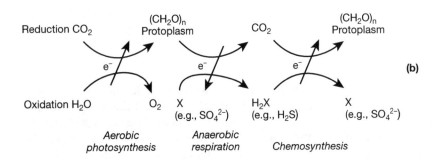

Figure 16–1 Oxidation–reduction reactions, showing (a) photosynthesis–respiration reactions under aerobic conditions; and (b) reactions in oxic and anoxic systems. The transfer of electrons (e^-) is indicated by arrows. *(Modified after Kortmann and Rich 1994.)*

ductivity at other trophic levels. Similarly, the redox state determines the extent that toxic trace metals, such as lead and mercury, are available for uptake by the biota. While inorganic phosphorus occurs virtually in only one oxidation state (PO_4^{3-})—and thus is not a redox element—its presence in the water column is nevertheless greatly influenced by the oxidation state of iron and abundance of ionic aluminum (Chapters 17, 19, and 27).

The redox potential expresses an environment's tendency to receive or supply electrons in the same way pH expresses the concentration of H^+ in solutions (Sec. 14.3). The redox potential (E_h) is estimated by measuring the electron flow with a pH meter having millivolt capacity. The electron flow is measured between a reference (standard) electrode (serving as a reference point of zero volts) and a platinum electrode. E_h is expressed as millivolts (mV) or volts (V). In other words, E_h represents the voltage necessary to prevent the flow of electrons between the environment and a reference electrode. E_h readings change with pH and are normally expressed as the potential at pH = 7. Alternately, the oxidation–reduction state of the water or sediment can be presented as the electrode potential (pE) which, representing the net balance of electron transfer reactions, provides a measure of the oxidizing or reducing intensity of the system. Positive potentials indicate that the system is relatively oxidizing, negative potentials indicate that it is relatively reducing.

In pure solutions, redox reactions can be described by paired electron transfers. A reductant is a compound that reacts by releasing electrons (an electron donor) and is oxidized, whereas an oxidant is a compound that takes up electrons (an electron acceptor) and becomes reduced.

$$\text{reductant}_1 + \text{oxidant}_2 \leftrightarrows \text{oxidant}_1 + \text{reductant}_2$$
EQ. 16.2

An example of a microbe-mediated oxidation reaction in nature involves the reduction of nitrate. Denitrifying bacteria are **facultative anaerobes**, and are able to switch from dissolved oxygen (DO) to NO_3 as an alternative electron acceptor in the oxidation of organic matter after the DO is depleted during aerobic respiration (Fig. 16–2 and Eqs. 3a and b in Table 16–1). The availability of NO_3 provides particular microbes with an energy source to maintain the cells that have already formed and for continued growth.

Natural waters and sediments are complex mixtures in which a large number of different organic and

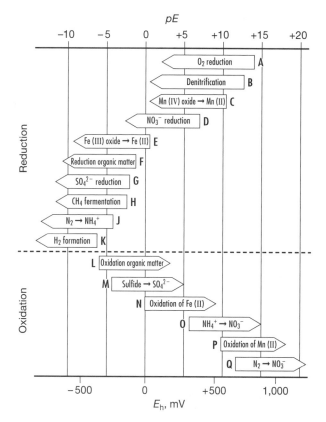

Figure 16–2 Biologically mediated redox processes (calculated for pH = 7) in the oxidation of organic matter (A–K, and L), nitrogen fixation (J and L), and the oxidation of reduced elements in the presence of oxygen (M–Q), which yields energy for the reduction of CO_2 in chemosynthetic processes. Overlapping E_h or pE ranges indicate that several redox systems operate at the same time. *(After Stumm and Baccini 1978.)*

inorganic redox reactions are taking place simultaneously. Many reactions in nature are neither complete or reversible—as opposed to reactions in pure solutions. In other words, the reactions have not reached equilibrium even though equilibrium conditions are assumed when redox values are presented. Therefore, the single E_h or pE value representing the oxidation state of water or sediment at any moment must be interpreted cautiously and the measurements are more accurately referred to as the **apparent redox potential**, or simply as **electrode potentials**, rather than by calling them **redox potentials**—which they are not (Mortimer 1971). In addition, measured potentials may be inaccurate or erroneous when the platinum electrodes become coated with an invisible organic or

inorganic film (Hayes et al. 1958). Low E_h sediments may also release reduced forms of elements to the overlying water even though the measured potential of the water itself is higher than the level of the reduced form. Together, these problems mean that the redox state of any particular element is usually inferred from the presence or absence of DO, or from the oxidation state of a particular measured redox element. According to Mortimer (1941, 1942) denitrification commences when about 4.0 mg O_2 l^{-1} remains in the water, but is virtually lacking (~0.1 mg l^{-1}) at the onset of the reduction of Fe^{3+} to Fe^{2+}. The disappearance of DO is associated with an E_h of ~200 mV.

In spite of its inaccuracies, measured electrode potentials have utility in qualitatively interpreting solubility changes for many important plant nutrients and trace metals, and in considering the geochemistry of mineral formation or solubilization in water and sediments. They also serve as an indicator (surrogate) of the degree of sediment oxygenation and the suitability of particular sediments for invertebrates sensitive to low DO conditions (hypoxia). Potentials integrate all sources of organic matter supply and provide an index of the biological activity within a system. The potentials and dissolved oxygen consumption rates (AHOD, VHOD; Sec. 15.5) are measures of trophic status and whole-system metabolism with great utility in aquatic management and research.

Hexagenia limbata (Ephemeridae, Ephemeroptera—a large burrowing benthic insect (mayfly) of the littoral and sublittoral zone of north temperate lakes, and an important fish-food organism—is normally present in the surficial sediments when the E_h is –120 mV (measured at 5 cm below the surface), and is always absent at E_h < –190 mV (Fig. 25–3). The maximum abundance (biomass) threshold is between E_h = –170 and –140 mV (Fig. 16–3). Above this threshold, factors other than degree of sediment oxygenation determine the abundance of *Hexagenia*. Lake eutrophication, resulting in an increased organic matter sedimentation rate, increased sediment decomposition rate, and a resulting reduction in DO leads to the decline or disappearance of *Hexagenia* and other oxygen-sensitive invertebrates. The low potentials reported mistakenly suggest that *Hexagenia* appears to thrive in anoxic sediments (Fig. 16–2), but this is so only because these particular E_h measurements served as surrogates, made at 5 cm—to improve voltage stability—thereby underestimating the E_h values in the upper few centimeters where the animals live. There, *Hexagenia* and other

Table 16–1 **Selected oxidation–reduction reactions and the maximum amount of free energy (ΔG_0^1) that can be derived from each, providing an indication of the amount available for work in cells. ΔG_0^1 represents the change in standard free energy when 1 mole of reactants (represented here by organic matter with an average oxidation state corresponding to glucose) is converted to 1 mole of product (at 25°C, at 1 atmosphere of pressure, and at a pH of 7). $kJmol^{-1}$ = 4.187 kcal.**

1. Oxidation of organic matter by aerobic heterotrophic microorganisms (using dissolved oxygen as a terminal electron acceptor) and the return of nutrients to the inorganic form.

   $(CH_2O)_{106}(NH_3)_{16}(H_3PO_4) + 138O_2 \rightarrow 106CO_2 + 16HNO_3 + H_3PO_4 + 122H_2O;$

   $\Delta G_0^1 = -3,190 \quad kJmol^{-1}.$

2. Oxidation of organic matter and the reduction of insoluble Mn(IV) to soluble Mn(II), catalyzed by facultative anaerobic microorganisms.

   $(CH_2O)_{106}(NH_3)_{16}(H_3PO_4) + 236MnO_2 + 472H^+ \rightarrow 236Mn^{2+} + 106CO_2 + 8N_2 + H_3PO_4 + 366H_2O;$

   $\Delta G_0^1 = -3,050 \quad kJmol^{-1}.$

3a. Oxidation of organic matter and the associated reduction of nitrate (denitrification), catalyzed by facultative anaerobic microorganisms and yielding NH_3 (NH_4^+) plus gaseous N_2 which is lost from the system.

   $(CH_2O)_{106}(NH_3)_{16}(H_3PO_4) + 94.4HNO_3 \rightarrow 106CO_2 + 55.2N_2 + H_3PO_4 + 177.2H_2O;$

   $\Delta G_0^1 = -3,030 \quad kJmol^{-1}.$

3b. $(CH_2O)_{106}(NH_3)_{16}(H_3PO_4) + 84.8HNO_3 \rightarrow 106CO_2 + 42.4N_2 + 16NH_3 + H_3PO_4 + 148.4H_2O;$

   $\Delta G_0^1 = -2,750 \quad kJmol^{-1}.$

4. Oxidation of organic matter and the associated reduction of insoluble Fe(III), catalyzed by facultative anaerobic microorganisms and yielding soluble Fe(II).

   $(CH_2O_{106}(HN_3)_{16}(H_3PO_4) + 212Fe_2O_3$ (or $424FeOOH) + 848H^+$

   $\rightarrow 424Fe^2 + 106CO_2 + 16NH_3 + H_3PO_4 + 530H_2O$ (or $742H_2O);$

   $\Delta G_0^1 = -1,410 \quad kJmol^{-1} (Fe_2O_3)$

   $\Delta G_0^1 = -1,330 \quad kJmol^{-1} (FeOOH).$

(continued)

burrowing larval insects are able to ventilate their burrows through undulating body movements that carry oxygenated water from the overlying hypolimnion into the burrows (Eriksen 1963) and remove waste products in the return flow.

Bacteria, Organic Matter, and Nutrient Cycling

Redox reactions in nature are biologically mediated (catalyzed) by microorganisms and the chemical sequence (Fig. 16–2, Table 16–1) is paralleled by an ecological succession of microbial taxa. Very few of these organisms have been studied in the laboratory and little is known about species assemblages in nature and how they change at different redox potentials. Nevertheless, systems exhibit a high redox potential

under well-oxygenated conditions because the dominating heterotrophic organisms have DO available as the terminal electron acceptor in the aerobic oxidation of organic matter (Fig. 16–1). The electrons themselves are produced in the metabolism (oxidation)[1] of reduced organic compounds to CO_2 (Eq. 1 in Table 16–1).

▲ Only a fraction (~1% to 50%) of the energy released in microbially mediated redox processes becomes available for use by the organisms for growth, with the balance either lost in respiration or stored in sediments. (Chapter 22). Furthermore, organisms can only carry out processes that are thermodynamically

---

[1]In fact, oxidation does not involve the removal of free electrons but of usually paired hydrogen atoms, in a process called dehydrogenation. H^+ cannot accumulate so oxidation is accompanied by the reduction of the H^+ acceptor and the release of energy.

Table 16–1   (continued)

5. Oxidation of organic matter and the associated reduction of sulfate, catalyzed by obligate anaerobes and yielding reduced sulfur.

   $(CH_2O)_{106}(NH_3)_{16}(H_3PO_4) + 53SO_4^{2-} \rightarrow 106CO_2 + 16NH_3 + 53S^{2-} + H_3PO_4 + 106H_2O$;

   $\Delta G_0^1 = -380$ kJmol^{-1}.

6. Fermentation of organic matter and the production of methane, catalyzed by obligate anaerobes.

   $(CH_2O)_{106}(NH_3)_{16}(H_3PO_4) \rightarrow 53CO_2 + 53CH_4 + 16NH_3 + H_3PO_4$;

   $\Delta G_0^1 = -350$ kJmol^{-1}.

7. Oxidation of selected reduced inorganic compounds in the presence of oxygen by chemosynthetic (chemolithotrophic) microbes.

   a. $CH_4 + 2O_2 \rightarrow CO_2 + 2H_2O$;

      $\Delta G_0^1 = -810$ kJmol^{-1}.

   b. $HS^- + 2O_2 \rightarrow SO_4^{2-} + H^+$;

      $\Delta G_0^1 = -797$ kJmol^{-1}.

   c. $2NH_4^+ + 3O_2 \rightarrow 2NO_2^- + 4H^+ + 2H_2O$;

      $\Delta G_0^1 = -275$ kJmol^{-1}.

   d. $2NO_2^- + O_2 \rightarrow 2NO_3^-$;

      $\Delta G_0^1 = -75.8$ kJmol^{-1}.

   e. $4Fe^{2+} + O_2 + 4H^+ \rightarrow 4Fe^{3+} + 2H_2O$;

      $\Delta G_0^1 = -4$ kJmol^{-1}.

Source: After Froelich et al. 1979, and Fenchel and Blackburn 1979.

possible and, therefore, only act as redox catalysts of particular reactions (Table 16–1, Fig. 16–2).

Aerobic environments normally have redox potentials greater than about +350 mV, but when nearly all the DO has been consumed other, more weakly oxidized, chemicals (Mn^{4+}, NO_3^-, Fe^{3+} and SO_4^{2-}) sequentially accept electrons from the organic matter being oxidized and become reduced (Fig. 16–2). For example, sulfate reduction is matched by reduced sulfur production (Fig. 16–2 and Eq. 5 in Table 16–1) which, in the presence of available reduced iron, may be precipitated as ferrous sulfide (FeS) (Sec. 19.3).[2] Both Fe and S in precipitated FeS are now in a re-

duced state and they cannot participate further as alternative electron acceptors in redox reactions.

In general, the oxidation of organic matter follows a sequence in which the oxidant yielding the greatest free-energy change per mole of organic matter is succeeded by the next most efficient oxidant (see Table 16–1), thereby yielding the maximum energy possible to the microbes. This process continues until all oxidants or all oxidizable organic matter is consumed. Last in the process are fermentation processes catalyzed by **obligate anaerobes** in which part of the organic molecule serves as the fuel (electron donor), and another part of the substance is used to combust it (oxidant), serving as the electron acceptor. In the anaerobic decomposition of organic contaminants (e.g., halogenated organic compounds or polychlorinated biphenyls) the microorganisms prefer those contaminants that can serve as either an electron donor or acceptor in redox reactions. The effectiveness of microbes in breaking down organic contaminants de-

[2]The reduction of SO_4^{2-} in anaerobic hypolimnia and sediments and its storage in sediments as FeS or organic sulfur consumes some of the H_2 produced, thereby removing H^+s from the system. This removal is an important buffering process in acidified waters and their sediments, allowing them to be less acidic than anticipated on the basis of the H^+ input (Chapter 27).

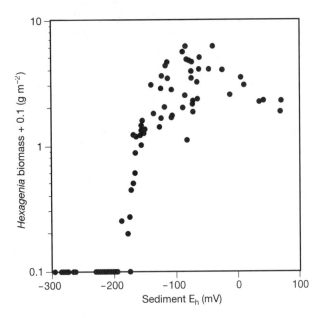

Figure 16–3 The biomoass of *Hexagenia limbata*, a large burrowing mayfly (Insecta) versus the oxidation–reduction potential measured at 5 cm below the sediment surface at 81 littoral and sublittoral stations in 12 Québec (CA) lakes. Note that 0.1 is added to each biomass value to allow 0 values to be plotted on the log-scale graph. *(After Rasmussen 1988.)*

pends, at least in part, on the presence of suitable non-contaminant counterpart electron acceptors or electron donors; and on the structure of the contaminants themselves, including the degree of chlorination (Chapter 28). In the oxidation process, complex organic compounds are converted to acetate, hydrogen (H_2), carbon dioxide, and, if less complete, to other organic compounds. Methanogenic bacteria use the acetate (CH_3COOH) and H_2 to reduce CO_2 to methane (CH_4), creating the typically elevated CH_4^+ levels found in anoxic freshwater sediments and hypolimnia (Fig. 18–11).

$$CH_3COOH \rightarrow CO_2 + CH_4 \qquad \text{EQ. 16.3}$$

$$CO_2 + 4H_2 \rightarrow CH_4 + 2H_2O, \qquad \text{EQ. 16.4}$$

and where CH_4 is available:

$$CH_4 + H_2O \rightarrow 2H_2 + CO_2. \qquad \text{EQ. 16.5}$$

At this point, the possibilities for microbial energy generation and growth have become exhausted. The highly reduced end products that have accumulated are metabolized to yield energy only when oxygenated conditions return (Eqs. 7a–e in Table 16–1, and Table 22–1).

Chemosynthetic Microbes

Other microbes, the **chemosynthetic (chemolithotrophic) bacteria**, are able to metabolize the reduced end products—including NH_4^+, S^-, Fe^{2+}, H_2, CO_2, CH_4 plus small amounts of reduced organic molecules—in energy-yielding reactions upon return of DO. This energy allows the reduction of CO_2 to organic matter (microbial protoplasm) and the return of the redox elements to a more oxidized form (Fig. 16–2 and Table 16–1).

While photosynthetic organisms use solar energy, chemosynthetic ones obtain their energy from reduced compounds and synthesize organic matter in the absence of light. Microbial protoplasm produced by catalyzing redox reactions serves as food for protozoa and other zooplankton in the *microbial food web* (Sec. 22.9). However, microbes do not catalyze all of the reduced inorganic compounds; a varying, but sometimes important, fraction is oxidized chemically upon contact with dissolved oxygen (Chapter 19).

The relative importance of different electron acceptors in the oxidation of organic matter depends on system morphometry and on the concentration of available redox elements. A lake with a large hypolimnion contains a greater mass of DO at the onset of stratification, and DO is able to serve longer as the principal electron acceptor than in a shallow lake with a small hypolimnion volume (Fig. 15–10). The availability of redox elements is important as well. For example, sulfate reduction plays a much more important role in the oxidation of organic matter in marine and estuarine waters and sediments than in their freshwater counterparts where SO_4 concentrations are much lower, with the result that the redox sequence in fresh water is shifted more quickly to fermentation processes and methane (CH_4) production (Table 16–2). In addition, where the pool of oxidized iron (Fe^{3+}) is exceptionally large—as in ancient, highly weathered soils underlying low latitude freshwater wetlands—it can serve as the principal electron acceptor in the anaerobic oxidation of organic matter (Eq. 4 in Table 16–1, and Roden and Wetzel 1996).

While the relative importance of particular electron acceptors in anaerobic respiration has been examined by many limnologists, only two studies appear to have examined all the principal ones. It is evident that

Table 16–2   **Relative importance of anoxic carbon mineralizing reactions during summer stratification in the hypolimnia of five lakes. NS = not significant; ND = not determined.**

Reaction	Percent Anoxic Carbon Metabolism				
	Vechten (NL)	Wintergreen (US)	Lake 226N (CA)	Lake 227 (CA)	Lake 223 (CA)
Mn^{4+} reduction	ND	NS	0.6	0.2	0.6
NO_3^- reduction	ND	NS	5	0.1	0.05
Fe^{3+} reduction	ND	NS	2.9	1.6	5.3
SO_4^{2-} reduction	ND	13	16	16	20
CH_4 production	70	87	75	82	74

Source: Kelly et al. 1988 and Capone and Kiene 1988.

methanogenesis (CH_4 production) is the primary route by which organic matter is respired in the anoxic hypolimnia of the few freshwater lakes that were examined, followed distantly by SO_4^{2-} (Table 16–2). Other redox elements played a minor role in these lakes, either because their concentrations were low (NO_3^- and SO_4^-) or because they were in particulate (insoluble) form in the water and sediments (Fe^{3+}, Mn^{4+}). In aerobic lake and river water, the oxidation of CH_4, Fe^{2+}, and other reduced compounds (including NH_4^+) derived from lake sediments and wetlands results in lower DO levels in the receiving waters (Sec. 15.4).

The scale at which research is carried out has a definitive impact on the conclusions that are drawn (Sec. 2.6). For example, among lakes varying greatly in trophic status (nutrient levels), but with less variation in the surface area and depth (morphometry) range scales, concentrations of critical plant nutrients or the resulting algal biomass are the best predictors of methane (and CO_2) production (Fig. 16–4). Conversely, where trophic status varies little among systems but morphometry varies greatly, morphometry is the best predictor of CH_4 (plus CO_2) released during decomposition of organic matter (den Heyer and Kalff 1998).

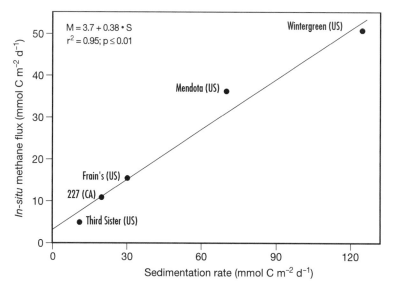

Figure 16–4   In-situ hypolimnetic methane flux rate as a function of the sedimentation rate of organic carbon into the anoxic hypolimnia of five North American freshwater lakes that vary greatly in both primary production and resulting sedimentation rates. Note the CH_4 flux is accompanied by a CO_2 flux not shown here (see Eqs. 16.3, 16.5, and Eq. 6 in Table 16–1). (*Modified after Kelly and Chynoweth 1981.*)

Highlights

- Oxidation–reduction reactions, or redox reactions, are largely mediated (catalyzed) in nature by microorganisms.
- Redox reactions are directly or indirectly responsible for most organic-matter oxidation, nutrient recycling, and energy flow from microbes to higher trophic levels.
- In contrast to reactions in pure solutions, in nature many redox reactions are neither complete nor reversible and redox potentials must be interpreted cautiously.
- Dissolved oxygen (DO) is the electron acceptor of choice in aerobic systems, resulting in lower levels of DO in the hypolimnia and sediments during metabolism (oxidation) of sedimenting and sedimented organic matter. ▲ When DO is exhausted, other electron acceptors (e.g., NO_3^-, SO_4^{2-}, organic matter) are reduced sequentially. The reduced compounds that are produced then serve as electron donors upon the return of DO, providing energy to yet other types of microorganisms.
- ▲ While the apparent redox potentials are usually inferred from the presence or absence of DO, measured potentials provide an indication of the oxidation state of elements and the types of microbes and other organisms that are dominant. Potentials are also useful for inferring the suitability of the sediment environment for aerobic invertebrates.

17

Phosphorus Concentrations and Cycling

17.1 Introduction

It is now widely recognized that the eutrophication of fresh water can be controlled by reducing the input of plant nutrients, especially phosphorus and nitrogen. Of these two elements, phosphorus is generally considered in shortest supply for algal growth in oligotrophic waters at midlatitude, characterized by a high N:P ratio (Sec. 8.1). Therefore, most nutrient-control programs have focused on it. The N:P ratio of the protoplasm of rapidly growing algal assemblages varies only moderately around the idealized Redfield ratio (Table 8–2); substantially lowering the phosphorus supply raises the N:P supply ratio above the element demand ratio and usually reduces the algal biomass. Another reason for the focus on phosphorus has been that the fixation of atmospheric nitrogen by cyanobacteria makes control of nitrogen more difficult (Chapter 18). Finally, phosphorus removal from wastewater through biological and chemical precipitation (Fig. 8–14) is readily feasible and less expensive than removal of nitrogen.

Phosphorus enters aquatic systems from the land and by direct atmospheric deposition on water surfaces (Tables 8–9 and 8–10). Catchment contributions are normally dominant, except where the catchment area is very small and also composed of a substrate low in phosphorus (Secs. 9.1–9.3). Phosphorus is typically high in demand relative to supply and phosphorus retention is high with little released to receiving waterways in well-watered and vegetated drainage basins in the temperate zone. The phosphorus that is released is primarily in the dissolved organic form, supplemented by organic phosphorus contained in particles washed from the land with additional dissolved phosphorus from the atmosphere. In poorly vegetated basins—including crop-growing areas—phosphorus is usually released to waterways while sorbed to soil particles, particularly during periods of storm runoff (Sec. 8.5).

17.2 The Classical Model of Phosphorus Cycling

According to the classical view of phosphorus cycling, based on very important research done before and after World War II by W. Einsele and W. Ohle in Germany, and C. H. Mortimer in England, PO_4^{3-} is strongly sorbed to iron oxyhydroxides (e.g., FeOOH) or precipitated as $FePO_4$ under aerobic conditions. Over time much of the PO_4^{3-} entering lakes and wetlands is removed from the water column after sorption to oxidized and highly insoluble FeOOH aggregates (flocs) and sedimenting organic particles, including algal cells. The aggregates continue to serve as effective sorption surfaces at the oxidized sediment surface and thus as an effective barrier to phosphate diffusion. The aggregates not only trap phosphate that is released following sediment decomposition of sedimented organic particles but, more importantly, serve as a barrier to PO_4^{3-} diffusing upward from the anoxic sediments below, preventing its return to the water column (see also Chapter 19). Mortimer (1941) determined that a sediment redox potential of ~200 mV separates the oxidizing (Fe^{3+} and associated

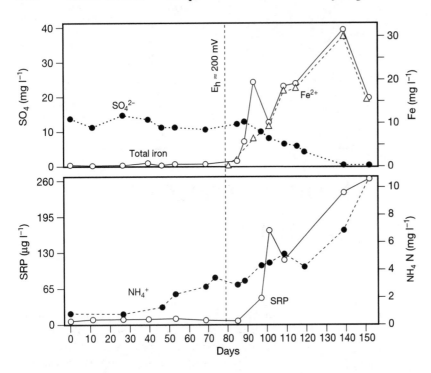

Figure 17–1 Variations in the chemical composition of water overlying the profundal sediments of Lake Windermere (GB). Sediments were examined in experimental sediment-water tanks over 152 days. Note the solubilization of iron and phosphorus at a redox potential of ~200 mV. *(Modified after Mortimer 1941).*

PO_4^{3-}; insoluble) from the reducing conditions (Fe^{2+} and PO_4^{3-}; soluble) (Fig. 17–1).

The ~200 mV redox boundary rises towards the sediment surface with increasing eutrophication or organic pollution, making more sedimented organic matter available for decomposition. When anoxic conditions reach the surface, $FeOOHPO_4$ complexes dissolve and PO_4^{3-} (plus Fe^{2+}) can diffuse into the hypolimnion. Interstitial concentrations of PO_4 are typically 5–20 times greater than in the overlying waters (Boström et al. 1982) and much diffusion to overlying waters is possible. The release of elements from sediments to the overlying waters is known as **internal loading**. This well-established term is widely used in limnology, but is a misnomer because the elements in question were already part of the aquatic system. The internal loading of reduced redox elements (Fe^{2+}, NH_4^+, NO_2^-, S^{2-}) from sediments imposes considerable oxygen demand in the overlying water where they serve as electron donors during microbe-mediated aerobic oxidation of organic matter (Chapters 16, 18, and 19).

Further Development of the Classical Model

▲ A few years after its articulation, Hasler and Einsele (1948) suggested that the iron-phosphate model was more complex than first proposed. They argued

that not only iron but also sulfate could affect the release of phosphorus from anoxic sediments. The microbial reduction of SO_4^{2-} (Table 16–1 and Fig. 16–2) yields sulfide (S^{2-}) which, upon the formation of highly insoluble FeS and FeS_2 (Sec. 19.3), can remove enough reduced iron from solution to allow phosphorus to escape adsorptive control. Waters high in sulfate are indeed characterized by low iron concentrations and thus by low Fe:P ratios, although the phosphorus release mechanism may only partly be the one postulated by Hasler and Einsele (Fig. 17–2, and Caraco et al. 1993). Nevertheless, the role of sulfur in anoxic phosphorus release is greatest in calcareous lakes containing much more SO_4^{2-} than Fe^{3+}. Conversely, the role of sulfur in affecting phosphorus release is small in lakes and wetlands in igneous and other basins receiving little sulfur from their catchments, which are typically characterized by a high Fe:S supply ratio (Chapter 19).

The classical model was greatly refined during the last few decades of the 20th century (Boström et al. 1982), incorporating important laboratory research by U. Tessenow (DE). Recent work on shallow Danish lakes determined that sediment phosphorus release under oxic conditions is negligible when the surficial sediment Fe:P ratio is > 15–20 (as mass), but is large when the ratio is < 10, presumably because insufficient iron is available to bind and precipitate the dissolved

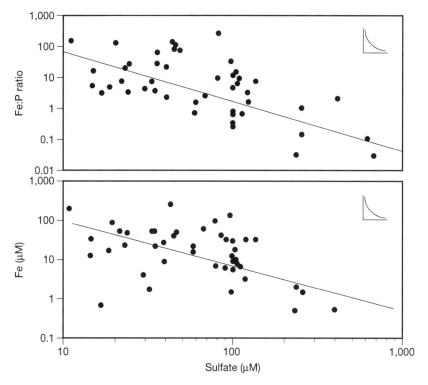

Figure 17–2 The among-lake relationship between sulfate and iron concentration and the negative impact of a high sulfate concentration on the Fe:P ratio in the bottom waters of US lakes. The data support the hypothesis that when sulfate is reduced to sulfide under anoxic conditions, iron sulfides are precipitated and the Fe:P ratio is lowered in the process. A low ratio provides fewer iron oxide flocs and a reduced P sorption. (*After Caraco et al. 1993.*)

phosphorus diffusing upward to the overlying water (Jensen et al. 1992). Other research has shown that sediment resuspension enhances phosphorus release in shallow eutrophic lakes (Hamilton and Mitchell 1988).

Although the Hasler and Einsele (1948) work on the role of sulfur suggested there was more to the release of PO_4^{3-} than iron binding, the Fe:P paradigm continued to dominate scientific thinking (see Boström et al. 1982). It was only when phosphorus cycling was examined in many additional lakes during the final three decades of the 20th century that it became widely accepted that some systems did not fit the classical model. That recognition led to a reevaluation of the model, which then led to a modified interpretation. When the data the original model was based upon were reviewed, they too revealed that new doubts are confirmed by at least some of the old data but not noted at the time. A new scientific paradigm is then born.[1]

---

[1]The Nobel laureate J. B. S. Haldane (1963) observed that there are normally four stages in the acceptance of a scientific idea: (1) This is worthless nonsense; (2) This is an interesting, but perverse, point of view; (3) This is true, but quite unimportant; (4) I always said so.

17.3 The Modern Model and Aerobic Phosphorus Release

In the classical model, phosphorus release from sediments is considered to be primarily a chemical process. Microorganisms play an indirect role by utilizing dissolved oxygen (DO), NO_3^-, SO_4^{2-}, Fe^{3+}, and Mn^{4+} as electron acceptors in organic matter oxidation, thereby affecting the solubility of chemical species (Table 16–1 and Fig. 16–2). However, more recent research is indicating that the release rate is significantly affected by decomposition processes involving microbes and thus by biological rather than chemical mechanisms as originally postulated by C. H. Mortimer and others (Prairie et al. 2001). The bacteria are directly involved by releasing **soluble reactive phosphorus** (SRP) in the water following cell lysis and upon release during anaerobic conditions of solubilized polyphosphate granules accumulated under aerobic conditions. Whatever the precise mechanisms, biological release is important—with somewhere between 10 and 75 percent of the potentially soluble sediment phosphorus not sorbed but held within the cells of microbes (Boström et al. 1988, Gächter and Meyer 1993). Furthermore, reduction of Fe(III) is not strictly chemically induced in the absence of DO. Reducing conditions alone are not enough to reduce Fe(III) to Fe(II), in part because the aggregates are stabilized by a coating of organic matter. Iron-reducing bacteria using Fe(III) as an electron acceptor in the oxidation of organic matter, are necessary (Lovley 1991, Eq. 4 in Table 16–1, and Chapter 19).

If the classical model were applicable, Fe(II) and PO_4^{3-} should be released simultaneously upon the chemical reduction of the Fe(III) complex. But Gächter et al. (1988) has shown that release patterns are out of phase. They also observed that sediments are less able to take up dissolved phosphorus upon sterilization with antibiotics—this would not be the case if the processes were strictly physical or chemical. Reexamination of the data collected by Mortimer (1942) in Estwaite Water (GB) showed that the PO_4^{2-} and Fe^{2+} release patterns had also been out of phase. Recently, Prairie et al. (2001) found the classical theory inadequate when they reported that no relationship existed between iron and phosphorus release in the anoxic hypolimnia of a series of Quebec (CA) lakes, and also showed that many oligotrophic lakes can go anoxic without phosphorus release. Release appears to depend not so much on hypolimnetic DO concentrations as on the balance between the supply of phosphorus in sedimenting particles and the—still poorly understood—retention capacity imposed by biological and chemical means. In practice, it does not really matter whether the released phosphorus had been part of an Fe complex or held by microorganisms. However, understanding how the phosphorus is held is crucial for consideration of underlying mechanisms that are not only interesting in themselves but may also lead to better predictive models.

▲ The recent reevaluation of the classical theory is important for several reasons:

1. It provides a useful reminder that correlations do not provide proof of a mechanistic link, even though a particularly mechanistic interpretation seems to be the most plausible explanation at the time.

2. It allows consideration of alternative hypotheses, explaining why some lakes with anoxic hypolimnia release little or no phosphorus (Caraco et al. 1991). It also allows the possibility of an improved explanation (prediction)—even in lakes that appear to fit the classical model.

3. The phosphorus released from inorganic $FeOOHPO_4$ (and $AlOOHPO_4$) complexes is inor-

ganic PO_4^{3-}, whereas phosphorus directly released by microbes is organic or colloidal polyphosphate phosphorus. Unfortunately, the standard colorimetric molybdate technique that is used for determination of PO_4^{3-}, uses reagents that hydrolyze (break down) labile organic phosphorus causing it to be measured as PO_4^{3-}. Yet, only one to 10 percent of the PO_4^{3-} may have been in this form prior to the chemical analyses (Rigler 1973). Consequently, hypolimnetic increases in dissolved phosphorus following the development of anoxia had always been thought to be in the PO_4^{3-} form, and thus in accord with the classical model—even though it may have been organic phosphorus released by microbes. Such analyses are now routinely reported as soluble reactive phosphorus (SRP).

4. The classical model pertains to the behavior of phosphorus in the hypolimnia rather than lakes as a whole. Large increases observed in the well-oxygenated mixed layer of shallow (unstratified) but productive lakes that cannot be attributed to external loading are not explained by the classical model. Nor is the classical model appropriate to explain why phosphorus release from well aerated and turbulent littoral zones is dependent on the phosphorus concentration gradient across the sediment–water interface (Boström et al. 1982).

Aerobic Phosphorus Release

Many mechanisms are responsible in varying degrees for aerobic phosphorus release. They range from those with physical causes to those with underlying biological causes. Their relative importance varies over time and space. Increases in internal loading in highly productive shallow lakes with an oxic sediment surface layer have been experimentally linked to increases in temperature (Søndergaard 1989)—suggesting that biological processes are probably involved—and to increases in pH resulting from high rates of photosynthesis (Fig. 17–3). Elevated pH appears to allow some of the phosphorus adsorbed to FeOOH flocs to be replaced by OH^-, with the phosphate released to the water. Microbial respiration and decay also release phosphorus and other elements into the interstitial water where it is subject to diffusion to the overlying water. Conversely, microbes play an indirect role in reducing phosphorus release by maintaining a sediment redox potential high enough to inhibit or reduce release from sediments. They do so by using elevated

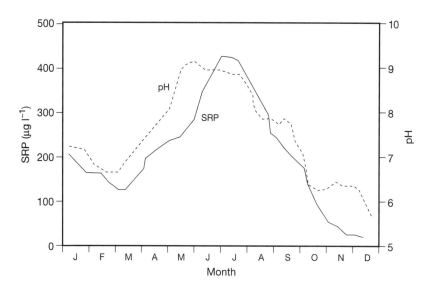

Figure 17–3 Seasonal variation in soluble reactive phosphorus (SRP) and pH in shallow (unstratified), hypertrophic Lake Glaningen (SE). Note that the correlation is no proof that increased pH is the cause of the changes in phosphorus flux from the sediments, although this is theoretically plausible. *(Modified from Ryding and Forsberg 1977, in Boström et al. 1982.)*

concentrations of NO_3^- and SO_4^{2-} at the sediment–water interface as electron acceptors in the oxidation of organic matter (Table 16–1).

Abundant benthic macroinvertebrates can also increase phosphorus release from oxic sediments through burrowing, filtering, feeding, and excreting activity at the sediment–water interface; these activities are collectively referred to as **bioturbation**. However, conflicting laboratory work suggests that the activity of macroinvertebrates increases the dissolved oxygen flux into sediments, thereby decreasing phosphorus release. Finally, a sediment coating of algae and bacteria (*biofilm*) may either increase or decrease phosphorus release depending on growing conditions.

Another group of mechanisms affecting phosphorus release are linked to turbulence induced by wind or gas bubbles. The turbulence enhances the transport of dissolved phosphorus from anoxic depths in the sediments directly to the overlying water, bypassing the phosphorus diffusion barrier imposed by ferric oxyhydroxide complexes at the oxic sediment surface (Kawai et al. 1985). Respiratory gas bubbles produced by microbes cause sediment turbulence that stimulates the release of SRP (Ryding and Forsberg 1976, and Ahlgren 1988).

While experiments under controlled conditions are required to identify and quantify the importance of particular mechanisms and processes, they are not geared for quantifying the relative or absolute importance of individual mechanisms in nature. For example, experimental work on quantifying the release of phosphorus is usually based on diffusion rates from profundal zone sediment cores brought to the laboratory. Even if laboratory conditions resemble the diffusion rates characterizing the nonturbulent profundal zone, they cannot be a surrogate (proxy) for release rates for shallow water sediments exposed to considerable turbulence.[2]

It is clear from whole-lake mass-balance calculations in shallow lakes that release rates per unit area from the shallow water sediments of deeper lakes are typically five to 10 times greater than those measured on profundal zone cores from the same lakes incubated under aerobic conditions in the laboratory (Enell and Löfgren 1988). The imbalance appears to reflect the combined effect in nature of diffusion, turbulence, bioturbation, and the release of phosphorus from littoral zones. The importance of turbulence in phosphorus release is suggested by a whole-lake mass-balance:core-determined release ratio that is much closer to the profundal sediment ratio (Enell and Löfgren 1988), indicating that under nonturbulent conditions the phosphorus transport out of deep water sediments is indeed attributable to molecular diffusion.

Models are at best an approximation of the real world and all models are, over time, subject to modification or qualification until the original model becomes unrecognizable or is totally rejected. The Einsele-

---

[2] "Attempts to infer process from patterns are fraught with the usual pitfalls associated with the interpretations of correlations. While an experimental approach goes far in side-stepping this pitfall experiments may be difficult, expensive, or impossible when the patterns and processes of interest apply to a system of large scale, such as entire lakes, catchments, or landscapes." (S. G. Fisher 1994)

Ohle-Mortimer phosphorus–iron model is no exception to this rule. It has continued to provide an important framework for ideas and data on phosphorus cycling and its link to the Fe cycle, but is less complete than its authors were able to surmise 60 years ago. Models are to be used, not unquestioningly believed.

▲ 17.4 The Mass-Balance Equation and Phosphorus Cycling

Aquatic systems are open systems with inputs and outputs. If one assumes that the input of substance M is constant and the input is instantly and completely mixed throughout the waterbody, then the steady state or equilibrium version of the relationship between inputs, losses, outputs, and the concentration of M can be expressed with a simple mass-balance model:

$$\frac{\Delta M}{\Delta t} = I - O - (S_{gross} - R). \qquad \text{EQ. 17.1}$$

Where $\Delta M/\Delta t$ = storage gain or loss of substance M over time t; I = external load; O = outflow loss; S_{gross} = loss to sediments; R = internal load; and $(S_{gross} - R)$ = net sedimentation (S_{net}).

Inland waters are subject to seasonal and diurnal changes in the absolute and relative importance of photosynthesis and respiration, as well as to changes in weather and climatic conditions and it is evident that they cannot be in a steady state, for any length of time.[3] But, since seasonal patterns repeat themselves more or less each year or growing season, equilibrium ($\Delta M/\Delta t = 0$) is assumed in steady state mass-balance modeling, which is usually done with a seasonal or annual perspective rather than with the hourly or daily dynamic reality.

The many phosphorus budgets now available, based on input-output modeling, have demonstrated both seasonal and annual variability in the rate of net phosphorus sedimentation. Mass-balance budgets for

phosphorus (and nitrogen)—made with applied goals in mind—pointed out the importance of sediments as a phosphorus source and has encouraged fundamental research on nutrient cycling; this in turn influences the interpretations made in more applied research.

Lakes with Oxic Hypolimnia

Temperate lakes with oxic hypolimnia are considerably different from those with anoxic ones. Lakes with oxic hypolimnia are generally deep and have thick hypolimnia able to hold a large mass of dissolved oxygen following overturn (Sec. 15.5). Their drainage ratios (CA:LA or CA:V) are normally relatively small, implying small water loading (Chapter 8), and thus a long water residence time (WRT), as well as small catchment-nutrient loading (see Table 17–1 and Chapter 8) compared to most lakes with anoxic hypolimnia.

Mass-balance measurements have shown that the fraction of the external phosphorus load supplied on an annual basis that is retained (R) and stored in the sediments rather than flushed out is about twice as large as for those with anoxic hypolimnia (Table 17–1). The fraction of phosphorus retained by lakes and wetlands declines exponentially with increasing water loading (m yr^{-1}) and increases with the WRT (yr) (Fig. 9–7)

$$R = \frac{1}{(1 + \sqrt{WRT})}, \qquad \text{EQ. 17.2}$$

and the phosphorus concentration in the water (not lost to sediments or outflow) is

$$P_{water} = \frac{P_{inflow}}{1 + \sqrt{WRT}}, \qquad \text{EQ. 17.3}$$

assuming that $P_{lake} = P_{outflow}$. This equation indicates that, among lakes, the P concentration of inflowing water rather than the actual P loading (kg yr^{-1}) is the single best indicator of the water concentration.

Deep lakes with a long WRT (> 10 yr) typically retain 70–90 percent of the incoming phosphorus. Progressively less of the incoming phosphorus is stored permanently in the sediments as WRT declines (Fig. 9–7). Short WRT systems are also usually shallow and less well-stratified or completely unstratified. Consequently, a greater fraction of the phosphorus input is retained in the epilimnia, where it is potentially available to primary producers and microbes, before being flushed out. A combination of larger nutrient loading to short WRT lakes and higher nutrient retention in the water column (Fig. 9–3 and Table 17–1) explains the typically higher nutrient concentrations, algal tur-

---

[3]The assumption that natural systems normally reach and maintain equilibrium has its roots in a world-view derived from Greek metaphysics, which proposes that nature must ultimately express an ordered reality. While most scientists now recognize that natural systems are not in a state of true equilibrium, systems are often assumed to be in equilibrium over the course of short-term or annual studies. Assuming equilibrium conditions simplifies experiments, and is also encouraged by the simplicity of equilibrium mathematics.

Table 17–1   **Measured phosphorus retention (R_{obs}) and external loading (L_{ext}, mg m^2 yr^{-1}), as well as the mass-balance calculated internal loading (L_{int}, mg m^2 yr^{-1}) and percent of total loading from sediments (%$_{int}$) in selected temperate zone lakes of known water residence time (WRT, yr) and mean depth ($\bar{z}$, m) at the time of reporting. Negative L_{int} rates reflect a net adsorption of phosphorus to the sediments (retention). Note the typically high external and internal P loading in the generally shallow lakes with anoxic hypolimnia and a short water residence time.**

Lake	L_{ext}	L_{int}	%$_{int}$	R_{obs}	WRT	$\bar{z}$
Oxic hypolimnia						
Cayuga (US)	550	58	10	0.51	9	55
Aegeri (CH)	160	−16	0	0.73	9	49
Turler (CH)	300	−22	0	0.69	2	14
Superior (CA, US)	30	−4	0	0.93	189	148
Michigan (US)	100	−11	0	0.83	31	84
Huron (CA, US)	70	−11	0	0.93	48	59
Ontario (CA, US)	680	−53	0	0.59	8	89
Zurich, upper (CH)	1,320	−288	0	0.75	5	50
Washington (US)	460	−6	0	0.55	3	33
Maggiore (CH, IT)	3,390	−1,353	0	0.64	4	177
Clear (CA)	40	1	2	0.73	8	13
Tahoe (US)	42	−4	0	0.90	714	303
Anoxic hypolimnia						
Chautauqua (US)	340	164	33	0.17	1.4	7
Saratoga (US)	1,600	−205	0	0.53	0.4	8
Chemung (US)	220	26	11	0.63	4.2	9
Shawaga 1967 (US)	653	58	8	0.49	0.7	6
Harriett (US)	710	355	33	0.01	0.8	9
Hallwiler (CH)	550	69	11	0.47	4	28
Baldegger (CH)	1,750	485	22	0.31	4.5	34
Esrom 1973 (DK)	600	78	12	0.67	17	12
Bergundasjöen 1973 (SE)	8,810	2,954	25	0.40	1	2
1974[1]	2,110	3,239	61	−0.80	1	2
1975[1]	410	1,281	76	−2.39	1	2
1976[1]	240	1,746	88	−6.54	1	2

[1]Years following sewage diversion.

Source: After Nürnberg 1984.

bidity, and fish production than is observed in deeper, longer WRT lakes nearby.

Stratified lakes receiving modest external phosphorus (and nitrogen) input, have low algal production and sedimentation rates. If the lakes also receive little organic matter from their catchments, there is little hypolimnetic respiration, with the result that the surface sediments remain oxic. The lakes therefore exhibit little or no internal loading (Table 17–1). Such lakes, in mass-balance parlance, serve as "sinks" rather than "sources" of phosphorus, with the fraction of externally supplied phosphorus that is retained indicated by the WRT. But even lakes and wetlands with internal loading are sinks, albeit to a much smaller extent.

Lakes with Anoxic Hypolimnia

Lakes with anoxic hypolimnia are typically shallow, ($\bar{z} < 10$ m) and have a relatively short WRT (< 2 yr), a high external phosphorus loading (Table 17–1), and a high algal biomass unless they are humic and suffer from an anoxia brought about by respiration of allochthonous organic matter. High water loading (q_s) in nonhumic lakes implies high nutrient input and a WRT too short for effective sedimentation and retention of most of the organic matter and particle-bound nutrients. The hypolimnetic sediments overlain by the typically shallow hypolimnia are able to hold only a small mass of dissolved oxygen—thereby facilitating anoxia, low redox potential, and resulting high internal phosphorus loading from both the anoxic and the oxic shallow-water sediments. A combination of high internal loading and a short WRT implies that a modest fraction of the externally derived phosphorus is retained by the system (Table 17–1).

How much of the internal load of a particular lake is attributable to release from anoxic sediments, and how much is released aerobically from a large littoral zone cannot be resolved by mass-balance modeling, but requires experimentation to measure the relevant rates and identify the mechanisms. However, it is clear from mass-balance calculations that the anoxic hypolimnetic sediments of eutrophic and hypertrophic lakes typically release ~10–30 mg P m^{-2} d^{-1} (Table 17–2). Work on hypertrophic Lake Søbygaard (DK) shows that the exceptionally large gross release rate of 100–200 mg P m^{-2} d^{-1} in summer is largely balanced by a sedimentation rate of 100–150 mg P m^{-2} d^{-1}, demonstrating the net release rate to be highly sensi-

tive to changes in the sedimentation rate. When sedimentation rates were low following periods of phytoplankton collapse, net release rates briefly increased to 100–200 mg P m^{-2} d^{-1} because the gross release rate was largely unaffected (Søndergaard et al. 1993).

Oxic versus Anoxic Internal Loading

Steady state mass-balance determinations made on eutrophic lakes cannot distinguish between phosphorus diffusing from anoxic hypolimnia into epilimnia via the thermocline, and the phosphorus released by shallow-water oxic sediments overlain by epilimnetic water. However, release rates based on lakes too shallow to stratify do provide a measure of oxic release rates. Net long-term release rates of 14–38 mg P m^{-2} d^{-1} in eutrophic but unstratified Lough Neagh (Northern Ireland, GB) (Stevens and Gibson 1976) and 8–30 m^{-2} d^{-1} in shallow ($\bar{z} = 1$ m) and hypertrophic Søbygaard (DK) show *oxic* release to be of the same magnitude as the *anoxic* release rate from hypolimnetic sediments of eutrophic lakes (Table 17–2).

There are major discrepancies between whole-lake measured sediment release rates obtained from mass-balance calculations measuring net release (sedimentation–release) and those obtained from profundal sediment cores in the laboratory measuring gross release (Table 17–3). Conclusions drawn on the basis of each of the two techniques should therefore be quite different. Possible reasons why oxic cores in the laboratory release disproportionately little phosphorus include (1) absence of natural turbulence which, by its removal of the released dissolved phosphorus from the sediment–water interface, maintain a high diffusion gradient; (2) absence in the laboratory of a continual supply of phosphorus containing organic matter sedimenting from the water column; (3) possible underestimation of bioturbation if the few cores that are analyzed happen to come from sites with little benthos; (4) inaccuracies in extrapolating (spatial scaling up) from the few cores analyzed to the behavior of the whole sediment surface; (5) problems in temporal scaling up from core studies carried out over days to the seasonal average values obtained from mass-balance studies; and (6) gross errors in the mass-balance estimates resulting from either poor hydrological records or phosphorus determinations that are not representative or are inaccurate. While experiments under controlled conditions are of great importance in studying mechanisms and processes that are confounded in nature, extrapolations from the laboratory to nature

Table 17–2 **Average daily phosphorus net release rates based on mass-balance calculations from the anoxic sediments of selected lakes during summer anoxia.**

Lake	Release Rate (mg P m^{-2} d^{-1})
Alderfen Broad (GB)	20
Baldegger (CH)	10
Bergundasjöen (SE)	25
Erie (CA, US)	7
Esrom (DK)	12
Mendota (US)	11
Norrviken (SE)	9
Rotsee (CH)	28
Shagawa (US)	12
Mean for 15 lakes (including the above)	14

Source: Modified from Nürnberg 1985.

Table 17–3 **Phosphorus release (mg P m^{-2} d^{-1}) from pelagic and oxic sediments of eutrophic lakes estimated from mass-balance computations and from laboratory experiments using intact cores.**

Lake	Mass Balance	Laboratory Cores	
	Aerobic	Aerobic	Anaerobic
Glaningen, SE	47	2	18
Ramsjön, SE	13	0.3	20
Ryssbysjön, SE	9	0.7	20

Source: Data from Ryding and Forsberg 1976.

should be made with the greatest caution because the simplifications made in experiments purposely reduce the complexity and reality well below that encountered in nature, and controlled experiments are usually carried out over short spatial and temporal scales and not over the seasonal whole-lake scale of most observational studies. Conversely, whole-system measurements based on mass-balance modeling (which is not error-free either) cannot resolve the underlying mechanisms and principal processes which require experimentation under controlled conditions. Research at particular spatial and temporal scales has both possibilities and limitations.[4]

17.5 Sediment Phosphorus Release and Phytoplankton Production
▲

In terms of phytoplankton production (Chapter 21), a unit of available phosphorus released by epilimnetic sediments is much more important than the same quantity released by anoxic hypolimnetic sediments. The former is released directly into the photic zone and is available for uptake and growth, whereas the latter is released into an aphotic hypolimnion and requires diffusion to the epilimnion or the entrainment of hypolimnetic water by storms before becoming accessible to plants (Figs. 17–4, 11–9, and Eq. 12.8).

High total phosphorus (TP) concentrations in anoxic hypolimnia yield steep phosphorus gradients between the epilimnia and hypolimnia that facilitate diffusion. Using the concentration gradient measured in a eutrophic Québec lake plus estimates of the eddy diffusion rates (Sec. 12.5), Nürnberg (1985) calculated the average vertical phosphorus transport to the photic zone as ~4 mg m^{-2} d^{-1}, which was equal to about 32 percent of the phosphorus accumulated in the hypolimnion during stratification. Equivalent work on two highly eutrophic Swiss lakes shows an even higher vertical transport rate of 33 and 20 mg m^{-2} d^{-1}, representing between 50 percent and 100 per-

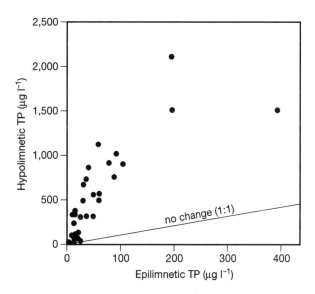

Figure 17–4 A comparison of hypolimnetic and epilimnetic TP concentrations for 36 worldwide lakes having anoxic hypolimnia just before fall turnover. Note the large increase in hypolimnetic TP, which is available to euphotic zone organisms following diffusion, the progressive erosion of the metalimnia, entrainment during storms, or following destratification. *(After Nürnberg and Peters 1984.)*

---

[4]Mechanisms are elusive in nature. In a rare and revealing example, two groups of scientists independently studied Shagawa Lake (US) over the same time period and came to totally different conclusions about the mechanism responsible for their observations. Clearly, both conclusions could not have been correct. The most plausible reason for the discrepancy is not experimental error but unwarranted assumptions in extrapolating findings made over particular time and spatial scales to longer periods and to the system as a whole.

cent of the phosphorus accumulated in the anoxic hypolimnion of one of the two lakes (Bloesch et al. 1977).

The vast majority of low humic lakes in largely undisturbed catchments in the temperate zone lack anoxic hypolimnia and typically experience little hypolimnetic sediment phosphorus release and the phosphorus supplied to the phytoplankton must be derived from the oxic shallow-water sediments and external inputs. Caraco et al. (1992) estimate that the euphotic zone phosphorus pool of oligotrophic Mirror Lake (US) would be totally exhausted in less than one month in summer without fresh inputs from shallow-water sediments, supplemented by phosphorus obtained via the atmosphere. But, as summer phosphorus concentrations in the open water of oligotrophic to mesotrophic lakes do not change very much, inputs must largely balance losses (Guy et al. 1994).[5]

Internal loading from sediments, nutrient recycling in the water column, and atmospheric input become progressively more important among lakes as the catchment to lake area (CA:LA) ratio declines and catchment derived nutrient loading also declines (Sec. 9.4). An extreme case is Lake Michigan (US, CA:LA = 1.6), which released about 837 mg P m^{-2} (27 mmol P m^{-2}) from its nearshore aerobic sediments following a large storm during spring turnover, a quantity 10 times greater than the annual input from all external sources (Brooks and Edgington 1994, and see Sec. 12.12). The inputs stimulate bacterial production before the onset of the spring algal bloom (Cottner 2000).

17.6   Phosphorus Control, Internal Loading, and Lake Management

Highly eutrophic lakes are unsightly, with low water transparency resulting from a very large algal biomass. This unsightly appearance is magnified when the phytoplankton community is dominated by scum-forming cyanobacteria (Sec. 21.6). If they are deep enough to stratify, the hypolimnia of such lakes are thin (Fig. 15–10), causing them to become readily anoxic as the result of high algal sedimentation and decomposition

---

[5]"Just how . . . lakes are able to support a crop of phytoplankton from May to July or August without any appreciable decrease in the soluble phosphorus of the upper water or only a slight one, is not known." (C. Juday et al. 1927)

rates. High diffusion rates plus entrainments during windy periods allow considerable internal loading to the euphotic zone (Fig. 11–9), supplemented by internal loading from oxic shallow-water sediments. Summer fish kills occur when the epilimnia turn temporarily anoxic during destratification events and large quantities of highly reduced inorganic and organic (oxygen-consuming) compounds enter the mixed layer (see Secs. 15.4 and 16.2).

Highly eutrophic lakes and slow-flowing lowland rivers rich in nutrients are turbid and are further characterized by abundant and unsightly algal and bacterial mats on stones and sediments in shallow water. Equally unsightly mats characterize nutrient-polluted shallow rivers. In response to public complaints major efforts have been made to reduce the external nutrient loading.

Phosphorus Remediation

The most widely used remediation techniques used to restore and manage lakes and rivers include (1) phosphorus, and sometimes nitrogen, removal (*abatement*) in sewage treatment plants before releasing treated wastewater into receiving rivers or lakes (Fig. 8–14); (2) diverting wastewater; (3) using natural or constructed wetlands to trap incoming phosphorus and enhance denitrification, thereby lowering the nutrient release into receiving rivers and lakes (Sec. 18.4); and (4) constructing well-vegetated buffer strips along waterways to trap particulate and dissolved phosphorus (and nitrogen) released by agriculture and deforested (clear cut) drainage basins (Chapter 18). Other less widely used and still experimental approaches include (5) controlling sediment phosphorus release (phosphorus inactivation) following the addition of aluminum sulfate (alum) or ferric chloride and its subsequent precipitation and storage in aerobic lake sediments as $AlPO_4$, $FePO_4$, or as $Fe(OOH)PO_4$ flocs (Fig. 17–5); (6) oxidizing phosphorus-rich organic sediments in lakes with lime ($Ca(OH)_2$) and the resulting precipitation of the insoluble mineral apatite [$Ca_{10}(PO_4)_5(OH)_2$], or injecting liquid calcium nitrate into sediments or the hypolimnia. Adding lime raises the pH and the resulting increased iron flocculation provides a sorptive surface for the phosphorus as long as the sediment surface remains aerobic. The proposed NO_3^- additions stimulate denitrification which raises the redox potential of the sediments or water (Chapter 18 and Sec. 16.2), allowing the Fe(III) flocs

Figure 17–5 Aluminum sulfate (Alum) additions to a eutrophic US lake. The precipitation of phosphorus as $AlPO_4$ from the water column and storage in aerobic sediments as $AlPO_4$, $FePO_4$ or $Fe(OOH)PO_4$ flocs results in a (temporarily) reduced phytoplankton biomass and increased water transparency. (*Photo courtesy of Sweetwater Technology, Division TeeMark Corporation.*)

to continue to serve as sorption sites for PO_4^-—but the electron acceptor (NO_3^-) might also enhance organic matter decomposition (Sec. 16.2) and thereby increase the potential for phosphorus release! Other approaches include (7) dredging nutrient-rich sediments—an expensive option—to reduce internal loading and increase the average underwater light climate (Sec. 10.7); (8) withdrawing hypolimnetic water high in phosphorus, iron, manganese, and ammonia. This technique, developed by the Polish limnologist P. Olszewski, only requires a pipe inserted into the hypolimnion to siphon the anoxic water to a site below the outflow level of reservoirs. This is not feasible in lakes and a pump can be used; (9) harvesting submerged macrophytes, as these interfere with water transportation, recreation, and river discharge causing

flooding (Chapter 24); (10) lowering the water level (*draw-downs*) in reservoirs to control macrophytes (Chapter 29) and thereby exposing fish that had been hiding among the plants to open-water predators; (11) *biomanipulation*, which most often involves reducing the abundance of zooplanktivorous fish, allowing a greater abundance of large filter-feeding zooplankton able to control the algae (Chapter 23) thus increasing the transparency; and (12) hypolimnetic aeration. Examples of nutrient abatement and lake aeration techniques are described below.

▲ Nutrient Abatement: Success and Failure

The diversion of wastewater to the sea from Lake Washington, Seattle (US), was an early success story that is still unsurpassed. The lake's rapid response following the 1963–1967 diversion is widely cited as evidence for the effectiveness of nutrient control programs in lake management (Fig. 17–6). Reasons for the rapid recovery include relatively rapid flushing (average WRT = 2.4 yrs) that quickly removed the high nutrient water, plus a suitable morphometry. The large volume of low-temperature hypolimnetic water (z_{max}, 60 m) held a large mass of DO at the onset of stratification. This large pool of DO prevented the development of widespread anoxia in the hypolimnion, even during the years of greatest wastewater input (Edmondson 1966, 1977). Furthermore, the lake received mostly readily divertible point-source loading of wastewater from an urban catchment rather than the more typical—and much more difficult to manage—nonpoint-source loadings from mainly agricultural drainage basins.

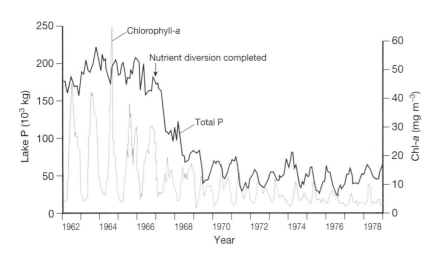

Figure 17–6 Total mass of phosphorus and chlorophyll-*a* concentration in near-surface water in Lake Washington, US (LA = 87.6 km², $\bar{z}$ = 33 m, average WRT = 2.4 yr), plotted monthly. Sewage diversion was started in 1963 with the diversion of about 28 percent of the effluent volume and was > 99 percent complete by March, 1967. (*Modified from Edmonson and Lehman 1981.*)

In accord with the classical model (Sec. 17.2), there was no net release of phosphorus from the oxic sediments prior to diversion with the sediments remaining a sink for phosphorus. Little hypolimnetic phosphorus was returned to the euphotic zone by internal loading before (and after) the diversion of the inflowing wastewater. Finally, Lake Washington (and some other mildly eutrophied lakes (Marsden 1989), have sediments whose sorptive capacity for phosphorus was never exceeded, allowing for effective retention of sedimented phosphorus.

The early, well-documented Lake Washington success story encouraged many nutrient abatement programs. Probably the most successful among them was the remediation of Lake Maggiore (CH, IT). It is an even deeper lake ($\bar{z}_{max} = 370$ m, WRT = 4 yrs) that had been only moderately eutrophied over a relatively short period and it also maintained an oxic hypolimnion (de Bernardi et al. 1996). However, a large outlay of taxpayer money has rarely been able to duplicate the dramatic success at Lakes Washington and Maggiore because the conditions for a quick recovery were not present. Lakes with anoxic hypolimnia that have accumulated large quantities of nutrients in their sediments over many years contain a very large reservoir of nutrients and organic matter. Such lakes respond much more slowly to a reduction in external nutrient loading and organic matter, and continue to exhibit very high sediment respiration (DO uptake) and high release for many years even after the external load is greatly reduced. In addition, nearly all the highly eutrophied lakes that are the most obvious candidates for nutrient abatement lie in nutrient-rich drainage basins that are farmed or located in urban areas (Table 17–1). A combination of large continuous internal loading from highly enriched sediments and

continued significant input from nonpoint sources—always difficult to control—may prevent or delay substantial improvement for many years after construction of expensive wastewater treatment facilities.

Lakes Norrviken (SE), Shagawa (US), and Søbygaard (DK) are three examples of highly eutrophic, shallow lakes that have received wastewater for many years before abatement. Wastewater diversion reduced the external load to Lake Norrviken by about 87 percent (Ahlgren 1980). However, the water quality in Lake Norrviken improved only modestly during the following 11 years, despite a fivefold reduction in phosphorus loading, with the average summer transparency only rising from 0.6 to 1.2 m, and average summer TP concentrations declining by about one-third as the result of continued internal loading from the nutrient-rich sediments (Fig. 17–7). The internal loading, together with significant input from the nutrient-rich catchment, maintained summer TP at a level sufficient (~95 µg l^{-1}) to allow a high summer algal biomass (~40 µg chl-a l^{-1}) to be maintained.

The modest response of Lake Norrviken ($\bar{z} = 5$ m, WRT = 1.3 yr) to a large external reduction in phosphorus loading is compounded by a simultaneous reduction in external nitrogen inputs. Reductions in nitrogen, plus the loss of nitrogen from the system by denitrification, appears to have made the lake—with a N:P ratio of ~15 by atoms (or 7 by mass)—possibly more nitrogen than phosphorus limited (Ahlgren 1980) and if so a poor test case for phosphorus reduction. Lake Norrviken is certainly not the only eutrophic lake examined where the interpretation is confounded by possible nitrogen limitation prior to or following treatment. In highly eutrophic lakes, the primary limiting nutrient is often nitrogen unless the

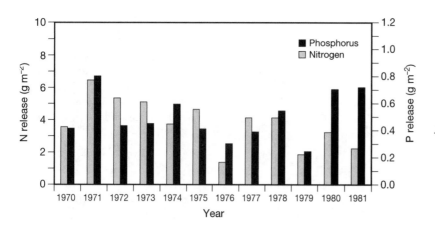

Figure 17–7 Summer net release of phosphorus and nitrogen from the sediments of eutrophic Lake Norrviken, SE (LA = 2.67 km², $\bar{z}$ = 5 m, average WRT = 1.3 yr), calculated for the period between the early summer minimum and late summer maximum concentration. *(Modified after Ahlgren 1988.)*

typically high algal biomass in the turbid lakes is constrained by light (Forsberg and Ryding 1980, and Chapter 21). The greatest possibility of nitrogen limitation exists when TP is high and the N:P ratio is reduced, as in Lake Norrviken (Fig. 8–2, and Downing and McCauley 1992).

Nutrient abatement was tried at Shagawa Lake in Minnesota (US, $\bar{z}$ = 6 m; WRT = 0.8 yr). A new, advanced wastewater treatment plant reduced the external phosphorus load by 75 percent (Larsen et al. 1979) and during the next 16 years was followed by a significant reduction in the spring phytoplankton biomass, but no change in the summer biomass. This was attributed to continued large internal loading from the sediments during warm summer months. Postrestoration summer TP levels in the epilimnion (~250 μg l^{-1}) remained three times larger than the reduced levels seen in spring and summer total nitrogen (TN) concentration (~35–800 μg l^{-1}) were double the levels encountered in spring. The summer N:P ratio of about 16 (by atoms, 7 by mass) suggests a near optimal nutrient supply ratio for growth (Table 8–2, Eq. 16.1) and possibly not a primary nitrogen limitation. Summer water transparency remained at about two meters, and scum-forming blue-green algae species continued to dominate the plankton (Wilson and Musick 1989).

Work on Lake Søbygaard proves that even very large percent reductions in phosphorus input are insufficient in hypertrophic systems that have received very high nutrient input for many years. Lake Søbygaard (DK, $\bar{z}$ = 1 m, summer WRT = 25–30 days) had received large amounts of poorly treated wastewater for many years, but external phosphorus loading was reduced by

80–90 percent in 1982. Since then, TP concentrations in the main inlet have usually been between 100 and 300 μg P l^{-1} while average summer outflow has been higher, at 500–1,000 μg l^{-1}. However, net internal loading has gradually declined from 8 g P m^{-2} yr^{-1} to 2 g P m^{-2} yr^{-1} between 1982 and 1990. The mean chlorophyll-*a* level has declined from about 800 μg l^{-1} in 1984 and 1985 to ~200–300 μg l^{-1} since 1987, and the lake is still characterized as hypertrophic (Chapter 21). Søndergaard et al. (1993) estimate—on the basis of a gradual decline in surficial sediment concentration—that sufficient stored phosphorus remains to support net release for another 10 years or so.

Few lakes have responded to abatement as well as Lakes Washington and Maggiore and few, if any, as poorly as Lake Søbygaard (Fig. 17–8) where the external loading (~2.5 g P m^{-2} yr^{-1}) from agricultural nonpoint sources remains far too high for a meaningful recovery. If a reduction in external and internal loading is insufficient to allow the biomass to be constrained by nutrients, high and unsightly algal turbidity will remain. Unfortunately the nutrient input from rich agricultural soils is frequently underestimated when remediations are planned (Cullen and Forsberg 1988). Forsberg (1985) notes that phosphorus input has to be reduced by at least 60–70 percent to increase the water transparency of hypertrophic Swedish lakes from 0.5 m to 2 m. However, an analysis of net annual sediment release in a variety of eutrophic European lakes typically shows the onset of a decline in TP concentrations—the first sign of recovery—within five years of implementing remediation measures (Sas 1989). Even so, external loading of these same lakes usually continues to be too high to reduce

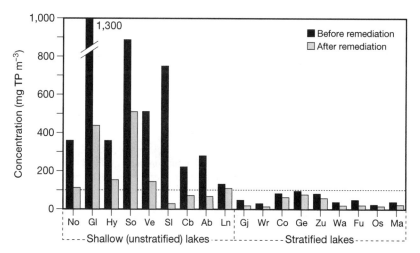

Figure 17–8 Phosphorus concentrations in 18 European lakes before and after remediation measures were undertaken. Note that the horizontal dashed line indicates the approximate phosphorus concentration (~100 mg P m^{-3}) below which the phytoplankton biomass responds to remediation, including macrozooplankton grazing and the water clarity shows an increase. *(Modified after Sas 1989.)*

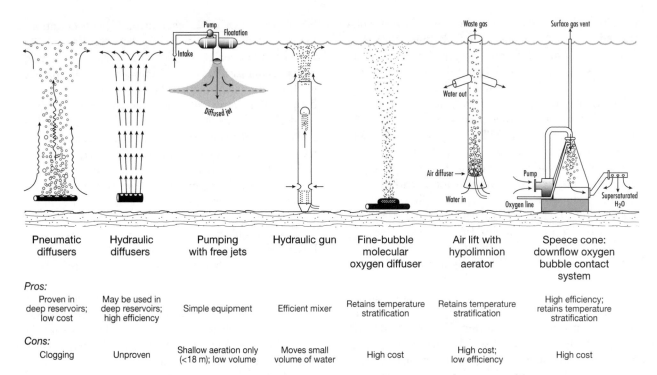

Pneumatic diffusers	Hydraulic diffusers	Pumping with free jets	Hydraulic gun	Fine-bubble molecular oxygen diffuser	Air lift with hypolimnion aerator	Speece cone: downflow oxygen bubble contact system
Pros:						
Proven in deep reservoirs; low cost	May be used in deep reservoirs; high efficiency	Simple equipment	Efficient mixer	Retains temperature stratification	Retains temperature stratification	High efficiency; retains temperature stratification
Cons:						
Clogging	Unproven	Shallow aeration only (<18 m); low volume	Moves small volume of water	High cost	High cost; low efficiency	High cost

Figure 17–9   A variety of re-aeration devices and their advantages and disadvantages. *(Largely after Johnson 1984, and Beutel and Horne 1999.)*

postrestoration TP to less than the 30 μg l^{-1} concentration often used to distinguish eutrophic from mesotrophic lakes (Fig. 17–8 and Sas 1989).

Lake Aeration

Aeration is sometimes used as a management tool where nutrient reduction is not feasible or insufficient, but aesthetic considerations, recreational fishing, or the need for improving the quality of raw water for drinking warrants the expense of aeration. Two simple approaches are used (Fig. 17–9). In the first technique, the whole water column is kept destratified by means of either circulation pumps or a bubbled-air system placed on the bottom (Fig. 17–10). High concentrations of reduced sulfur, methane (CH_4), ammonia, phosphorus, iron, and manganese—characteristic of the anoxic hypolimnia of eutrophic lakes—are lowered, while odorous toxic H_2S is oxidized to SO_4^{2-} and the quality of raw water withdrawn from aerated potable reservoirs is increased.

The downside to artificially circulating the whole water column is that the temperature is homogenized and no cold-water fish habitat is available in higher latitude lakes and reservoirs. Also, a decline in algal

biomass was noted in less than half the 23 studies reviewed by Cooke et al. 1993. Nor is aeration invariably successful in reducing the high relative abundance of the large scum-forming and odor-producing blue-green algal species (Sec. 21.2) or concentrations of available phosphorus. In fact, transparency declined following artificial circulation in over 50 percent of reported case studies (R. A. Pastorak et al. 1982, in Kortmann et al. 1994).[6]

In the second approach only the hypolimnion is aerated leaving the epilimnion and metalimnion intact (Fig. 17–9). Internal nutrient loading to the hypolimnion is greatly reduced where hypolimnetic DO concentrations are raised to about 3.5 mg l^{-1} (Gächter and Wuest 1993). Unfortunately, total phosphorus concentrations in the epilimnia of many European and

---

[6]"More than 40 years of research have yielded variable conclusions about the effects of lake aeration on water quality. For instance, hypolimnetic DO concentrations usually increase . . . , whereas total P (TP), NH_4^+, and chlorophyll-*a* (chla) may not change, or they may increase or decrease () . . . In general, most aeration studies are compromised by limited data sets. . . . , application of other manipulations prior to or during the study (), and changes in external loading during the study ()." (E. E. Prepas and J. M. Burke 1997)

Figure 17–10   Whole water column aeration (oxygenation) of a Japanese lake, in the process destratifying the water column. *(Photo courtesy of Pearson Education/PH College.)*

North American lakes were hardly affected by hypolimnetic aeration. Continued or periodically high nutrient loading from agricultural or urban areas often prevents a decline in sedimentation of organic matter, thereby maintaining a high DO consumption rate at the sediment surface and continued high phosphorus release (Gächter and Wehrli 1998).

Work in the United States has ascribed the failure in reducing high internal phosphorus loading to aerators with insufficient capacity (Kortmann et al. 1994). Furthermore, the aerobic or facultative bacteria (Chapters 16 and 22) replacing the anaerobic species have a greater capacity to oxidize the large accumulated pool of highly reduced organic and inorganic compounds in the hypolimnia than had been assumed when the aerators were purchased, thereby lowering DO levels (Kortmann and Rich 1994).

There has been a great deal of work on aeration, but it is evident that much remains to be learned about aeration processes and their effect on the chemistry and biota of aquatic systems.

17.7   The Empirical Modeling of Phosphorus

The phosphorus loading concept implies that there is a quantifiable relationship between the amount of phosphorus entering a waterbody and the response to this input that can be measured by water transparency,

phytoplankton biomass (Fig. 8–16), and energy flow (Fig. 17–11). E. Naumann surmised more than 75 years ago (Chapter 2) that increased nutrient loading from the land would increase algal productivity. Phosphorus loading was qualitatively modeled first by D.S. Rawson in 1939, and used quantitatively by an engineer named C. N. Sawyer (1947), who concluded that nuisance summer blooms of algae in Wisconsin lakes result if critical levels of inorganic nitrogen and phosphorus (300 μg N l^{-1} and 10 μg P l^{-1}) were exceeded during spring overturn. However, R. A. Vollenweider was the first to formulate drainage-basin loading criteria for phosphorus and nitrogen allowing oligotrophic waters to be distinguished from eutrophic waters (Vollenweider 1968). The criteria he developed have their conceptual origin in a study of the input–output mass-balance of phosphorus and nitrogen.

Under equilibrium conditions, there is a relationship between nutrient input/output and within-lake concentrations. The relationship changes systematically with loading (g m^{-2} yr^{-1}) and water residence time (WRT) (Eq. 17.3)—WRT determines the extent incoming nutrients are sedimented (retained) rather than flushed (Sec. 9.6). Vollenweider (1976) and others subsequently refined the loading relationships defined in 1968. The resulting empirical models, or steady-state *statistical models*, are widely used today in the management of lakes and research, and are central to modern theories of lake eutrophication (Reckhow and Chapra 2001). The models are based on data collected from many lakes and describe the behavior of the *average* lake receiving a particular nutrient load, but they provide no real insight into underlying mechanisms or changes over time and space. The relationship between nutrient loading and in-lake concentration is obtained and the 95 percent confidence limits are typically determined. The models are geared toward predicting how *groups* of lakes are expected to respond to nutrient addition or abatement. But loading data are cumbersome to obtain and most empirical models instead consider the relationship between in-lake phosphorus concentrations and the biotic parameters of interest (Fig. 8–16).

Short-term dynamic reality is commonly explored with *dynamic models* that mathematically and conceptually accommodate changes over time and space, and underlying mechanisms (Sec. 17.8). All but the simplest of such models are sufficiently complex and system specific and are used more for interpolation rather than extrapolation (prediction). But dynamic models were not developed to cope with the stochastic

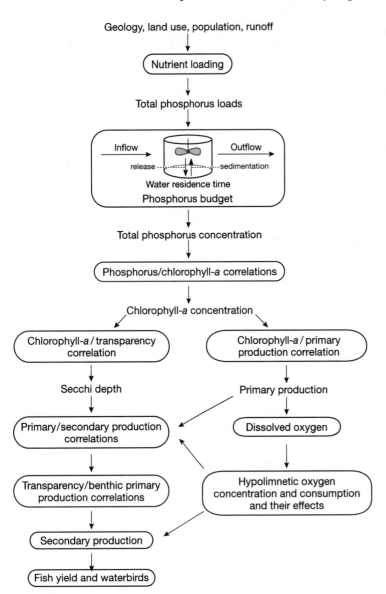

Figure 17–11 Schematic diagram of empirical phosphorus loading models for lakes and reservoirs and their role in research and lake management. *(Modified after Ahlgren et al. 1988.)*

(random) variation imposed by factors such as storms or disease, nor with the long-term behavior of systems, for which yet other model types are more appropriate (Straškraba et al. 1993).

Lakes and Their Catchments

The nutrient loading concept and resulting empirical models are important not only in lake management and research but also because their development forced limnologists to consider lakes and wetlands and their catchments as units, with the catchments and atmosphere included as external sources of nutrients.

Empirical eutrophication models remind limnologists that lakes (and wetlands) are open systems with inputs and outputs. The utility of incorporating mean depth (a surrogate for volume and water residence time or water loading) reminds limnologists that the response of lakes to nutrients is modified by lake and catchment morphometry and by hydrology. Modeling efforts drew attention to the importance and feasibility of making testable predictions and the development of these models quickly led to the realization that the models are lake type and climate specific. Humic lakes and lakes exhibiting a low N:P ratio (with reduced probability of primary phosphorus limitation), as well as lakes having high inorganic turbidity (light limitation) are often excluded as they do not fit the lake type of greatest interest. Nor are the most commonly used empirical models applicable to eutrophic lakes with anoxic hypolimnia, which—in addition to external inputs—often experience internal phosphorus loading (Sec. 17.4).

The following equation development was derived from Ahlgren et al. 1988. Readers not concerned with model derivations should skip to Section 17.8.

▲ The basic mass-balance model for total phosphorus can be written as:

$$\Delta\left(\frac{PV}{\Delta t}\right) = M - PV \cdot \frac{Q}{V} - S. \qquad \text{EQ. 17.4}$$

See Table 17–4 for symbols and units. By considering a lake to be in equilibrium on an annual basis $[\Delta(PV/\Delta t) = 0]$, the amount of phosphorus retained (equals input minus output) is equal to the mass in the water minus the net sedimentation (S). However, Vollenweider concluded that it would be more logical to consider retention as a function of the in-lake mass or concentration:

$$\Delta\left(\frac{PV}{\Delta t}\right) = M - PQ - \sigma P, \qquad \text{EQ. 17.5}$$

where the sedimentation coefficient σ is the fraction of the in-lake mass of P removed annually by sedimentation via all processes other than outflow.

At steady state, Eq. 17.5 is reduced to

$$Pv = \left(\frac{M}{Q/V + \sigma}\right), \qquad \text{EQ. 17.6}$$

and substituting the flushing rate ρ_w for Q/v:

$$PV = \left(\frac{M}{\rho_w + \sigma}\right). \qquad \text{EQ. 17.7}$$

To yield the within-lake concentration, the mass in Eq. 17.7 is divided by lake volume (v):

$$P = \left(\frac{M/V}{\rho_w + \sigma}\right). \qquad \text{EQ. 17.8}$$

M/V is equivalent to $L_p/\bar{z}$, and substitution in Eq. 17.8 produces

$$P = \left(\frac{L_p/\bar{z}}{\rho_w + \sigma}\right), \qquad \text{EQ. 17.9}$$

which is the basic equation underlying most phosphorus loading models. This equation is important in both theoretical and applied limnology because it provides a quantitative basis for understanding why some lakes are oligotrophic and others eutrophic. Equation 17.9 also provides a partial explanation for the tendency of deep lakes to be more oligotrophic than shallow ones, and that rapidly flushed lakes retain less phosphorus and fewer particles of all kinds because a higher fraction of them leave via the outflow than in lakes that flush more slowly. When both the numera-

Table 17–4 **Symbols and units used in the phosphorus mass-balance models.**

P = TP = in-lake total phosphorus concentration (mg m^{-3})
P_i = inflow mean total phosphorus concentration (mg m^{-3})
L_p = annual P loading (mg m^{-2} yr^{-1})
V = lake volume (m^3)
PV = total mass of phosphorus in lake (mg)
t = time
M = annual input mass of phosphorus (mg yr^{-1})
O = annual outflow mass of phosphorus (mg yr^{-1})
Q = discharge of the outlet of lake (m^3 yr^{-1})
q_s = water discharge height (m yr^{-1})
S = annual net sedimentation of phosphorus (mg yr^{-1})
σ = sedimentation coefficient (yr^{-1})
v = apparent P settling velocity (m yr^{-1})
$\bar{z}$ = lake mean depth (m)
ρ_w = Q/V = flushing rate (yr^{-1})
τ_w = $1/\rho_w$ (WRT, yr) or WRT
R_p = P retention coefficient

Source: After Ahlgren et al. 1988.

tor and denominator are multiplied by ρ_w (the flushing rate) the equation becomes

$$P = \frac{L_p}{q_s} \cdot \frac{\rho_w}{(\rho_w + \sigma)}. \qquad \text{EQ. 17.10}$$

Where $\rho_w/(\rho_w + \sigma) = 1 - R_p$, with R_p as the retention coefficient is acting on the average inflow concentration. Equation 17.10 is difficult to apply because the sedimentation coefficient (σ, yr^{-1}) is not a constant, but is affected by (1) phosphorus loading and lake phosphorus content; (2) its uptake by organisms and adsorption to particles; and (3) apparent settling velocity (v) of the particles. The velocity is affected by particle size, stratification, turbulence, food-web structure, etc. (Secs. 20.4 and 21.6).

Both Vollenweider (1976) and Larsen and Mercier (1976) found that the flushing rate (ρ_w) was a good surrogate for the changing sedimentation coefficient (σ). Substituting $1/\tau_w$ for σ and rearranging Eq. 17.10 yields the widely used equation

$$P = \frac{L_p/q_s}{(1 + \sqrt{\tau_w})}. \qquad \text{EQ. 17.11}$$

Equation 17.11 states that the in-lake phosphorus concentration is a function of the phosphorus loading corrected for the water residence time (flushing). However, further work has shown that not only the phosphorus loading rate but also the in-lake phosphorus concentration affect the sedimentation of phosphorus. Relatively deep, slowly flushed lakes receive proportionately little water and little phosphorus from undisturbed drainage basins and they have a low load:lake content ratio. In these lakes, the net phosphorus sedimentation (retention) is largely a function of in-lake concentration, reflecting efficient retention (efficient recycling) in the water column. Conversely, rapidly flushed lakes draining nutrient-rich catchments receive considerable amounts of phosphorus from the land. They have a high load:lake content ratio, and sedimentation (retention) is largely a function of external loading. Prairie (1989) estimated that an average 25 percent of the P load and 18 percent of the lake P content of 120 lakes examined sedimented (retained) annually. Reliable estimates of P retention must consider both P load and lake P content, with retention related to τ_w.

Using the *critical phosphorus concentration* (10 μg l^{-1}), which Vollenweider used to separate oligotrophic from mesotrophic lakes (30 μg l^{-1} separates mesotrophic from eutrophic lakes) and Eq. 17.11, it is evident that the *critical loading* (L_c) beyond which the

average oligotrophic lake is expected to become mesotrophic is

$$L_c = \frac{10 \cdot q_s}{(1 + \sqrt{\tau_w})}, \qquad \text{EQ. 17.12}$$

which is the well-known form of Vollenweider's 1976 loading model.

Many modifications and elaborations of the phosphorus loading versus phosphorus concentration models exist; some use flushing rate (ρ_w) or its inverse, the annual water loading rate (q_s, m yr^{-1})—which is the product of $\bar{z}$ and the flushing rate—rather than the water residence time (τ_w, yr). For example, Canfield and Bachmann (1981) examined a large data set comprised of North American natural lakes and reservoirs and found, as computed by Ahlgren et al. (1988), that

$$P = \frac{L_p/q_s}{(1 + 0.129\,\rho_w^{-0.451}\,P_i^{0.549})}, \qquad \text{EQ. 17.13}$$

where P_i = mean inflow concentration and describes observed within-lake phosphorus concentration better than the original Vollenweider formulation (Eq. 17.11) that was based on only 30 mostly larger European and North American lakes. R. A. Vollenweider and J. Kerekes (OECD, 1982) subsequently examined the applicability of Eq. 17.11 by using a much larger data set comprising north temperate zone lakes. Using regression analysis, they concluded that the following modification of the original equation (Eq. 17.11) best describes the overall dependence of lake phosphorus concentration on external loading and water residence time for their selection of low-humic and probably phosphorus-limited (inorganic N:SRP > 10 as mass) lakes,

$$P = 1.55\left[\frac{L_p/q_s}{(1 + \sqrt{\tau_w})}\right]^{0.82}. \qquad \text{EQ. 17.14}$$

The utility of empirical models for predicting the average within-system nutrient (or algal biomass) concentration of any waterbody that was or was not included in the data set used in developing the equation is dependent on how similar the system is to the average system of the data set used. The phosphorus retention coefficient (R_p) is best estimated from both loading (L_p) and lake content (PV)

$$P_p = L_p + \frac{PV}{L_p} \qquad \text{EQ. 17.15}$$

After substitution, rearrangement, and using 25 percent of the load and 18 percent of the lake content sedimented (Prairie 1989), the retention coefficient is

$$R_p = \frac{0.25 + 0.18\tau}{1 + 0.18\tau}. \qquad \text{EQ. 17.16}$$

This equation states that the dependence of R_p on WRT (τ) is attributable to sedimentation of both the load and lake content.

Equations 17.11 and 17.14 are easily rearranged to estimate the phosphorus loading. Such a rearrangement is useful for confirming the reliability of estimated phosphorus loadings that are based on direct measurements or export coefficients (Table 8–8). The utility of this exercise rests upon having measured the parameters of the equations with little error.

For most limnologists, the prediction of phosphorus loading or in-lake phosphorus concentration is only a stepping stone towards predicting, for example, aspects of phosphorus dynamics, algal biomass or primary production, dissolved oxygen consumption rates, secondary production, or fish yields (Fig. 17–11). Equivalent work on small lotic systems suggests that when in-stream total phosphorus concentration is kept below 30 μg P l^{-1}, and total nitrogen concentration is kept below 350 μg N l^{-1}, the benthic algal biomass is not likely to exceed nuisance levels of 100 mg chl-*a* m^{-2} (Dodds et al. 1997).

17.8   The Dynamic Modeling of Phosphorus

There must be a trade-off between the simplicity and generality of models, and the precision of predictions.[7] Empirical statistical models such as those described by Eqs. 17.11, 17.13, and 17.15 have the desired simplicity and considerable generality. However, the best-known models are not truly general, instead they are restricted to a particular set of nonhumic freshwater lakes in the temperate zone with typically oxic hypolimnia where the N:P ratio suggests a primary phosphorus limitation. Furthermore, models make statements about the *average* behavior of large "populations" of waters, and their application to individual systems may result in large errors.

In empirical models, the natural complexity is reduced to annual or seasonal averages of loading and concentration. Steady-state empirical models also say nothing about the effect of phosphorus over time and space within particular lakes of interest. But the largest

shortcoming of empirical modeling is that without a cause and effect structure, it can only reliably forecast under conditions similar to those reflected in the data set that was used to construct the models. The problem of multiple causes and effects—which can be plausibly explained by different mechanisms or combinations of mechanisms—retard progress in ecology and limnology, and sets limits upon the utility of empirical and dynamic modeling. Furthermore, individual scientists with roots in different paradigms tend to hold strong views about which mechanisms or processes are believed most important.[8]

There is a need for more flexible, powerful models. The desired models should not only be able to predict future states but also provide explanations about why the system of interest behaves as it does based on both the mechanisms and processes that are employed by biological or chemical species or groups of species. Lake specific answers would be obtained. This need has been partially met by the development of *dynamic simulation models* with varying complexity (Sec. 2.6), complementing the empirical models discussed above. But the dynamic models have their limitations. In the case of phosphorus modeling, little is known about the processes and mechanisms to be modeled, some necessary data are lacking, and it is unknown which environmental factors exert the strongest influence on different phosphorus compartments over time and space. Three dynamic models are discussed below.

▲   EAWAG Model

Dynamic eutrophication models available today differ greatly in complexity depending on the goals (generality versus precision) and interest and background of the creators. An example of a simple dynamic phosphorus model is the EAWAG model, developed at the Swiss Federal Institute for Environmental Science and Technology [EAWAG] (Fig. 17–12). This model is simple because it uses only soluble reactive phosphorus (SRP), particulate phosphorus (PP), and dissolved oxygen concentration as the *state* or *predictor variables* evaluated over time. Fixed rate constants (*model parameters*) for photosynthesis, respiration, mineralization,

---

[7]"Any model remains an intellectual plaything of limited impact . . . unless it can be tested by and verified by experiment, or by field observation, or both." (C. H. Mortimer 1975)

[8]Individual bias was first recognized in 1897 by T. C. Chamberlin when he wrote, "There is the then imminent danger of an unconscious selection and of a magnifying of phenomena that fall into harmony with the theory and support it and an unconscious neglect of phenomena that fail that coincidence."

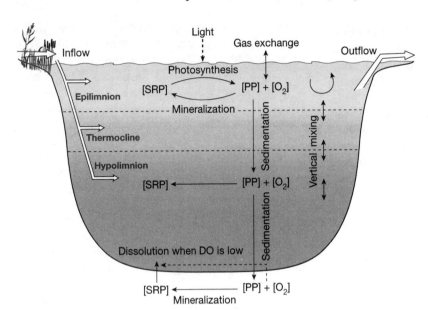

Figure 17–12 Schematic diagram of a simple dynamic model. The EAWAG model is based on a limited number of required measurements. *(Modified after R. Gächter et al. 1983, in Ahlgren et al. 1988.)*

sedimentation, sediment oxygen consumption, sediment phosphorus release, flushing, and in-lake transport between the epilimnia, metalimnia, and hypolimnia were either measured or obtained from published data. The experimentally determined rate constants were obtained in one region of Switzerland—characterized by a particular climate, geology, and lake morphometry—and the resulting model is therefore a regional model, but it can be "recalibrated" for work elsewhere.

The EAWAG model is simple from a dynamic modeling perspective; it employs many fixed (nondynamic) constants, some of which are obtained from steady-state statistical (empirical) models. However, data requirements for predicting phosphorus concentrations are much larger for the EAWAG dynamic model than for empirical models. This particular disadvantage is acceptable to those who value the insights to be gained by considering aspects of the dynamics that are based on a mechanistic underpinning. This underpinning allows simple (simplistic) dynamic models to be seen as a stepping stone toward more comprehensive models. Whether EAWAG or another simplified dynamic model with empirical aspects will allow for better predictions than is possible with equivalent regionally developed but far simpler empirical models is not well resolved (Ahlgren 1988).

It is important to keep in mind that simplicity, generality, and predictive power are not the only criteria considered to be important in modeling. While predictive power is certainly a long-term goal of most

scientists favoring dynamic simulation modeling, the primary short-term goal is usually an exploration to discover gaps that can form the basis for research and more comprehensive models. Indeed, simulations point to outcomes that are rarely imaginable on the basis of common sense and they provide new explanations for events observed in nature (Scheffer and Beets 1994).

Lake Glumsoe Model

The EAWAG model is interesting because it explicitly explores the possibilities of embedding empirical aspects within a simple dynamic model, thereby benefitting from an explicit consideration of processes and mechanisms. Much more complex than the EAWAG model for phosphorus, but of only moderate complexity, is the ecosystem model originally developed for the highly eutrophic Lake Glumsoe in Denmark (Fig. 17–13). It describes the behavior of three trophic levels (phytoplankton, zooplankton, and fish) as a function of the phosphorus supply. Instead of the two phosphorus state variables used in the EAWAG model, the Glumsoe model uses a large number of simultaneous equations that include eight phosphorus variables, another nine nonphosphorus state variables, plus 19 parameters describing the lake, and 28 fixed rate constants to define the phosphorus, nitrogen, and energy fluxes between compartments. The 28 fixed rate constants have to be determined—an enormous and almost impossible task—or drawn from the litera-

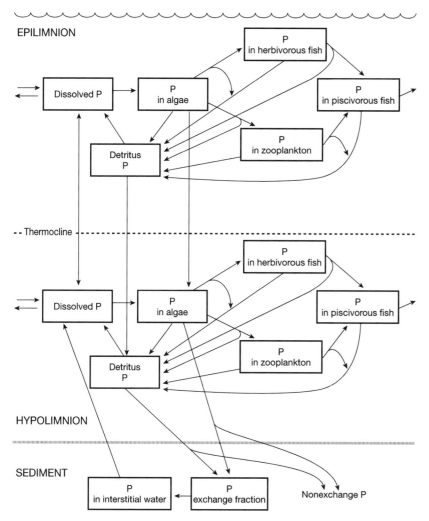

Figure 17–13 Schematic diagram of the Glumsoe model of phophorus cycling. Note its complexity relative to Fig. 17–12, and its simplicity relative to Fig. 17–14. *(Modified after Jørgensen et al. 1986.)*

ture with their relevance to the system in question uncertain. The relatively complex Glumsoe model and other such models are abstractions and a simplification of reality. Some of the particular constants selected from the laboratory and field literature require adjustment (calibration, tuning) to make the model results fit observed lake behavior. Sophisticated techniques are available for this purpose and good fits can be obtained, but since adjusted constants have no known biological reality the models are not very useful for predicting the behavior of the same lake after a change in nutrient loading or fish introduction, or for predicting the behavior of other lakes that inevitably differ physically, chemically, and biologically from the lake on which the model was originally based. Moreover, the same good fit between observed and modeled data can be produced by manipulating different sets of constants or state variables.

PCLOOS Model

A final, more recent example of an interesting, but considerably more complex, dynamic simulation model is the PCLOOS model developed by Janse et al. (1992). It attempts to deal with both water and sediment processes and addresses not only phosphorus but also carbon fluxes (Fig. 17–14). This model has been partly calibrated and used to simulate (interpolate) what should happen in a Dutch lake following a reduction in external phosphorus loading.

Recent Developments

Most of the dynamic modeling done today has less grandiose goals than the ecosystem-level modeling discussed above. The limitations of modeling complex systems is better understood and most current dy-

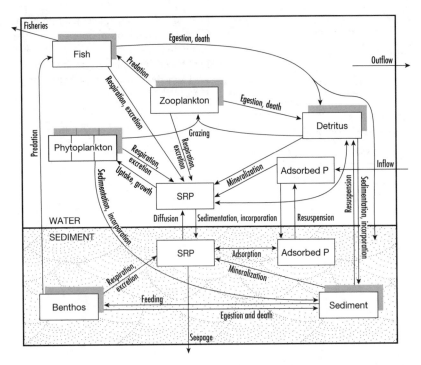

Figure 17–14 Overview of the PCLOOS model of phosphorus and carbon flow in aquatic systems. Compartments modeled in two units are indicated by two boxes, one behind the other. The phytoplankton compartment is composed of three functional groups: cyanobacteria, diatoms, and green algae. The arrows denote both carbon and phosphorus fluxes. *(Modified from Janse et al. 1992.)*

namic modeling exercises focus on more readily testable components of the earlier ecosystem models.[9]

The complexity of even relatively simple dynamic models is sufficiently large that consideration of their underlying assumptions is beyond the scope of a general limnology text. However, the importance of developing both simple empirical and more complex simulation models is without question. Empirical and simulation models combined with other modeling approaches complement one another. They make it possible to develop relatively simple hybrid models that combine empirical modeling based on field survey data with dynamic modeling of population responses to changes in environmental factors, based on laboratory experiments, to produce population level forecasts (Shuter and Regier 1989). But even hybrid models are not easy to present (explain) in a general

textbook. For all their limitations, simple empirical models provide a much needed pattern (generality) and are an important starting point in research. They also play a crucially important role in aquatic management, not because they make precise predictions but because the rate of environmental degradation is so large that there is no alternative to making management decisions based on imprecise empirical models. See Scheffer and Beets 1994 for an excellent and easy-to-understand discussion of the possibilities and limitations of modeling approaches.

Highlights

- Phosphorus is the element usually in shortest supply for algal growth in oligotrophic and mesotrophic waters characterized by a N:P supply ratio that is much higher than the demand ratio of algal protoplasm.
- According to the classical $FePO_4$ model, sedimenting and sedimented iron oxyhydroxide flocs (aggregates) strongly sorb PO_4^{3-}. Sedimented aggregates serve as an effective phosphate diffusion barrier to sediment soluble reactive phosphorus (SRP) as long as the aggregates remain intact under oxic conditions.
- When DO concentration approaches zero and the sediment surface redox potential declines to below ~200 mV, the aggregates dissolve and the elements diffuse into the overlying water.

---

[9]"Science can be viewed as the process of building successively "better" descriptive and predictive models of the world. But how does one define "better?" In the past, scientists have tended to narrow their questions in order to achieve higher accuracy. This leads to [dynamic] models with low articulation [few components] but high descriptive accuracy. They say much about little. More recently scientists have begun to take a "systems view" that looks at phenomena more comprehensively. This strategy leads to highly articulated [many component] models with low accuracy. These models say little about much." (R. Costanza and F. H. Sklar 1985)

- Microbes play a major direct and indirect role in phosphorus release from sediment, modifying the classical view of sediment release as a primarily chemical process. Phosphorus is not always released following the dissolution of iron aggregates.
- ▲ In shallow eutrophic lakes, physical disturbance of the oxic sediments allows dissolved phosphorus to escape to the overlying water, bypassing the diffusion barrier.
- ▲ The fraction of incoming phosphorus that is retained by lakes and wetlands and stored in the sediments increases with water residence time.
- ▲ As the catchment area to lake area ratio declines, sediment release, nutrient recycling in the water column and the atmosphere become increasingly important as sources of phosphorus for algal and microbial growth.
- The principal phosphorus (and nitrogen) remediation (abatement) techniques involve (1) nutrient removal from wastewater in sewage treatment plants or wetlands before release; (2) diversion of wastewater from rivers and lakes; and (3) the maintenance or construction of well-vegetated buffer strips along waterways to trap dissolved and particulate phosphorus released as the result of particular land-use practices.
- ▲ The recovery of lakes following phosphorus abatement is quickest in rapidly flushed lakes where the nutrient and organic matter inputs had been small enough to allow an oxic hypolimnion to be maintained. Recovery is slowest in those shallow lakes with an anoxic hypolimnion that had received and stored large quantities of nutrients and organic matter in their sediment over many years, allowing them to serve as an ongoing source of phosphorus to the overlying water. Recovery is impossible where catchment exports remain high.
- There is no "best" modeling approach. The most appropriate models change with the temporal and spatial scales at which questions are posed.

18

Nitrogen Cycling

18.1 Introduction: The Atmosphere, the Land, and the Water

Nitrogen plays a central role in inland waters. Nitrogen and phosphorus are commonly the elements in greatest demand by plants and the heterotrophic microbes relative to supply (Table 8–1). The nitrogen supply is therefore of great importance in determining the primary productivity of aquatic systems and microbial recycling of organic matter.

Nitrogen, in contrast to phosphorus, exists in a variety of oxidized and reduced forms (oxidation states) that allow it to serve as an electron donor and receiver in a host of oxidation–reduction reactions (Table 16–1) of central importance in nutrient cycling and biogeochemistry.

Ionic and organic nitrogen are supplied to aquatic systems from drainage basins and the atmosphere. Total nitrogen (TN) in precipitation ranges from a low of ~100 kg km^{-2} yr^{-1} in polar regions to over 2,000 kg km^{-2} yr^{-1} in or downwind from industrial and agricultural regions in the north temperate zone (Table 8–10). Boring et al. (1988) reported a similar range over primarily forested regions of the United States also indicating long distance transport from regions of emission.

Atmospheric nitrogen inputs to the land have approximately doubled globally—and even more locally—as the result of human activity, with air masses carrying the emitted nitrogen over long distances be-

fore deposition (Vitousek et al. 1997). In the northeastern US, the atmospheric deposition of nitrogen on the landscape, primarily originating from fossil fuel combustion, currently exceeds all other individual nitrogen inputs from fertilizer, imported food and feed, and nitrogen fixation by crops (Jaworski et al. 1997). A 12-year record of TN deposition in Ontario (CA), downwind from major US source regions, showed that about half the *bulk precipitation*—precipitation collected in open containers (Chapter 27)—to be in the form of NO_3^-, a third as NH_4^+, and the balance as total organic nitrogen (TON) (Molot and Dillon 1993). More limited data suggest that a similar $NO_3:NH_4^+$ ratio characterizes the wet tropics (Downing et al. 1999) whereas measurements in central Europe indicate a slightly greater deposition of NH_4^+ than NO_3^-, mostly in summer (Kopáček et al. 1997). The ratio is further skewed toward NH_4^+ in western Europe as the result of enormous emisions from livestock wastes (Chapter 27).

Long-term Increase in Nitrogen Deposition

An indication of the widely reported long-term increase in nitrogen deposition from the atmosphere is provided by nitrate measurements made in remote Lake Superior (CA, US). The lake is characterized by an unusually small catchment area to lake area (CA:LA) ratio (1.6) and thin, nutrient-poor boreal for-

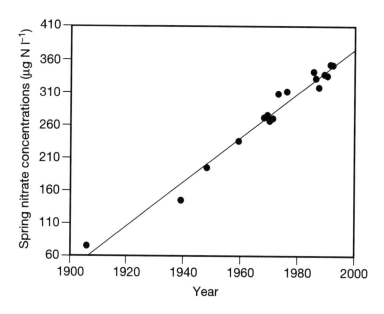

Figure 18-1   Spring nitrate concentrations in remote Lake Superior (CA, US), a lake surrounded by a disproportionately small drainage basin composed of hard (igneous) rock, (CA:LA = 1.6). *(Data from Bennett 1986, and Williams and Kuntz 1999.)*

est soils that release little nitrogen. Yet, spring nitrate concentrations increased nearly sixfold between 1906 and 1992. Concentrations have been increasing at a rate of ~2% yr^{-1}, indicating that it doubles every 34 years, with no sign of slowing down (Fig. 18-1). The rate of increase is almost identical to that observed in the Mississippi River since 1965 (Vitousek et al. 1997). This suggests that most of the nitrogen in Lake Superior, as in the Mississippi River which drains the rich agricultural land of the US Midwest, is derived from upwind agricultural sources in the same region. In northwestern Europe, the rate of increase has been even higher than in Lake Superior. D. F. Brakke (1988) in Vitousek et al. (1997) reported that nitrate concentrations doubled in remote Norwegian lakes over less than a decade. The rapid increase indicates either an atmospheric supply rate growing at twice the rate seen in Lake Superior or nitrogen saturation of the typically thin soils, with some Norwegian catchments now serving as a source rather than as a sink for atmospheric nitrogen (Chapter 27).

Work on other remote boreal forest lakes in Ontario indicates that most of the nitrogen increase in remote Lake Superior is indeed derived via the atmosphere from distant sources (Molot and Dillon 1993). The research shows atmospheric contributions to rise systematically with a decline in the CA:LA ratio. Their model suggests that about three-quarters of the nitrate measured in Lake Superior is derived from dis-

tant sources via the atmosphere and only one-quarter is from the catchment.

The atmosphere is important not only as a source of nitrogen but of other materials. This is evident from work on lead and PCBs, showing that—with the near absence of catchment sources—about 97% of lead and ~90 percent PCBs deposited in the sediments of Lake Superior arrive via the atmosphere from sources far away (Table 28-3). These findings are also supported by research on acidifying precipitation, which exerts negative effects on waterways far removed from industrial sources (Chapter 27). Naturally, the atmosphere contributes proportionally less to lakes and wetlands in large catchments and agricultural catchments releasing large quantities of nitrogen to waterways.[1] Even so, high atmospheric nitrogen inputs into remote lakes reduce the possibility of a primary nitrogen, rather than phosphorus, limitation. Thus, it is believed that raised atmospheric inputs of nitrogen are responsible for shifting an originally nitrogen limited Lake Tahoe (US) to a primarily phosphorus limited lake (Goldman 2000).

---

[1]"Using relatively undisturbed areas as references, Howarth et al. (1996) estimate that riverine total N fluxes for most of the temperate regions surrounding the North Atlantic ocean may have increased from pre-industrial times by 2- to 20- fold. For the North Sea region, the N increase may have been 6- to 20- fold." (Vitousek et al. 1997.)

▲ Nitrogen Retention on the Land

Well-vegetated remote catchments effectively retain most of the nitrogen (and phosphorus) deposited on the land from the atmosphere (Table 9–2), exporting only a modest fraction of the two elements to waterways. Vitousek and Howarth (1991) argue for a primary nitrogen limitation in undisturbed temperate zone forested ecosystems. Their conclusion is supported by nitrogen mass-balances for the oligotrophic forest lakes mentioned above (Molot and Dillon 1993), which show that about two-thirds (67%) of the atmospheric TN deposited is stored or denitrified on land, 12 percent is denitrified within lakes, four percent is stored in lake sediments, and only 17 percent is exported from lakes by outflowing rivers. Moreover, the N:P ratio in streams draining the basins is generally lower than it is in precipitation, pointing to preferential retention of N on land unless the nitrogen is released to the atmosphere as a gas. Conversely, N:P ratios in lake outflows is slightly higher than input ratios, pointing to preferential P retention by lakes. The two different N:P ratios support the idea that P-limited temperate zone lakes can coexist with N-limited forest drainage basins in relatively pristine regions (Molot and Dillon 1993).

The N:P ratio in streams leaving forested drainage basins is much higher than the equivalent ratio in most streams draining nutrient-rich agricultural catchments in the temperate zone (Table 8–9). These lower ratios support experimental evidence for reduced phosphorus limitation in most nutrient-rich lakes (Chapter 21). There also appear to be climatic differences in the retention of nitrogen; limited evidence points to much lower nitrogen retention by tropical moist forests than by temperate counterparts, with high nitrogen-flux rivers draining most tropical catchments (Downing et al. 1999). Conversely, arid-zone lakes appear to receive exceptionally little nitrogen from their poorly vegetated drainage basins via inflowing rivers (Galat and Verdin 1988).

18.2   Nitrogen Transformation Processes

Nitrogen fixed by photosynthetic or heterotrophic microbes becomes part of the *particulate organic nitrogen* (PON) pool, as does the dissolved NO_3^- taken up by photosynthetic organisms in a process called **assimilative nitrate reduction** (Fig. 18–2). In this process, energy captured photosynthetically is used to reduce oxidized NO_3^- or NO_2^- to the reduced nitrogen of the protoplasm.

In the absence of dissolved oxygen (DO), oxidized nitrogen (NO_3^- and NO_2^-) serves as a final electron acceptor in the oxidation of organic matter by facultative heterotrophic microorganisms at the oxic–anoxic interface, the sediment surface, and at the oxycline present in some metalimnia. The oxidized forms of nitrogen are reduced in a **denitrification** or **dissimilative nitrate reduction** sequence, yielding gaseous N_2, as well as gaseous N_2O, an important greenhouse gas emitted into the global atmosphere.

Another fraction of NO_3^- is reduced to NH_4^+ in **dissimilative ammonia production** (Fig. 18–2, and Eq. 3b in Table 16–1). However, the largest source of NH_4^+ results from the breakdown and mineralization of organic matter by both aerobic and anaerobic bacteria in a process called **ammonification** (Fig. 18–2).[2] Ammonification commences with detrital particles in the photic zone and continues during sedimentation and after particles arrive at the sediment surface. Large quantities of total inorganic nitrogen (TIN) are released into the water column from the sediments in eutrophic lakes and wetlands (Table 8.8). For example, Höhener and Gächter (1994) estimated a sediment release rate ranging 16–31 mgN m^{-2} d^{-1} during an eleven-summer study of the effects of hypolimnetic aeration on the water quality of a eutrophic Swiss lake.

▲ Large quantities of hypolimnetic NH_4^+ and other nutrients can be returned to epilimnia during entrainment of hypolimnetic water during storms (Fig. 11–9 and Chapter 12). Two summer days of strong winds associated with a cold front resulted in a 1.2 m lowering (erosion) of the thermocline of Lake Mendota (US) and the entrainment of about 0.33 g m^{-2} of NH_4^+ from the anaerobic hypolimnion plus about one-third as much total phosphorus. The internal loading of the epilimnion was followed within six days by about a threefold increase in the algal biomass (Stauffer and Lee 1973). Much of the nitrogen supplied was presumably taken up as NH_4^+ during **assimilative ammonium reduction**. However, in the presence of DO

---

[2]Under high pH conditions in warm (hyper) eutrophic waters there can be a rapid shift from NH_4^+ to toxic NH_3, leading to decreases in fish culture trout growth, and possibly severe damage to fish stocks.

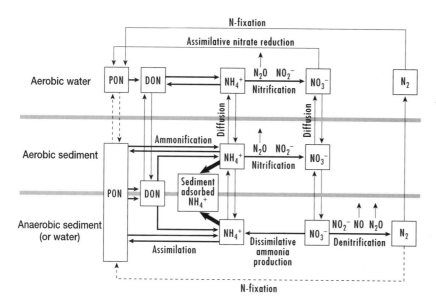

Figure 18–2 Nitrogen transformations near the sediment–water interface. Not shown are inputs from the land and atmosphere or stream outflows. Minor pathways are also not shown. (*Modified from Kamp-Nielson and Anderson 1977.*)

some of the NH_4^+ will have been nitrified (to NO_2^- and NO_3^-) by chemosynthetic bacteria (Fig. 18–2, Table 16–1) or chemically oxidized (Chapters 16 and 22).

18.3   Nitrification

Nitrification, the biological oxidation of NH_4^+ usually to NO_3, is catalyzed by a variety of microorganisms who thereby obtain energy for their metabolism (Tables 16–1 and 22–1). It is biologically-mediated conversion, not chemical oxidation, that is primarily responsible for the high ratio of $NO_3:NH_4^+$ observed in the well-oxygenated epilimnia of unpolluted clear-water lakes, and partially responsible for the high ratio that is observed in streams draining well-aerated agricultural soils. Conversely, upland streams draining forested areas are dominated by NH_4^+ released by poorly drained forest soils. The importance of nitrification is important in nitrogen cycling because the oxidized forms of nitrogen that are produced (primarily NO_3^- and NO_2^-) can then partake in denitrification reactions resulting in loss to the atmosphere of gaseous N_2 (Sec. 18.4).

The overall nitrification reaction,

$$NH_4^+ + 2O_2 \rightarrow NO_3^- + H_2O + 2H^+ \quad \text{EQ. 18.1}$$

shows that two moles of DO are needed for the oxidation of each mole of NH_4^+ to NO_3^-. Expressed differ-

ently, ~4 mg DO are required to oxidize 1 mg NH_4^+ nitrogen. Nitrification therefore exerts large demands on the pool of DO stored in hypolimnia (Table 18–1), in ice-covered lakes and rivers (Sec. 15.6), and downstream from sewage treatment plant outfalls releasing large quantities of NH_4^+ and organic matter (Fig. 15–7).

▲   Nitrification Sites

The principal site for nitrification is the interface (oxycline) between oxic and anoxic waters or sediments. The anoxic zone is characterized by a relatively high NH_4^+ supply rate, resulting from ammonification in the sediments. This process has been examined primarily in profundal sediments and in water-column oxyclines, but an equally important nitrification environment is provided by the little-studied littoral zone of lakes; the land–water plus hyporheic ecotones of lotic systems (Fig. 8–7); and wetlands. In all of these ecosystems, a combination of high organic matter production and decomposition, high summer temperatures and a high DO supply rate provides an ideal sediment environment for nitrification of the NH_4^+ produced during decomposition, yielding the high NO_3^- concentrations that are of concern in human health issues. However, NH_4^+ oxidation will terminate at the NO_2 or N_2O stage under nearly anoxic conditions (Downes 1988). Nitrification stops altogether under anoxic conditions, but the pertinent facultative bacteria remain present,

Table 18–1 **The contribution of water column plus sediment nitrification to DO consumption in hypolimnia, based on the assumption that all nitrification was attributable to chemosynthetic bacteria.**

Lake	Depth (m)	Oxygen Consumed (%)
Grasmere (GB)	10–15	25–35
	15–20	25–35
Mendota (US)	12	63
Blelham Tarn (GB)	6–10	up to 100
Lake Erie (CA, US)	hypolimnion	1
Lake Ontario (CA, US)	hypolimnion	7–14
Lake Ontario (CA, US)	hypolimnion	40
Lake St. George (CA)	ice covered	71

Source: Hall 1986, and Knowles and Lean 1987.

ready to recommence nitrification when DO is reintroduced.

The depth to which nitrification is possible within sediments ranges from zero to several centimeters. It is determined by the DO levels of the overlying water and by the thickness of the *diffusive boundary layer* (DBL), which helps determine the rate of DO diffusion into the sediments (Sec. 24.7). Therefore, the observed rate of nitrification (and the coupled denitrification) is principally a function of three substrates: the NH_4^+ (or NO_2^-) pool available, the DO supply rate, the CO_2 available and the water temperature with its effect on rates of metabolism (Prosser 1986). CO_2 is important because nitrifying microbes are chemosynthetic autotrophs—organisms that reduce CO_2 into organic carbon with the energy they obtained from oxidation of NH_4^+ (Table 16–1, and Eqs. 7c and d).

There have been many technical and interpretational problems associated with benthic chambers lowered onto the sediment surface, discouraging their use in determining *in situ* nitrification and denitrification rates.[3] The nitrification process has therefore been primarily studied in sediments manipulated in the laboratory, and sometimes in relatively undisturbed hypolimnetic sediment cores taken to the

laboratory (this applies equally to denitrification, see Sec. 18.4).

Environmental Factors Affecting Nitrification

The high NO_2^- and N_2O (a by-product) concentrations observed in the hypolimnion of a New Zealand lake (Fig. 18–3) were attributed to termination of nitrification at the NO_2^- stage under nearly anoxic conditions ($< \sim 0.2$ mg $O_2 l^{-1}$). The processes and mechanisms responsible for observed concentrations of the various nitrogen species change over time and place and nitrification rates are not only affected by the availability of the required substrates and modified by temperature but also by pH, with rates severely reduced at pH < 5 in acidified waters. Thus, high NH_4^+ and NO_2 concentrations in the well-oxygenated epilimnion of Lake Orta (IT), at a time when the pH had been lowered to about four by industrial effluents, were attributed to termination of the nitrification process at the NO_2^- stage (Mosello et al. 1986). Another example is when nitrification in two experimentally acidified Canadian lakes became blocked when the pH was lowered to between 5.4 and 5.7 (Rudd et al. 1988). This blocking allowed summer epilimnetic NH_4^+ concentrations to increase more than sixfold from < 18 to 108 μg l^{-1}.

Nitrifying Organisms

Nitrification is carried out by two groups of bacteria: the ammonium oxidizing bacteria (e.g., Nitrosomonas) which oxidize NH_4^+ to NO_2^-), and by nitrite

---

[3]"Benthic chambers placed on the sediments and used to measure sediment release rates must be calibrated with independent methods to ensure that the processes observed within the chamber [reduced turbulence and diffusion effects] reflect the processes taking place outside. Total inorganic nitrogen (TIN) fluxes measured with benthic chambers in 1985 overestimated observed accumulation rates of TIN by the hypolimnetic mass balance approach by up to 7-fold." (Höhener and Gächter 1994.)

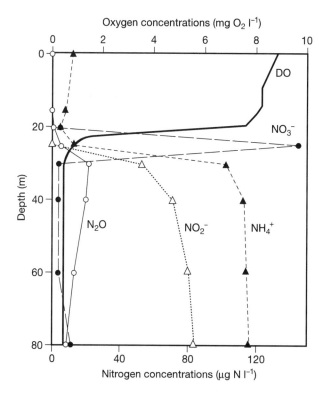

Figure 18–3 Nitrate (NO_3^-), nitrite (NO_2^-), ammonium (NH_4^+), nitrous oxide (N_2O) and dissolved oxygen (DO) concentrations in Lake Rotoiti, (NZ) on 11 April 1985. Note the importance of the oxycline at 20–25 m as the principal site for nitrification and denitrification, reflected in large concentration changes with depth in the principal forms of ionic nitrogen. *(Modified after Downes 1988.)*

oxidizing bacteria (e.g., *Nitrobacter*). The former is better adapted to low DO concentrations, but both are chemoautotrophic and able to couple the energy released by the oxidation of organic matter to fixation of CO_2 needed to satisfy their carbon demand.

18.4   Denitrification

Denitrification is a bacterially mediated process of dissimilatory reduction of nitrogen oxides (NO_3^- and NO_2^-) first to gaseous nitrous oxides (NO and N_2O) and then dinitrogen gas (N_2) (Fig. 18–2). The process is carried out by many heterotrophic, facultative anaerobic bacteria (Knowles 1982) and fungi at oxic–anoxic interfaces in lakes, rivers, and wetlands (Table 8–8). The microbes use NO_3^- or NO_2^- as the

terminal electron acceptor in the oxidation of organic matter (Fig. 16–2, and Eqs. 3a and b in Table 16–1). Denitrification and nitrification are closely coupled, but denitrification is the process responsible for the loss of fixed nitrogen to the atmosphere, primarily as N_2 but some as N_2O.[4] Additional nitrate is lost through **assimilative reduction** into microbial and algal protoplasm (Fig. 18–2). Recognition of the close relationship between nitrification and denitrification is used to promote nitrogen removal rates in wastewater treatment plants (Fig. 8–14).

Denitrification and the Cycling of Other Elements

Some facultative anaerobic bacteria carrying out denitrification are species in the genus *Pseudomonas*, but denitrifiers are also found among species of *Achromobacter*, *Bacillus*, and *Micrococcus* (Keeney 1973). Because they are facultative anaerobes, they can all use NO_3^- as the terminal electron acceptor in respiration when the DO supply becomes limiting.

With glucose as the organic substrate, oxic and anoxic processes can be compared directly (Keeney 1973).

$$\text{Oxic: } C_6H_{12}O_6 + 6O_2 \rightarrow 6CO_2 + 6H_2O \qquad \text{EQ. 18.2}$$

$$\text{Anoxic: } C_6H_{12}O_6 + 4HNO_3 \qquad \text{EQ. 18.3}$$

$$\rightarrow 6CO_2 + 6H_2O + 2N_2 + 4H_2O$$

Like other elemental cycles, the nitrogen cycle does not occur in isolation from other cycles. For example, the sulfur and nitrogen cycles are linked because NO_3^- can be denitrified while sulfur is oxidized:

$$5S + 6KNO_3 + 2CaCO_3 \qquad \text{EQ. 18.4}$$

$$\rightarrow 3K_2SO_4 + 2CaSO_4 + 2CO_2 + 3N_2$$

Similarly, the nitrogen and phosphorus cycles are coupled, with NO_3^- serving as an important electron acceptor maintaining the redox potential at a level high enough to prevent the reduction and solubilization of

---

[4]W. M. Lewis, Jr. (2000) hypothesizes that tropical lakes may exhibit primary nitrogen limitations more often than their temperate zone counterparts (Chapter 21) as the result of postulated higher denitrifications rates in the yearround warm waters, characterized by lower DO concentrations and greater probability of the anoxic conditions required for denitrification (Chapter 15).

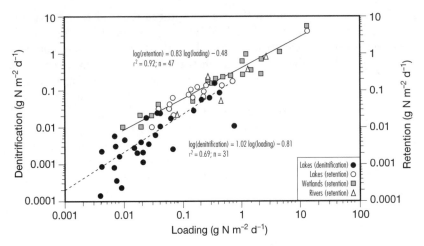

Figure 18–4 Relationship between nitrogen loading to largely temperate zone inland waters and denitrification (determined by mass-balance) as well as between nitrogen loading and nitrogen retention, (input minus output) which includes denitrification. Note that the denitrification rate does not change with loading (slope = 1.0) over the range examined, but retention declines (slope = 0.8). The discrepancy appears to be the outcome of examining results over different range scales, with the retention data set also including highly eutrophic systems. (*Saunders and Kalff 2001a, and Fleischer and Stibe 1991.*)

iron phosphate complexes until it is exhausted. As long as the complexes remain intact, profundal sediment release of PO_4^{-3} is prevented in eutrophic lakes (Sec. 17.2 and Chapter 19). Empirical evidence for this link was provided by Andersen (1982), who reported a positive correlation between high concentrations of NO_3^- in the overlying water and low P release from sediments in 31 shallow, highly eutrophic Danish lakes. He interpreted the low PO_4^{3-} release rate to be the result of a high NO_3^- diffusion rate—facilitated by turbulence—into the sediments. It has been proposed that injecting NO_3^- into anaerobic sediments of eutrophic lakes might be a useful management tool to prevent P release (W. Ripl in Björk et al. 1979).

There are two ways of determining denitrification rates, each with associated advantages and disadvantages. The mass-balance approach yields denitrification rates indirectly from seasonal or annual differences between measured whole-system inputs and outputs of nitrogen. Its greatest advantage lies in providing an integrated whole-system measure:

$$\text{Denitrification (mg N m}^2\text{ period}^{-1}) = \text{(terrestrial}$$
$$+ \text{atmospheric N input}$$
$$+ \text{aquatic N fixation)}$$
$$- \text{(outflow N}$$
$$+ \text{sediment stored N}$$
$$+ \Delta\text{water N)}$$

EQ. 18.5

The disadvantage of the mass-balance approach is the considerable number of measurements that must be made to obtain denitrification by difference. Consequently, any errors in individual input and output measurements are attributed to denitrification. Finally, a steady-state mass-balance model is a black-box modeling approach (Table 2–2), one that considers only inputs and outputs and provides no insight into the underlying mechanisms or processes that yield the rates observed. The alternative involves experimental determinations of denitrification in a limited number of sediment cores taken to the laboratory, providing needed mechanistic insights but that are difficult to extrapolate to nature.

From the perspective of aquatic management, it does not really matter whether the nitrogen retained (input–output) is stored in the sediments or denitrified and lost to the atmosphere. Either way it reduces undesirable high levels of NO_3 in drinking water supplies and reduces its eutrophication potential.[5]

Denitrification, Nitrogen Release to Waterways, and Environmental Factors

Nitrogen loading (input) allows a first estimate of the outputs from aquatic systems, albeit it only over a large range scale (Fig. 18–4). But, the slope of the line in the figure (0.83) shows that eutrophic systems, in-

---

[5]High nitrate concentrations (> 10 mg l^{-1} as N) in drinking water drawn from wells or from nitrate-rich surface waters causes methoglobinemia (MetHb) in formula-fed infants less than 6 months, following the reduction of water nitrate in the body to nitrite. Maximum concentration allowed in drinking water are typically < 10 mg l^{-1} as N. MetHb is a condition resulting from the conversion of hemoglobin (Hb) to MetHb, which is unable to transport DO. The result is the characteristic "blue baby syndrome."

cluding wetlands, are proportionally less effective in preventing incoming nitrogen from leaving via the outflow than is the case for oligotrophic ones. Eutrophic systems usually have larger drainage basins and are shallow, allowing for rapid flushing. The resulting reduced sedimentation of the organic matter containing nitrogen (and phosphorus) is ultimately responsible for the proportional decrease in retention (Fig. 18–4, Windolf et al. 1996). Even so, unless flushing rates and nitrogen loadings are exeptionally high, wetlands and shallow lakes remain effective nitrogen traps (Table 8–8) and of great utility in the treatment of livestock waste and sewage effluent.

▲ Nitrogen release rates from catchments to receiving rivers and nitrate concentrations in the rivers rise with increasing human density—with human density as a proxy for organic waste production, soil quality, and agricultural activity (Fig. 18–5). However, where the nitrogen plus phosphorus supply rate to waterways is very high, as in rich agricultural areas, the receiving rivers, lakes, and estuaries exhibit extreme eutrophication. For example, the export of plant nutrients from the intensively farmed drainage basin of the Mississippi River to the Gulf of Mexico is so high that the estuary experiences periodic sediment anoxia, toxic algal blooms, and fish kills (Downing et al. 1999).

Denitrification in shallow-water sediments is enhanced by the presence of submerged and emergent macrophytes. Macrophytes trap sedimenting organic particles and are able to translocate oxygen produced in the leaves to the roots. It diffuses from the roots into the surrounding sediments, allowing a linked nitrification and denitrification in the root zone and overlying sediments. Christensen and Sørensen (1986) compared denitrification in sediments with and without vegetation cover provided by the submerged perennial macrophyte *Littorella uniflora*. Plants reportedly accounted for as much as 50–70 percent of the annual sediment denitrification, but sediments trapped by them probably made an important (but unmeasured) contribution as well, compared to little or no sedimentation of nitrogen containing particles outside the macrophyte bed. Within macrophyte beds and wetlands denitrification rates are highest in shallow, warm water, declining systematically and rapidly with increasing depth (Saunders and Kalff 2001b).

The stems of emergent macrophytes serve as an important conduit not only for the downward transport of DO but also for the upward flux of N_2 and N_2O produced in the root zone and lost to the atmosphere (Fig. 18–6). Work on emergent macrophytes

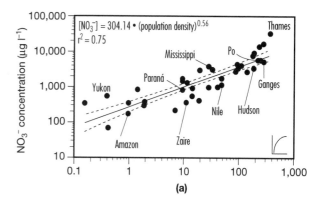

(a)

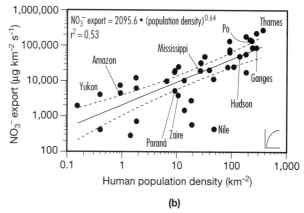

(b)

Figure 18–5   Effect of human population density and associated agricultural activities in large river catchments on (a) the mean annual nitrogen concentration in the rivers; and (b) the specific export of nitrate nitrogen from the catchments to the sea. Dashed lines show the 95 percent confidence intervals around the mean. *(Modified after Cole et al. 1993a.)*

(Chapter 24) in an experimental system showed that about one-quarter of NH_4^+ fertilizer added to the root zone was lost as N_2 via the plants stems within a month, following nitrification in the root zone, diffusion of the NO_3^- produced back to the anaerobic zone, denitrification and transport (Reddy et al. 1989).

18.5   Nitrogen Fixation: Rates and Process

On earth, nitrogen is overwhelmingly present as molecular N_2 and fixed forms of nitrogen are subject to depletion by denitrification. The nitrogen available to the global biota would decline were it not for biologi-

Wetland plant

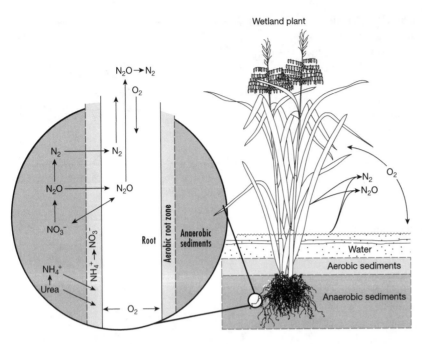

Figure 18–6   Schematic presentation of nitrification–denitrification in the root zone of rice and other emergent aquatic macrophytes. *(After Reddy et al. 1989.)*

cal nitrogen fixation offsetting the losses. Nitrogen fixed terrestrially on well-vegetated drainage basins is largely retained or denitrified (Sec. 18.1). When there is only modest terrestrial export, substantial planktonic nitrogen fixation occurs within lakes and wetlands (Sec. 18.6) receiving streamwater with an unfavorable (low) N:P supply ratio relative to the demand ratio of the algae (see Table 8–2 and Sec. 21.13). Although nitrogen fixation rates increase as the N:P ratio declines (Fig. 18–10), rates are low compared to the denitrification rates measured in highly eutrophic waters (Seitzinger 1988) normally characterized by low N:P ratios (Sec. 21.13). The shortfall has to be offset by nitrogen inputs from catchments and the atmosphere.

Biological nitrogen fixation is an enzyme-catalyzed process by which a wide variety of prokaryotic organisms—both freeliving and symbiotic—reduce atmospheric nitrogen (N_2) to ammonia (NH_3). The pertinent organisms include some of the photoautotrophic cyanobacteria (blue-green algae), as well as a variety of aerobic and anaerobic heterotrophic and chemoautotrophic (chemosynthetic) bacteria (Howarth et al. 1988). The principal cyanobacteria capable of nitrogen fixation belong to filamentous genera that have specialized, heavy-walled cells called **heterocysts** providing the anoxic conditions required (Fig. 18–7, and Sec. 21.2). The heterocystous

freshwater genera include, amongst others, *Anabaena, Aphanizomenon, Gloeotrichia, Nodularia, Cylindrospermum, Mastigocladus,* and *Nostoc* (Paerl 1990). The first two are normally planktonic, whereas the others grow primarily on surfaces (*periphyton*, Chapter 24).

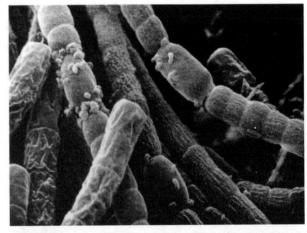

Figure 18–7   An electron micrograph showing strands of filamentous cyanobacteria with heterocysts growing on an underwater leaf of a freshwater macrophyte and showing heterotrophic bacteria on the leaf and filaments. *(Courtesy of American Society of Limnology & Oceanography/ASLO.)*

Nitrogen Fixation and Environmental Factors

▲ Nitrogen fixing cyanobacteria require the energy of the sun to reduce N_2 to the organic nitrogen of protoplasm. Nitrogen fixation therefore varies diurnally and typically declines with depth in lakes and wetlands (Fig. 18–8). N_2 fixation is an expensive process, the energy used for fixation will not be available for growth. Consequently, cyanobacteria able to fix nitrogen normally stop when NH_4^+ or NO_3^- become readily available (Carr and Whitton 1982) or when the irradiance is low. The fixed nitrogen is in due course recycled by leakage of extracellular nitrogen from the cells, ammonification upon death (Fig. 18–2), and animal predation on the nitrogen-fixing organisms.

In a field study, Viner (1985) assessed the relative importance of environmental factors and the presence of heterocysts in predicting the daytime N_2 fixation rate in Lake Rotongaro (NZ). He found that light climate and abundance of heterocysts explained 86 percent of the variation in observed fixation rates (Fig. 18–9). However, irradiance is expected to be much less important at the scale of a lake district over which irradiance (and temperature) would vary relatively little,

but NH_4^+ and NO_3^- concentrations and the N:P loading ratio (Fig. 18–10) vary greatly.

▲ 18.6   Nitrogen Fixation Rates: Plankton vs Littoral Zone

Nitrogen fixation rates have been very largely examined in the plankton and the littoral rates have received little attention. However, the abundance of nitrogen fixing cyanobacteria growing on surfaces in oligotrophic lakes and wetlands suggests that nitrogen fixation is probable, even when pelagic zone rates are not measurable. The experimental addition of phosphorus to oligotrophic boreal forest and arctic lakes to lower the N:P supply ratio led to substantial nitrogen fixation in the littoral zone, even though little or no fixation was measured in the plankton before or after the fertilizations took place (Table 18–2, p. 281). It appears that the disproportionally high denitrification rates measured in shallow water sediments (Sec. 18.4) yields a low N:P supply ratio that stimulates littoral zone nitrogen fixation.

The relative contribution of the littoral zone to whole-lake nitrogen fixation is not only a function of

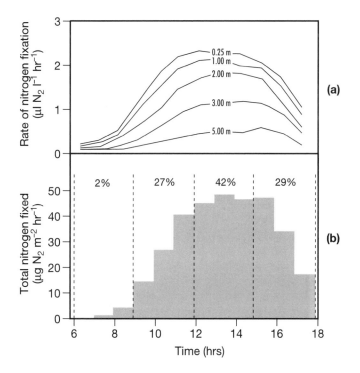

Figure 18–8   Nitrogen fixation (a) at different depths on 12 September 1975 and (b) per unit surface area on 25 November 1975 in Rietvlei Dam, ZA. (*After Ashton 1979.*)

Variable	Percent contribution to total variance explained			
Incident photosynthetically available radiation (PAR)	40	56	86	95
Extinction coefficient	16			
Heterocysts per m^2	30			96
NH$_4^+$ + NO$_3^-$ concentration	9			
Temperature	1			

Figure 18–9 Analysis of variance for the regression of the rate of daytime N$_2$ fixation on other variables in a seasonal study of a New Zealand lake. Note that the role of NH$_4^+$ plus NO$_3^-$ was a minor one here probably because concentrations varied only modestly over time (temporal scale) in the single lake. Conversely irradiance, which varies greatly on a day to day basis, exerted a major influence at these particular temporal and spatial scales. *(From Viner 1985.)*

the N:P supply ratio and the light climate but also of system morphometry. Work in both the littoral and open water of a small English Lake showed littoral fixation rates to be about three times higher on a whole lake basis than in the pelagic zone (Jones and Simon 1981). These high littoral rates were the result of a very high specific fixation rate (mg N m^{-2} d^{-1}) and a large littoral zone (m^2). Consequently, littoral zone contributions to nitrogen fixation and other metabolic

processes is large in shallow systems, but minor in steep-sided lakes dominated by their pelagic zones. Contradictory conclusions that have been published in the literature are often the result of work having been carried out over different spatial and temporal scales (Sec. 2.6)

18.7 Forms and Quantities of Nitrogen in Inland Waters

The enduring work on the phosphorus cycle carried out by W. Einsele, W. Ohle, and C. H. Mortimer (Sec. 17.2) was complemented in the early 1940s by equally important work on nitrogen cycling by C. H. Mortimer. His laboratory and field research continues to provide the conceptual foundation for interpreting the forms, concentrations, and seasonal cycles of inorganic nitrogen in water and sediments (Fig. 17–1).

In well-oxygenated epilimnia of nonhumic lakes, the combined inorganic nitrogen is largely present as NO$_3^-$, its most oxidized form. Ammonium (NH$_4^+$), produced upon the death or decomposition of organisms or excretion by animals, will typically be low because the ammonium produced is readily oxidized first to NO$_2^-$ and then to NO$_3^-$. Plants preferentially take up the reduced NH$_4^+$ rather than the oxidized NO$_3^-$, further contributing to lower NH$_4^+$ concentrations.

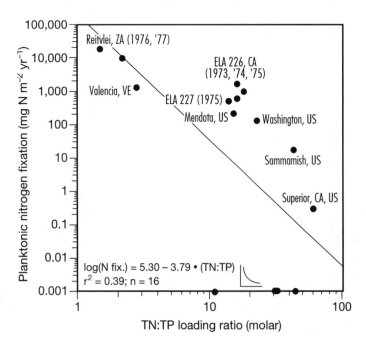

Figure 18–10 The relationship between planktonic nitrogen fixation and the TN:TP loading ratio showing large nitrogen fixation rates when the TN:TP supply rates are unfavorable relative to the demand ratio of the nitrogen-fixing organisms. Selected lakes are named. The baseline data points represent fixation rates below detection in ELA lakes 227 and 261. *(Data from Howarth et al. 1988.)*

Table 18–2   **Nitrogen fixation rates by littoral zone algae growing on macrophytes plus the contributions of nitrogen fixation to whole-system nitrogen loading. Note the importance of fixation in systems fertilized only with phosphorus, yielding a low N:P ratio, and that little or no nitrogen fixation occurred when additional nitrogen was supplied. ND = not determined.**

Lake	Year	Maximum Fixation (mg N m^{-2} hr^{-1})	Percentage of Total N Loading	Fertilization
Hymenjaure (SE)	1973	0.40	17	P
	1974	1.66	48	P
	1975	0.05	< 1	P and N
Stugsjön (SE)	1974	0.09	6	ND
	1975	0.17	7	ND
Crater (US)	1979	0.10	ND	ND
Tahoe (US)	1979	0.14	ND	ND
226NE (CA)	1976	1.87	ND	P and N
P & N (CA)	1981	0.26	5	P and N
Far (CA)	1981	0.95	28	P
Spring (CA)	1981	0.56	16	ND

Source: After Bergmann and Welch 1990.

But NH_4^+ dominates in humic systems, characterized by low redox potentials and reduced DO levels. Moreover, high summer concentrations of NH_4^+ in streams and epilimnia of nonhumic lakes are almost always attributable to the input of wastewater from sewage treatment plants or livestock. Large rivers receiving considerable NH_4^+ and NO_3 input from their terrestrial and aerial catchments have NO_3 concentrations and terrestrial NO_3 export coefficients (μg km^{-2} s^{-1}) correlated with the human population density (Fig.

18–5). However, total nitrogen in streams and rivers draining undisturbed forests is mostly organic nitrogen, with the total concentration and the fraction composed of NO_3^- rising with increasing human disturbance (Howarth et al. 1996). The dominance of dissolved organic nitrogen in the total dissolved nitrogen pool of oligotrophic waters was demonstrated as early as the 1920s by Birge and Juday and their coworkers. For a trophic classification of nitrogen, phosphorus, and chl-*a*, see Table 18–3.

Table 18–3   **Summer near-surface average nutrient concentrations (μg l^{-1}) for classifying lakes and rivers into different trophic state categories.**

	Trophic State	Inorganic N	Total N	Total P	TN:TP	Chl-*a* Suspended	Chl-*a* Benthic
Lakes	Oligotrophic	< 200	< 350	< 10	~35	< 3.5	—
	Mesotrophic	200–400	350–650	10–30	~25	3.5–9	—
	Eutrophic	300–650	650–1200	30–100	~14	9–25	—
	Hypertrophic	500–1500	> 1200	> 100	~12	> 25	—
Rivers	Oligotrophic	—	<< 700	< 25	~28	< 10	< 20
	Mesotrophic	—	700–1500	25–75	~22	10–30	20–70
	Eutrophic	—	> 1500	> 75	~20	> 30	> 70

Source: After Vollenweider 1968, Forsberg and Ryding 1980, and Dodds et al. 1997.

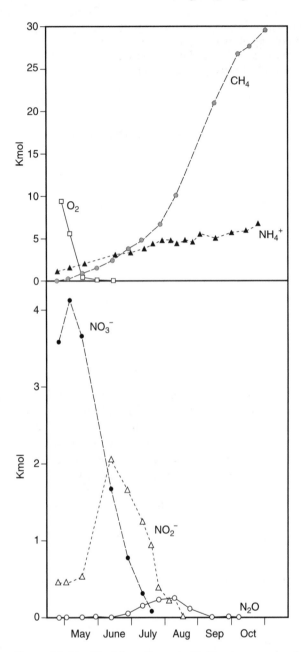

Figure 18–11 Total hypolimnetic (8–15 m) contents of inorganic forms of combined nitrogen, methane, and dissolved oxygen in Lake St. George (CA) in 1982. *(Modified from Bedard and Knowles 1991.)*

▲   Forms of Nitrogen and Their Cycling

The combined inorganic ionic species (thus excluding N_2) are known together as **dissolved inorganic nitrogen** (DIN) or as **total inorganic nitrogen** (TIN). The other operationally defined fractions are **dissolved organic nitrogen** (DON) and **particulate organic nitrogen** (PON).[6]

Winter DIN concentrations in unpolluted temperate lakes are often higher than during summer as the result of a relatively low photosynthetic rate (demand) in relation to supply rate at latitudes where the soil does not freeze or freezing is of short duration, and where much of the annual runoff occurs during the winter months.

Hypolimnetic waters normally contain elevated levels of DIN because of diffusion of NH_4^+ from the sediments combined with release from decomposing sedimenting particles in the hypolimnetic water column and a low net uptake of nitrogen by organisms in the aphotic zone. Consequently, hypolimnetic NH_4^+ concentrations increase over time (Fig. 18–11). When DO becomes depleted the NO_3, followed by NO_2, become major electron acceptors in the oxidation of organic matter, and their concentrations decline during the resulting denitrification process. Methane (CH_4), an end product of the anoxic decomposition of organic matter (Eq. 16.3), rises in step with the amount of NH_4^+ liberated (Fig. 18–11). The introduction of DO during fall overturn allows the oxidation of NH_4^+ to NO_3^- and CH_4 to CO_2 (Fig. 18–11 and Eqs. 7a, c, d, in Table 16–1), returning concentrations to the prestratification state.

Highlights

- Nitrogen and phosphorus are the elements normally in highest demand by aquatic plants and microbes relative to supply from drainage basins and the atmosphere.
- The probability of a primary nitrogen limitation by phytoplankton is greatest where the N:P supply ratio is well below the algal demand ratio of ~16:1 as atoms.
- Nitrogen is supplied to aquatic systems from the drainage basins and the atmosphere, and can be produced *in situ* through nitrogen fixation.
- Nitrogen in lakes and wetlands that is not lost via the outflow is largely denitrified rather than stored in

---

[6]A *dissolved* substance is operationally defined as one that passes a membrane filter with an average pore size of ~0.20-0.45 μm, with the fraction retained on filters considered *particulate*. However, some very fine particles plus colloidal materials pass the filters and are included in the dissolved fraction even though they are not dissolved. The role of filtration in changing the relative importance of the dissolved and particulate fractions, skewed as the result of cell breakage, cell leakage, or coagulation, has not been resolved.

sediments. The relative importance of denitrification rises with increasing water residence time and decreasing water depth. Nitrogen lost to the atmosphere through denitrification is no longer available to the biota, making wetlands and shallow lakes particularly useful for treating wastewater.

- Nitrogen cycling affects other nutrient cycles (e.g., phosphorus and sulfur).
- Human activity has raised—and continues to raise—the export of nitrogen from the land to waterways and, via the atmosphere to aquatic systems far from regions of industrial and agricultural activity.

19

Iron, Manganese, and Sulfur

19.1 Introduction

Iron and manganese are essential micronutrients for plants and iron may occasionally control algal production in inland waters. Reduced inorganic sulfur concentrations, together with irradiance, determine the production rates of photosynthetic sulfur bacteria below the oxycline in transparent lakes (Sec. 22.10).

The solubility of Fe and Mn in the ionic (not complexed) form is very low in pure solutions. The calculated solubility of ionic iron in well-oxygenated waters that are not exceptionally acidic is less than 10 μg l^{-1} (Mill 1980).[1] Yet total iron concentrations measured in filtered river water are about 1,000 times higher, indicating a sorption or complexation with organic colloids and other forms of dissolved organic matter in lakes, rivers, and wetlands, greatly affecting metal behavior and concentration. Since there is little iron in ionic form, it can occasionally limit algal growth (Sec. 21.13).

The physical, chemical, and biological interactions of iron and manganese with other elements and organic matter are well-recognized. Iron plays a particularly important role in determining the concentration, solubility, and flux rates of phosphate, sulfur, trace metals, and dissolved organic carbon (DOC) in waterways. Furthermore, alkalinity is generated in acidified lakes upon the microbial reduction of oxidized iron and sulfate, thereby buffering aquatic sediments against acidification (Tables 27–1 and 16–1). In lakes, buffering is permanent if the reduced elements are not reoxidized during the next period of destratification and reoxygenation. The iron and manganese in aquatic systems is derived from drainage basins, primarily as oxides.

19.2 Iron Cycling

The classic work of C. H. Mortimer (1941–1942) on iron (and phosphorus, see Chapter 17) cycling in Esthwaite Water (GB) provides the basis for our present understanding of the iron cycle in low humic (low DOC) nonacidified waters.

In the presence of dissolved oxygen (DO) and the absence of much dissolved organic matter, ferric iron [Fe^{3+} or Fe(III)] forms a number of insoluble oxides and oxyhydroxides (FeOOH). The aggregates (flocs) that are formed settle onto the sediments, also the primary reason why well-oxygenated stream sediments are frequently covered by a rusty brown layer of sedimented Fe(OH)$_3$ carried in as reduced iron (Fe^{2+}) in groundwater.[2] Ferric iron becomes subject to microbially and chemically mediated reduction in lakes, wetlands, and the hyporheic zone of lotic systems once the DO, oxidized manganese (Mn^{4+}), and nitrate have been

---

[1] The solubility of Fe and most other metals under oxic conditions is particularly elevated under the very low pH conditions encountered in highly acidic volcanic lakes and acidified streams draining acidic coal mine effluents (Sec. 14.5 and Chapter 27).

[2] Fe (and Mn) oxides have a large surface area per unit weight and their surface hydroxyl groups have a high affinity (adsorption capacity) for a variety of trace elements and dissolved organic matter. Oxidized surfaces also facilitate oxidation of sorbed elements through redox reactions.

utilized as terminal electron acceptors during the microbially mediated oxidation of organic matter (Eqs. 2 and 4 in Table 16–1). DO consumption rates (respiration) per unit volume are highest in surficial sediments and these become anoxic prior to deoxygenation of the overlying water. The chemical oxidation and reduction of iron are greatly dependent on pH, with the oxidation rate of Fe(II) increasing by a factor of ~100 per unit pH rise (Davison and Seed 1983).

Sediment Release, Transport and Precipitation

In the absence of reduced sulfur, the Fe(II) produced under anoxic conditions is soluble and able to diffuse from the sediment into the water column. The Fe(II) in the hypolimnia is partly derived through diffusion from sediments on the slopes of the basins and is transported laterally by horizontal eddy diffusion plus hypolimnetic currents (Secs. 12.6 and 12.12) at much greater rates than is possible through vertical diffusion (Sec. 12.5). This view differs from the belief held when Mortimer did his seminal research. Then, noth-

ing was known about relative diffusion rates and it was thought that hypolimnetic Fe (Mn, etc.) concentrations were the result of only vertical eddy diffusion from sediments. In addition, analyses of profundal zone sediment cores taken at different depths in the hypolimnion of a eutrophic Swiss lake show horizontal transport to result in Fe loss from shallow hypolimnetic sediments and deposition (accumulation), following oxidation in the deepest portion by a process called **geochemical focusing** (Schaller et al. 1997).

Inputs of Fe(II) from sediments (internal loading) can be high, occasionally yielding concentrations well above 1,000 µg l^{-1} in the anoxic hypolimnia of noncalcareous (low bicarbonate) lakes (Table 19–1). However, the reduced Fe(II) is rapidly chemically and biologically oxidized by bacteria in the genus *Siderocapsa*, for example, following contact with DO in the water-column oxycline and at the sediment surface during overturn (Fig. 19–1). The aggregates that form sink onto oxidized sediments but dissolve when anoxic conditions return. Measurements have shown that more than 90 percent of the recycled Fe originated from the top 1 cm of the sediment surface and a

Table 19–1 **Chemical conditions in the anoxic water or anoxic sediment–water interface of selected lakes at the end of summer. Subscripts t and s stand for total and soluble, respectively. ND = no data. Note inter-annual variations as well as differences in concentrations and the ratios between lakes and between calcareous (high bicarbonate) and noncalcareous (low bicarbonate) lakes.**

Lake	Date	Sampling Depth (m)	Depth Above Sediments (m)	Concentration (mg m^{-3} or µg l^{-1})				
				Fe$_t$	Fe$_s$	Mn$_t$	Mn$_s$	S^{2-}
Calcareous								
Mendota (US)	9/77	24.0	0.5	150	60	625	625	+
	9/78	23.8	0.4	80	40	490	490	3.6
	9/79	23.8	0.6	40	ND	500	500	3.3
Rostherne (GB)	10/81	29.0	0.05	130	< 3	3,300	3,320	< 0.01
Baldegg (CH)	9/77	65.0	1.0	40	35	390	390	3.0+
Monona (US)	9/78	17.2	0.1	~20	~20	300	290	5.3
Noncalcareous								
Sebasticook (US)	summer/ 80	17.2	0.2	893	ND	8,352	ND	ND
Esthwaite (GB)	9/80	~15.0	0.1	1,720	ND	600	ND	ND
Shagawa (US)	7/77	12.5	0.5	7,700	ND	930	ND	ND
Lake 227 (CA)	9/79	10.0	~0.3	10,050	ND	ND	ND	ND

Source: After Stauffer 1987.

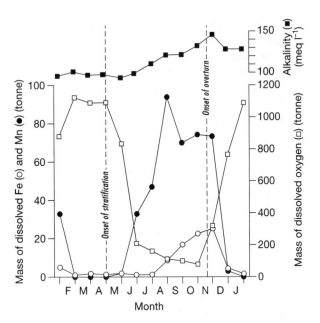

Figure 19–1 Seasonal changes in the mass of dissolved iron, dissolved manganese, and dissolved oxygen in the bottom waters (15–32 m) of Lake Sammamish, US. Note (1) the release of dissolved Mn from sediments when the upper part of the hypolimnion still contains considerable dissolved oxygen; (2) a more modest release of dissolved Fe when virtually the whole hypolimnion is anoxic; (3) the rapid precipitation and decline of previously dissolved Mn and Fe following overturn and reoxygenation of the water column; and (4) a seasonal increase in alkalinity following the consumption of H^+ during the reduction of the oxidized elements that are able to serve as electron acceptors (Chapters 16 and 27). *(Modified from Balisteri et al. 1992.)*

similarly high fraction of the sediment flux of NH_4^+, CH_4, and dissolved inorganic carbon (DIC) (see Cook 1984).

The precipitation rate of Fe(III) and other trace metals (including aluminum, see Chapter 27) declines among systems with increasing DOC because of sorption to dissolved organic matter.[3] Consequently, much more iron is retained in oxic water columns than is predicted on the basis of research on iron solubility in pure water. Recent work has demonstrated that there is light-induced (photochemical) reduction of Fe(III) to Fe(II) in surface waters adding to the tiny pool of

soluble Fe available for uptake by algae in oxygenated waters (Emmenegger et al. 1998).

▲ Iron Aggregates and Their Retention

The retention within stratifying lakes of both externally and internally loaded iron is generally high (60–99%) because Fe(III) aggregates are sedimented and prevented from leaving the systems through flushing during overturn periods (Table 19–2). The retention of Fe (and other metals) rises as the water residence time increases (Fig. 19–2). However, whole-system acidification experiments have shown an increased dissolution of sedimented Fe with declining pH, resulting in increased water concentrations.

A fraction of the sedimented iron (and Mn) may not be subject to resolubilization during the next period of anoxia because some has been converted to much less soluble mineral forms. Mineral formation—and sometimes further transformation to yet other mineral forms—is known as **diagenesis**; the term is sometimes also applied to the transformation of organic compounds from one form to another in microbial respiration. Transformed crystals of a precipitate are generally less soluble than the original crystals making diagenesis an effective way to reduce or prevent solubilization when environmental conditions once again become conducive to solubilization.[4]

Sedimenting Fe(III) aggregates range in diameter from about 0.05 μm to 0.5 μm (Davison and Tipping 1984), but are usually between 0.20 μm and 0.35 μm. However, the aggregates contain much more than just Fe. For example, in Esthwaite water (GB) the flocs contain 30–40 percent by weight each of Fe and organic matter, with the balance contributed by phosphorus, nitrogen, manganese, silica, sulfur, calcium, and magnesium (Tipping et al. 1981). In addition, adsorption of organic matter makes aggregates effective sorption sites for trace metals (Sec. 19.4), thereby facilitating their movement from the water to the sediment.

---

[3]The brown-red color of DOC in iron-rich areas is enhanced by the sorption of Fe(III) to DOC. Water color will therefore be a poor indicator of the DOC concentration among climatic/geological regions that differ in iron and the fraction of DOC that is colorless (Chapter 22).

[4]Much of the research on diagenesis has been done on meromictic lakes where the hypolimnion and sediments are permanently anoxic, and on saline lakes. The latter are characterized by vigorous diagenesis during periods of high evaporation when the solubility product of different ion pairs is sequentially exceeded (Sec. 13.7). The degree of diagenesis is frequently such that only a fraction of the precipitated crystals are resolubilized when more dilute runoff water enters the lakes or wetlands during the following rainy season.

Table 19–2 **Iron budgets for selected lakes. External and internal inputs as well as outputs (mg m^{-2} yr^{-1}) and water residence time (WRT, yr). ND = no data.**

Lake	Input External	Input Internal	Input Total	Output	Retention External (%)	Retention Total (%)	WRT
Harp (CA)	1,266	0	1,266	456	64	64	3.2
Rawson (CA)	647	0	647	94	85	85	6.4
Oneida (US)	974	0	974	386	60	60	0.7
Mendota (US)	12,800	0	12,800	130	99	99	5.5
Esthwaite (GB)	20,000	2,700	22,700	6,400	68	72	0.3
Finjasjoen (SE)	6,500	8,514	15,014	5,455	16	64	0.3
Lake 227 (CA)	1,431	8,932	10,363	ND	ND	ND	2.1
Blue Chalk (CA)	269	1,043	1,311	81	68	94	5.5

Source: After Nürnberg and Dillon 1993.

19.3   Iron and Sulfur

Inorganic sulfur is the single most important ionic determinant of the solubility and cycling of iron and most trace metals. When SO_4^{2-} is reduced during the microbial oxidation (respiration) of organic matter:

$$2(CH_2O) + SO_4^{2-} \rightarrow S^{2-} + 2CO_2 + 2H_2O; \quad EQ.\ 19.1$$

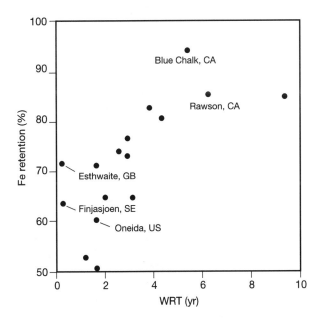

Figure 19–2 Total iron retention versus water residence time (WRT, yr) for 17 temperate zone lakes. *(Modified after Nürnberg and Dillon 1993.)*

any gaseous S^{2-} or the bisulfide ion (HS$^-$) that is produced combines with Fe (as well as with trace metals and occasionally Mn) to form almost insoluble precipitates under anoxic conditions. See Eqs. 4 and 5 in Table 16–1 for a more complete reaction.

The Fe:SO$_4$ Supply Ratio

This ratio is the principal determinant of the extent to which reduced Fe (and most trace metals, see Sec. 19.6) is solubilized under anoxic conditions. High levels of dissolved iron in anoxic hypolimnia are found only where the Fe:SO$_4$ supply ratio is high, as in low conductivity boreal forest lakes receiving disproportionately little SO_4^{2-} from their drainage basins (e.g., Lake 227 and Blue Chalk, Table 19–2). When anoxic conditions develop, the Fe(II) that is liberated precipitates as FeS when the solubility product is exceeded, in the presence of enough sulfide, giving sediments a characteristic black color. Conversely, a high external Fe:SO$_4$ loading ratio allows little or no H$_2$S production because any HS$^-$ is precipitated as FeS. Such systems contain high levels of residual dissolved iron. High concentrations are also typical in the anoxic hypolimnia of eutrophic lakes where reduced Fe and associated phosphorus are liberated after dissolution of the FeOOHP complexes in sediments under anoxic conditions (Table 19–2; Fig. 17–2).

Where the terrestrial Fe:SO$_4$ supply ratio is low, as in the calcium carbonate-rich catchments of midwestern North America, northern Germany, or the

Ukraine, virtually all Fe is precipitated as FeS and anoxic hypolimnia typically contain little Fe (e.g., Lake Mendota, Table 19–1) but considerable H_2S. Where the lakes are also sufficiently transparent, a layer of photosynthetic sulfur bacteria develops just below the oxycline, utilizing the reduced sulfur (Sec. 22.10). Calcareous lakes retain not only most Fe but also retain a substantial fraction of the Mn supply (Sec. 19.5), primarily as insoluble $MnCO_3$. Outflowing rivers from such lakes consequently carry little Fe and Mn, and downstream lakes and wetlands are characterized by exceptionally low Fe:S and Mn:S supply ratios.

▲ Sulfur Diagenesis

Where the climate and morphometry allow an extended period of lake or wetland anoxia and much diagenesis, sediments acquire a more stable, less soluble precipitate in the form of ferrous disulfide (FeS_2), known as the mineral pyrite (Capone and Kiene 1988). There is growing evidence that pyrite can form much more quickly than once thought feasible. Even so, much of the reduced sulfur is not stored permanently as FeS or FeS_2 minerals but rather as organic sulfur compounds, which are less readily reoxidized following reoxygenation than inorganic forms. The organic sulfur is formed directly during microbially mediated reduction of SO_4^{2-}, and abiotically by reactions of reduced sulfur with sediment organic matter (Rudd et al. 1986, Urban 1994). In some saline systems, most of the sulfur may be stored as gypsum ($CaSO_4$) or anhydride.

▲ 19.4   Iron and Organic Matter

Iron and most other metals become adsorbed to sedimenting inorganic aggregates (Sec. 19.2) and also bond with colloidal[5] and dissolved organic molecules known as **organic ligands**. Many ligands are humic-acid polymers that release H^+ ions (protons) and are negatively-charged hydrophobic ("water fearing") molecules (Sec. 8.8). The molecules form metal–

---

[5]Intermediate-sized colloids falling between the operationally defined "particulate" (retained by 0.2 or 0.45 μm membrane filters) and "soluble" phase have been disregarded by aquatic scientists because they are difficult to separate and analyze (Allan 1986). In some lakes colloid carbon levels are much higher than concentrations of particulate organic carbon (Burnison and Leppard 1983).

organic complexes by ion exchange, surface adsorption, or chelation and the small colloidal Fe(III) particles that are formed are subject to little sedimentation. The existence of Fe(III) aggregates in anoxic hypolimnia has been attributed to their stabilization by organic matter and to insufficient reducing conditions at the sediment–water interface during the stratification period (Sigg et al. 1991). It is the bonding of metals to organic ligands that most confounds our ability to base an interpretation of metal behavior in inland waters on either measured redox potentials in nature or on the behavior of metals in pure inorganic solution in the laboratory. The bonding of Fe (and most trace metals) to dissolved organic matter (DOM) allows higher metal levels in both oxidized waters (redox potential $E_h > {\sim}200$ mV) at nearly neutral pH and in acidic waters than is expected from thermodynamic considerations (Urban et al. 1990, Nürnberg and Dillon 1993). The increased solubility of oxidized iron (and aluminium) in acidified waters at pH < 5 allows the additional bonding with DOM and floc formation (sedimentation) responsible for the characteristically increased clarity of lakes following acidification (Chapter 27).

19.5   The Manganese Cycle

Manganese oxides [Mn(III), Mn(IV)] become reduced at higher redox potentials than Fe^{3+} and SO_4^{2-} (Fig. 16–2), with the result that appreciable levels of dissolved Mn(II) appear in the hypolimnia or in hypoxic groundwater at a time when Fe remains in the insoluble hydroxide form. Conversely, manganese oxidation and precipitation commence when Fe remains dissolved (Fig. 19–1). Manganese is precipitated primarily as relatively large (1–5 μm) Mn oxide aggregates, but also as $MnCO_3$ in calcareous waters low in DOC. In comparison to Fe(III), little oxidized Mn is bound to organic matter (Urban et al. 1990). Inorganic flocs may contain large amounts of Ca, but Mg, Si, P, S, Cl, K, and Ba have been found as well (Tipping et al. 1984). Following microbially mediated oxidation, most Mn(II) is precipitated as Mn(III) or Mn(IV) oxyhydroxide flocs and not as MnS because little reduced sulfur is normally available at the relatively high E_h at which Mn(II) is formed (Table 19–1). A larger fraction of precipitated Mn oxide flocs than precipitated Fe (as FeS or FeS_2) is solubilized in the anoxic hypolimnia of calcareous lakes. Solubilized Mn (or Fe) is also subject to geochemical focusing and accumulation in deep water sediments (Sec. 19.2).

▲ Observation shows that the high concentrations of reduced iron in anoxic hypolimnia are derived from the sediments, but this is not necessarily the case for manganese. Most Mn(II) accumulating in anoxic hypolimnia or wetlands appears to be the result of a reductive dissolution of sedimenting particles derived from catchments (Eq. 2 in Table 16–1, Urban et al. 1990, Balistrieri et al. 1992), but not all reduced (dissolved) Mn is in the ionic Mn(II) form. Work on a Japanese lake (Yagi, 1988) shows that nearly half was in a colloidal form instead.

Another important difference between the Fe and Mn cycles is the slower oxidation of reduced manganese at pH > 6–7, following its lateral plus vertical transport from anoxic sediments toward the well-oxygenated water above. A fraction of the iron is rapidly (hours to days) reoxidized chemically and then precipitated (removed) at a neutral pH range, but the much slower reoxidation of Mn(II) is apparently slowly (hours to months) catalyzed by bacteria, among which *Metallogenium spp.* is particularly important. As the result of slower reoxidation, a considerable fraction of soluble Mn can be flushed out of low bicarbonate lakes during overturn (Davison and Tipping 1984), and from wetlands with anoxic sediments (but see Fig. 19–1). Mn and Fe concentrations in inland waters show an inverse correlation with pH, and the sediments of recently acidified systems are a net source of them to the water (Fig. 27–8 and Urban et al. 1990). At the same time, lowering the pH during acidification increases the relative stability of Mn^{2+} ions, thereby suppressing the formation of Mn(III) and Mn(IV) oxyhydroxides and their precipitation.

19.6 Iron, Manganese, and Trace Metals

The cycling of Fe and Mn within aquatic systems (Fig. 19–3) can be conceptualized as "wheels" in which the oxidized forms are precipitated as metal oxide, as a salt crystal, or coprecipitated with dissolved organic matter. The two elements become partly or mostly resolubilized under reducing conditions, except for Fe in the presence of reduced sulfur when insoluble FeS is formed.

▲ Sedimenting FeOOH and MnOOH aggregates scavenge redox-sensitive trace metals and arsenic from oxygenated waters that are not highly acidic (Gunkel and Sztraka 1986, Kuhn et al. 1994). If the sediment surface is oxic, the trace element solubility in the surficial sediment pore water is low (Tessier et al. 1985), and the possibility of a significant return of the elements to the water column is small. When sediment respiration rates are high, anoxic conditions develop rapidly at the sediment surface of lakes and wetlands, even when overlying waters remain well oxygenated. The soluble Fe(II) and Mn(II) that is released then allows Fe(II), Mn(II) and, presumably trace metals to be recorded in the water column under oxygenated conditions until the reduced forms have become biologically and chemically oxidized there and precipitate.

However, the return of reduced trace metals to the overlying water is small where scavenged elements are sedimented into anoxic hypolimnia or sediments containing excess S^{2-}. Soluble metals (Me) such as Zn, Cd, Pb, and Hg are then precipitated as MeS or sorbed to iron sulfide in the sediments. Conversely, anoxic hypolimnia may contain relatively high trace metal concentrations where the Me:S supply ratio is high

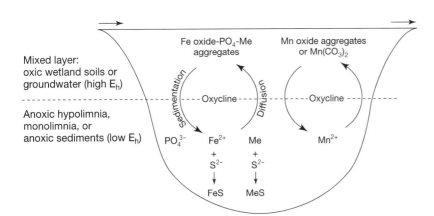

Figure 19–3 A schematic diagram of the iron, manganese, and trace metal (Me) wheels showing the sedimentation of aggregates (flocs) under oxic conditions in lakes, wetlands, groundwater, and in the hyporheic zone located below the surficial sediments of many lotic systems (see Fig. 8–7). Sedimentation of aggregates is followed by their dissolution in anoxic sediments and upward diffusion toward oxygenated water.

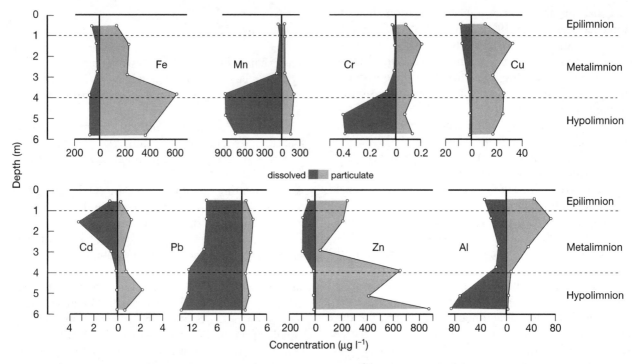

Figure 19–4 Vertical distribution during stratification of iron, manganese, and trace metals in Schäfer Lake, a 0.045 km² polluted urban lake in Berlin, Germany, on 13 July 1983. The fraction of dissolved metal is indicated by the quantities to the left of the vertical lines, and particulate metal is to the right. An oxycline was present in the metalimnion between 3 m and 4 m. Note the high hypolimnetic concentration of dissolved (filtrable) Mn and low concentration of dissolved Fe, the variation in concentration and solubility among trace metals, and the differences among trace metals in the dissolved and particulate fractions in the epilimnion and hypolimnion. *(After Gunkel and Sztraka, 1986.)*

(Fig. 19–4). Finally, the solubility of trace metals too increases as pH declines in acidified waters (Fig. 27–8).

Human Activity and Element Cycling

Human activity regionally affects the cycling of all the elements mentioned above through agricultural and industrial activity, acidifying precipitation containing trace metals, and urban waste incinerators releasing trace metals and organic contaminants that are deposited on waters and in drainage basins (Chapters 27 and 28). For example, acidifying precipitation increasing the NO_3^- loading to aquatic systems allows the NO_3^- to serve as an important electron acceptor in the oxidation of organic matter by microbes (Eq. 3 in Table 16–1) in sediments, thereby stabilizing the E_h at ~100 mV (Chapter 16) and preventing the reduction of Fe(III) to Fe(II). Atmospheric SO_2 or SO_4^2 and H^+ deposited on poorly buffered catchments lower the

typically high Fe:S ratio and pH of drainage basins and inland waters, thereby increasing the solubility of Fe, Mn, and trace metals in soils and their export to waterways (Chapter 27).

The rudiments of the Fe, Mn, and S cycles were appreciated by Mortimer (1942), and partially understood even earlier by S. Yoshimura (1931) and associates in Japan, but the gradual quantification of the cycles plus their coupling to many trace metals is largely based on research during the last three decades.

Highlights

- Following their export from drainage basins, the cycling of iron (Fe) and manganese (Mn) in aquatic systems can be conceptualized as "wheels," in which the oxidized forms [Fe(III), Mn(III), (IV)] are precipitated as metal oxides, salt crystals, or coprecipitated with organic matter as inorganic-organic aggregates (flocs). The elements are partly solubilized as Fe(III) and Mn(II) under reducing conditions in anoxic hypolimnia, anoxic wetlands, and in the anoxic portion of the hyporheic zone of lotic systems.

- The resolubilization of reduced iron [Fe(II)], most trace metals (Me), and occasionally Mn(II) under anoxic conditions is low where reduced inorganic sulfur (S^{2-}, HS^-) concentrations are high and insoluble FeS, FeS_2, MnS, and MeS is formed.

- Fe and Mn cycling in aquatic systems is no longer seen as only the outcome of chemical oxidation and reductions. There has been recent recognition of the important direct and indirect roles microbes play in redox reactions.

- ▲ Fe and Mn are of great importance through their physical, chemical, and biological interactions with sulfur, other elements, and organic matter affecting the concentrations of phosphate, nitrate, and most trace metals available to the biota. The very low solubility of Fe under well-oxygenated conditions in euphotic zones occasionally limits algal growth.

- The onset of anoxic conditions and the removal of an Fe oxide diffusion barrier at the oxic–anoxic sediment interface results in high concentrations of soluble Fe and Mn in anoxic hypolimnia and slowly flushed anoxic wetlands.

- Sulfur plays an important role in Fe, trace metal, and sometimes Mn cycling with reduced ionic sulfur forming insoluble precipitates under anoxic conditions, preventing their release into the overlying water.

- ▲ Humans affect the regional cycling of elements through agricultural and industrial activity; by greatly increasing the nitrate, sulfate, phosphate, trace metal, and hydrogen ions deposition on water and their drainage basins; and by allowing increased dissolved oxygen consumption rates in organically enriched waterways.

C H A P T E R

20

Particle Sedimentation and Sediments

20.1   Introduction

A major determinant of the sediments suitability as a physical and nutrient substrate for the biota is its physical and chemical structure. In shallow water, algae grow on sediments and the rooted macrophytes obtain nutrients from them (Chapter 24). Heterotrophic bacteria and invertebrates live on and within the spaces between sediment particles, while fish use sediments for feeding and nesting (Chapter 26). Sediment respiration, dominated by microbes (Sec. 25.7), largely determines whether the sediment surface will be oxic or anoxic and whether the apparent redox potential is high or low (Chapter 16). This has a major impact on the suitability of the sediments for invertebrates requiring dissolved oxygen (Chapter 25), and the extent that the sediments can serve as a source or sink of nutrients and trace metals (Chapters 15 to 19, and 28) to the water column.

The likelihood of sedimented particles plus adsorbed nutrients and contaminants remaining at the site of sedimentation, resuspending and then transporting and depositing elsewhere, or flushing from the system is a function of system morphometry (size, shape, and depth) which determines exposure to winds and wind-induced turbulence (Chapter 12). Lake morphometry (depth) also influences the length of time available for organic particles to decompose in the water column rather than become sedimented. The sediment texture and chemical composition in all inland waters is modified by the sediment biota. However, using the same metaphor as before, the sediment biota operate and interact on a "stage" whose size, shape, and suitability for the "actors" is determined beyond the shoreline and, in lakes, is also greatly affected by decomposition in the water column above the sediments. The preserved remains of some of the "actors" are increasingly providing insights into the history of lakes, quiescent bays of rivers, and wetlands as well as their response over time to changes in climate and human disturbance.

20.2   Origin and Distribution of Sediments

Lake and wetland sediments are usually overwhelmingly derived via inflowing rivers from drainage basins. The soils, the vegetation cover, the drainage ratio (CA:LA ratio), and the catchment slope determine not only the particle supply rate but also the extent to which the supply to aquatic systems is composed of inorganic particles rather than organic matter (Chapter 8).

An important fraction of catchment-derived phosphorus, iron, manganese, and inorganic nitrogen exported to aquatic systems as NH_4^+ travels adsorbed to organic and inorganic particles (Likens 1984, Prairie and Kalff 1988 and Chapters 8, and 17 to 19). There is, therefore, a correlation between the quantities of particulate matter exported from the land and the quantities of phosphorus, nitrogen, iron, and other plant nutrients received by inland waters.

In any climatic or geologic zone, the shape and size of an aquatic system is a function of catchment morphometry (Chapter 7). The water residence time of

Table 20–1   Correlation matrix of organic content (OC), water content (WC), bulk density (BD), maximum depth (z_{max}), degraded chlorophyll-*a* (chl-*a*), lake surface area (LA), predicted inorganic sedimentation rate (ISR$_{pred}$, mm yr^{-1}), and predicted organic sedimentation rate (OSR$_{pred}$, mm yr^{-1}) in the profundal zone of North American lakes. All terms are logarithmically transformed except for WC. $P < 0.001$. Note that the phytoplankton organic matter content (as degraded chl-*a*) is not reflected in the sediment attributes of the generally oligotrophic lakes, implying that the sediment organic matter is overwhelmingly of allochthonous origin and that the within lake primary production is either largely respired in the water columns or overwhelmed by the supply of allochthonous organic matter. With increasing lake area and associated increase in maximum depth ($r = 0.81$) the among-lake organic and water content of the profundal sediments declines ($r = -0.75$ and -0.81, respectively). The organic content of the sediments is best predicted from the inorganic matter sedimentation rate ($r = 0.98$, $r^2 = 0.96$).

	OC	WC	BD	z_{max}	chl-*a*	LA	ISR$_{pred}$	OSR$_{pred}$
OC	—	0.87	−0.86	−0.59	—	−0.75	−0.73	−0.60
WC	0.87	—	−0.88	−0.61	—	−0.81	−0.65	−0.53
BD	−0.86	−0.88	—	0.49	—	0.68	0.73	0.67
z_{max}	−0.59	−0.61	0.49	—	—	0.81	0.27[1]	—
CHL-*a*	—	—	—	—	—	—	—	—
LA	−0.75	−0.81	0.68	0.81	—	—	0.45	0.35[2]
ISR$_{pred}$	−0.73	−0.65	0.73	0.27[1]	—	0.45	—	0.98
OSR$_{pred}$	−0.60	−0.53	0.67	—	—	0.35[2]	0.98	—

[1]$P < 0.05$.
[2]$P < 0.01$.

Source: From Rowan et al. 1992b.

lakes is linked to both lake and catchment attributes, including erosion, so it is not surprising that lake attributes are correlated (covary) with profundal sediment characteristics (Table 20–1). The advantage of covariation is the ability to develop simple empirical models capable of predicting a variety of sediment attributes from easily measured morphometric attributes (Table 20–2). The associated disadvantage is that covariation confounds interpretations of "cause" and "effect."

System Morphometry, Particle Distribution, and Biological Activity

System and catchment morphometry determine where most biological activity occurs in aquatic systems. In shallow transparent lakes, and even more so in streams and wetlands, the biota and its metabolism are overwhelmingly associated with the sediments. Except in highly eutrophic lakes a shallow water column sets limits on the areal planktonic production possible (mgCm^{-2}d^{-1}) facilitates the loss of particles to the sediments but also facilitates resuspension. Conversely, profundal sediment metabolism is low in deep lakes, where a high fraction of the organic matter is decomposed by microbes in the water column or consumed by zooplankton and fish, and relatively few organic particles reach the profundal sediments. This is reflected in a negative relationship between the sediment organic content and lake depth ($r = 0.59$, Table 20–1). Even in deep lakes, a much higher fraction of settling organic matter reaches the shallow water than deep water sediments; this is evident from a much higher shallow-water sediment respiration rate (CO_2 and CH_4 release; den Heyer and Kalff 1998).

Direct evidence for the dependence by the sediment biota on events in the water column include an observation by Lellák (1966) that the biomass of the macroinvertebrate benthos of Czech floodplain ponds was higher in years when the macrozooplankton in the overlying water were reduced, presumably because more food particles reached the sediments. A number of studies have quantified the effect of interannual variation in diatom sedimentation on populations of

Table 20–2 **Regression equations to predict bulk density (BD, g ml⁻¹), water content (WC, % wet wt.), organic matter (OM, % dry wt.), chlorophyll-*a* (chl-*a*, mg m⁻³) and zoobenthic biomass (ZB, g m⁻²) of profundal sediments from sediment load (SL, tonnes yr⁻¹), water residence time (WRT, yr) and sediment retention (SR, tonnes yr⁻¹) for a variety of temperate lakes.**

Equation	R^2	S.E. est.	n
$\log(BD) = 0.00739 \log(SL) - 0.00542 \log(WRT) + 0.0129$	0.78	0.0087	38
$WC = -2.370 \log(SL) + 2.224 \log(WRT) + 96.931$	0.80	2.709	38
$\log(OM) = -0.094 \log(SL) + 0.066 \log(WRT) + 1.628$	0.77	0.114	38
$\log(chl\text{-}a) = 0.204 \log(SR) + 0.170$	0.54	0.231	38
$\log(ZB) = 0.114 \log(SL) - 0.342 \log(WRT) + 0.430$	0.50	0.387	27

Source: From Rowan and Kalff 1991.

benthic invertebrates (e.g., Johnson and Wiederholm 1992). Deeper lakes tend to have lower benthos:zooplankton biomass ratios than shallow lakes (Table 25–4), presumably the result of greater organic particle utilization in the water column.

Small, relatively low-density particles in shallow water are readily resuspended by turbulence and transported by currents in all aquatic systems, but are particularly abundant in the water column of shallow lowland lakes at mid and low latitudes where a high proportion or all of bottom sediments are subject to wind-induced turbulence and resuspension. The current velocities required to resuspend and transport particles in lakes are primarily determined by lake size

(fetch) and wind speed (Fig. 20–1). In deep lakes, fine (low density) particles ultimately settle in the deepest water and are there little subject to resuspension.

▲ Sediment Distribution in Lakes

The process by which catchment-derived sediment particles that enter lakes are sedimented, resuspended and transported toward low-energy sites of permanent deposition in the profundal zone is known as **sediment focusing**. Sediment distribution in deeper lakes can be conveniently subdivided into three zones based on differences in their potential for resuspension. The two principal zones are the

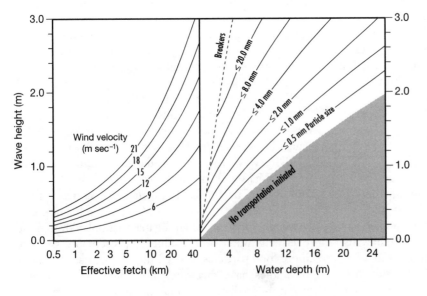

Figure 20–1 The relationship between effective fetch and wave height as well as between wave height and bottom dynamics in wind exposed littoral zones based on work on Lake Vättern (SE). The *effective fetch*, a surrogate for wind energy, is based on the wind reaching a particular location from a 42° angle to each side of the principal wind direction. (*Modified after J. O. Norrman 1964, in Håkanson and Jansson 1983.*)

zone of sediment erosion (ZSE), characterized periodically by high turbulence and dominated by coarse-grained inorganic sediments; and the zone of sediment accumulation (ZSA) a zone of low turbulence dominated by fine inorganic particles (fine silt 2–32 μm; clays < 2 μm diam.) and organic particles of similar low density (Håkanson and Jansson 1983). The two zones are separated by a transition zone, the zone of discontinuous sediment accumulation (ZDA) also known as the zone of transportation, where sediment accumulation is interrupted by rare periods of resuspension and transport during major storm events. While the surficial sediments in this zone are indistinguishable from those of the ZSA, the thickness of this layer is smaller.

Separating the ZSE and the ZDA is a narrow band, normally only a few meters wide, where the transition of coarse to fine sediments takes place. This particular transition zone is known as the deposition boundary depth (DBD) (Fig. 12–3b). The position of the DBD is a predictable function of maximum fetch (F, km) or exposure (km^2), which serve as surrogates for the wave energy (turbulence) experienced, and underwater slope (%) (Rowan et al. 1992a).

$$\text{DBD (m)} = -0.107 + 0.742 \log(F) \quad \text{EQ. 20.1}$$
$$+ 0.0653 \text{ slope}$$

The DBD is found at < 3 m in small wind-protected lakes (F < 1 km) with a shallow underwater slope. In such lakes, the thermocline is sometimes below the DBD (Sec. 11.7). Conversely, the DBD is located at 40 m or more, well below the thermocline in very large temperate lakes (fetch > 200 km).

The annually largest storms determine the depth to which sediment resuspension is possible, but underwater slope affects the stability of the deposited sediments and therefore the DBD at particular sites. Furthermore, a steep slope allows currents to impact the sediments more directly and powerfully than shallow slopes. Therefore, the DBD frequently occurs at a depth greater than is computed on the basis of only wave-energy theory, but is predictable when underwater slope is also considered (Eq. 20.1, and Blais and Kalff 1995). The steeper the slope, the smaller the zone of accumulation. Being able to predict the depth beyond which only fine particles can accumulate (the DBD) is important for determining where plant nutrients, organic matter, and contaminants (primarily adsorbed to fine particles) are deposited and stored (Rowan and Kalff 1991).

Fine Sediment Accumulation in Shallow Water

The profundal zone is not the only site of long-term sediment accumulation in deep lakes. The required conditions are also present in quiescent bays and macrophyte beds. The plant biomass—or better yet, the amount of space occupied by the biomass per unit volume of water, the *biomass density* (Sec. 24.8)—is a good predictor of how much the turbulence is reduced and how much fine sediment can accumulate (Fig. 20–2). Average specific sedimentation rates (mm m^{-2} yr^{-1}) are many times higher in dense macrophyte beds than in the profundal zone (Benoy and Kalff 1999). Thus macrophyte beds permit portions of the littoral zone of lakes and rivers bordered by wetlands to serve as net "sinks" rather than "sources" of particles and associated nutrients and contaminants (James and Barko 1990), allowing them to be highly productive habitats (chs. 21 and 26).

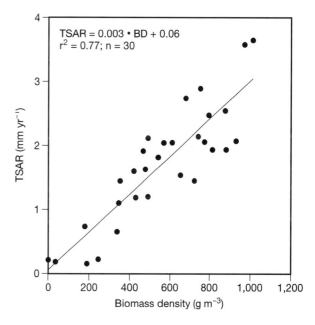

Figure 20–2 The relationship between the biomass density, a measure of how the submerged macrophyte biomass is distributed in the water column, and the total sediment accumulation rate in the littoral zone of Lake Memphremagog (CA, US) (LA = 102 km^2, $\bar{z}$ = 16 m). Note that sediment accumulation is highest at sites characterized by a dense cover of low-growing plants (high biomass density), and lowest where the plants fill little of the water volume above the sediments (Sec. 24.8); and that net sediment accumulation is restricted to sites with overwintering plants because the intercept is not significantly different from zero. (*Benoy and*

The sediments of lotic systems show great variation in particle size as a function of typically enormous spatial and temporal variations in turbulence (Fig. 20–3, and Sand-Jensen and Mebus 1996), which has a decisive influence on sediment accumulation and the distribution of benthic plants and animals (Fig. 20–4). Fine sediments in lotic systems accumulate temporarily at low-energy sites between and behind boulders or within seasonal macrophyte beds, more permanently in bordering wetlands, and permanently in reservoirs.

20.3   Sedimentation and Sediment Traps

Lakes and reservoirs deep enough to have a profundal zone of sediment transport and accumulation serve as traps for fine sedimenting particles. Rates of sediment accumulation interest three different groups of limnologists with different goals who also happen to work at quite different temporal and spatial scales (Fig. 2–6). The group interested in net whole-system accu-

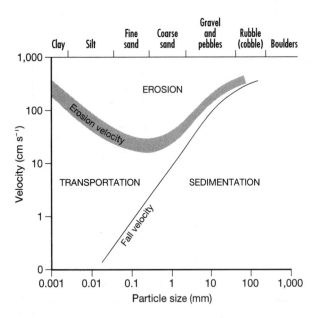

Figure 20–3 Relation of mean current velocity in water at least 1 m deep to the size of mineral grains that can be eroded from a stream bed of material in low turbidity streams. Below a velocity sufficient for the erosion of grains of a given size (shown as a gray band), grains continue to be transported. Deposition occurs at lower velocities than required for erosion of a particle of a given size. Velocities have to be higher to effect erosion in turbid (higher density) systems. (*After M. Morisawa 1968, in Allan 1995.*)

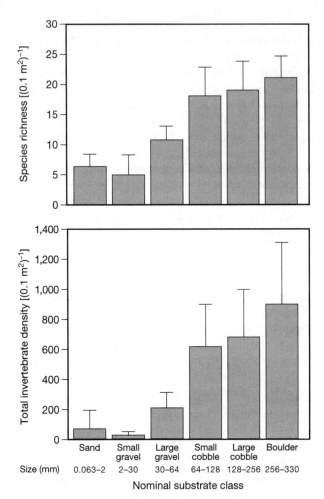

Figure 20–4 Species richness and invertebrate density at different sites in the Mohaka River (NZ) draining pasture land on 2 February 1988, under a narrow range of current velocities (~0.4 m s^{-1}) and depths and under conditions of baseflow. Extensions represent approximately 95 percent confidence intervals. Note, the wide variation in substrate size combined with little variation in environmental conditions prevent a confounding of the effect of substrate size on the biota and effects linked to time and variation in environmental conditions. (*After Quinn and Hickey 1990.*)

mulation rates over seasonal or annual scales usually determines *retention* with mass-balance calculations (input–output). A second group, the paleolimnologists,[1] focus on the long-term net accumulation of par-

---

[1]Paleolimnology is the study of the history of inland waters through sediments. Paleolimnologists are interested in reconstructing and interpreting past communities and environmental conditions from physical, chemical, and biological information contained in sediment profiles from cores collected in the ZSA.

ticles (years to centuries or millenia), using dated sediment cores taken at a limited number of deep water sites (Secs. 20.5 and 20.6). Until recently, more common than either of these studies were short-term sediment trap investigations (days to months, but sometimes carried out over an annual period), for example, of the loss rate of organic matter or species from the pelagic zone, or sediment accumulation rates. Some of the particles may be subsequently resuspended and others respired in the sediments or water column; therefore sediment trap data cannot be equated with *net* accumulation rates.

Traps and Trapping

Sediment traps are suspended singly or in series at one or more depths in the pelagic zone (Fig. 20–5). The containers have traditionally varied greatly in width, size, and shape and were suspended in a variety of ways at a variety of depths for different periods of time. Preservatives were added in some studies to inhibit bacterial decomposition or zooplankton grazing within the traps. Preservatives have their own problems because they can't resolve whether the macrozooplankton that were collected sank or swam into the traps.

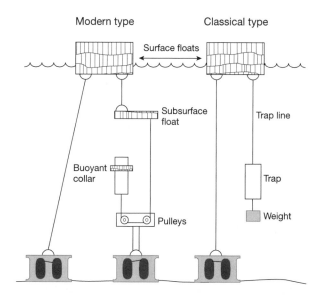

Figure 20–5 Schematic diagrams of a modern and classical type of sediment trap configuration. The modern version is little affected by motions imparted by waves on the surface float or the trap line affecting the quantity and quality of material collected in the trap immediately below. *(Modified after Dillon et al. 1990.)*

Traps designed and suspended differently collect quite different amounts of material. This was not recognized 30+ years ago. Then, sediment traps were used to examine sedimentation rate changes over time or depth in the profundal zone of single lakes; any type of trap permits that. The fact that the measured rates were relative rather than absolute was finally recognized in the early 1980s when interlake comparisons became more common. Laboratory and field studies at that time determined the optimum size and shape of traps. It is now widely accepted that simple tubes without adornments (e.g., funnels on top) with a large *aspect ratio* (height–internal diameter) of > 5:1 (10:1 for turbulent waters) yield the best results. Modern traps are suspended from subsurface floats minimizing the up-and-down movement associated with surface floats and preventing winds from moving the traps out of a vertical position. A flotation collar keeps the traps—suspended at one or more depths—upright (Fig. 20–5, and Bloesch 1996).

The sedimentation rate is determined as follows

$$F = \text{flux (g dry wt m}^{-2}\text{d}^{-1}) \qquad \text{EQ. 20.2}$$

$$= \frac{\text{subsample dry wt. (mg)} \cdot \text{total sample vol. (cm}^3)}{10 \cdot \text{subsample vol. (cm}^3) \cdot \text{trap area (cm}^2) \cdot \text{period (d)}}$$

Sedimentation Patterns

Trap studies carried out in the pelagic portion of oligotrophic dimictic and monomictic lakes exhibit a frequently noted bimodal pattern of sedimentation, with typically highest deposition rates during overturn periods. Sedimentation in temperate lakes is usually maximal in summer, but polymictic lakes have great temporal variation in sedimentation.

Depending on the system examined, seasonal variation in sedimentation can be attributed to variation in (1) phytoplankton production (Chapter 21); (2) allochthonous particles carried by inflowing rivers (Sec. 8.8); (3) calcite precipitation (Sec. 14.6); (4) organic aggregate formation through turbulence and microbial action, with the sedimenting aggregates known as **lake snow**; or (5) resuspension of the bottom material followed by resedimentation.[2] Semiarid zone reservoirs and lakes in crop-growing areas, with exposed soil receiving high sediment input from their poorly

---

[2]"Sedimentation rates measured at 5 m above the bottom at both [Lake Ontario] stations were extremely variable with time and the inshore rates at times exceeded the offshore rates by a factor of 50." (F. Rosa 1985)

vegetated catchments (Sec. 8.8), exhibit peak sediment loads during periods of maximum runoff.

Sediment Resuspension

Even the best-designed traps set in the profundal zone collect much more material on an annual basis than is produced in the euphotic zone and derived from the drainage basin; it has become evident that periodic resuspension and subsequent resedimentation of the previously sedimented material is responsible for greater trapping (*overtrapping*) by profundal sediment traps. The traps collect not only particles sedimented from the euphotic zone (*primary flux* or *new sedimentation*) but also sedimented particles that have been temporarily resuspended and are resettling (*secondary flux*). Profundal traps typically measure *gross* rather than *net* sedimentation on an annual basis. However, during periods of stratification (reduced sediment resuspension) and negligible sediment loading from the catchment, the same traps provide a measure approximating net sedimentation of the then-dominant primary flux of autochthonously-produced particles.

The importance of resuspension first became evident to algal ecologists examining trapped particles under the microscope. For example, Simola (1981)

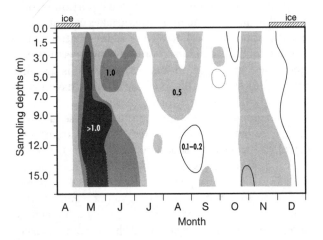

Figure 20–6 Profundal sedimentation rate (g dry wt. m^{-2} d^{-1}) in Lake Lovojärvi, FI (LA = 5.2 ha, $\bar{z}$ = 7.7 m, z_{max} = 17.5 m) during 1974. Fifty percent or more of the diatoms collected in the profundal zone during March, late September, and early December were organisms transported from the littoral sediments to open water prior to sedimentation. The identified shaded isopleths and unshaded areas represent different rates of sedimentation. *(Modified after Simola 1981.)*

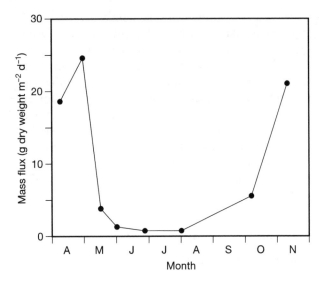

Figure 20–7 Temporal mass flux of particles collected in sediment traps suspended at a depth of 30 m at a 45-m-deep station in southeastern Lake Michigan (US, LA = 58,016 km^2, $\bar{z}$ = 84 m). Note the resuspension and sediment focusing in spring and autumn when the lake is isothermal, and low sedimentation rates during the stratified period when primary production (particle production) rates are greatest. *(Modified after Gardner et al. 1989.)*

noted that the seston collected in pelagic zone traps during the overturn period after ice-out was primarily composed of suspended benthic diatoms (*periphyton*, Sec. 24.9) from the littoral zone rather than phytoplankton (Fig. 20–6). He noted a second smaller peak after a storm during fall overturn. Simola calculated that greater than 50 percent of the organic matter reaching the undisturbed profundal sediments of his small wind-protected (partially meromictic) lake was derived from littoral and allochthonous sources rather than phytoplankton primary production.

Principal sites of resuspension are lake specific. In large, wind exposed Lake Michigan (US), resuspension occurs in both the littoral and profundal zones; the highest resuspension rates are from above ~15 m, and occur during spring and fall overturn periods (Fig. 20–7). However, resuspension is not only a function of turbulence over sediments but also of the properties of those sediments (particle size, shape, and density). Resuspension is reduced by microbial mats—often several millimeters thick—formed over profundal sediments, and by a film of living algae and bacterial filaments in shallow water. Invertebrate activity and fish-feeding reduces the mechanical coherence (cohesion) of these mats, permitting increased resuspension of the sediments.[3] For example, large populations of carp feeding in wetlands uproot large quantities of macrophyte vegetation and can resuspend enough sediment to greatly increase turbidity.

Quantification of Sediment Resuspension

▲ The quantitative importance of resuspended material is rarely known because only a small fraction of sediment-trap studies have had their findings corrected for resuspension. Those that have done so used a variety of sediment tracers, such as the short-lived [7]Be (Berillium, half-life 53.1 d) or the long-lived [137]Cs (Cesium, half-life 30 yr) isotopes. As [7]Be is quickly removed by decay, its concentration is much lower in older sediments than in fresh material. Hence, resuspended particles are characterized by low [7]Be. Conversely, ^{137}Cs is accumulated in sediments and resuspended sediments have a higher activity than newly sedimented particles (see Sec. 20.6). Other tracers used to determine resuspension are based on the difference between the stable carbon isotope ratio of plankton to sediment-derived particles, as well as on differences in sediment and plankton pigment ratios for example, and the organic matter or aluminium content of resuspended sediments versus that in primary sedimenting material.

The importance of resuspension has been clearly demonstrated by a comparison of sequential sediment trap data with whole-lake mass-balance (input–output) determinations of the fate of nutrients (Chapter 17). A four-year mass-balance comparison made on a small Precambrian Shield lake makes the point (Dillon et al. 1990). Gross sedimentation rates of phosphorus, nitrogen and iron were, respectively, 17, 3 and 31 times greater than those calculated from whole-lake input–output measurements. But mass-balance determinations are not error free either (Sec. 5.2), their accuracy depends on the quality of the discharge data and the frequency and quality of the nutrient determinations. However, it is unlikely that the mass-balance work by an expert team of scientists missed more than 90 percent of the incoming phosphorus and iron.

---

[3]"In most lakes, the hypolimnetic sediments are cohesive, not only because of the dominant small [and cohesive] particles, but also because of various biological processes. For example, the sulfur bacterium *Beggiatoa* can form large mats in the transition zones of low oxygen concentrations in eutrophic lakes (). Further, benthic activities (e.g., tubes formed by oligochaetes and chironomids) may consolidate large parts of the sediments." (Bloesch 1995)

20.4 Sinking Velocities and Sedimentation Rates

An accurate estimation of rates of sedimentation loss is very important for determining the rate of nutrient and contaminant removal from the water to the sediments. It is equally important in determining the time available for algal cells and other organic particles to decompose, thereby permitting recycling of nutrients and contaminants in the water column. This, in turn, determines the rate at which organic particles are deposited onto the sediment surface for subsequent consumption, microbial decomposition, or storage.

The rate at which organic particles derived from the euphotic zone are deposited on the sediment surface ($g\,m^{-2}\,d^{-1}$) along with particle quality is a deciding factor in determining the abundance and activity of sediment microbes (Sec. 22.11), dissolved oxygen concentrations, and suitability of the sediments as a habitat and feeding site for invertebrates and their fish predators (Chapters 25 and 26).

▲ The Sedimentation Process

Although the sedimentary process is theoretically well understood, there remains a need for good data to model water-column and sediment processes, including horizontal variation and changes in sedimentation with depth. The paucity of data together with temporal changes in particle size and density plus methodological problems with traps, confound easy interpretations and the development of good predictive models.

It may seem surprising that sedimentation takes place at all in turbulent environments. In 1851, G. G. Stokes examined the settling rate of small (< 500 μm), nearly spherical inorganic particles in nonturbulent water and articulated an equation that has guided the thinking of limnologists and oceanographers about algal sedimentation. The Stokes equation describes settling rate as a function of particle density, particle radius, plus the density and viscosity of the water. The effect of radius on the settling of organic particles is evident from Table 20–3 and explains, for example, why lakes and reservoirs receiving large quantities of clay-sized particles during periods of high surface runoff remain turbid for long periods unless rapidly flushed.

The modified Stokes equation (Hutchinson 1967) for algae is:

Table 20–3 **Settling rate of inorganic particles of the same density and different diameters under laminar-flow conditions.**

Particle Type	Particle Diameter (μm)	Settling Rate (m d⁻¹)
Fine sand	125–250	950–2,246
Very fine sand	62–125	225–950
Coarse silt	31–62	57–225
Medium silt	16–31	16–57
Fine silt	8–16	4–16
Very fine silt	4–8	1–4
Coarse clay	1–4	0.1–1
Fine clay	0.1	0.001

Source: Largely after R. A. Ferrara and A. Hildick-Smith 1982, in Cooke et al. 1995.

$$V = \left(\frac{2}{9}\right) \frac{g \cdot r^2 (D - D^1)}{n \cdot \phi} \qquad \text{EQ. 20.3}$$

Where V = sinking velocity, g = gravitational acceleration (980 cm sec⁻¹), r = radius of a sphere of identical volume to the algae, D = density of the algae, D^1 = density of the water, n = coefficient of viscosity of the water, ϕ = coefficient of form resistance, a measure of cell deviation from a sphere of the same volume and density. Equation 20.3 is widely used to predict particle sedimentation rates, but strictly applies only under nonturbulent (laminar) conditions. Sinking fluxes occur even though the mixed layer is largely turbulent, resulting from laminar flow or nearly laminar conditions existing at the interface between the epilimnion and metalimnion and at the sediment–water interface, allowing settling particles to escape a turbulent water column in lakes and rivers. The loss rate is a function of the number of particles and their probability of entering the nonturbulent layer. The probability of sedimentation increases with increasing settling rate (excess density) and decreasing thickness of the mixed layer (Eq. 21.4). Consequently, sinking fluxes are the product of particle concentration and particle sinking rates just above a boundary layer. Sinking fluxes in the mixed layer are increased upon flocculation of dissolved organic matter, and coagulation of small particles following their collision, thereby increasing particle size (density) and sinking rates of the lake snow produced (Weilenmann et al. 1989).

Sinking Rates in Nature

Estimated sinking velocities based on the Stokes equation suggest that 100 μm spherical and nonspherical organic particles (form resistance factor 1.5; excess density 0.05) at 4–10°C sink at about 17 m d⁻¹ and 11 m d⁻¹, respectively, in nonturbulent water (compare with Table 20–3). Measured sinking velocities in nature of (1) rapidly sinking large diatoms (Chapter 21); (2) crystals of calcite (Chapter 14); (3) fecal pellets of copepods or large (3–20 mm) organic aggregates; and (4) inorganic-organic aggregates, such as iron and manganese oxides plus sorbed organic matter and trace metals (Chapter 19), are roughly similar to those based on the Stokes equation for particles of similar size and density ($\geq$ 10 m d⁻¹). However, most organic and inorganic particles in inland waters are smaller than 10 μm.

Both theory and empirical observation of metalimnetic and hypolimnetic settling velocities indicate the sinking rate of 10 μm particles is typically ~0.25 m d⁻¹ (also Sec. 21.6) for particulate organic carbon (POC), particulate phosphorus and nitrogen, and pigments. At that rate, a 10 μm particle takes 40 days to sink 10 m and 200 days to settle 50 m, leaving ample time for decomposition and changes in the C:P and C:N ratios of the particles (Baines and Pace 1994).

The sinking velocity of larger phytoplankton (> ~70 μm) varies greatly, depending on whether the organisms are able to control their buoyancy and on their physiological condition (Sec. 21.6). Large diatoms may have sinking velocities as low as those of a 10 μm particle during periods of rapid growth, but they typically sink as rapidly as > ~6 m d⁻¹ during senescence and following death (Horn and Horn 1993). In contrast, larger particles (Hutchinson 1967), including zooplankton fecal pellets may sink > 100 m d⁻¹. Macrozooplankton contribution to sedimentation is not only direct through sinking fecal pellets or aggregates and dead bodies but also indirect by predation on small algae (Chapter 23), sometimes causing a shift to larger (inedible) forms with a higher sedimentation rate (Bloesch and Bürgi 1989).

Daily sinking losses of seston from the epilimnia range from 0.5 percent to 20 percent, suggesting that half the particles are lost from the mixed layer over 10–200 days (Baines and Pace 1994). The wide range is attributable to differences in size and type of phytoplankton and other particles available for sedimentation (Bloesch and Sturm 1986), and to differences in mixed-layer thickness.

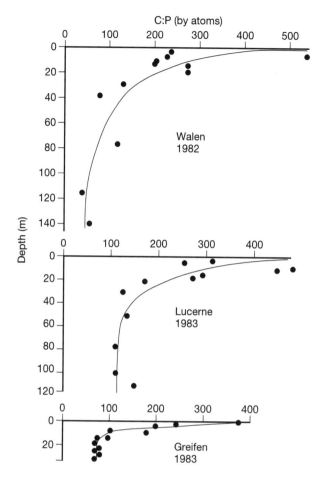

Figure 20–8 Average particulate C:P profiles in the water column of three Swiss lakes during the period when sestonic C:P ratios were maximal. *(After Gächter and Bloesch 1985.)*

Nutrient Recycling

Hypolimnetic traps set at different depths show a progressive decline with depth in organic matter trapped as the result of decomposition—primarily by bacteria attached to the sinking particles (Fig. 20–8). Not only do the attached microbes thereby lower the C:P ratio by respiring some of the carbon, bacteria also acquire phosphorus from the water when the particles have either a higher seston C:P ratio (Uehlinger 1986) or higher N:P supply ratio than the equivalent bacterial demand ratio for growth (Sec. 22.5). Laboratory work has shown that under phosphorus-limiting conditions, phosphorus is acquired and retained by bacteria on sinking particles whereas excess carbon and nitrogen are liberated. Conversely, sinking particles are dispro-

portionately rich in nitrogen whenever the microbial biota is primarily nitrogen limited (high C:N) (Tezuka 1990).[4]

Only a modest fraction of the particles produced in the epilimnia leave the mixed layer by sedimentation. Comparisons of primary production rates in epilimnia with trap-measured losses of particles and their nutrients confirm that epilimnia are of great importance for nutrient recycling and retention (Baines and Pace 1994). In the primarily small oligotrophic and mesotrophic lakes that were examined, between 15 percent and 30 percent of the primary production and phosphorus was typically lost by sedimentation from the epilimnia. In other words, 70–85 percent of the primary production is decomposed and the nutrients are recycled in the mixed layer, not lost to sedimentation. Rapid nutrient recycling means that only modest quantities of nutrients need be supplied from beyond the shoreline (Chapter 8) or returned from the hypolimnia by entrainment (Chapter 12) to maintain epilimnetic production in nutrient-poor lakes. The rapid recycling of autochthonously produced particles in epilimnia and the possibility for decomposition in the hypolimnia implies that only a minor fraction of euphotic zone production, dominated by large particles (Bloesch and Uehlinger 1990), reaches the profundal sediments in deep lakes to nourish microbes (Chapter 22) and sediment invertebrates (Chapter 25) living there. The extent to which the metabolism of the profundal benthic biota of oligotrophic lakes is dependent on the supply of terrestrially produced particles is not well resolved in lakes, but is known to be important in forest streams (Sec. 8.2).

The relative importance of the mixed layer in recycling organic and plant nutrients is reduced in shallow unstratified lakes where particles are rapidly sedimented on calm days only to be partially resuspended during periods of turbulence. The distinction between planktonic and benthic processes is nebulous in shallow waters.

---

[4]Laboratory work under controlled (simplified) conditions is of great importance, not only to discern underlying mechanisms and processes that provide plausible or probable explanations for observations made under uncontrolled conditions in nature but also to provide new hypotheses (theories) that can be explored in nature, and to determine the physiological limits beyond which organisms cannot function.

20.5   The Sediment Record

Profundal sediments in the zone of sediment accumulation (Sec. 20.2) contain an invaluable record of past conditions. Lakes, wetlands, quiescent bays and floodplains of rivers, and the sediments of former—dessicated or infilled lake basins (*paleolakes*)—contain a system history. Paleolimnologists can infer a great deal about the past when sediments collected with coring devices are examined and dated.

Coring Devices

Most coring devices are either *gravity corers* or *piston corers*. Gravity corers, widely used for collecting cores in soft, relatively recent sediments, consist of a weighted tube with a valve mechanism that closes at the upper end once the core is taken, preventing loss of the core when the device is hauled to the surface (Fig. 20–9). Piston corers are designed to take rela-

tively long cores (several meters) when the core tube is driven into the sediments. A sliding piston inside the tube moves up, eliminating the hydrostatic pressure created while the core tube is filled, and offsetting the tendency of sediment to be pushed down as the core tube is driven in. A third device, used to collect sediments with a high water content, is not a tube. It is a probe filled with dry ice and alcohol (~ −70°C) that freezes the adjacent sediment before the *freeze sampler* (with adhering sediment) is pulled to the surface for subsampling. Cores and freeze samples are subsampled—usually by slicing them into layers—to isolate specific time periods (Smol and Glew 1992). Deep-coring ancient lakes involves an offshore drilling platform and coring device similar to those used to drill for oil in the sea. A 1,400 m-long core taken in Lake Biwa (JP) has brought three million years of the history of the lake and region to light (Horie 1987).

Figure 20–9   A sediment gravity corer with a core visible in the transparent core tube. (*Photo courtesy of J. Smol.*)

Transfer Functions and Hindcasting

The single greatest advance made by paleolimnologists in recent decades—and one of the most important advances in limnology as a whole during that period—was the development of *transfer functions* allowing the fossil record to be quantified and interpreted in terms of important environmental variables such as pH, total phosphorus, salinity, DOC, aluminium, trace elements, temperature (climate), and profundal zone DO. To accomplish this, a *calibration data set*, also known as a *training* or *reference data set*, is created for a series of waters (usually > 40) in a particular geographic region. This is done by first quantifying the biological indicators under study (e.g., diatoms, chrysophytes, invertebrates, or fish scales) in recently deposited sediments (usually the top 1 cm) and statistically linking the abundance of organismal remains to the present-day chemical or physical variable (Fig. 20–10). In the second step, equations based on the quantitative relationships currently observed are used to infer past conditions (*hindcasting*) based upon abundance of the same species in dated cores. It is assumed that the relationship between species composition and environmental conditions in the calibration data set is constant over time and relevant to conditions when the fossils were deposited.

▲ Hindcasting saw major progress during the 1980s when the ability to distinguish between lakes and wetlands that had been acidified as the result of upwind

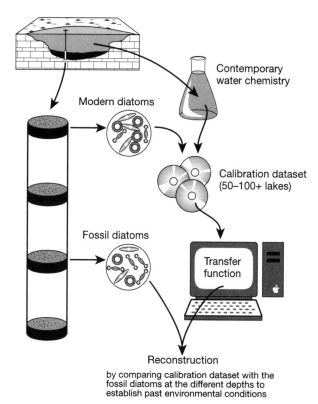

Figure 20–10   Outline of steps to be taken to create a calibration dataset, and its use in reconstructing the environmental conditions encountered by, for example, fossil diatoms in dated sediment cores. *(After Hall and Smol, 1999.)*

sources of acidic emissions and those that were naturally acidic (Sec. 27.3) became of great importance. Other major paleolimnological advances made during and since this period greatly enhanced our understanding of the history and development of lakes and wetlands. These advances included improved techniques for coring, sediment dating (Sec. 20.6), statistical analyses, and a clearer understanding of the environmental factors controlling organism and sediment accumulation. These improvements have allowed major insights into past physical, chemical, and biological conditions and helped debunk an old notion that lakes change systematically from more oligotrophic to more eutrophic during their developmental history. They also showed the importance of long-term fluctuations in climatic conditions (decades to centuries or longer) on inland waters and their biota; data that are difficult to derive from typical short-term research projects in limnology (see Fig. 20–11).

Successful attempts to reconstruct the historical acidity of lakes were followed by equally important reconstructions of past trophic and climatic conditions, salinity, temperature, dissolved oxygen concentrations in hypolimnia, changes in contaminant emissions and land use over time and even the historical population size of salmon stocks returning to lakes to spawn and die (Finney et al. 2000). The preserved mouthparts of certain *Diptera* (Insecta) species, varying in sensitivity to hypoxic and anoxic conditions, have been used as indicators of DO conditions in the past, and also reveal the importance of fish predation. The most commonly used organisms are diatoms, most of which preserve well. Others include chrysophyte algae (Chapter 21), cladocerans and some rotifers, chironomids and chaoborids (Chapter 23), and mollusc shells (see Anderson 1993, and Charles and Smol 1994). Pollen, spores, and the seeds of aquatic plants have been used in studies of water-level and climate change, but rarely for water chemistry reconstruction.[5]

Diagenesis

▲ Reliable dating of the sediments is essential for inferring past limnological conditions and sediment accumulation rates. Stratigraphic markers, such as sediment layers deposited at a known time, are not always present and sedimentation rates are often so small (< 1 mm yr^{-1}) that even core slices one or two centimeters thick encompass more than 10 or 20 years (Table 20–4). There are still other problems even when sediment markers are available. One major difficulty is *diagenesis*, postdepositional changes in any sediment variable by any mechanism, including the poor or differential preservation of organisms which confounds what is observed (Sec. 19.2).[6] Gases produced

---

[5]"While many limnologists are, perhaps, still unaware of the possible benefits of the sediment record, paleolimnologists also must accept the temporal and ecological resolution of their results and, hence, its limitations. For paleolimnologists, it is important then that they match the interpretation of the biotic remains in the sediments to the temporal resolution of the profile and the dominant ecological processes that are observable or relevant at that time scale." (N. J. Anderson 1995)

[6] • "Differential preservation of the organism remains is a major problem . . . particularly hampering comparisons between different lakes or even between different levels within a single sediment sequence." (H. Simola 2000)

• "In paleolimnology . . . diagenesis is at best an annoyance and at worst a nightmare, weakening nearly every generalization that investigators may wish to make." (M.W. Binford et al. 1983)

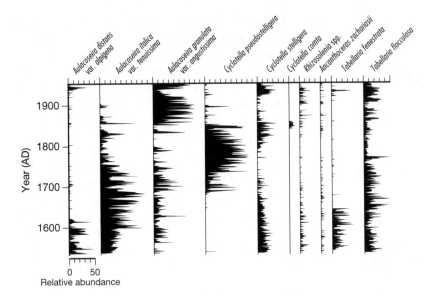

Figure 20–11 Year-to-year stratigraphy of the relative abundance of 10 plankton diatoms in Lake Lovojärvi, FI (LA = 0.05 km^2, $\bar{z}$ = 7.7 m) from 1539 to 1956 obtained with a freeze sampler. Note that (1) there are species which have maintained their frequency in the community throughout the 418 year period (*A. zachariasii*); (2) other species have changed in frequency, replacing each other as the most dominant species in the community (*A. italica* var. *tenuissima*, *A. granulata* var. *angustissima*, and *C. pseudostelligera*); and (3) this replacement may occur gradually (*A. italica* var. *tenuissima* to *C. pseudostelligera*) or rapidly (*C. pseudostelligera* to *A. granulata* var. *angustissima*). (*After Simola et al. 1990*.)

Table 20–4 **Net accumulation rates in the profundal zone of selected lakes, determined by radioisotope dating of the sediments. Ranges probably represent either sites differing in underwater slope or cores taken from within the zone of discontinuous sedimentation. Variation in the mg cm^{-2} yr^{-1}:mm yr^{-1} ratios probably reflect differences in the water content and thus in the organic or inorganic nature of the sediments at the different locations. ND = no data.**

	Accumulation Rate	
Lake	**mm yr^{-1}**	**mg m^{-2} yr^{-1}**
Baikal, south basin (RU)	~0.1–0.15	ND
Superior (CA, US)	0.1	7–25
Tanganyika (KE, TZ, UG)	0.3–0.5	ND
Hidden (AU)	0.2–1.7	ND
Lagodiville (FR)	0.7	13
Michigan (US)	0.1–4.0	10–20
Tahoe (US)	1.0	21
Huron (CA, US)	1.0–1.1	21 and 51
Shinji (JP)	1.2–2.7	13–71
Pavin (FR)	1.3	13
Windermere (GB)	2.4	60
Washington (US)	3.8	ND
Macquarie (AU)	1.0–6.0	ND
Mendota (US)	6.0	18
Tansa (IN)	~40	~2,800

Source: Krishnaswami and Lal 1978, Johnson 1984, and Timms 1992.

during decomposition, such as methane, diffuse and bubble upward when cores are raised to the surface, potentially disturbing the sediment distribution. Burrowing animals also disturb the upper sediments, a process known as *bioturbation* (Sec. 17.3), making it possible for microfossils, algal cells, and zooplankton resting eggs to be moved to layers other than where they were originally deposited. Furthermore, the possibility that some material originally deposited in a zone of discontinuous sediment accumulation was resuspended long after the original deposition and then added to the sediment profile in the zone of sediment accumulation can cause difficulties in stratigraphic analyses.

The evaluation of core slices is highly time consuming and few studies collect more than one from the deepest portion of a lake, assuming the single core to represent fairly the whole zone of sediment accumulation (see Battarbee 1991). Yet it is evident, from the advances in paleolimnology based on transfer functions and single cores, that the effect of these uncertainties have been sufficiently small to avoid confounding patterns at the among-lake scale. For example, among lakes varying greatly in pH, it is apparent that the effect of pH on the biota at a particular interval scale (Sec. 2.6) is much larger than the variation imposed by uncertainties. But empirical patterns based on many water bodies, exhibiting considerable data scatter around the line of best fit, make for much uncertainty when used to predict the behavior of any one individual system in both paleolimnology and eutrophication research (Sec. 21.14). Minor problems at one scale may be major at another.

▲ 20.6 Dating Sediments

The known decay rate of particular long-lived radioactive isotopes released into the atmosphere allow them to serve as useful markers for dating sediments. The isotope returns to the earth's surface by precipitation or adsorbed to dust (aerosol) particles. Once the marker enters a lake or wetland, the isotope is either sorbed to or incorporated into particles and transported to the sediments. The age of a sediment core slice is determined by comparing an isotope's current specific activity (activity per unit mass) with the value recorded at a specific depth in a core. The isotope's decay rate (or half-life: the time required for radio activity to decay in half) must be known, and a constant isotope supply rate and no diagenesis is assumed (Krishnaswami and Lal 1978). When dating is done at several depths in a core, it is possible to calculate sedimentation rates over time.

One isotope in wide use is radioactive carbon (^{14}C), with a half-life of 5,730 ±40 years. The long half-life makes it particularly useful for dating organic matter deposited as far back as 40,000 years BP. But as the radioactive decay is only 1.25 percent over a 100-year period it is too small for accurate dating of recent (< 400 yrs) sediments.

Lead-210 as a Dating Tool

The isotope of choice for dating recent sediments (deposited during the last ~150 yrs of unprecedented human impact) is lead 210 (^{210}Pb), a natural isotope released into the atmosphere through radioactive decay of radon-222 (^{222}Rn) gas escaping from the earth's crust.[7] Decay (daughter) products of ^{222}Rn, including ^{210}Pb, are deposited on aquatic systems by precipitation or dry-fall, providing a continuous flux of ^{210}Pb onto land and water. Upon entering a lake or wetland, the isotope is quickly adsorbed to particles and is accumulated within weeks in the sediments where it decays with a half-life of 22.3 years. As pointed out above, recent sediment profiles can be dated by comparing the

^{210}Pb activity at depth relative to that at the surface, assuming no postdeposition changes. However, the background or *supported* ^{210}Pb from catchment-derived ^{222}Rn must be subtracted first, leaving the atmospherically derived *unsupported* ^{210}Pb for dating (Krishnaswami and Lal 1978, Battarbee 1991, and Schelske et al. 1994).

Biological and Chemical Changes over Time

When several layers within the recent sediment profile are dated it becomes possible to determine changes in sedimentation rates over time (Table 20–4), allowing estimates of short-term and long-term changes in the abundance of aquatic species whose remains were preserved in the sediments (Fig. 20–11). Similarly, changes in contaminant deposition rates over recent decades (Sec. 28.5), changes in the pH of lakes or wetlands during the last century and a half (Sec. 27.2), or past climatic conditions reflected in either the identifiable pollen grains of terrestrial plants or the remains of the aquatic biota deposited and preserved in the sediments, can be determined.

A recent reconstruction of drought patterns in the Great Plains region of North America was based on an estimation of the salinity optima of modern diatom assemblages preserved in the *surficial* sediments of 53 local lakes covering a wide salinity gradient. Then a transfer function (Sec. 20.5), linking present salinity and modern diatom assemblages, was developed and used to infer past salinities based on the diatom assemblages observed in dated cores (^{14}C, ^{210}Pb) from a single lake, Moon Lake (US). The results indicated that the extreme droughts of the 1930s and the 1890s (the Dust Bowl periods) were not exceptional (Fig. 20–12). Extreme drought periods were more common before 1200 AD, probably because the same atmospheric circulation patterns that produce drought conditions today were more frequent and persistent (Laird et al. 1996).

Cesium-137 and Other Markers

Another useful marker is the radioisotope cesium-137 (^{137}Cs), deposited after the onset of atmospheric nuclear bomb testing in 1954. The principal marker is the peak fallout recorded during 1963 when testing was at its height, after which fallout declined rapidly to background levels. The 1986 explosion of a nuclear powerplant at Chernobyl (UA) followed by the release and widespread distribution of ^{137}Cs over parts of Europe, along with other isotopes, provides a recent

---

[7]Radon-222 is a noble gas in the decay chain of Uranium-238, which has a 4.5×10^9 yr half-life and was present during the formation of the earth. For the complete decay series see Krishnaswami and Lal 1978.

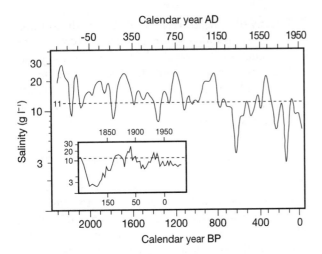

Figure 20–12   A 2,300 year record of the average salinity of saline Moon Lake (North Dakota, US), inferred from the relative abundance of particular diatom taxa in the upper 4 m of a core taken at a ~13 m deep site. The inset expanded scale shows average salinity changes over the last two centuries. Note the wide divergence in chemical conditions (~3–30 g l^{-1}) from the long-term average of 11 g l^{-1}. The fluctuations in salinity, runoff, nutrient loading, lake area, and depth must have affected all aspects of the lake biota. Also note that the smoothed line hides much additional intra-year and among-year variation. (*Modified after Laird et al. 1996.*)

sediment marker in the affected countries. The 1963 ^{137}Cs peek is readily detected in north temperate zone lake cores, and its half-life is long enough (30 yrs) to remain detectable today. It therefore provides a marker for an examination of events during the last 50

years of greatest human impact on inland waters and their catchments (Fig. 20–13). Recent work shows ^{137}Cs to be useful only in higher alkalinity (moderate to high conductivity) lakes, where postdepositional mobility is negligible. Post-depositional mobility prevents it from being useful in low HCO_3^- (low conductivity) lakes or wetlands, including those impacted by acidifying precipitation (Blais et al. 1995).

Simpler, less expensive core markers than radioactive tracers include (1) clearly defined sediment layers deposited during a recorded massive storm and resulting sediment runoff or the construction of a road for which records are available, (2) historically dated mountain slides, volcanic eruptions, or forest fires, (3) the rapid rise in *stable lead* (^{208}Pb) in cores associated with the onset of coal burning at the time of industrialization, (4) the first recorded presence of ragweed pollen (*Ambrosia spp.*) in north temperate zone sediment profiles representing the recorded time of forest clearing for agriculture purposes, allowing the weed to invade (Blais et al. 1995), and (5) the first appearance in sediments of specific chemicals, for example detergents such as surfactants or fluorescent whitening agents, whose date of introduction is known. Each marker provides a single date.

Prior to advances made during the last three decades, at a time when studies of single lakes were the rule, paleolimnologists worked primarily on the anoxic profundal sediments of meromictic lakes where sediments most clearly exhibit annually deposited laminations, known as *varves*. Lengthy periods of sediment anoxia and the absence of bioturbation by invertebrates requiring dissolved oxygen help prevent

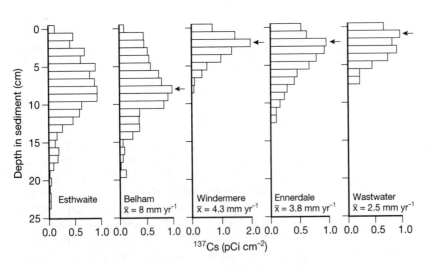

Figure 20–13   Depth profiles of ^{137}Cs deposited in five English lakes that were cored between 1970 and 1973. Note (1) the average sedimentation rates range threefold in one lake district from 2.5 mm yr^{-1} in Wastwater to 8.0 mm yr^{-1} in Blelham; (2) the arrows mark the year of maximum ^{137}Cs fallout; and (3) the apparent mobility of ^{137}Cs in Esthwaite water cores, eliminating its use as a date marker. (*Modified after Pennington et al. 1973.*)

disturbance (diagenesis). Varved sediments are particularly suitable for historical monitoring studies because they provide direct calendar dating, based on a pronounced seasonality in the type of material that is sedimented (Fig. 14–7). In paleolimnology too scientists differ as to what they consider accomplished and accomplishable (e.g., Smol 1990 and Battarbee 1991).

▲ ## 20.7 Profundal Sediment Characteristics

There is debate about whether the large among-lake variation in organic matter content (percent of dry weight) and water content (percent of wet weight) of profundal sediments is the outcome of differences in autochthonous primary production or of differences in catchment-derived organic and inorganic sediment supply rate. Inorganic sediments dilute the percent organic matter content (Rowan et al. 1992b). Evidence exists for both scenarios. Productive lakes in areas of low inorganic sediment loading also experiencing relatively low inputs of allochthonously derived organic matter exhibit surficial sediments dominated by autochthonously produced organic matter (Rybak 1969, and Pennington 1978). Conversely, allochthonously derived organic matter dominates the sediments of oligotrophic temperate lakes (Wissmar et al. 1977) which as a result do not show a link between water and sediment concentrations of algae pigments (degraded chl-*a*, Table 20–1). Among oligotrophic lakes, the largely allochthonously derived sediment *organic matter content* (percent of dry weight) is primarily determined by considerable variability in the supply rate of inorganic sediments from drainage basins, diluting the organic matter (Tables 20–1 and 20–2).

Sediments and Nutrient Cycling

The fate of plant nutrients captured by recently sedimented particles and their attached bacteria depends on the sediment environment. For example, phosphorus is typically strongly retained by aerobic profundal sediments. In eutrophic lakes, phosphorus and DOC adsorbed to Fe(III) flocs (Chapters 17 and 19) or sedimented $CaCO_3$ crystals (Sec. 14.6) is subject to release to the overlying water when anoxic conditions develop, the result of high rates of sediment microbial respiration (Sec. 17.2). Additional nutrients are released by the microbes themselves (Sec. 17.3). In the case of nitrogen, microbially mediated denitrification at the oxic–anoxic interface of well oxygenated littoral and profundal sediments is responsible for the loss of nitrogen to the atmosphere from both lakes and wetlands (Sec. 18.4), thereby lowering the sediment N:P ratio and raising the C:N ratio above that of arriving particles (Dillon et al. 1990). It is evident that understanding nutrient cycling between sediments and water requires information on the physics (turbulence, diffusion rates), chemistry, and metabolism of the biota.

Highlights

- The physical and chemical structures of sediments are the principle determinants of their suitability as a source of nutrients to the sediment biota, the overlying water, and as a physical habitat for the biota.
- In shallow, transparent lakes and even more so in streams and wetlands, the biota is overwhelmingly associated with the sediments. A shallow water column can limit the areal planktonic production and facilitates both the loss of particles to the sediments and their resuspension.
- ▲ Deeper lakes exposed to wind are characterized by two principal sediment zones, a shallow water zone of sediment erosion (ZSE) dominated by coarse (dense) particles, and a largely nonturbulent deeper zone of sediment accumulation (ZSA), characterized by fine particles received from the overlying water and the ZSE. Quiescent bays and macrophyte beds are shallow water areas where fine sediments and their adsorbed nutrients accumulate.
- Sediment traps continue to be widely used in lakes for measuring the loss rate of particulate matter and planktonic algal species from the mixed layer to the hypolimnia and sediments.
- The rate at which particles are deposited on the sediment surface, together with organic particle quality, is of deciding importance in determining the productivity of the sediment biota in all inland waters.
- ▲ While the sinking velocity of large (dense) particles is high (≥ 10 m d^{-1}), most particles are small (< 10 μm) and sinking velocities are low, allowing considerable time for decomposition and nutrient recycling within the mixed layer. Only a modest fraction of autochthonously produced organic particles leave the mixed layer through sedimentation.

- ▲ A variety of radioisotopes are available for dating sediment cores, with ^{210}Pb the isotope of choice for dating sediments deposited during the last ~150 years of unprecedented human impact. Simpler, less expensive core markers—providing only a single date—include the deposition of an identifiable sediment layer or a pollutant whose date of introduction is known.

- ▲ Paleolimnological advances during recent decades allow the sediment fossil record to be quantified and interpreted in terms of environmental variables such as pH, total phosphorus, salinity, DOC, temperature, hydrology, and profundal dissolved oxygen concentration.

21

The Phytoplankton

21.1 Introduction

The algae in the open water of lakes and large rivers—the **phytoplankton**—are by far the best-studied of all biotic groupings in inland waters and oceans. Tens of thousands of papers have been published on the physiological ecology and physiology of phytoplankton species in the laboratory and field, the seasonal dynamics of individual species, and the apparent causes of their waxing and waning. The community structure and the productivity of phytoplankton assemblages in relation to environmental factors and biological interactions have also received a great deal of attention. An enormous amount is now known about phytoplankton, and for a modest number of species some broad generalities are possible at the species level in nature. Yet, predictions at the species level are hampered by the hundreds of species present in individual waters, most of whom have received little study, and the host of possible interactions among them. Predictions are further hampered by interactions between species and the changing physical/ chemical environment, and by interactions between the phytoplankton and their temporally and spatially changing predators. Even so, well-studied temperate zone lakes tend to have both a recurring and characteristic assemblage of dominant species at particular times of the year. Such seasonal regularities continue to stimulate phytoplankton ecologists to examine the mechanisms, processes, and environmental factors determining the composition of the communities observed (Fig. 21–1). However, the apparent regularity of the biomass cycle and dominant species in individual systems is confounded in some years by large shifts in both composition and timing of biomass peaks, providing a reminder of the difficulties that must be surmounted before predictions can be made about individual species.

Algal Classification

The phytoplankton community is composed of several groups of algae and one major group of photosynthetic bacteria, the **cyanobacteria**. The latter were known and classified as the *Cyanophyta* or **blue-green algae** by algal systematists before it was widely appreciated that the group is much more closely related to the bacteria. But, since photosynthetic cyanobacteria functionally resemble the remaining phytoplankton groups, they are typically discussed as a component of the phytoplankton rather than of the true bacteria.

The traditional classification of major groups of aerobic (oxygenic) and photosynthetic plants based on morphology or pigmentation is increasingly giving way to one based on evolutionary (molecular-based) relationships.[1] This began with the recognition that blue-green algae, Division *Cyanophyta*, belong to the Empire Bacteria (*Prokaryota*) and not to the Empire *Eukaryota*. The prokaryotic cyanobacteria lack,

---

[1] The hierarchy of biological classification is basically as follows: Domain, Kingdom, Subkingdom, Phylum (animals and bacteria) or Division (plants), Subdivision, Class, Order, Family, Genus, Species, Subspecies (variety).

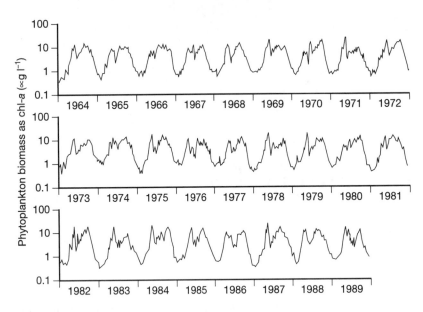

Figure 21–1 The phytoplankton biomass in the euphotic zone (1 m) of the North Basin of Lake Windermere, GB (LA = 8 km², z̄ = 25 m) sampled weekly or bi-weekly (in winter) from 1964 to 1989. Note that (1) the data are plotted on a logarithmic scale and the seasonal and inter-year variability is much higher than immediately apparent; (2) there is a regular seasonal pattern, albeit with appreciable inter-year variation; and (3) there is both a spring and a late summer biomass maximum in most years. *(After Talling 1993.)*

among other features, both a distinct membrane-bound nucleus and distinct pigment-containing organelles (plastids). The remaining (eukaryotic) algae all share membrane-bound nuclei and plastids. The eukaryotic algae are composed of a number of often distantly related Divisions; their taxonomy is complex and has been treated in various ways by systematists. However, there are only six Divisions commonly encountered in inland waters (Sec. 21.2). A seventh—the *Rhodophyta* (red algae)—can be important, mostly in oligotrophic streams where they grow attached to the substrate. The different major groups of algae have traditionally been recognized at the Division level (e.g., *Chlorophyta* = green algae) or Class level (e.g., *Bacillariophyceae* = diatoms). For the sake of simplicity, they have all been given Division status here, but see Graham and Wilcox 2000.

The number of freshwater algal species is enormous. In Quebec alone, some 3,000 taxa of freshwater phytoplankton and benthic algae have been identified (Poulin et al. 1995), with 440 taxa considered planktonic, indicating that the majority of species are primarily benthic. The term **taxa** (singular taxon) refers to both species and subspecies or varieties. Padisák (1992) reported 417 taxa, about 30 percent of which were primarily benthic, in a single shallow Hungarian lake. Fortunately, from the point of view of the ecologist, most of the species identified in waterways are rare, many appearing only after storms in lakes or floods in rivers which resuspend benthic, littoral, and wetland forms into the plankton.

Benthic vs Planktonic Algae

Planktonic algae that spend a portion of their life cycle on substrates are known as **meroplankton** to distinguish them from true phytoplankton, also known as **holoplankton** or **euplankton**, which are found in the plankton only. The importance of meroplankton is particularly evident in shallow lakes or the shallow portions of deep lakes where they readily sink onto the sediments or macrophytes, only to be resuspended during windy periods (Fig. 21–2). Alternatively, meroplankton species emerge from a sediment resting phase when favorable conditions return. The ability of many planktonic plant (and animal) species to survive in shallow or deep water sediments and to provide an inoculum when favorable conditions return is of great importance in structuring plankton communities and also plays a role in the transport of protoplasm-bound plant nutrients from the sediments to the euphotic zone, a role which has not been thoroughly investigated. Most Divisions of freshwater phytoplankton have species with benthic resting stages.

Those algae normally associated with shallow water sediments or its vegetation are known as **periphyton** or benthic algae (Chapter 24). When they become detached, they are called **tychoplankton**, or simply meroplankton. An examination of the algae of the pelagic zone of Lake Constance (AU, CH, DE) showed that nearly half of the 190 species identified were tychoplankton (Schweizer 1997). Other work on

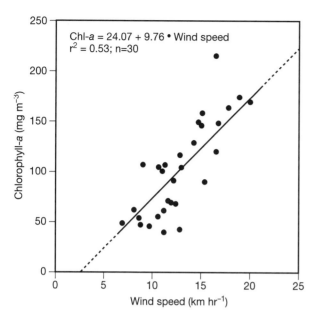

Figure 21–2 Relationship between average daily wind speed and algal biomass as chl-*a* in Lake Apopka, Florida, US (LA = 125 km^2, $\bar{z}$ = 1.7 m., z_{SD} = 20 cm.) which reflects the role of turbulence in resuspending sedimented algae plus sediments on high wind days. Note that (1) extrapolations beyond the data range should always be done with great caution as the relationship may be different. Thus, the model suggests, surely incorrectly, that there be zero chl-*a* at wind-speeds of around 2.5 km hr^{-1} and that chl-*a* concentrations keep rising with increasing windspeed above 20 km hr^{-1}; and (2) the highly eutrophic lake switched from a macrophyte-dominated to a phytplankton-dominated system following eutrophication and finally a hurricane (see Sec. 24.11). *(From Schelske et al. 1995.)*

two small forest lakes (US) showed that about one-third of the species were recruited from the sediments, with the recruitment of the different species changing spatially and temporally. Most species were recruited from shallow sediments (< 4 m), and often increased the total phytoplankton abundance by 10–50 percent d^{-1} (Hansson 1996). Similarly, virtually all floating algae in rapidly flushed streams are tychoplankton and originate from the substrate or the land–water eco-tone. In contrast, many of the algae in the plankton of larger, more slowly flushed rivers (stream order > ~5; Sec. 8.2) are true phytoplankton found in both slow-flowing rivers and lakes. The suspended algae of lotic systems are commonly referred to as **potamoplank-ton**, but the true potamoplankton are only those that reproduce within the water column.

The different designations for algae, referring to their originating habitat, acknowledges that the true plankton (euplankton) form a continuum with those inhabiting the sediments or living on macrophytes and also with the potamoplankton of rivers. The traditional view that the phytoplankton community of aquatic systems can be understood in isolation from the community living or resting on substrates is un-tenable, particularly so in shallow systems.

Species Richness

The number of algal species seen in aquatic systems is determined partly by the physical/chemical/biologi-cal environment and partly by the taxonomic compe-tence and persistence of the investigators. Species richness is further determined by how well the col-lected samples were preserved and the techniques used to examine them. An experienced phytoplankton ecologist using a traditional light microscope will probably encounter somewhere between 70 and 200 species during an annual study of the plankton of a particular lake, most of which would be uncommon. However, a long-term examination of the plankton by an algal taxonomist would probably yield > 400 taxa in temperate zone freshwater lakes, but lower num-bers in saline lakes. While dominant species change seasonally, Scandinavian studies show that a mere six to eight species contribute more than 90 percent of the biomass at any time in a single lake (Willén 1976). This implies that the vast majority of species present are rare. However, species that are rare at one time may be abundant during another season or in another lake or river where conditions differ. Finally, abun-dance is not a good measure of growth rate or the im-portance of a species in energy flow. Some relatively rare small forms grow exceptionally fast but show lit-tle change in abundance because they are removed equally fast by predators. Conversely, large motile or buoyancy-regulating cells or colonies may suffer little loss and therefore be abundant even though they grow slowly (Sec. 21.2).

21.2   Species Composition and Phylogenetic Generalities

Enormous effort has gone into determination of the algal species composition of lakes and rivers, and a considerable number of excellent descriptions of species distribution over time in individual systems are available. Yet, the extensive and impressive amount of information collected over the last 80–100 years has not enabled the development of models able to predict, even roughly, the species composition of individual systems over time and space.[2] The difficulties encountered have long encouraged research on more aggregated scales (Genus, Division), yielding the generalities outlined below. However, these are too broad to be considered interesting (useful) research by many limnologists working at less aggregated scales. What at first glance might seem to be modest accomplishments is not attributable to a lack of thoughtful, high-quality research but rather to the large numbers of species and the daily, seasonally, and spatially changing physical, chemical, and biological environment determining growth and loss rates in individual species. It is the imbalance between the magnitude of the relative growth and loss rates of individual species which determines their contribution to the community biomass over time.

While a priori prediction of species composition is a goal, individual lakes exhibit sufficiently large seasonal and interyear repetition in the abundance of dominant species and genera (Fig. 21–1) to have encouraged the development of schemes describing typical seasonal or midsummer patterns for particular lake types. The among-lake patterns are nearly always described as a function of nutrient richness. However, growing recognition that the stability of the water column exerts a major influence on the light climate experienced (Figs. 10–12 and 12–5), and thereby on the relative growth rates and abundance of the different species, has in recent years encouraged the development of promising schemes which include water column stability (light climate), nutrient richness, and flushing time (Reynolds 1997).

The Intermediate Disturbance Hypothesis

The phytoplankton of any waterbody are composed of many species, indicating that interactions for resources are usually insufficient to allow large-scale elimination through competitive exclusions. This shows that species are not in an equilibrium (*steady state*). It is clear that externally produced disturbances

are sufficiently common to interrupt or reverse progress towards a steady state. Frequent disturbances exclude all but a few rapidly growing pioneer species. Maximum diversity is therefore hypothesized to be the result of intermediate levels of disturbance. The **Intermediate Disturbance Hypothesis** (IDH) of J. H. Connell (1978) is an attempt to reconcile the more recent nonequilibrium view with the earlier equilibrium view by arguing that the observed organizational state of communities is the outcome of progress toward equilibrium since the last disturbance. Figure 21–3 shows conceptually how the timescale of disturbance exerts an impact on the community structure of different types of lakes (Reynolds 1993). When the timespan between disturbances lengthens, a successional sequence of dominant species develops. The intermediate disturbance hypothesis is one of a series of conceptual schemes to predict and understand changes that have been observed at the species or genus level of phylogenetic organization. A number of better quantified broad generalities have emerged at the Division level, the most aggregated scale of algal systematics.

Division Cyanophyta: cyanobacteria, ~1,350 freshwater species

Large colonies, clusters of intertwined cells, or single filaments (Figs. 21–4 and 18–7) of cyanobacteria frequently dominate the algal biomass of nutrient-rich (high biomass) temperate lakes (Fig. 21–5) and slowly flowing rivers in summer. They also dominate in ice covered polar lakes and polar meltwater streams that freeze solid in winter. At lower latitudes the large cyanobacteria may dominate slow-flowing rivers and lakes for much of the year, or even year-round (Fig. 21–6).

Moreover, the large cyanobacteria predominate among the benthic algae of hot springs and antarctic pools. In addition, tiny cyanobacteria (< 2 μm) (Sec. 21.3) commonly dominate the algal biomass in oligotrophic lakes, and yet other cyanobacteria dominate the algal biomass in poorly illuminated metalimnia. All this variation is a tribute to an extraordinary structural and functional heterogeneity among the cyanobacteria. They are proof that many factors (light, nutrients, CO_2, pH, temperature, turbulence, competition, selective grazing by predators) determine the dominance of particular species in all Divi-

---

[2]"The forces controlling species composition, diversity, dynamics, and stability of an ecosystem remain one of the major mysteries of modern science." (D. Tilman 1996)

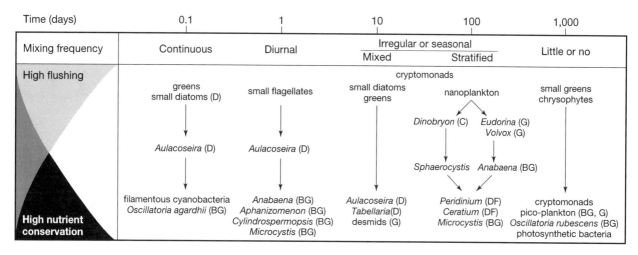

Figure 21–3 A conceptual matrix of dominant phytoplankton associations in relation to the time available to achieve equilibrium (vertical axis) against the frequency of mixing events (horizontal axis) and the observed rate of flushing and associated nutrient input (vertical axis). The algae mentioned frequently represent only one of a group of species. Divisions (phyla) are given as follows: D = diatoms; BG = blue-green algae; G = green algae; C = chrysophyta; DF = dinoflagellates. The arrows show possible or probable successional sequences unless stopped, or setback by a new disturbance. Disturbances include not only the frequency and regularity of water column mixing and resulting light climate changes, changes in flushing, and nutrient input, but also variation in nutrient input from the profundal zone or sediments and variation in abundance of predators. *(Modified after Reynolds 1993.)*

sions.[3] However, the relative importance of the large cyanobacteria species in summer, and year-round in low latitude lakes, typically increases as nutrient concentrations rise (Fig. 21–5); the water TP or TN concentrations serve as a surrogate, or proxy, for community biomass (chl-*a*). Species able to fix nitrogen often dominate in systems with disproportionately little nitrogen (TN:TP < 10 by mass) (Fig. 18–10).

▲ The large cyanobacteria contribute an average of > 70 percent of the summer phytoplankton biomass in hypertrophic (> ~100–1,000 μg TP l[−1]) lowland lakes in the temperate zone (Fig. 21–5). The mechanisms responsible for their dominance have not been resolved because the importance of particular mechanisms changes over time and space; these include nutrient availability, pH/CO_2 conditions, reduced respiration rates, or simply a lower loss rate (by grazing, sinking, or disease) than experienced by other groups of algae. Thus, a lower loss rate rather than a higher

growth rate was responsible for the observed shift from a diatom to a blue-green algal dominance in a small lake (Knoechel and Kalff 1975). More recent research in Florida similarly concludes that the dominance of large cyanobacteria was the result of the other phytoplankton groups suffering disproportionately higher losses (Agustí et al. 1990).

Large cyanobacteria usually decline disproportionally following periodic overturns when the $z_{eu}:z_{mix}$ ratio declines and the algae spend increasing amounts of time in the dark (Fig. 10–12). The occasional dominance of the large cyanobacteria in oligotrophic lakes and the frequent high abundance of photosynthetic picocyanobacteria (< 2 μm) in them, further precludes a simple but adequate explanation for the success of the cyanobacteria based on a few environmental variables.

Cyanobacteria and Aquatic Management

The large cyanobacteria have received a great deal of attention in fundamental research and the management of eutrophic and hypertrophic lakes and rivers.

---

[3]"There is no evidence that any one factor is of overriding importance in determining the abundance of a particular species." (G. E. Fogg 1965)

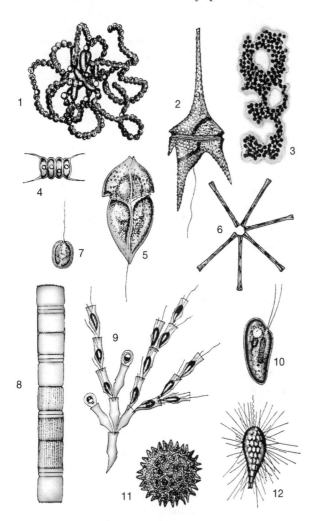

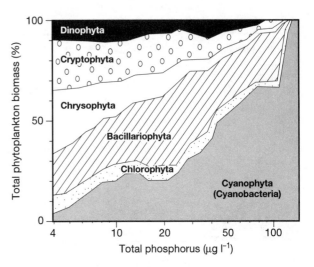

Figure 21–5 Changes in temperate zone phytoplankton community structure, excluding the picophytoplankton, with changes in water column total phosphorus in primarily stratified freshwater lakes. As total phosphorus serves as a surrogate for algal biomass (chl-*a*), the fraction of cyanobacteria also rises with increasing algal biomass (see also Fig. 21–6). Not shown is a not infrequent shift from a cyanobacteria to a Chlorophyta dominance in hypertrophic (> 500 µg TP l⁻¹) unstratified lakes and sewage lagoons (see Jensen et al. 1994). *(Modified from Watson et al. 1997.)*

Figure 21–4 Selected phytoplankton. (1) cyanobacterium cluster: *Anabaena flos-aquae*, (2) dinoflagellate: *Ceratium hirundinella*, (3) cyanobacterium colony: *Microcystis flos-aquae*, (4) green algae colony: *Scenedesmus quadricauda*, (5) dinoflagellate: *Gymnodinium helveticum*, (6) diatom: *Asterionella formosa*, (7) chrysophyte: *Chrysococcus rufescens*, (8) filamentous diatom: *Aulacoseira islandica*, (9) chrysophyte colony: *Dinobryon divergens*, (10) cryptomonad: *Cryptomonas obovata*, (11) green alga (desmid): *Pediastrum boryanum*, (12) chrysophyte: *Mallomonas caudata*. Not to scale.

The highly visible **blooms** or **outburst** (high biomass) of large filamentous or colonial cyanobacteria (especially in the genera *Aphanizomenon*, *Anabaena*, and *Microcystis*; Fig. 21–4), which produce surface scums on nutrient-rich lakes on calm days, are not only aesthetically displeasing but furthermore interfere with the functioning of water supply plants requiring potable water from the epilimnia or rivers. Sand filters are

commonly used in the plants to remove the algal particles from the water. When abundant, the large cyanobacteria clog the filter beds more frequently and, together with their metabolites and those produced by other algae and fungi, impart an unpleasant earthy/musty taste and smell to the water, which in the case of the cyanobacteria is apparently the result of cell decomposition (Bierman et al. 1983). Concentrations of even a few nanograms per liter are readily noted, with the affected water then requiring special and expensive treatment to make it acceptable to consumers (Wnorowski 1992).[4]

Some strains of the large cyanobacteria (as well as some dinoflagellates and diatoms) produce toxins that can cause human illness and death as well as the death of cattle,

---

[4]Chlorination of water supplies is commonly used to prevent bacterial growth in the water distribution systems and to kill microbes before releasing treated wastewater (Sec. 8.5), and also to pretreat raw water supplies obtained from eutrophic surface waters. The chlorination causes cell lysis and the release of odor- and taste-producing substances that make their removal even more difficult.

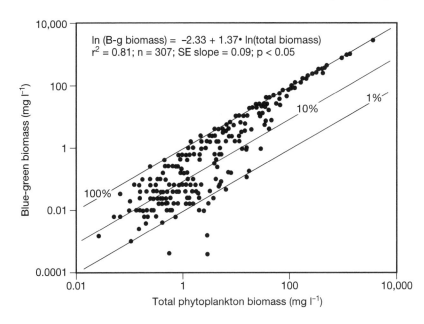

Figure 21–6 Relationship between blue-green algal biomass (wet wt) and total phytoplankton biomass (wet wt) in 165 Florida (US) lakes. The solid lines show the lines for which the biomass of blue-green algae equals 100 percent, 10 percent, and 1 percent of the total phytoplankton biomass. Note that (1) the cyanobacteria make up an increasing fraction of the total biomass as the total biomass increases, and (2) while the cyanobacteria may represent any fraction of the total algal biomass when the total is below about 50 mg l^{-1}, they represent nearly 100 percent of the total in hypertrophic lakes (total biomass > 100 mg l^{-1}). *(After Canfield et al. 1989.)*

wildlife, and dogs when present in high concentrations. Some of the strains exert little known **allelopathic** (toxic and inhibitory) effects on zooplankton and other competing phytoplankton.[5] Toxic algae almost certainly have caused poisonings of fish and waterfowl, but hard proof appears to be lacking. The cyanobacteria toxins include peptides, alkaloids, organophosphorus compounds and lipopolysaccharides, some of which affect the liver system (*hepatotoxins*) while others are *neurotoxins*. Recent studies in Europe and North America have demonstrated that 25–75 percent of blooms produced by toxic strains encountered in eutrophic lakes are toxic to humans (Sivonen et al. 1990, Bell and Codd 1994). Hepatotoxic heptapeptides, often referred to as *cyanoginosins*, or *microcystins* when produced by the genus *Microcystis*, are usually the causal agents (Lindholm and Meriluoto 1991). In the temperate zone, the presence of toxins is most often linked to a high biomass of the species *Microcystis aeruginosa*. Other types of toxins, including neurotoxins (e.g., anatoxins), are most commonly re-

leased when blooms come to an end (senescence). See Chorus and Bertram, 1999.

Resting Stages

The relative abundance of cyanobacteria (and other phytoplankton species) is determined by seasonally changing growth and loss processes (Secs. 21.6, 7, 8, and 11), as well as by the size of the innoculum available to colonize and compete with other species for the limited nutrient supply and light available. All Divisions contain species that have either resting stages or vegetative cells able to over-winter in the water or sediments. Among common bloom-forming cyanobacteria, the filamentous genera *Anabaena* and *Aphanizomenon* can produce thick-walled resting cells (cysts), called **akinetes**. In contrast, the common colonial genus *Microcystis* (Fig. 21–4) lacks such cells and survives unfavorable conditions in a vegetative form in or on the sediments.

The sudden appearance of meroplankton can contribute significantly to the phytoplankton community biomass (Fig. 21–2) and modify its composition (Schelske et al. 1995, Hansson 1996). In one shallow, highly eutrophic Swedish lake, the resting-cell biomass of living *Microcystis* colonies in the sediments was greater (per m^{-2}) throughout the year than in the water column during the summer bloom (Boström et al. 1989). Finally, meroplankton recruited from the more nutrient-rich sediments normally contain high levels of nutrients and may contribute significantly to

---

[5]"The benefits to the organisms of producing such poisonous chemicals is, at the time of writing, still an unexplained paradox." (Reynolds 1997)

"An overall synthesis of interactions between toxic phytoplankton and their grazers is elusive because blooms and grazer interactions are situation-specific. Many contradictions are due to a variety of toxins which may have different physiological effects on consumers and to differences in toxin potency or intracellular concentrations." (J. T. Turner and P. A. Tester 1997)

the internal nutrient load of the mixed layer (Barbiero and Welch 1992, and Chapter 17). The meroplankton serve as a reminder that the biota of the benthos, the littoral zone of lakes and rivers, and the pelagic zone are more closely linked than is usually recognized.

Division Chlorophyta: Green algae, ~2,400 freshwater species

The **Chlorophyta** is composed of unicellular flagellated and nonflagellated cells, colonies, and filaments (Fig. 21–4), plus the macroscopic charophyceans—a group of ancient macrophytes (Chapter 24). The microscopic green algae make a major contribution (typically 40–60%) to the planktonic species richness in freshwater lakes, but only rarely contribute much of the biomass of oligotrophic and mesotrophic temperate zone lakes (Fig. 21–5). The phytoplankton species richness in a tropical wetland was also numerically dominated by chlorophytes but again they contributed only modestly to the biomass (de Oliveira and Calheiros 2000). In contrast, the biomass contribution of the green algae is frequently large in highly polluted polymictic lakes, wastewater lagoons, and nutrient-rich farm ponds with total phosphorus concentrations > ~500 µg l⁻¹ (Happey-Wood 1988, Spodniewska 1983, and Jensen et al. 1994), as well as in nutrient-rich floodplain lakes and rivers everywhere. Highly saline shallow lakes and wetlands too are frequently dominated by green algae, mainly *Dunaliella spp.*, whose photoprotective pigment gives the water a pinkish color when the algae are abundant. The green alga *Chlamydomonas nivalis*, which contains a red photoprotective pigment, are a common snow alga at high latitudes and altitudes. Furthermore, benthic forms (periphyton) often dominate the inorganic substrate of eutrophic rivers and the littoral zone of freshwater lakes (Chételat et al. 1999). It is unresolved why the green algae typically contribute more to the algal biomass of most tropical and subtropical freshwater lakes than to their north temperate zone counterparts of similar trophic status (Fig. 21–7).

Division Euglenophyta: ~1,020 freshwater species

The **Euglenophyta** are characterized by a modest number of small and medium-sized flagellated species that normally make a negligible contribution to the phytoplankton biomass of stratifying lakes (Fig. 21–5). The group makes its greatest contribution to the bio-

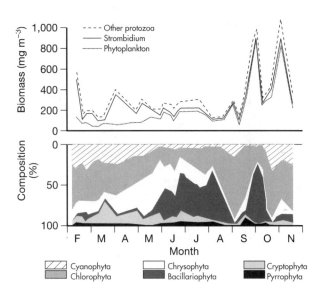

Figure 21–7   Seasonal cycle of phytoplankton and protozoan abundance and phytoplankton composition in the northern portion of Lake Tanganyika (TZ, ZR) in 1975. Note (1) the high relative importance of the Chlorophyta, and (2) the intermittently large biomass of *Strombidium*, a protozoan containing algal chloroplasts, and other ciliate protozoans. *(After Hecky and Kling 1987.)*

mass of small, highly eutrophic ponds where they often codominate with chlorophytes. Together they give such waters a brilliant grass-green color that differs from the dull tone characterizing surface scums of cyanobacteria. The euglenoids are often abundant among the littoral zone and wetland vegetation and are swept into rivers and lakes following storms (de Oliveira and Calheiros 2000).

Division Bacillariophyta: Diatoms, ~5,000 freshwater species

The **Bacillariophyta** commonly dominate the species composition of the plankton during periods of overturn in wind-exposed (turbulent) lakes with a sufficient supply of available dissolved silica to build their external skeletons (frustules) (Kalff and Watson 1986). Planktonic diatoms usually supply the highest fraction of the growing season community biomass in mesotrophic systems (~10–30 µg TP l⁻¹), contributing proportionally less to the biomass in both nutrient-poor and nutrient-rich temperate lakes (Fig. 21–5). Moreover, diatoms play a major role in both tropical (Fig. 21–7) and ultra-oligotrophic polar lakes (Fig. 21–8). Their contribution to the community biomass is, if anything, even larger in the potamoplankton

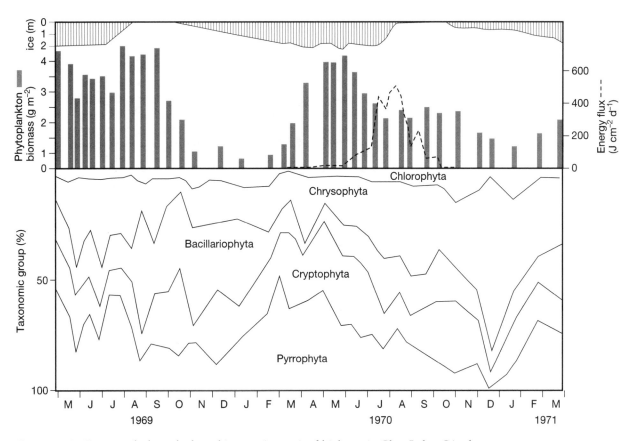

Figure 21–8 Integrated phytoplankton biomass (wet wt) of high arctic Char Lake, CA, the percentage of the biomass belonging to the principal taxonomic groupings, and the photosynthetically available energy flux entering the water column. *(Modified from Kalff et al. 1975.)*

(Rojo et al. 1994) and among the benthic algae (periphyton) of lotic systems (Chételat et al. 1999).

Although diatoms vary greatly in size, with the smallest unicells having a diameter of about 2 μm and ranging to 2 mm (i.e., by four orders of magnitude), most species are large (Fig. 21–4) and readily collected with traditional plankton nets (Table 21–1). Heavy silicate frustules and the absence of flagella make all but the smallest species prone to sedimentation (Sec. 21.6) and disappearance from shallow mixed layers following the onset of stratification (Fig. 21–9, and Nixdorf 1994) or exhaustion of the silica supply. Sedimented diatoms can remain viable in the sediments for many years, ready to resume photosynthesis within hours of their return to the euphotic zone (Sicko-Goad et al. 1986). Living diatoms have been collected many centimeters below the sediment surface of lakes. Some of these cells (cysts) appear to have survived sometimes in excess of 100 years (Stockner and Lund 1970). They represent a large *seed bank*, analogous to the seed

banks of many terrestrial plants and aquatic macrophytes. Such banks allow immediate recolonization when suitable environmental conditions return, without the need for a much slower recolonization by dispersal from other inland waters or growth based on a small number of cells that had remained in the water column. Large diatoms are rarely important in wind-protected lakes or after the development of an ice cover, but the smallest forms can maintain a presence even in ice-covered (low turbulent) lakes (Fig. 21–8).

Division Chrysophyta: Golden-brown algae, ~450 freshwater species

This group consists of a relatively few species of small single-celled flagellates and flagellated colonies (Fig. 21–4). **Chrysophyta** typically contribute only modestly to the species richness and little to the phytoplankton biomass of eutrophic lakes (Fig. 21–5). However, chrysophytes frequently dominate, or

Table 21–1 **A division of phytoplankton, bacterioplankton, and protozooplankton on the basis of unit size.**[1]

Grouping	Maximum Diameter (D) or Length (L) (μm)	Phytoplankton Attributes
Femtoplankton	< 0.2 (D)	Consists of very small bacteria and viruses.
Picoplankton	0.2–2 (D)	Contains the smallest phytoplankton (~0.5–2 μm). All but the smallest are subject to significant predation by small rotifers, protozoa, and by some of the filter-feeding daphnid crustaceans; experience negligible sinking rates. Very high potential growth rates of the larger forms.
Nanoplankton	2–30 (D or L)	Many, often flagellated phytoplankton; principal food of the macrozoo plankton and microzooplankton; very low sinking rates. High potential and realized growth rates.
Microplankton	30–200 (D or L)	Small microplankton (< ~70 μm). Subject to some macrozooplankton grazing and prone to moderate sinking when nonmotile or lacking buoyancy control. Moderate potential growth rates. Large microplankton (> ~70 μm). Retained by traditional ~70 μm mesh size plankton nets (*netplankton*). Highly prone to sinking in the absence of buoyancy control; principal food of pelagic and benthic zone omnivorous fish as well as sediment microbes. Moderate to low potential growth rates.
Mesoplankton	200–20,000 (L)	Large cells and colonies. For attributes, see large microplankton.
Macroplankton	> 20,000 (L)	Large free-floating plants such as duckweed (*Lemna spp.*) in ponds; and the notorious water hyacinth (*Eichornia*) and waterfern (*Salvinia*) in tropical and subtropical lakes and slowly flowing rivers (see Fig. 15-1 and Chapter 24). Lowest potential growth rates.

[1]The present division differs from the one proposed by Sieburth et al. (1978) by expanding by 10 μm the maximum nanoplankton size range to include those organisms most subject to substantial predation by freshwater crustacean zooplankton.

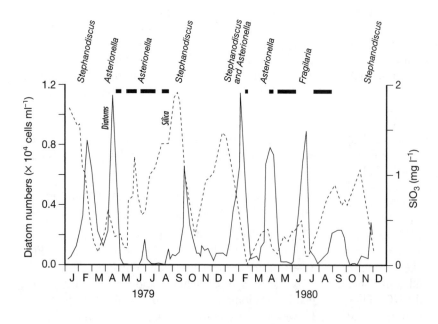

Figure 21–9 Seasonal variation in diatom abundance (cells, solid line) and dominant genera in a shallow (LA = 12.6 km², $\bar{z}$ = 10 m, $z_{\max}$ = 34 m) polymictic reservoir (Rutland water, GB), their decline during periods of stratification and sedimentation (thick bars at top of graph), and a decline in dissolved silica (SiO₃, dashed line) following periods of high diatom growth. (*Modified after Ferguson and Harper 1982.*)

codominate with the cryptophytes, the community biomass and species abundance in oligotrophic clear water and humic (dystrophic) lakes at all latitudes (Fig. 21–10). This division contains pigmented species closely related to nonpigmented forms (Protozoa).

Division Cryptophyta: ~100 freshwater species

The **Cryptophyta** are a second group of small or medium-sized flagellates (Fig. 21–4). In terms of community biomass, they make their greatest proportional biomass contribution in oligotrophic and mesotrophic temperate lakes (Fig. 21–5), but are also common in oligotrophic polar lakes (Fig. 21–8) and tropical waters, including tropical wetlands where they can dominate the biomass (de Oliveira and Calheiros 2000). This division is composed of relatively few species and thus contributes little to the phytoplankton species richness. Single-celled cryptophyte and chrysophyte flagellates, together with the smaller species of dinoflagellates of similar size, are usually the principal food of the larger rotifers and crustacean zooplankton. These same three groups of flagellates contain a number of species that are **mixotrophic** (more than one mode of nutrition), able to prey on bacteria and the smallest algae.

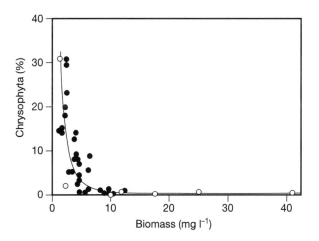

Figure 21–10 The relationship between average total phytoplankton biomass and the percentage of the total contributed by Chrysophyta in 27 Ontario, CA, lakes (●) and seven tropical lakes (○). The Chrysophyta contain a considerable number of species able to complement their nutritional needs by engulfing picoplankton-sized particles (*phagotrophy*). The plot suggests a disproportionate importance of phagotrophy in oligotrophic lakes. (*After Kalff and Watson 1986.*)

Division Pyrophyta: Dinoflagellates, ~220 freshwater species

The **Pyrophyta** typically contribute only a small fraction of either species number or growing season community biomass of temperate lakes (Fig. 21–5). The principal exception, and one poorly understood, is that the motile dinoflagellates (rather than large cyanobacteria with buoyancy control) occasionally dominate or codominate the summer biomass of stably stratified, eutrophic lakes (Fig. 21–3). Some species lack chlorophyll and live heterotrophically, engulfing small algae and bacteria. Dinoflagellates (Fig. 21–4) produce cysts that sink into the sediments and return to the plankton when favorable conditions return. Dinoflagellates often replace chlorophyte and cryptophyte flagellates in small, well-stratified lakes following acidification (Sec. 27.9).

21.3 Algal Size and Activity: Small Cells vs Large Cells

Cell size is a more important determinant of the (maximum) growth rate over a large size range than phylogenetic position. Cell size is also an important determinant of the pathways of material and energy flow in aquatic systems. An increasing number of limnologists have begun to consider the structure of algal communities and their functioning on the basis of organismal size and form rather than phylogeny.

During the first half of the 20th century, the size-based phytoplankton literature was largely based on those algae captured with the relatively coarse mesh plankton nets available at the time. A major technical advance was made in the 1930s with the development of what became known as the **Utermöhl inverted microscope**, which has its optic elements facing upward (inverted) from below the microscope stage. The algae in preserved water samples are allowed to sediment onto the glass bottoms of settling tubes. The glass bottom plate is next separated from the overlying settling tube, placed on the microscope stage, and examined via the upward facing optics. The Utermöhl microscope made it possible to quantify and identify the small algae (as well as the larger forms), showing their importance in the plankton (Wetzel and Likens 2000).

The larger organisms retained by traditional plankton nets have been known as the **netplankton**. Those too small to be quantitatively retained by the nets (< ~64–70 μm), but readily enumerated under the Utermöhl microscope, became known as **nanoplankton**. A very wide range of smaller mesh sizes available today has caused the original division to break down, and cells or colonies with a minimum dimension of < 30 μm, < 20 μm, and even < 10 μm are sometimes referred to as nanoplankton.

The next major technological advance was made in the 1970s with the application of cell-surface and DNA staining fluorescent dyes that allowed phytoplankton and bacterioplankton organisms as small as < 1 μm to be recognized and counted with the aid of fluorescence microscopes. Their use revealed large numbers of tiny phytoplankton and protozoans that previously had been largely overlooked. Following Sieberth et al. (1978) those algae, bacteria, and protozoa with a maximum dimension between 2 μm and 0.2 μm are now known as **picoplankton**. Those forms between 2 and 30 μm continue to be known as *nanoplankton* and the larger phytoplankton as **microplankton** (Table 21–1).

Size and Function

The most common picophytoplankton are the near-spherical, rod-shaped prokaryotic cyanobacteria with cell dimensions of 0.4–1 μm (Caron et al. 1985, and Fahnenstiel et al. 1986), and the slightly larger, smallest eukaryotic algae in the Division Chlorophyta (Sec. 21.2). The picophytoplankton are proportionately most abundant in highly transparent (oligotrophic) lakes when they may dominate the phytoplankton biomass with numbers up to > 10^6 cells ml^{-1}. The picophytoplankton are of much smaller relative significance in the biomass of eutrophic waters (Fig. 21–11).

At the opposite end of the phytoplankton size spectrum lie colonies of *Microcystis spp.*, colonial cyanobacteria, with a colony diameter up to 570 μm (Fig. 21–4). The resulting phytoplankton size range of individual unit volumes thus covers seven orders of magnitude. If, as Reynolds (1984a) points out, a *Synechococcus* cell of picoplankton size (< 10 μm^3) is equated with the size of a chicken egg, the *Microcystis* colony (~10^8 μm^3) would be the size of a small house.

The enormous range in volume of the phytoplankton and the large range in surface area to volume ratio (SAV) of phytoplankton cells (Fig. 21–12) allow for (1) great differences in the maximum growth rates possible, (2) major differences in the rate at which phytoplankton sink out of the euphotic zone, and (3) the ease and rate by which they are consumed by macrozooplankton and herbivorous fish. These three size-

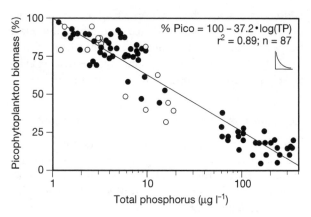

Figure 21–11 The importance of nutrient concentration, as represented by total phosphorus in predicting the relative (%) contribution of picophytoplankton to the total algal biomass (mg C m^{-3}) of lakes (●) and seas (○). Note the decline in the relative contribution of the picoplankton with increasing total phosphorus or its surrogate, algal biomass. *(After J. Cullen and J. Kalff, unpublished.)*

linked effects have a major impact not only on the phytoplankton size structure and species composition but also on the rate of nutrient turnover in aquatic systems and whether most of the nutrient recycling occurs in the mixed layer rather than in the hypolimnia or sediments (Sec. 20.4).

Algal Size, Growth, and Loss Rates

The small phytoplankton greatly dominate the average summer community biomass numerically in oligotrophic and mesotrophic lakes, but are increasingly replaced by algae of microplankton size at TP concentrations greater than ~40 μg l^{-1} with the result that eutrophic lakes in summer are usually dominated by microphytoplankton (Fig. 21–13).[6] The increasing dominance of the microplankton biomass is attained despite a higher growth potential of the algal nanoplankton and picoplankton (Fig. 21–12a).

The high potential and realized growth rate in nature of small algae is commonly attributed to a disproportionately large surface area and an associated large number of nutrient uptake sites per unit volume. A disproportionately thin layer of nonturbulent water

---

[6]Conditions experienced by temperate zone algae in summer resemble to a considerable extent those experienced by low latitude algae year-round. It is consequently tempting, and sometimes necessary, to equate the two and extrapolate from the abundant temperate zone summer database to the less well-studied inland waters at low

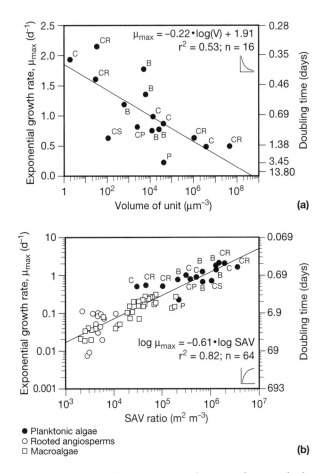

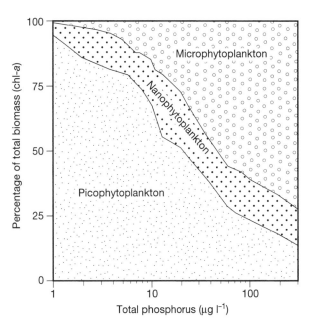

Figure 21–12 (a) Maximum growth rates of some planktonic algal species readily grown in culture and belonging to different divisions (C: Cyanophyta; CR: Chlorophyta; B: Bacillariophyta; CS: Chrysophyta; CP: Cryptophyta; P: Pyrophyta) in continuously light-saturated cultures at 20°C, plotted against their mean volumes. (b) Maximum growth rates of planktonic algae, macroalgae, and rooted macrophytes plotted against their surface area to volume ratios (SAV). Note the importance of cell volume (*V*) and the surface area to volume ratio in predicting maximum growth rates of algae and other aquatic plants, independent of their phylogenetic position, when examined over a wide range of *V* and SAV. *(Figure a after Reynolds 1989; Figure b data from Reynolds 1989, and Nielsen and Sand-Jensen 1990.)*

Figure 21–13 Relationship between total phosphorus and the average fraction of the total algal biomass (chl-*a*) during the growing season contributed by picoplankton (< 2 μm, derived from Fig. 21–11), nanophytoplankton (2–35 or 50 μm), and microplankton (> 35 or 50 μm), computed with the locally weighted sequential smoothing (LOWESS) technique. Note (1) the great dominance (> 75%) of picoplankton at TP concentration < ~8 μg l⁻¹, and their rapid decline with increasing trophic state, (2) the intriguing relative constancy of the nanophytoplankton at all trophic levels; and (3) the dominance (> 50%) of the large microphytoplankton in highly eutrophic systems (> ~40 μg TP l⁻¹). *(Data from Watson and McCauley 1992.)*

around small particles further facilitates the flux of materials into and out of the cells (Sec. 24.7). Even where the relative contribution of the nanoplankton plus picophytoplankton to the community biomass is somewhat smaller, as in mesotrophic lakes, the high growth rates of small forms allow them to dominate primary production (Fig. 21–14).

The primary production of the nanoplankton community appears to rise much more rapidly than the biomass with increasing trophic status (Fig. 21–15), suggesting that the nanoplankton suffer disproportionately high loss rates compared to the microplankton, whose biomass (Fig. 21–13) and community production (Fig. 21–14) appear to rise in step with increasing trophic status. The result is a typical microplankton dominance in eutrophic lakes in the summer (Fig. 21–13). The nanoplankton plus picoplankton size categories, (< 30 μm, Table 21–1) encompass organisms of a size most readily consumed by

latitudes. Caution is necessary because the stability of stratification in the tropics is typically lower (Chapters 11 and 12), allowing more entrainment of water and associated nutrients into tropical epilimnia from their (warm) hypolimnia than in the temperate zone (Sec. 11.6). Moreover, tropical algae experiencing a "perpetual summer," can take advantage of nutrients entering aquatic systems year-round, whereas this is not necessarily the case in the temperate zone.

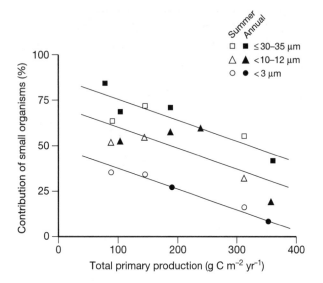

Figure 21–14 The relationship between total primary production and the relative contribution of small (≤ 35 μm) phytoplankton-size fractions in temperate European lakes. Squares = fraction ≤ 30–35 μm; triangles = fraction ≤ 10–12m; circles = fraction ≤ 3 μm. Filled marks indicate annual production and open marks represent summer production. Note the decline in the relative contribution of nano- and picophytoplankton (≤ 30–35 μm) to primary production (g C m⁻²) with increasing trophic state and the important contribution of picophytoplankton (defined here as ≤ 3 μm) in the most oligotrophic lakes. *(After Uehlinger and Bloesch 1989.)*

inland water protozoans, rotifers, and crustacean zooplankton, are known as *"edible algae."* The relative constancy of the nanoplankton's contribution to the community biomass over a wide range of trophic states must be attributable to community loss rates remaining a relatively constant function of growth rates (Fig. 21–13). Losses through zooplankton predation (Chapter 23) have received much attention, but less is known about other sources of mortality (viral cell lysis, autolysis, disease).

▲ Whatever the fate of the phytoplankton production in nature, the close relationship between cell size and growth rate in both the laboratory and nature, as well as between cell size, cell form, and sinking (Chapter 20), indicates that over a *large* size range, the influence of cell volume on growth rate is much stronger than differences in growth linked to the particular taxonomic grouping (Fig. 21–12a). Given the importance of unit size in determining growth rates, and the importance of size plus form (shape) of units in determining the fate of the organic matter that is produced,

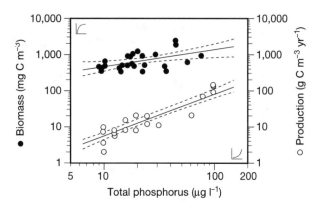

Figure 21–15 Relationship between the biomass (●) of the "edible fraction" (nanoplankton) of the phytoplankton and total phosphorus concentration and between nanoplankton production (○) and total phosphorus. Lines give best-fit regression and 95 percent confidence intervals. Note that, while the production of the edible phytoplankton rises with trophic state the biomass rises much less, indicating an increased loss rate of nanoplankton in more eutrophic lakes. *(Modified from Watson and McCauley 1988.)*

there is good justification for the use of size and form in phytoplankton ecology whenever generalities are sought about the functioning and structure of phytoplankton communities. This does not mean that the more traditional aggregation scheme—based on taxon—is unimportant. The considerable scatter around the line of best fit between maximum growth rate and cell size (Fig. 21–12) shows that variables other than size play a role in determining growth rates among species that differ little in size. Unfortunately, the relative importance of phylogeny in determining growth and loss rates in nature remains obscure.

Small organisms, characterized by high maximum growth and loss rates are sometimes referred to as **r-selected species**. Situated at the opposite end of the continuum are the **k-selected species**, dominated by larger organisms with a reduced maximum growth rate; these species must experience a reduced loss rate to be able to maintain their abundance.

21.4   Seasonal Biomass Cycles: A Conceptual Model

Winter and Early Spring

Dimictic and temperate zone monomictic lakes exhibit an annual minimum phytoplankton biomass in midwinter. The small amount of solar radiation reaching the water surface is, in dimictic systems, further reduced by ice and snow cover. Winter conditions may not be much more favorable in deep ice-free (monomictic) lakes in which turbulence carries the phytoplankton throughout the water column, yielding a highly unfavorable average light climate ($z_{eu} \ll z_{mix}$) (Fig. 10–12). The principal determinant of the low winter phytoplankton biomass and production (Fig. 21–16) is the low irradiance reaching the water surface. However, in temperate zone monomictic lakes, a significant winter biomass or bloom of large cyanobacteria that became established during the previous autumn is sometimes maintained, although without appreciable net growth.

Spring Bloom

Rapidly rising levels of solar energy reaching inland waters in spring allow for increased rates of primary production and a rising algal community biomass, even in ice-covered systems (Figs. 21–8 and 21–16). Rising levels of solar energy first raise the z_{eu}:z_{mix} ratio (Fig. 10–12) and water temperature in the shallow littoral zones, allowing for increased rates of photosynthesis and growth there. The progressively increasing irradiance then gradually raises the z_{eu}:z_{mix} ratio in the

deeper mixed zone. Large diatoms (e.g., *Asterionella formosa*) are favored under conditions of deep mixing ($z_{eu} \ll z_{mix}$). However, it is the onset of stratification (Sec. 11.4) that suddenly and greatly reduces the z_{mix} in deep lakes. The resulting quick rise in the z_{eu}:z_{mix} ratio (light climate) and temperature sparks the onset of a period of rapid phytoplankton growth, the **spring bloom**, utilizing nutrients washed in from the drainage basin or recycled in the water column during the autumn and winter. Rapidly growing flagellates usually dominate first (Fig. 21–16).

A shallower mixed layer (z_{mix}) in wind-protected lakes favors an earlier rise in the z_{eu}:z_{mix} ratio and onset of spring bloom than in wind-exposed ones. A high albedo of the snow cover and the high light-extinction coefficient of the snow plus ice delays the spring bloom in dimictic compared to monomictic lakes for periods of weeks to months (Chapter 10). Lakes with large littoral zones may receive an important innoculum of meroplankton (Sec. 21.1, and Fig. 22–7) that helps explain the frequently very rapid increase in the phytoplankton community biomass, even before the onset of stratification (Lund 1954).

The spring phytoplankton increase in temperate lakes is aided by a macrozooplankton development rate that greatly lags the phytoplankton growth rate

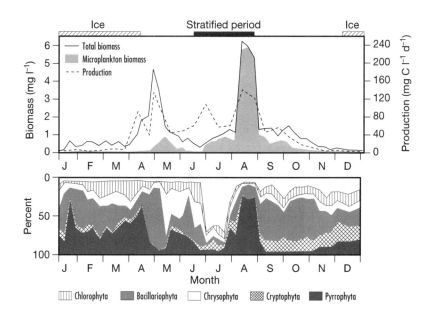

Figure 21–16 Phytoplankton production (dashed line), biomass (solid line), and the relative abundance of the different algal divisions in the 0–10 m water column of Lake Erken, SE (LA = 24.7 km², $\bar{z}$ = 9 m, CA:LA = 5.7) in 1960. The contribution of the nanoplankton (clear area) and microplankton (> ~70 μm, shaded area) to the biomass is shown. Note the importance of the nanoplankton, dominated by small diatoms and dinoflagelates, in the biomass and in production under ice and during spring bloom (April–May), followed by a period of high nanoplankton production but low biomass (clear-water phase) in June. Large blue-green algae (microplankton > ~70 μm) dominated the biomass in July, until the biomass and production of large dinoflagellates peaked in August. Production declined gradually during the autumn circulation period and the diatom-dominated community became once again increasingly composed of nanoplankton. (*After Nauwerck 1963.*)

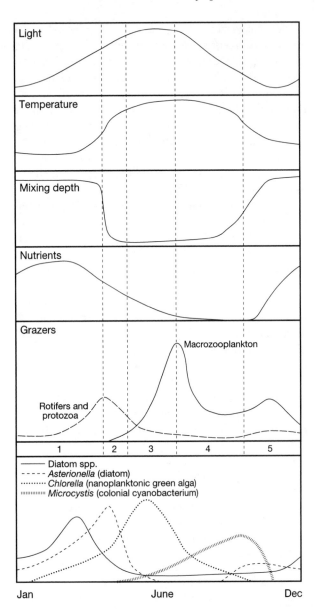

where both light ($z_{eu} \geq z_{mix}$) and nutrient conditions are favorable.

Late Spring to Summer

The onset of summer stratification has major implications for the phytoplankton community. Contact between the mixed layer and profundal sediments is reduced in stratifying lakes and thus reduces the possibilities for nutrient recycling (primarily P, N, and Si) from deep waters and their sediments to the euphotic zone (Chapter 12). In addition, external inputs from catchments normally decline in the continental portions of the north temperate zone, the result of seasonal reductions in river discharge (Fig. 8–4). A combination of higher temperatures and abundant food supply allows the herbivorous macrozooplankton growth rate to increase and their community predation rate to catch up with nanoplankton growth rates (Chapter 23), bringing about a decline in the edible algae. Moreover, the onset of stable stratification makes the large diatoms subject to increased loss by sedimentation from the now much thinner mixed layer, or to a reduction in diatom growth rate where the silica supply becomes exhausted. Whatever the mechanism, the spring phytoplankton community—commonly dominated by rapidly growing r-selected species and diatoms adapted to low light—are replaced by slower growing k-selected species (large colonial or filamentous cyanobacteria with buoyancy control or large dinoflagellates) better able to resist sinking and predation by filter-feeding zooplankton following the onset of a stable stratification. However, in years when stratification stability is particularly low and the light climate variable, the small r-selected algae may continue to dominate (Fig. 12–6).

The stereotypical spring bloom comes to an end as the result of increased macrozooplankton community growth and predation rates in the rapidly warming epilimnia or to an increased sedimentation loss rate of dominant large diatoms and their incorporated nutrients from the thinner mixed layer (Sec. 20.4). The resulting period of relatively or absolutely low algal biomass and increased transparency is known as the *clear-water phase* (Chapter 23). This phase is not very evident in oligotrophic lakes (chl-*a* < 7 mg m^{-3}), which are too nutrient poor to allow a biomass increase large enough to be readily detected (Fig. 21–18). Nor is a spring clear-water phase normally seen in highly eutrophic lakes where high nutrient supply rates and typically shallow water columns allow ready resuspension of sedimented microplankton—some having

Figure 21–17 Diagrammatic representation of the seasonal pattern in environmental conditions and planktonic biota of a hypothetical mesotrophic to moderately eutrophic monomictic lake in the north temperate zone. (*Modified after Reynolds 1990.*)

during periods of low water temperature (Fig. 21–17). Ciliate protozoans and rotifers may be abundant during the cold-water period, (Fig. 23–11) but they are only capable of removing the smallest algae. Predation losses of the nanoplankton due to herbivorous macrozooplankton thus tend to be small early in the season. The phytoplankton community biomass is, therefore, able to rise to a distinct spring maximum in those lakes

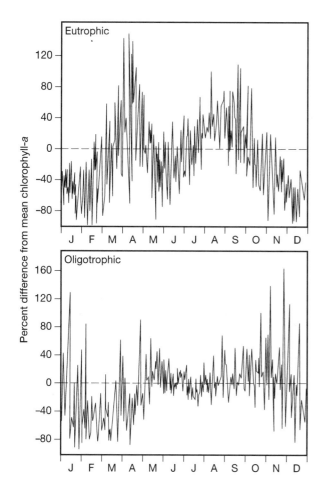

Figure 21–18 The pattern of chl-*a* change in 20 eutrophic (> 7 mg chl-*a* m⁻³) and 31 oligotrophic (< 7 mg chl-*a* m⁻³) temperate zone lakes. The time series for each lake year was scaled by expressing every measured chl-*a* concentration as the percent difference from the annual or seasonal mean for that lake. Scaled concentrations for each day were averaged to reduce scatter. The horizontal line represents the mean. Positive differences (the area above the line) indicate concentrations greater than the mean whereas concentrations less than the mean fall below the line. Note that a maximum is clearly visible only in the eutrophic lakes. *(From Marshall and Peters 1989.)*

buoyancy control—which are also minimally subject to zooplankton predation (Sec. 21.3).

Predation Control

The realization that macrozooplankton predation can make an important contribution to the establishment of a clear-water phase came later than the recognition of the importance of nutrients in determining phytoplankton biomass cycles. This is not surprising be-

cause algal culture research had long pointed to the importance of nutrients in phytoplankton growth, but such work had not also contributed to knowledge of the importance of herbivores in regulating the phytoplankton species composition and community biomass. Even today relatively little is known about the feeding habits and food selection in nature of the larger, well-studied crustacean zooplankton, and even less is known about the selective feeding by smaller forms (Chapter 23). When abundant, the larger *Daphnia* species in particular have a major impact on the phytoplankton community biomass or its composition (Fig. 21–19). Less is known about the role of zooplanktivorous fish, and in particular the role of larval fish and older young-of-the-year fish (YOY or 0⁺, Chapter 26) virtually all of whom feed on macrozooplankton (see Luecke et al. 1990, Jeppesen et al. 1997, and Chapter 23). A modest number of aquatic organ-

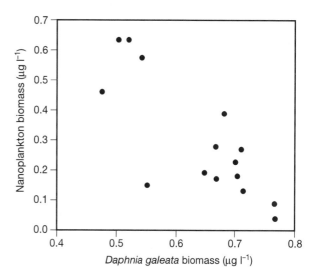

Figure 21–19 The relationship between the annual average wet weight of the principal large daphnid, *Daphnia galeata* and the nanophytoplankton biomass in the Saidenbach Reservoir (DE) between 1975 and 1990. Note that the demonstration of the pattern is aided by the focus on a single lake, where the morphometry is a constant and interannual variation in nutrient conditions are modest in comparison with such variations in inter-lake studies. The data scatter is further reduced by only considering annual averages rather than presenting individual measurements. Even so, considerable scatter, not attributable to *D. galeata*, remains. Whereas the research points to a cause and effect relationship, the correlation itself cannot provide the proof. Another (unstudied) variable correlating with the *Daphnia* abundance (e.g., fish predation) could be ultimately responsible for the changes in the nanoplankton biomass. *(After Horn and Horn 1995.)*

isms have a disproportionate impact on the structure and functioning of communities; they are recognized as **keystone species**.

Resources vs Predation

Unfortunately, the recognition that both top-down (predation) and bottom-up (resources) processes play a role in structuring communities does not help decide the relative importance of each over time and space. The literature is filled with studies that examined one factor, not both.[7] There is, however, a slowly growing recognition that top-down effects are relatively important in oligotrophic and mesotrophic lakes where a high fraction of the phytoplankton are small enough to be consumed by herbivorous macrozooplankton (Fig. 21–13), and periodically in eutrophic systems where predators of the macrozooplankton are rare enough to allow the macrozooplankton to greatly reduce the abundance of their phytoplankton prey.

The best evidence for the importance of top-down effects comes from the study of *single* systems in which the seasonal variation in nutrients is typically modest compared to a much larger seasonal variation imposed by predation. Conversely, bottom-up effects are particularly evident in nutrient-rich systems where most of the summer algal community is of microplankton size, too large to be consumed by zooplankton predators (Fig. 21–13). Furthermore, and not surprisingly, bottom-up effects are best evident in *among*-systems studies that range widely in nutrients (resources), but where predation effects vary less systematically. It should be realized that the two types of effects are closely coupled, with top-down effects resulting in a release of nutrients. The often-articulated dichotomy between the two types of effects is unwarranted. Studies carried out over different spatial or temporal scales by scientists adhering to different paradigms (who therefore focus on different aspects) inevitably reach different conclusions (Sec. 2.6).

It is widely accepted today that resources (nutrients, light) primarily determine both the among-system community production rates and maximum algal biomass attainable in inland waters, with temper-

ature controlling the maximum rate of growth and respiration. But, much debate and disagreement remains between individuals and groups carrying out research over a variety of temporal, spatial, and range scales concerning the relative importance of resource availability versus biotic interactions (predation/competition/disease) in structuring communities and determining their productivity.

Different mechanisms may be responsible for particular results, and the identified "causes" change with the temporal and spatial scales examined and the background beliefs of the investigators. *Proximal causes*, usually examined over short temporal and spatial scales in flasks, may be a resource (nutrient) limitation or high sedimentation rate of the most abundant species. Investigators with an interest in zooplankton and measuring their interaction with their prey over short periods tend to conclude that grazing effects are important. However, more *distal causes*, typically operating over seasonal or annual time scales, may entail a reduced predation pressure on the zooplankton resulting from a low predation exerted by a small cohort of young-of-the-year (YOY) zooplanktivorous fish, in turn caused by unfavorable conditions during spawning, egg development, or after hatching (Chapter 26). Alternatively, high mortality imposed by piscivorous fish could be responsible for sufficiently reducing the zooplanktivorous fish, allowing the macrozooplankton to be abundant. *Ultimate causes*, which usually reflect longer time scales, may include the nutrient loading from catchments which is much affected by land use plus year- to decade-long cycles in runoff (Fig. 5–15).

The undoubtedly correct position that top-down control is superimposed on the potential productivity determined by nutrient supply (Carpenter et al. 1985) is insufficient to reconcile the findings and views of individuals working at different scales. None of the principal determinants of the algal biomass and community structure in aquatic systems is constant over time or space, and all impacts manifest themselves on the algal community biomass with lags.

Autumn

Phytoplankton fall maxima are not uncommon, but among temperate zone lakes maxima are restricted to lakes combining a favorable $z_{eu}:z_{mix}$ ratio and sufficient nutrient supply to allow a noticeable biomass increase. A little-examined alternate scenario proposes that the cause of fall blooms is because of a reduction in top-down control of the algae, resulting from increased

---

[7]"Cataloging the outcome of single-factor studies is not synthesis. Ecologists tend to champion their favorite ecological factor (indeed some have made careers doing so) but collecting examples . . . and weighing their relative importance by the number of manuscripts in support of each, tells us little about how the world works." (M. D. Hunter and P. W. Price 1992)

predation pressure exerted on the macrozooplankton by rapidly growing YOY fish, or because of the negative effect of lowered temperatures on macrozooplankton growth. At higher latitudes, aquatic systems suffer from quickly shortening days that rapidly reduce the irradiance received at their surfaces. This reduction, together with a rapid rise in the thickness of the mixed layers resulting from thermocline deepening (Chapter 11), produces an unfavorable light climate (Sec. 10.11) for primary production even when nutrients are ample.

Tropical Lakes

The generalized picture painted above does not apply to low latitude lakes, where the variation imposed at higher latitudes by strong, regular seasonality in light and heat input is greatly reduced (Lewis 2000). Here the biomass cycles are best linked to often substantial seasonal or inter-year variability in runoff and associated nutrient supply, as well as to seasonal and short-term changes in water column stability (Sec. 12.5). Stability is affected by changes in density currents resulting from changes in river discharge and turbidity, wind speed, and net heat input (Sec. 11.4). Periodic reductions in stability allow for thickening of the mixed layer and increased recycling of nutrients from the frequently anoxic or hypoxic hypolimnia, and short-term variations in phytoplankton primary production, community biomass, and herbivorous zooplankton abundance.

The often considerable variation in timing and duration of wet seasons at low latitudes creates large inter-year variation in the direct nutrient input from rain (Chapters 4 and 8) and the quantity of nutrients (and particles) exported from catchments by rivers. The result is a much greater interannual variation in seasonal phytoplankton biomass cycle (Fig. 21–20) than is exerted by these same variables in the temperate zone. Other qualitative differences between mid and low latitude waters include the disproportionally great importance of omnivorous fish species at low latitudes who are able to feed on zooplankton, phytoplankton, detritus, or on benthic plants and animals. Moreover, a greater seasonal and interannual difference in river discharge and evaporation from water surfaces at low latitudes affects the size of littoral zones and associated wetlands that serve as feeding, breeding, and nursery areas for fish, waterbirds, and invertebrates (Chapters 24 and 26).

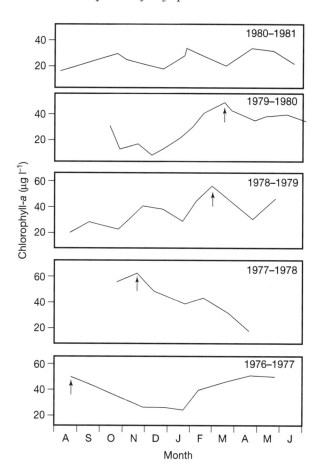

Figure 21–20 Seasonal changes in average chlorophyll-*a* concentrations in Lake Naivasha, KE (LA = 147 km², $\bar{z}$ = ~7 m) between 1976 and 1981. Note the absence of a regular seasonal pattern with distinct annual maxima (at arrow) in August, December, March, April, or lacking altogether (1980–1981). *(Modified after Kalff and Watson 1986.)*

▲ 21.5 The Composition of Phytoplankton Cells

All the components of living protoplasm, except lipids, have a density greater than fresh water (~1,000 gl⁻¹). The fat and oil content of algal cells usually accounts for only about two to 20 percent of the algal dry weight, therefore phytoplankton are slightly heavier than water and sink (Smayda 1970). The diatoms have silica deposits in their cell walls and are more dense than other algae; the larger species are particularly subject to sinking and removal from the euphotic zone of stratified lakes during periods of low turbulence (Sec. 20.4).

Cell Carbon

The organic carbon content of primarily cultured algae typically averages between 40 percent and 50 percent of their dry weight (Duarte 1992) (~15 percent of wet weight). The carbon content (C, pg) in nature is usually estimated from the volume (V, μm^3) of cells by equations, such as Rocha and Duncan's (1985), based on cultured freshwater algae:

$$\ln(C) = -1.952 + 0.996 \ln(V) \qquad \text{EQ. 21.1}$$

Precise carbon determinations must be made under laboratory conditions because water samples collected in nature contain nonliving particulate organic matter (*detritus*). Free-living bacteria and bacteria attached to detritus also cannot be readily separated from the algae by means of filtration or centrifugation. The detrital component greatly dominates the phytoplankton in oligotrophic and mesotrophic waters (Uehlinger and Bloesch 1989), but is small in highly eutrophic systems characterized by a disproportionately large living biomass (Harris 1986). In addition, the ratio of living-cell carbon to detritus changes seasonally within aquatic systems. Most estimates of living carbon are therefore based on recalculations from the necessarily imprecise measurements of seston carbon, cell volume, or indirectly from chlorophyll-*a* or the adenosine triphosphate (ATP) content of cells.

Much effort has been expended on the determination of carbon content in algae because it potentially provides a convenient measure of the community biomass (standing stock) in nature, and because carbon content is the basic measurement of primary production and energy flow in food webs (e.g., mg C m^{-2} t^{-1} or mg C m^{-3} t^{-1}). Yet such measurements remain far from unambiguous (Rocha and Duncan 1985) because of the large impact of methods and assumptions on results. For example, Ahlgren (1983) employed four widely used methods to estimate and compare the phytoplankton carbon content (g m^{-3}) of particulate matter (*seston*) samples from Lake Norrviken, SE, on 21–23 days during the spring (April) and fall (October). The estimated average carbon content ranged *fourfold* between 0.8 and 3.3 (g m^{-3}), depending on the technique used. Second, estimates of the *in situ* specific growth rates (mg C mg C^{-1} t^{-1}) varied about *fivefold* (half an order of magnitude), between a P:B ratio of about 0.04 and 0.22. It is evident that growth rates obtained by using different techniques for carbon con-

tent cannot and should not be equated and compared without question.

Chlorophyll-*a*

Biomass determinations from cell volume determined under the microscope are cumbersome and imprecise, therefore a large majority of aquatic studies have used chlorophyll-*a*, the dominant phytoplankton pigment, as an index or surrogate for the algal biomass. But Nicholls and Dillon (1978), Desortová (1981), Canfield et al. (1985), and others have shown much seasonal and intersystem variability in the chlorophyll-*a*: carbon ratio. This variability is in part the result of a wide range in chl-*a* content per cell within and among species resulting from different growing conditions (light, nutrients, temperature) and quantities of accessory pigments (e.g., xanthophylls in diatoms or phycobillins in cyanobacteria) that absorb light in other portions of the PAR spectrum and transfer energy to chlorophyll-*a* (Table 21–2). However, both high variability and a sometimes exceptionally high chl-*a*: biomass ratio may be an artifact if the picoplankton biomass is not measured during routine phytoplankton countings, whereas their contribution to the chl-*a* is captured in analyses.

Despite these reservations, chlorophyll-*a* determinations are widely and profitably used in research and management as a surrogate for community biomass. The estimates are easy to obtain and accurate enough for surveys and studies in which the chl-*a* interval scale (Sec. 2.6) is large. Chlorophyll-*a* plus other pigments are also detectable by means of satellite-borne sensors that, after *ground truthing* (calibration with field measurements), are widely used to estimate, the surficial algal and emergent macrophyte biomass in large lakes and oceans. Selected models predicting chl-*a* from algal biomass are presented in Table 21–2.

Nitrogen, Phosphorus, and Other Elements

Under optimal growing conditions in the laboratory, nitrogen contributes about one to nine percent ($\bar{x}$ = 5.5 ± 2.5), and phosphorus about 0.2–2.0 percent ($\bar{x}$ = 1.14 ± 1.41) of the dry weight of phytoplankton (Duarte 1992), with the geometric mean of N and P concentrations yielding an N:P ratio of 16:1 by atoms (7:1 by mass). The average atomic ratio is equal to the estimated 106C:16N:1P protoplasmic ratio of rapidly growing phytoplankton in nature (Sec. 16.1). The

Table 21–2 **Selected models of the relationship between phytoplankton fresh weight (B mg l^{1-}) and chlorophyll-*a* (µg l^{-1}).**

Equation	Correlation Coefficient	n	Locality and Comments
chl-*a* = 7.94 + 8.03·B	0.93	63	Lake Norrviken, SE; hypertrophic, cyanobacteria dominating
chl-*a* = 3.25 + 15.06·B	0.98	23	Lake Norrviken, SE; cryptomonads dominating
chl-*a* = 12.63 + 3.55·B	0.90	25	Holland Marsh, Ontario, CA; eutrophic
chl-*a* = 1.58 + 4.97·B	0.79	197	Czech reservoirs; individual measurements
chl-*a* = −1.69 + 6.38·B	0.87	11	Czech reservoirs; annual averages
chl-*a* = −10.42 + 3.41·B	0.79	13	Lake Naivasha, KE; monthly measurements
log (chl-*a*) = 0.55 + 0.50· log (B)	0.82	165	165 Florida lakes, US; single spring and summer mean values for each lake

Source: From Ahlgren 1970, Nicholls 1976, and Desortová 1981; modified after Kalff and Watson 1986, and Canfield et al. 1985.

protoplasm ratio is assumed to also represent the algal uptake (demand) ratio.

The uptake ratio, or *Redfield ratio* (Eq. 16.1), is sometimes used to provide a first estimate (after chemical analysis of the algal material) of whether P or N is most likely to be growth limiting (Healey and Hendzel 1979). Furthermore, by assuming the algal C:N:P ratio to be roughly constant, the amount of carbon fixed in primary production experiments provides an estimate of how much nitrogen and phosphorus was taken up from the environment at the same time. But the Redfield ratio is an idealized one, with different algal assemblages exhibiting considerable variation in nutrient concentrations and thus in the C:N:P ratio. While making assumptions is unavoidable in research and management, the effect of a particular assumption (e.g., P:B) on conclusions should be explored in a *sensitivity* or *simulation analysis*. This involves substituting a range of plausible values around the value selected and examining their impact on the conclusions drawn.

Much of the variation in cellular N and P concentrations in algae is the result of the organisms' ability to store some N and considerably more P, enough to satisfy the growth requirements for several cell divisions during periods when the uptake rate is higher than its immediate requirement (Duarte 1992). The excess uptake is commonly referred to as **luxury uptake** and the excess stored is known as **surplus** or **storage phosphorus** or **nitrogen**. Stored phosphorus or nitrogen is available for growth when the supply:demand ratio becomes unfavorable. When N and P (and Si, in the case of diatoms) becomes in such

short supply in the environment that growth is reduced after exhaustion of stored nutrients, the phytoplankton can subsequently accumulate considerable carbon during photosynthesis. The result is an elevation of cellular C:P or C:N ratios which commonly result in the release of some or all of the excess DOC into the water (Sec. 21.9).

Not only algal growth rates are reduced by a high C:P:N supply ratio; the herbivorous zooplankton too exhibit reduced growth rates when the protoplasmic C:P ratio of their prey is unsuitably high. It seems that, as the C:P supply rises above a threshold of ~300:1 (by atoms), the phosphorus content of the prey is increasingly too low for good growth of *Daphnia*. Where the inorganic nutrient content is ample, the nutritionally limiting factor shifts to the quality of the algal carbon (biochemical composition) (Chapter 23).

▲ 21.6 Algal Sedimentation and Buoyancy Control

The community biomass of lentic systems is the outcome of not only community growth rate but also of the rate algae are lost from the water column by sedimentation, predation, and other causes of death.

Particle Sedimentation

The density of phytoplankton protoplasm (diatoms excluded) is about two to nine percent greater than the density of their fresh water medium, whereas the diatoms with their silica-containing walls have densities

10–20 percent greater than fresh water (Reynolds 1984b). The excess density causes both nonmotile living and dead phytoplankton to sink.

The widely used modified Stoke's equation for estimating sinking rates (Eq. 20.3) suggests that nonmotile phytoplankton lacking buoyancy control (see below) have sinking rates that increase as a function of their radius squared, resulting in more rapid sinking of large cells. Sinking also increases with increasing cell density and varies inversely with the viscosity of the medium, allowing more rapid sinking in warm than cold water. Finally, sinking rates vary inversely with increasing "form resistance," indicating that protruberances reduce sinking speed. Moreover, such structural features make the cells appear larger to herbivores, making them less prone to predation.

Assuming a single mixing event and no growth or losses other than by sinking, the abundance at the end of the interval (N_t) is a function of the sinking velocity (V, m d^{-1}) and the thickness of the mixed layer z_{mix}:

$$N_t = N_0 \left(\frac{1 - V}{z_{mix}} \right)^{t-1} \qquad \text{EQ. 21.2}$$

Maintaining an abundance in the mixed layer requires not so much a low sinking velocity as a velocity that is low with respect to the thickness of the mixed depth (Fig. 21–21). However, the Stoke's equation applies only to cells subject to nonturbulent (laminar) flow. But turbulence in the mixed layer will not appreciably affect the sinking rate across the base of the mixed layer because laminar-flow conditions apply there, making the sinking loss of cells from a mixed layer independent of turbulence (Reynolds 1989, Horn and Horn 1993).

While the sinking velocity *among* nonmotile species is determined principally by the variables included in the Stoke's equation (Sec. 20.4, and Bloesch and Burns 1980), the variation in sinking velocity of any one species with a characteristic size and shape is a function of the physiological condition of the individual cells or cell units. For example, healthy diatoms have a sinking velocity that is about half that of dead or senescing cells, providing an explanation for the observation (Sommer 1984) that sinking velocities in hypolimnia are usually greater than the velocities of the more vigorous cells present just below the bottom of the euphotic zone.

Sinking Velocities and Their Effect

Most of the nonmotile larger microplankton sink at rates of between 0.2–2.0 m d^{-1} but with much variation among species, and much variation for single species over time (Sec. 20.4). While the average sinking velocity of filaments of living *Aulacoseira italica* (a large diatom, Fig. 21–4) in a German reservoir averaged 1.2 m d^{-1} the velocities ranged between 0.5 m d^{-1} and 4 m d^{-1} (Horn and Horn 1993). In contrast, very few nanoplankton and picoplankton are collected in traps set just below epilimnia, indicating that their sinking rates are negligible and implying that their organic matter and associated plant nutrients are almost totally recycled within mixed layers rather than lost to hypolimnia (Sec. 20.4).

A high loss rate from a euphotic zone must be offset by high growth rates if a species is not to disappear. A high growth rate is particularly essential for heavy diatoms in wind-protected lakes. Such lakes typically have thin epilimnia (e.g., 2–3 m) (Sec. 11.7) from

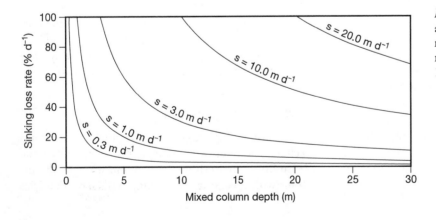

Figure 21–21 Specific sinking loss rates as a function of mixed column depth for nonmotile organisms with given sinking rates using the equation $r_s = (s \times h_m)^{-1} \, 100$. *(Modified after Reynolds 1989.)*

which diatoms in particular can be quickly lost through sedimentation during a few calm days (Eq. 21.2 and Fig. 21–9). As a consequence, the large diatoms contribute appreciably to the biomass of such lakes only during periods of overturn. However, unless diatoms exhaust the silica supply they commonly remain an important component of the biomass in lakes with a thick z_{mix}, as well as in shallow lakes where sedimented cells are resuspended at intervals (Figs. 21–2). In most eutrophic lakes and many large eutrophic rivers, the summer replacement of large diatoms by large filamentous or colonial cyanobacteria appears not to be the outcome of an inferior growth rate or nutrient acquisition rate of the diatoms but rather the result of a higher sinking loss rate than for large cyanobacteria containing gas vacuoles (Knoechel and Kalff 1975).

It is evident that sedimentation losses in lakes are much greater for large nonmotile and nonbuoyant algae than for small phytoplankton, many of which are motile as well. Conversely, losses imposed by zooplankton grazing are much greater for small species than large species (Sec. 23.10). Nanoplankton and picoplankton dominate oligotrophic lakes (Figs. 21–10, 11 and 13) and algal losses through sedimentation from mixed layers are expected to be proportionally lower on an annual basis than from eutrophic lakes dominated by large forms subject to little predation by herbivorous macrozooplankton. Predation and sedimentation have different effects on organic matter and nutrient cycling. Sedimentation involves stripping the material from the mixed layer and much decomposition in the profundal zone and sediments, whereas grazing entails regeneration (recycling) of the nutrients within the water column.

Buoyancy Control

Buoyancy in large nonmotile microplankton is partially controlled by changes in intracellular gas vacuole formation and changes in polysaccharide storage (*ballast*). Gas vacuole production—restricted to prokaryotic organisms—plays an important role in the buoyancy regulation of filamentous and colonial blue-green algae (cyanobacteria). The gas-filled structures have a much lower density than the surrounding water. Whether such a cyanobacterial cell or colony will be positively buoyant and rise, negatively buoyant and sink, or neutrally buoyant under nonturbulent conditions is determined by the fraction of the cell occupied by gas-filled vacuoles and the amount of photosynthate laid down as ballast, which affects the

density (Fig. 21–22). Whatever the precise mechanisms employed, buoyancy-controlled vertical movements in an upward direction of between about 0.5 m hr^{-1} and 3 m hr^{-1} have been recorded for large cyanobacteria under conditions of little or no turbulence (Paerl 1988). The rates are of the same magnitude (1–2.5 m h^{-1}) as the vertical migration of large flagellated phytoplankton (Sommer 1988).

Floatation and Surface Blooms

Scums or surface blooms of large blue-green algal colonies or filaments are particularly evident in nutrient-rich lakes in summer following a period of calm sunny weather.

The development of unsightly cyanobacterial surface scum (blooms) plus accumulation on downwind shores was responsible in the 1970s and 1980s for arousing the awareness of the public in many economically developed countries about the need to reduce the nutrient loading of lakes and slowly flowing lowland rivers.

Buoyancy control by large cyanobacteria or the motility of flagellated algae is insufficient to overcome the turbulence generated in the mixed layer by light winds. Even modest winds of 3–5 m s^{-1} produce enough turbulence in the mixed layer (vertical diffusion coefficient, $K_v > 0.01$ cm s^{-1} (Sec. 12.5), to exceed the movement of most phytoplankton by at least an order of magnitude (Reynolds 1984b). Even so, both buoyancy control and motility confer a considerable competitive advantage in the mixed layer whenever turbulence is low, and in metalimnia, allowing the organisms to track the most favorable light and nutrient conditions.

▲ 21.7 Parasitism and Disease

The observed phytoplankton biomass is the outcome of a host of growth and loss processes changing over time and space (Kalff and Knoechel 1978). The loss processes include not only species-specific sinking and predation (Chapter 23) but also species-specific rates of infection and death by pathogens. The latter have received relatively little attention from limnologists. Among the most important pathogens are viruses (Sec. 22.8), lysing bacteria, parasitic protozoa, and fungal parasites. Most pathogens appear to be specific to a particular species at a particular time. Among the

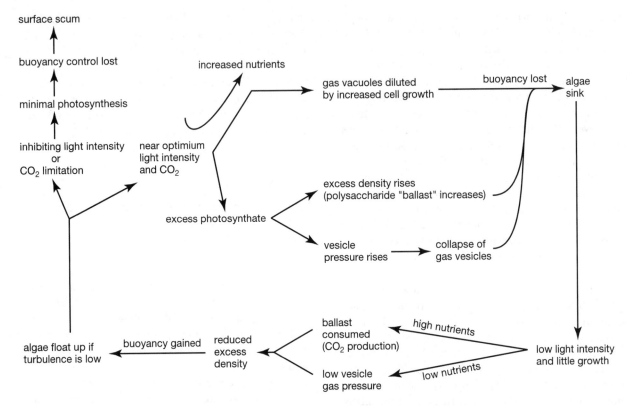

Figure 21–22   Conceptual model of the relationship between environmental factors, biosynthesis, and buoyancy in large cyanobacteria. More recent studies pay greater attention to the role of polysaccharide ballast than turgor pressure in buoyancy control. *(Modified after Reynolds and Walsby 1975.)*

protozoans, those species that extract the cell contents of their host are recognized as parasites; while others at the opposite end of the pathogen–predator continuum engulf their prey and are considered to be true microzooplankton predators.

Best-studied are the uniflagellate *Chytridiales* (*Chytridomycetes*)—known as **chytrids**—and their parasitic interaction with diatoms. These parasitic fungi have a life-cycle stage in which large numbers of free-swimming flagellated bodies, known as **zoospores**, are formed and released from the parent body or **sporangium**. The zoospores settle on phytoplankton cells, penetrate the cell with their flagellum and transfer host protoplasm to the externally attached zoospores. The zoospores grow and become sporangi which in turn form new zoospores (van Donk 1989). In the course of an epidemic there is an increase first of recently encysted zoospores, followed by a rise in sporangia and the death of the algal cells that bear them. Infection is light-dependent and related to the

photosynthetic conditions experienced by the host, the effect of temperature on the parasite, as well as by the host's density (Bruning 1991). Detailed work on a Japanese lake (Lake Suwa) suggests that high chytrid-induced diatom mortality is uncommon there because it requires not only an infrequently observed high diatom infection rate (> 30%) but also temperatures high enough to permit a high chytrid growth rate (Kudoh and Takahashi 1990). The difficulty in converting the percentage of parasitized cells into death rates is that the time between infection and cell death is generally unknown in nature. Even so, parasitism is an important loss mechanism which requires more investigation.[8]

---

[8]"Although parasitism is likely to be just as important as other loss processes, many of the dynamic models of phytoplankton-biomass changes have incorporated factors for the rate of removal by herbivores and for sedimentation, but losses due to parasite attacks have not been considered." (van Donk 1989)

21.8   Photosynthesis, Light, and Temperature

Experiments with nutrient-sufficient algal cultures has repeatedly shown a similar photosynthesis–irradiance (P–I) relationship. The P–I curve is described by a rectangular hyperbola, where the asymptote represents the **maximum photosynthetic rate** (P_{max}) while the initial linear slope is known as the alpha (α) region. The specific photosynthetic rate—the rate per unit biomass or chl-*a*—rises linearly with increasing irradiance over the low irradiance portion of the slope (α) where the algae are **light limited**. Photosynthetic rates are controlled by photochemical reactions in this region that are largely independent of temperature. The steepness of the light-limited slope (α) is determined by the nutrient, light, and temperature history of the culture, as well as by species-specific attributes. At higher intensities, the algae become increasingly **light saturated**. The onset of light saturation (I_k) is described by the point on the line where the slope becomes nonlinear (Fig. 21.23). The onset of light saturation in nature generally falls between 0.14 and 0.72 μmol m^{-2} h^{-1}. At saturating intensities biochemical (enzymatic) rather than photochemical reactions are rate-limiting and regulated by temperature (Davison 1991). At superoptimal intensities the photosynthetic rate (slope β) declines below the maximum observed

(P_{max}) as the algae become progressively more **photoinhibited**. Photoinhibition in nature commonly commences at about three times I_k (Harris 1978).

Although the effect of temperature on the P–I response is easily demonstrated under nearly optimal growing conditions in the laboratory, it appears that constraints imposed by low irradiance, usually encountered during circulation in the mixed layer, or nutrient limitations are the principal determinants of growth in oligotrophic systems (Fee et al. 1987, and Markager et al. 1999). However, where light and nutrient conditions are favorable, as in shallow nutrient-rich lakes, streams, and wetlands, the effect of temperature is readily demonstrated (Jewson 1976). Furthermore, temperature is the principal determinant of respiration rates and thus of the fraction of the photosynthetically produced organic carbon that is lost in respiration and not available for growth (Sec. 21.9).

▲ 21.9   Photosynthesis, Respiration, and Growth

The relationship between the photosynthetic process, the organic matter fixed, the dissolved oxygen (DO) evolved, and the growth of phytoplankton is not straightforward; nor are measures of primary production simple to interpret.

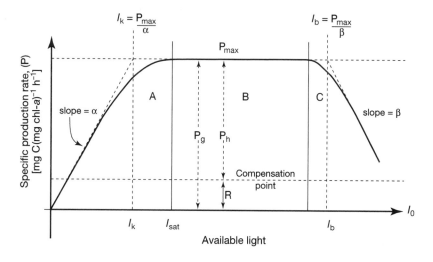

Figure 21–23 A generalized photosythesis–irradiance (P–I) curve. Useful parameters include α, the light-limited initial slope serving as an index of the efficiency with which quanta of light energy are utilized (quantum energy); I_k, an index of the onset of light-saturated photosynthesis reflecting the photoadaptive state of the particular species or community; P_{max}, the light-saturated rate of photosynthesis and a measure of the photosynthetic capacity of the cells under the particular nutrient and temperature conditions encountered; and I_b, the parameter indicating the onset of light inhibition. Zones A, B, and C represent the zones of light limitation, light saturation, and light inhibition, respectively. P_g and P_n represent gross and net photosynthesis, respectively, with $P_n = P_g - R$ (respiration). I_{SAT} is the flux at which photosynthesis becomes light-saturated (onset P_{max}).

The overall and greatly simplified photosynthetic equation:

$$6CO_2 + 12H_2O \xrightarrow{\text{light}} C_6H_{12}O_6 + 6H_2O + 6O_2$$

EQ. 21.3

shows that if only simple carbohydrates were produced, each mole of CO_2 taken up would result in the release of one mole of oxygen. In that case, the **photosynthetic quotient** (PQ)—defined as $+O_2/-CO_2$ by volume—is 1.0 and the **respiratory quotient** (RQ = $+CO_2/-O_2$) would be unity as well. However, most recorded PQs range between one and two (commonly ~1.2). A PQ higher than 1.0 tells us that algae typically synthesize a mixture of carbohydrates, fats, and proteins rather than only carbohydrates.

Excess carbon is excreted when the rate of carbon fixation temporarily exceeds the rate at which it can be either converted to protoplasm or stored as carbohydrates in the limited physical space available—when growth rates and photosynthetic rates are not balanced. An analysis of marine and fresh water studies shows that an average of ~13 percent of the photosynthate is excreted, but that the **extracellular release** (ECR – $\mu g\ C\ l^{-1}\ hr^{-1}$) varies widely between three and 41 percent among systems (Baines and Pace 1991). The carbon lost by ECR is naturally not available for algal growth and measures of carbon uptake therefore overestimate the amount of carbon that is assimilated and converted to protoplasm. While the released carbon is not available to the plants that produced it, it plays an important role in the nutrition of aquatic bacteria (Chapter 22). A second means by which photosynthate becomes unavailable for growth is through **photorespiration**, a significant loss process when the photosynthetic system is light saturated (at P_{max}) or photoinhibited (Harris 1978, Raven and Beardall 1981). The terms "carbon fixation," "production," and "photosynthesis" are here, and commonly elsewhere, used interchangeably but they do not have the same meaning. *Photosynthesis* refers to the process only, *gross production* to the rate of carbon fixation before respiration is taken into account, and *net production* refers to the gross production rate minus respiration. *Production* makes no statement as to whether gross or net production is measured. *Carbon fixation* refers simply to the rate carbon is fixed per period, usually measured by the uptake of $^{14}CO_2$ during photosynthesis, but without claims as to its fate. The term *primary production* refers to the rate at which energy is stored by photosynthetic activity in the form of organic substances which can be used as food (E. P. Odum 1959). It is used interchangeably with *primary productivity*, which once referred to yield rather than rate. For techniques used to measure primary production from dissolved oxygen (DO) evolution or the uptake of radiocarbon ($H^{14}CO_3$) see Wetzel and Likens 2000.

Carbon Assimilation and Respiration

While the ratio of carbon assimilation to respiration has been shown to be greatest at low temperatures, respiration becomes increasingly more significant as the temperature increases (Jewson 1976). The implication is that a considerably higher photosynthetic rate is needed for a specified amount of growth (net production) at high temperatures. This conclusion is of considerable significance in interpreting phytoplankton growth under poor light conditions in cool metalimnia and in snow and ice-covered lakes. The same low light flux would be insufficient in warmer water to compensate for a much higher respiration rate there.

The gross carbon-fixation rate at the depth of maximum photosynthesis (P_{max}, Figs. 21–23 and 21–24) normally greatly exceeds the algal respiration rate, but an appreciable fraction of the phytoplankton spend time under light-limited conditions in the deeper portions of the mixed layer. Thus, the phytoplankton community in Lunzer Lake, AT, was light-limited during incubation below about 2 m, whereas it was light-limited at all depths (except possibly at the very surface) in highly eutrophic Zeller Lake (AT, Fig. 21–24). At some depth in the mixed layer, the gross carbon-fixation rate (P) of the average cell is reduced or even declines to zero. However, respiration (R) is unaffected and the depth-integrated net production (growth) for the mixed layer as a whole may be modest, even though the rates of net carbon fixation are high at the depth where photosynthesis is maximal (P_{max}). Furthermore, for growth the carbon that is fixed during the day has to exceed the sum total of daytime and nighttime respiration.

21.10 Primary Production in Nature

Production Patterns and Their Modeling

The photosynthesis–irradiance relationship (P–I) observed for a particular species in the laboratory (Fig.

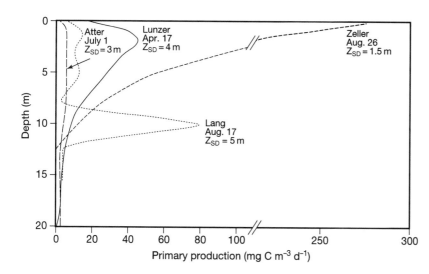

Figure 21–24 Primary production in four Austrian alpine lakes in 1960 as measured by the in situ uptake of $^{14}CO_2$. Oligotrophic Atter Lake (z_{SD} = 3 m) does not show a production maximum or the pronounced surface inhibition shown by mesotrophic Lunzer Lake. Transparent Lang Lake (z_{SD} = 5 m) exhibits a metalimnetic phytoplankton maximum, while highly eutrophic Zeller Lake (z_{SD} = 1.5 m) has a sufficiently high biomass and resulting light extinction to exhibit the maximum photosynthesis at the surface. *(Derived from Findenegg 1964.)*

21–24). The P–I curve characterizing a phytoplankton community in the mixed layer (Fig. 21–25c) can be combined with the observed light extinction (Fig. 21–25b), incoming irradiance (Fig. 21–25a), and vertical distribution of the algal biomass (Fig. 21–25e–h) to conceptually explain the observed vertical carbon-fixation profiles (Fig. 21–25i–l).

▲ Talling (1957) developed an important hybrid mechanistic-empirical model, describing water column integrated rates of photosynthesis based on the impressive early modeling of the P–I relationship by E. L. Smith (1936). The simple Talling model describes the water column integrated photosynthetic rates during the incubation period as a function of P–I curve attributes (P_{max}, I_k), irradiance entering the water column (I_o), the light extinction coefficient (k_{min}), and an assumed homogeneously mixed algal biomass (B) as

$$\Sigma\bar{A} = \frac{P_{max} B}{1.33 \, k_{min}} (\ln I_o - \ln 0.5 \, I_k) \quad \text{EQ. 21.4}$$

where $\Sigma\bar{A}$ = hourly rate of areal photosynthesis (mg O_2 m^{-2} h^{-1}) or carbon uptake (mgC m^{-2} hr^{-1}), P_{max} = photosynthetic capacity at saturation [mgO_2 chl-a^{-1} hr^{-1} or mgC chl-a^{-1} hr^{-1}, B = phytoplankton community biomass (as mg chl-a m^{-3}), k_{min} = vertical extinction coefficient of the most penetrating spectral region, or k_{PAR} (ln units m^{-1}), I_0 = surface flux of PAR (W m^{-2}),

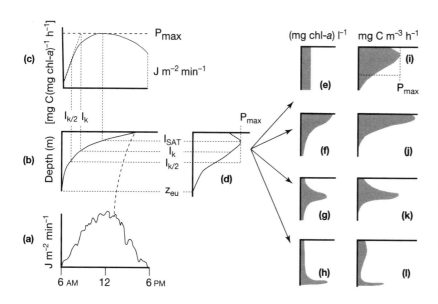

Figure 21–25 The surface irradiance (a), the light extinction with depth (b), and the P–I curve (c) which allows the specific primary production (μg C mg chl-$a^{-1}h^{-1}$) to be estimated as a function of the estimated light climate (d). The specific primary production (mg C chl-a^{-1} t^{-1}) multiplied by the algal biomass (as chl-a) at each depth (e–h) yields the different photosynthetic profiles shown (i–l, mg C $m^{-3}h^{-1}$). The first profile is common in well-mixed epilimnia of mesotrophic lakes, the second in eutrophic lakes with surface blooms, the third in those lakes with a metalimnetic biomass maximum, and the fourth in shallow transparent lakes and streams with most of the biomass on the sediments. *(Modified after Capblancq 1982.)*

k_{PAR} (ln units m^{-1}), I_0 = surface flux of PAR (W m^{-2}), and I_k = irradiance characterizing the onset of light saturation.

The simple Talling model accurately predicts *in situ* values of $\Sigma\bar{A}$ (Jones 1977), despite the fact that in contrast to more sophisticated recent modeling efforts (see Prézelin et al. 1991), it does not account for possible light inhibition near the surface on bright days.

Empirical relationships between community algal biomass (as chl-*a*) and areal rates (per m^2) of photosynthesis and production (Table 21–3) are based on a roughly constant relationship between production per unit biomass in the mixed layer and irradiance, thereby allowing the areal production to be approximated from the measured biomass. The community biomass further serves as a surrogate for the maximum volumetric primary production [A_{max} mg C (or mg O$_2$) m^3 t^{-1}] at the depth of **saturating (optimal) irradiance** (I_{sat}). Volumetric production at I_{sat} rises approximately linearly with increasing chl-*a* to a maximum at the highest phytoplankton chl-*a* levels encountered (~1,000 mg m^3). Yet, higher volumetric rates are noted in benthic microalgal mats where the cells are more tightly packed, albeit in a thin layer (Fig. 21–26).

Daily and Annual Production Rates

Areal rates of annual phytoplankton production in lakes vary over about three orders of magnitude—between the ~4 g C m^{-2} yr^{-1} recorded in an ultraoligotrophic high arctic lake and measurements made in a hypereutrophic (hypertrophic) subtropical reservoir (~ 5,700 g C m^{-2} yr^{-1}) (Table 21–4). The true range is even larger but annual data are lacking for the most ultra-oligotrophic antarctic lakes. Measurements taken during summer in one such lake (Lake Vanda, Vincent and Vincent 1982) show daily rates to be only about half as large as those measured in the least productive polar lake (Char Lake), suggesting an annual rate of only about 2 g C m^{-2} yr^{-1} in ultra-oligotrophic antarctic lakes. But, as most of the primary production in

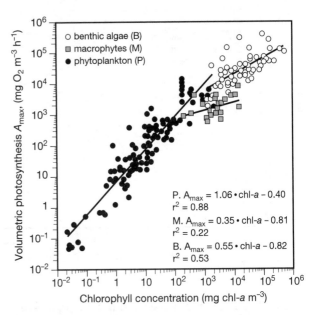

Figure 21–26 Maximum volumetric rates of primary production (A_{max}) as a function of the chlorophyll-*a* concentration in phytoplankton communities, macrophyte stands and benthic algal mats. Note the importance of chl-*a* as a predictor of photosynthesis over a large range (interval) scale, whereas chl-*a* has no predicitve power over a narrow range scale (~10fold), over which variables other than chl-*a* determine rates of photosynthesis (Sec. 2.6). (*After Krause-Jensen and Sand-Jensen 1998.*)

highly transparent lakes is carried out by benthic plants (Fig. 24–17), the variation in areal primary production (phytoplankton plus benthic plants) is much smaller than for the phytoplankton alone. The relative contribution of phytoplankton to whole system production increases with increasing turbidity because the benthic algae are increasingly outcompeted for light (Chapter 24).

The exceptionally high areal rates of *daily* phytoplankton primary production in hypertrophic low lati-

Table 21–3 **Relationships between hourly (*A*) and daily (*ΣA*) rates of integrated primary production (mg C m^{-2} t^{-1}) and chlorophyll-*a* concentration (mg chl-*a* m^{-2}) in epilimnia.**

Equation	r^2	Location
$A = 0.85 \cdot$ chl-*a* + 9.46	0.58	Cooking Lake (CA)
$\Sigma A = 0.14 \cdot$ chl-*a*	0.91	Lake Chad (CM, NE, NG, TD)
$A = 1.20 \cdot$ chl-*a* + 17	0.76	Lake Constance (AT, DE, CH)

Source: After Cabrera and Montecino 1984.

Table 21–4 **Phytoplankton primary production for selected lakes, listed from lowest to highest annual production. Results obtained with the dissolved oxygen technique were divided by three to obtain the carbon equivalent. S = saline. Note the (1) highest daily and annual production rates are, with few exceptions, encountered at low latitudes; (2) large seasonal variation in primary production, even at low latitudes; and (3) low and high photosynthetic capacity, at a combination of low temperature/low nutrients and high temperature/high nutrients, respectively. ND = not determined.**

Lake	Latitude	Maximum Photosynthetic Capacity, P_{max}; [max. mg C (mg chl-a)$^{-1}$ hr^{-1}]	Maximum Rate of Fixation, A_{max} (mg C m^{-3} hr^{-1})	Range of Daily Integral Fixation, ΣA (mg C m^{-2} d^{-1})	Annual Production, $\Sigma\Sigma A$ (g C m^{-2} yr^{-1})
Char, CA	74°42'N	0.7	1	0–70	4
Hakojärvi, FI	61°15'N	0.6	6	10–77	5
Watts, AQ	68°36'S	17.3	6	0–340	10
Meretta, C	74°42'N	2.7	10	0–800	11
Aleknagik, US	59°20'N	3.0	2	73–207	13
George, US	43°31'N	ND	31	25–227	16–28
Pink, AU (S)	38°02'S	3.7	19	5–18	24
Port Bielh, FR	42°50'N	2.2	5	0.1–242	25
Neusiedler, AT	47°50'N	7.0	87	10–733	41
Biwa, JP	35°00'N	3.7	37	6–390	60–90
Little Manitou, CA, (S)	51°48'N	5.3	146	0–1,188	70
Trummen, SE	56°52'N	3.3	433	500–2,433	180
Castanho, BR	3°00'S	15.7	200	~1,900	290
Humboldt, CA (S)	52°09'N	8.5	433	333–1,3333	680
Maggiore, CH, IT	45°57'N	6.3	38	233–1,800	365
Leven, GB	56°10'N	6.7	343	133–7,000	340–620
Corangamite, AU (S)	38°05'S	8.0	367	187–3,333	760
Zeekoe, ZA	~34°05'S	7.9	1,524	1,200–4,350	860–1,000
Chilwa, MW (S)	15°30'S	ND	ND	1,800–4,333	730–1,300
Søbygaard, DK	~56°15'N	9.2	3,700	3,000–12,000	766–1,862
Red Rock, AU (S)	38°05'S	40.0	6,333	60–17,667	2,200
Aranguadi, ET (S)	9°39'N	6.0	10,000	~19,000	ND
Hartbeespoort Dam, ZA	25°45'S	14.3	5,916	400–30,900	5,712

Source: Data from Westlake 1980, Heath 1988, Hammer 1986, Harding 1997, and Robarts and Zohary 1984.

measured in ultra-oligotrophic polar lakes. The very low annual phytoplankton production of polar waters is usually the result of a severe nutrient limitation, a short growing season, and low water temperatures. The addition of sewage wastewater to one such lake (Meretta, CA) raised annual production threefold over that in the adjacent but unfertilized Char Lake (Table 21–4). The substantial nutrient addition was insufficient to raise the annual production more, probably because of low water temperatures. Low temperatures (usually < 4°C) severely limit primary production at

optimal light intensities in the water column (I_{sat}), but do not prevent a slow accumulation of a substantial community biomass which, combined with the water color and modest irradiance, contributes to a low *effective light climate* (Sec. 10.7). Together, low temperatures and a poor light climate allow an only moderate response to increased nutrient supply. However, low primary production rates (Table 21–4) are not restricted to polar regions.

The most productive systems on an annual basis are shallow, extremely nutrient-rich lowland lakes at low latitudes. The combination of a variable but year-round high irradiance and shallow water column's yields a high effective light climate. This, combined with high water temperatures and high nutrient levels, yields a very high biomass and results in high areal production rates. The highest production rates recorded at the depth of optimal irradiance (I_{sat}) are the product of a high **photosynthetic capacity**—also known as the **assimilation number** [mg C (mg chl-a)$^{-1}$ hr^{-1}]—and an exceptionally high biomass (Table 21–4).

Hypertrophic Lakes

Certain shallow saline lakes are among the most productive lakes globally, but Hartbeespoort, a South African freshwater reservoir, appears to be the most productive lake on record (Table 21–4). A combination of high year-round irradiance, high temperatures, large wastewater loading into a shallow mixed layer permit the extraordinarily high depth-integrated production rates observed.

High water temperatures in inland waters at low latitudes allow a high rate of microbial decomposition and recycling of nutrients contained within the organic matter respired, making them quickly available for uptake and growth by the algae. In the temperate zone, the ideal growing conditions are most closely approached in shallow but hypertrophic lakes (e.g., Lake Søbygaard, DK) in summer (Table 21–4). Even so, lower water temperatures and a resulting lower P_{max} appear to be responsible for distinctly smaller *daily* rates (Table 21–4) than for their tropical counterparts (Osborne 1991, Jeppesen et al. 1997a). On an *annual* basis, the lower annual depth-integrated production rates in hypertrophic temperate zone and high latitude lakes are primarily attributable to a combination of a long winter period of low irradiance and low water temperatures.

Photosynthetic Efficiency

The efficiency (%) of light utilization, the **photosynthetic efficiency**, of aquatic plants in whole systems can be obtained by dividing the photosynthetically stored energy on an areal basis (g C m^{-2} d^{-1}) by the photosynthetically available energy (PAR, µmol m^{-2} d^{-1}) after converting to comparable energy units. Computed phytoplankton efficiencies are typically well below one percent, but reach to between one and four percent in shallow eutrophic systems (Harding

1997). The one percent value based on phytoplankton is a large underestimate of whole system efficiencies in clear oligotrophic lakes where most of the production is contributed by benthic plants (Fig. 24–17). Even so, light scattering and absorption in water columns is large enough to reduce the photosynthetic efficiency of highly productive aquatic systems to well below that encountered on land.

21.11 Production:Biomass (P:B) Ratios and Specific Growth Rates in Nature

The P:B ratio is defined as the ratio of carbon fixed (P) to phytoplankton biomass, expressed as carbon (B) per unit time. It provides a useful measure of the rate at which algal carbon turns over. The P:B complements the photosynthetic capacity which expresses the ability or potential to fix carbon per unit of chlorophyll-a [mg C (mg chl-a)$^{-1}$ t^{-1}]. From summer measurements of the P:B and P:chl-a ratios in mixed layers some important generalities have emerged:

1. The average P:B ratio in mixed layers is high in transparent oligotrophic systems (Fig. 21–27), and is low within and among aquatic systems when the community biomass is high (eutrophic waters). Low ratios in eutrophic waters imply a low productivity per unit biomass or chlorophyll-a (Tereshenkova 1985). Low average specific growth rates further imply that loss rates must also be low to allow the high biomass to be maintained.

2. The normally high P:B observed in nutrient-poor transparent waters commonly dominated by small species (Fig. 21–27) implies a rapid recycling of both organic matter and the nutrients in high demand. Rapid recycling further requires an efficient coupling between the phytoplankton, their zooplankton grazers, and associated microbes.

3. A reduced community P:B under conditions of high phytoplankton biomass is primarily attributable to a poor effective light climate, which is the result of **self shading**. Light limits the growth rate of the individuals and the community, not nutrients. Superimposed is the reduced growth potential of the typically large-sized phytoplankton species dominating most eutrophic lakes and slowly flowing rivers (see Fig. 21–13). While the community P:B declines with increasing self shad-

Table 21–5 The prediction of summer chlorophyll-*a* (μg l^{-1}) from spring or summer total phosphorus (μg l^{-1}) and/or total nitrogen (μg l^{-1}) in a variety of rivers and lakes. Note the (1) different amounts of chl-*a* predicted per unit TP or TN; (2) variation in the fraction of the variance (r^2) explained; (3) difference in the chl-*a* predicted per unit TP in stratified versus unstratified lakes and the better predictions possible in low TP (< 30 μg l^{-1}) than high TP (> 30 μg l^{-1}) lakes; and (4) large variation in the sample size (n) and the nutrient concentration range examined (not shown).

Equation	Location	Model	r^2	n	Authors
lotic					
1.	North temperate zone	log (chl-*a*) = −1.65 + 1.99 · log (TP) − 0.28 · [log (TP)]^2	0.67	292	Van Nieuwenhuyse and Jones 1996
2.	Eastern Canada	log (chl-*a*) = −0.26 + 0.73 · log (TP)	0.76	31	Basu and Pick 1996
lentic					
3.	northern and western Europe and northern North America[1]	log (chl-*a*) = −0.55 + 0.96 · log (TP)	0.77	77	OECD 1982
4.	northern and western Europe and northern North America	log (chl-*a*) = −0.39 + 0.87 · log (TP)	0.69	133	Prairie et al. 1989
5.	northern and western Europe and northern North America	log (chl-*a*) = −3.13 + 1.44 · log (TN)	0.69	133	Prairie et al. 1989
6.	northern and western Europe and northern North America	log (chl-*a*) = −2.21 + 0.52 · log (TP) + 0.84·log (TN)	0.81	133	Prairie et al. 1989
7.	midwestern United States	log (chl-*a*) = −1.09 + 1.46 · log (TP)	0.90	143	Jones and Bachmann 1976
8.	largely from Japan and North America[1]	log (chl-*a*) = −1.13 + 1.58 · log (TP)	0.95	56	Dillon and Rigler 1974
9.	Alberta, CA; nonstratified	log (chl-*a*) = −0.68 + 1.25 · log (TP)	0.69	25	Riley and Prepas 1985
10.	Alberta, CA; stratified	log (chl-*a*) = −0.56 + 1.02 · log (TP)	0.64	31	Riley and Prepas 1985
11.	Florida, US	log (chl-*a*) = −0.15 + 0.74 · log (TP)	0.59	223	Canfield 1983
12.	Florida, US	log (chl-*a*) = −2.99 + 1.38 · log (TN)	0.77	223	Canfield 1983
13.	Florida, US	log (chl-*a*) = −2.49 + 0.27 · log (TP) + 1.06 · log (TN)	0.81	223	Canfield 1983
14.	Argentina; all lakes	log (chl-*a*) = −1.94 + 1.08 · log (TP)	0.78	97	Quirós 1990
15.	Argentina; lakes TP≤ 30	log (chl-*a*) = −2.60 + 1.44 · log (TP)	0.75	57	Quirós 1990
16.	Argentina; lakes TP>30	log (chl-*a*) = −2.03 + 1.06 · log (TP)	0.44	40	Quirós 1990
17.	North Island, NZ	log (chl-*a*) = −1.13 + 1.35 · log (TP)	0.84	21	Pridmore et al. 1985
18.	North Island, NZ	log (chl-*a*) = −2.56 + 1.22 · log (TN)	0.73	16	Pridmore et al. 1985

[1]Lakes with inorganic N:PO$_4$P ≥ 10.

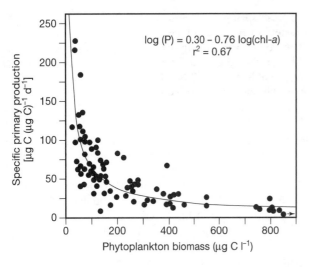

Figure 21–27 Specific primary production, also known as the *activity coefficient*, as a function of phytoplankton biomass in the euphotic zone of the Kličava Reservoir (Czech Republic) determined over 24 hours between March and October over a 7–10 year period. Note that the primary production per unit biomass typically declines with increasing biomass, with the biomass serving also as a surrogate for the nutrient concentrations. The commonly elevated specific production in transparent oligotrophic systems reflects dominance by rapidly growing picoplankton and nanoplankton (see Fig. 21–12). The low community growth rates at high biomass (high nutrients) usually appear to be the result of dominance by slower growing microplankton and poorer light conditions created by self shading. The specific production, especialy during periods of low biomass, is overestimated as the picoplankton biomass was not considered. *(Modified after Javornický 1979, in Westlake 1980.)*

slowly flowing rivers (see Fig. 21–13). While the community P:B declines with increasing self shading, the areal community production (mg C m^{-2} d^{-1}) continues to rise with increasing nutrient richness but at a progressively slower rate. The fascinating outcome is that the most productive lakes on an areal basis are inhabited by, on average, the slowest growing phytoplankton, characterized by low average P:B and P:chl-*a* ratios and able to minimize losses rather than maximize growth.

4. An among-system examination of the P:B of larger microplankton (> ~70 μm) and nanoplankton (2–30 μm) shows no change in the average P:B of the microplankton with increasing levels of critical nutrients (Watson and McCauley 1988). But as the nanoplankton P:B appears to increase (Fig. 21–15), nutrient-rich lakes should become dominated by

small algae. The fact that they are usually dominated in summer by large microplankton (Fig. 21–13) indicates a disproportionately large loss rate of nano- and picophytoplankton in eutrophic systems, thought to be largely attributable to zooplankton predation supplemented by other less well-investigated loss processes. The, at the same time, disproportionately small loss rate to predation and sedimentation of dominating large cyanobacteria with buoyancy control and by large dinoflagellates contributes to a typical summer dominance of eutrophic lakes by large algae.

▲ Specific Growth Rates in Nature

Observed changes in population density are the net outcome of growth in the water column plus inputs from lake or inflowing stream sediments and from upstream lakes, minus the losses resulting from sedimentation, predation, disease, physiological death, and export via the outflow. Determination of the different gain and loss terms is difficult, uncertain, and highly time consuming. The observed population growth rate of assumed exponentially growing organisms (μ) in nature is commonly considered to be the measured increase in the biomass (B) over a time interval ($t_2 - t_1$), during which losses are assumed to be negligible.

$$\mu = \frac{\ln B_{t_2} - \ln B_{t_1}}{t_2 - t_1}, \qquad \text{EQ. 21.5}$$

where μ is an **exponential growth constant** and the **generation** or **doubling time** (t_g), the time needed for the population to double, is

$$\frac{\ln 2}{\mu} = \frac{0.693}{\mu} \qquad \text{EQ. 21.6}$$

The spatially and temporally changing growth and loss rates experienced by individual species is reflected by changes in species composition and community biomass over time. Periods of stability reflect a balance between the growth rate and the sum of the various loss rates.[9]

Maximum specific growth rates in both surface waters and in the laboratory, where loss rates can be

---

[9]Growth and loss rate estimates and their interpretation among all organisms, are greatly influenced by the sampling interval or time scale. For rapidly growing algae that also suffer high loss rates, a *weekly* sampling may encompass seven or more generations that experienced temporally and spatially large variations in *daily* growth and loss rates. Growth rate discrepancies reported are attributed to many different causes, but almost never linked to differences in time scale.

reduced to zero, range from about 0.1–0.6 per day for large microplankton (> 70 μm) to about 1.5–2.0 day^{-1} for small flagellates during periods when losses are minimal (Fig. 21–12 and Harris 1986). Abundance changes of picophytoplankton in nature suggest maximum growth rates ranging from 2–5 d^{-1}, unless the highest values are affected by inputs from sediments or littoral zones rich in picophytoplankton, resulting in overestimation of the highest growth rates reported (Happey-Wood 1991). Maximum phytoplankton growth rates are much higher than for macrophytes with their low surface area to volume ratio (Fig. 21–12 and Chapter 24).

21.12   Limiting Nutrients and Eutrophication

A growing concern in the early 1960s in North America and parts of western Europe about the effects of **eutrophication** (nutrient enrichment) in recreational lakes culminated in a major conference in 1967 (NAS 1969). The consensus was that phosphorus and nitrogen (but mostly phosphorus) were the two elements responsible for summer algal blooms observed in previously oligotrophic lakes in the temperate zone currently receiving wastewater or runoff from fertilized fields.[10] At about the same time, Sakamoto (1966) demonstrated a link between phosphorus and nitrogen concentrations and the resulting algal biomass in Japanese lakes (Fig. 21–28). The next major step was taken by Vollenweider (1968) who produced simple empirical equations (models) linking the export of phosphorus from drainage basins to the phosphorus concentration of lake water, followed by linking water P concentrations to the phytoplankton biomass (Fig. 8–16 and Sec. 17.7). The Vollenweider models and other similar models have had a major impact on lake management and the direction of fundamental research on nutrient cycling (Chapters 17 and 18) and the effect of nutrients and nutrient ratios on species succession.

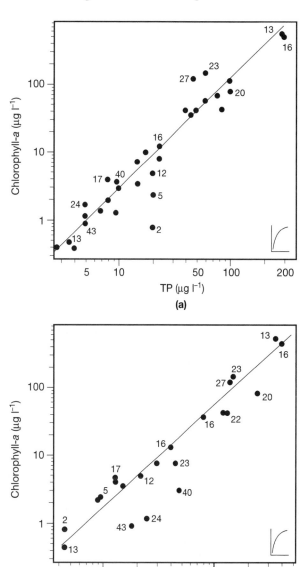

Figure 21–28 Relationship between chlorophyll-*a* content and (a) total phosphorus (TP), or (b) total nitrogen (TN) in surface and near-surface waters of Japanese lakes in May, June, and July. Numbers next to points represent the N:P ratio (by mass). Note that lakes with exceptionally low and exceptionally high N:P ratios are outliers in the TP–chl-*a* and TN–chl-*a* relationships, respectively. (*Modified from Sakamoto 1966.*)

---

[10]Early evidence for the importance of phosphorus as a limiting element was obtained experimentally by W. R. G. Atkins (1923) in England, C. Juday et al. (1926) in the United States, and by H. Fischer (working on fish culture, 1924) in Germany. Fischer reported a marked increase in carp production when ponds were fertilized with groundup, phosphorus-rich slag or superphosphate, whereas the addition of nitrogen and potassium without phosphorus raised the yield much less.

In the early 1970s, it quickly became evident that nearly half of the phosphorus exported to waterways in populated areas of economically developed countries was derived from phosphorus added as a water softener to household detergents. The use of detergents had rapidly increased during the first several decades following World War II as increasing affluence allowed many more households to purchase washing machines and dishwashers. Government efforts to encourage manufacturers to reduce the phosphorus content of detergents were naturally resisted. Manufacturers first argued that CO_2 derived from the microbial oxidation of organic matter in sewage, not phosphorus, was the fertilizing agent and, furthermore, that there was no safe and effective alternative to the use of phosphorus as a water softener.

The evidence most instrumental in persuading law-makers that phosphorus controls had to be instituted was not the solid scientific evidence that had been accumulated since the 1920s but rather the simple whole-lake nutrient enrichment experiments carried out in the **Experimental Lakes Area** (ELA) of northwest Ontario (CA). There, D. W. Schindler and coworkers added combinations of (1) inorganic nitrogen plus phosphorus, (2) nitrogen alone, (3) nitrogen plus organic carbon (as sucrose), and (4) sucrose alone to small oligotrophic Precambrian Shield lakes; the organic carbon was provided as a source of organic matter for the heterotrophic bacteria and thus as a source of additional respiratory CO_2 available to the phytoplankton. They then compared the whole-system responses to changes in **reference lakes** (similar nearby lakes) or to the other half of lakes partitioned with plastic curtains.

The different additions showed, abetted by dramatic aerial photographs (Fig. 21–29), that neither nitrogen or carbon alone, nor their combination, appreciably increased the algal biomass of the oligotrophic lakes, but phosphorus additions were necessary for production of a high algal biomass (Schindler et al. 1973). The research permitted two important additional conclusions. First, phytoplankton nitrogen fixation is stimulated when a low N:P ratio is created—< 22:1 by moles or < 10:1 by mass (Flett et al. 1980, and Chapter 18). This particular finding showed that a natural source of nitrogen was potentially available through nitrogen fixation. Second, the addition of nitrogen plus phosphorus is sufficient to produce a high algal biomass, without requiring the addition of CO_2 or other elements. Thus, the algae were able to utilize atmospheric CO_2

Figure 21–29 Lake 226, Experimental Lakes Area (CA), showing a surface cyanobacterial algae bloom (*Anabaena spiroides*) in the far basin fertilized with phosphorus, nitrogen, and carbon. No bloom developed in the near basin, fertilized with the same quantities of nitrogen and carbon, but no phosphorus. The two basins were separated using a reinforced vinyl sea curtain. *(Photo courtesy of D. S. Schindler.)*

after its diffusion into the low DIC lakes, and could do so without the external CO_2 provided through the decomposition of organic matter in wastewater or experimental sucrose additions. However, the same work also showed that the *rate* of primary production can be constrained by shortages of DIC, thereby reducing the rate at which the community biomass increased.

The findings at ELA, and subsequently elsewhere in lakes and rivers, yielded the still relevant conclusion that phosphorus control should be the focus of nutrient abatement in the temperate zone, at least where unpolluted systems are characterized by a high N:P ratio. Conversely, large-scale nitrogen removal requires the creation of anoxic conditions in the treatment process to maximize denitrification (Fig. 8–14 and Sec. 18.4). Other research on reducing phosphorus export to waterways, involving land management in agricultural basins that minimizes nonpoint source losses, shows nitrogen export is reduced concomitantly. Finally, limnological research has demonstrated that, although some aquatic systems partially offset nitrogen reduction by nitrogen fixation, over the longer term reduction in phosphorus is a more reliable way to lower the algal biomass.

These findings, together with a large number of much more difficult to interpret short-term bioassay studies in flasks plus attempts to model phosphorus dynamically (Sec. 17.8), drew most scientific and public attention to phosphorus and its behavior. The resulting **phosphorus paradigm** caused the role of nitrogen and other nutrients, the importance of lake morphometry (Fee 1979) and predation as codeterminants of algal biomass, phytoplankton production, and community structure, to receive disproportionately little attention.[11] Nor was it widely recognized that additions of phosphorus alone to whole lakes are usually insufficient to appreciably raise the algal biomass (Fee 1979 and Elser et al. 1990). To raise the algal biomass both phosphorus and nitrogen are normally required.

21.13   Nitrogen vs Phosphorus

The important and influential whole-lake fertilization experiments at ELA, subsequently supported by findings elsewhere, persuaded legislators in a number of political jurisdictions to impose limits on the phosphorus levels allowed in detergents. Indeed the central goal of the early whole-lake fertilization experiments had been to demonstrate the need for aquatic management to impose controls on phosphorus release to north temperate zone waterways. That goal naturally drew attention away from nitrogen's importance in determining phytoplankton biomass and community composition.

Collection and analysis of data from the different whole-lake and limnocorral fertilization experiments in the temperate zone literature (Elser et al. 1990)

showed that in 80 percent of the *lake years* (number of lakes × number of years) examined, both nitrogen and phosphorus had to be added together to elicit a significant increase in algal production and biomass. Addition of just phosphorus or nitrogen to the generally oligotrophic systems yielded a response in only six percent of the lake years evaluated. The conclusion that addition of both elements is required and sufficient to greatly raise the algal biomass in the vast majority of oligotrophic lakes indicates a relatively close balance between the supply of phosphorus and nitrogen in nature (the N:P ratio). That balance implies that a relatively small addition of phosphorus alone, or wastewater disproportionately rich in phosphorus, to high N:P systems is enough to shift algal communities from a primarily P-limited to an N-limited state. Other work on lotic systems allows the same conclusion.

The typically high N:P supply ratio of drainage water from well-watered, unpolluted mid and high latitude catchments is well above the N:P demand ratio of phytoplankton protoplasm (Table 8–2), suggesting that phosphorus is the nutrient most commonly limiting algal growth (Sec. 8.1). However, there is considerable variation within and among species in laboratory-determined optimal N:P demand ratios, therefore a supply ratio of > 16:1 by atoms (7:1 by mass) is assumed to reflect a phosphorus limitation, and a ratio of < 10:1 by atoms (7:1 by mass) reflects a nitrogen limitation (see Fig. 21–28 and Smith 1982). Intermediate values indicate an approximately balanced growth.[12]

Nitrogen Limitation

The low N:P supply ratios needed to yield N-limited algae in well-watered temperate regions are encountered where aquatic systems receive runoff or groundwater from fertilized soils or wastewater (Fig. 21–30 and Table 8–2). Modest inputs of relatively concentrated and characteristically low N:P wastewater from urban areas and livestock (between 14:1 and 6:1 by atoms) is sufficient to not only substantially raise absolute nutrient concentrations but also to lower the N:P ratio of the receiving waters, making nitrogen limitation plausible.

---

[11]As we discussed in Chapter 2, science is subject to changes in fashion. During paradigm periods, most scientists working on a particular type of problem are part of a consensus that a certain approach or a particular set of issues, is the most significant and offer a solution to the problems perceived to be most important at the time. Examples in limnology include lake classification in the 1920s and 1930s, (Chapter 2), trophic levels and energy flow in the 1950s and 1960s (Chapter 2), and phosphorus in the 1970s and 1980s (Chapter 17). The phosphorus paradigm is currently not so much rejected as modified by a renewal of interest in the role of nitrogen, through work on the effects of zooplankton and fish predation on nutrient levels and phytoplankton community structure—the top-down paradigm (Sec. 23.11), and an explosion of research on the role of microbes (Chapter 22) and microzooplankton (Chapter 23) in determining both algal production and the flow of organic matter and nutrients in aquatic systems.

---

[12]A proper nutrient supply ratio is based on the *available* nutrient pools rather than on the *total* pools of phosphorus and nitrogen (TP and TN) that also include little available fractions. Yet, total phosphorus is the best longer-term predictor of biomass (chl-*a*) produced (Peters and Bergmann 1982). This is presumably because TP best reflects the size of the nutrient pool that becomes available as the result of decomposition.

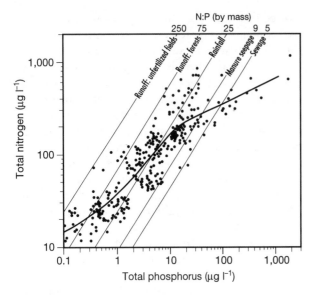

Figure 21–30 Relationship between mean summer TN and TP concentrations in epilimnetic waters of largely temperate zone waters. The average trend in the data (solid line) was estimated using a LOcally WEighted Sequential Smoothing (LOWESS) technique. The thin black lines show N:P ratios of selected potential nutrient sources to the lakes. Multiply by 2.21 to obtain molar (atomic) N:P ratios. Note the typically high N:P ratio in runoff from unfertilized fields and forests and the progressive decline in the N:P ratio towards highly eutrophic (high TP) systems, where the ratio increasingly resembles manure seepage and wastewater. *(After Downing and McCauley 1992.)*

▲ The slope of the relationship between TN and TP concentrations is typically smaller than unity (Fig. 21–30) and it is evident that the TN:TP ratio declines among aquatic systems as TP levels rise. The likelihood of an N:P supply ratio unable to satisfy the algal demand ratio for nitrogen is greatest for eutrophic systems (TP > ~30–100 µg l^{-1}) (Downing and McCauley 1992) and for oligotrophic semiarid zone systems that receive exceptionally little fixed nitrogen from their poorly vegetated catchments or the atmosphere (Sec. 18.1).

The importance of TN in predicting algal biomass in highly eutrophic systems was demonstrated by Prairie et al. (1989) who showed the correlation between TP and chl-*a* concentrations to be greatly influenced by TN at high phosphorus concentrations (> ~60 µg P l^{-1}), suggesting a primary nitrogen limitation. The effect of nitrogen was insignificant at low TP concentrations (oligotrophic systems). Although

nitrogen limitation becomes more probable in highly eutrophic systems (Fig. 21–30), the self shading produced by a high algal biomass increases the possibility of light limitation, not nutrient limitation, in turbid waters. Finally, the N:P ratio declines sharply with increasing salinity in North American lakes, suggesting that TN is the limiting nutrient in more highly saline (semi-arid zone) waters (Chow-Frazer 1991).

The literature shows that the possibility of a primary nitrogen limitation is higher in oligotrophic waters at lower latitudes than in the temperate zone. Low N:P districts are found in subtropical Florida (US), where many lakes and streams are located on marine sediments high in phosphorus (Canfield 1983). Semiarid zone tropical and temperate reservoirs (Sec. 29.6) and streams, as well as a variety of tropical South American lakes (Setaro and Melack 1984) have N:P supply ratios suggesting primary nitrogen limitation. Lebo et al. (1994) and Lewis (2000) hypothesize that low latitude lakes may be more prone to nitrogen limitation as the result of year-round high water temperatures and a resulting high denitrification at the oxic–anoxic interface of the typically anoxic hypolimnia of tropical lakes (Secs. 15.2 and 18.4) . However, other low latitude lakes appear to be either phosphorus limited or have an N:P supply ratio close to the algae demand ratio (Talling and Lemoalle 1998).

Other Limiting Factors

Phosphorus and nitrogen were the only two nutrients necessary to greatly increase growth rates and community biomass in ~80 percent of the oligotrophic freshwater lake fertilization experiments in which both elements were added (Elser et al. 1990), but there was no growth response in the remaining 20 percent of the lake years examined. This lack of response may have been attributable to a nutrient limitation by elements other than nitrogen and phosphorus. Short-term bioassays in flasks have variously pointed to Fe, Co, Mn, Mo, or S as the most probable candidates. A limitation by one or more of these elements is most plausible where the N:P supply ratio approaches the demand ratio of the plants (Forsberg and Ryding 1980, Healey and Hendzel 1980). An iron limitation has been demonstrated in the central Pacific Ocean and an occasional iron limitation in Lake Erie (CA, US) is suggested by bioassay experiments (Twiss et al. 2000). Other possible reasons for the lack of response to the addition of N and P include primary light limitation in humic or turbid lakes, or an abundance of large herbivores able to prevent an algal biomass response. Even so, algal species differ somewhat in their N:P demand ratio and diatoms are further dependent on a silica availability. Consequently, changes in nutrient ratios in nature can be expected to affect the species composition even when the community biomass is constrained by phosphorus and nitrogen.

▲ 21.14 Empirical Nutrient–Phytoplankton Relationships

The external nutrient supply which plays such a central role in predicting the growing-season algal community biomass and production of inland waters is usually overwhelmingly obtained from drainage basins via streams and rivers (Fig. 5–15). Exceptions include lakes and wetlands with a particularly low drainage ratio (CA:LA) receiving most nutrients via the atmosphere, or low latitude lakes with higher ratios but subject to so much catchment and stream evapotranspiration that the receiving lakes and wetlands are primarily dependent on direct precipitation on the water surface (Chapter 8 and Sec. 9.5). Finally, wetlands and those lakes located in exceptionally deep, porous drainage basins receive most of their water and a significant portion of their nutrient supply from groundwater rather than from inflowing streams.

The importance of drainage basins was recognized by E. Naumann (SE) at the beginning of the 20th century (Chapter 2); qualified by Pearson (GB); and formalized by D. S. Rawson (CA), whose findings stimulated R. Vollenweider and others to develop simple quantitative models linking the phytoplankton biomass to the *external supply* of total phosphorus (Sec. 17.7). But most empirical models only link average TP or TN *concentrations* in streams or lakes to the average algal biomass (chl-*a*) because concentration measurements are much simpler than obtaining nutrient loading estimates from the land and atmosphere (Fig. 21–28). See Table 21–5 for a small sample of the many empirical nutrient–biomass models.

The equations and plots in Table 21–5 predict widely varying amounts of chlorophyll-*a* per unit of nutrient. Differences in the regression slopes, the intercepts, and the data scatter have received considerable attention, but remain poorly resolved. However, it is evident that lakes and rivers differ too much among regions for there to be a single empirical model characterizing the relationship everywhere. Considerable among-system variation in the relationship between the total phosphorus concentration (or its covariate, total nitrogen concentration, Table 9–1) and the algal biomass (as chl-*a*) is partially the result of differences in temporal, spatial, and range (interval) scales. Differences in climate, lake morphometry, nutrient supply, and biological interactions add much additional variation. This is particularly evident from Argentinean work in which groups of lakes differing in their trophic status, animal biota, and sediment respiration yielded a variety of lake type specific models (Tables 21–5 and 21–6). Some models (Table 21–5) are based on individual data points, others on summer data only, while yet others represent annual averages.[13] Possibilities and limitations of steady-state regression models such as those in Tables 21–5 and 21–6 are discussed in Sec. 17.7.

The TP–chl-*a* Relationship and Aquatic Management

Empirical total phosphorus–chlorophyll-*a* relationships are widely used in research and aquatic management, but several precautions should be noted:

1. The relationships are only applicable to primarily phosphorus-deficient systems and should not be used if the phytoplankton yield is largely constrained by some other factor (e.g., light, nitrogen, flushing).

2. The relationships are not precise, as is evident from the scatter around the regression slopes (Figs. 8–16 and 21–28). The empirical models reflect the *average* behavior of the systems examined and cannot be used to predict how changes in phosphorus loading or phosphorus concentration will change the algal biomass in particular waterbodies. If individual waterbodies are to be managed, available system-specific information should

---

[13]Laboratory and among-system field studies have long appreciated the power of resolution gained by allowing the variable(s) of interest, whether it is lake or river size, nutrient supply in chemostats, fish predation in limnocorrals, or nutrient concentrations among lakes or rivers, to vary greatly while holding other environmental variables relatively or absolutely constant. Studies further differ in spatial and temporal scale and it is not at all surprising that there is a lot of disagreement about the generality of interpretations made and conclusions drawn.

Table 21–6 **Factors affecting the influence of total phosphorus (TP, µg l⁻¹) on algal biomass (as chlorophyll-*a*, µg l⁻¹) in Argentinean lakes and reservoirs. MASI = mean macrozooplankton body weight (><3.6 µg dry wt); DO_b = dissolved oxygen at 0.5 m above deep water sediments (><2 mg l⁻¹); %SS = frequency of sampling zooplanktivorous silverside (fish); SE = standard error of the slope, a measure of the data scatter; all equations P < 0.001.**

Equation	Limits	TP range	Equation	r^2	SE	n
1.	all data	1–1,288	log (chl-*a*) = −1.943 + log 1.08 (TP)	0.78	0.06	97
2.	MASI≤ 3.6	1–1,288	log (chl-*a*) = −1.879 + log 1.12 (TP)	0.82	0.06	75
3.	MASI>36	2–398	log (chl-*a*) = −2.646 + log 1.10 (TP)	0.75	0.14	22
4.	%SS>0	1–1,288	log (chl-*a*) = −1.432 + log 1.04 (TP)	0.80	0.07	58
5.	%SS=0	2–350	log (chl-*a*) = −1.916 + log 0.84 (TP)	0.74	0.08	39
6.	DO_b≤ 2	15–398	log (chl-*a*) = −0.338 + log 0.89 (TP)	0.80	0.13	14
7.	DO_b>2	1–1,288	log (chl-*a*) = −2.045 + log 1.06 (TP)	0.80	0.06	83

Source: Quirós 1990.

be obtained and given preference. Management decisions should never be based only on TP–chl-*a* relationships.

3. The best relationships for management (predictive) purposes are regionally derived models, which normally best represent local climate, hydrology, geology, land use, system morphometry, and the biota of the waters to be managed. A particular model cannot be expected to be applicable to waters different from those used to develop the model.

4. Aquatic management aimed at increasing the transparency through phosphorus effluent abatement is likely to be successful if lake and river TP concentrations can be reduced to well below 100 µg P l⁻¹, and is most quickly effective in lakes that have not received high nutrient inputs for many years (Sec. 17.6), but is unlikely to have an effect on chlorophyll-*a* concentrations in hypertrophic systems (TP > 100 µg l⁻¹). There, reductions in nitrogen loading may bring about large reductions in algal biomass (McCauley et al. 1989).

5. The probability that *biomanipulation* (reduction) of zooplanktivorous fish, allowing enhanced macrozooplankton predation on (small) phytoplankton and resulting in increased water transparency, increases if the water TP concentration is < 50–100 µg l⁻¹ (Sec. 23.8).

6. Shallow systems can be dominated by either macrophytes or phytoplankton over a wide range of nutrient concentrations (< 50– > 150 µg P l⁻¹), but the probability of algal dominance rises with

increasing TP (Chapter 24). The benthic algal biomass in lotic systems is generally kept below nuisance levels (~100 mg chl-*a* m⁻²) when water TP concentrations are maintained below 30 µg l⁻¹ (and TN < 350 µg l⁻¹) (Dodds et al. 1997).

▲ 21.15   The Maximum Phytoplankton Biomass

The highest biomass encountered in specific eutrophic lakes or slowly flowing nutrient-rich rivers is of great importance in management because public complaints about algal blooms pertain to summer maxima, not the average algal biomass. Biomass maxima attract scientists interested in the mechanisms, processes, and photosynthetic rates that permit very large standing crops in inland waters.

The highest standing crops, or **photosynthetic cover**, reach ~300 to ~1,000 mg chl-*a* m⁻² and are primarily encountered in warm hypertrophic freshwater lagoons and shallow saline lakes at mid and low latitudes receiving high irradiance and not subject to much sediment resuspension (Talling et al. 1973, Osborne 1991, and Jeppesen et al. 1997a). These lakes typically have a mean depth of less than < ~1–1.5 m or, like the Hartbeespoort reservoir (Table 21–4), are stably stratified with a thin z_{mix} yielding the required favorable z_{eu}:z_{mix} ratio. Higher values (up to ~2,000 µg l⁻¹, Fig. 21–26, Talling et al. 1973) probably represent a local downwind accumulation of algae in lakes or resuspension of recently sedimented organisms. Few inland waters have photosynthetic covers exceeding 100–200 µg chl-a m^{-2}. This is attributable to one or more of the following reasons: (1) insufficient nutrients; (2) low temperatures; (3) insufficient incoming irradiance; (4) unfavorable z_{eu}:z_{mix} ratio; (5) high planktonic or benthic herbivory, or (6) flushing rapid enough to prevent algal biomass accumulation.

A high community biomass is usually dominated by large colonies or filaments of cyanobacteria (Figs. 21–4, 21–5 and 21–29) during periods of stable stratification. Laboratory research has shown that the maximum biomass attainable increases with increasing unit size (Agustí and Kalff 1989). Packing the biomass in larger units reduces light extinction (Fig. 10–6) and increases the z_{eu}:z_{mix} ratio, thereby allowing an increase in community biomass and depth-integrated production.

Not all highly eutrophic systems are dominated by large cyanobacteria during the warm season. Hypertrophic freshwater lakes and wastewater lagoons with total phosphorus (TP) concentrations > 500 – 1,000 µg P l^{-1}—which are also invariably shallow (unstratified)—may be dominated in summer by a nanoplanktonic chlorophyte (*Chlorophyta*) biomass or a mixture of chlorophytes and large cyanobacteria.

The reason for a lack of dominance by large cyanobacteria is much discussed, but unresolved. Explanations include frequent water column mixing, favoring nanoplankton over the slower-growing large cyanobacteria, or low macrozooplankton predation pressure on small algae resulting from an abundance of zooplanktivorous fish. Physiological mechanisms have been invoked as well, such as the superior ability of small cells in hypertrophic systems to take up the frequently limited supply of CO_2 (Jensen et al. 1994). Determinants of a particular phytoplankton size and species structure have not been resolved because quite different mechanisms or processes—operating over different scales and each manifesting itself with a lag—can bring about a particular community.

Prediction of the maximum algal biomass to be expected is of considerable importance in management. Jones et al. (1979) examined a 50-lake data set from western Europe and North America and concluded that the maximum chl-a concentration observed is an average of 1.7 times the mean summer chl-a.

$$\text{Max. chl-}a = 1.7 \cdot \bar{x} \text{ summer chl-}a + 0.2.$$
$$r^2 = 0.58$$

EQ. 21.7

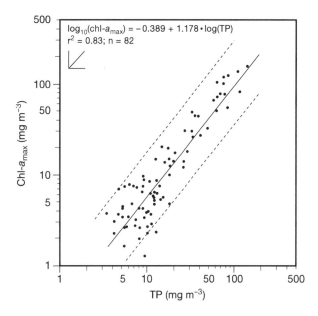

Figure 21–31 Relationship between the maximum observed chlorophyll-a concentration (chl-a_{max}) and the growing season mean total phosphorus concentration in the euphotic zone of 82 European, North American, and New Zealand lakes. Dashed lines represent the 95 percent confidence interval for an individual point. *(After Pridmore & McBride 1984.)*

The discrepancy between the maximum biomass predicted from Eq. 21.7 and the observed biomass which was particularly large in the highest chlorophyll-a lakes—the ones of greatest concern. Such lakes tend to be shallow and subject to much resuspension, thereby greatly increasing the variation in the water column algal biomass over time (Fig. 21–2). A superior approach was taken by Pridmore and McBride (1984) who developed an empirical relationship between the observed maximum chl-a and the mean total phosphorus of the water (Fig. 21–31).

Highlights

- Freshwater algae consist of a very large number of species (> 10,000), a minority of which are found in the plankton. An experienced phytoplankton ecologist would identify between 70 and 200 species in the plankton of a single lake during a year-long study, while a taxonomic study might reveal ~400. A modest number of species contribute most of the community biomass. Species richness is reduced in saline waters.
- Some species in the plankton live part of the time on or in substrates and are not true plankton.

- The phytoplankton range in size from as small as < 1 μm in diameter, with organisms < 2 μm defined as picophytoplankton. The algal picophytoplankton, together with those in the next size category, the nanophytoplankton (2–30 μm), have the highest maximum growth rates, are of a size readily eaten by a variety of freshwater invertebrates, and have a density low enough to make them little subject to sedimentation. Large phytoplankton, the microplankton, range in size from > 30 μm to colonies of ~600 μm in diameter. They are characterized by lower maximum growth rates and, when lacking flagella (motility) or buoyancy control, are most subject to sedimentation.

- Growth rate and community biomass in temperate and polar waters is primarily constrained by light limitation in winter, while nutrient limitation usually limits community production and biomass in all but highly turbid waters at the same latitudes in summer, and year-round at lower latitudes. Superimposed are biomass oscillations and species fluctuations produced by predation, sedimentation, disease, or flushing that help structure the community.

- ▲ The specific growth rate [mg C g C^{-1} (or mg chl-*a*) d^{-1}] in the laboratory and communities in nature, is light-limited under low irradiance. Growth rates increase to a maximum at an intermediate intensity encountered somewhere in the upper water column or shallow-water sediments, and become photoinhibited yet closer to the surface.

- ▲ The annual depth-integrated phytoplankton rates of primary production ranges about three orders of magnitude from a low of < 4 g C m^{-2} yr^{-1} in ultra-oliogotrophic polar lakes to a high of 6000 g C m^{-2} yr^{-1} in a hypertrophic low latitude lake. Ultra-oligotrophic lakes experience a very short growing season, low temperatures, and very low nutrient inputs from their drainage basins. Annual production rates (g C m^{-2} yr^{-1}) are lower than the highest daily rates measured in highly nutrient-rich (hypertrophic) tropical lakes that also experience high temperatures and high irradiance year-round.

- ▲ The primary production (P) to phytoplankton biomass (B) ratio (P:B), expressed as carbon, is a useful measure of specific growth rates. The P:B ratio in the mixed layer of lakes and slowly flowing lowland rivers tends to be low when the algal biomass is high, yielding a poor effective light climate for photosynthesis. Low organismal growth rates demand low loss rates if a species is not to disappear. The typically high P:B ratios seen in nutrient-poor water, characterized by a low community biomass of typically small algae and high water transparency, imply a rapid individual growth rate, a high loss rate, and rapid recycling of organic matter and the nutrients in highest demand.

- ▲ The two nutrients in highest demand relative to their supply are phosphorus and nitrogen. Experimental additions to oligotrophic freshwater lakes and streams show that only combined shortages of P and N constrain the community biomass in virtually all environments that are not light-limited, but the *rate* of biomass accumulation can be constrained by a DIC shortage, especially in low DIC lakes. Other elements (e.g., silica for diatoms) also affect algal species composition.

- A primary phosphorus limitation is common in well-watered temperate regions, characterized by an N:P supply ratio (>~16:1 by atoms) well above the demand ratio of the algae. A primary nitrogen limitation is most probable in agricultural and urban areas where the runoff is both rich in nutrients and characterized by a low N:P supply ratio (below the demand ratio of the algae). A primary nitrogen limitation outside the temperate zone is most likely in regions where the surface geology is rich in phosphorus, in low latitude semiarid catchments where the soils release disproportionally little N, and in some other tropical lakes.

- ▲ Public complaints about algal blooms in the temperate zone pertain largely to the summer maxima rather than the average biomass. Highest standing crops or community biomass, typically dominated by large cyanobacteria reaching ~300–1,000 μg chl-*a* m^{-2} are encountered in shallow ($\bar{z}$ < 1.5 m) nutrient-rich waters during periods of high irradiance plus high temperature.

22

The Bacteria

22.1 Introduction

The small size (mostly < 1 μm diameter) and morphological diversity—much less than the protozoa or algae—has greatly affected the development of aquatic microbiology. Most of the heterotrophic prokaryotes are too small to be seen or identified under the light microscope and early work had to be based on culturing of microbes. That work, started during the last decade or so of the 19th century, was an outgrowth of concern about the importance of disease-causing sewage bacteria in lakes and rivers from which towns drew their drinking water.

The early emphasis on culture work was most useful in the development of culture media that enabled researchers to detect and count wastewater bacteria in drinking water, and identify a variety of sewage and other pathogenic species that could be cultured on the media developed. The emphasis on culturing bacteria led directly to important work in microbial physiology, work that then contributed much to the development of conceptual ideas about the roles of microbes in nutrient cycling. All of these accomplishments were, from an ecological perspective, somewhat offset by the associated emphasis on laboratory rather than field research, and a disregard for the vast majority of heterotrophic bacteria that could not, and still cannot, be grown on available culture media. Those forms that can be cultured frequently change size, shape, and physiology under different growing conditions, indicating that traditional systematic and physiological characteristics of species are not very useful in describing planktonic bacteria (Hobbie 1988).

Difficulties in enumerating the abundance of aquatic bacteria in nature—a prerequisite for work on their distribution and dynamics—have been gradually overcome. Major technical breakthroughs during the last two decades or so make it possible to obtain reasonably reliable bacterial abundance and biomass measurements—measurements that reveal the bacterial abundance in fresh water to be about three orders of magnitude greater (Bird and Kalff 1984) than had been believed on the basis of the traditional culture studies. The much higher abundances revealed by the recent advances in technique have shown that the bacterial biomass contributes between about five and 15 percent of the total planktonic biomass in fresh water (del Giorgio and Gasol 1995). All this greatly stimulated fairly successful efforts to measure bacterial production rates and to assess the importance of planktonic bacteria in nutrient cycling and food webs. The outcome has been a veritable explosion of papers on different aspects of microbial ecology during the last two decades. The new counting techniques not only showed that planktonic bacteria are unexpectedly abundant but also that they are much smaller than those typically seen in culture. Genetic-fingerprinting techniques, based on sequences of ribosomal RNA, have begun to make it possible to distinguish between phylogenetic groups of microbes (Subclasses, phylotypes) in natural communities on the basis of nucleic-acid sequences. It is evident from the many new rRNA sequences that have been found that the vast majority of bacterioplankton species are *not* represented in cultures.

The techniques used in microbial ecology continue to evolve very rapidly and too little time has

349

passed to be certain whether some of the emerging generalities are fact, artifact, or a mixture of the two. As a consequence, the present chapter is much more speculative than any of the others. Many of the recent findings and ecological interpretations made will probably not stand the test of time because experiments are still overwhelmingly restricted to laboratory and field studies carried out in small containers (<< 1 l), over short periods (< 1 day) on single factors, making it difficult to extrapolate the findings to nature (see Fig. 1–2).

While the role of heterotrophic bacteria in the utilization and decomposition (liberation) of organic matter has long been recognized in a qualitative sense, the more recent work is starting to allow some quantification as well. Much of the work is being done in oceans and estuaries and this chapter will, more than any other, draw from the marine literature. To do so assumes that the principal constraints on bacterial growth do not differ materially between marine and fresh waters. While this conclusion appears to be generally true, two major differences have become apparent. First, planktonic freshwater bacteria encounter a substantial supply of organic carbon produced on drainage basins and littoral zones (*allochthonous production*), while oceanic bacteria are overwhelmingly dependent on carbon produced in situ (*autochthonous production*). Second, and probably as a consequence, lake plankton contains more heterotrophic biomass per unit autotrophic biomass (Sec. 22.7).

Recent developments in microbial ecology are having ramifications elsewhere in the aquatic sciences. The fluorescent-staining techniques that have revolutionized the enumeration of heterotrophic bacteria (Hobbie et al. 1977) now also permit the picophytoplankton (< 2 μm) to be readily seen (Chapter 21). Their previously unrecognized abundance has led to a small revolution in phytoplankton ecology, raising questions about some of the interpretations made during the last 70 years in studies when researchers were unaware of either their existence or their abundance. Moreover, the recent developments in microbial ecology have pointed to the important role the heterotrophic bacteria play in determining phytoplankton abundances and production, resulting from the ability of the bacteria to compete successfully with the phytoplankton for inorganic nutrients that are in short supply. Recent research on bacterial biomass and production has also raised many questions about the fate of all this production, leading to much experimental work on the consumers of heterotrophic bacteria. The

identification of the protozoans (mostly flagellates and ciliates) as major predators has allowed an explosion of research on protozoan ecology (Chapter 23). The research has demonstrated that the traditional division between botany and zoology is unwarranted—certain protozoa are able to both photosynthesize and graze on bacteria. In addition to the heterotrophic bacteria, there are two other groups of aquatic bacteria. The first group—the cyanobacteria (blue-green algae)—were discussed in Chapter 21, the second group—the green and purple photosynthetic bacteria of anoxic environments—will be dealt with in Sec. 22.10.

▲   22.2   From Past to Present

The early microbiological studies, starting in 1889 and using organic-rich growth (culture) media that we now know greatly underestimate the variety and abundance of microbes present, laid the groundwork for the second phase of development. This occurred when L. M. Snow and E. B. Fred, working with Birge and Juday, in Wisconsin, developed the first microscope technique allowing enumeration of (larger) bacteria under the light microscope (after concentrating and staining them). The new microscope method, independently developed in the former USSR a few years later, showed the bacterial abundance in unpolluted water to be roughly ten thousand times greater than the numbers obtained by traditional culture methods (A. S. Razumov 1932, in Kuznetsov 1970).

The new microscope techniques were perfected and employed in the former USSR between the two World Wars, when interest in aquatic microbial ecology languished somewhat in the rest of Europe and in North America. The isolation of the former USSR, as a result of language and politics, meant that the abundant and excellent work carried out over a 30–40 year period in both aquatic and soil microbiology was largely unappreciated or hidden from the view of Western scientists.

Following World War II, Soviet scientists not only developed radioisotopic techniques in an attempt to measure heterotrophic bacterial respiration but also explored the use of fluorescent dyes to stain bacteria (Gorlenko et al. 1983). Starting in the 1960s and 1970s, renewed interest by western scientists in the planktonic microbes led to the development of convenient fluorescent-staining techniques that now allow routine determination of bacterial abundances. In the modern technique, the bacteria are first collected on filters, then stained with a fluorescent dye, and finally observed using an epifluorescent microscope (Francisco et al. 1973, Hobbie et al. 1977). The staining techniques have revealed that the vast majority of planktonic bacteria are too small (0.1–1.0 μm diameter) to have been visible with the old staining techniques employing nonfluorescent dyes. Counts obtained with the earlier dyes appear to have yielded abundances that are twofold lower in oligotrophic lakes and sevenfold lower in mesotrophic lakes than those obtained by fluorescence microscopy. The underestimates in more eutrophic lakes were much larger (~twentyfold; Bird and Kalff 1984). Unfortunately, there is no simple conversion factor to reconcile old and new data, primarily because the size distribution—and thus the visibility—of bacteria under the light microscope greatly varies between systems. The reconciliation of data collected in the past, using different techniques, with modern data is not restricted to microbial ecology, but is common whenever major technical advances are made. For example, much of the older (pre-1960) phytoplankton literature did not take the nanoplankton into account, whereas the more recent literature has largely overlooked the picophytoplankton in analyses and interpretations. In another example, nearly all of the pre-1960 determinations of pH were made with dyes, and the results are not easily convertible into modern measurements made with pH meters. As a result, it has been difficult to compare present levels of acidification with pre-acidification dye data (Chapter 27). A last example is the major improvement in technique and laboratory procedures that have made the earlier (pre-1980) literature on trace-metal levels in waters impossible to interpret (Chapter 28).

22.3 Bacterial Size, Form, and Metabolism

Until the development of easily performed epifluorence techniques for counting, most aquatic bacteria were thought to be > 1 μm short rods (cylinders), and the average bacterial cell in aerobic waters was believed to have a volume about an order of magnitude greater than the ~0.03 μm³ volume (range 0.01–0.2 μm³) determined with fluorescence techniques by, for example, Cole et al. (1993). It is now evident that cocci (spheres or near-spheres), ranging from 0.1–0.6 μm in diameter, dominate most waters but with a variable fraction of vibrio (curved cells), rod, colony, and fila-

mentous forms (5–~55 μm); rods arranged in chains (5–100 μm) are also present. While it makes physiological sense for smaller organisms with a more favorable surface to volume ratio (Sec. 21.3) to be favored under conditions when substrates are limiting, there is no systematic relationship between bacterial size and trophic status (Cole et al. 1993).

▲ The observed size distribution of planktonic bacteria may be principally due to selective predation of the larger bacteria by filter feeding zooplankton and, when abundant, mussels, rather than substrate availability. For example, experimental manipulation of enclosures (mesocosms) led to the development of a community of small cocci and rods that were typical of the lake itself when the large water flea *Daphnia* dominated. Grazing by *Daphnia* on large bacteria and protozoans prevented the latter from restricting the abundance of small bacteria (~80% of biomass < 1 μm). Conversely, when the protozoans (protists) were allowed to dominate, their grazing led to rapid reduction of smaller cells, and an increase of the large filamentous bacteria that were resistant to protozoan grazing (~90% of biomass > 3 μm) (Jürgens et al. 1994). When abundant, grazers can structure bacterial communities and bring about changes in the growth rate and taxonomic composition of the assemblages favoring rapidly reproducing and/or grazing-resistant species (Jürgens et al. 1999).

Metabolic Types

The impossibility of identifying often superficially similar species by light microscopy, and the impossibility of culturing more than a miniscule percentage (≤ 1%) of the organisms has long led microbiologists to emphasize metabolic type (Table 22–1) rather than the taxonomy of the organisms collected. The principle metabolic division is between **autotrophic** bacteria who obtain all or virtually all cell carbon required for biosynthesis by reducing CO_2, and **heterotrophic** bacteria who obtain all such carbon from reduced organic substances. A second division, based on the source of energy used, distinguishes between **chemosynthetic** bacteria (**chemoautotrophs**)—autotrophic forms that utilize energy obtained from energy-yielding (exergonic) *chemical* reactions to reduce CO_2 to organic matter—and the **photosynthetic** bacteria (**phototrophs**). These bacteria use *light* energy for the reduction of CO_2 to organic carbon (Table 22–1). A third division is based on the source of electrons for growth. Organisms can be either **organotrophs**

Table 22–1 **Classification of major groups of bacteria according to their metabolism. (For more detailed terminology see Sec. 22.3; for more on redox reactions, see Chapter 16.)**

Major Metabolic Types	Electron Donor	Electron Acceptor	Carbon Source	End Products (other than cell carbon)	Organism
(1) Photosynthetic autotrophs					
(a) Cyano-bacteria (blue-green algae)	light, H_2O	H_2O	CO_2	O_2	green plants (aerobic)
(b) Photosyn-thetic bacteria	light, H_2S, S, $S_2O_3^{2-}$, H_2	H_2O	CO_2	S, SO_4^{2-}, H_2O	green and pur-ple sulfur bacteria (anaerobic)
(2) Chemosynthetic autotrophs	H_2S, S°, $S_2O_3^{2-}$, NH_3, NO_2^-, Fe^{2+}, H_2, S, Mn^{2+}	O_2, NO_3^-, CO_2	CO_2	S°, SO_4^{2-}, NO_3^-, Fe^{3+}, H_2O, N_2, S, CH_4, Mn^{4+}	nitrifying bac-teria, color-less sulfur bacteria, methanogens (aerobic and anaerobic)
(3) Photosynthetic heterotrophs	light, organic substances (sugars, alcohols, acids)	H_2O	organic substances	H_2O	nonsulfur purple bacteria (anaerobic)
(4) Heterotrophs (selected types only)					
(a) majority of microorgan-isms and all animals	organic substances	O_2	organic substances	organic acids, alcohols, etc.	heterotrophic bacteria and animals (aerobic)
(b) denitrifers	organic substances	NO_3^-	organic substances	N_2, NO_2^-, NH_3	denitrifying bac-teria (anaero-bic)
(c) sulfate reducers	primarily organic substances	SO_3^{2-}, $S_2O_3^{2-}$ ($S_4O_6^{2-}$), NO_3^-	organic substances	H_2S ($S_2O_3^{2-}$), N_2	sulfate reducing bacteria (anaerobic)
(d) fermenters	organic substances	organic substances	organic substances	H_2, CO_2, organic acids, NH_3, CH_4, H_2S	fermentation bacteria (anaerobic)

Source: Modified after Gorlenko et al. 1983.

who use electrons from organic matter, or **lithotrophs** which use electrons from inorganic compounds such as sulfide, hydrogen, or water. Thus, algae, cyanobacteria, and macrophytes are **photolitho-** **trophs**, heterotrophic bacteria are **chemoor-ganoheterotrophs**, and photosynthetic bacteria are **photolithoautotrophs**, although some are **photo-organoheterotrophs** and some can even be **pho-**

toorganoautotrophs. These convenient groupings based on broad metabolic types or size categories are, unfortunately, also an impediment to further insight because it is becoming clear from exciting molecular work (using the DNA coding for 16 SrRNA sequences) that the microbes vary enough to allow groupings of bacteria (phylotypes); and that predators select not only on the basis of size but also phylotype (Jürgens et al. 1999). Other molecular work shows seasonal and ecological variability in the pool of morphologically similar bacteria (Pernthaler et al. 1998).

22.4 Abundance, Biomass, and Distribution

The planktonic bacterial abundance of free-living cells normally oscillates between 10^5 and 10^6 cells ml^{-1} in all but ultra-oligotrophic or hypertrophic lakes (Table 22–2). Values as high as 10^8 cells ml^{-1} have been noted in shallow, but hypertrophic African saline lakes. Numbers approaching 10^8 ml^{-1} have also been recorded in temperate and antarctic fresh waters (see Pedrós-Alió and Guerrero 1991). Even so, the abundance range normally encountered among temperate zone lakes is surprisingly narrow and, within single lakes, cell numbers usually vary only by a factor of five to ten annually (Fig. 22–1), with lowest numbers in hypolimnia and in winter. However, a mere twofold annual variation was noted in a polymictic tropical African lake and a Central American lake experiencing weak seasonality (Gebre-Mariam and Taylor 1989, and Erikson et al. 1998). Abundances in north temperate lotic systems are typically of the same magnitude as in lakes (10^6 ml^{-1}) (Basu and Pick 1997).

The annual variation in bacterial abundance in temperate lakes is much smaller than the equivalent variation for phytoplankton abundance or biomass, upon which the heterotrophic bacteria depend as an important source of low molecular weight organic carbon and associated inorganic nutrients. The reason for the relatively modest annual variation in bacterial abundance within individual lakes is not well-resolved. The four determinants of variation considered to be most important are (1) inorganic nutrients and organic carbon availability (Sec. 22.6); (2) the consumption of bacteria by predators (Sec. 22.7), including predatory bacteria and viral infections (Sec. 22.8); (3) variation in size and sinking rates; and (4) variation in the fraction of the bacteria that are metabolically active.

▲ The apparent metabolically active bacteria of the plankton are usually a modest fraction of the among-system total abundance (15–30%), with the active fraction varying about twofold in summer. The apparent metabolically active fraction in lakes appears to be much higher than the 12–16 percent active fraction noted in oligotrophic streams, groundwater, and the open oceans, but is considerably lower than in estuaries where about half the bacteria are apparently active, provided the technique used gives a reliable measure of activity (Fig. 22–2). In single systems (e.g., Lake Rodó, UY), the number of active bacteria changes seasonally in step with the total quantity and temperature, and at times is close to 100 percent (Sommaruga and Conde 1997).

An interesting exception to the typically modest annual variability in bacterial abundance in individual lakes are the Varzéa lakes of the Amazon region of South America (Sec. 6.5). Those lakes exhibit two orders of magnitude annual variation in bacterial abundance related to their flushing rate. During the dry season water levels are at their lowest and the lakes contain ~10^7 cells ml^{-1}, but the abundance declines to ~10^5 cells ml^{-1} during the high-water phase when the water residence time is short and bacterioplankton are rapidly flushed (Pedrós-Alió and Guerrero 1991).

The bacterioplankton of most transparent lakes are usually small (0.2–1 μm; 0.02–0.1 μm^3), and largely free-living. Bacteria attached to particles gen-

Table 22–2 **The growing season abundance of heterotrophic planktonic bacteria in the mixed layer of low humic lakes.**

Trophic status	Chlorophyll-*a* (μg l^{-1})	Abundance (cells × 10^6ml^{-1})
Oligotrophic	< 3	< 1.7[1]
Mesotrophic	3–7	1.7–6.5
Eutrophic	7–40	> 6.5

[1]Oligotrophic humic (brown-water) lakes usually have two to three times higher abundance (Johansson 1983).

Source: After Bird and Kalff 1984, and Forsberg and Ryding 1980.

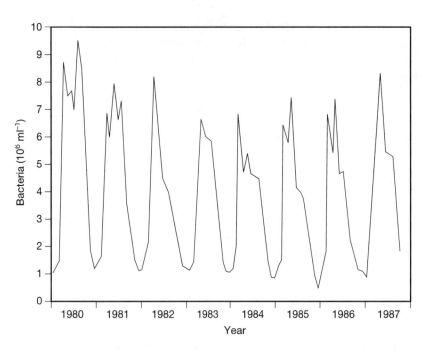

Figure 22–1 Monthly averages of bacterial numbers observed from 1980 to 1987 in the 0–10 m layer at a pelagic station in the upper arm of Lake Constance (AT, CH, DE). *(After Güde 1990.)*

erally contribute a small percentage of the total number in relatively transparent oligotrophic and mesotrophic lakes that are low in particles (Fig. 22–3). However, they can dominate the free-living bacteria in particle-rich systems (Lind and Dávalos-Lind 1991). The attached bacteria tend to be larger (0.05–0.35 μm^3) and they contribute proportionally more to the community biomass (Fig. 22–3). The attached bacte-

ria are most abundant toward the end of algal blooms when there is much algal senescence and detritus, and when bacterial production is high (Fig. 22–3). While the larger bacteria (1.0–3.0 μm; 0.05–0.38 μm^3) typically contribute no more than about a quarter of the total number, their contribution to the total bacterial biomass may be double that—their relative abundance seems particularly great in humic lakes (Pinel-Alloul and Letarte 1993) and anoxic hypolimnia (Cole et al. 1993) that are also low in zooplankton predators. The larger bacteria (> 1.0 μm) contribute disproportionately to the total bacterioplankton production (Fig. 22–3), probably because the growth rate of the smallest bacteria is low (Bird and Kalff 1993) with a high fraction of them inactive (dormant). The plankton also contains modest numbers of large thread-like bacteria (> 3 μm) with a cell size of 5.0–8.4 μm^3 (Pernthaler et al. 1996) containing 50–100 times more carbon than the average-sized bacterium.

The free-living bacteria differ so vastly in volume and activity that bacterial numbers are not very good indicators of bacterial biomass, production, or energy flow in food webs (Sec. 22.9). Yet most studies still report only the abundance, probably because size determinations are time consuming and the fluorescent stains that are used frequently produce a halo effect around the cells making precise microscope measurements difficult. It is, furthermore, insufficiently recognized that very small organisms vary as much in size as larger organisms.

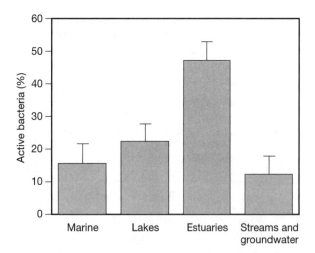

Figure 22–2 Average and standard deviation of the percent active bacteria in aquatic systems as measured by the bacterial uptake of a vital stain. *(Modified after del Giorgio and Scarborough 1995.)*

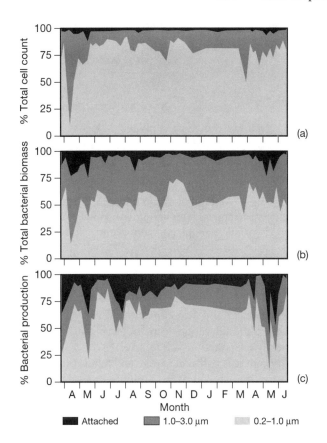

Figure 22–3   The relative abundance (a) and biomass (b) of two size classes of free-living plus the attached heterotrophic bacterial production (c) as a function of the total in the surface water (0–3 m) of Lake Constance (AT, CH, DE). Note that the bacteria attached to particles are rare but large and contribute disproportionately to the community production. *(Modified from Simon 1987.)*

Within single temperate lakes the planktonic bacteria exhibit a fairly regular seasonal cycle, with their highest numbers normally occurring in summer (Fig. 22–1). Abundances appear to be more closely linked to discharge (Sabater et al. 1993) or trophic status (Basu and Pick 1997) in the small number of lotic systems evaluated. There is, however, no consistent pattern in the relationship between algal biomass (chl-*a*) and bacterial abundance or biomass in individual lakes. Some plankton studies have shown that bacterial peaks follow algal peaks, an expected pattern if bacteria are constrained by resource limitation. Other studies show that algal and bacterial peaks coincide, and yet others show no relationship between the two peaks, a pattern expected when grazing or other losses determine bacterial abundance. Interpretations are surely confounded by sampling intervals too large and too variable to allow unambiguous interpretation (see interval scale, Sec. 2.6).

22.5   Heterotrophic Bacterial Abundance and Environmental Factors

Among aquatic systems, the bacterial abundance is closely related to both inorganic nutrient levels (Fig. 22–4) and the usually linked algal biomass (chl-*a*). This coupling is evident from the high correlations between bacterial abundance on one hand and the chl-*a*, total phosphorus, and light extinction on the other (Table 22–3).

Total Phosphorus and Chlorophyll-*a*

The classical view of bacterial–phytoplankton interactions is one in which the heterotrophic bacteria of the plankton are largely dependent on the organic matter produced by the phytoplankton, with phytoplankton production itself constrained by a shortage of phosphorus, nitrogen, or light. Correlations between bacterial abundance and phytoplankton biomass in low humic systems (Table 22–3) fit the above interpretation. However, the view is undergoing revision. The recognition that the bacteria—characterized by a tenfold lower C:P ratio (~10:1 as mass) than the

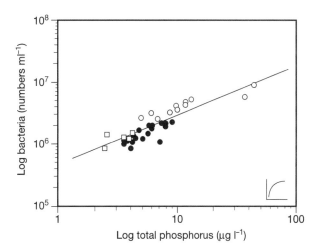

Figure 22–4   Log-log plot of the average bacteria numbers and total phosphorus for warm, monomictic Hobiton Lake, British Columbia, CA (□), subarctic Yukon lakes, CA (●), and dimictic lakes in eastern Canada (○). (r^2= 0.81, n= 35, p< 0.01). *(After Hardy et al. 1986.)*

Table 22–3 **Simple Pearson correlation coefficients between epilimnetic bacterial abundance (log B), algal biomass (chl-*a*), total phosphorus (TP), and limnological variables in 23 lakes (CA and US), each visited once in summer. NS = not significant.**

Factor	log (B) r^2	n
log(chl-*a*)	0.71***	23
log(TP)	0.76***	23
Sampling depth	NS	23
Temperature	0.55**	23
Conductivity	NS	23
Light extinction coefficient (k_d)	0.68***	23
Alkalinity	NS	23
pH	NS	23
PO_4^{3-} uptake constant	NS	23
Mean depth	−0.62**	18
log(surface area)	−0.47*	22
log(catchment area)	NS	23
DOC	0.55**	22

*P<0.05
**P<0.01
***P<0.001

Source: From Currie 1990.

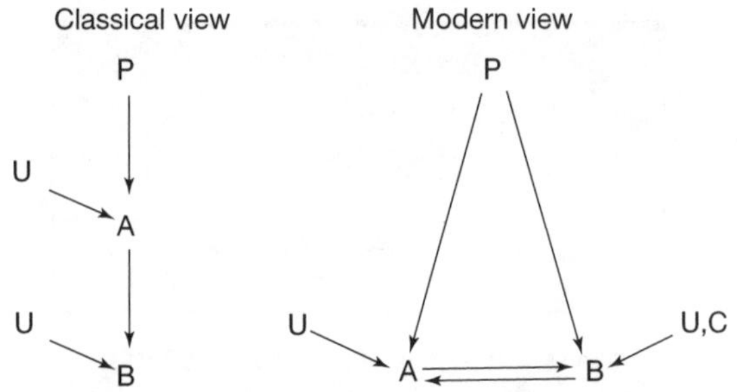

Figure 22–5 Conceptual models describing the possible impact of algae (A), phosphorus or nitrogen (P), and other environmental factors (U) on bacterioplankton (B) in inland waters. U includes temperature, predators, and disease, while C represents allochthonous organic matter. *(Modified after Currie 1990.)*

phytoplankton (Fagerbakke et al. 1996)—compete successfully with algae for limiting inorganic nutrients when the organic matter supply is not their principal constraining factor. This scenario is most plausible in oligotrophic clearwater and humic systems characterized by a high C:P or C:N ratio (see Fig. 24–11). Currie (1990) suggests, based on earlier laboratory experiments, that the lower correlation observed between bacterial abundance and algal biomass (chl-*a*) than between bacterial abundance and total phosphorus (TP) indicates competition between the two groups of organisms for phosphorus (Table 22–3).

Furthermore, at any one concentration of chl-*a*, significant additional variance in bacterial abundance is statistically explained by TP (Currie 1990).

Although it is still speculative and needs more corroboration, the above research points to a scenario in which not only phytoplankton but bacteria too are positively affected by phosphorus (Fig. 22–5). Currie (1990) and others have interpreted these and other similar data to mean that heterotrophic bacteria are able to out-compete algae for phosphorus when dissolved organic carbon (the energy source) is abundant (high C:P ratio), and phosphorus is more limiting than carbon. But bacterial growth becomes simultaneously limited by C and P when the C:P ratio is lowered as the result of bacterial respiration of excess carbon, or resulting from an increase in external or internal loading of phosphorus. If and when the C:P supply ratio declines even further, to below the bacterial demand ratio, the microbes are expected to be constrained by a carbon limitation (Hessen et al. 1994).

Among-system correlations between bacterial abundance (or biomass) and measures of the resource supply (TP, chl-*a*) indeed point to (but cannot prove)

Table 22–4 **Models describing the relationship between bacterial abundance (B, 10^6 cells ml^{-1}) and total phosphorus (TP, µg l^{-1}), chl-*a* (µg l^{-1}) and temperature (°C) based on data collected in US and CA lakes on single dates in summer.**

Equation	Model	Adjusted r^2	n
1	log (B) = 6.2 + 0.41 · log (TP)	0.56	23
2	log (B) = 6.5 + 0.29 · log (chl-*a*)	0.45	23
3	log (B) = 5.8 + 0.39 · log (TP) + 0.021 · temp.	0.81	23

Source: From Currie 1990.

Table 22–5 **Regression equations describing the relationship between chlorophyll-*a* (mg m^{-3}) and the abundance of free-living bacteria (B, 10^6 cells ml^{-1}), determined by epifluorescence techniques in aquatic systems.**[1]

Equation	Model	r^2	n	Location
1	log (B) = 5.445 + 0.630 · log (chl-*a*)	0.92	23	Japanese lakes
2	log (B) = 5.911 + 0.763 · log (chl-*a*)	0.90	20	freshwater literature
3	log (B) = 6.277 + 0.569 · log (chl-*a*)	0.66	13	Quebec lakes
4	log (B) = 5.835 + 0.736 · log (chl-*a*)	0.79	19	marine literature
5	log (B) = 5.960 + 0.524 · log (chl-*a*)	0.75	35	fresh and marine waters
6	log (B) = 6.50 + 0.29 · log (chl-*a*)	0.45	23	CA-US lakes
7	log (B) = 6.62 + 0.45 · log (chl-*a*)	0.55	31	Ontario and Quebec rivers
8	log (B) = 8.83 + 0.58 · log (chl-*a*)	0.63	93	African reservoirs

[1]Additional equations and references can be obtained via anonymous ftp at ftp.icm.csic.es/pub/gasol.

Source: After Aizaki et al. 1981, Bird and Kalff 1984, Cole et al. 1988, Currie 1990, Basu and Pick 1997, Bouvy et al. 1998.

the importance of resources in determining bacterial abundance (Table 22–5, Fig. 22–4) and bacterial production (Sec. 22.7); but the slopes of the regression equations describing the relationship between chl-*a* and bacterial abundance (Table 22–5) are well below unity (0.29–0.76). This shows that, among systems, bacterial abundance in both lakes and rivers rises at a considerably slower rate than algal biomass, pointing to either (or both) an increasingly smaller bacterial ability to compete with algae for resources in more eutrophic systems or disproportionately higher loss rates (Sec. 22.9). Higher light extinction in bacteria-rich lakes (Table 22–3) serves as a surrogate for the higher algal biomass (turbidity) there, while the mean depth and its correlate, surface area, are indicative of more rapid flushing and higher nutrient inputs (see Fig. 9–3).

Organic Matter

The dissolved organic carbon (DOC) concentration varies with lake type (Table 22–6). The principal organic carbon source for microbes in inland waters depends on the extent that the catchment is a larger source of available organic carbon than of inorganic nutrients.[1] Systems receiving water from well-drained agricultural drainage basins releasing relatively little dissolved organic carbon but disproportionately large amounts of inorganic nutrients are dominated by autochthonously produced organic carbon (e.g., Lake Esrom, DK) (Table 22–7). Conversely, systems dominated by forested or poorly drained catchments releasing large quantities of DOC with high C:P (Fig. 8–17) and C:N release ratios and little autochthonous primary production are dominated by allochthonously produced organic carbon (e.g., Wood's Lake, US). Small lotic systems, more tightly tied to the land than lakes, tend to be dominated by allochthonous carbon, particularly in forested or wetland regions.

Table 22–6 **Concentration of dissolved organic carbon (mg C l^{-1}) in different lake types. (See also Table 8–12).**

Lake Type	DOC Concentration	
	mean	range
Oligotrophic	2[1]	1–3
Mesotrophic	3[1]	2–4
Eutrophic	10[2]	3–34
Dystrophic	30[3]	20–~90

[1]Low-humic lakes with DOC primarily of allochthonous origin.

[2]Intermediate-humic lakes with DOC primarily of autochthonous origin.

[3]High-humic lakes with DOC overwhelmingly allochthonous.

Source: Modified after Thurman 1985.

---

[1]Note that organic matter produced in the littoral zone of a lake or wetland (Chapter 24) and subsequently exported to the pelagic zone is as much an allochthonous source to the plankton as is organic matter from the drainage basin. However, the littoral zone organic matter becomes an autochthonous source of organic matter at the scale at which the system as a whole is the unit of study.

Table 22–7 **Sources of organic carbon, as a percentage of the total input for selected temperate zone lakes, values are rounded off and therefore do not total 100 percent. Note the relative importance of autochthonous carbon sources in the low-humic fresh water and above all in Pyramid Lake and in eutrophic Lake Esrom, both of which receive disproportionately little organic carbon from their catchments. Highly dystrophic (humic) Wood's Lake is an example of a system receiving much allochthonous carbon. The relative availability of allochthonous organic carbon is poorly understood. The trophic designations are: oligotrophic (O), mesotrophic (M), eutrophic (E), dystrophic (D), saline (S).**

Lake	Paäjärvi FI	Lawrence US	Mirror US	Wood's CA	Pyramid US	Esrom DK
Autochthonous						
phytoplankton	59	19	65	—	96	88
periphyton	2	18	2	6	—	5
macrophytes	7	39	2	3	0	5
Subtotal	68	76	69	9	96	98
Allochthonous						
fluvial	—	11	15	90	4	2
atmospheric (leaf fall)	—	—	10	1	0	0
Subtotal	33	11	25	91	4	2
Mean depth (m)	14	6	6	1	59	12
Tropic status	D	M	O	D	S	E

Source: From Galat 1986.

▲ Dissolved Organic Carbon and Its Availability

Dissolved organic carbon (DOC) is a mixture of compounds with more than 50 percent present as large molecular-weight humic and fulvic acids collectively referred to as *humic substances*. The remainder consists of neutral acids and compounds of smaller molecular weight, such as mono- and polysaccharides as well as amino acids, that are much more available to microbes. The terrestrial DOM exported by streams and groundwater can be operationally divided into two pools: an old (>> 40 years) soil-water pool of extensively catchment-recycled organic matter carried to streams by groundwater, and a pool of more recent and probably more microbially labile (available) DOC derived from leaf litter or the surficial soil layer that is exported as surface and subsurface runoff during periods of high runoff (Schiff et al. 1997). The old soil-water pool contains considerable amounts of lignin-like substances, phenolic polymers unique to vascular plants, that are widely believed to be rela-

tively resistant (refractory, recalcitrant, nonlabile) to further decomposition. The recent DOC pool contains more smaller molecular weight sugars, amino acids, peptides, and other simple compounds leached from recently produced plant detritus. In all inland waters, the allochthonous pool is supplemented by both simple and more complex recently produced compounds of algal and macrophyte origin.

There are very few well-quantified generalizations about the composition of natural DOC and its availability to microbes. The number of compounds is vast and the composition of DOC changes over time and space. The techniques used to measure DOC fractions and microbial metabolism are not always easy to interpret and continue to evolve in ways that make it difficult to compare the results of differing techniques, making different assumptions.

Both DOC and POC fractions (the latter after solubilization by extracellular enzymes deployed on bacterial cell surfaces, *ectoenzymes*) are utilized to some extent by the heterotrophic bacteria. Ultraviolet radiation also degrades high molecular-weight organic compounds (humic substances) into lower molecular-weight molecules that appear to be disproportionately available as carbon and energy sources. In the process, nutrients are released (Moran and Zepp 1997).

The efficiency by which the organic carbon substrate is converted to biomass is reflected in the fraction of the substrate that is used for growth rather than for respiration. Phrased alternatively, the **conversion efficiency** or **growth efficiency** (yield) is the bacterial biomass produced per unit organic carbon utilized (Table 22–8, and del Giorgio and Cole 1998).

There is much variation in the growth efficiency of different substrates (Table 22–8), in part because the substrates differ in the fraction of carbon readily available.[2] Growth efficiency is further affected by the C:N and C:P ratio of the organic matter. When a limiting inorganic nutrient is experimentally added—and the C:N or C:P ratio is thereby lowered—more of the formerly unavailable POC or DOC becomes available and the conversion efficiency increases (Table 22–8). Growth efficiency rises as phytoplankton production rises, reflecting an increased supply of high quality carbon with relatively low C:P and C:N ratios (del Giorgio and Cole 1998).

Allochthonous vs Autochthonous Carbon in Microbial Metabolism

Even low-humic systems receive considerable amounts of organic carbon, mostly as dissolved organic carbon (DOC) from their drainage basins (Sec. 8.8) with additional amounts from littoral zones. The allochthonous input of decomposable organic matter is sufficient to lower the photosynthesis to respiration (P:R) ratio of most streams (Table 8–5) and the epilimnia of many low-humic oligotrophic lakes to well below unity (del Giorgio and Peters 1993). Planktonic bacteria in high-humic lakes, characterized by high light extinction, low dissolved nutrient levels, and a re-

Table 22–8 **Estimates of growth efficiency (conversion efficiency, BCE) of bacteria in utilizing particulate and dissolved organic substrates. (See also Table 22–3.)**

Substrate	Yield (%)
DOC from a humic lake	4–5
DOC from Amazon River (BR)	3–46
DOC from Ogeechee River, Georgia (US)	31
DOC from leaf leachate	10
POC from *Spartina* (emergent macrophyte)	10
POC from *Juncus* leaves (emergent macrophyte)	2.5
POC from natural seston	9
POC from phytoplankton	14–24
POC from phytoplankton	58
lignocellulose from *Carex* (emergent macrophyte)	29
lignocellulose from same *Carex* plus added NH_4^+	54
Excreted organic carbon (fresh water literature)	31–75

Source: From Pomeroy and Wiebe 1988, Tranvik and Höfle 1987, and del Giogio and Cole 1998.

sulting low autotrophic primary production, are overwhelmingly nourished by allochthonous carbon, with the supply rate linked to high periods of stream inflow (Table 22–7, and Bergström and Jansson 2000). In shallow clearwater lakes dominated by lake and wetland macrophytes, some of the required organic carbon is derived from littoral zones (Table 22–7, and Coveney and Wetzel, 1995). Even so, a high fraction of the DOC received from well-vegetated drainage basins is thought to be quite resistant (recalcitrant) to further microbial decomposition in aquatic systems because of a typically lengthy exposure to microbial degradation and transformation on land or in wetlands. But even if only a very small fraction of the large pool of allochthonously DOC supplied is available to aquatic microbes, the quantities substantially add to the pool of presumably more recently synthesized and more readily available autochthonous DOC provided by the phytoplankton and benthic plants.

Another explanation for the apparent recalcitrance of much of the allochthonous organic matter involves, as mentioned, not so much the molecular structure of the DOC itself as the C:P or C:N ratio. Streams and

---

[2]The quantity and quality of DOC is not only of great importance in limnology but also in environmental toxicology (Chapter 28) and public health. Chlorination of drinking water and treated sewage wastewater prior to its release to waterways serves to kill intestinal bacteria. Unfortunately, the chlorine, together with the wastewater DOC and natural DOC, allows the bacteria in the receiving waterways to synthesize haloacetic acids of concern and trihalomethane precursors. The subsequent conversion of precursors to trihalomethanes (e.g., trichloromethane or chloroform) with carcinogenic, and possibly mutagenic, properties provides a health risk where concentrations are elevated and the water is a source of drinking water. (Palmstrom et al. 1988)

groundwater draining undisturbed, and well-vegetated drainage basins usually exhibit ratios vastly higher than the demand ratio of the bacterial cells.

The extent to which allochthonous organic matter is utilized by planktonic microbes has become an important issue in limnology. Does allochthonous organic matter make an appreciable contribution to the observed planktonic bacterial biomass and production? This is suggested by the commonly elevated ratio of bacterial to phytoplankton carbon in unproductive lakes (Fig. 22–6). Furthermore, the observation of systematically higher bacterial biomass per unit algal biomass in inland waters than oceans fits the notion of allochthonous organic carbon substantially supplementing the autochthonously produced pool in inland waters (Simon et al. 1992). In addition, the observed greater community biomass of heterotrophic than autotrophic organisms in oligotrophic lakes suggests that at least a portion of the freshwater energy demands in the plankton is satisfied by organic carbon derived from drainage basins, and the seldom-considered benthic plants (del Giorgio and Gasol 1995).

Temperature and Lake Morphometry

Bacterial biomass tends to be greater in warm than cool epilimnia (Eq. 22.1 and 3, Table 22–4), and temperature has a positive effect on bacterial growth

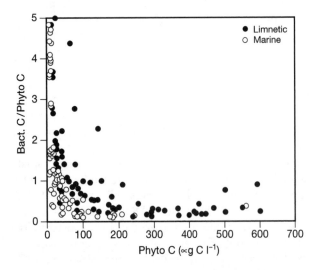

Figure 22–6 The relative importance of bacterial organic carbon (Bact. C) to phytoplankton carbon (Phyto. C) in freshwater and marine systems of increasing trophic state (phytoplankton biomass). *(After Simon et al. 1982.)*

(White et al. 1991). Laboratory research has repeatedly demonstrated that higher temperatures (up to some definite limit) increase metabolic rates, thereby increasing the rate organic matter is recycled. But lake temperatures covary with lake depth in among-lake studies and it is unclear to what extent the higher metabolism is a physiological response to temperature or the result of environmental factors that are linked to water depth (Table 22–3). Shallow lakes that are also warm tend to be more nutrient-rich than deep lakes (Fig. 9–1), and allow a higher rate of resuspension of sedimented and littoral zone bacteria into the water column, providing another explanation for observed higher bacterial abundances. In addition, water temperature at higher latitudes covaries with a host of seasonal effects, such as zooplankton grazing, primary production, and hydrology.

22.6   Resource Limitation vs Grazing Control of Bacterial Abundance

Disagreements in the literature about the relative importance of resource availability versus predation control in determining both abundance and community structure of not only the bacterioplankton but also at other trophic levels, are often the outcome of examinations over different temporal and spatial scales (Sec. 2.6). The correlation between algal biomass (chl-*a*) and bacterial abundance at a time-averaged scale (Table 22–5) suggests that resource availability is an important determinant of bacterial abundance among lakes that exhibit considerable variability in algal biomass and bacterial abundance. Bacterial production rates too are an important function of resource availability because bacterial abundance and community production are correlated (Sec. 22.7); but the bacterial abundance to chl-*a* ratio declines with increasing chl-*a*, a surrogate for resource availability. Based on a simple food web model, Sanders et al. (1992) argue that bottom-up control (food supply) is most important in regulating bacterial abundances in oligotrophic systems, and top-down control is most important in eutrophic environments (Sec. 22.11) where the bacterial/algal biomass ratio is lowest (Fig. 22–6).

Gasol et al. (1995) noted that the impact of grazers is less apparent *among* lakes varying several orders of magnitude in resource availability than within *single* lakes. Individual systems exhibit a more, modest annual variation in resource availability, whereas the seasonal variation in abundance and impact of bacterial grazers typically varies by one or two orders of magnitude. Consequently, the seasonal impact of grazers is more readily evident at the within-system scale of inquiry than the among-system scale (Sec. 2.6).

Work on phytoplankton-based food chains suggests that phytoplankton community production is usually most strongly constrained by resource limitation (nutrients, light), while their zooplankton or fish predators are most commonly controlled by organisms feeding on them. Little is known about which type of control governs the heterotrophic bacteria, partly because of uncertainies in determining the production rates and carbon content of bacterial cells in nature. Work by Psenner and Sommaruga (1992) indicates rapid shifts between bottom-up and top-down control, indicating, once again, the importance of the time and spatial (within vs among) scale of studies on the conclusions drawn (Sec. 2.6).

22.7 Heterotrophic Bacteria: Production, Losses, and System Contribution

The large number of bacterial production measurements now available show, despite technical uncertainties, that among systems varying greatly in resource availability production rates increase with increasing algal biomass (Table 22–9) and phytoplankton production (Cole et al. 1988). But resource availability is less seasonally variable in single systems (smaller range scale) and the equivalent pattern is not as easily discerned, requiring experimental manipulations (Psenner and Sommaruga 1992).

Empirical models developed to predict bacterial production (Table 22–9) allow some interesting observations. First, the slopes of Equations 1, 2, and 3 in the table are not significantly different from unity (0.81–1.22), thus it appears that bacterial abundance and production rise in step among aquatic systems, with the result that production per cell apparently does not change systematically with trophic status. Second, the relationship between abundance and production is much weaker in freshwater than marine systems (Eqs. 2 and 3 in Table 22–9). In inland waters, bacteria are in much closer contact with their drainage basins, littoral zones, and the sediments as sources of organic carbon and plant nutrients than is the case for their open-ocean counterparts. Furthermore, catchments greatly vary not only in size but also in the timing, quantity, and quality of organic matter and nutrients exported, further weakening the strong link between autochthonous primary production or algal biomass (chl-*a*) and the bacterial production that is observed in open oceans.

▲ There is now abundant evidence that protozoan and other grazers selectively remove the larger, most metabolically active bacteria (Šimek et al. 1999), modifying the size distribution, taxonomic composition, and lowering the growth rate per cell as a result (Pernthaler et al. 1996). Selective grazing induces contrasting survival strategies among bacterial strains, involving either shifts to a high division rate of the

Table 22–9 **Relationships among systems between heterotrophic bacterial production (BP, μg C l^{-1} d^{-1}), bacterial abundance (A, 10^9 cells l^{-1}), phytoplankton biomass (chl-*a*, μg l^{-1}) and temperature (temp., °C) in euphotic zones.**

Equation	Model	r^2	n	Habitat
1	log (BP) = 0.89 + 1.22 · log (A)	0.68	700	fresh water and marine
2	log (BP) = 0.95 + 0.81 · log (A)	0.21	275	fresh water
3	log (BP) = 0.71 + 1.12 · log (A)	0.77	207	marine
4	log (BP) = 0.52 + 0.52 · log (A) + 0.041 · temp.	0.37	275	fresh water
5	log (BP) = 0.90 + 0.49 · log (chl-*a*)	0.20	219	fresh water

*Additional equations and references can be obtained via anonymous ftp at ftp.icm.csic.es/pub/gasol.

Source: From White et al. 1991.

survivors or the development of grazing resistance through an increase in cell size (Pernthaler et al. 1997). As a group, very small bacterial cells appear to be little grazed. They grow slowly and many of the organisms apparently are dormant (Bird and Kalff 1993). For more on grazing, see Sec. 23.10.

System Contribution

Bacterial abundance increases with trophic status, but at a progressively slower rate. Note the low slopes (<< 1) in Table 22–5. The great importance of bacterial carbon relative to phytoplankton carbon also declines with trophic status (Fig. 22–6). This suggests a proportionally smaller role for the planktonic bacteria in lake metabolism of more eutrophic systems because bacterial production per unit biomass (specific production) does not increase sufficiently, or increase at all, with increasing trophic status to offset the proportionally smaller bacterial biomass in eutrophic systems (White et al. 1991). Smaller, less diverse data sets also suggest that specific production and respiration do not change—or may even decline—with increasing bacterial abundance (Lind et al. 1997, Cimbleris and Kalff 1998, Sommaruga and Robarts 1997). Consequently, still-limited evidence points to a dominant role played by bacteria in the plankton of highly oligotrophic inland waters, but a progressively smaller role in more productive inland waters. However, the role of bacteria in the system as a whole would not decline if, as expected, reduced water column production is offset by enhanced microbial activity in the sediments. Eutrophic lakes are typically shallow (Chapter 7), and a shallow water column allows rapid particle sedimentation on calm days (Fig. 21–21). This permits more of the particles to reach the sediments, and presumably yields a disproportionately larger role for sediment bacteria in shallow unstratified lakes with well-oxygenated sediments. At the same time, greater possibilities for resuspension point to a close connection in shallow waters between planktonic, epiphytic, and sediment bacteria (Sec. 22.11). Epiphytic bacteria in macrophytes beds are abundant and exhibit areal production rates well above those of the bacterioplankton (Theil-Neilsen and Søndergaard 1999). The higher rates of algal and bacterial production near shore (Fig. 22–7) probably represent organisms washed out of the macrophyte beds or derived from littoral sediments.

Gross heterotrophic bacterial production, which includes respiration, appears to vary from less than 10

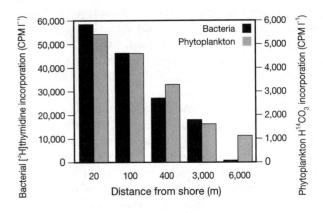

Figure 22–7 Transects of phytoplankton primary production and bacterial production (thymidine incorporation), expressed as counts per minute (CPM) of tracer taken up at a 2 m depth, from the littoral to the central portion of the upper basin of Lake Constance (AT, CH, DE). *(After Güde 1990.)*

percent of the water column integrated primary production in (shallow) eutrophic lakes to an extrapolated 50–60 percent in highly oligotrophic clearwater systems (Cole et al. 1988). The percentages are, encouragingly, reflected in the bacterial: phytoplankton biomass ratio (Fig. 22–6).

▲ 22.8 Viruses

Viruses are extremely abundant. Direct counts, by electron microscopy, show between 10^5 and 10^8 viral particles ml^{-1} in planktonic environments (Fig. 22–8, and Bergh et al. 1989). Most of the particles are small (< 70 nm or 0.070 μm in diameter), with a very small fraction larger than 100 nm (Maranger and Bird 1995). The few freshwater production rates that have been measured range between ~10^9 and 10^{10} particles $l^{-1}day^{-1}$ (Weinbauer and Höfle 1998). How frequently the **virioplankton** component of the **femtoplankton** (< 0.2 μm) (Table 21–1) imposes a major loss rate on the bacteria and algal populations—through cell rupture (*lysis*) following infection—remains obscure. There are indications that somewhere between 10 percent and 20 percent of the heterotrophic bacteria (but fewer of the algae?) in the sea may be lysed (ruptured) daily during the warmer portions of the year (Cottrell and Suttle 1995). Viral lysis in the oxic epilimnion of a German lake accounted for eight to 42 percent of the summed lysis plus heterotrophic nanoflagellate (HNF) grazing mortality, and the majority of bacterial production in the anoxic hy-

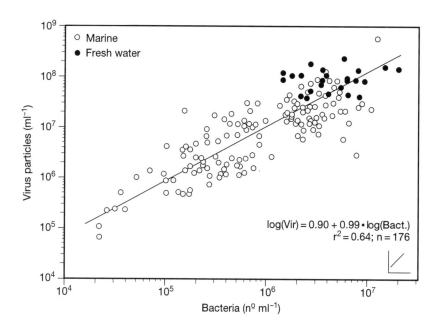

Figure 22–8 Relationship between viral and bacterial abundance in freshwater and marine habitats. The line represents the regression equation describing the overall relationship. Note that the relationship was insignificant for the freshwater lakes, which ranged over only one order of magnitude. Over this moderate range in bacterial abundance, differences in techniques used or assumptions (other than bacterial abundance) determine viral abundance. (See Section 2.8 regarding the importance of the range scale examined in correlation analysis and in inferring cause and effect.) (*After Maranger and Bird 1995.*)

polimnion, harboring relatively few grazers, was removed by viruses. Interestingly, estimates of bacterial production in the two lake layers show it to be totally or almost totally balanced by the summed mortalities (Weinbauer and Höfle 1998). Work elsewhere suggests that resource supply and **bacterivory** (predation on bacteria), not lysis, control bacterioplankton abundance at the among-system scale (Pedrós-Alió et al. 2000). Clearly, no broad generalities have yet emerged explaining the importance of the virioplankton, but it is apparent that lysis can be important over particular spatial and temporal scales. Viruses can impose not only a substantial bacterial mortality but further impact the bacterial community growth and composition through the release of DOC from bacterial cells that is then reused by other bacteria (Fuhrman 1999).

Bacterial and viral abundance are correlated among systems with typically an order of magnitude greater viral abundance. But no pattern has emerged for the relatively few eutrophic fresh waters that have been examined, covering a narrow range scale (Fig. 22–8 and Sec. 2.6). Other significant (positive) relationships have similarly been found between viral abundance and chlorophyll-*a* concentration, total phosphorus, and bacterial production (Maranger and Bird 1995). However, it remains unclear whether the viruses seen on electron micrographs are bacterial or algal viruses. It is similarly largely unclear what determines the observed changes in viral abundance, and how much viruses affect the abundance and community structure of bacteria (and algae) by selective infestation (predation) rather than through carbon and nutrient recycling.

22.9   The Microbial Food Web

The importance of heterotrophic bacteria in the decomposition of organic matter has long been recognized (Kuznetsov 1970). It has also long been known that the inorganic nutrients that are liberated would be available for primary production in the water column or at the sediment surface, which in turn would sustain the other trophic levels in the food chain. Since bacteria can only directly utilize dissolved material, it was thought that the microbes could not be important consumers of particulate organic matter (POM). The dissolved organic matter (DOM) was also thought to be present in low concentrations in the oligotrophic lakes best-studied and it was assumed that planktonic bacteria did not grow rapidly.

The idea that bacterial production might be an important energy source for zooplankton and benthic invertebrates, and via them for higher trophic levels, was not widely appreciated until recent decades when their great abundance became recognized.[3] In terms of food

[3]This insight had come much earlier to water-pollution biologists, working outside the mainstream of microbial ecology, on the self-purification of rivers with distance below pollution sources. (Kuznetsov 1970)

web structure and energy flow, the heterotrophic bacteria and phytoplankton occupy the same trophic level, both nourish the primary consumers (Jones 1992). From a food web perspective, heterotrophic bacterial production can be considered as primary production in that it converts dissolved to particulate matter or synthesizes organic matter by chemosynthetic autotrophy. The current widespread recognition of the role of aquatic bacteria in food webs (Fig. 22–9) had to await technical developments, largely accomplished during the last two decades.

Bacteria are, as we have seen, imperfect transformers of organic substrates (Sec. 22.5). Much of the gross production is lost in respiration. How much is lost rather than converted into bacterial protoplasm depends on the quality of the organic substrate and availability of the required inorganic nutrients. The bacterial conversion (or growth) efficiency (BCE or BGE) of substrate into bacterial biomass production plateaus near 50 percent in the plankton of highly eutrophic water but can be as high as 60–75 percent when the bacteria utilize recent plant production (Table 22–8). In that case, at least half of the measured gross bacterial production is potentially available to bacterial grazers, and via them to higher trophic levels. Conversely, in highly oligotrophic inland water systems where the terrestrially derived substrates are of poor quality, the conversion efficiencies can be as low as a few percent (Table 22–8, and del Giorgio and Cole 1998).

Correct conversion (growth) efficiencies are needed to accurately estimate the importance of bacteria in food webs (see below), but such determinations are far from simple even under experimental conditions (see del Giorgio and Cole 1998). Most field studies, therefore, must assume a conversion factor whose appropriateness to a particular study is unknown. As the growth efficiency is expressed in terms of carbon, information on the carbon content of the bacteria in nature must also be available. This information is generally not available and a literature value has to be assumed. Unfortunately, bacteria appear to vary two- to fivefold in their carbon content, and the particular volume: carbon conversion factor that is assumed has a major effect on the calculated growth efficiency. Finally, it is not clear that the bacteria thriving under the experimental conditions used to determine the BCE are the same as those who dominate in situ assemblages.

Bacterial Grazers

The recent literature indicates that the principal grazers of the planktonic bacteria are the protozoa, chief among them the *heterotrophic nanoflagellates* (HNF). The HNF are usually < 12 μm in their longest dimen-

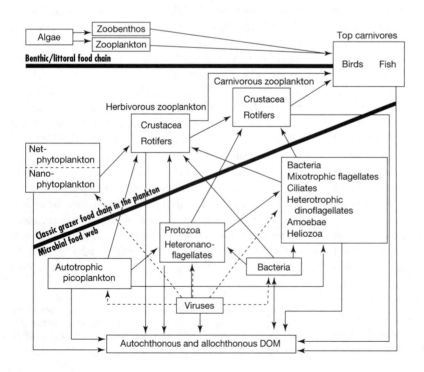

Figure 22–9 The contemporary view of microbial food web structure (below the lower thick line) in relation to the "classic" grazer food chain (above the thick line) in plankton. There is also a little-studied benthic/littoral food chain that subsidizes the classic grazer food chain and through turbulence is coupled (not shown) to the microbial food web. Full lines and arrows indicate feeding interactions, broken arrows indicate viral infections. The pool of dissolved organic matter (DOM) used as a substrate by the bacteria is replenished by various release processes (excretion, exudation, cell lysis, sloppy feeding) from each compartment and from the catchment. *(Modified from Weisse and Stockner 1993.)*

sion (Chapter 23, and Weisse and Müller 1990). Their conversion efficiency averages ~30 percent (range 10–53%; Nagata 1988). The sometimes highly abundant and somewhat larger small ciliates appear to graze primarily on larger bacteria and the HNFs (Fig. 22–9). The protozoans are in turn preyed upon by *microzooplankton* (< 200 μm) such as rotifers and small crustaceans, or by *macrozooplankton* (> 200 μm) such as the larger filter-feeding cladocerans (Chapter 23). Some of the large herbivorous cladocerans can and do directly graze larger bacteria.

▲ The bacterial filtering efficiency of larger cladocerans is typically low, but the water volume filtered or "cleared" by the individuals is exceptionally large (100–3,000 μl animal^{-1} hr^{-1}; Porter et al. 1983) compared to typically ~1.5 nl animal^{-1} hr^{-1} for individual HNF (Fukami et al. 1991). The much less abundant bacterivorous cladocerans therefore can have a major impact only when the animals are abundant (Vaqué and Pace 1992). Abundant large cladocerans can filter the entire volume of shallow lakes in less than a day. However, the observation that the bacteria: HNF ratio is typically highest when large cladocerans are abundant suggests that the principal role of the cladocerans is the removal of the HNF predators rather than bacteria (Gasol et al. 1995).

Despite an individually small water- (and particle-) filtering rate of only ~10–100 bacteria per hour, HNFs are frequently abundant enough to clear a significant fraction of the epilimnia of oligotrophic lakes daily. In Lake Constance (AT, CH, DE), the HNF remove an average of about 50 percent of the annual bacterial production. The generally larger but less abundant ciliates clear considerably more water (1–10 μl animal^{-1} hr^{-1}), in the process typically removing 10–125 HNF ml^{-1} hr^{-1} (Weisse and Müller, 1990) plus a potentially large number of bacteria (Šimek and Straškrabová 1992). But the limited available evidence suggests that, on a per animal basis, the ciliates are probably more important as grazers of picoplankton, nanoplankton, and HNFs than of bacteria (Šimek et al. 1995).

Not only does the abundance of protozoans (*protists*) vary seasonally within and among aquatic systems but measures of the impact of their predation on bacteria appear to be confounded by the wide variety of techniques in use, and yielding different results (Fig. 22–10). Work assembled by Sanders et al. (1989) indicates that bacterivory and bacterial production are linked: the grazers that were examined were able to remove, on average, almost 100 percent of the bacterial production (Fig. 22–11).

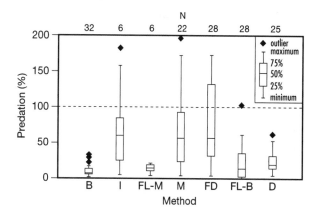

Figure 22–10   Box-whisker plot of the mean percentage of bacterial standing stock removed daily by protist predation according to the method used in the freshwater and marine literature. B: fluorescent beads; I: protists inhibitors; FL-M: minicells (stained bacteria); M: radiolabelled minicells; FD: filtration and dilution; FL-B: fluorescently labelled bacteria; D: fluorescent particle disappearance. The number of individual observations (N) are given above the plot. The central line of each box is the median of the distribution, and the box limits are the 25 per cent quartiles of the data. The whiskers (vertical lines) cover the entire data range, except extreme observations (◆), some of which are off the scale in groups I, M and FD. (*Modified after Vaqué et al. 1994.*)

The questions raised during the last decade about the fate of the bacterial production have encouraged more work on the ecology of the protozoa and their role in the predation of planktonic bacteria, and also revealed that some algae are important bacteria grazers. Photosynthetic flagellates, some phylogenetically closely related to the HNFs (Chapter 23), are occasionally significant predators (Fig. 22–12). Many of the pigmented flagellates, which typically dominate the phytoplankton biomass of oligotrophic lakes (Chapter 21), may well be the dominant group of bacterial grazers there (Bird and Kalff 1987, Berninger et al. 1992). Organisms such as these, with more than one mode of nutrition, are known as **mixotrophs**. Those mixotrophs feeding on particles are known as **phagotrophs** (Chapter 21). The existence and abundance of phagotrophic algae in the plankton complicates deciphering the energy flow pathways within the microbial loop (Fig. 22–9) and wreaks havoc on the trophic level concept. The common occurrence of greenish ciliates, able to photosynthesize by means of chloroplasts, often sequestered from their algal prey (Fig. 21–7, and Sanders 1991), further confounds the notion of distinct trophic levels.

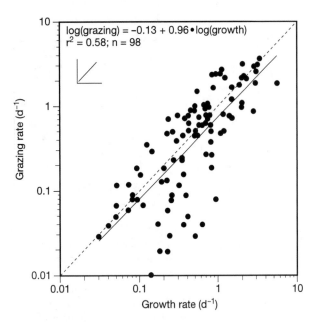

$$\log(\text{grazing}) = -0.13 + 0.96 \cdot \log(\text{growth})$$
$$r^2 = 0.58; \, n = 98$$

Figure 22–11 The relationship between bacterial production (growth) and community bacterivory (grazing) in freshwater and marine systems. The dashed line represents a 1:1 relationship and the solid line represents the best-fit equation (shown). The considerable data scatter reflects either an expected short-term imbalance between production and grazing or inaccuracies in the methods used for determining both production and grazing rates. *(Modified after Sanders et al. 1989.)*

The **food chain**, or more appropriately, the **food web** into which heterotrophic bacterial production (and picophytoplankton production) flows primarily to the protozoans and the often closely related phagotrophic algae, is known as the **microbial loop** (Azam et al. 1983) when the discussion pertains to the flow of organic carbon from a higher to lower trophic level within the microbial food web. The loop complements the classical food web leading from nanoplankton to the macrozooplankton and up (Fig. 22–9).

Energy Flow in Food Webs

Much research in lakes is underway on different aspects of the microbial food web. Not surprisingly, it turns out to be considerably more complex than originally envisaged, requiring the aggregation of organisms into functional groups for modeling purposes.[4]

Small HNF are consumed by larger ones before they are eaten by one or more size classes of ciliates and by some rotifers (Chapter 23). Rapid progress in deciphering microbial food webs is slowed by lack of information on the autecology of the protozoans[5] (and other small organisms in the water and sediments), and by the practical limitation of having to sample them at intervals much longer than their doubling time in nature (Fig. 2–2). Quantification of the energy flow in planktonic food webs is also slowed by the presence of phagotrophs and allochthonous carbon inputs from drainage basins and the littoral zone. Not only is the plankton of oligotrophic systems, dominated by heterotrophic bacteria rather than algae (Fig. 22–6), but the plankton community biomass of all heterotrophic plus autotrophic organisms is also dominated by heterotrophic organisms when chl-*a* concentrations decline below the 2–5 μg chl-*a* l^{-1} (del Giorgio and Gasol 1995). The authors propose a compensating energy subsidy from outside the plankton. A drainage basin DOC subsidy dominates energy flow to the microbes in humic systems (Bergström and Jansson 2000), characterized by high light extinction and low rates of benthic primary production, and contributed significantly (13–43%) to whole system respiration in four humic US lakes (Cole et al. 2000). Transparent oligotrophic lakes receiving relatively little allochthonous organic matter from their catchments are dominated by benthic rather than planktonic primary production (Fig. 24–17) that can subsidize planktonic food webs.

The complexities of the microbial loop may not matter very much from the point of view of energy flow to higher trophic levels, unless a high fraction of bacterial carbon flows directly into the large filter-feeding cladocerans fed upon by fish (Wylie and Currie 1991). The reason for this lies in the loss of energy by respiration at each level (step) in the food web. In other words, it depends on the conversion efficiency of the bacterivores able to convert the bacterial protoplasm into biomass. Assuming a conversion efficiency as high as 25–50 percent between bacteria and their

---

[4]"If we go too far in pooling organisms together, we will lose all the natural history that ultimately drives evolution of both organisms and communities. If we stop short and retain many groups, we will

be confronted with a complexity that will escape analysis." (C. Pedrós-Alió 1994)

[5]"Many of the uncertainties about food webs result from our lack of knowledge of natural history of both microorganisms and metazoans . . . Yet many people view natural history as 19th century descriptive biology, not worthy of serious consideration in the age of genetic engineering. The fact is that much of our modeling and other efforts at quantification are limited by our ignorance of life histories, feeding strategies, parasitic and predatory interactions, and the like." (Pomeroy and Wiebe 1988)

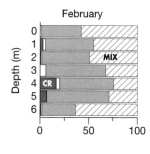
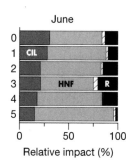
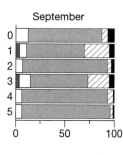

Figure 22–12 Relative grazing impact on bacteria by different planktonic bacterivores in eutrophic and monomictic Lake Ogelthorpe, Georgia, US (LA = 30 ha; z_{max} = 8.5 m). Crustaceans (CR), ciliates (CIL), heterotrophic nanoflagellates (HNF), mixotrophic algae (MIX), and rotifers (R). Note the modest grazing contribution of the crustaceans. (*Modified after Stockner and Porter 1988.*)

predators, as well as between the different predators, virtually all the bacterial carbon that is produced dissipates in only a few feeding steps. Even if the conversion efficiency from bacteria to the HNFs were 50 percent, and the HNFs were consumed directly by ciliates rather than other HNFs, nearly all the bacterial carbon in the plankton will be lost through respiration in a three-step food chain, with the ciliates receiving about 12 percent and the macrozooplankton a mere six percent of the bacterial energy (Pomeroy and Wiebe 1988). Three to five steps is more probable in the microbial loop, and it is evident that little would remain for transfer to macrozooplankton and from there to zooplanktivorous fish (fish CE ~5–10%). Only when there is a major one-step transfer from bacteria to large bacterivorous macrozooplankton is it likely that appreciable amounts of bacterial carbon becomes available to zooplanktivorous fish (Wylie and Currie 1991). The possibility of such a one-step transfer to higher trophic levels is much greater in lotic systems and the sediment of lakes and wetlands. There, filter-feeding insect larvae such as common black flies (Simuliidae) effectively filter bacteria (and algae) from the water in streams, and are particularly abundant below lake outlets. Macrobenthic invertebrates feeding on bacteria-coated detritus also receive much of their energy from bacteria (see Sec. 25.2, and Jónasson 1972). They, along with species that feed on fungi-coated particles, thereby short-circuit the microbial loop by eliminating several trophic transfers and increasing the microbial energy available to macroinvertebrates and their fish predators (Meyer 1990).

Although the literature is far from convincing, it appears that in the absence of unusually heavy grazing pressure on bacteria by macrozooplankton or benthic macroinvertebrates, the principal role of microbial food webs lies, as originally thought, in the degradation (respiration) of organic matter. The inorganic nutrients recycled in the water column play a critical role in sustaining not only planktonic but also benthic primary production and ultimately the fisheries at the top of the aquatic food chain (Fig. 22–9).

In the jargon of energy-flow studies, the microbial food web appears to be primarily an energy *sink* in the plankton but, in the process, a crucially important nutrient *link*. Where the microbial food web in the plankton serves as a loop, it is largely responsible for recycling critical nutrients within the euphotic zone (Nakano 1994). In the process, it greatly reduces losses by sedimentation of particulate nitrogen and phosphorus from the site of primary production and the water column.

Little is known not only about bacterial loss rates but about loss rates at all trophic levels, the temporally and spatially changing importance of grazing by different predators, particle sedimentation, or disease and viral infections which together determine the observed bacterial loss rates. However, a comparison of the modest number of experimental studies in which both bacterial plankton production and grazing losses were measured (Fig. 22–11) indicate that predation is frequently the major loss factor for free-living bacteria, with bacterial production and loss rates of the same magnitude over the 12–24 hour time scale of the studies.[6] Even so, the existence of systematic seasonal and short-term oscillations in bacterial abundance and community biomass points to periodic imbalances between bacterial growth and loss rates (Psenner and Sommaruga 1992). Aquatic scientists have traditionally spent more time and effort, at all trophic levels, measuring production rates than the much more difficult to obtain sum total of the different loss rates encountered. Unfortunately, a prediction of population or community abundance and community structure requires more than a better understanding of the

---

[6]"Numerous processes occurring in the microbial loop are similarly difficult to evaluate because the shortest time in which a measurement can be made is longer than the time over which a process occurs." (Frost et al. 1988)

losses experienced in the plankton. It further requires much more research incorporating both the littoral zone and sediments into food webs, and a consideration of the energy subsidies from drainage basins.

▲ 22.10 Photosynthetic Bacteria

The heterotrophic bacteria of the plankton normally require dissolved oxygen and need no light, but the **green** and **purple photosynthetic bacteria** typically require both light and anoxic conditions (Table 22–10). The photosynthetic bacteria are therefore found in abundance only at the bottom of illuminated oxyclines in either the water column or sediments. Their bright pigmentation, larger size, and interesting physiology/biochemistry, plus easy identification and culture, have long made photosynthetic bacteria the object of physiologists' attention. But, the specialized growing conditions of the bacteria—often well below the lake surface in meromictic lakes—had until re-

cently, drawn relatively few microbial ecologists to these bacteria.

Although the green and purple bacteria (Table 22–10) share a prokaryotic cellular architecture and the ability to convert light energy into a chemically usable form with the cyanobacteria (blue-green algae; Sec. 21.1, and Chapters 16 and 21), they differ fundamentally from them in other ways. Virtually all cyanobacteria (and the eukaryotic algae) evolve molecular oxygen as a photosynthetic by-product whereas photosynthetic bacteria do not (Table 22–1). With a few exceptions, photosynthetic bacteria require anoxia because the synthesis of their bacteriochlorophylls is repressed under oxic conditions. The anoxygenic photosynthesis of green and purple bacteria, in contrast to the oxygenic photosynthesis of the eukaryotic algae and cyanobacteria, is primarily dependent on the availability of reduced forms of sulfur. Carbon dioxide is reduced to bacterial biomass in paired redox reactions when reduced sulfur is oxidized in the light (Chapter 16). Hydrogen sulfide (H_2S), elemental sul-

Table 22–10 **Selected genera of anoxygenic photosynthetic bacteria. The number in parentheses indicates the number of species recognized in each genus.**

Group	Morphology
Purple Bacteria	
Sulfur Bacteria (Chromatiaceae and Ectothiorhodospiraceae)	
Amoebobacter (2)	cocci embedded in slime; gas vacuoles
Chromatium (11)	large or small rods
Lamprocystis (1)	large cocci or ovoids; gas vacuoles
Thiocapsa (2)	small cocci
Thiopedia (1)	small cocci arranged in sheets; gas vacuoles
Ectothiorhodospira (4)	small spirilla; do not store sulfur inside cells; common in hypersaline lakes
Nonsulfur Bacteria (Rhodospirillaceae)	
Rhodopseudomonas (8)	rods, dividing by budding
Rhodospirillum (6)	large or small spirilla
Green Bacteria	
Sulfur Bacteria (Chlorobiaceae)	
Chlorobium (5)	small rods or vibrios
Pelodictyon (3)	rods or vibrios, some forming a three dimensional net; gas vacuoles
Gliding Bacteria (Chloroflexaceae)	
Chloroflexus (2)	multicellular filaments up to 100 μm long
Oscillochloris (1)	large filaments up to 2,500 μm long; gas vacuoles

Source: Modified after Madigan 1988.

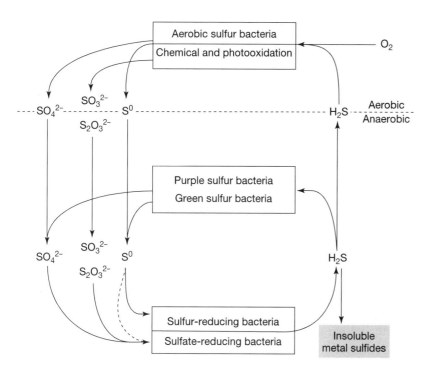

fur ($S°$), and thiosulfate ($S_2O_3^{-2}$) are the most common electron donors (Fig. 22–13). If sulfide concentrations are high, elemental sulfur will be formed as the first step in the oxidation of sulfate, as was shown experimentally by C. B. van Niel more than 60 years ago (see Madigan 1988).

$$2H_2S + CO_2 \xrightarrow{\text{light}} CH_2O + H_2O + 2S°$$

EQ. 22.1

But when concentrations are low, the sulfide is oxidized directly to sulfate during the reduction of CO_2 to biomass.[7]

$$H_2S + 2CO_2 + 2H_2O \xrightarrow{\text{light}} 2CH_2O + H_2SO_4$$

EQ. 22.2

Photosynthetic bacterial cells or colonies are much larger than the very small heterotrophic bacteria of the oxic water column (Sec. 22.3). Most can readily be

seen under the light microscope and are therefore much better known taxonomically than the heterotrophic bacteria of the plankton. In the late 1800s and the first several decades of the 1900s, sulfur bacteria became the object of study of several outstanding schools of bacterial physiologists, mostly in the Netherlands, France, and former Soviet Union (USSR). But only the soviet schools and their "offspring" in eastern Europe achieved excellence in the microbial ecology of inland waters (and soils) during the 20th century, as was mentioned in Sec. 22.2. As a consequence of historical developments, much more has become known about the taxonomy, physiology, and physiological ecology of the sulfur bacteria than about the very diverse heterotrophic bacteria of the plankton, which are now receiving so much attention from microbial ecologists.

Bacterial Types

Photosynthetic species have been divided into two broad groupings: the purple bacteria and the green bacteria. The purple sulfur bacteria of the family *Chromatiaceae* (Table 22–10) and the green sulfur bacteria of the family *Chlorobiaceae* are obligate anaerobic and phototrophic, primarily using reduced sulfur compounds as electron donors (Tables 22–1 and

---

[7]"The role of phototrophic bacteria as sulfide consumers is probably more important than their contribution to primary production per se; hydrogen sulfide is a highly poisonous substance for most forms of aquatic life and for many bacteria. Sulfide oxidation generates nontoxic species of sulfur, allowing the upper layers of a stratified lake to remain oxic and thus suitable for eukaryotic microorganisms and aquatic macroorganisms." (Madigan 1988)

16–1). The globules of sulfur (S°, produced as the intermediate oxidation product during photosynthesis) are characteristically formed outside the cells of the green bacteria, but arise inside the cell of purple bacteria. The presence or absence of the refractive S° globules, together with the different photosynthetic pigments, serves as an easy diagnostic tool. However, some species of purple sulfur bacteria tolerate low dissolved oxygen (DO) conditions and then function as chemosynthetic autotrophs (Table 22–1), showing them to be metabolically flexible and allowing them to grow, albeit more slowly, in the presence of DO and absence of light.

Distribution

Purple sulfur bacteria usually dominate closer to the water surface than green bacteria. The green bacteria have exceptionally efficient light-harvesting units (*chlorosomes*) allowing them to grow photosynthetically deeper in the water column at light fluxes as low as about 0.3 μmol m^{-2} s^{-1}, (Pfennig 1989). There, they benefit from higher concentrations of H$_2$S at the chemocline. However, both groups have light compensation levels well below those of the photosynthetic algal community frequently growing just above them in the oxygenated portion in the metalimnia of transparent lakes (Fig. 22–14).

Many of the purple sulfur bacteria have flagellae, allowing them to migrate vertically in chemoclines in response to phytotaxic or chemotaxic cues. Most of the planktonic species of green sulfur bacteria, and some of the purple sulfur bacteria, contain gas vacuoles that impart buoyancy control and allow them to position themselves in the nonturbulent metalimnia to optimize light and substrate availability. Daytime depletion of S^{2-} stimulates downward migration, followed by upward movement in the morning after nighttime respiration and diffusion has replenished the S^{2-} supply, and light has once again become the principal factor limiting photosynthesis (van Gemerden et al. 1985).

Water samples from the oxycline (chemocline) of transparent lakes with the appropriate combination of light and sulfide frequently contain a thin dense layer or **plate** of photosynthetic sulfur bacteria that also includes sulfate-reducing and other heterotrophic bacteria. Depending on the dominant pigmented species the plates may be pink, pink-red, brown-red, purple-pink or various shades of green (Fig. 22–15); the color is imparted by a variety of carotenoid pigments rather

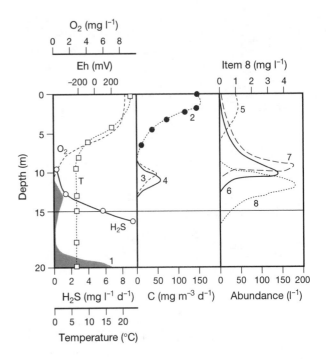

Figure 22–14 Microbiological characteristics of Lake Belovod, RU. 1 = rate of sulfate reduction, 2 = algal photosynthesis, 3 = bacterial photosynthesis, 4 = chemosynthesis, 5 = algal abundance, 6 = Protozoa, 7 = Cladocera, 8 = community biomass of purple sulfur bacteria. *(Modified after Gorlenko et al. 1983.)*

than by particular bacterial chlorophylls (bact. chl-*a*, *b*, *c*, *d*, *e*). Most purple bacteria contain only bact. chl–*a*, whereas the green sulfur bacteria have either bact. chl-*c*, *d*, or *e* as the major chlorophyll but also contain a small amount of bact. chl-*a*. However, it is the bacterial chl-*a* which plays the central role in converting light energy into ATP in both groups.

A brightly colored plate of photosynthetic bacteria at depth is rarely evident from the surface except during destratification, when turbulence at the bottom of the miximolimnion (Sec. 11.19) can sweep the upper portion of the plate into the mixed layer, giving rise to designations such as "pink lake." Shallow wastewater lagoons that are highly wind protected and become anoxic a short distance below the water surface because of enormous respiration rates, may look pink or rose-red in summer. Salt-evaporation pans used for winning salt too are often similarly colored. The pans are most commonly dominated by red-colored species of the motile family *Ectothiorhodospiraceae* (Table 22–10), but can be dominated by similarly pigmented algae (Sec. 21.2) and nonphotosynthetic bacteria. The

steep or wind-protected shoreline, are candidates for
meromixis (Secs. 7.4 and 11.10). The probability of
sufficient irradiance reaching the bottom of the oxycline to allow the development of a bacterial plate is a
function of drainage basin size and steepness as well as
geology. Only barren or small well-vegetated catchments with a moderate to high drainage basin slope
have sufficiently modest DOC and inorganic nutrient
loading to allow the high transparency (low extinction
coefficient, k_d) required to permit bacterial photosynthesis below the oxycline (Chapter 10). Catchment geology determines the Fe:S loading ratio, which in turn
helps determine whether there will be free reduced
sulfur available for oxidation below the oxycline
(Sec. 19.3).

Among meromictic lakes with bacterial plates—
implying the availability of reduced sulfur—observed
production rates are primarily a function of the large
among-system variation in light availability. Highest
production is therefore seen in lakes in which the oxycline is found close to the water surface (Fig. 22–16).
The quantities of reduced sulfur are apparently less of
a limiting factor; Montesinos and Van Gemerden
(1986) reported that low-sulfide lakes (< 1 mg S l^{-1})
were not outliers in a light versus photosynthesis plot.
The most productive lake on record is Solar Lake (IL),
a shallow (z_{max} = 4.5 m) saline pool combining high

Figure 22–15   A plate of pink photosynthetic bacteria from
below the oxycline at ~7 m in meromictic Lake Mahoney,
British Columbia (CA). *(Photo courtesy of T. G. Northcote.)*

salt level of the evaporation ponds is normally so high
that the DO solubility is negligible (Sec. 15.2) allowing photosynthetic bacteria to grow virtually to the
water surface. Similarly, anoxic hot springs and estuarine mud flats containing large quantities of reduced
sulfur often have a surface layer of cyanobacteria, followed first by a layer of purple bacteria overlying a
layer of green bacteria. Plates are usually dominated
by a single species, but additional species exhibit maxima at particular depths within the overall plate
(Gorlenko et al. 1983).

Growth and Loss Rates

The potential for bacterial photosynthesis is partially a
function of lake and catchment morphometry at the
among-lake scale of inquiry. Only lakes with a large
"relative depth" (maximum depth–surface area) and a

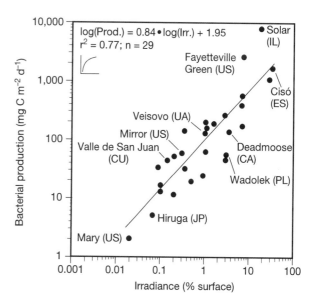

Figure 22–16   The relationship between the irradiance at
the oxic–anoxic interface and the areal production of the
photosynthetic bacteria growing there. *(Modified after Montesinos and Van Gemerden 1986.)*

irradiance with a high sulfide supply and high temperature (45–50°C). The lowest bacterial production rates are encountered in water and sediments where the bottom of the oxycline is found at a depth where the light climate is poor (Fig. 22–16).

Bacterial (and phytoplankton) primary production in metalimnia declines as the biomass of phytoplankton or other particles in the mixed layer above the bacteria increases, thereby increasing the light extinction. Thus, the bacterial plate of an Australian reservoir disappears temporarily during algal blooms and is completely lacking during years when floods greatly raise the inorganic turbidity (Banens 1990).

Photosynthetic bacterial production rates are, with few exceptions, relatively low, expressed as a per unit area of lake-surface basin, and equally low when expressed on a per cell basis and compared to the equivalent phytoplankton production per cell in the mixed layer (Table 22–11). Even so, the bacterial abundance and biomass (bacterial chl-*a*) may be large, although concentrated in a narrow band at the depth where the combination of light and the H_2S supply is optimal. Overmann et al. (1991) counted as many as 4×10^8 cells ml^{-1} of large purple sulfur bacteria in a 10-cm-thick bacterial plate in Mahoney Lake (CA). Hall and Northcote (1990) recorded concentrations of 780–1,050 mg bacterial chl-*a* m^{-2} in the same plate (Fig. 22–15). It is evident that the "photosynthetic cover" (mg chl-*a* m^{-2}) can be of the same magnitude as the highest values recorded for phytoplankton in hypereutrophic lakes (Sec. 21.15).

A combination of a usually modest bacterial primary production rate (mg C m^{-2} d^{-1}) and high biomass indicates a typically very low growth rate per cell within bacterial plates. Measured doubling times of between two and four months in Mahoney Lake show that growth rates are exceedingly low regardless of the assumptions made to obtain the estimated rates. Depth-integrated average doubling times of purple sulfur bacteria were also slow in a small well-studied Spanish pool (Lake Cisó) during the stratification period (Pedrós-Alió and Guerrero 1993), but varied seasonally and with depth within the plate. Doubling times in the upper portion of the plate averaged a mere 36 hours, but the bottom layer was inactive and largely nonviable (Pedrós-Alió and Guerrero 1993).

If the bacterial growth rates are normally low, the loss rates resulting from sedimentation, senescence, viralysis, disease, grazing, and washout must also be low for the community to maintain itself. Unfortunately, loss rates have received relatively little attention. High numbers of ciliated protozoa, daphnids, and copepods have been seen at the upper surface or just above bacterial plates. Daphnids and copepods found with pink, purple, or red guts filled with photosynthetic bacteria, plus grazing experiments, confirm that macrozooplankton can and do feed in anoxic waters containing toxic H_2S for at least short periods (Sorokin 1970, van Gemerden and Mas 1995). Some protozoans even live permanently under anoxic conditions (Chapter 23), and predatory bacteria can play an important role as well (van Gemerden and Mas 1995).

Table 22–11 **Contribution of anoxygenic photosynthesis to the total annual photosynthetic production of lakes containing phototrophic bacteria. Note that these lakes are probably biased toward those having substantial bacterial plates and high bacterial production.**

Lake	Production (g C m^{-2} yr^{-1})		% of total
	algae	bacteria	
Cisó (ES)	25	250	92
Fayetteville Green (US)	51	239	83
Smith-Hole (US)	35	35	50
Waldsea (CA)	38	32	46
Vilar (ES)	188	84	31
Solar (IL)	76	49	31
Deadmoose (CA)	69	14	17
Banyoles III (ES)	116	18	14
Big Soda (US)	300	50	10
Knaack (US)	342	17	5
Haruna (JP)	85	4	4
Vechten (NL)	192	7	3

Source: See Montesinos and van Gemerden 1986.

Nevertheless, the typically low bacterial production rates and long cell-doubling times indicate that grazing and other losses must be low during the growing season.

Even though noticeable plates of photosynthetic bacteria occur in only a small fraction of lakes and their contributions to whole-system energy flow is usually modest, the work on their physiology and ecology drew attention to the importance of bacteria in redox reactions, organic matter decomposition, and nutrient cycling. It is now evident that both the heterotrophic plus photosynthetic bacteria dominate nutrient recycling through paired redox reactions (Sec. 16.2). In the process they mediate the links between the carbon, nitrogen, phosphorus, sulfur, iron, and trace-metal cycles, and greatly affect the productivity and biogeochemistry of aquatic systems.

▲ 22.11 Heterotrophic Sediment Bacteria

The technical problems associated with counting bacteria and the determination of microbial production in the sediments are much greater than the already formidable difficulties encountered in the plankton. How the "rain" of autochthonous and allochthonous particulate organic matter derived from the upper waters and the abundance of "piggy-backed" bacteria help determine the abundance and production of the profundal sediment microbial community remains virtually unknown. The sediment–water interface, where the arriving particulate organic matter is primarily mineralized (Sec. 19.2), is difficult to sample. The intense microbial metabolism in sediment bacterial mats depletes electron acceptors such as DO and NO_3 within millimeters of the sediment surface (Chapter 16). Just below this layer, iron, manganese, and sulfate serve as electron acceptors in the oxidation of organic matter in the anoxic sediments and then dissolve, affecting the recycling of the four elements and, indirectly, phosphorus (Chapters 16 and 19). Water-column aggregates composed of organic and inorganic matter averaged 5.5 mm in diameter (range < 3–20 mm) in Lake Constance (AT, CH, DE), and were densely colonized with $5–80 \times 10^6$ bact. agg^{-1} (Grossant and Simon 1998), with the bacteria generally three times larger than free-living bacteria. However, the organic/inorganic aggregates, known as *lake snow*, containing living algae, detritus, and zooplank-

ton remains are also heavily colonized by heterotrophic bacteria being transported to the sediment. Once sedimented, the water-column bacteria become an addition to the sediment community (Sec. 22.7). Turbulence resuspends microbes from the sediments or carries them into the open water from the land–water interface or via inflowing streams. Consequently, the planktonic bacterial community of inland waters forms a continuum with their benthic, littoral zone, and catchment-derived counterparts, but the scientists studying the microbes tend to restrict their attention to a single habitat.

The sediment bacterial abundance, expressed per unit volume of surface sediment is, depending on the water column thickness, typically between two and 1,000 times greater per unit volume ($0.1–25 \times 10^9$ cells ml^{-1}) in the surficial sediments than in the water column above (Schallenberg and Kalff 1993).[8] Their size is also typically several times larger (~0.1–0.2 μm^3 $cell^{-1}$) than the free-living bacteria of the water column. Organic matter and associated nutrients play a major role in determining sediment-production rates (Fig. 22–17 and Boström et al. 1989), with the fraction of the incoming organic matter sedimented and retained, and potentially available to sediment bacteria, a function of the water residence time (Fig. 9–8).

Whatever the reasons, the single best indicator of bacterial abundance (BA) is the sediment-water content (with highest numbers at intermediate water-content levels; Fig. 22–18) while bacterial heterotrophic production in rivers and profundal sediments is, among systems, a function of the organic content of sediments (Fig. 22–17) and flushing rate (FR) of lakes (FR = 1/WRT; Schallenberg and Kalff 1993). Increased flushing allows greater export of both organic matter and inorganic nutrients from catchments (Fig. 9–3), normally reflected in enhanced phytoplankton (Fig. 9–4), and bacterioplankton production rates (Cimbleris and Kalff, unpbl. data). Indeed, the positive correlation between bacterial abundance and flushing rate points to the importance of allochthonous resources in determining sediment bacterial abundances among systems and, indirectly,

---

[8]Sediment bacteriology has its roots in soil microbiology, where abundances have traditionally been expressed per unit dry mass (cells g dry wt^{-1}). While this is not unreasonable for well-drained agricultural soils with a low and almost constant water content, its (still common) application to sediments varying greatly in percent of solids makes their interpretations much more difficult than doing so on the basis of water volume, including solids.

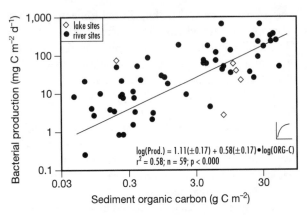

Figure 22–17 Relationship between heterotrophic bacterial production and surficial sediment organic carbon (ORG-C). Note the high proportion of river samples and the high scatter around the line of best fit, indicating the impact of other unmeasured factors on bacterial production, or problems with the bacterial production techniques in sediments. Factors other than sediment organic carbon content dominate the bacterial production where the organic carbon range varies less than about ten fold. *(After Sander and Kalff 1993.)*

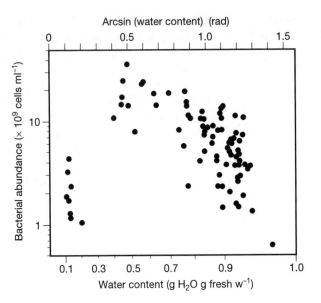

Figure 22–18 Relationship between sediment bacterial abundance and sediment water content. Water content in grams water per gram fresh mass was arcsine transformed. *(After Schallenberg and Kalff 1993.)*

the abundance of sediment macroinvertebrates (Eq. 5 in Table 20–2).

Experimental work in laboratory mesocosms has shown that diatom additions to the sediments—simulating a major sedimentation event of high quality organic matter—resulted in increased bacterial production within two hours, and a tenfold increase within 24 hours (Goedkoop et al. 1997). But such experiments are not aimed at predicting the importance over time and space of particle decomposition (respiration) in the sediment versus the water column.

The relative importance of sediments as a site for organic matter decomposition and bacterial production increases with decreasing water depth. Rapid sedimentation of particles from shallow water columns on calm days (Sec. 20.4 and Fig. 21–21) decreases the time available for bacterial decomposition in the water column of lakes. Furthermore, a comparison of water-column bacterial production ($\mu g\ C\ m^{-2}\ d^{-1}$) with measures of bacterial sediment production ($mg\ C\ m^{-2}\ d^{-1}$) in a set of relatively deep oligotrophic lakes (Climberis and Kalff unpubl. data) indicates a one order of magnitude to several orders of magnitude higher sediment rate. The apparent uncoupling of water column and sediment rates provides a suggestion that the profundal sediment bacteria may be primarily nourished by littoral zone primary production plus large quantities of typically high C:P and C:N organic matter from the

well-vegetated drainage basins that once in the sediments, allow an increased BGE as the result of elevated levels of recycled P and N there. Other recent work on macrophyte dominated lakes shows that phytoplankton production in the shallow water column is insufficient to satisfy sediment bacterial carbon demands without also considering carbon inputs from the macrophytes (Coveney and Wetzel 1995, and Kirschner and Velimizov 1999). It is apparent that understanding the dynamics of planktonic and sediment bacterial communities requires consideration of events in the water column, the littoral zone, and the drainage basin.

Highlights

- The heterotrophic bacteria of the plankton are dominated by small forms (0.2–1.0 μm) that also exhibit only modest morphological diversity, preventing their identification under the microscope.
- Recent, or fairly recent, developments in (1) staining heterotrophic bacteria, facilitating their enumeration; (2) techniques for measuring bacterial production; and (3) bacterial identification in nature, using molecular techniques; plus (4) techniques for measuring grazing rates by predators is allowing rapid progress in aquatic microbiology.

- The abundance of free-living bacteria in water columns usually oscillates between 10^5 and 10^6 cells per ml^{-1}, except in ultra-oligotrophic and hypertrophic waters.
- Bacterial abundance and production increase among systems as the resource availability increases, but at a progressively slower rate, pointing to disproportionally higher loss rates in eutrophic systems.
- Heterotrophic bacteria compete with algae for limiting nutrients (e.g., phosphorus and nitrogen).
- The principal bacterial predators are the typically abundant heterotrophic nanoflagellates (HNFs), small ciliates, and phagotrophic algal flagellates, but certain macrozooplankton and macrobenthic invertebrates are major consumers when abundant.
- ▲ Typically high bacteria:phytoplankton biomass ratios and autotrophic:heterotrophic biomass ratios in the plankton of oligotrophic lakes points to energy (organic matter) subsidies from the littoral zone and drainage basins.

- ▲ The C:P and C:N ratios of organic matter are an important determinant of organic matter availability to microbes and their conversion (growth) efficiency.
- ▲ Very recent studies on aquatic viruses indicate that they can be important as a source of microbial mortality, but also release microbe-derived organic matter to the microbial community.
- ▲ The organic matter utilized in microbial food webs is often largely respired within the microbial loop, rather than available as microbial production to higher trophic levels. In the process plant nutrients in high demand are recycled.
- ▲ Communities of purple and green photosynthetic bacteria commonly form a dense layer at the bottom of the oxycline in lakes and sediments receiving both a sufficient irradiance and a supply of reduced sulfur.
- ▲ The sediment bacterial community in the surficial sediments is, depending on water column thickness, typically between two and 1,000 times greater than in the water column above.

23

Zooplankton

23.1 Introduction

The inland water zooplankton, or animal plankton, range in size from small protozoan flagellates less than 2 μm in their longest dimension to large crustaceans of several centimeters. The **macrozooplankton**, larger than 200 μm and comprised principally of crustaceans, have been the subject of many limnological studies during the last 100 years. **Microzooplankton**, those animal organisms smaller than 200 μm include the rotifers, the smallest instars (larval stages) of copepods, as well as the protozoans. In practice, the microzooplankton designation is frequently applied to the rotifers alone, a large group of animals that rank second to crustaceans in the amount of attention accorded them by biologists concerned with the zooplankton. For both taxonomic and analytical reasons protozoans or **protozooplankton** are usually, as in this text, considered separately from the other animal plankton.

Crustacean zooplankton and all but the smallest rotifers can be quantitatively sampled with traditional plankton nets and traps. Rotifers are also easily preserved and can be readily examined with light microscopes under low magnification. This is not the case for the much smaller ciliated and flagellated protozoans that must be collected, preserved, and examined with techniques that are generally foreign to those interested in larger organisms. The techniques required for the delicate protozoans are more similar to techniques needed for the study of phytoplankton and are quite different from those employed to examine larger

zooplankton. As a consequence, the protozoans have received little attention from aquatic scientists until recently, although in 1920, H. Lohmann (DE) proposed an important role for colorless flagellates in pelagic food webs. The neglect the protozoans have suffered has been alleviated with a surge of interest in the ecology of these organisms when their abundance and importance as predators of heterotrophic bacteria was recognized (Chapter 22). The recent attention accorded to protozoans has led to a near revolution in ideas about the role of the zooplankton in aquatic systems. It has similarly become widely recognized during the last 30 years or so that planktivorous fish and other invertebrate predators, when abundant, have major direct and indirect effects (top-down effects) on zooplankton abundance and community structure. This recognition changed earlier beliefs based upon the idea that the availability of resources (bottom-up effects) and competition for these resources among zooplankton species were the principal determinants of abundance, distribution and community structure of animal plankton in nature.

Apart from the zooplankton groups mentioned above, there are a few species of insect larvae (e.g., *Chaoborus* species and other midges, Chironomidae), jellyfish, and larval clams (e.g., *Dreissena*, the zebra mussels) that may at times be an important component of the animal plankton. Recently hatched larval fish can also be considered part of the plankton for the first weeks of their life—until they become strong enough swimmers to determine their position in the water column regardless of turbulence, at which point they become part of the **nekton** rather than plankton.[1]

▲ 23.2 Zooplankton Sampling

The earliest zooplankton (and phytoplankton) biologists, working before and shortly after the turn of the 20th century, used nets made of silk bolting cloth (available for sieving flour) to collect semiquantitative samples. The finest cloth available had mesh openings of ~60–70 μm, fine enough to retain virtually all the adult crustaceans and all but the smallest rotifers. The nets retained few protozoa and those were usually destroyed or grossly distorted by the formalin and alcohol preservatives used for the larger organisms. Even today, most zooplankton limnologists interested in crustaceans and rotifers continue to use nets with ~60–70 μm meshes and make primarily vertical hauls. However, researchers interested in rotifers increasingly use nets with mesh openings of about 35 μm, but even fine nets underestimate the abundance of the smallest rotifers and considerably underestimate the number of eggs of most species, which can pass through nets with a 10 μm mesh. Protozoans are best collected with sampling bottles or pumps rather than nets.

All nets, new or old, provide resistance to the flow of water through them, with the result that less water passes through plankton nets than is surmised from the area of its mouth and the distance it is hauled. As a consequence, the abundance and biomass of organisms is underestimated. To minimize this investigators frequently use nets with a small ratio of opening to net length to reduce the resistance to flow. Nevertheless, the backpressure produced during horizontal towing or vertical hauling creates a shock wave ahead of any net that may allow the strongest-swimming macrozooplankton or fish larvae to evade capture, resulting in an underestimation of their abundance. But a flow meter installed inside the mouth of the net will at least obtain an accurate measure of the volume of water filtered.

Traps have been developed to evade these problems. A simple modern trap is the Schindler-Patalas

Figure 23–1 A Schindler-Patalas trap for collecting zooplankton. The two hinged doors swing upward when the trap is lowered and close at the desired depth when the hauling-up commences. The organisms contained within the volume (e.g., 30 l) are collected in the attached net. *(Photo courtesy of J. Kalff.)*

trap (Fig. 23–1) made of transparent plastic that provides little warning to the light-sensitive zooplankton. Traps and motorized pumps with a tube intake collect significantly more macrozooplankton than either hauled nets or opaque water-sampling bottles (Table 23–1). Echolocators (also used to determine fish abundance, Chapter 26) and light-beam devices (optical plankton counters) that measure the reduction in light transmission brought about by zooplankton assemblages at different depths as they are lowered, have the important advantage of revealing the spatial and temporal distribution of the macrozooplankton zooplankton. But these methods require calibration with trap- or pump-collected samples. Despite the development of a variety of sampling methods, nets remain the col-

---

[1]The distinction between the strong-swimming **nekton** and the **plankton**, with their weak locomotory power is actually arbitrary. The collective term for all particulate matter in the water column is **seston**—composed of **bioseston** (plankton and nekton), **abioseston** of inorganic origin, and **tripton** of organic origin.

Table 23–1   **Relative effectiveness of several zooplankton sampling devices, with numbers caught by the transparent Schindler-Patalas trap given an index number of 100 plus the 95 percent confidence limits for a vertical series of 10 replicates.[1]**

Species	Sampling Devices			
	28-l Schindler-Patalas trap	Opaque 9-l Water-Sampling Bottle	13-cm Clark-Bumpus Metered Tow Net with 94 μm Mesh	30-cm Metered Tow Net with 76 μm Mesh
Holopedium gibberum	100±18	65±15	71±18	65±24
Daphnia sp.	100±14	62±18	74±13	60±30
Leptodora kindtii	100±26	54±41	35±24	59±33
Diaptomus leptopus	100±21	60±22	54±23	49±21
Diaptomus minutus	100±18	105±25	68±12	86±31
Mean	100±19	69±24	60±18	64±28

[1]The relative effectiveness of pumps relative to nets is similar to the Schindler-Patales trap to net ratio for macrozooplankton (Pace 1986).

Source: Modified after Schindler 1969.

lection device of choice in half the recent studies (McQueen and Yan 1993).

Estimates of zooplankton (and phytoplankton) abundance collected in lakes and rivers, with whatever device, are subject to additional inaccuracies as a result of inappropriate preservatives, inappropriate subsampling of the material collected, and errors incurred in counting and measuring the organisms. But more serious than any of these difficulties in obtaining accurate estimates of abundance is the heterogeneous horizontal and vertical distribution, or *patchiness* of the zooplankton (Fig. 23–2). Patchiness can impart errors of unknown magnitude when making vertical net or trap hauls at the often single deep-water lake station sampled while midchannel or midbasin sampling of lotic and lentic systems will underestimate the organisms in well-vegetated littoral areas.

The heterogeneous distribution of lake zooplankton appears in part determined by (1) system morphometry (lake depth and shape); (2) the configuration of inflows and outflows; (3) prevailing winds; (4) current patterns; (5) upwellings; (6) competition for food resources between zooplankton species; (7) predators (Urabe 1990, Gliwicz and Rykowska 1992); and (8) there is also a diurnal component as the result of vertical migration in deeper lakes, and horizontal or transversal migrations between littoral zone weed beds and the open water in shallow lakes containing predators but no low-light hypolimnion to hide in during the daytime (Sec. 23.14).

Good and frequent abundance determinations are prerequisites for accurate production estimates and linking abundance patterns to environmental factors in all organisms, and the sampling problems described

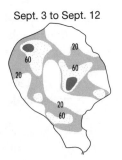
Sept. 3 to Sept. 12

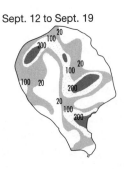
Sept. 12 to Sept. 19

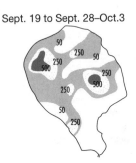
Sept. 19 to Sept. 28–Oct.3

Figure 23–2 Distributions of *Bosmina obtusirostris* during September 1968 in subarctic Lake Latnjajaure, SE (LA = 0.73 km², $\bar{z}$ = 17 m). Based on trap catches (individuals per day) at different depths. *(Modified from Nauwerck 1978.)*

above are great enough to allow only a resolution of substantial differences observed within and among studies.

Zooplankton research has, as a consequence of historical developments, been overwhelmingly dominated by studies of the relatively large planktonic crustaceans, mostly the *Daphnia* species (water fleas, Fig. 23–3). Their fascinating diurnal migrations and frequent seasonal changes in size and shape have particularly intrigued limnologists (Sec. 23.14). Their large size and ability (when abundant) to filter and remove a large fraction of the small phytoplankton and protozoa has made them *keystone* species—species with a disproportionately large impact on community production structure.

The ease with which many crustaceans (and rotifers) can be cultured has generated a large volume of literature on the metabolism and growth of particular species in the laboratory. A much smaller, usually quite separate, group of scientists has been interested in the

rotifers; fewer researchers have addressed the protozoans. The zooplankton literature reflects a larger concern with individual species, or with a small group of species within a single systematic subdivision, using a variety of techniques for collecting, preserving, and counting. Traditional zooplankton research makes generalizations about zooplankton communities as a whole much more difficult than for their phytoplankton counterparts.

23.3 Protozoa, Rotifers, and Crustaceans

Protozoa

Among the unicellular protozoa (Kingdom Protista), the heterotrophic nanoflagellates (HNFs) (Fig. 23–4) are the major consumers of free-living bacteria (Chapter 22), picophytoplankton (Finlay et al. 1988), and

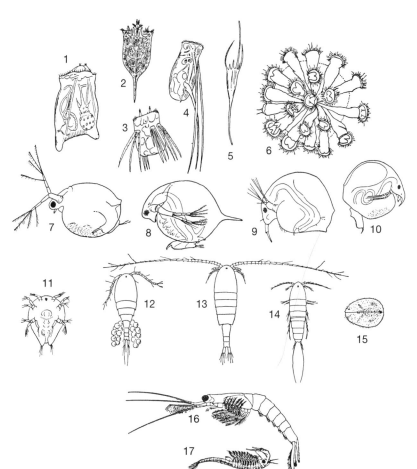

Figure 23–3 Selected zooplankton, not drawn to scale. Rotifers: (1) *Asplanchna*, (2) *Keratella*, (3) *Polyarthra*, (4) *Filinia*, (5) *Kellicottia*, (6) colony of *Conochilus*; Cladocerans: (7) *Ceriodaphnia*, (8) *Daphnia*, (9) *Bosmina*, (10) *Chydorus*; Copepods: (11) cyclopoid copepod: *Nauplius* larva, (12) cyclopoid copepod: *Cyclops*, female, (13) calanoid copepod: *Diaptomus*, (14) harpacticoid copepod (primarily benthic): *Canthocamptus*; other selected Crustaceans: (15) Ostracods: *Cypridopsis* (benthic), (16) *Mysis* (benthic-planktonic), (17) *Eubranchipus* (littoral). (*After Needham and Needham 1962.*)

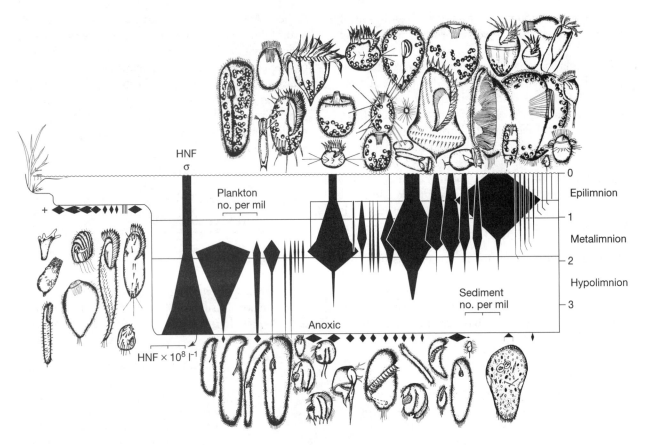

Figure 23–4   The species and abundance of protozoa in Priest Pot Pond (GB) on 24 June 1987.
(Modified after Finley et al. 1988.)

other smaller HNF species. The abundant HNFs ($\sim10^5$–10^8 l^{-1} or higher in highly eutrophic lotic and lentic systems) range in size from about 1.0 to about 20 μm (Fig. 23–5). They include nonpigmented species within four orders of Phytomastigophora (cryptomonads, dinoflagellates, euglenoids, and chrysophytes) that structurally have very closely related pigmented counterparts in the phytoplankton (Chapter 21). In addition, there are two orders of Zoomastigophora (choanoflagellates and kinetoplastids) that are zooflagellates in the strict sense of the word. With the HNFs so closely related to their algal counterparts, it is not surprising that some of the photosynthetic flagellates are able to engulf picoplankton-sized particles in a process known as *phagotrophy* (Bird and Kalff 1986, 1987, and Olrik 1998).

The second group of protozoa, the ciliates, are larger in size ($\sim$8 μm to $\sim$300 μm) but are also less abundant ($\sim10^2$–10^4 l^{-1}) (Fig. 23–4). They are somewhat more easily preserved and counted and are better

known than HNFs, but their role in food webs remains poorly resolved. Even less is known about the other types of protozoans. While the smallest planktonic ciliates (8–20 μm) may be primarily picoplanktivorous (Jürgens and Šimek 2000), the larger ciliates appear to feed on somewhat larger particles, the size of HNFs and small nanophytoplankton (Finlay et al. 1988).

Among the ciliates, those containing captured chloroplasts from ingested algae or those containing more permanent symbiotic green algae (*zoochlorellae*) are common. Finlay et al. (1988) noted that about one-third of the planktonic ciliates in a eutrophic Priest Pot Pond (Fig. 23–4) contained zoochlorellae. Some were so green they are presumed to contribute importantly to primary production. Such organisms, having more than one mode of nutrition and known as *mixotrophic*, can dominate the phytoplankton and zooplankton biomass of tropical Lake Tanganyika (ZR, TZ) (Fig. 21–7) and have been shown to contribute

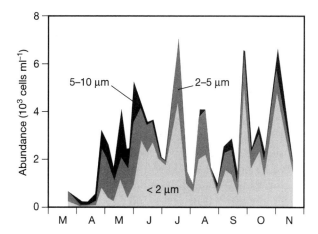

Figure 23–5 The seasonal abundance of heterotrophic fla-gellates (HNFs) with a maximum length of < 2, 2–5, and 5–10 μm from between zero and eight meters in the upper arm of Lake Constance (AT, DE, CH). Note that HNFs < 2 μm numerically dominate most of the year. However, they contribute negligibly to the HNF biomass (not shown), which is dominated by the 5–10 μm organisms, including when the community biomass is greatest between April and June. *(After Weisse and Müller 1990.)*

substantially to planktonic photosynthesis in two Aus-tralian lakes (Laybourn-Parry 1997).

Some of the larger ciliates, as well as some small (1–5 μm) HNFs, are abundant in anoxic hypolimnia and the underlying sediments (Fig. 23–4) where they live anaerobically and feed on heterotrophic bacteria (Sec. 22.9), nanoflagellates (Sec. 21.2), and colonies of purple photosynthetic bacteria (Sec. 22.10). Some species of planktonic flagellates and ciliates are nor-mally not free-living but are attached to large phyto-plankton (e.g., diatoms), rotifers, or crustacean zooplankton. Organisms growing externally attached to other organisms are known as *epibionts.*

Among the other groups of protozoans are two or-ders of amoebae that are primarily associated with the sediments and littoral aquatic vegetation, and large numbers of meroplanktonic species that are periodi-cally swept into the plankton of lakes, rivers, and wetlands.

Rotifers

Rotifers (Phylum Rotifera) (Fig. 23–3), typically an order of magnitude less abundant than protozoans, are the most important soft-bodied *metazoan* (multicellu-lar) invertebrates in the plankton. Their name comes from the apparently rotating wheel of cilia above the lorica, known as the *corona*, used for locomotion and sweeping food particles towards the mouth. The mouth is generally anterior and the digestive tract contains a set of jaws (*trophi*) to grasp food particles and crush them. The many unusual and beautiful shapes of rotifers, combined with sufficient size to allow ready examination under the light microscope, has attracted many investigators since the 19th century (Koste and Holloway 1993).

Relatively few (~100) but ubiquitous rotifer species that are totally planktonic have been described world-wide. A much larger number (~300) are periphytic or are associated with sediments and the vegetation of lit-toral zones. Somewhere between one-third and one-half of all the planktonic species described are present in single lakes (Nauwerck 1963, Pauli 1990), and many are also found in lotic systems and wetlands. Some species occur year-round while others are ephemeral, occurring during short periods of the year. The most abundant planktonic taxa in a particular waterbody also tend to be the most abundant taxa in other similar water-bodies of the same lake district or region.

Planktonic rotifers (and cladoceran crustaceans) have a very short life cycle under favorable conditions of temperature, food, and photoperiod. Egg develop-ment typically takes from about three days at 10°C to as little as one day at 25°C (Fig. 23–6, and Hutchinson 1967). The young reach maturity in a matter of days under optimal conditions and each female produces up to about two dozen young during a one-to-three week lifetime. As a consequence, rotifers produce many generations each year and can rapidly increase in abundance under favorable food and temperature conditions. The short generation time of smaller ro-tifers allows them to be disproportionately abundant in rapidly flushed streams and reservoirs (Basu and Pick 1996 and Sec. 29.4). Invertebrates producing more than one generation per year are known as *mul-tivoltine*, those completing two life cycles per year are *bivoltine*, and those that produce one generation per year are known as *univoltine*. In addition, species re-quiring two years to complete a life cycle are known as *semivoltine*, and those with a longer life cycle are *mero-voltine*. A univoltine species at midlatitudes may be univoltine or even merovoltine at high latitudes.

An important reason rotifers (and cladocerans) can be multivoltine is because they produce unfertil-ized but diploid (2N) eggs yielding only females through parthenogenesis. In most species the eggs are

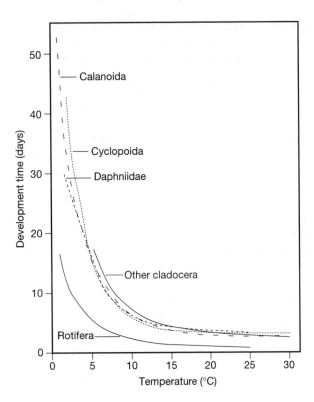

Figure 23–6   Duration of egg development for Daphniidae, other Cladocera; Cyclopoida, Calanoida and Rotifera as a function of temperature. Note the short development time of the much smaller rotifers. *(After Bottrell et al. 1976.)*

released into the water, but a few species breed them internally. In species that can produce males, the nonsexual reproduction—also known as **amictic** or **parthenogenic** reproduction—ends when unfavorable conditions are on the way. The haploid (N) eggs and the males then produced allow sexual or **mictic** reproduction. The fertilized eggs of rotifers (calanoid copepods and cladocerans) are usually thick-walled resting eggs that sink to the sediments to await the return of more favorable conditions. The eggs can survive desiccation in ephemeral bodies of water and can be aerially transported along with sediment particles to other waterbodies.

Rotifers differ not only greatly in size (over three orders of magnitude) and form but also in feeding behavior. Most common species belonging to the genera *Keratella, Brachionis, Filinia,* and *Conochilus* (Fig. 23–3) are omnivorous, feeding on picoplankton, small flagellates, and small ciliates (< 20 μm). Other rotifers, such as species in the genus *Polyarthra* and *Synchaeta,*

tend to select particles up to about 40 μm but prey on somewhat larger protozoans as well (Arndt 1993).

Crustacean Zooplankton

The first group, the **cladocerans** (Suborder Cladocera), are normally covered by a hard chitinous cover that is known as the *carapace*. Respiration in these, and all other crustacean zooplankton, is through either the body surface or via gills. The two large second antennae are responsible for giving the cladocerans their common name—water fleas—and are used for rowing through the water (Figs. 23–3 and 23–7).

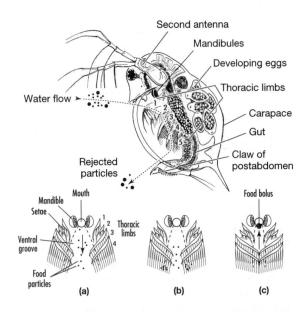

Figure 23–7   Diagrammatic representation of feeding in the cladoceran *Daphnia*. The thoracic legs flip back and forth several times per second, moving water through the space between the legs (food groove), as shown in (a). The first two legs and setae on the carapace are used to eject large or distasteful particles. Particles of food (mostly bacteria and algae) stick to the long setae (b) of legs three and four, due to electrostatic attraction. As the legs move (c), they brush against one another, and comb the collected food and mucous into a ball (bolus) which is moved toward the mouth (d). At this point, the food bolus can be rejected if it is distasteful, or if the gut is too full. The claw of the postabdomen is used to flick the bolus out of the food groove. If the food bolus is accepted, it is chewed by the mandibles and swallowed. Because of the small size of the filters and the rapid movement of the legs, Reynolds number considerations suggest that these legs cannot act as sieve filters, but must act as electrostatic filters. *(Modified after Russell-Hunter 1969. Description courtesy of S. Dodson)*

Most cladocerans are filter-feeders on suspended living and detrital particles.[2] Their thoracic limbs are covered by a network of hairs (*setae*), which in turn are covered with a network of closely spaced (a few μm) *setules*. Particles are captured electrostatically by the comb of setae and setules when the legs move (Fig. 23–7). Rejected particles are removed via the lower carapace. The distance between the setules increases with each molt and varies considerably between species. As a result the size of the smallest particles that can be collected varies with the molt within as well as among species, making body size and taxonomic position poor indicators of whether picoplankton are a major component of their diet (Geller and Müller 1981). Finally, it is increasingly evident that even the filter-feeders feed somewhat selectively. Consequently, the capability of filtering a certain-sized particle does not also mean that particles of that size are selected in the proportion that they occur in the seston. Rather, the efficiency ('selectivity') of removal is a complex function of the availability, size, shape, and taste or nutritional quality of seston particles, whose absolute and relative abundance changes over time and space.

Not all cladocerans are filter-feeders. There are a number of large carnivorous genera that grasp their prey. Among them are one or more species of the genera *Polyphemus*, *Leptodora*, and the omnivorous spiny water flea (*Bythotrephes*).

The cladocerans, like rotifers, are parthenogenic most of the time with diploid (2N) amictic eggs hatching into miniature adults that typically mature in three to six weeks. Under suitable conditions allowing frequent molts, the eggs produce the next generation within a few weeks. The lifetime of cladocerans is longer (one to three months in the laboratory) than rotifers and each cladoceran may produce many offspring. Haploid (N) eggs and males are produced in most species when unfavorable conditions announce themselves, including abundant predators. The fertilized resting eggs (2N), covered by resistant cases, known as *ephippia* (singular: ephippium), settle to the sediments to await the return of more favorable conditions. A frequently large pool of *diapausing* (or *resting*) eggs[3] in lakes explains the sudden reappearance and abundance of cladocerans (and rotifers) in spring or when rains fill ephemeral bodies of water. At other times rapid increases reflect the high reproductive potential associated with asexual reproduction.

Copepods

Population growth in the second of the two major groups of crustacean zooplankton, the **copepods** (Class Copepoda), is more regulated by longevity and survival rates than by egg production, as in the cladocerans. Copepods (Fig. 23–3) are dependent on sexual reproduction and—with only the female half of the population able to produce young—the potential for rapid reproduction and rapid changes in population size of adult copepods is greatly reduced. A long development time, necessitating lengthy experiments, and a complex life history with difficult to distinguish early larval stages has resulted in much less experimental work on copepods than cladocerans. Similar to cladocerans and rotifers, copepod egg development time is almost totally determined by water temperature (Fig. 23–6). After hatching, the young molt 11 times before becoming adults. During the first five or six molts the juveniles are known as *nauplii* (singular: *nauplius*) which, under the microscope, superficially resemble tiny spiders more than they resemble the adult form (Fig. 23–3). During the remaining five molts the immatures are known as *copepodites*. The latter resemble the adult, also called the copepodite VI

[2]Cladocerans are present and may be important filter-feeders in saline lakes of low to moderate salinity, but the most important and best-known filter-feeding crustacean is the brine shrimp *Artemia salinas or closely related species* (class Branchiopoda), which occur worldwide in fishless saline lakes and tolerates high salinities (> 300‰). It has been widely introduced for aquaculture purposes and by managers of solar ponds used for harvesting salt (after precipitation upon evaporation) as grazers to control the phytoplankton biomass. *Artemia*'s ability to synthesize hemoglobin allows it to function in high-salinity waters characterized by very low dissolved oxygen concentrations (Sec. 15.2). Other important filter-feeding organisms in moderately saline lakes and wetlands (mostly < 30‰) are the tilapias, a group of fishes (Chapter 26). The best-known birds restricted to salt lakes are the herbivorous filter-feeding flamingos which feed primarily on filamentous cyanobacteria (e.g., *Oscillatoria platensis*) and small invertebrates (Fig. 14–2).

[3]The diapausing eggs of individual species of daphnids, copepods, and rotifers are found in large numbers in lake sediments (~10^4–10^6 m²). Eggs survive for decades or longer as determined from associated dated organic matter in core slices. Viable copepod eggs are apparently deposited as long as 330 years ago. Resting eggs preserve a large egg bank that maintains species richness and allows for genetic variation favoring hatching and survival of those strains best adapted to particular conditions. (Hairston 1996)

stage. One or more of the copepodite stages may over-
winter or survive otherwise unfavorable conditions by
entering into a diapause (or resting) phase in the sediments, followed by a sometimes sudden return to the
plankton when conditions improve.

▲ The number of copepod generations per year is
lower than for the rotifers and cladocerans, ranging
from ~15 (generation time 20–30 days) in tropical hypereutrophic Lake George (UG) characterized by a
year-round "summer", to one generation per year in
high arctic oligotrophic Char lake (CA). As for the
eggs, development time beyond the egg stage is, as in
other crustaceans, largely a function of temperature
(Nauwerck 1963). But where temperature varies relatively little, the role of food is predominant. Thus *Cyclops scutifer* is univoltine in central Scandinavia (Fig.
23–8), but semivoltine or merovolitine in the arctic—
taking as much as three years to mature under the low
temperature and low food availability conditions that
characterize artic Swedish lakes (Morgan 1980).
Moreover, development time has a genetic basis, different species develop at somewhat different rates
under optimal conditions.

Among the three orders of copepods, **cyclopoid
copepods** are generally predatory (carnivorous) on
other zooplankton, fish larvae, and tadpoles, but relatively few species of animal can be conveniently pigeonholed as to diet and the cyclopoid copepods in
fact feed also on algae, bacteria, and detritus. The second group of copepods, the **calanoid copepods** (e.g.,
Diaptomus) change their diet with age, sex, season, and
food availability. The calanoid copepods, long considered to be herbivorous, are now known to be omnivorous, feeding on ciliates and rotifers as well as on
algae, bacteria, and detritus. The calanoid and cyclopoid copepods are disproportionately abundant in
oligotrophic and dystrophic waters at all latitudes.
The third group, the **harpecticoid copepods** are primarily benthic (Fig. 23–3).

Other Crustaceans

A third major group of inland water crustaceans are
the **ostracods** (Class Ostracoda), which are primarily
benthic rather than planktonic and are frequently very
common in sediments and on macrophytes but occasionally swept into the plankton. Although the cladocerans and copepods dominate the crustacean
zooplankton in abundance and in numbers of species,
fairy and clam shrimps (Class Branchiopoda) can be

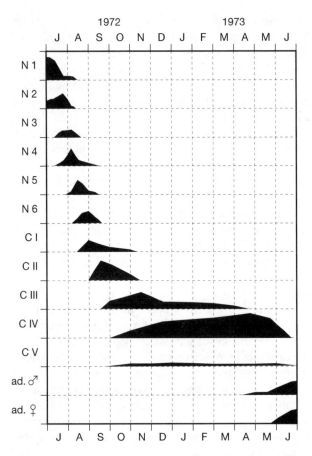

Figure 23–8   The succession in the frequency of the different instars in the 1972–1973 generation of the univoltine cyclopoid copepod *Cyclops scutifer* in Lake Øvre Heimdalsvatn,
NO (LA = 0.8 km², $\bar{z}$ = 4.7 m). Note the six naupliar stages
(N1–6), followed by five copepodid stages (CI–V) and the
adult (reproductive) stage. *(After Larsson 1978.)*

common especially in shallow saline waters (Fig.
23–3). The last class of inland water crustaceans to be
mentioned are the **malacostracans** (Class Malacos-
traca), which include the mysids. Among the mysids,
are the oppossum shrimp *Mysis relicta* (Fig. 23.3).

▲ 23.4 Species Richness
and Its Prediction

Well-studied mesotrophic freshwater lakes and lowland rivers at temperate latitudes contain between 50
and 100 planktonic zooplankton species, protozoa ex-
cepted (Morgan 1980, and Kobayashi et al. 1998). An
average of some 100 species, greatly dominated by ro-

tifers, characterize Brazilian rivers and floodplain lakes (Rocha et al. 1995). There is an abundant littoral zone community which has been well-studied in Lake Ladoga (RU) where it is composed of 141 taxa—68 rotifers, 50 cladocerans, and 23 copepods (I. V. Telesh 1996, in Raspopov 1996). Invertebrate species abundance is greatly reduced in saline inland waters (Fig. 13–9). Thus Green (1993) reports just three rotifer species each in two African saline lakes, a tiny fraction of the planktonic species noted in temperate and tropical freshwater lakes. The exact number of species recorded is, as in the algae (Chapter 21), a function of the intensity of the investigation, the taxonomic competence of the investigator, and the goals of the inquiry (species composition versus abundance or biomass), and is affected by whether genetic rather than morphological differences are being examined.

The temperate and tropical zooplankton community in individual freshwater lakes is annually dominated by one to three cyclopoid copepods, one calanoid copepod, typically three to 10 cladocerans, and a similar number of rotifers, with the most abundant species contributing between 60–90 percent of the total number of individuals in each of the three taxonomic groupings (Morgan 1980). More limited information suggests that typically a half-dozen ciliates and presumably a similar number of heterotrophic nanoflagellates dominate the protozoan community.

The principal predictor, albeit a weak one, of crustacean and rotifer species abundance is lake area (Fig. 23–9). The pattern is widely explained by postulating a larger number of habitats in larger systems, with individual species best-adapted to specific habitats. Habitat differences exist not only between shallow and deep waters but also between warm and cold water, illuminated and dark regions, zones of low and high turbulence, as well as between lake regions of low and high dissolved oxygen. Even open-ocean habitats, which probably closely resemble the pelagic zone of large lakes, can be partitioned into five or more subhabitats (McGowan and Walker 1979). The explanation, the **habitat diversity hypothesis**, is but one of a number of theories originally proposed to explain the relationship between species abundance and the surface area of oceanic islands. Freshwater habitats can be visualized as islands surrounded by land (see Brönmark 1985).

The second best determinant of crustacean zooplankton species abundance is annual phytoplankton primary production (resource availability). The number of species is reduced in highly oligotrophic lakes

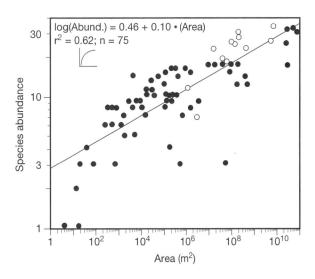

Figure 23–9   The relationship between freshwater lake surface area and the number of planktonic crustacean zooplankton species in 66 North American lakes (filled circles), and the number of species of planktonic rotifers in 12 lakes from England and Africa (open circles). Note the variation in species abundance at a given lake surface area, indicating that factors other than lake surface area also affect species abundance. Other (probably important) factors include resource availability, temperature, predation, and system morphometry. (*Data from Dodson 1992, and Green 1993.*)

and peaks at relatively low rates of planktonic primary production and may decline at higher rates of primary production ($> \sim 180$ mgC m^{-2} yr^{-1}) (Dodson et al. 2000). Finally, the number of crustacean zooplankton species has been shown to be weakly a function of the number of lakes within 20 km, suggesting a possible role for waterfowl, wind, flowing water, and fish in the distribution of the organisms (Dodson 1992).

The determinants of species abundance change, as do other limnological attributes, with the scale of studies. In individual lake districts over which lake area and productivity vary relatively little, other variables become important as the principal predictors. Thus, the single best predictor of crustacean zooplankton species abundance in 60 generally small boreal forest lakes with nearly neutral pH and varying little in primary production, was the maximum depth—with depth probably a surrogate measure of habitat diversity (Keller and Conlon 1994). But, if the lakes selected had instead varied relatively little in depth, size, and trophic state but had ranged widely in pH, the latter would emerge as the statistically most significant variable (Fig. 27–10).

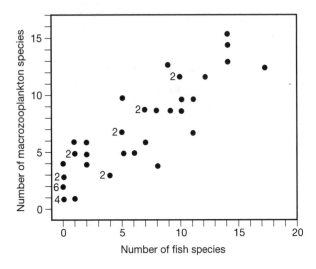

Figure 23–10 Relation between the number of macrozooplankton and fish species in 47 Ontario Lakes. Numerals indicate coincident points. *(After Sprules 1975.)*

The contribution of predation to species abundance is clearly important, but is only readily demonstrated experimentally (Sec. 23.7). The features determining zooplankton species abundance over a particular range scale generally also determine the species abundance of other systematic groupings (Figs. 23–10 and 27–10).

23.5   Seasonal Cycles

There are fewer quantified generalities that can be made about zooplankton than about phytoplankton community biomass cycles. Not only are zooplankton (and phytoplankton) cycles subject to interyear and decade or longer variation in weather but also to system-specific cycles in phytoplankton resource production. Moreover, they are affected by the size distribution of their algal and bacterial prey, as well as disease and *selective* predation by fish and invertebrate predators. All of these impacts vary in their relative intensity over time and space, thus they prevent a simple interpretation of the reasons for the abundance and relative distribution of a species or its biomass. But an additional difficulty is that zooplankton cycles usually refer only to the abundance or biomass of selected species of crustaceans or rotifers rather than the zooplankton community as a whole. Nevertheless, some broad generalities describing zooplankton succession and seasonal cycles have emerged that have conceptual value, at least for the particular system type and region of the temperate zone for which they were developed. Such generalizations are of limited utility for studies carried out in a different climatic zone or even in the same climatic zone if the waterbodies differ substantially from those used to create the generalities in such factors as the relative size of littoral zones, water residence time (flushing rate), hypolimnetic oxygen concentration, or the abundance of zooplankton predators.

Lake Erken

One of the best and most complete studies of a zooplankton community is the classical study of dimictic mesotrophic Lake Erken, SE, by Nauwerck (1963) who, except for the heterotrophic nanoflagellates, examined the entire zooplankton community (Fig. 23–11). The five-month period of ice cover (mid-December to mid-April, but shorter in recent decades) is a period of low species abundance and community biomass. The few remaining cladoceran species (*Daphnia*, *Bosmina*, and *Ceriodaphnia*) congregate near the profundal sediments and gradually decline in abundance as the winter progresses. The copepodite stages of the *Cyclops spp.* diapause in the sediment. As the days start to lengthen and phytoplankton photosynthesis recommences just below the snow and ice cover (Fig. 21–16), a community of nanophytoplankton, ciliates, and rotifers develop in the surface waters, dominated by the ciliate *Bursaria truncatella*.

Phytoplankton production and nanophytoplankton biomass rises quickly after ice-out and the onset of spring circulation, accompanied by a rapid increase in ciliate biomass (primarily *Bursaria*, Fig. 23–11). Following the collapse of the nanophytoplankton and their presumed ciliate predators during the early spring overturn period, the zooplankton biomass briefly returns to winter levels and the community during that period is dominated by the large predacious rotifer *Asplanchna priodonta*. Subsequently, the community becomes increasingly dominated by cyclopoid copepodites after their return to the water column from the sediments, as well as by the calanoid copepod *Eudiaptomus graciloides*, following the rapid development of its overwintering eggs and nauplii. After the onset of stratification, filter-feeding species continue to dominate during the zooplankton biomass maximum. When the *Eudiaptomus* population crashes in July it is suc-

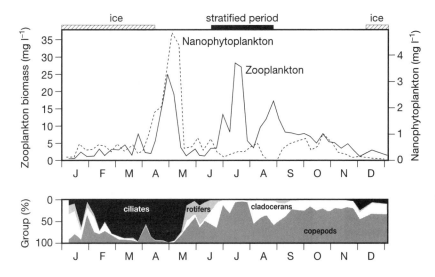

Figure 23–11 The seasonal biomass cycle of zooplankton (HNFs excluded) and nanophytoplankton (wet wt), as well as the relative importance of the different taxonomic groupings of zooplankton (%), HNF excluded, in the surficial 20 meters of Lake Erken, SE (SA = 27.4 km², z̄ = 9 m) in 1960. Note the overwhelming under-ice dominance of ciliates in the spring, the collapse of the spring nanophytoplankton biomass following the ciliate maximum in April, the July peak of copepods, and the relative importance of the filter-feeding cladocerans in August and September at a time when the small ("edible") phytoplankton virtually disappeared, presumably through grazing. *(Modified from Nauwerck 1963.)*

ceeded by a late summer cyclopoid copepod–daphnid assemblage that remains abundant during and following lake destratification (Fig. 23–11).

The decline of the filter-feeding daphnids in September is accompanied not only by an absolute increase in their nanophytoplankton prey but also in their relative contribution to the phytoplankton biomass (Fig. 21–16). The daphnid decline yields a crustacean zooplankton community dominated (80%) by copepodite copepods (mostly *Diaptomus*), with the balance made up of cladocerans (*Daphnia* and *Bosmina*, Fig. 23–3) whose abundance and biomass gradually decline in step with the decline in nanophytoplankton biomass and water temperature. The details of the Lake Erken seasonal cycle are naturally lake and year specific. However, a long period of ice cover and a relatively short period of stratification imposes a pattern on the biomass cycle that has counterparts in dimictic lakes elsewhere.

Dimictic and Monomictic Lakes

Relatively large well-studied warm monomictic lakes of the temperate zone, such as Lake Constance (AT, CH, DE), Lake Geneva (CH, FR), Lake Windermere (GB), Lake Ontario (CA, US), and Lake Washington (US), lack winter snow and ice cover, and the associated high albedo and light extinction (Sec. 10.3). The lakes are also usually located at a somewhat lower latitude than dimictic lakes (Sec. 11.2) and receive higher winter irradiance. They experience both higher water

temperature and higher winter primary productivity. Warm monomictic lakes at higher latitudes further differ from their dimictic counterparts because they receive much of their water and associated nutrients during fall and winter periods of high runoff (Fig. 8–4). As a result they typically maintain a higher winter plankton biomass than seen in dimictic lakes of similar trophic status (trophy). Monomictic Lake Constance, whose plankton has been exceptionally well-studied, experiences sufficient warming to allow the onset of stratification in April, when dimictic Lake Erken has traditionally been ice covered (Fig. 23–11). The early onset of stratification in Lake Constance, which reduces the depth of mixing (z_{mix}), improves the light climate (z_{eu}:z_{mix}, Fig. 10–12) and raises water temperatures (Sec. 10.11), allowing elevated levels of phytoplankton and bacterial production long before this becomes possible in Lake Erken and other dimictic lakes.

The increased algal and bacterial biomass (Fig. 22–1) together with higher water temperatures allow much earlier egg development as well as higher growth rates of protozoans (Fig. 23–5), rotifers, and crustacean zooplankton in Lake Constance and other warm monomictic lakes of the temperate zone, than is possible in their dimictic counterparts with a lengthy period of ice cover. Longer autumnal days and resulting greater daily irradiance further extends the growing season of temperate zone monomictic lakes, allowing more generations of zooplankton and potentially permitting a greater role for zooplankton in structuring phytoplankton communities than at higher latitudes. The stereotypical seasonal algal–zooplankton sequence seen in Lake Constance (AT, CH, DE) is also broadly evident in a number of other large, monomictic lakes in central Europe (Sommer et al. 1986).

Clear-water Phase

▲ Macrozooplankton grazing in mesotrophic lakes, above all by large filter-feeding *Daphnia*, at times brings about a springtime **clear-water phase** characterized by high transparency. The transparency can be further enhanced by increased sedimentation of diatoms if the clear-water phase occurs at the onset of stratification. A clear-water phase in Lake Constance first became evident when eutrophication allowed the soluble reactive phosphorus (SRP) concentration to exceed 20 µg P l^{-1} (Tilzer et al. 1990), at a time when the 'critical' concentration of large daphnids first exceeded 2–4 mg dry wt. m^{-2}. This occurred not only in Lake Constance but in most of the 10 lakes examined (Lampert 1985).

While the grazing activity of the large filter-feeding zooplankton can contribute appreciably to the production of a spring (or summer) clear-water phase, the daphnids are certainly not the only contributors to the development of such a phase. For example, a multiyear study of monomictic Lake Geneva (CH, FR) noted that the *Daphnia* biomass was able to explain only 35 percent of the among-year variation in magnitude of the phase (Balvay et al. 1990). This implies that the remaining 65 percent must have been attributable to other causes. Prominent among them are probably increased particle (diatom) loss by sedimentation following the establishment of a temperature stratification (Secs. 21.2 and 21.6), plus reduced turbulent transport of deep-water nutrients into the euphotic zone (Sec. 12.11). An abundance of large-bodied filter-feeding zooplankton is not a prerequisite for the development of a clear-water phase. In monomictic Lake Michigan (US), the spring algal biomass decline and the resulting increase in transparency is instead tightly coupled to the onset of stratification, (Lehman and Cáceres 1993), making a physical/chemical explanation more plausible than one based on zooplankton grazing.

Two not-mutually-exclusive hypotheses have been proposed to explain the common late spring to early summer decline of daphnid abundance in mesotropic to eutrophic lakes. One theory attributes the decline to over-exploitation of the edible algae (resource limitation hypothesis, bottom-up control) allowing a shift in the phytoplankton community to one dominated by large inedible species. Alternatively, increased water temperatures plus associated large increases in the abundance of planktivorous fish and the larvae of other species or invertebrate predators (e.g., *Chaoborus*) allow sufficiently enhanced consumption of larger daphnids and other macrozooplankton to greatly reduce their numbers in summer (predation limitation hypothesis, top-down control) (Luecke et al. 1990).

Low Latitude Pattern

The pronounced seasonality seen at mid and high latitudes becomes increasingly muted at lower latitudes (Fig. 23–12). More modest seasonal changes in irradiance and temperature, together with greater seasonal variability in rainfall, river discharge, and wind speed, affect water column stability, mixed-layer light climate, external and internal nutrient loading, primary production (Chapters 21 and 24), fish reproduction (Chapter 26) and, ultimately, zooplankton population dynamics.

Superimposed on seasonal cycles are longer-term climatic cycles ranging from years to decades and even centuries (Sec. 5.4) imposing further patterns on the

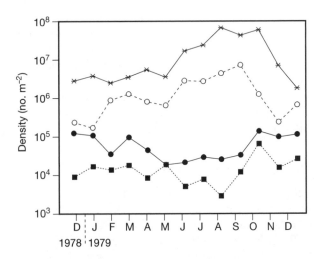

Figure 23–12 Density per square meter of protozoans (✕), rotifers (○), copepod nauplii (●), and macrozooplankton (■) in monomictic and mesotrophic Lake Oglethorpe (LA = 0.3 km², z̄ = 2.3 m, WRT = 80 d), Georgia (US) at ~34°N. *(After Pace and Orcutt 1981.)*

biota of inland waters at all latitudes. Longer-term patterns are often recorded best in the sediments (Sec. 20.5).

▲ 23.6   Long-term Variation in Zooplankton Abundance

Multiyear studies, of which there are few (Fig. 1–1), show a commonly large interannual variability in total crustacean zooplankton abundance (and benthos, too). Coefficients of variation close to 100 percent for the summer months are common (Evans and Sell 1983). In addition, there may be more year-to-year variation in zooplankton abundance over a relatively few years than over long periods. More longer-term data sets are badly needed to shed light on longer-term variation and its causes.

A comparison of the species composition of crustacean zooplankton in Lake Mendota (Wisconsin, US) in the 1970s, with equivalent work by F. A. Birge in the 1890s, shows more interyear variation in the abundance of major species during each study period than between the two periods (Pedrós Alió and Brock 1985). Among the very long-term studies are those of the crustacean zooplankton of the Bratsk Reservoir below Lakes Baikal and in Dalnee (RU). Bratsk reveals about an eightfold maximum variation and Lake Dalnee about a fivefold maximum variation in the annual average biomass of the crustacean zooplankton community (Morgan 1980), and an equally large variation in the contribution of the dominant species. The major interannual changes were noted over periods of a few years and there was no long-term trend, as might be expected if the lakes had been subject to eutrophication or systematic human interference. The large interannual and decade-long periods of variability can have many different causes, including changing predation pressure imposed by fish or invertebrate predators (Fig. 23–13) and changes in the zooplankton food supply as the result of interannual changes in runoff and nutrient loading (Fig. 5–14) or water column stability (Fig. 12–6), and duration of ice cover linked to weather systems originating over the Atlantic or Pacific Oceans (Sec. 5.4). Other plausible mechanisms include changes in competitive interactions between the crustacean species examined on the one hand and the unmeasured rotifers and protozoa on the other, or a changing combination of some or all of the above.

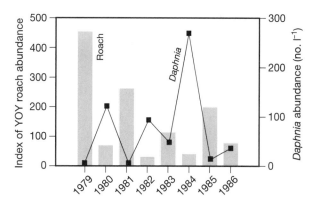

Figure 23–13 Summer variation in young-of-the-year (YOY) zooplanktivorous fish density (*Rutilus rutilus*) and Daphnia abundance in a small English lake. Note that there is a clearly negative relationship between fish density and *Daphnia*, testifying to the important effect of fish predation on large (and vulnerable) zooplankton. (*After Townsend 1991.*)

The proximal biological causes that bring about seasonal and interyear changes in resource availability or predator density may have more ultimate causes linked to intraannual and interannual changes in weather and runoff, as well as to decade-long or longer climatic cycles. Indeed, proximal and ultimate causes simply represent different time and spatial scales of study.

Water Column Stability and Zooplankton

Evidence for climate-linked cycles is provided by a 40-year data set of biweekly crustacean zooplankton samples collected in Lake Windermere (GB). The data show an increase in biomass in the 1970s but no changes in species composition. The biomass increase is attributed to eutrophication and the resulting greater food availability (Fig. 23–14). More pertinent here is the existence of a roughly 10-year cycle in zooplankton biomass. This cycle coincides with the approximately 10-year periodicity in sea surface temperature in the region, linked to the North Atlantic Oscillation (Sec. 5.4). Within the 10-year cycle is a shorter-term pattern of low macrozooplankton biomass after a warm June, and high summer biomass after a cool June. The interdecade variation in the June water temperature is, as noted by the researchers, too small to have had a direct effect on the survival and

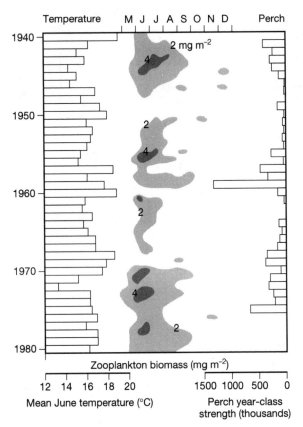

Figure 23–14 The contour diagram shows the annual and decadal long variation in the biomass of crustacean zooplankton in the north basin of Lake Windermere, GB (LA = 8 km²; z̄ = 25 m) between 1940 and 1980. The left-hand histogram shows the year-to-year variations in mean June temperature based on weekly measurements at a central station in the north basin. The right-hand histogram shows year-to-year variations in the year-class strength of perch between 1940 and 1976. In 1976 the perch polulations were dramatically depleted by fungal disease. *(Modified after George and Harris 1985.)*

reproductive success of the zooplankton. However, the interannual differences were sufficient to have brought about changes in water column stability affecting the timing of phytoplankton succession and thereby the availability of suitably sized food particles for the zooplankton. George and Harris (1985) provide some evidence that the crop of edible nanoplankton, which succeeds the early spring diatom bloom following the onset of stratification, is more likely to coincide with the period of rapid *Daphnia* production in cool years when the phytoplankton succession is slowed. Intermittant mixing in cool years may disrupt the succession toward a summer community of large

(inedible) algae, allowing for continued abundance of edible-sized algae and a series of *Daphnia* generations between June and October (Fig. 12–6, and George et al. 1990). In contrast, warm Junes are associated with early water column stratification and the development of an early nanoplankton crop prior to the period of rapid *Daphnia* production. Matching or mismatching between zooplankton production and food availability appears to have an important effect on zooplankton abundance in much the same way that many fish biologists believe that the magnitude of young-of-the-year (YOY) fish stocks is determined by match or mismatch with its zooplankton food supply (Chapter 26, and George and Harris 1985).

The changes in resource availability brought about by eutrophication and changes in weather and climate explain 35 percent of the year-to-year variation in the Lake Windermere zooplankton biomass. The year-class strength of perch, the dominant planktivorous-benthivorous fish (Fig. 23–14), statistically explained an additional 6.5 percent of the variation. While the findings are impressive indeed, more than half of the total variation in zooplankton biomass remained unexplained.[4]

Fish and Zooplankton

A third and last example of a long-term study is the *dynamic modeling* of data collected over 14 years, in Lake Mendota (Wisconsin, US). Here too, the analysis points to the role of planktivory by fish in determining the biomass and community structure of the macrozooplankton. An order of magnitude change in interyear planktivory rates on *Daphnia* was attributed to first a waxing (growing) and then a waning effect of a single strong year-class of cisco (*Coregonus aztedi*), a pelagic fish that remains planktivorous throughout its life (Rudstam et al. 1993). Years with a low *modeled* (not measured) planktivory rate were characterized by a

---

[4]High-quality long-term data are of the greatest importance in distinguishing short-term fluctuations from long-term trends, and in linking patterns in the biota to environmental factors. Long-term data are rare because the needed long-term commitment does not fit the philosophy of most organizations funding research and it is frequently not profitable in terms of the flow of publications which is used as a measure of the success of scientists. However, the long-term collection of either poor-quality data, data of little relevance, or high-quality data with poorly documented or changing methodology is not only wasteful of scarce resources but useless in terms of discerning useful patterns, and for the management of aquatic systems.

high biomass of larger *Daphnia spp.* as well as by their earlier development in spring and longer presence in summer. The simulations projected the effect of the planktivory to have been important in winter, spring, and late summer but not during the early summer period. A bloom of large *Daphnia* in early summer appears to have permitted sufficient removal of algal biomass to allow a clear-water phase. In contrast to the findings of several biomanipulation experiments (Rudstam et al. 1993), the modeled (inferred) high planktivory rate on *Daphnia* did not lead to reduction in overall macrozooplankton biomass, apparently because the modeled cyclopoid copepod biomass increased during the years of highest calculated fish predation on the daphnids. The importance attributed to selective fish predation on zooplankton community structure is supported by German research that showed a compensatory increase in cyclopoid copepods in years when daphnids were few (Horn and Horn 1995).

23.7   Top-down Control of Zooplankton

Hrbaçék (1958) was the first to demonstrate that the manipulation of planktivorous fish stocks can have a dramatic effect on zooplankton abundance, biomass, and community structure, during manipulation of Czech Republic fish ponds. A second decisive experiment involved the introduction of zooplanktivorous fish to a United States lake, which led to a dramatic shift from large-bodied crustacean zooplankton (length ≥1 mm) to smaller species less subject to fish predation (Fig. 23–15). Since that time, there has been a veritable explosion of research on the effect of fish feeding on the aquatic biota (Northcote 1988), and in particular, on the size-selective predation of fish on zooplankton and benthos.

Removals of zooplanktivorous fish, brought about experimentally by netting, poisoning, and naturally

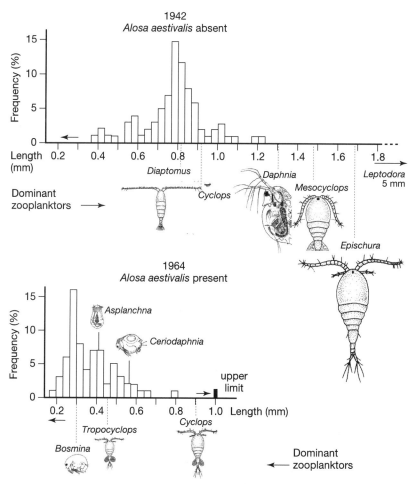

Figure 23–15 The composition of the crustacean macrozooplankton of Crystal Lake, Connecticut (US), before 1942. After 1964 a population of alewife (*Alosa aestavalis*), a zooplanktivorous fish, became well established. Each square of the histograms indicates that at least one percent of the total sample counted was within the particular size range. The specimens depicted represent the mean size (length from posterior base line to the anterior end) of the smallest mature instar. The arrows indicate the position of the smallest mature instar of each dominant species in relation to the histograms. The predacious rotifer *Asplancha* is the only noncrustacean included in the study. Note the reduction in zooplankton size and changes in dominant species with the introduction of the planktivorous alewife. (*Modified after Brooks and Dodson 1965.*)

by winter kills resulting from under-ice anoxia, and piscivorous fish additions to lakes, have unambiguously demonstrated the importance of fish in structuring the macrozooplankton community and, indirectly, the phytoplankton community (Chapter 26). The resulting changes in transparency have been shown to indirectly affect the development of submerged macrophytes (Chapter 24, and Scheffer 1998).

A high abundance of zooplanktivorous fish greatly reduces the abundance of the large zooplankton upon which they preferentially feed (Fig. 23–16). Mills et al. (1987) suggest that the size distribution of macrozooplankton in lakes provides a useful index of the structure (zooplanktivorous versus piscivorous) of fish communities. The large filter-feeding zooplankton species selectively preyed upon by zooplanktivorous fish are replaced by smaller forms that are apparently less able to exert the same predation pressure on the phytoplankton community. The phytoplankton are then able to increase their community biomass and production. However, selective zooplankton feeding on small phytoplankton does not always reduce the community phytoplankton biomass but may shift the algal community to one dominated by large cells and colonies (e.g., cyanobacteria) little subject to grazing. Even so, a major (> 75%) experimental reduction in zooplanktivorous fish allowed more large daphnids as well as increased water transparency (Meijer et al. 1999). Whether the proximal cause is reduced fish predation on the zooplankton, as generally believed, or is at least in part attributable to a reduction in nutrient recycling by the fish, is not well resolved.

▲ Experimental Manipulations

A growing number of limnocorral (mesocosm) experiments, in which predators can be readily and severely manipulated or excluded, convincingly demonstrate a major possible direct or indirect impact of top-down control on plankton communities (Fig. 23–17).[5] Although the enclosure results are convincing, they can

---

[5]"Ecologists are still confused about the distinction between 'pulse experiments' in which one briefly applies a perturbation and then watches the system relax, and the much more common 'press experiment' in which one applies a sustained perturbation [as in limnocorrals and whole lake manipulations]. These two types of FEs [Field Experiments] have very different interpretations. For example, many of the species interactions revealed by press experiments are not direct effects, as the experimenter often believes, but chains of indirect effects." (Diamond 1986)

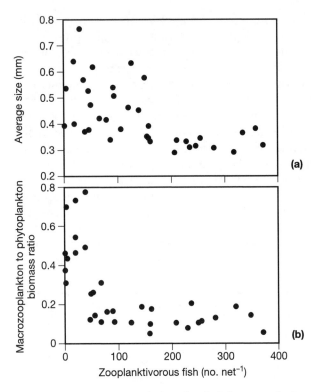

Figure 23–16 The average length of cladocerans during summer (a) and the macrozooplankton to phytoplankton biomass ratio (b) as a function of the catch per unit effort (CPUE) of zooplanktivorous fish in multiple mesh-sized gill nets set in shallow unstratified but highly eutrophic Danish lakes. Each point represents one lake. Note the fourfold variation in fish density in a set of lakes lying in one climatic zone and varying little in depth (morphometry) and trophic state, allows the pattern to emerge. The considerable scatter indicates that sampling limitations or more than predator abundance determines zooplankton size and the zooplankton to phytoplankton ratio. (*After Jeppesen et al. 1997.*)

provide no insight about the importance of fish-imposed top-down control in natural systems where the abundance of YOY or older zooplanktonivorous fish is either unknown or typically varies much less than in enclosure experiments (Chapter 26). Where zooplanktivorous fish are rare or absent, invertebrate planktivores (e.g., predacious crustaceans or *Chaoborus*, Sec. 23.9) usually dominate. These predators feed preferentially on smaller macrozooplankton and rotifers, and communities can become dominated by large zooplankton species (Hanarato and Yasumo 1989).

The successful but often severe ("sledgehammer") manipulation or elimination of fish stocks in limno-

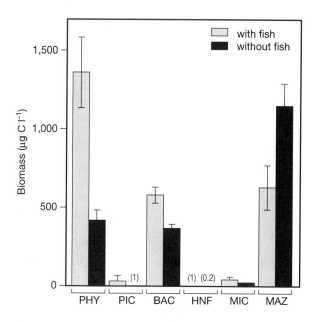

Figure 23–17 Average biomass from April to September of phytoplankton (PHY), picoalgae (PIC), bacteria (BAC), heterotrophic flagellates (HNF), rotifers (MIC), and macrozooplankton (MAZ) in duplicated enclosures with and without zooplanktivorous fish. Bars represent deviation between replicate enclosures. *(After Christoffersen et al. 1993.)*

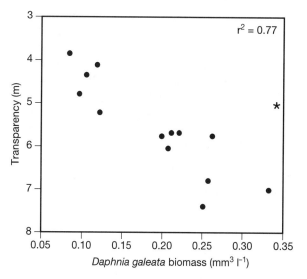

Figure 23–18 The among-year relationship between average summer biomass of the principle herbivore, *Daphnia galeata* and water transparency (z_{SD}), in Saidenbach reservoir, DE (LA = 1.5 km²; $\bar{z}$ = 15 m). A roughly threefold increase in *Daphnia* biomass was associated with a 3 m increase in transparency. The asterisk marks a year with high flushing rates and therefore high inorganic turbidity, which was excluded from the analysis. *(After Horn and Horn 1995.)*

corrals[6] and a number of largely small lakes has led to proposals that reducing planktivorous fish might be a useful management tool for increasing water transparency (Shapiro 1990) in lakes where reductions of nutrient inputs are difficult to achieve (Sec. 23.8).

It is evident that a major experimental reduction (> 75%) of the typically abundant zooplanktivorous fish in shallow eutrophic lakes allows large populations of large filter-feeding daphnids to develop (see also Fig. 23–16. Through their high community filtering rate they reduce the algal biomass enough to greatly increase water clarity (Fig. 23–18, and Meijer et al. 1999). The top-down effects of fish predation on macrozooplankton and macrozooplankton predation

on the phytoplankton is frequently referred to as a *trophic cascade*. However, the effects of predator-mediated interactions weaken with every step down the food chain (McQueen et al. 1986). As a result, top-down effects are far from clear-cut between piscivorous fish at the top of a food web and phytoplankton at the bottom. Furthermore, direct effects are not always easy to distinguish from indirect effects and can be confounded by the effect of increases in other zooplankton predators (e.g., *Chaoborus*).

▲ 23.8 Biomanipulation and Lake Management

Water Clarity

Aquatic management in the temperate zone is primarily concerned with the end result, usually water clarity,[7] and not with the importance of the different mechanisms that determined the desired transparency.

---

[6]Limnocorral manipulations are normally strong and qualitiative (e.g., fertilized versus not fertilized, fish versus no fish) and sustained to allow a clear response over the controls. Such strong manipulations have yielded valuable information on possible processes and mechanisms but at the same time suffer from (1) not knowing whether the effects are direct effects as the experimenter believes, or chains of indirect effects (e.g. reduced nutrient recycling in the absence of fish or changes in the flow of carbon and nutrients between phytoplankton and bacteria); and (2) an inability to identify the dominant mechanisms and processes operating in nature.

[7]High water clarity is not the desired goal in lake management in portions of the world where fish yield (kg ha⁻¹ yr⁻¹) for human consumption is the principal concern (Chapter 26).

Even so, the identification and quantification of mechanisms is the first step toward the development of models that better predict impacts on aquatic systems. When abundant, filter-feeding macrozooplankton commonly play a key role in reducing the phytoplankton (and protozoan) biomass.

Biomanipulations of temperate zone lakes have been proposed as a management tool for increasing water clarity in eutrophic systems in which a sufficient reduction in internal or external nutrient loading cannot be brought about. Many biomanipulation experiments have been carried out or are in progress.

Fish Manipulations

The effects of biomanipulations on water clarity, most commonly through reduction of the zooplanktivorous fish, have most successfully been demonstrated in enclosures and shallow small lakes but less successfully in larger deeper lakes (van der Vlugt et al. 1992, Scheffer 1998).[8] Fish in small shallow systems are obviously much easier to manipulate strongly enough to yield a large response that exceeds the normally large intrayear and interyear phytoplankton variability through all causes. However, strong fish manipulations of larger or deeper systems is not only expensive or impossible but are often opposed by lake users. Fishermen (in Europe) may resist either major reductions of planktivorous fish (by netting or poisoning) or projected decreases following an increase in piscivorous fish (by stocking). Furthermore, biomanipulations to improve water quality over the long-term require large and sustained fish manipulations to maintain the desired nonequilibrium between predator and prey (Kitchell 1992). Sustained biomanipulation is not only expensive but uncertain because of the notorious interyear variability in the number of young-of-the-year (YOY) fish produced by the managed fish stocks (Sissenwine 1984), and the enhanced food availability and survival of the YOY (Chapter 26).

---

[8]"Most defenders of biomanipulation as an effective tool for reducing algal biomass tacitly presume the manageability of fish stock control in lakes and reservoirs. The presumption becomes more and more fragile with the increasing size of the water body, as planktivorous fish cannot be removed either by poisoning, or by draining. Any substantial lowering of the impact of planktivory, by enhancing piscivory via artificial addition of predatory fish, requires a massive stocking programme of adult fish predators. The cost-effectiveness of much stocking is often questionable, especially because of the uncertainty of its stability." (Seda and Kubečka 1997)

Finally, as mentioned, high zooplankton grazing pressure on edible algae may not result in long-term increased water clarity but rather replacement by large inedible algae.[9] For more see Kufel et al. 1997.

There is vigorous and stimulating debate by proponents and skeptics as to the possibility of interpreting whole-lake biomanipulation experiments or their utility (de Melo et al. 1992, and Carpenter and Kitchell 1992). The 'truth', as usual, lies in the middle. Biomanipulation is probably less useful as a *management* tool than suggested by its strongest proponents, but can be useful when combined with substantial nutrient reductions, especially in small lakes. Benndorf (1990) proposes an areal phosphorus loading threshold of ~0.6 g TP m^{-2} yr^{-1} below which top-down effects can be demonstrated (in phosphorus-limited lakes). Jeppesen et al. (1997) suggest fish manipulations as a supplement to loading reductions in shallow eutrophic lakes when lake TP concentrations are lowered to the 50–100 μg TP l^{-1} range, the range over which chlorophyll-*a* concentrations in Danish lakes respond to changes in TP.

Invertebrate Predators

The abundance of large-bodied filter-feeding zooplankton in a particular lake is the outcome of the resources available (above all nanophytoplankton and protozoans) and planktivorous fish abundance. It is also determined by the abundance of predacious cladocerans such as *Leptodora* and *Bythotrephes*, or their predacious cyclopoid copepod counterparts (Gliwicz et al. 1978), or the abundance of large omnivorous shrimps (e.g., *Mysis relicta*); all of which feed selectively on larger daphnids. Probably even more important in structuring zooplankton communities in many lakes, is the abundance of larval insect predators in the genus *Chaoborus* (Sec. 23.9), which themselves are a source of fish food. Finally, the abundance of zooplankton is negatively affected by a large abundance of filter-feeding molluscs and other invertebrates at the sediment surface of shallow lakes. These compete with the macrozooplankton for algal and protozoan food particles. An example is the increasing water clarity and reduction in macrozooplankton in Lake Erie following the invasion of zebra mussels (Sec. 25.6).

---

[9]"Currently, the unpredictability of biomanipulation as a management tool makes investments hazardous and may cause social resistance to the development of biological solutions to environmental problems." (A. Persson and L. A. Hansson 1999)

While the debate about the practical utility of bio-manipulation experiments in lake management continues, the experiments have been invaluable in exploring the mechanisms and processes involved in top-down impacts on the structuring of planktonic and littoral communities. Furthermore, the experiments are raising important questions about nutrient cycling, food-web structure, and the role of microbial foodwebs (Sec. 22.9).

Increasing Fish Food: Introducing *Mysis relicta*

Fisheries biologists have introduced the opossum shrimp *Mysis relicta* to many north temperate lakes where it did not previously exist in an attempt to increase the food available to salmonids (lake trout and salmon). This attempt at lake management has had major unforeseen consequences. Predation by *Mysis* on larger cladocerans has led to a virtual disappearance of large daphnids,[10] resulting in a major modification of the macrozooplankton community and an increase in the abundance of smaller less or nonpredated zooplankton (Langeland 1988). Declines in large herbivorious zooplankton typically result in increased phytoplankton biomass.

Fish communities have also been affected by *Mysis* introductions. Its introduction in Lake Tahoe (US) led to a rapid and unexpected decline of the kokanee salmon (*Oncorhynchus nerka*), which fed primarily on the previously abundant large cladocerans. Furthermore, the expected increase in the catch of lake trout (*Salvelinus namaycush*) did not materialize even though the latter fed increasingly on *Mysis* (Goldman 1981). Similarly, a decline in the char and whitefish fisheries in Scandinavian lakes is at least partly attributable to the effect of *Mysis* introductions (Lasenby et al. 1986). The introductions increase the length of the food chain by one step, with the macrozooplankton energy now mainly flowing to the fish via the mysids. With a zooplankton growth efficiency (food intake/growth × 100) of only ~15–50 percent (Sec. 23.10) the introduction of *Mysis* resulted in a decrease rather than the expected increase in the yield of planktivorous fish. The lengthening of food chains has the further unfortunate

result of elevating the levels of persistent contaminants in fish, such as salmon or lake trout, at the top of the food chain (Sec. 28.8).

The Great Lakes of North America: A Natural Biomanipulation Experiment

The Great Lakes of North America are particularly well-studied large-scale natural (unintended) biomanipulation experiments. Long-term phytoplankton records and fish catch statistics together with some 30 years of crustacean zooplankton data place Lake Michigan and the other Laurentian Great Lakes among the best-studied lakes worldwide. Major perturbation of the food-web structure were first noted following the introduction of the sea lamprey eel (*Petromyzon marinus*) in the 1830s, with the predator feeding on large piscivorous fish (Fig. 23–19). The next major change in the food-web structure of the lakes was the introduction of pacific salmon species and the invasion of the alewife (*Alosa pseudoharengus*), a zooplanktivorous fish, followed during the last decade or so by the introduction of two predacious cladocerans, the spiny water flea (*Bythotrephes longimanus*), filter-feeding zebra mussels (*Dreissena spp.*, Sec. 25.6); plus several fish species, a New Zealand mollusc, and the fishhook water flea (*Cercopagis pengoi*) in 1998.

Many of the recent invaders originally came from the Caspian Sea region of the former USSR (Sec. 5.8) and 'escaped' after canals were constructed linking the region with central and western Europe, or as the result of transplants into newly created reservoirs elsewhere in the former USSR. However, invasions of the somewhat salinity-tolerant species into western Europe (Ketelaars et al. 1998) and the Laurentian Great Lakes is attributed primarily to ships releasing ballast waters with their stowaways, taken on elsewhere. But not all invaders arrive in ballast water. Otherwise introduced species include the common carp (*Cyprinus carpio*), eurasian water milfoil (*Myriophyllum spicatum*, a submerged macrophyte in the littoral zone and wetlands, see Chapter 24), and purple loosestrife (*Lythrum salicaria*, an emergent macrophyte in wetlands).[11]

Investigations into the effects of these introductions have allowed a number of somewhat uncertain conclusions as to their individual effects on plankton

---

[10]Mysids are omnivorous and exhibit two distinct feeding modes: Large food items are picked up using the thoracic appendages, while algae and other suspended particles are filtered from the incoming water current produced by the same appendages.

[11]In subtropical regions (e.g. Florida, US) many tropical plants and animals originally imported for the aquarium trade or to decorate ponds have escaped and thrived, with some a major threat to the native biota.

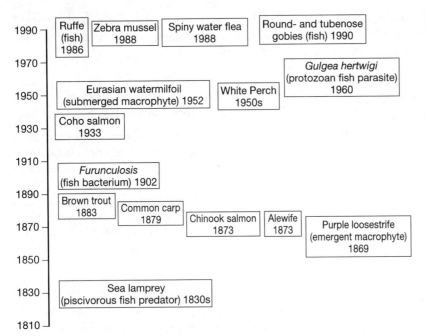

Figure 23–19 A timeline of those of the 140 nonindigenous species introductions considered to have substantial impacts on the Great Lakes. (*Modified after Mills et al. 1993.*)

and lake function. The effect of introductions to the Laurentian Great Lakes is inconclusive not because the research has been deficient but because the effect of each introduction has cascaded through the food web of the lakes, in turn setting up new cascades, with the time scales over which these indirect effects operate are poorly known. This, together with the impossibility of clearly distinguishing direct from indirect effects, the effect of one introduction from the next, effects from longer-term climatic cycles, as well as from chemical and shoreline changes brought about by human activity. All this, and exacerbated by the absence of controls, has largely confounded unambiguous interpretations of the effect of the introductions on the zooplankton and other components of the food web (Mills et al. 1987, and Lehman and Cáceres 1993).

An exception to the difficulty of linking cause and effect is the 'sledgehammer' effect of the recent introduction of zebra mussels (Sec. 25.6) on the crustacean zooplankton of shallow Lake Erie. Huge populations of the benthic mussels rapidly filter and remove the small algae and bacteria also consumed by filter-feeding macrozooplankton. They further consume the rotifers, ciliates, and nauplii that were normally eaten by omnivorous copepods. The contribution of the macrozooplankton to whole-lake secondary production has declined in step with the increase in benthic production (Johannson et al. 2000).

Even though the effects of introductions are typically difficult to resolve, the abundant research in the North American Great Lakes, western Europe, Florida, New Zealand, and elsewhere has greatly contributed to an appreciation of the complexity of planktonic and littoral food webs, and the important role of the zooplankton therein.

23.9   *Chaoborus:* The Phantom Midge

Fish, predacious zooplankton, and filter-feeding zebra mussels are not the only predators of zooplankton, nor are they necessarily the most important. In mesotrophic and eutrophic lakes and ponds, nonbiting flies in the genus *Chaoborus* (Family Chaoboridae, Order Diptera) are often present in sufficient numbers to have a major impact, primarily on intermediate-sized species of macrozooplankton (Hillbricht-Ilkowska et al. 1975).

The phantom midges are known as such because of their transparent larval stages (Fig. 25–3). The two-pigmented gas bladders change their density upon expansion and contraction, allowing the larvae to move up and down the water column. The bladders serve as the targets for echolocators, making it possible to follow their daily migrations. Four aquatic stages cover over 98 percent of the duration of their life cycle. Body length increases about fourfold between the first and fourth instar stages. In highly oligotrophic, food-limited temperate zone lakes there may only be one generation of *Chaoborus* produced in two years. Conversely, development can be completed in six to seven weeks at summer temperatures in warm food-rich lakes (Muttkowski 1918).

Following pupation, the motile pupa rises to the surface to metamorphose and emerges from the water surface as a nonbiting mosquito-like fly. The fertilized females deposit their eggs over a brief period (< 6 days). A large number of eggs are laid in rafts near the shore (~500 eggs female^{-1}). This, together with a high abundance of midges, particularly in lakes lacking much fish predation, may result in the emergence of ~100,000 first instar larvae per square meter (Neill 1985). Although these suffer from enormous mortality, the maximum densities of all instar stages combined often range between about 5,000 m^{-2} and a record 130,000 m^{-2} in a fish-free system (Xie et al. 1998). In Lake Victoria (KE, TZ, UG), *Chaoborus* larvae appear to be the most common insect larvae in the lake,[12] followed by the chironomids (Chapter 25, and MacDonald 1956); together they produce enormous mating swarms (Fig. 23–20).

The first and second instar larvae, typically < 200 μm in width, inhabit the lower depths of the epilimnion, preying on nanoplankton, large protozoans, very small crustaceans, and noncolonial rotifers (Neill 1988). In contrast, the much longer (13–18 mm) third and fourth instars exhibit pronounced diel vertical migrations but are benthic during the day and hide in hypoxic or anoxic sediments in lakes with fish predators. At night they feed on plankton, acting as ambush predators on large rotifers, intermediate-sized *Daphnia* in the epilimnion, and other similar-sized macrozooplankton (Moore 1988). During this period they

Figure 23–20   A mating swarm of lake flies (chaborids and chironomids) on Lake Victoria (KE, TZ, UG). *(Photo courtesy of G. Kling.)*

also replenish their dissolved oxygen supply. Their preference for intermediate-sized zooplankton (including early molts of larger species) has been postulated as a trade-off between an increased encounter rate of the more rapidly swimming large cladocerans and reduced handling efficiency of large prey (Pastorak 1981), but may be more simply a function of the *gape width* (mouth diameter) setting limits on the size of ingestible particles. While the early instars (I and II) are too small to be eaten by fish, they are consumed by invertebrate predators such as predatory copepods (Fischer and Frost 1996). However, the abundance of the larger instars III and IV is strongly constrained by fish predation and chaborids may be absent from lakes with high fish predation pressure (Lamontagne and Schindler 1994). When abundant, *Chaoborus* competes with zooplanktivorous fish for prey. A reduction in fish predators during lake acidification, commonly results in an increase in *Chaoborus* abundance (Chapter 27 and Locke and Sprules 1993).

Chaoborus instars III and IV typically exhibit no diurnal migration in lakes and ponds that lack zooplanktivorous fish, but will commence migrations following the

---

[12]"The periodic appearances of lake-fly swarms [involving synchronous pupation and adult emergence] is well known on many [African] lakes. What appear to be gigantic clouds of smoke, sometimes of over fifty meters in height and occasionally more than a kilometre in length, rise from the surface of the lake. The majority of the swarms on Lake Victoria appear a few days after the new moons. They are mainly composed of chaborids, especially *Chaoborus anomalus*, with a lesser number of chironomids of the genera *Chironomus, Tanypus* and *Procladius*. The water seems to boil with struggling fish devouring the pupae as they rise to the surface to emerge; the swarms are often followed by flocks of white-winged black tern, kites and other birds taking advantage of the abundant feast . . . The swarms are often blown onto the land by onshore winds . . . whether this is a significant drain on the lake's productivity remains to be determined." (Beadle 1981).

introduction of fish (Dawidowicz et al. 1990). Vertical migrations may be extensive. The fourth instar larvae migrate up to 200 m between the deep hypoxic (low oxygen) layer and the upper waters of Lake Malawi (MM, MZ, TZ) at a calculated rate of about 25 m h^{-1} upward and 40 m h^{-1} downward (Sec. 23.14, and Irvine 1997).

Chaoborus predation has a direct impact not only on the mid-sized macrozooplankton and earlier molts of larger species but indirectly on the size and food web structure of zooplankton and fish communities. This was demonstrated when large daphnids were experimentally removed from enclosures in a fishless lake. The resulting reduced competition for food resources allowed a large increase in the abundance of small rotifers and small cladoceran species. The increased abundance of the small zooplankton provided such an abundance of food particles that many more of the early *Chaoborus* instars survived and reached the fourth instar stage (Neill 1984). This particular experiment is mentioned for two reasons. First, as a reminder that the role of competition for resources should not be overlooked despite recent emphasis in the literature on predation as the major force in structuring zooplankton communities, (Neill 1984, Yan et al. 1985). Second, despite the emphasis in the literature on adult forms or late developmental stages of particular zooplankton species, the abundance of zooplankton (insects, fish, etc.) will not be predictable without sufficient information on the survival and growth of the early developmental stages.

▲ 23.10   Zooplankton Feeding

The recognition that growth of a zooplankton population or community requires both enough food of sufficient quality and appropriate size range to be readily eaten has led to important research, mostly in the laboratory, on different aspects of the filtering, assimilation, and respiration rates of primarily macrozooplankton. This work has made a major contribution to appreciation of the physiological and behavioral ecology of zooplankton and their energy requirements for growth and reproduction. The *filtering rate* (F) or *clearance rate* (ml animal^{-1} d^{-1}) is calculated from the volume of water cleansed by the average individual per unit of time:

$$F = \frac{\ln(C_o) - \ln(C_t)}{t} \times \frac{W}{N} \qquad \text{EQ. 23.1}$$

Where C_o = initial concentrations of food organisms (no. ml^{-1}), C_t = the final concentration of food organisms (no. ml^{-1}), W = water volume (ml), N = number of animals, and t = duration of experiment.

The amount of food consumed, the *grazing rate* (G) or *ingestion rate* per animal per unit of time, is the product of the filtering rate (F) and the mean food concentration (C):

$$G = F \times \frac{C_o + C_t}{2} \qquad \text{EQ. 23.2}$$

normally expressed in terms of the energy content, wet or dry weight, or carbon content.

Russian researchers report that the gross growth efficiency (food intake/growth × 100) of macrozooplankton varies between 15 percent and 32 percent in nature (Winberg 1972). But the gross efficiency will naturally range beyond these values when food quality is either very high or very low (Richman and Dodson 1983). While the gross growth efficiency of both protozoans and metazoan animals commonly varies widely around a mean of 20–30 percent (Straile 1997), the growth efficiency is a function of the C:P ratio of the particles consumed (Elser et al. 2000), which is normally unknown in nature. Consequently, a particular reported value must be assumed, leaving much uncertainty as to the fraction of the seston consumed, assimilated, and used for growth rather than respired.

Filtering Rates and Particle Size

Laboratory measurements show a systematic increase in daphnid filtering rate with temperature to an optimum range beyond which the filtering rate declines rapidly (Horn 1981), and a similar increase with increasing body size (Knoechel and Holtby 1986, and Jürgens et al. 1996). Although larger zooplankton have a higher filtering rate than small ones (Sec. 22.9), the filtering rate per unit biomass (the *mass-specific filtering rate*) declines with increasing organismal size— regardless of taxonomic position—when examined over a wide body-size range (Peters and Downing 1984). Filtering and grazing rates also depend on food concentration (Horn 1981) and quality.

Most direct measures of filtering (grazing) have been determined in beakers, either by adding unlabelled algae or other food particles and measuring their decline some hours later or, by adding ^{32}P or ^{14}C labelled food particles and allowing the zooplankton to graze for a few minutes before collecting the animals and determining their radioactivity. To facilitate the analysis, the numbers of animals used in laboratory experiments are usually considerably higher than observed in nature. This may result in (unrecognized) grazing rate reductions to less than half those measured under more natural densities (Fig. 23–21).

Probably the most elegant and least artificial grazing studies are those using the *Haney in situ grazing chamber* (Haney 1973), based on a device developed by Z. M. Gliwicz (PL). The clear plastic chamber (resembling a Schindler-Patalas trap, Fig. 23–1), is lowered and when closed entraps a known volume of water and its organisms. Closing the doors opens a small piston which releases radio-labelled cells (bacteria, algae, or yeast) of a suitable size. After a few minutes of feeding, but before radio-labelled food particles are defecated, the chamber is hauled to the surface and the zooplankton are collected. From the calculated radioactivity per unit volume of chamber at the outset, the feeding time, and the radioactivity of the zooplankton after feeding, the grazing rate of individual zooplankton species or the community as a whole can be determined (Knoechel and Holtby 1986). Feeding studies based on providing highly palatable particles of an optimal size (~5–10 μm) in Haney chambers or in the laboratory will naturally yield a higher grazing rate on those particles than on the whole natural spectrum of

nanoplankton (2–30 μm) present in nature, and will further yield a rate that even more greatly overestimates the grazing on natural assemblages, composed not only of picoplankton and nanoplankton but also on the little (if at all) consumed larger microplankton (> ~30 μm) (James and Forsyth 1990, and Cyr and Pace 1993).

Grazing Impacts on the Prey

The frequently modest whole-system grazing rates in nature (ml animal^{-1} d^{-1} × no. of animals) measured, even on edible-sized particles (< ~30 μm), suggest that most of the time the principal role of macrozooplankton is to structure the prey community rather than act as the principal determinant of the biomass of phytoplankton, protozooplankton, and bacterioplankton communities (Fig. 23–22).

In both terrestrial and aquatic communities, the loss rates imposed by herbivores is primarily imposed on the most rapidly growing prey organisms rather than being experienced equally by all potential food particles (Fig. 24–10). This is probably because the high-quality food, rich in the minerals, nutrients, proteins, and fatty acids required for rapid growth, is disproportionately present in fast-growing prey organisms (see Fig. 24–11). Recognition of the importance of selective feeding has provided support for the hypothesis that the abundance of microphytoplankton in aquatic systems is partly attributable to larger phytoplankton being little subject to zooplankton grazing.

▲ 23.11   Nutrient Cycling and Zooplankton

Zooplankton affect the phytoplankton community biomass and structure directly by selective grazing as well as indirectly through the release (recycling) of critically needed mineral nutrients. The selective effect of each mechanism and the effect of the two processes combined on the phytoplankton community structure and abundance remains poorly known. Field manipulations of grazers yield changes in the phytoplankton community that at times are better explained by changes observed in algal growth rates linked to mineral nutrients than by grazer-induced mortality (Sterner 1989). Predators exert not only a top-down but also a bottom-up effect on their prey because they recycle nutrients, making them available to the primary producers. These complexities are increased

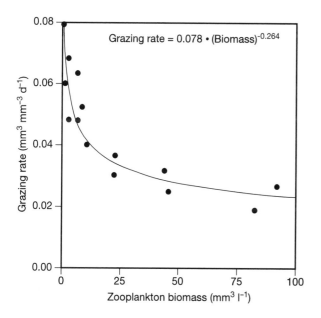

Figure 23–21 A laboratory-determined relationship between macrozooplankton biomass (*Daphnia hyalina*) and the species-specific grazing rate, showing a pronounced decline in grazing with increased crowding. *(After Horn 1981.)*

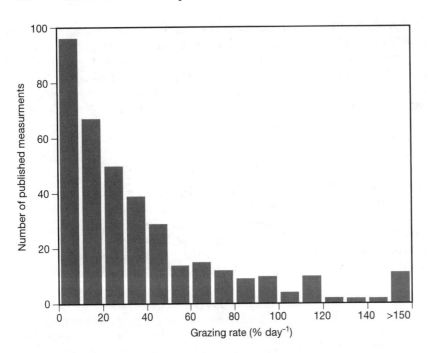

Figure 23–22 Published frequency distribution (n = 369) of macrozooplankton grazing rates on nanoplankton (< 35 μm) in epilimnia during the main growing season in lakes of the Northern Hemisphere, New Zealand, and year-round in South African lakes. Grazing rates were measured on single radioactively labelled food particles, radioactively labelled algal assemblages, and by the changing abundance over time of algae in water samples. The last technique normally yielded the lowest grazing rates. Note that 51 percent of all the observations had community grazing rates smaller than $0.25 \ d^{-1}$. Thirty experimentally measured grazing rates showed that crustaceans grazed from 2–21 percent per day (mean = 8%) of the chlorophyll in the algal (< 35 μm) community. (*After Cyr and Pace 1993.*)

when it is recognized that herbivores (and bacteria too) normally have lower and less variable N:P (Table 8–2) and C:P protoplasmic ratios (*stoichiometry*) than their phytoplankton prey (Hessen and Andersen 1992, and Elser et al. 2000). Consequently, when the phytoplankton prey provides the predator with a higher protoplasmic C:N:P supply ratio than its demand ratio, the zooplankton predator will—over its short lifetime at least—disproportionately retain the phosphorous acquired and release the excess nitrogen, producing an even higher and less favorable C:P and N:P supply ratio for phytoplankton production. Although food quality affects zooplankton species composition and abundance, it is less clear whether periodically high macrozooplankton abundances exerts an impact on phytoplankton and protozoan production over longer time scales, or in the presence of substantial fish predation. Even so, the different macrozooplankton groups differ in their protoplasmic C:P and C:N demand ratios (see Elser et al. 2000), and in the required essential fatty acids and amino acid levels supplied by their prey (Sec. 21.5).

Food Quality

The importance of a balanced diet has long been recognized in fish and livestock production. Zooplankton species composition and growth may be constrained by the C:N:P ratio of their food and associated

changes in the biochemical composition of the consumers and their prey, including changes in algal cellwall thickness and digestibility. Recent work indicates that biochemical composition affects *Daphnia* growth, but only when the C:P supply is low enough for there to be no phosphorus limitation (Boersma 2000). While prey quality and quantity as well as predation and disease are important proximal determinants of zooplankton production and abundance, more ultimate causes include the geology and land use. These determine the mineral nutrient and organic matter supplied from drainage basins as well as the morphometry of aquatic systems, which determines flushing, habitat availability, and depth of the water column.

Daphnia species, which have a particularly low protoplasmic N:P ratio (demand ratio, ~14:1 by atoms) for growth and egg production, dominate in eutrophic systems characterized by a low seston N:P ratio (supply ratio). Conversely, calanoid copepods have a slightly higher nitrogen content (% N), but greatly reduced phosphorus content (% P)—thus they have a higher protoplasmic N:P (~30–50:1 by atoms) ratio. They tend to be proportionately more common in oligotrophic waters with a characteristically high seston N:P ratio (Elser and Urabe 1999). Work on two Dutch lakes showed that food quality is indeed predictively linked to macrozooplankton growth and abundance, with *Daphnia* abundance declining systematically when the seston C:P ratio increased from

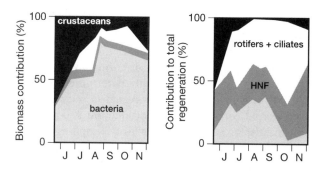

Figure 23–23 Contribution of zooplankton and bacteria to the biomass and ammonium (NH_4^+) regeneration in mesotrophic Lake Kizaki, JP (LA = 1.4 km², $\bar{z}$ = 29 m) in 1992. Note (1) the disproportionately small (relative to their biomass) bacterial contribution to nutrient regeneration, presumably because the majority of planktonic bacteria were metabolically inactive (dormant) (see Figure 22–2); and (2) the disproportionately large contribution of the HNF and other small noncrustacean zooplankton to nutrient regeneration. (*Modified after Haga et al. 1995.*)

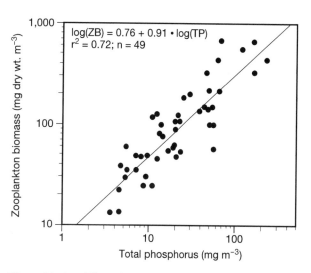

Figure 23–24 The relationship between total phosphorus concentration (TP) and macrozooplankton biomass (ZB) in 49 lakes, 47 of which are in the north temperate zone. (*After Hanson and Peters 1984.*)

250 to 500 (as atoms), even though most are selective feeders and do not simply consume bulk seston (De-Mott and Gulatti 1999). Zooplankton and other heterotrophic organisms (including bacteria) who consume and assimilate food disproportionately high in carbon will have a low gross conversion efficiency and excrete or respire the excess carbon.

Small organisms tend to have a higher metabolic rate per unit biomass than larger organisms, therefore it is not surprising that the protozoa contribute more—and the larger crustaceans less—to nutrient recycling than is suggested by their contribution to the community biomass (Fig. 23–23). Conversely, macrozooplankton abundance and growth rate are more important as a determinant of the survival and growth rate of YOY fish, and of those fish that are zooplanktivorous as adults (Chapter 26).

23.12   Resource Availability and Zooplankton Biomass

Eutrophic lakes tend to have a higher zooplankton biomass than oligotrophic ones. This conclusion applies equally to the protozoans (Porter et al. 1985), the rotifers (Pauli 1990), and the macrozooplankton (McCauley and Kalff 1981), thereby providing empirical evidence for the widely accepted recognition of the importance of resource (food) availability on zoo-

plankton abundance and community biomass. As the phytoplankton biomass increases among systems the zooplankton biomass increases too but at a lower rate (log-log slopes < 1).

The basis for smaller macrozooplankton to phytoplankton biomass ratios in eutrophic lakes is poorly resolved primarily because predator–prey interactions change with the temporal and spatial scale examined (Sec. 2.6), and because more than species interactions determine the abundance of populations and communities. Physical/chemical variables among lakes that are linked with resource availability and, correlate with the macrozooplankton biomass and production (see below) include temperature (+), water depth (–), the water residence time of systems (+) (Yan 1986), and the pH of the water (–). Total phosphorus (Fig 23–24) and nitrogen are other predictors of macrozooplankton biomass (+). The limited work on protozoan zooplankton shows that they too are correlated with resource availability (bacterial and algal abundance, +), water temperature (+), and the abundance of their presumed cladoceran predators (–) (Gasol and Vaqué 1993).

23.13   Zooplankton Production

Most available zooplankton production measurements were obtained during the life of the International Biological Programme (IBP) of the late 1960s and early 1970s. A concerted effort was then made to obtain information on productivity and energy flows in both aquatic and terrestrial systems.

Secondary production by herbivores cannot be defined, as in primary production, on the basis of the synthesis of organic molecules. Instead, production represents the growth of an individual, a population, or a community. In zooplankton (benthos and fish) ecology, production rates have usually been measured for individual populations of particular interest. Then the rates obtained for the different species are summed and spatially and temporally scaled up to produce a measure of the production of the dominant species.

Production rate determinations are time consuming. This, along with the simplifying assumptions that must be made to obtain the rates are important reasons why relatively few such measurements are being reported today. Instead, differences in secondary production rates are usually inferred from differences in the biomass of the groups of organisms considered. Where absolute community rates are desired, these are most frequently computed on the basis of the body size of the organisms and their abundance, using literature-derived average metabolic rates obtained for similar species.[13]

▲ Production Measurements

When there is neither recruitment to the population or death during the interval, production (P_R) is the total increase in biomass:

$$P_R = (B_t - B_o) = W_t N_t - W_o N_o \qquad \text{EQ. 23.3}$$

where B = biomass, W = weight of an individual, N = number of individuals, and the subscripts t = beginning and o = end of the time interval. The average weight is usually obtained from length measurements that can be converted to weight by means of a predetermined length–weight relationship (Langeland 1982).[14] However, the simplifying assumptions are un-

realistic and birth as well as death rates have to be measured. While this is time consuming it is relatively straightforward for species such as the copepods, with clearly identifiable life stages or **cohorts** whose abundance over time can be readily determined (see Fig. 23–8). This is trickier for zooplankton such as cladocerans and rotifers where reproduction is continuous during the growing season, with overlapping generations that are difficult to distinguish.

Where the life stages cannot be distinguished because of continuous reproduction, one of three basic methods is employed to obtain production rates. In the first method, laboratory-determined growth-rate data are combined with population estimates made in the field to compute production. In the second approach, population estimates are combined with laboratory or field-obtained physiological measurements, such as the ratio between production and assimilation (assimilation = ingestion – egestion), in which the ratio describes population maintenance costs. Widely used is a variant of the third method in which production (P_R) is determined as the product of the abundance measured and an experimentally or literature-determined *turn-over time*, the time required for a population biomass (B) in steady state to replace itself (P/B):

$$P_R = \frac{P_R}{B} \times B \qquad \text{EQ. 23.4}$$

The various methods used in the literature for calculating production rates make different assumptions. This is usually not a major issue in individual studies where the primary goal is to show and explain changes over time and space, using a standard methodology. It is a much more serious issue when among-system generalities are sought. To evaluate the impact of assumptions made in computing production rates, Andrew (1983) applied four different methods to a *single* population of *Daphnia* in nature. Even though the methods used to collect and process the *Daphnia* were identical, the computed production ranged fourfold between 13 and 51 g dry wt. m^{-2} yr^{-1}. These results do not even account for the large impact in the literature of

---

[13]"Much of the mathematical-appearing literature on production consists merely of restating natural history observations in terms of symbols and subjecting them to simple algebraic manipulations. Not that there is anything wrong with this; it clarifies the ideas and proves a uniform basis for comparable calculation. Nevertheless, we have little mathematical theory behind studies of secondary production; what there is has mainly to do with population dynamics." (W. T. Edmondson 1974)

[14]Equations describing the nonlinear relationship between the sizes (weights) of plants and animals (*W*) and another of their characteris-

tics (e.g., length, *y*) are called **power formulae** ($Y = aW^b$), where *a* and *b* are constants. Because *Y* and *W* increase at different rates, power formulae involving body size relations are usually referred to as **allometric equations** (Greek, allos = other; metros = measure). Transforming (converting) the observed values to their logarithms provides a convenient linearization for ease of computation as $\log_{10}(Y) = \log_{10}(a) + \log_{10}(W)$. (Peters 1983)

among-study variation imposed by differences in collecting methodology or sampling devices (Table 23–1), including net mesh size, frequency and time of sampling, the number of replicate samples taken, or preservation and counting procedures. Consequently, it is unclear how much among-system variation reported in zooplankton biomass and production for inland waters is real rather than methodological. Whatever the uncertainties, they are so large that few direct measurements of zooplankton production and benthic animal production have been made in recent years. Instead, organismal abundance is used as a proxy for population production. But where production estimates are needed these are usually obtained from measures of population biomass, plus the individual weight of organisms and water temperature. (See Downing and Rigler 1984.)

Determinants of Production

In a comprehensive empirical study Plante and Downing (1989) assembled data on 137 invertebrate populations (including zooplankton, benthic insects, annelids, and molluscs) and found that secondary production rates (P, g m^{-2} yr) were positively related to the mean annual population biomass (B, g m^{-2}) and water temperature (T, °C), and negatively linked to the maximum individual weight of the organisms (W_m, mg dry wt.):

$$\log (P) = 0.06 + 0.79 \cdot \log (B)$$
$$- 0.16 \cdot \log (W_m) + 0.05 \cdot T \qquad \text{EQ. 23.5}$$
$$R^2 = 0.79; F = 165; P < 0.001; n = 137$$

The single most important correlate of production is indeed population biomass (Fig. 23–25), which alone accounts for 63 percent of the 79 percent variation accounted for by Eq. 23.5). Population biomass serves as a useful surrogate for production in all groups of organisms larger than bacteria when the weight (size) of individuals and the water temperature are known. The multiple regression model (Eq. 23.5) summarizes several major attributes of zooplankton and benthic invertebrates (Chapter 25). First, with increasing total biomass population production increases systematically, but at a progressively slower rate as indicated by a slope coefficient of less than one (0.79). Consequently, the annual production per unit biomass, or per organism (known as specific production) declines with the increasing weight of individuals. Second, production rises as organismal size declines (–0.16). Third, the positive effect of temperature on production confirms many findings in the literature showing that egg development, growth rates, and feeding rates normally increase with temperature. The model, based on temperate zone data, predicts the observed higher growth rates noted at lower latitudes.

The 21 percent variation in invertebrate production that is not explained by the model (Eq. 23.5) could be the result of, for example, differences in pH (Plante and Downing 1989), with invertebrate population production rates sometimes reduced in acidified lakes (Sec. 27.9). Population production rates are further affected by the water residence time; higher production rates are linked to a decreasing WRT (Plante and Downing 1989). A shorter WRT reflects a larger input of both inorganic nutrients and organic matter

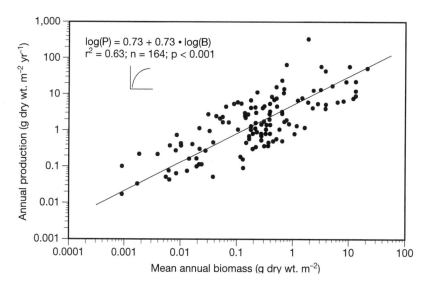

Figure 23–25 Relationship between the mean annual biomass (B) of aquatic invertebrate populations and the annual production (P). Some of the variations in production at any one biomass are attributable to differences in the weight of individual organisms and temperature (see Eq. 23.5), food quantity and quality, and uncertainties in the biomass and production measurements. *(Modified after Plante and Downing 1989.)*

from drainage basins (Fig. 9–3), and proportionately reduced sedimentation of food particles. A third variable, not evaluated in Eq. 23.5, is system morphometry (Yan 1986). However, the single largest source of unexplained variation in the model may well be imprecision in both population biomass and production estimates from the literature used to build the empirical model.

Zooplankton production rates are normally higher in lakes with a deep rather than shallow thermocline (Plante and Downing 1989), possibly because thermocline thickness serves as a surrogate for lake size and depth where more of the organic matter in a thick water column is available to the zooplankton, whereas in a shallow water column more of the sedimenting particles are consumed in the sediments following a much shorter water column residence time (See Table 25–4). An alternate but equally testable hypothesis is that lakes with a shallow thermocline are typically small and, more importantly, located in small catchments that export relatively little plant nutrients and organic matter (Sec. 9.5).

The principal predictors of zooplankton production always change with the hierarchical, spatial, and temporal scales over which studies are executed (Sec. 2.6). But that does not reduce the utility of empirical models such as Eq. 23.5. Where time and resources do not allow a direct determination, such models allow a first estimate of species and community production. Equally important, they permit a very useful a priori prediction of the magnitude of expected production rates before undertaking a specific study. In addition, empirical models set useful limits on what can be expected in nature. The limits are important for conceptualization and development of dynamic models of ecosystem components and whole ecosystems (ecosystem models). Empirical models also raise useful questions concerning the differences noted between observed production rates and those predicted. Finally, predictor variables and computed coefficients stimulate questions about the underlying processes and mechanisms that serve as a basis for explanatory studies.

▲   ## 23.14   Diel Migration and Cyclomorphosis

Diel Migration

The pronounced diel vertical migration (DVM) of macrozooplankton and distinct seasonal changes in the morphology of successive generations, known as

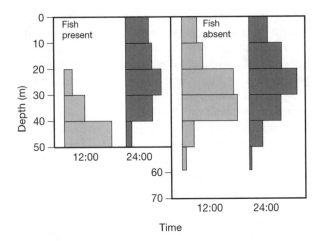

Figure 23–26   The midday and midnight depth distribution (percentage of the total population) of *Cyclops abyssorum* in August 1985 in two Polish mountain lakes: Lake Morskie Oko, containing fish; and Lake Czarny nad Morskim, lacking fish. Note the more pronounced vertical migration in the lake with fish. *(After Gliwicz 1986.)*

cyclomorphosis, of crustacean zooplankton and rotifer species has intrigued and attracted limnologists and oceanographers for close to a century, but comprehensive explanations remain elusive.[15]

Research on vertical migration (Fig. 23–26) has given rise to a wide variety of explanatory hypotheses. They consider both proximal causes (e.g., food, respiration) of the behavior as well as the ultimate or adaptive value of migrations (e.g., predator avoidance). Ideas have continued to evolve and today few limnologists would suggest that any single factor is responsible. McLaren (1974) produced metabolic and demographic hypotheses, arguing that zooplankton migration accrues an energy bonus and the organisms spend a portion of each day in cold water where respiration is reduced. This reduction allows more energy to be expended on growth and reproduction. The synchronized periodicity, normally (but not always) downward during early morning, and upward to the epilimnion after sunset, should at the same time provide the macrozooplankton a daytime zone refuge from visual predators (Sec. 10.12). The metabolic hypothesis was further developed by Geller (1986) and other temperate zone scientists who proposed that the

---

[15]"After more than 150 years of field observations and 100 years of laboratory research, diel migrations remain an enigma." (Huntley 1985)

organisms obtain their energy bonus by being able to feed in the warm food-rich epilimnion without appreciably increasing their cold-water determined respiration rate. The metabolic hypothesis ties migration to energy conservation during periods of assumed food limitation, but is of little relevance to explaining migrations observed in the tropics where water column temperature changes are a few degrees at best. More recently, a number of researchers have united the food limitation and predator avoidance hypotheses and argue that migration is the result of a necessary compromise between the two (Lampert 1989, and Dini and Carpenter 1992).

The environmental cues that trigger migration usually involve changes in light intensity (Table 10–2) and the release of **infochemicals** by a variety of predators. Experimental work on limnocorrals with and without fish or *Chaoborus* predators, also varying in food availability and depth, definitely shows greater diel vertical migration of macrozooplankton in enclosures with predators, and reduced migration in deep enclosures without abundant food. Other experimental work has demonstrated that particular chemical cues provided by the excretory products of predators, known as **kairomones**, stimulate migration (Nesbitt et al. 1996), as do **alarm substances**, cues provided by injured or partially eaten prey organisms of the same species.

Work on copepods demonstrated that vertical migration is more pronounced in lakes with long-established fish populations than in those in which zooplanktivorous fish were introduced more recently (Gliwicz 1986a). This particular finding provides evidence that migration is not only a short-term phenotypic response to the presence of predators but also involves natural selection and a genotypic response involving two or more genotypes.

The present consensus is that predator avoidance is the most important causal factor inducing migrations, with food availability of secondary importance. The predation avoidance hypothesis is further supported by observations of increased diel migration *Daphnia* in more transparent lakes (Dodson 1990), in which the animals spend the daylight hours in the sediments or at depth where they cannot be seen by predators. Those lakes showing little diurnal migrations of *Daphnia* were either highly oligotrophic (little food) or had a minimal population of zooplanktivorous fish. In addition to affecting migrations, infochemicals induce a variety of morphological and life-history changes, including the onset of male and ephippial (resting) egg production.

Horizontal Migration

Diel horizontal migration (DHM) in shallow lakes (lacking hypolimnia) containing fish, with a daytime aggregation of pelagic zooplankton in macrophyte beds and nighttime movement into the pelagic zone, provides even more evidence that predator avoidance is an important reason for migrations (Fig. 23–27). Laboratory experiments have shown that the presence of fish or introducing water in which fish had lived initiated migration of *Daphnia magna* toward plant beds (Lauridsen and Lodge 1996). Russian research (Kiselyev 1980, in Grigorovich et al. 1998) reports evening migrations of cladocerans along the bottom slope from the deep profundal zone to the littoral zone, followed by a return migration to the profundal zone in early morning.

Full moonlight appears to reduce the extent of *Daphnia* migration toward the surface by about 2 m (Dodson 1990). This may not be sufficient to minimize predation by visual predators. For example, the sudden rise of the full moon some hours after sunset and the darkness ascent of zooplankton in an African reservoir suddenly made them visible to sardines in surface waters; the fish inflicted major mortality on

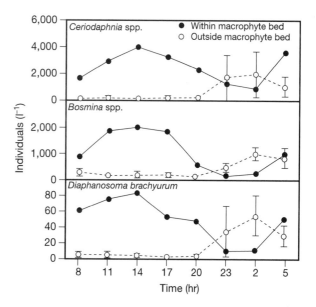

Figure 23–27   Diel variations in the abundance of various cladocerans in a 2-m exclosure (open to small fish and zooplankton) containing dense coverage of submerged macrophytes at a reference station in Lake Stigsholm, DK (LA = 0.01 km², z̄ = 0.8 m, z_{max} = 1.2 m). (*After Lauridsen et al. 1996, in Jeppesen et al. 1997.*)

the crustaceans (Gliwicz 1986b). The extent of migration is furthermore a function of organismal size (Peters et al. 1994). The large opposum shrimp (*Mysis relicta*, Fig. 23-3) in ultra-oligotrophic Lake Tahoe may move 900–1,000 m vertically during 24 hours, possibly a record for a freshwater zooplankton (Goldman, 1981).

Cyclomorphosis

This term, describing seasonal changes in generational morphology, was coined by R. Lauterborn (DE) in 1904. Cyclomorphosis has been observed in phytoplankton (dinoflagellates, and colony formation in unicellular green algae), zooplankton (rotifers, copepods, and cladocerans) and fish, but has been investigated best in daphnids. These express it most strongly as seasonal changes in head shape (helmet size) and the length of the carapace spine or dorsal crest (Jacobs 1987). In rotifers it is expressed primarily by changes in spine length. Strictly speaking, cyclomorphosis entails the production of different phenotypes by the same genotype. However, genetic studies have revealed that there *is* a genetic component to cyclomorphosis (Black 1980). Even morphologically very similar plankton organisms may seasonally show distinct changes in genotype, thereby confounding the interpretation of cyclomorphism.

There are abundant laboratory studies (Havel 1987) linking cyclomorphism to temperature, food, light, turbulence, and water soluble organic substances that serve as infochemicals released by potential predators. Intriguingly, experimentally induced or enhanced morphological changes seen during the various developmental (ontogenetic) stages in the laboratory are usually less extreme than what is observed in nature. This implies that either laboratory conditions reduce the number or strength of identified inducing factors or the existence in nature of yet unidentified inducing factors (Jacobs 1987).

Many adaptive explanations for cyclomorphosis have been proposed over the last century. One of the earliest theories was that sinking rate is reduced as a result of increased surface area or enhancement of oxygen diffusion during periods of low oxygen concentration. More modern explanations include decreased visibility to predators or the increased difficulty experienced by predators in handling zooplankton with longer spines or an unusual body shape. For example, posterior spine development in laboratory clones of the rotifer *Keratella cochlearis* is en-

hanced by a water-soluble factor released by two different species of cyclopoid copepods and a predatory rotifer (Stemberger and Gilbert 1984). The spined form of this rotifer is less susceptible to predation by the inducing species. A difficulty confounding adaptive interpretations of cyclomorphosis on growth and survival is that morphological changes are accompanied by changes in physiology and behavior.

Probably the most widely accepted hypothesis today is that seasonally changing selective predation by temporarily abundant predators act as a selective force for the subsequent cyclomorphosis that is observed. This is supported by observations of clone cultures showing much variation in morphology even in the absence of predators.[16]

Highlights

- Inland-water zooplankton range in size from small protozoan flagellates less than 2 μm in their longest dimension to large crustaceans several centimeters in length.
- Inland-water zooplankton consist primarily of protozoans, rotifers, and crustaceans. Abundance and individual growth rates are highest among small protozoans and lowest among large crustaceans.
- Well-studied freshwater lakes and rivers contain 50–100 planktonic zooplankton (protozoans included), plus a roughly equal number of littoral species. Species abundance declines with increasing salinity. Species richness increases with system size, attributable to increased habitat variation. The trophic status serves as an indicator of resource availability.
- ▲ Long-term studies of crustacean zooplankton in unpolluted lakes reveal large interannual and decade-long variability in zooplankton abundance, species composition, and community biomass. The variation has been primarily attributed to changing predation pressure imposed by fish or invertebrates, and indirectly to changes in nutrient loading, flushing rates, water column stability, or the duration of an ice cover

---

[16]Broad generalizations in biological limnology reflect the behavior of many species. No such general interpretations have emerged from the studies on vertical migration and cyclomorphosis of individual zooplankton species. This failure appears, the result of continuing attempts to draw broad generalities from the study of single rather than many species. In other words, the failure appears to stem from attempts to extrapolate from the specific to the general without the identification of a broad generality that unites the diverse observations made.

which can be linked to interannual variations in weather and decades-long climatic cycles.

• A high abundance of zooplanktivorous fish, including young-of-the-year, greatly reduces the abundance of large zooplankton on which they preferentially feed. The reduction in large filter-feeding herbivorous zooplankton typically allows an increase in phytoplankton biomass or a shift to an algal community dominated by large cells and colonies little subject to predation.

• ▲ Biomanipulations, primarily through reductions of zooplanktivorous fish, have been proposed as a management tool in temperate zone lakes for increasing water clarity through enhanced removal of phytoplankton by the resulting abundance of large filter-feeding zooplankton.

• Eutrophic waters have a higher protozoan, rotifer, and crustacean biomass and production than oligotrophic waters, providing empirical evidence for the widely accepted recognition of the importance of resource (food) availability on zooplankton abundance and community biomass.

• ▲ Predators exert both top-down and bottom-up effects on their prey because they also recycle nutri-ents and thus affect the C:N:P supply ratio of the phytoplankton.

• Secondary production by invertebrates represents the growth of an individual, a population, or a community.

• Secondary production rates of zooplankton (and benthic insects) is, among populations, best predicted on the basis of the population biomass (+), individual weight of organisms (–), and water temperature (+).

• ▲ The frequently modest whole-system grazing rates in nature suggest that the principal role of macrozooplankton is to structure the prey community rather than determine its biomass.

• ▲ The present consensus is that predator avoidance is the most important causal factor inducing diel migrations, with food availability of secondary importance.

• ▲ While comprehensive explanations remain elusive, the most widely accepted hypothesis for seasonal changes in zooplankton morphology (cyclomorphosis) is that seasonally changing predation by temporarily abundant predators superimposes a selective force on the already substantial variation in morphology present in the absence of predators.

24

Benthic Plants

24.1   Introduction

The influence of large aquatic plants or **macrophytes** on the dynamics of the shallow water (littoral) zone of lakes is generally very large. Since F. E. Eggleton's (1931) pronouncements on the subject, some textbooks go so far as to define the *littoral zone* on the basis of the area covered, or potentially covered, by macrophytes. However, the littoral zone encompasses the fraction of sediments and overlying water column sufficiently well-illuminated to allow not only macrophyte growth, but also the growth of the sediment algae to the bottom of the euphotic zone. The macrophytes are often the principal primary producers not only in the littoral zone of lakes (Wetzel 1983) but also in shallow rivers (Sand-Jensen 1997), and they dominate wetlands. Macrophyte abundance and biomass determine the distribution and abundance of the **periphyton**, the algae attached to substrates (Sec. 24.9).

The macrophyte community can be subdivided in a variety of ways, depending on the goal of the researcher. The broadest taxonomy-based subdivision is between the **macroalgae** (Characeae; Chlorophyta) and large filamentous or sheet-forming algae in several divisions, the **non-vascular plants** composed of liverworts, bog mosses (*Sphagnum*) and true mosses (bryophytes), and the **vascular plants** (angiosperms). Alternatively large plants can be subdivided on the basis of habitat as **emergent** (*helophytes*), **free-floating**, and **submerged macrophytes** (*hydrophytes*). The emergent species usually grow only partly submerged to a depth of 1–2 m, and not infrequently grow above the waterline of lakes, wetlands, and rivers with only

their roots located in wet or damp soils. The emergent and submerged macrophyte categories form a continuum, with some species having both emergent and submerged forms. The continuum also extends from emergent to terrestrial vegetation with species that grow both in water and on land (**amphibious plants**) and others that grow on land and occasionally in water (**secondary water plants**).

The submerged forms, which are usually but not always rooted (Fig. 24–1), are often subdivided on the basis of their growth form into tall **canopy-producing** species that reach the water surface sometime during the growing season with most of the biomass near the top of the stand (e.g., water lilies), **erect** species that reach some distance into the water column with a roughly uniform biomass distribution along their length, and low-growing **bottom-dwelling** species that have most of their biomass near the substratum.

The emergent littoral vegetation of rivers and lakes help reduce shoreline erosion through their dampening effect on wave energy, and also serves as a major wildlife habitat. The submerged forms—together with the submerged portions of emergent species—trap particles and associated nutrients, with the plants and sediments forming an important substrate for bacteria and periphyton (Sec. 24.9). In addition, macrophytes serve as habitat for substrate-associated invertebrates (*zoobenthos*) feeding on periphyton, detritus and associated microorganisms, and their zoobenthic predators (Chapter 25). Macrophytes also provide a daytime refuge for pelagic zooplankton in shallow lakes (Fig. 23–27). Moreover, macrophytes provide a habitat for the feeding, breeding, and hiding

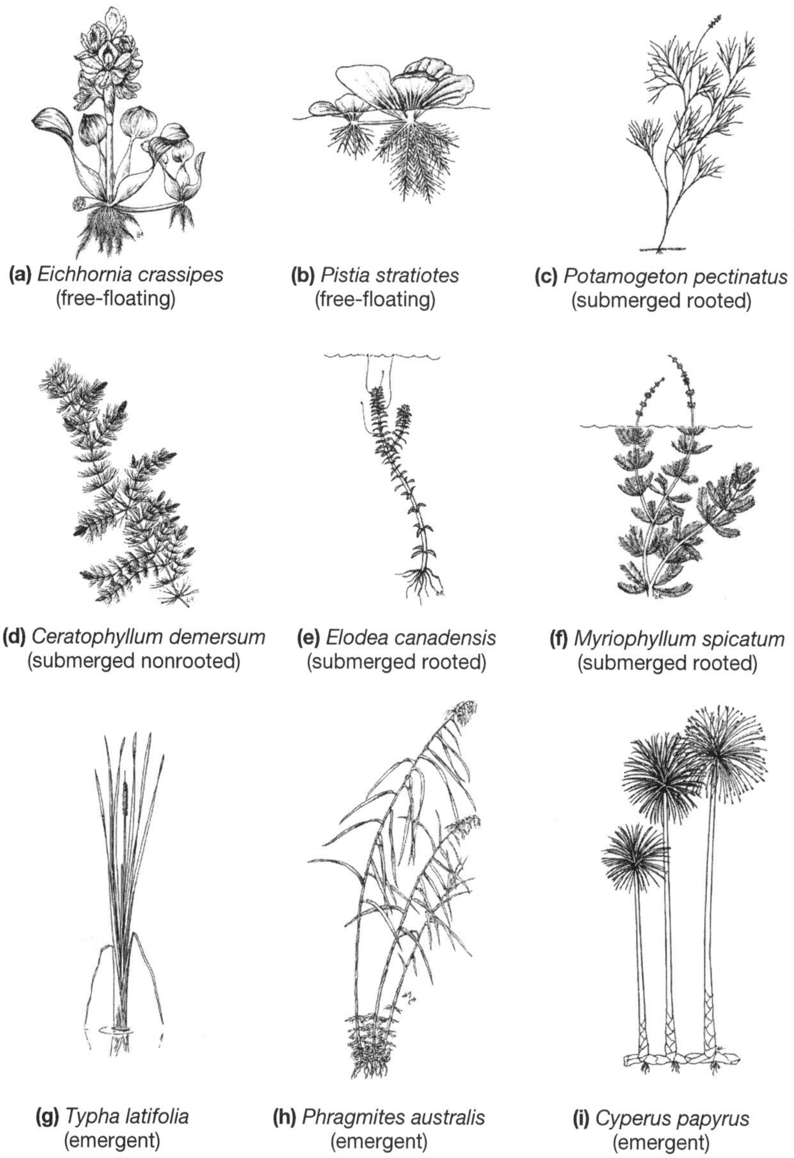

(a) *Eichhornia crassipes*
(free-floating)

(b) *Pistia stratiotes*
(free-floating)

(c) *Potamogeton pectinatus*
(submerged rooted)

(d) *Ceratophyllum demersum*
(submerged nonrooted)

(e) *Elodea canadensis*
(submerged rooted)

(f) *Myriophyllum spicatum*
(submerged rooted)

(g) *Typha latifolia*
(emergent)

(h) *Phragmites australis*
(emergent)

(i) *Cyperus papyrus*
(emergent)

Figure 24–1 Some species of free-floating, submerged, and emergent macrophytes. Not to scale. *(From a variety of sources.)*

of littoral fish and for pelagic or riverine fish species feeding in shallow water. Finally, the macrophyte-dominated littoral zone provides a habitat for waterfowl, songbirds, amphibians, reptiles, and mammals.

Where abundant, the submerged macrophytes are important suppliers of organic matter to inland waters, and their decomposition can have a major effect on dissolved oxygen concentrations and the cycling of nutrients and contaminants. The canopy-forming sub-

merged macrophytes of lakes interfere with fishing, boating, and swimming when dense and are often considered visually unattractive by property owners, prompting demands for their removal (Sec. 24.11). Dense beds of submerged macrophytes in lotic systems impair discharge and thereby increase the potential for flooding and interfere with boat transportation.

Among free-floating macrophytes, the small duckweeds (e.g., *Lemna spp.*) are probably the best known.

They cover the surface of nutrient-rich wind-protected ponds and ditches in temperate and tropical regions. More important from a management perspective are larger species, such as the water hyacinth (*Eichhornia crassipes*) (Fig. 24–1a), the nile cabbage (*Pistia stratiotes*) (Fig. 24–1b), and aquatic ferns (e.g., *Salvinia molesta*) which sometimes cover part or even the whole surface of nutrient-rich lakes and reservoirs (See Fig. 15–1), slowly flowing rivers, or wetlands at lower latitudes. They interfere with water transport and reduce the underwater light climate, negatively affecting the phytoplankton and submerged macrophytes. Free-floating macrophyte species compete with phytoplankton for nutrients because their roots are suspended in the water; and only manage this successfully in nutrient-rich waters sufficiently protected from the wind to prevent the plants from being washed ashore.

24.2   Wetlands and Their Utilization

Wetlands are waterlogged landscapes that cover about 8.6 million km^2 or 6.4 percent of Earth's land area (Mitsch and Gosselink 1993); this is about three times the area of lakes. The greatest extent of wetlands is found in the boreal forest regions of North America and Eurasia, and around the equator (see Fig. 4–1).

A definition of wetlands that satisfies everyone has not been developed because the suitability of a definition depends on the goal and the field of interest of its creators. Some focus on soil development, others on aspects of the hydrology, and yet others emphasize the vegetation or biota in general. Here **wetlands** are defined as a place sufficiently saturated with water to enable plants characteristic of wet soils to grow (P. E. Greeson et al. 1978, in Howard-Williams 1983). Wetlands and lakes form part of a continuum, with the deeper portions of large wetlands considered as lakes with a large littoral zone.

There is no universal classification scheme for wetlands. The available schemes were created for particular purposes and address higher latitude wetlands characterized by their particular climatic, hydrological, geological, and vegetation conditions (Morant 1983). Selective terms and types of wetlands derived for the north temperate zone are presented in Table 24–1. Workable definitions of wetlands and wetland types are important to wetland scientists and are equally important to assist managers and regulators in wetland management. They need clear and legally binding definitions to prevent or control the draining of wetlands for agriculture or urban development, as well as preventing the large-scale cutting of forested wetlands. The channelization (straightening) of winding lowland rivers separates rivers from the wetlands on their floodplains, encouraging drainage for agriculture (Sec. 4.3). The conversion of deepwater wetlands in southern and southeast Asia to a monoculture of rice, and the conversion of wetlands (shallow lakes) for aquaculture are the major threats to wetlands in that area. The value of wetlands for flood control, water storage, and water purification has been estimated at US ~\$15,000 ha^{-1} yr^{-1} (Sec. 1.1). This does not even consider their value as a habitat for fish and wildlife, recreation, or maintaining biodiversity. In less economically developed countries at low latitudes, wetlands are the home of millions of people dependent on them for their livelihood and for maintaining their culture.[1]

▲ About half the wetland area in the contiguous United States has disappeared since 1780, mostly because they were drained for agricultural purposes. The impact has been greatest in the lower Mississippi valley and the prairie pothole (slough) region—a major waterfowl breeding area in the northern portions of the prairie region of the United States and adjacent Canada (Mitsch and Gosselink 1993). In some states such as California, Ohio, and Iowa, less than 10 percent of the original wetlands remain; virtually all freshwater wetlands have been lost near major urban centers worldwide. The losses are even greater in western Europe where wetlands have been drained since the Middle Ages. Of those remaining in the Netherlands and Germany in 1950, more than half were drained by 1980 (Gleick 1993). Not only the wetlands are lost, the birds, mammals, and fish disappear with them.

Population growth in South and East Asia stimulates dam building and draining or converting wetlands for agricultural or aquacultural purposes. By 1988 about half the original wetland area in a number of southeast Asian countries had been drained or greatly modified (Gleick 1993)—in populated Java (ID) only some 6 percent remained—and conversions have probably been accelerating. The diversion of in-

---

[1]"The principle forces driving government policies such as those on land use, on agricultural subsidies and on water pricing, are economic. There is therefore an urgent need to effect detailed analyses which can demonstrate the multiple public benefits of wetland conservation, as opposed to the public cost and more limited private benefits of wetland conversion." P. J. Dugan 1994.

Table 24–1 **Selected wetland terms and types.**

Bog	Peat-accumulating wetland dominated by mosses. Bogwater is acidic (typical pH ~3–5), has a low or negative alkalinity, and is nutrient poor.
Fen	Peat-accumulating wetland receiving some higher alkalinity groundwater from a mineral substrate. As a result the water has a higher nutrient content, a circumneutral pH (6–7), and supports a marsh-like vegetation. Usually dominated by sedges.
Marsh	Permanently or periodically inundated wetland characterized by nutrient-rich water and subject to seasonal fluctuations. The emergent vegetation is adapted to saturated soils and submergent macrophytes are present in deeper water. In European terminology, a marsh must have a mineral substrate and does not accumulate peat.
Mire	European, synonymous with any peat-accumulating wetland (bog, fen).
Playa	Shallow ephemeral ponds or lagoons in semiarid regions exhibiting appreciable seasonal changes in water level and an elevated salinity.
Slough	Marsh or shallow lake system in the northern prairie regions of the United States and adjacent Canada.
Swamp	Wetlands dominated by trees and shrubs (North America), a forested fen or reed-dominated wetland (Europe), and a tree- or reed-dominated wetland (Africa). Temperate zone swamps are fed by nutrient-rich groundwater from primarily mineral sediments.
Wet meadow	Grassland or savannah with waterlogged soil near the surface but without standing water for most of the year.
Open water	Deeper portions of wetlands and the shallow-water zone (littoral zone) of lakes and rivers, typically inhabited by submerged macrophytes.

Source: Largely after Mitsch and Gosselink 1993.

flowing stream water for irrigation purposes or the extensive use of wetlands for livestock grazing or fodder (hay) are less overt ways of destroying or degrading wetlands. Even so, vast and largely undisturbed freshwater wetland areas remain, especially in the sparsely populated regions of the boreal forest region of Eurasia and North America, as well as in the internal and coastal deltas of South American, African, and New Guinean river basins (Table 6–3).

Wetland Rehabilitation and Wastewater Treatment

Legislation is now in place in several countries insisting or recommending that there be no additional net loss of wetlands and encouraging the restoration (*rehabilitation, reconstruction*) of degraded wetlands—or even the creation of new ones (*mitigation wetlands*, US) (Mitsch and Gosselink 1993). However, little attention is being given to recreating the appropriate hydrological conditions and establishing the appropriate plant communities (Bedford 1996). Furthermore, the success of wetland rehabilitation and creation is rarely evaluated, but where this is done the results have been variable, particularly when the hydrologic regime has

been changed. In any case, the degree to which restored wetlands resemble the original is largely unknown. Restoration of an early successional marsh or reed swamp is obviously easier than recreating a forested wetland. Peatlands were established over thousands of years and cannot be returned to their original state at all (Gorham 1996).

Better recognition of the importance of wetlands and growing interest on the part of ecologists, biogeochemists, and legislators has spurred rapid developments in wetland ecology over recent decades. Biogeochemists have become interested in wetlands and their local and global role as a source and sink in carbon, nitrogen, and sulfur cycling (Gorham 1996). The role of wetlands in the transformation of mercury to a more toxic form has encouraged important research in boreal forest wetlands and their associated lakes (Sec. 28.9, and St Louis et al. 1994).

There is growing interest in using natural and artificial wetlands for treating wastewater effluents. Construction costs vary widely but construction and maintenance costs compare favorably with equivalent costs for more traditional wastewater treatment facilities (Mitsch and Gosselink 1993). Whether treating the wastewater in a wetland is appropriate depends partly on the climate, partly on the volume of water to be treated, and partly on the size and depth of available wetlands in the area. A host of factors influence nutrient retention. Among them are the wastewater loading rate ($m^3 \ m^{-2} \ yr^{-1}$) and seasonally changing water residence time or 'flow through.' Retention is further influenced by the pH and sorptive capacity of the soils (Sec. 17.6) and the condition of the vegetation. Finally, temperature has an effect on waste retention through its effect on the metabolism of microbes, macrophytes, and periphyton.

Effective wetlands for wastewater treatment must allow large quantities of nitrogen to be lost through denitrification, with a release of N_2 and N_2O to the atmosphere (Secs. 8.4 and 18.4). Effective wetlands should also have a high long-term sorptive capacity for phosphorus (Sec. 17.6). Whereas nitrogen retention (retention = input − stream output) is often high (Fig. 18–4), long-term phosphorus retention is typically lower and quite variable and dependent on the amount of phosphorus sorbing extractable aluminium and iron in the sediments (Sec. 17.2).

The remainder of this chapter discusses the emergent and submerged macrophytes based largely on the well-developed lake literature. However, the information is equally relevant to aquatic plants in lotic systems and wetlands.

24.3 Macrophyte Distribution and Species Richness

Emergent macrophytes form a fringe around many lakes, border most lowland rivers, and dominate wetlands. They grow at sites where the underwater slope is shallow and they are protected from wave-induced turbulence, allowing plants to root and preventing uprooting. Drainage basins in well-watered regions with slopes low enough to experience impeded (slowed) drainage typically have extensive wetlands surrounding the local lakes and rivers, making it difficult to decide where a particular wetland ends and the lake or river begins. This is an even greater difficulty at low latitudes where large seasonal and interannual differences in rainfall and runoff bring about large changes in the size of the wetlands (Fig. 24–2). The littoral zone can therefore be seen as an interface or transition zone (ecotone, Sec. 8.3) between the drainage basin and the open water.

The emergent community of the freshwater littoral zone is stereotypically dominated by the *Phrag-*

Figure 24–2 A seasonal swamp, showing emergent and floating level macrophytes, in the Okavango wetlands, an internal delta (max. 10,000 km², z̄ = ~1m) in Botswana, southern Africa, where ~96 percent of the seasonally inflowing river water is lost by evapotranspiration. *(Photo by J. Thorsell/Courtesy of Ramsar.)*

mites reedbeds (Fig. 24–1h) of Eurasia and parts of Africa, the cattail (*Typha spp.*) (Fig. 24–1g) and bullrush (*Scirpus spp.*) marshes of temperate zone North America, and the papyrus (*Cyperus papyrus*) swamps (Fig. 24–1i) of tropical Africa and western Asia. The vegetation is less characteristically dominated by one or two species in South America, and South and East Asia. Emergent macrophytes are lacking only at high latitudes where a short, cold growing season, including frost during the growing season and probable ice damage to the roots, makes conditions unsuitable.

Both the emergent and submerged community of lakes and wetlands in polar and boreal forest regions are dominated by mosses, but submerged angiosperms with surface floating leaves (e.g., pondweeds, *Potamogeton spp.*, Fig. 24–3a) are typical in the humic (high light extinction) lakes and wetlands of the boreal zone. Submerged macrophytes dominate when the transparency is high. Aquatic systems lack an emergent community altogether where the land–water interface is unsuitably steep.

Species Richness

The number of macrophyte species typically increases with lake size (Fig. 24–4), probably because larger lakes not only have a larger littoral zone but also because they tend to have a greater variety of habitats as a result of more variation in depth, exposure, sediment type, and underwater slope. Other variables shown to

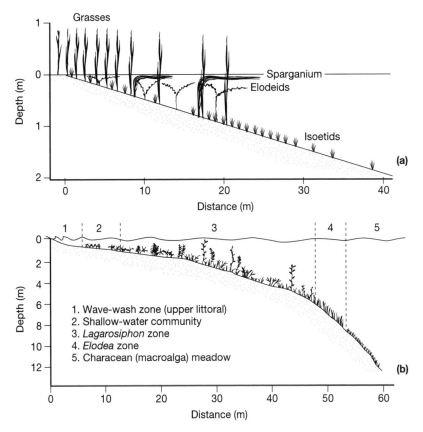

Figure 24–3 Emergent and submerged macrophytes in (a) Lake Suomunjärvi (LA = 6.4 km^2), a brown-water Finnish lake and (b) Lake Taupo (LA = 616 km^2), a transparent New Zealand lake. Note the very different maximum depth of colonization, growth forms, underwater slopes and near-shore turbulence that allows macrophyte growth up to and above the water line in the Finnish lake but not in the larger and more wind/wave swept shore of Lake Taupo. *(Part a from Toivonen and Lappalainen 1980; part b modified from Howard-Williams and Vincent 1983.)*

have a positive or negative influence on species richness, shown by work on Scandinavian lakes and wetlands, are altitude and latitude (–), abrupt water level variations (–), alkalinity (+), pH (+), turbidity (–), and salinity (–). Under severe climate conditions (≥ 7 month ice cover) angiosperms are replaced by moss species. In any lake most species are present only in shallow water, but some extend to a considerable depth (Nichols 1992). The dozen or so (range 0–45) submerged species noted in the average Scandinavian lake are but a small fraction of the 339 taxa (192 angiosperms, 129 bryophytes, and 18 charophytes) recorded in the region (Rørslett 1991). A total of some 1,000 submerged freshwater and estuarine species have been recorded worldwide (Cook 1996).

▲ The relationship between pH and species abundance is weak. It is confounded by system size because large lakes and wetlands typically have more species (Fig. 24–4). Similarly, elevated salinity does not always reduce species abundance because the response is species specific and further modified by species interactions. Experimental work on emergent and submerged freshwater macrophytes growing under

brackish conditions in coastal zones has shown that seed germination and seedling survival decline when the salinity (TDS) rises above about 16‰ (~25,000 μS cm^{-1}) (Ungar 1974). Above this concentration ionic and osmotic adjustments to rising salinity appear to become increasingly difficult. Species adapted to low salinity are gradually displaced at higher salinities in coastal waters by marine species, and by a variety of high-salinity adapted species, known as *halophytes*, in inland saline lakes and wetlands. However some macroalgae (Characeae) and angiosperms, such as *Ruppia spp.* and the closely related *Potamogeton pectinatus* (widgeon grass, Fig. 24–1c), are found both in low salinity fresh waters and waters with a salinity of >> 50‰ (Hammer 1986), indicating that the disappearance of freshwater species need not be a direct physiological effect of elevated salinity.[2]

---

[2] "As salinity increases rooted macrophytes become fewer in species and eventually disappear. At high salinities benthic algae may occur but in many hypersaline lakes there is, in effect, no littoral zone in the normal sense in spite of adequate light and the whole lake is pelagic." Hammer 1986.

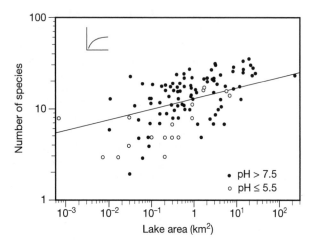

Figure 24–4 Observed submerged species richness as a function of lake area (km²) and pH, a measure of alkalinity. The solid line indicates the predicted species richness for Scandinavian and Finnish lakes. Acidic lakes (median pH 5.5) are shown by open circles. Lakes having a median pH ≥ 7.5 are plotted with solid circles. Note (1) most of the acidic low alkalinity lakes shown are small and the larger ones did not exhibit a particularly impoverished flora compared to the alkaline lakes; and (2) the effect of lake size is most evident only over a large range scale. Other unmeasured determinants (e.g., turbidity, mean depth) dominate among lakes varying less than about tenfold in lake area. *(Modified from Rørslett 1991.)*

▲ 24.4 Macrophyte Biomass and Its Determinants

Macrophyte biomass or community structure is most commonly examined at the scale of individual quadrants in selected shallow water weed beds or along depth transects perpendicular to shore. The information obtained on species composition, community structure, biomass, or productivity of the small scale quadrats is normally expressed per m². Stevenson (1988), in a review based on quadrat studies, reported a peak underwater biomass of 91–529 g dry wt. m⁻² for selected temperate lake systems. Data collected along transects are occasionally extrapolated (scaled up) from the inevitably small area examined to the whole of the littoral zone (g m⁻² of littoral) for a comparison of, for example, production per unit area of the macrophyte and phytoplankton communities. One such comparison (Table 24–2) further demonstrates how conclusions change with the spatial scale selected. Less common are studies that explore the influence

and importance of macrophytes biomass production at the scale of whole systems (tonnes lake⁻¹ or gm⁻² of lake surface yr⁻¹). We will begin this discussion at the whole-lake scale before addressing the much more abundant findings at more site-specific spatial scales within and among inland waters.

The area covered by macrophytes (m² lake⁻¹) and the total macrophytes biomass (tonnes lake⁻¹) represent somewhat different aspects of how macrophytes influence aquatic systems. The total biomass particularly affects primary productivity, nutrient dynamics, and the oxygen balance of a system, whereas the area covered by the plants has its greatest impact on the distribution of the periphyton, the invertebrates, and the abundance, distribution, and species composition of fish. Together, whole-system biomass and the area covered by plants determine how rapidly wind-induced turbulence is reduced in the littoral zone. The magnitude of the reduction in turbulence in turn affects particle sedimentation rates and the fluxes of gases, nutrients, contaminants, and organic matter into and out of littoral zones (Secs. 12.7 and 20.2).

The most extensive comparative analysis of emergent and submerged macrophytes among systems to date is an analysis of 139 primarily temperate zone lakes of generally low water color (Duarte et al. 1986). The study identified lake morphometry as the single best predictor of the area covered (ha) and the biomass (tonne dry wt.) of *emergent* macrophytes among lakes. On average, emergent plant biomass and cover increase linearly with lake size and decline as the littoral zone becomes steeper (Eqs. 1–6 in Table 24–3). About seven percent of the lake area turns out to be covered by emergent macrophytes, regardless of lake size. In contrast, both the area covered and the whole-system biomass of *submerged* macrophytes declines proportionately as lake size increases (Eqs. 7–12 in Table 24–3). Therefore, the relative importance of submerged macrophytes on a whole-lake-system scale declines (Sec. 24.10). The same study showed lake area and incident irradiance to be the best predictors of both submerged macrophyte cover and biomass. It is not surprising that larger lakes typically have an absolutely larger total biomass. But the *rate* of biomass increase also begins to decline with increasing surface area because large lakes tend to be deeper as well (Table 9–1) and a progressively a smaller fraction of the area receives sufficient sediment illumination to support plant growth (Eq. 10 in Table 24–3). The im-

Table 24–2 **Primary production by various components of Swartvlei Lake (ZA). Values in brackets are subtotals. Note (1) the epibenthic periphyton was not examined; and (2) a particular rate of primary production in the littoral zone need not be reflected in a similar role when the lake as a whole is considered, an issue of scale.**

	Net Production g C m^{-2} yr^{-1} (littoral)	Area Occupied % (whole lake)	Net Production g C m^{-2} yr^{-1} (whole lake)	% of Net Production (whole lake)
Potamogeton pectinatus (subtotal)	910	7.8	71.8	(52.4)
Characeae (subtotal)	970	19.7	19.2	(14.0)
Submerged macrophytes (total)	1,880	27.5	91.0	66.4
Emergent macrophytes (*Phragmites/Scirpus*)	690	2.4	16.5	12.0
Epiphytic periphyton (*Cladophora*)	130	16.4	10.5	7.6
Phytoplankton	27	69.9	19.0	13.8

Source: Modified after Howard-Williams and Allanson 1981.

pact of macrophytes, also tends to decline as the salinity rises (Eqs. 2, 5, and 12 in Table 24–3). It is therefore evident that submerged plants exert their greatest influence on small transparent freshwater lakes, with their greatest impact at lower latitudes where the annual irradiance received at the water surface is higher (Fig. 5–10). Small wonder that such lakes have been favored for the study of submerged macrophyte communities (see Sec. 24.10).

Table 24–3 **Models relating the total area colonized by emergent (A_e) and submerged (A_s) macrophytes, the percent area colonized (%A_e and %A_s), and the total emerged (B_e) and submerged (B_s) biomass as a function of environmental factors. A = lake area (ha), $I_{\bar{z}}$ = underwater irradiance (at mean depth), S = salinity indicated by a dummy variable of 1 or 0 for a conductivity <> 1,000 μS cm^{-1}, z_{max} = maximum depth, and L_{ic} = length of ice cover (days).**

Equation	r^2	n	P	SE est.
(1) $\ln(A_e) = 1.0 \cdot \ln(A) - 2.6$	0.87	60	< 0.0001	1.33
(2) $\ln(A_e) = 0.72 \cdot \ln(A) - 0.69 \cdot \ln(S) + 0.72$	0.90	60	< 0.0001	1.28
(3) %$A_e = 2.81 \cdot \ln(A) - 0.21(I_z) - 6.58 \cdot \ln(z_{max})$	0.41	53	< 0.0001	11.59
(4) $\ln(B_e) = 1.1 \cdot \ln(A) - 1.4$	0.70	36	< 0.0001	2.53
(5) $\ln(B_e) = 0.57 \cdot \ln(A) - 1.34(S) + 4.77$	0.76	32	< 0.0001	2.47
(6) $\ln(B_e) = 1.17 \cdot \ln(A_e) + 0.89$	0.91	28	< 0.0001	1.45
(7) $\ln(A_s) = 0.91 + \ln(A) - 1.1$	0.80	76	< 0.0001	1.47
(8) $\ln(A_s) = 0.94 \cdot \ln(A) + 0.85 \cdot \ln(I_z) - 3.7$	0.89	51	< 0.0001	1.05
(9) $\ln(B_s) = 0.89 \cdot \ln(A) - 1.42$	0.59	70	< 0.0001	3.04
(10) $\ln(B_s) = 0.95 \cdot \ln(A) + 1.12 \cdot \ln(I_z) - 4.6$	0.66	56	< 0.0001	2.14
(11) $\ln(B_s) = 0.99 \cdot \ln(A_s) + 0.37$	0.84	51	< 0.0001	1.26
(12) %$A_s = 1.4(I_z) + 0.07(L_{ic}) - 24(S) - 0.90$	0.60	55	< 0.0001	11.20

Source: After Duarte et al. 1986.

Biomass and Production

Most biomass and production data (Table 24–4) in the literature address only above-ground components, thereby excluding roots or *rhizomes* (root tubers). Reported values are therefore usually underestimated by 15–25 percent or more. The shoot:root biomass ratio varies greatly among species and growth forms; bottom-dwelling angiosperms have the lowest ratio (~1–3) and taller erect and canopy-producing angiosperm species generate the highest ratios (2–7). Angiosperms growing in nutrient-poor environments have disproportionately large root systems and a reduced shoot:root ratio (Sand-Jensen and Søndergaard 1979). Charophytes and bryophytes have little rhizoids, whereas the angiosperm *Ceratophyllum demersum* (coontail) lacks holdfast structures altogether (Fig. 24–1d). (Stevenson 1988, and Middelboe and Markager 1997).

The difficulty of sampling roots means that the precision of root measurements is particularly low.

Table 24–4 **Maximum production (g dry wt. m^{-2} yr^{-1}) of aquatic plant communities (upper 10% of values) estimated with a variety of techniques. g m^{-2} yr × 10^{-2} = tonne ha^{-1} yr^{-1}; 1 tonne = 1,000 kg. For differences between C$_3$ and C$_4$ plants consult a plant physiology text.**

Plant Type	g dry wt. m^{-2} yr^{-1}
Submerged macrophytes, temperate fresh waters	500–1,000
Submerged macrophytes, tropical fresh waters	2,000
Phytoplankton	1,500–3,000
Terrestrial plants (forests, pastures, crops)	2,000–8,500
Emergent floating C$_3$ macrophytes (subtropical fresh waters, *Eichhornia crassipes*)	4,000–6,000
Emergent rooted C$_3$ macrophytes, fresh waters, esp. *Phragmites australis, Typha spp.*	5,000–7,000
Emergent rooted C$_4$ macrophytes, tropical fresh waters	6,000–9,000

Source: Modified after Westlake 1982.

However, even the measurement precision of the above-ground components greatly varies depending on the particular sampling technique used, the plant biomass (Table 24–5), and the inevitable assumptions that must be made in the computations of production from changes in biomass (Table 24–6). Most production calculations are based only on seasonal changes in the above-ground standing crop (biomass) and do not consider losses experienced during the growing season. Where losses are considered, they are often obtained by selecting a generic loss estimate from the literature. Other seasonal and longer-term studies of macrophytes (invertebrates and fish) most often compute the production on the basis of biomass estimates multiplied by a production:biomass (P:B) ratio, also derived from the literature. Fortunately, such ratios apparently span a narrow range for submerged angiosperms (1.2–1.5) and macroalgae (1.7–1.9) (Howard Williams et al. 1986). In studies carried out over short time scales (hours), researchers typically attempt to obtain direct measurements of underwater production by placing whole plants or sprigs in containers, incubating them for several hours, and measuring the in situ photosynthetic rates (carbon uptake or oxygen evolution).

It is evident, despite imprecisions in biomass and production determinations, that under optimal conditions the associated community biomass (g m^{-2}) and maximum productivity (g C m^{-2} yr^{-1}) of temperate zone *emergent* macrophytes is higher than for their equivalent *submerged* counterparts (Table 24–4). This

Table 24–5 **Number of replicate samples needed for various sampler sizes and levels of aquatic macrophyte biomass in order that the standard error (SE) of replicate samples average 20 percent of the mean standing biomass.**

Sampler Size (cm^2)	Macrophyte Biomass (g dry wt. m^{-2})				
	3.2	10	32	100	320
100	42	26	16	9	6
316	35	21	13	8	5
1,000	29	18	11	7	4
3,162	25	15	9	6	3
10,000	21	12	8	5	3

Source: After Downing and Anderson 1985.

Table 24–6   **Examples of estimates of the annual primary production of (A) the emergent macrophyte *Phragmites australis* in western and central Europe employing some of the different assumptions commonly made, and their effect on the production determined; and (B) the emergent macrophyte of *Typha* (cattail) in Lake Chilwa, with the standard error. Biomass values are means from fifteen stands.**

(A) Method	Annual Primary Production (g dry wt. m^{-2} yr^{-1})	Relative to Best Estimate (%)
Maximum shoot biomass	1,143	0.55
Maximum shoot biomass + estimate of leaves lost	1,160	0.56
Maximum biomass – minimum biomass	1,600	0.77
Maximum shoot biomass + (maximum underground root, age 4 yrs)	1,718	0.82
Turnover estimate based on regularly measured changes in biomass	2,085	1.00
Maximum shoot biomass + leaves lost + (maximum – minimum root biomass)	2,235	1.07

(B) Method	Biomass (g dry wt. m^{-2}) and Annual Production (g dry wt. m^{-2} yr^{-1}
Maximum shoot biomass	2,537 ± 87
Minimum shoot biomass	−1,122 ± 170
Annual increment	1,415 (turnover value 0.558)
Estimated leaves lost	+ 165
Total production	1,580
Mean ash content	− 129 (8.2%)
Total organic production	1,451
Total littoral swamp production over 552 km^2 (tonnes yr^{-1})	8 × 10^5

Source: (A) Modified from Westlake 1982; (B) from Howard-Williams and Lenton 1975.

is due to the higher illumination of emergent species compared to submerged plants, and slow diffusion rates of dissolved inorganic carbon from the water into the plants. But emergent macrophytes at lower latitudes are not necessarily more productive on an annual basis. They commonly experience substantial seasonal or interannual variation in water level that leave the emergent plants stranded along desiccated shorelines for periods of months or even years. However, low latitude submerged macrophyte communities in systems subject to little water level variation do not experience seasonal die-back which, together with higher, more evenly spaced irradiance (Fig. 5–10) and higher temperatures, allows a much greater biomass and annual production than is possible in the temperate zone.

24.5 Submerged Macrophyte Distribution: Light and Lake Morphometry

▲

The effect of light flux on macrophyte distribution is easy to demonstrate and was first noted in Finland some 60 years ago by L. Maristo (1941). Both the greatest average depth of colonization and the average water depth (along shore to deep-water transects) where the biomass is maximal are a function of transparency (Fig. 24–5).

Angiosperms have a higher light-compensation level for sustained survival than macroalgae (Characeae), and angiosperms disappear from oligotrophic north temperate waters at depths where

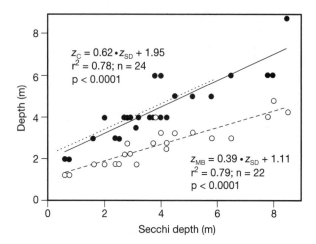

Figure 24–5 Relationships between maximum depth of colonization (z_c) and the depth of maximum biomass (z_{MB}) of submerged macrophytes and water transparency, in 25 Québec, Vermont, and New York lakes. The fine dashed line represents the average z_c for Wisconsin (US) lakes (Nicholls et al. 1990). *(Modified after Duarte and Kalff 1990.)*

usually less than about 1,800 J cm² (or 21% of incident PAR) is received over the growing season. Macroalgae (Characeae) survive and grow at greater depths where at least 1,200 J cm⁻² (or 10% of incident PAR) are received over the growing season (Chambers and Kalff 1985). But in more turbid European waters with little or no winter ice cover and associated light extinction, macrophytes are found down to about two percent of PAR (Middelboe and Markager 1997). The most deeply growing macrophytes in highly transparent lakes are true mosses (bryophytes). Growth form also dictates the minimum light requirement of aquatic plants (Middelboe and Markager 1997). Independent of the light climate, macrophytes are absent where the light flux is ample but the underwater slope is too steep.

The maximum depth where angiosperms can grow is determined primarily by irradiance rather than by hydrostatic pressure, as was once believed. The maximum ranges from a few centimeters along highly turbid wind-protected shorelines to about 18 m in highly transparent fresh waters. Angiosperms are found to 50 m in yet more transparent marine habitats (Middlboe and Markager 1997). Macroalgae (Characeae) and mosses (Bryophyta) extend much deeper because they are more tolerant of shade; in highly transparent monomictic Lake Tahoe (California, US) the macroalgae extended to as much as 60 m and the mosses to

153 m in the 1960s, with secchi transparences of ~30 m (Frantz and Cordone 1967).

The minimum depth of colonization by submerged macrophytes increases as turbulence (exposure) increases (Chambers 1987) and, at higher latitudes, seems to be linked to ice thickness and ice scouring of the rooting-zone sediment (Welch and Kalff 1974). Seasonal changes in water level and wave exposure are the principal determinants of how close to shore even short species of submerged macrophytes can grow at low latitudes and in reservoirs everywhere.

Underwater Slope and Turbidity

Turbulence, rather than irradiance, is the primary determinant of biomass in relatively transparent mid and low latitude streams and rivers, with the biomass declining at high water discharge and water depth (Chambers et al. 1991). Whereas the maximum depth of colonization and depth of maximum biomass in lakes are roughly predictable from the transparency

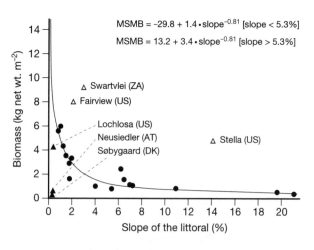

Figure 24–6 The relationship between underwater slope and the maximum submerged macrophyte biomass (MSMB) encountered along transects perpendicular to the shore. This model was developed in relatively transparent (z_{SD} = 4m) Lake Memphremagog (CA, US) which is characterized by a wide variety of slopes. The model was subsequently used to examine its predictive power in other temperate lakes in Europe and North America (●), in subtropical lakes (△), and in highly turbid lakes (▲). Note (1) the good fit of the relatively transparent temperate zone lakes to the model; (2) a lower than predicted biomass in turbid lakes; and (3) a higher than predicted biomass at lower latitudes characterized by a longer or year-round growing season. *(Modified after Duarte and Kalff 1986.)*

(Fig. 24–5), the biomass itself (g m^{-2}) is not a function of transparency but rather of the underwater slope at depths where irradiance is not the primary limiting factor. There the rooted biomass is greatest where the site-specific slope is lowest (< ~5%) (Fig. 24–6). Rooted plants are unable to grow where the slope is greater than about 15–20 percent (Duarte and Kalff 1990) regardless of the transparency, presumably as a result of sediment slumping (erosion), greater direct wave impact, poor rooting conditions, and no long-term accumulation of nutrient-rich fine sediment. The importance of underwater slope in predicting maximum biomass is not lake specific because the predictive models (Fig. 24–6) developed for a single lake could predict the maximum biomass in other temperate zone lakes. However, the same models are poor predictors of maximum biomass in turbid or humic lakes (z_{SD} depth < 2 m) where the actual biomass is lower than predicted, and in low latitude lakes where the maximum biomass observed is much higher (Fig. 24–6).

Latitude

Low latitude lakes typically receive greater annual irradiance. Plants there have a year-long growing season and experience higher water temperatures. The resulting good growing conditions—in systems not suffering from large water level variations—explain not only the larger maximum biomass at a particular underwater slope but also the greater depth to which rooted macrophytes can grow (Fig. 24–7). The greater maximum biomass and greater depth of colonization at low latitudes implies that submerged macrophytes can be expected to play a more important role in aquatic systems than at higher latitudes. Therefore, it is not really surprising that public demands for macrophyte control in the United States come primarily from southern rather than northern states (Sec. 24.11). A second reason is that southern (US) lakes are located on ancient well-eroded drainage basins (Sec. 5.5) and are often shallow in comparison with their glaciated counterparts which also experience a shorter growing season.

The seasonal or annual light flux a lake receives is a major determinant of both the maximum depth of macrophyte colonization and the depth at which the maximum biomass is found (Figs. 24–5 and 24–7). Consequently, the maximum depth of macrophyte colonization declines among lakes, and the deeper portions of wetlands, as the turbidity increases—

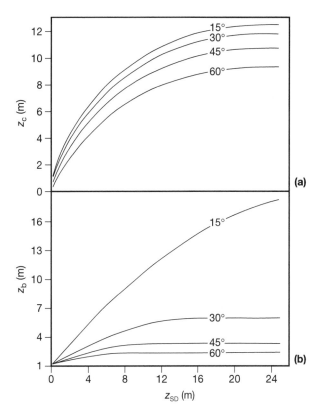

Figure 24–7   Predicted changes in (a) the maximum depth of colonization (z_c), and (b) the depth of the maximum biomass of macrophytes (z_b) as a function of the transparency (z_{SD}) and latitude. Note that the models predict a systematic increase in z_c and z_b with declining latitude at any particular transparency. *(After Duarte and Kalff 1987.)*

caused by high phytoplankton biomass or its surrogate, water nutrient concentration. In shallow Danish lakes with > 150 μg P l^{-1} and a transparency of $z_{SD} \leq 1$ m, an appreciable submerged macrophyte cover is typically lacking (Fig. 24–8).

The great sensitivity of macrophytes to changes in irradiance, is most evident from work on individual lakes. Examples include long-term reductions in the depth of colonization in response to increased turbidity (light extinction) (Sec. 24.10 and Ozimek and Kowalczewski 1984), and the invasion and rapid expansion of macrophytes in a biomanipulated lake following a sharp decline in algal turbidity upon a reduction in zooplanktivorous fish (van Donk et al. 1990). The biomanipulation enabled an abundant community of herbivorous macrozooplankton to greatly reduce the algal biomass (Sec. 23.7).

Over several years, changes in water level and sediment suspension (turbidity) can have an important

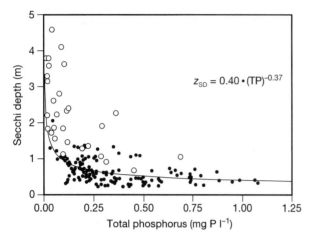

Figure 24–8 Secchi depth in relation to lake water total phosphorus in shallow Danish freshwater lakes. Open circles represent lakes with a substantial submerged macrophyte cover (> 30%); filled circles represent lakes where the macrophyte cover is modest (< 30%) or unknown. Each point represents one lake and is a time-weighted average of all data collected between May and October. Note (1) the exceptional nutrient richness of many of the lakes; (2) relatively few of the lakes with > ~100 μg l[1] and a transparency lower than ~1 m have appreciable macrophyte covers; (3) nutrient rich lakes with an extensive macrophyte cover have an elevated transparency for a given TP concentration; (4) clear or turbid water can exist at the same phosphorus concentration over a wide range of phosphorus concentrations; and (5) the pattern among the shallow lakes is not confounded by much variation in underwater slope. *(Modified from P. Kristensen et al. 1988, in Jeppesen et al. 1990.)*

impact on macrophyte distribution (Fig. 24–9). Over the short term, increases in light extinction in a New Zealand lake, resulting from river sediment inputs, reduced the maximum depth of colonization from 6.7 m to 1.8 m. It was five months before sedimentation and flushing of the suspended particles allowed recovery to

commence (Johnstone and Robinson 1987). This particular recovery was quick because the plants regrew from storage organs (*rhizomes* or *turions*) that had survived in the sediments; this is not possible where conditions have been unfavorable for years and the plants must invade again.

24.6   Submerged Macrophyte Distributions and Plant Nutrients

The role of nutrients in determining macrophyte growth became an issue of scientific and management interest in the late 1970s when the importance of phosphorus, and sometimes nitrogen, in determining phytoplankton biomass and production had been convincingly demonstrated. Even today there are relatively few well-substantiated reports of inorganic nutrient limitation of submerged macrophyte growth in fresh water (Barko et al. 1986, Anderson and Kalff 1986), and there seem to be no such reports for emergent macrophytes. This paucity stands in contrast to the common occurrence of phytoplankton nutrient limitation in nature (Secs. 17.6 and 21.12). There are several reasons for the discrepancy. The phytoplankton biomass, which is transported through the mixed layer, is primarily constrained during the growing season by nutrients rather than light in transparent oligotrophic lakes. This is evident from pronounced increases in phytoplankton production and algal biomass (Fig. 8–16) after the experimental addition of fertilizer or wastewater to previously oligotrophic lakes.

Not only are macrophytes, with their roots in relatively nutrient-rich sediments, less likely to be nutrient-limited than the phytoplankton but the relatively large and slow-growing macrophytes (Fig. 21–12 and Table 24–7) also have much lower nitrogen and phosphorus requirement per unit carbon (biomass) fixed.

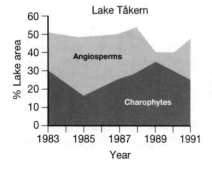

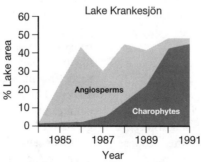

Figure 24–9 Percent of lake area covered by submerged macrophytes in two Swedish lakes, Tåkern (area outside reed beds = 31.3 km²; $\bar{z}$ = 1.0 m) and Lake Krankesjön (area outside reed beds = 2.9 km²; $\bar{z}$ = 1.5 m). *(Modified after Blindow 1992.)*

Table 24–7 **In situ mean growth rates and ranges of phytoplankton communities, benthic macroalgae, and selected species of slowly and more rapidly growing rooted macrophytes in temperate habitats during the growing season (April through October).**

Community and Species	Growth rate (day^{-1})	
	Mean	Range
Phytoplankton		
Lake Windermere, (GB)		0.02–0.39
Lake Castle, (US)	0.33	0.07–0.65
Lake Erken, SE	0.36	0.08–1.15
Lake Wingra, US	0.86	0.21–1.76
Lake Bysjön, SE	1.01	0.30–1.78
Periphyton		
Cladophora spp.	0.16	0.03–0.27
Rooted macrophytes		
Ranunculus peltatus	0.10	0.07–0.12
Potamogeton pectinatus	0.09	0.04–0.17
Potamogeton crispus	0.07	0.05–0.09
Sparganium emersum	0.03	0.02–0.06
Littorella uniflora	0.006	
Lobelia dortmanna	0.004	

Source: After Sand-Jensen and Borum 1991.

Macrophytes (mean tissue composition ~500C:24N:1P by atoms) need larger quantities of structural carbon to support stems and branches than the much smaller, more rapidly growing phytoplankton (~110C:16N:1P) (Sec. 16.1 and Duarte 1992). Consequently, a nutrient limitation of growth and community biomass is less likely, or less severe, for sediment-bound macrophytes than for phytoplankton (Sand-Jensen and Borum 1991).

It is now widely recognized that submerged macrophytes can take up inorganic nutrients from both the sediments in which they are rooted and via the leaves from the water column above, but the rooted macrophytes typically obtain most of their phosphorus and nitrogen from the sediments. Explanations include generally much higher interstitial nutrient levels to be tapped, and in oligotrophic takes, a poor competitive ability for water column nutrients with the phytoplankton and bacteria characterized by a much larger surface area to volume ratio (Fig. 21–12). Therefore it is not surprising that rooted macrophytes, particularly in oligotrophic lakes, obtain most or all of their phosphorus and nitrogen from the

relatively nutrient-rich sediments (Best and Mantai 1978, Denny 1980, and Carignan 1982). For the same reason, the free-floating macrophytes with their roots dangling in the water and the submerged but not rooted *Ceratophyllum demersum* (Fig. 24–1d) are only abundant in eutrophic systems where the water nutrient levels are high enough to exceed the nutrient demands of the kinetically more efficient bacteria and phytoplankton. A significant inorganic nutrient limitation is equally unlikely in lowland rivers rich in nutrients and CO_2 (Sand-Jensen 1997).

24.7 Submerged Macrophyte Distribution and Dissolved Inorganic Carbon (DIC): A Physiological Exploration

The photosynthetic rate of submerged macrophytes growing close to the water's surface in dense macrophyte beds may at times be primarily constrained not by irradiance, temperature, or (sediment) nutrients but by a shortage of DIC (Spence 1967, Hutchinson 1975, and Barko et al. 1986). Removal of CO_2 from the water by rapidly photosynthesizing plants, not matched by the dissociation rate of HCO_3^- and the release of CO_2, can bring about a CO_2 limitation under conditions of high pH (Fig. 14–1 and Sec. 14.2). Furthermore, diffusion of CO_2 from the air into the water is slow and unable to compensate quickly when the wind velocity is low and the water's surface is calm.

The ease of transport of available CO_2 or HCO_3^- into macrophytes is hampered by a thick **diffusive boundary layer** (DBL) around the plants. The diffusion rate of gases into and out of all plants is partly a function of the concentration gradient and partly a function of the thickness of the DBL. Within dense beds, the typically thick DBL is the result of reduced turbulence and the high dynamic viscosity of water (Chapter 3). Viscous forces dominate over turbulent forces at a small distance from the plant or sediment surface (Sec. 12.2). As a result, the surface film of water sticks to plant and sediment surfaces and does not partake in the water circulation. The outcome is a relatively low exchange rate of gases in solution between plants and the surrounding water, impeding rapid growth. The DBL ranges from about 5–10 μm for phytoplankton under well-stirred conditions to several 100 μm for submerged macrophytes in slow-flowing rivers, to DBLs of several millimeters for benthic algae or sediments under conditions of low turbulence (Madsen and Sand-Jensen 1991). Plants with thick epiphyte layers have proportionately thicker DBLs resulting from reduced turbulence (eddy diffusion) at leaf surfaces. With measured diffusion coefficients about four orders of magnitude lower in water than air, submerged macrophytes face a much greater risk of CO_2 depletion than the aerial portions of emergent macrophytes or phytoplankton.

The absence of sufficient turbulence to facilitate diffusion can be an impediment to macrophyte growth and affect species composition. Lake sites with high wind exposure are usually dominated by small compact (rosette-like) species with basal growth, stiff leaves, and a strong root system. In contrast, species with long, slender, highly flexible stems usually dominate shallow river channels or shorelines of sufficiently low turbulence to allow rooting and prevent damage to the stems (Fig. 24–3a). Macrophytes with floating leaves, such as water lilies, are restricted to low turbulence waters. Erect and canopy-forming species with apical growth commonly dominate in deeper waters under relatively sheltered conditions (Fig. 24–3).

Inorganic Carbon Limitation

The difficulty of experimentally determining where, when, and how much submerged macrophyte photosynthesis is constrained by a shortage of CO_2 (and not another factor) is compounded by the existence of DIC sources other than dissolved CO_2. These other sources include (1) sediment CO_2 taken up via the roots; (2) CO_2 uptake from the atmosphere by the leaves of floating-leaved forms or canopy species; (3) reutilization of respired CO_2 within the plant; (4) CO_2 obtained from the dissociation of HCO_3^- at the plant surface which also yields largely insoluble CO_3^{2-} and thus precipitates $CaCO_3$ crystals (marl) on the leaves (Sec. 14.6); and (5) direct utilization of HCO_3^- by energy-driven uptake mechanisms. The ability of many macrophytes to produce transport proteins, including the externally released carbonic anhydrase enzyme allowing them to use HCO_3^-, is important. In dense beds of rapidly photosynthesizing plants at pH > 7, bicarbonate concentrations are commonly four to 140 times greater than the available pool of free CO_2 (Fig. 14–1 and Duarte et al. 1994). Carbon dioxide uptake via the roots contributes a negligible fraction (< 1.5%) of the CO_2 uptake by most angiosperms. This was clearly evident from the low radioactivity of the leaves after radiocarbon $H^{14}CO_3^-$ was added to the root zone of plants growing in nature (Loczy et al. 1983).

The only submerged macrophytes known to use much sediment-derived CO_2 are the small rosette species, including isoetids (Fig. 24–3a). They have stems with large longitudinal channels (*lacunae*) that extend from the root to the shoot. The lacunae facilitate CO_2 diffusion upward from the roots during the day, with oxygen diffusing in the opposite direction. It may be more than a coincidence that rosette species often dominate the submerged macrophyte community in softwater oligotrophic lakes of the boreal forest region that are characterized by a small DIC pool in the water.

Although short-term experiments under controlled conditions in flasks or chambers placed over a few plants can evaluate the use of different carbon sources and the particular uptake mechanisms used, such experiments are not designed to evaluate the absolute and relative importance of DIC limitations over time and space in nature. One reason is the difficulty of scaling up experiments on a few shoots or plants under these conditions (Nielsen and Sand-Jensen 1991) to the behavior of entire weedbeds or whole rivers, lakes, or wetlands.[3] Another is that HCO_3^- concentration—the most common DIC surrogate used—correlates with many other environmental factors. Finally, there are environmental factors other than CO_2 that affect the productivity of a population or community in nature.

24.8 Plant Size, Community Structure, and Function

Plant Size

Plant and animal size (weight) provides a first measure of not only the maximum abundance possible but also growth rates under *optimal* conditions when examined over a wide size range (Figs. 2–4, 21–12, and 26–8). The maximum growth rate of macrophytes, characterized by a low surface area to volume ratio (SA:V) and therefore a small number of nutrient uptake and gas

---

[3]The importance of environmental factors other than the one that is under study at any one time—with the others held more or less constant—is of little concern when the goal is to identify mechanisms or particular processes. Nor is it a concern in nature whenever variation is relatively modest in all variables other than the one examined (e.g., Figs. 24–6 and 24–8). But it is a major problem when the goal is to determine the relative or absolute importance of a particular environmental factor or mechanism in nature, let alone its importance over time and space, where there is also considerable variation in other variables.

exchange sites per unit volume, is much lower than the SA:V of the much smaller phytoplankton or attached algae (Fig. 21–12). Furthermore, rooted macrophytes have to produce and maintain an elaborate system for nutrient uptake from sediments (roots), translocation, and structural support (stems). The result is natural macrophyte growth rates that are typically an order of magnitude smaller than for the phytoplankton (Table 24–7 and Fig. 21–12b).

Predation on Macrophytes

The observed lower growth rate of macrophytes in nature, the resulting lower biomass doubling rate (longer biomass turnover time), and a higher necessary investment in structural carbon (Sec. 24.6) gives the macrophytes a lower protein content and a much higher C:N:P ratio than the element demand ratio of the herbivores. The higher structural carbon content appears to be responsible for making macrophytes as a group less desirable to herbivores and bacteria than phytoplankton or periphyton. This interpretation is supported empirically by a typically lower fraction of macrophytes consumed by herbivores (Fig. 24–10).

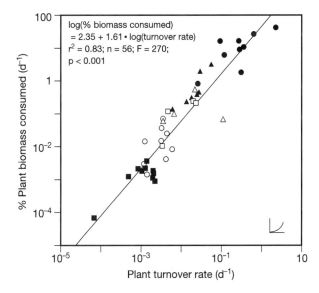

Figure 24–10 The relationship between plant turnover rate and the per cent of the biomass consumed daily by herbivores. Symbols represent phytoplankton (●), macroalgae (▲), sea grasses (□), wetland and freshwater macrophytes (△), grasses (○), and shrubs and trees (■). Note the systematic increase in the percent plant biomass consumed with increasing plant turnover rate. (*After Cebrián and Duarte 1994.*)

▲ Reduced macrophyte loss rates to herbivores implies that a larger fraction of the production remains available for decomposition as detritus and for storage in the sediments. The decomposition rate of macrophyte detritus at summer temperatures is also much slower (mean half-life, 58 days) than for phytoplankton (mean half-life, 17 days) (Enríquez et al. 1993) or attached algae. In addition, much higher C:N and C:P ratios of macrophytes than phytoplankton or periphyton retard microbial decomposition of macrophytes, which have C:N:P supply ratios much higher than the demand ratio for bacterial and fungal growth (Fig. 24–11). The large fraction of macrophyte production that is not directly consumed by herbivores but instead flows along the detritus pathway of decomposition implies slower decomposition and a primarily sediment-based decomposition for most macrophyte production. Consequently, the energy flow from littoral invertebrates to fish is based mainly on periphyton and phytoplankton.

Community Structure and Macrophyte Growth Form

Dense macrophyte beds with a closed canopy are composed of relatively large plants, but few of them per m^{-2}. In contrast, less dense quadrats tend to have a larger number of small plants (Fig. 24–12) in both aquatic and terrestrial habitats (Duarte and Kalff 1987). The principal reason for this pattern is a decline in the amount of light available per plant as the density increases, leading to increased *self-shading* and mortality. The amount of light available per plant is furthermore a major determinant of the weight of individual plants (Barko and Smart 1981). As a result, individual plant weights are greater in transparent than turbid lakes and rivers.

The fraction of the water column occupied by macrophytes is a function of the growth form of the species present. Canopy-forming species fill more of the water column than bottom-dwelling or erect species (Sec. 24.1). A convenient and quantitative way to describe plant growth form is the **biomass density** (BD): the plant biomass per unit volume of water occupied (g dry wt. m^{-3}). Species such as *Myriophyllum* (Fig. 24–1f) and those among the *Potamogeton spp.* (Fig. 24–1c) whose flowers must reach the surface for pollination have most of their biomass near the surface and disproportionately little in the rest of the water column when fully grown. Such species have a low BD and are responsible for public demands for macrophyte control because their biomass is most likely to

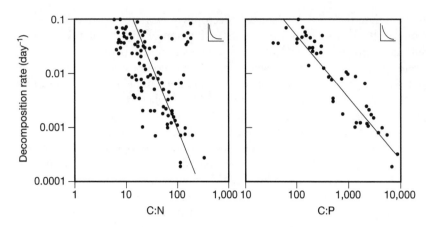

Figure 24–11 The relationship between the primarily microbial decomposition rate (k, day^{-1}) and the initial C:N and C:P atomic ratios of substrates. Note that decomposition rates are typically highest at the lowest C:N and C:P ratios which most closely approach the demand ratio of the microbial protoplasm. The data scatter must be attributable in part to the effect on microbial activity of the not reported temperature and, in case of C:N ratio, to a probably primary phosphorus limitation. *(After Enríquez et al. 1993.)*

interfere with fishing, recreation, and water transport. In contrast, low growing (high BD) species either lack flowers (e.g., *Chara spp.* and *Isoetes spp.*) or have underwater pollination (e.g., *Ruppia maritima* and *Ceratophyllum demersum*). Common species such as *Elodea canadensis* and the closely related *Hydrilla verticillata*, which is restricted to somewhat lower latitudes, share an intermediate growth form (Fig. 24–13).

Differences in plant growth form has a major impact on community functioning. Competition for light (self-shading) within stands is high among low growing species characterized by a high BD. Such species dominate only in highly transparent waters. Conversely, low BD species allow much greater light penetration into stands. Canopy formation makes it possible for plants to compete successfully for light with the phytoplankton and other macrophytes growing below. But canopy formation comes at a price. The close proximity of much of the biomass to the surface make low BD species—with their often delicate and fragile stems—more vulnerable to physical damage by waves or currents at exposed sites and to herbivory by aquatic birds. Low BD species in slow flowing rivers typically grow only in protected sites, whereas species with flexible ribbon-like leaves (e.g., *Sparganium vallisneria*, Fig. 24–3a) are able to grow at more exposed sites.

Among sites varying little in transparency, both experimental and observational data show that the predominant growth form is indicative of sediment nutrient conditions. Communities composed of low growing (high BD) species usually dominate on infertile sediments that characterize highly transparent nutrient-poor (oligotrophic) waters, whereas low BD species are favored on more fertile sediments (Chambers 1987). In yet more eutrophic lakes, subject to a rapid light extinction, only low and intermediate BD species (e.g., *Elodea* and *Ceratophyllum*) are able to thrive but only in the shore zone of wind-protected waters.

More than growth form determines success. Species also differ in attributes such as the specific leaf area (cm^2 mg dry wt.$^{-1}$), chlorophyll-*a* content, photosynthetic potential, (Fig. 24–14) and the CO_2 uptake compensation point (Nielsen 1993). These differences presumably favor different communities under different conditions.

24.9   Attached Algae

Algae growing attached to substrates are known as *periphyton*. Those growing on other plants, principally macrophytes, are known as **epiphytic periphyton** or

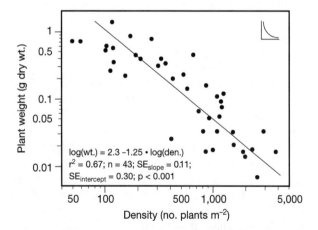

Figure 24–12 Relationship between submerged macrophyte weight and density measured at the end of the growing season. The line represents the regression equation. *(After Duarte and Kalff 1987.)*

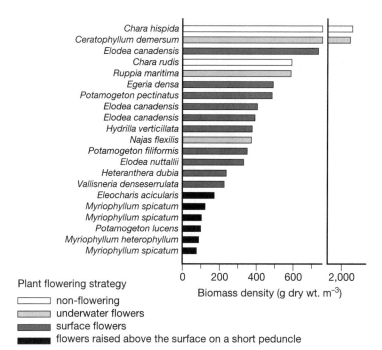

Figure 24–13   Values of biomass density for mature stands of different species. Average density was used when several values were reported in different studies. *(After Duarte and Kalff 1990.)*

Plant flowering strategy
- non-flowering
- underwater flowers
- surface flowers
- flowers raised above the surface on a short peduncle

epiphytes, whereas those growing as a greenish or brownish scum on stones, wood or sediments, are collectively referred to as **epibenthic periphyton**. The latter are frequently divided into **epilithic periphyton** which grow on stones, **epipelic periphyton** which grow on sediments, and *epixylic periphyton* growing on wood. The periphyton community, together with its associated bacteria, fungi, protozoans, and metazoans has long been known as **aufwucks** (German, meaning "to grow upon"), but the term is rarely used today and has been replaced by **biofilm**. A last group of nonplanktonic algae, the **metaphyton**, are not directly attached to substrata but are derived from, and associated with substrata in areas protected from

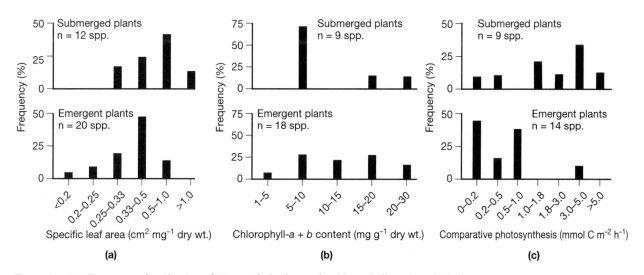

Figure 24–14   Frequency distribution of (a) specific leaf area, (b) chlorophyll-*a* + *b*, and (c) photosynthesis of emergent and submerged macrophyte species under standard laboratory conditions. *(Modified after Nielsen 1993.)*

waves and currents. They are typically composed of clouds of primarily filamentous green algae entangled among macrophytes or trapped between the sediment and water surface near shore. The epilithic, epiphytic and epixylic periphyton are totally or overwhelmingly dependent on nutrients from the water column and under favorable light conditions exhibit large increases in production when nutrient levels are raised (Sec. 24.10). In contrast, the epipelic periphyton can tap DIC and nutrients diffusing from the sediments where interstitial nutrient concentrations are usually orders of magnitude greater than in the water column above (Sec. 17.5) and, therefore, are not highly nutrient-limited (Blumenshine et al. 1997).

The periphyton and macrophytes have received vastly less attention from limnologists than the phytoplankton. This is partly the result of sampling difficulties, including a heterogeneous periphyton distribution in nature.[4] But sampling is further impeded by the difficulty of separating the periphyton from the substrates on which they grow. Until recently, most periphyton studies have minimized this problem by suspending glass slides among plants or placing objects (bricks, cleaned stones, plastic strips,) to be colonized on the sediments (Fig. 8–9) in the process sacrificing some reality for simplicity and ease of sampling.[5]

The epiphytic periphyton can be subdivided into two types of communities. The *tightly attached* community is operationally defined as one that cannot be removed when the collected submerged macrophytes are shaken in a closed container. The *loosely attached*

species, including planktonic forms sedimented onto the macrophytes and the sediments below, are readily dislodged during storms and then returned to the plankton where they are known as *meroplankton* (Sec. 21.1). Macrophyte primary production measurements in lakes and wetlands include the tightly attached periphyton plus probably a small fraction of the loosely attached periphyton community that did not become detached from the macrophytes during the manipulations. The periphyton in rapidly flowing streams or wave-swept lake shorelines lack a loosely attached periphyton community.

▲ The periphyton in inland waters at all latitudes are typically dominated by a variety of diatoms, green algae, and cyanobacteria. The absolute and relative abundance of each group changes seasonally and is linked to seasonally changing nutrient supplies, light conditions, scouring in streams during floods (Stevenson et al. 1996), flushing, external nutrient supply rates in wetlands, and predation.

The epibenthic periphyton growing on stones along the shore of polluted rivers and lakes are often dominated by firmly attached bright green and highly visible filaments of *Cladophora spp.* (Chlorophyta) supplemented by other large filamentous species of green algae (e.g., *Ulothrix*, *Spirogyra*, and *Oedogonium*). A localized dense green cover of filamentous algae on stones at or just below the waterline serves as a telltale indicator of local (point source) wastewater pollution.

The large epibenthic or epilithic filamentous genera (> 10,000 μm^3) become increasingly abundant as nutrient levels rise and are not usually grazed by freshwater invertebrates, but the biomass of more readily grazed smaller forms does not increase with trophic status—presumably due to high predation losses (Cattaneo 1987). The smaller epibenthic and epilithic forms are a well-documented source of food for herbivorous invertebrates, ranging from insects (e.g., Chironomids) to snails (gastropods), oligochaetes, and littoral zooplankton (Chapters 23 and 25, and Steinman 1996). These animals in turn are the principal resource for littoral fish and waterfowl.

The effect of predation has been very clearly demonstrated in exclosures covered with screens of progressively smaller mesh size that allowed only certain grazers to enter. The experiments yielded quite different epiphyte assemblages on the artificial macrophytes commonly used in aquaria. The different-sized grazers were clearly able to dramatically change the structure of the periphyton communities (Fig. 24–15), but importantly, they did not change the periphyton

---

[4]Much less is known about the ecology and physiology of rooted or otherwise attached plants than about their phytoplankton counterparts, even though macrophytes usually dominate shallow lakes, slow flowing rivers, canals, and wetlands, while benthic algal production is higher than phytoplankton production in many shallow lakes. Sand-Jensen and Borum (1991) attribute the neglect to (1) the historically strong tradition of phytoplankton ecology, and (2) the relative ease by which the phytoplankton biomass (chl-*a*) and production can be measured in situ, and easy examination of the physiology of phytoplankton in culture. This is contrasted with the heterogenous distribution of macrophytes and periphyton and the greater difficulties encountered in working with them in nature and the laboratory.

[5]"In an [eight week in situ] experiment comparing substrate colonization [using glass plates and sterilized rocks at eight m in Lake Tahoe, US] with the natural epilithic periphyton community, artificial substrate methods underestimated productivity by as much as 95%. The species composition of the periphyton. . . . was quite different from that of the natural sublittoral epilithic community." Loeb 1981.

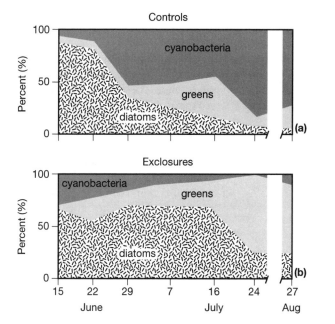

Figure 24–15 Relative importance of diatoms, green algae, and blue-green algae in Lake Memphremagog (CA, US) during 1979, expressed as percent of total epiphyte volume of communities growing on (a) plastic plants at control sites, and (b) on plastic plants in nearby screened (100 μm nylon mesh) exclosures that prevented the entry of snail and insect herbivores. Note the seasonally increasing importance of large blue-green algal epiphytes in nature (controls), while in the absence of macrobenthic grazers (exclosures) the diatoms remained dominant for much of the season, only to be replaced by green algae in July. *(After Cattaneo 1983.)*

community biomass (Cattaneo 1983). However, the observational literature variously reports effects, no effects, and even increases in algal biomass following predation (Steinman 1996) which indicates that impacts differ with the type, the relative abundance of predators and prey, and the temporal, spatial, and range scales of the research.

24.10   Eutrophication and Benthic Plants

It is apparent from the often very high biomass of submerged macrophytes at suitable sites in highly transparent lakes, shallow rivers, or wetlands that rooted macrophytes are able to out-compete the sparse phytoplankton of the open water zone whenever low nutrient levels, rapid flushing in rivers, or heavy predation limits areal rates of phytoplankton primary

production and their biomass. A growth and biomass limitation by nutrients is much less likely for rooted macrophytes that are able to tap the normally orders-of-magnitude-larger pools of sediment nutrients (Carignan and Kalff 1980, Sand-Jensen and Borum 1991). A film of epipelic algae capable of accessing nutrients diffusing from sediments below are similarly less likely to be nutrient-limited than phytoplankton in the water above (Fig. 24–16, and Blumenshine et al. 1997). The films further stabilize the sediments. But lake-eutrophication studies show that when the nutrient supply to previously oligotrophic lakes increases, the phytoplankton community with its high growth potential (Table 24–7) is increasingly able to out-compete the relatively slow-growing submerged macrophytes and the deeper growing epibenthic algae for light (Fig. 24–17). Without sufficient light, nutrient enrichment has little or no effect on growth.

The negative effect of eutrophication on benthic plants is primarily the result of increased light extinction brought about by a now elevated phytoplankton biomass, but is secondarily the outcome of increased light attenuation by colored dissolved organic compounds released by the increased phytoplankton biomass (Sec. 10.4). Epiphytic and epibenthic periphyton are further affected by the shading produced by macrophytes and the self-shading within periphyton biofilms (Fig. 24–16, and Lowe 1996). Therefore, it is not surprising that only where primary light limitation is lacking—near the shoreline of lakes, in shallow lotic systems and wetlands, and for epiphytes on macrophytes growing just below the surface—has eutrophication been followed by an increase in periphyton production and biomass. Moreover, macrophytes growing close to the water surface are negatively affected by the light attenuation produced by the biofilm of epiphytic periphyton, which are largely dependent on nutrients from the water and thus able to respond quickly to added nutrients where light conditions are favorable (Fig. 24–17, and Moeller et al. 1988). However, little is known about the environmental conditions within the periphyton biofilms.[6]

---

[6]"The environmental conditions and the response of organisms are different in open water and on solid surfaces. Nevertheless, there has been a tendency to characterize the physio-chemical environment of benthic plants by measurements in open waters and to believe that major growth-regulating variables for phytoplankton would have a similar importance to benthic plants." Sand-Jensen and Borum (1991).

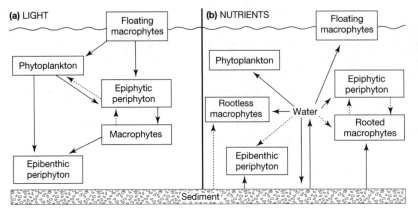

Figure 24–16 Conceptual models of (a) shading effects among aquatic phototrophic communities of phytoplankton, epibenthic periphyton, and rooted macrophytes with their associated epiphytic algae. Arrows point to functional groups that suffer in competition for light. Arrows with dashed lines show shading effects of normally more modest importance. (b) Utilization of the nutrient sources in the open waters and in the sediment. Primary importance of a particular nutrient source is shown by a full line, and secondary importance by a dashed line. *(Modified from Sand-Jensen and Borum 1991.)*

The response of benthic plants to eutrophication is dependent on the scale. The growth of periphyton in shallow water is enhanced where nutrients, not light, are limiting primary production. The response is negative in deep water where eutrophication increases the previous light limitation. Correct values obtained for the absolute and relative growth rates of macrophytes and periphyton in near-shore quadrats are incorrect when extrapolated to other depths. Measurements made in macrophyte beds along a transect perpendicular to shore can reveal the effect of light on benthic plants growing at different depths, but research at this scale is insufficient to reveal the impact of benthic plants on the system as a whole without knowing the system bathymetry and the distribution of the plants over time and space. As we have seen, causes (mechanisms) identified and conclusions drawn at one scale may be incorrect at others (Sec. 2.6).

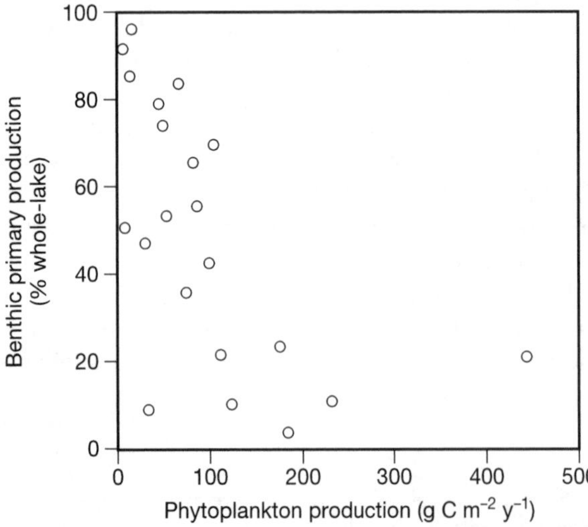

Long-term Effects of Eutrophication

The net effect of increased light extinction is the disappearance of macrophytes and periphyton from the deeper portions of their former habitat. Among the macrophytes, low growing (high BD) species inhabiting deeper waters with their biomass furthest from the water surface are affected first. Best able to cope are the canopy-forming (low BD) species such as Eurasian water milfoil (*Myriophyllum spicatum*) or the ubiquitous sago pondweed (*Potamogeton pectinatus*), which are able to cope relatively well with turbidity by having numerous starch-filled root storage organs that allow them to send a shoot towards the surface before (or despite) the development of a phytoplankton bloom or elevated inorganic turbidity.

▲ In highly eutrophic waters submerged macrophytes and periphyton will at best be present in a narrow band along the shoreline and dominated by species with a low or intermediate BD able to cope with a shallow water column (e.g., *Ceratophyllum demersum*, coontail or *Elodea*-like species, Fig. 24–1). The disappearance of epibenthic algae and submerged macrophytes from the deeper littoral zone proportionately reduces the importance of the plant-associated epiphyton and its predators (Fig. 25–7). The partial or total loss of a macrophyte and epibenthic periphyton vegetation cover commonly leads to increased sediment resuspension and the possibility

Figure 24–17 The percent contribution of periphyton and submerged macrophytes to whole-lake primary production as a function of phytoplankton production (trophic status). *(From Y. Vadeboncoeur and J. Vander Zanden, unpubl. data.)*

of increased nutrient diffusion from the formerly vegetated sediment surface, thereby further increasing light extinction and nutrient release.

Indirect effects of the loss of benthic plants other than increased sediment resuspension and linked nutrient release from the former benthic plant-occupied zone include elevated sediment respiration resulting from an enhanced phytoplankton sedimentation rate. The increased respiration lowers sediment and hypolimnetic oxygen concentrations, thereby enhancing the return of nutrients to the water column (Sec. 17.2). The additional nutrient loading allows further increases in the algal biomass, yet further accelerating the loss of benthic plants (Duarte et al. 1994). The net result will not only be a shift away from a primary production by sediment-associated plants (Fig. 24–17) but will also change the rate and principal site of organic matter decomposition.

Benthic Plants and Energy Flow

Decomposition of the relatively high C:P and C:N macrophyte tissue is slow (Sec. 24.8 and Fig. 24–11) compared to the decomposition of more rapidly growing protein-rich and low C:P phytoplankton and periphyton with their high growth potential. As a result, a much smaller fraction of the macrophyte production is consumed by herbivores (Fig. 24–10). Thus, Cyr and Pace (1993) note that the median value for the percentage of phytoplankton and periphyton productivity that is removed by herbivores in the **grazing pathway** of energy and material flow is much higher (~80%) than for the submerged and emergent macrophytes (~30%). Conversely, a much greater fraction of macrophyte production is consumed (decomposed) *after* death by microbes and benthic invertebrates in what is referred to as the **detritus pathway** of energy and material flow. However, strong manipulation experiments in the literature suggest that the importance of macrophyte grazers (waterfowl, muskrats, fish, crayfish, and other invertebrates) in reducing macrophyte biomass and community productivity may be underestimated at times (Lodge 1991, and Newman 1991). For example, Jupp and Spence (1977) report that *Potamogeton filiformes* in exclosures totally protected from waterfowl grazing became up to 5.3 times larger than unprotected plants. Furthermore, the introduction of the crayfish *Orconectes rusticus* in some Wisconsin (US) lakes led to a major decline of submerged angiosperms and associated invertebrate communities (Lodge and Lorman 1987).

The impact of resident plus migrating waterfowl and some fish feeding on macrophyte shoots has received little attention from limnologists (Søndergaard et al. 1996) but can be important (van Donk et al 1994).

While strong manipulations yield strong responses on the part of the biota these are unfortunately difficult to extrapolate to less strongly or differently impacted natural systems. Nevertheless, it is evident that eutrophication shifts whole-lake primary production away from benthic plants and littoral zones toward phytoplankton communities (Fig. 24–17), and to fish communities dominated by zooplanktivorous fish (Fig. 26–16).

Lake Mikolajskie: A Thirty-year Chronicle

Probably the best long-term study of the effect of eutrophication on littoral zones has been carried out over some 30 years on Lake Mikolajskie (PL). The principal effects reported in a synthesis of the work (Pieczyńska et al. 1988) include a 37 percent decline in the area covered by submerged macrophytes, but a much larger decline (90%) in the total biomass between 1963 and 1980—a period when the average transparency declined from about 3.0 m to 1.1 m. In one of the two basins, the maximum depth of macrophyte colonization declined by 2 m over a mere nine years. There was a major shift in the dominant species, from low-growing Characeae in 1963 toward a more diverse community with a significant fraction of the biomass now composed of several species of taller (intermediate BD) *Potamogeton spp.* and *Elodea canadensis*. More recently, the community became dominated by the low BD *Myriophyllum spicatum* (Kowalczewski and Ozimek 1993), a species that is widespread in Europe and North America (Fig. 24–1f). There was also a 66 percent decrease in the macrophyte biomass (g m^{-2}) within the area of the lake where the plants were able to maintain themselves. The decline in the area covered and biomass m^{-2} led to a striking summer increase in the periphyton biomass, dominated by epibenthic mats of the filamentous *Cladophora* and *Vaucheria* species in the portion of the littoral zone vacated by the slower-growing submerged macrophytes. The remaining submerged macrophytes became covered by the same filamentous species acting as epiphytes. The summer epiphyton coatings have exceeded the macrophyte biomass. No major change occured in the density of macrobenthic invertebrates per m^{-2} of the plant-inhabited portion of the littoral zone. Even so, the large decrease in the total area colonized by submerged macrophytes and the large decrease in the plant biomass available for colonization in the areas still covered by macrophytes resulted in approximately an 85 percent decline in the whole-lake abundance of epibenthic macroinvertebrates between 1971 and 1980. The magnitude of the invertebrate decline varied among macrophyte species. Algal mats became a more important habitat for invertebrates but at a lower density than on the macrophytes they replaced. There was no significant change over 21 years in the density (stems m^{-2}) or dry weight (g m^{-2}) of the dominant emergent macrophyte, *Phragmites australis* (Fig. 24–1h), except in those areas directly impacted by human activity (shoreline modifications, water skiing, or proximity to sewage outfalls).

24.11   Lake Management and Macrophytes

The restoration of shallow eutrophic lakes through reductions in external nutrient loading is notoriously difficult, primarily because the sediments contain a large pool of nutrients accumulated over many years that continue to be released for sometimes a decade or more (Sec. 17.6). Where the restoration is effective and water clarity increases sufficiently there is typically a rapid macrophyte recolonization of the area vacated during the previous period of enhanced eutrophication. But the initial recolonization is not necessarily by the species originally present (Blindow 1992). The subsequent species succession is most rapid in lakes that had maintained a small macrophyte community capable of providing a substantial inoculum (van Donk et al. 1993), or in water where the macrophyte loss was recent enough for the sediments to contain still-viable seeds or rooting structures. In other lakes macrophyte recovery has been incomplete or lacking altogether.

Delays in submerged macrophyte recovery appears to have various causes: (1) continued high sediment-nutrient release or still sufficient external or internal nutrient loading allowing the maintenance of a high algal biomass (turbidity); (2) high turbidity resulting from wind-induced resuspension of fine particulate matter delaying or preventing recolonization (Meijer et al. 1990); (3) waterfowl and coot grazing on the recolonizing macrophytes (Søndergaard et al. 1996); (4) sediment resuspension and the resulting enhanced internal nutrient loading by bottom-feeding fish such as common carp or bream (Tátrai et al., 1997); and (5) slow changes in the often large zooplanktivorous fish stocks allowing continued high predation on the macrozooplankton, and the resulting absence of top-down control of the phytoplankton by macrozooplankton.

Macrophyte Restoration

▲ It is evident from restoration research on six shallow Danish lakes, and from considerable research elsewhere, that the phytoplankton biomass and community structure is highly dependent on nutrient levels. Reducing zooplanktivorous fish stocks in the lakes to allow more herbivorous zooplankton was only effective as a manipulation tool below a threshold of 50–100 μg TP l^{-1} (Jeppesen et al. 1997). The authors propose that piscivorous fish stocking to control zooplanktivorous fish only be entertained as a remediation tool below the TP threshold range: where the phytoplankton biomass declines in step with the TP. Even a modest increase in the transparency of typically very shallow ($\bar{z}$ = 1.5 m) unstratified Danish lakes greatly increases the likelihood of offshore macrophyte recolonization. An extensive macrophyte cover is conducive to inducing a further shift from a turbid stage to a clear-water stage (Scheffer 1998).

After biomanipulation, rapid shifts between a clear-water state dominated by macrophytes, and a turbid state characterized by either high algal biomass or high inorganic turbidity have been observed, suggesting the existence of two alternate vegetation equilibria (stable states) in nutrient-rich (> ~100 μg TP l^{-1}) shallow west European lakes (see Fig. 24–8, Scheffer 1998, and Jeppensen et al. 1990). These nutrient-rich lakes teeter between phytoplankton dominance of the biomass and production, and a clear-water phase when macrophytes dominate and the energy and material flow is shifted toward the detritus pathway rather than the grazing pathway.

The reasons for a switching from a macrophyte to an algae-dominated system, or vice versa, are poorly understood and subject to much speculation, but minor changes in wind speed (resuspension) and hydrology (nutrient loading), on decadal time scales, or changes in zooplanktivorous fish stocks probably tip the balance. Exclosure experiments have shown that waterfowl grazing can prevent or retard the development of macrophyte beds as well. A variety of mechanisms have been postulated as being responsible for enhanced water clarity following the establishment of macrophytes (Søndergaard and Moss 1999). Among them are (1) increased sedimentation and lowered resuspension as the result of reduced turbulence in macrophyte beds; (2) macrophyte shading of phytoplankton; (3) macrophyte release of compounds that reduce phytoplankton growth (*allelopathic compounds*); (4) sufficient uptake of nutrients by macrophytes and epiphytes to reduce nutrient availability to the phytoplankton; and (5) increased grazing by large pelagic zooplankton who use the macrophyte beds as a daytime refuge from fish predators (Fig. 23–27), supplemented by plant-associated littoral zooplankton (e.g., chydorids). Enclosure experiments have shown that the projected high zooplankton grazing rates are most plausible when the zooplanktivorous fish abundance is modest ($< 3m^{-2}$) and more than 20 percent of the water volume is occupied (infested) by macrophytes which provide a refuge for fish (Fig. 24–18). However the most plausible cause inevitably changes with the time and spatial scale investigated (Sec. 2.6).

The replacement of dense blooms of phytoplankton by dense beds of canopy-forming macrophytes is a mixed blessing in aquatic management because macrophytes interfere with swimming, boating, and fishing. The particular trade-off is considered worth exploring intensively in countries such as the Netherlands, Denmark, Poland, and parts of England where the nonpoint nutrient supply from agriculture is high enough that even where urban wastewater treatment is good, shallow lakes remain turbid. But the trade-off is not acceptable in, for example, the south central and southern United States where hypertrophy is less common, where the growing season is much longer (Fig. 24–7), the plant biomass accumulation large and where there is little or no winter die-back. The plant cover in the typically shallow waterways there is so high to engender demands for macrophyte control. In cooler higher latitude western Europe, calls for macrophyte control (by weed cutting) are primarily to prevent the flooding that results when dense beds of macrophytes impede drainage in slowly flowing lowland streams and rivers, and to facilitate water transport in canals and rivers.

Macrophyte Control

There are three basic approaches used, singly or in combination, for managing (controlling) aquatic weed problems: mechanical harvesting and removal of submerged macrophytes, herbicide treatment, and biological control. A fourth method, used routinely in steep-sided fish ponds everywhere, is to add enough fertilizers to maintain phytoplankton blooms that produce a sufficiently high light extinction to prevent macrophyte development. However, the fertilization approach to macrophyte control is unsuitable in the

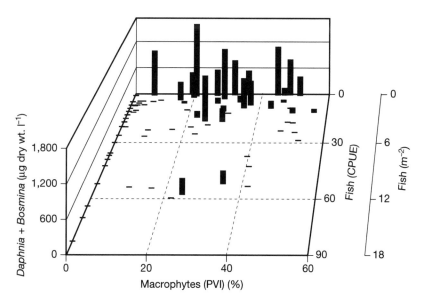

Figure 24–18 Biomass of the dominant pelagic cladocerans (*Daphnia* + *Bosmina*) versus the catch per unit effort (CPUE) of 0+ and 1+ roach and three-spined sticklebacks, and fraction of water volume infested by macrophytes (PVI) in Lake Stigsholm (SE, LA = 0.01 km², $\bar{z}$ = 0.8 m). Enclosure experiments involved manipulation of plants (mainly *Potamogeton spp.*) and fish density. Note that a high macrozooplankton biomass occured at an intermediate (> ~20%) to high PVI, but only when the zooplanktivorous fish abundance was modest. (*After Jeppesen et al. 1997.*)

management of multiple-use lakes (Nichols 1991). Each of the first three management approaches above has advantages and associated disadvantages.

Mechanical Harvesting

Mechanical harvesting is expensive, must be repeated at regular intervals during a (long) growing season, and typically has little or no long-term effect on the biomass or lake area covered by plants. Furthermore, only relatively small areas can be cut by a single machine and costs are high (Table 24–8). In larger lakes only selective macrophyte harvesting is practical, cutting channels to facilitate boating and fishing, creating cruising lanes for large piscivorous fish otherwise unable to penetrate and feed in the dense foliage, and removing the shoots of canopy forming species to encourage the growth of low-growing (high BD) species (Fig. 24–19).

Harvesting plant biomass, once thought to be an effective way to reduce lake nutrient levels, is rarely used for that purpose today because the nutrients removed are typically a minor fraction of the nutrient loading received from the drainage basins. Mechanical harvesting may actually encourage the spread of nuisance species because many are able to propagate from even small fragments produced during the cutting. Harvesting also removes very large numbers of small fish and invertebrates. The effects of such removal are virtually unknown but the impact of small fish removal need not be negative, particularly where dense macrophyte beds prevent natural harvesting by the piscivorous predators of the frequently overabundant and slow-growing zooplanktivorous fish (Chapter 26).

Other mechanical ways to reduce an overabundance of submerged macrophytes include water level fluctuations in reservoirs (Sec. 29.6). *Drawdown* of the water level may indeed eliminate macrophytes but, depending on the timing and duration, may simply select for species able to cope. Fiberglass screens are sometimes used to prevent plant growth over small areas at docking or swimming sites, but the sedimentation of particulate matter onto the screens provides a macrophyte rooting substrate after a few years.

Herbicide Treatment

Herbicide treatment is the most commonly used control method in the United States. Herbicides, which are toxins, are easy to apply, usually effective, and act rapidly. Arsenic was used with varying effectiveness in the 1920s and 1930s. The herbicide 2-4-D was developed in the 1940s and is still used on land and, for controlling Eurasian milfoil (*Myrophyllum spicatum*). Many other types of herbicides have been developed; they differ in chemical formulation, method and timing of application, mode of action, and persistence.

Most modern herbicides degrade in less than a month but others last longer. Some are broad-spectrum herbicides that also kill algae. Most are selective for macrophytes and some are moderately selective for particular species. Most herbicides in use are **contact herbicides** that act upon contact and kill through interference with photosynthesis. While the aboveground portions of the plants are killed the roots are not, allowing rapid regrowth.

The **systemic herbicides** are a second type that are not only absorbed by the leaves but are also translocated to the roots. They are potentially more effective than contact herbicides, but tend to suffer from poor translocation. When they are used at higher concentrations they act as a contact herbicide instead. The dying plants can, when dense, cause anoxia following their sudden death and decomposition. Fish kills are the result and sediment phosphorus is commonly released (Sec. 17.2).

One major disadvantage of all herbicides is that sufficient contact time is required for absorption before the herbicide is flushed away or diluted for the treatments to be effective. Therefore only small areas can be treated at any time. The required elevated concentrations are expensive (Table 24–8) and treatments

Table 24–8 **Typical 1991 cost ranges (US$) per treatment per hectare for mechanical harvesting, herbicide treatment, and biological control (grass carp) for aquatic plant management in the United States.**

Treatment	Midwest	Florida
Mechanical harvesting[1]	400–1,100	900–4,200
Herbicides	600–1,100	400–1,000
Grass carp[2]	200	70–100

[1]Excludes capital cost of equipment; high values refer to densest infestations.

[2]Costs vary with the number of fish required and their longevity.

Source: From Cooke et al 1993.

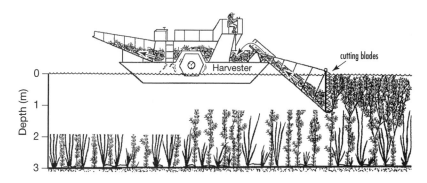

Figure 24–19 General or selective harvesting of macrophytes, with the selective harvesting goal to remove a canopy forming species (here eurasian milfoil, *Myriophyllum spicatum*) in order to stimulate the growth of low growing species, (wild celery, *Vallisneria americana*). Note the cutting blades and the transport of the plants into the boat. *(After Engel 1987.)*

must be repeated after regrowth. The biggest issue of all is the increasing unacceptability of chemical controls and the addition of biocides (toxins) to waterways. Therefore, the informative lake management book by Cooke et al. (1993) does not discuss herbicides, which is appropriate in my view.

Biological Control

The best-studied and, in the United States, most successful method of biological control is obtained by stocking chinese grass carp, also known as white amur (*Ctenopharyngodon idella*, Cyprinidae), a plant eating (*phytophagous*) fish native to northeast Asia. Its use has been contentious because it was introduced into North America without any studies of its beneficial and harmful effects. Grass carp were initially thought to be incapable of reproduction under local conditions but this turned out to be incorrect. Today the fish is totally banned from some North American jurisdictions whereas others only allow the use of the specially bred sterile (triploid) version. The popularity of grass carp is high because they are generally effective and provide long-term control of plants (fish life-span, ~8 years).

Fish stocking rates and costs (Table 24–8) vary because fish feeding rates are a function of water temperature and vary further as a function of the growth rates and density of the plants to be controlled. The fish, not surprisingly, prefer some plant species over others and while a particular density may remove all submerged and emergent macrophytes in one lake it may simply encourage the growth of less palatable species in others, unless the fish is much more densely stocked. The total removal of macrophytes improves certain types of recreational activities, but it also entails the loss of habitat for many species, including fish, water birds, and other wildlife, and it may enhance shoreline erosion in wind-exposed lakes. Finally, the selective reduction of macrophyte biomass, a commonly desired management goal, is difficult to attain because excess fish are difficult to recapture.

The effect of macrophyte control on fish species composition and sport fishing is poorly understood and is greatly in need of fundamental research. Fish species other than grass carp used for biocontrol include the silver carp (*Hypophthalmichthys molitrix*), the bighead carp (*H. nobilis*) and members of the tropical tilapias (Family Cichlidae, Chapter 26). The latter have been used to prevent macrophytes from clogging California irrigation canals (Opuszynski and Shireman 1995).

Not only phytophagous fish but also phytophagous insects have been used for macrophyte control, often with only modest success because few phytophagus predators feed on a single plant species, and because the abundance of the predators is determined by more than food availability. One of the most successful control experiments was conducted over a number of years and yielded a large reduction in the surface area of Louisiana (US) waterways covered by the floating exotic *Eichhornia crassipes* (Fig. 24–1a) from 500,000 to 122,000 ha. This was accomplished following the testing, importation, culture, and release of a beetle native to Argentina that feeds only on this particular plant. A second, even greater success was the virtual elimination of another floating macrophyte *Salvinia molesta* (invasive water fern; Fig. 15–1) from the Sepik River floodplain system of Papua New Guinea using beetles, previously used successfully for the same purpose in Australia (Thomas and Room 1986). The *Salvinia* infestations had greatly interfered with crucial water transportation and fishing by the local people, leading to the abandonment of villages. *Salvinia* control was achieved about two years after introduction of the beetle, which as adults feed on *Salvinia* buds whereas the larvae attack both buds and rhizomes.

Highlights

- Large aquatic plants can be subdivided on the basis of habitat between the amphibious, emergent, free-floating, and submerged forms.
- Periphyton include the benthic algae associated with macrophytes and those growing on sediment or rock substrates.
- ▲ Emergent macrophytes extend from above the shoreline to a depth of roughly 1–2 m, while the submergent angiosperms extend down to a depth where ~2–10 percent of the incoming irradiance (PAR) penetrates, unless the underwater slope is unsuitably steep or the turbulence too great. Macroalgae and mosses extend much deeper.
- Macrophytes and periphyton dominate the phytoplankton in shallow transparent waterways where the rooted macrophytes plus the periphyton growing on fine nutrient-rich sediments are able to tap sediment nutrient concentrations much higher than those available to the phytoplankton. But in nutrient-rich waters the phytoplankton outcompete (shade) the bottom-living plants for light.
- Globally, wetlands cover an area about three times that of lakes and include the fringing littoral zones of rivers and lakes. The historically extensive wetlands have been largely drained for agricultural purposes in densely populated agricultural areas and thus are lost as (1) a crucially important feeding, breeding, and hiding area for fish and wildlife and thus the maintenance of high biodiversity; (2) for water purification; (3) water storage and flow regulation; and (4) a home for fishermen and seasonal farmers, particularly in the economically developing world.
- ▲ During periods of high photosynthesis, the photosynthetic rate of macrophytes may be constrained by a slow diffusion rate of dissolved inorganic carbon from the water into the plants.
- Macrophyte growth rates under optimal conditions are typically an order of magnitude smaller than for the much smaller phytoplankton and periphyton.
- ▲ Macrophytes have a much higher structural carbon content for stems and therefore a lower protein and nutrient (especially P and N) content and lower desirability to most herbivores and heterotrophic bacteria than the phytoplankton and periphyton.
- Eutrophication shifts whole lake primary production away from benthic plant and littoral zones toward phytoplankton communities and zooplanktivorous fish.
- ▲ Macrophyte beds are an important refuge from piscivorous fish for both young-of-the-year fish and small fish species, and a partial refuge for macrozooplankton and benthic invertebrates from their fish predators.
- ▲ Dense beds of macrophytes interfere with river discharge, water transport, boating, and fishing, and also prevent piscivorous fish from feeding efficiently in the dense foliage. Three management approaches are mechanical harvesting, herbicide treatment, and biological control using macrophyte-eating fish or insect predators.

25

Zoobenthos

25.1   Introduction

The **zoobenthos** is the animal community living in association with the substrate–water interface. The fauna of the sediment surface, the **epifauna**, and those of the surficial sediments, the **infauna**, have been investigated best. In addition, there is diverse fauna associated with the macrophytic vegetation, the *epiphytic* fauna, and a second one, the *hyporheic* fauna, living well below the substrate surface in the permeable gravel of rivers. The hyporheic fauna is part of a continuum that ranges from the fauna of the sediment-surface of lakes to the *floodplain* fauna of the groundwater adjacent to many lotic systems (Sec. 8.3).

The zoobenthic community contains organisms ranging in size from the poorly investigated protozoans to large clams and crayfish. Nearly all investigators have routinely used relatively coarse sieves (pore size 400–1,000 μm, but most commonly 400–500 μm) to separate the larger forms—the **macrobenthos**, including most insects—from soft sediment substrates. They use a similar mesh size to collect stream organisms removed from the substrate. The use of coarse sieves and nets facilitates sampling and quantification but at the cost of greatly underestimating the **meiobenthos** (< ~400–100 μm) which is composed primarily of rotifers, copepods, young chironomids, small oligochaetes, and nematodes. The **microbenthos** (< 100 μm)—composed of small species including protozoans and juveniles of larger forms—are completely overlooked. The oversight is not serious when the biomass of macrobenthic animals consumed by fish and water birds is at issue (Sec. 25.5), but it is very serious when density or species estimates are needed (Fig. 25–1). Researchers working on Mirror Lake (New Hampshire, US) noted that of more than a million individuals m^{-2}, > 98 percent of the individuals and two-thirds of the typically small species would have been overlooked if only the macrobenthos and meiobenthos (there > 250 μm) had been investigated (Strayer 1985). About half the animals were found in the top centimeter. Unfortunately, the literature is largely based on the macrobenthos which hinders synthesis.[1] Sieves are unsuitable for sampling the **megabenthos** (> 1000 μm), which includes crayfish and large bivalve molluscs, that require different sampling techniques.

Size and Energy Flow

Normalized size spectra indicate that, similar to plankton, the biomass of unicellular (algae, protozoa) and multicellular (metazoan) benthic organisms are roughly equal in all the logarithmic size classes examined (Fig. 25–2). This implies that the metabolism of the benthos is greatly underestimated by considering only organisms of macrobenthic size and weight. But

---

[1] "I would like to make a plea for limnologists to choose their sieve mesh sizes more carefully. Historically, the mesh sizes used have been highly variable . . . The reason . . . is that the 'ideal' mesh size (i.e., one that retains the most animals and the least sediment) varies with the sediment type and the objectives of the study. While this approach has been useful for individual scientists, it has also left the field of limnology with a body of 'quantitative' data that is very difficult to interpret." (Strayer 1985.)

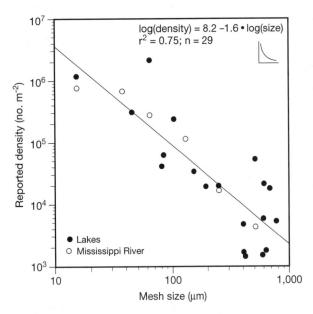

Figure 25–1 The effect of the mesh size used to screen sediment samples on the apparent density of the zoobenthos in aquatic systems. *(Modified after Strayer 1985.)*

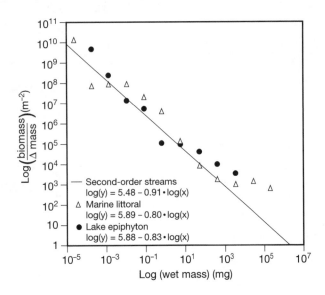

Figure 25–2 Normalized community biomass spectra in benthic communities. The line represents the average spectrum observed in second-order Laurentian streams (CA), with a broader generality indicated by equivalent data from other littoral assemblages. The pattern shows the biomass to be roughly equal in all logarithmic size classes. *(After Cattaneo 1993.)*

the regularity also indicates that, like the plankton, it may be possible to use size-based community descriptions together with information on the systematic changes in metabolism with size (Fig. 2–4 and Eq. 23.5) to estimate rates of community production and respiration or contaminant flow in food webs (Cattaneo 1993). Strayer (1991) argues for the existence of a meiofaunal "loop," analogous to the microbial loop (Sec. 22.9), within which most of the energy is respired, thereby allowing only a minor fraction of the zoobenthic energy to flow to higher trophic levels.

Sampling the Benthos

The best-investigated group is the insects (Fig. 25–3). In streams, shallow littoral sites, and wetlands, insects and other macroinvertebrates are usually sampled within quadrats placed on the sediments. In lotic systems the epifauna and infauna are typically captured with a net attached to the downstream side of quadrats. The nets are coarse to facilitate water flow and the passage of fine particles. While coarse nets (and sieves) are appropriate for collecting the macrobenthos they do not capture the early larval stages of most macrobenthic species quantitatively. Consequently, the abundance and distribution of the early life-history stages remains poorly known (Hynes 1970).

Emergence traps placed on the water surface are used to collect insect species upon their emergence as flying adults. Some other techniques include light traps to attract night-flying insects and daytime sweep-netting of flying adults to complement the sediment sampling of the larvae. Quite different techniques are required to sample the hyporheic fauna or the much less abundant megabenthos (>1000 μm).

Stream and littoral zone sediments are highly heterogeneous, requiring many replicate samples for quantitative studies (Table 1–1). In work on lakes this has led to a focus on the biota of the homogeneous particle size *zone of sediment accumulation* (Sec. 20.2), and also encouraged research on the more easily sampled planktonic macroinvertebrates but at the price of neglecting the productive littoral zone.

25.2   Taxonomic Distribution, Species Richness, and Abundance

The number of reported species depends not only on the taxonomic skills and interests of investigators and the temporal (days, years) and spatial (m², whole-

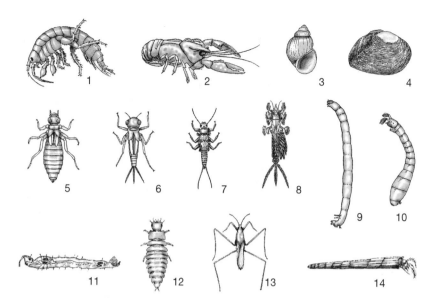

Figure 25–3 Selected benthic organisms from lentic and lotic waters, not drawn to scale. Crustaceans: (1) *Gammarus* (scud, Amphipoda), (2) *Cambarus* (crayfish, Decapoda); molluscs: (3) *Campeloma* (snail), (4) *Unio* (bivalve); larval insects: (5) *Gomphus* (dragonfly, Odonata), (6) *Hyponeura* (damselfly, Zygoptera), (7) *Acroneuria* (stonefly, Plecoptera), (8) *Hexagenia* (mayfly, Ephemeroptera), (9) *Chironomus* (midge, Diptera), (10) *Simulium* (blackfly, Diptera), (11) *Chaoborus* (Diptera), (12) *Amphizoa* (beetle, Coleoptera), (13) *Gerris* (water strider, Hemiptera), (14) *Triaenodes* (cased caddisfly, Trichoptera). (*After Needham and Needham 1962.*)

system) scales examined but also on the research goals. For example, detailed taxonomic work on the insects of two American streams revealed 300 and 350 insect species (see Allan 1995). A much more comprehensive, nearly two-decade-long study of a German stream (Breitenbach) with a small pond on it reported 476 insect species plus 568 other metazoan organisms for a total 1,044 species (see Allan 1995). The true flies (Order Diptera) contributed 74 percent of all insect species. The largest group of dipterans belong to the Family Chironomidae—the midges (152 spp.)—a group of nonbiting mosquito-like flies that form mostly small but sometimes enormous mating clouds over water or nearby land (Fig. 23–20). Their species richness is tremendous in lakes as well. In the shallow well-studied Rybinsk reservoir (RU), 135 chironomid species have been identified. Other species-rich groupings in the German stream included the Phylum Nematoda (125 spp.), Rotatoria or rotifers (106 spp.), the Annelida or oligochaetes (56 spp.), and Tubellaria (50 spp.). Much less diverse were the Crustacea (24 spp.), the Hydracarina or water mites (22 spp.), and the Mollusca (12 spp.).

Both aquatic and terrestrial research has repeatedly demonstrated that relatively few species dominate numerically or in terms of community biomass at any one time, with the remainder rare. Thus, earlier work on the same German stream by J. Illies (1971) showed that of the 148 insect species trapped following emergence the 15 most abundant species contributed 80 percent of the total number of individuals (Allan 1995).

▲ 25.3 Life-History Aspects

The enormous number of species whose life histories are often poorly known[2] and possessing considerable flexibility in rates of development and diet preclude broad generalities based on systematics.[3] However, some general conclusions have been drawn. Many temperate zone insects have a single generation per year (univoltine) but development may take two years under poor growing conditions, for example in poorly oxygenated hypolimnia or where resources are limiting in highly oligotrophic waters. Life cycles that typically take one year at temperate latitudes may, as in the zooplankton (Sec. 23.3), cover two or three years for the same or closely related species at higher latitudes. Among species with more than one generation per year at temperate latitudes are the multivoltine small crustaceans, while some of the lotic midges (Chironomidae) and a number of blackfly species (Family Simuliidae) have two generations per year (bivoltine). Others, including some leeches, snails, crayfish, and clams commonly have a two-year life cycle in the temperate zone. The number of generations is not always larger at lower latitudes where

---

[2]"We know very few of the life histories of the ecologically very important Chironomidae. This is mainly because of our inability to identify the larvae." (Hynes 1970.)

[3]"In the present state of our knowledge of the life histories of benthic animals we are almost overwhelmed by facts, and it is difficult to discern any clear over-all pattern." (Hynes 1970.)

stream discharges vary greatly seasonally, where water courses may dry up altogether (Resh and Rosenberg 1984) and periods of anoxia are more common. Many benthic invertebrates have resting (diapause) stages, allowing eggs, larvae, or adults to survive unfavorable conditions.

Feeding Groups

The feeding mechanism of primary consumers can be grouped into four broad functional categories: filter-feeding, deposit feeding and collecting, scraping, and shredding, while those categorized as predators feed on primary consumers or other predators (Fig. 8–5). Insects that filter-feed use silken nets or modified body parts to filter algae, bacteria, and detritus from the water column, primarily in flowing water. When abundant, bivalves (molluscs) are important filter-feeders in lakes and rivers (Sec. 25.6). Where turbulence is low and fine organic particles are accumulated, as in the hypolimnia or at the land–water interface, the benthos is greatly dominated by deposit feeders and collectors browsing the sediment surface or burrowing in soft sediment to feed on bacteria-rich particles at the oxic–anoxic sediment interface (Sec. 25.4). Scrapers dominate in shallow flowing water and turbulent littoral zones where they feed on the biofilm of benthic algae, fungi, and the associated bacteria. In streams, shredders consume large detrital particles, such as fallen leaves and their attached fungi and bacteria, or consume macrophyte detritus in littoral zones and wetlands. Only crayfish and a relatively few insect species feed on living macrophytes.

The functional classification is far from perfect when it is applied to the many omnivorous species or species that change feeding category with age or size. Thus, omnivorous stoneflies (Order Plecoptera, Fig. 25–3) ingest more stream periphyton and detritus when small, and more animal prey when large (Allan 1995).

25.4   Lake Morphometry, Substrate Characteristics, and the Zoobenthos

The Littoral Zone

Littoral zones—often dominated by macrophytes—are widely recognized as the most productive regions of lakes (Chapter 24, and Kajak and Hillbricht-

Ilkowska 1972). Littoral zones also have the largest number of animal species, contain the highest animal biomass and density, and have the highest secondary production (Brinkhurst 1974). The littoral zone of lakes has been subdivided into seven zones (Wetzel 1983) but here we have reduced them to three. The **upper littoral** zone reaches from the shoreline sprayed by waves in wind-exposed lakes to where the emergent macrophytes (if present) disappear (~1–2 m), the **middle littoral** extends down from the lower edge of the upper littoral zone to the depth where rooted submerged macrophytes disappear, and the **lower littoral** or **sublittoral zone** extends from the lower edge of the middle littoral zone to the bottom of the euphotic zone (Sec. 10.7). The sublittoral zone is, during the growing season, occupied by benthic algae. Below the sublittoral zone lies the profundal zone (Fig. 1–5). However, from a zoobenthic perspective the bottom of the shallow water zone is best indicated by the *Deposition Boundary Depth (DBD)* separating the zones of fine and coarse sediment (Sec. 20.2) or, alternatively, by the depth of the thermocline which separates the warm food-rich mixed layer from the much colder deep waters at higher latitudes.

Some of the best long-term work on lake macrobenthos has been done in Lake Esrom (DK). Jónasson (1972) summarizes this work, started in the 1930s by K. Berg and colleagues, and provides an introduction to the abundant research carried out in Germany and elsewhere in Europe between the Great Wars.

About 90 percent of the total of > 500 macrobenthic species found in north European lakes, such as Esrom, inhabit the productive and warm littoral zone. Roughly 30 percent of the species are found in the narrow, turbulent, and often wave-swept upper littoral zone. Most are herbivorous (algivorous), feeding on benthic periphyton and associated bacteria. Many are adapted to cope with high turbulence by having flattened or streamlined bodies, or big claws or suckers for attachment. More than half of the 155 species found in the upper littoral zone of wave-exposed shores are also characteristic of rivers. (H. Ehrenberg 1957, in Jónasson 1978).

The largest number of macrobenthic species occupy the structurally diverse and often macrophyte-dominated middle littoral zone where they live on the macrophytes or in the organic sediments below. Even in a lake such as Mirror Lake (US), which has little macrophyte cover, the greatest species richness is in the area of maximum macrophyte cover at 1 m (Fig. 25–4). In the less structurally complex sublittoral zone

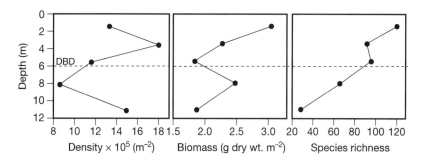

Figure 25–4 Mean abundance, biomass, and species richness of macrobenthic, meiobenthic, and microbenthic metazoans in oligotrophic Mirror Lake, US (LA = 15 ha, $\bar{z}$ = 5.8 m) as a function of depth and the observed as well as predicted depth of the depositional boundary depth (DBD, Eq. 20.1), separating the zones of coarse and fine sediments. Note that (1) macrophytes are scarce in the lake; and (2) the water overlying the deep water sediments (9.5–11m) becomes anoxic for a period in both winter and summer. *(Modified from Strayer 1985.)*

there are no macrophytes to dampen turbulence, and periodic fine-sediment resuspension and transport leaves behind a base of coarse sediments, which in Lake Esrom is occupied by a mere 50 macrobenthic species.

The algivores, bacterivores, and detritivores of the sublittoral zone and their invertebrate predators are dependent in part on the quantity and quality of organic matter obtained from the biofilm of periphyton and their associated bacteria growing on the relatively immobile coarse (sandy) substratum that is sometimes partially covered by wood debris. But other important food sources include the phytoplankton plus bacteria-coated detrital particles raining down from the water column above, along with littoral and terrestrially derived organic matter washed out of the littoral zone. The relatively firm sublittoral substrate in Lake Esrom and other wind-exposed deep lakes is typically dominated by large detritivores and algivores such as the isopod *Asellus*, the amphipod *Gammarus* (Fig. 25–3), and large mussels including zebra mussels (Sec. 25.6).

Turbulence, Substrates, and the Biota

Turbulence has a major effect on the abundance and composition of the benthos through its effect on sediment characteristics. It also serves as a proxy for the oxygenation of surficial sediments, lake stratification, and water temperature. The only lentic study in which the effect of turbulence on invertebrate biomass and distribution was explicitly investigated (Rasmussen and Rowan 1997) showed that epilimnetic depositional sites (ED) of sediment and organic matter accumulation in macrophyte beds contained an average of

more than twice the zoobenthic biomass of nearby sites without macrophytes, characterized by coarse sediments and little organic matter (epilimnetic nondepositional sites (EN) of high turbulence) (Fig. 25–5). Superimposed on the substrate effects are the influence of temperature and the thickness of the overlying water column, with thicker and warmer

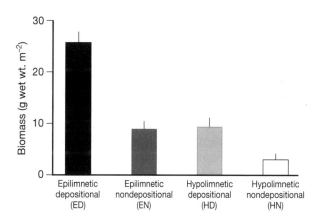

Figure 25–5 Mean macrozoobenthic biomass (±1SE) in Lake Memphremagog (CA, US) in relation to whether sample sites were depositional (high in organic content, below the depositional boundary layer (DBD), see Sec. 20.2) or erosional (low in organic content, above the DBD) as well as to whether they were epilimnetic (littoral and warm) or hypolimnetic (cool). Tukey multiple ranges test shows: ED > EN and HD > HN; $p < 0.01$. Note that the demonstration of the pattern was aided by the near absence in this single system of potentially confounding variation in trophy, temperature, climate, etc., allowing the difference between depositional and nondepositional areas to emerge clearly. *(After Rasmussen and Rowan 1997.)*

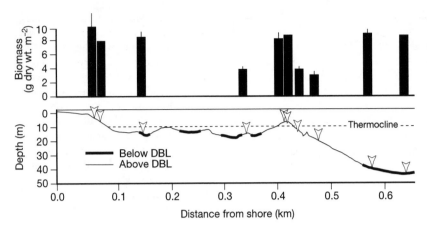

Figure 25–6 Zoobenthic biomass (± 1SE) in relation to depositional conditions and temperature (18°C above and 6°C below the thermocline) along transect 3 in Lake Memphremagog (CA, US). Arrowheads show location of sampling station. Thick line indicates depositional sites below the depositional boundary depth (DBD). Note (1) the highest biomass is in shallow depositional quadrats above the thermocline and at depositional sites below the thermocline; (2) the lowest biomass is found at sites of erosion (nondeposition, coarse sediment) below the thermocline. *(After Rasmussen and Rowan 1997.)*

water columns permitting greater respiration of sedimenting organic matter, thereby reducing the quantity and quality of the material arriving at the sediment surface. Epilimnetic depositional sites (ED) apparently have the largest resource base, allowing for a greater biomass than at equivalent well-oxygenated depositional sites in the hypolimnion (HD, Fig. 25–5). The benthos there experiences an on average smaller supply of high-quality sedimenting food particles and much lower water temperatures, but see Figure 25–6.

Epilimnetic depositional sites are found only in low-sloped areas where turbulence is typically dampened by macrophyte beds, or in small protected bays of large lakes. The dampening effect rises with increasing macrophyte biomass, which increases as the underwater slope declines (Fig. 24–6). Macrophyte beds reduce turbulence, in the process enhancing net sedimentation (Fig. 20–2), thereby creating a more varied food rich habitat for the benthos, providing an explanation for the elevated invertebrate biomass in the ED. Slope, through its effect on macrophyte development (Fig. 24–6) and sedimentation, is a good predictor of the zoobenthic biomass (Rasmussen and Kalff 1987).

The impact of sediment characteristics and temperature on the zoobenthic biomass is scale-dependent. They are most evident at the scale of individual systems that have much spatial variation in sediment characteristics and temperature but little or no variation in trophic status, turbidity, and climate. The same investigation carried out at the among-system scale,

over which both trophic status and morphometry vary greatly, shows that both variables are important predictors of the zoobenthic biomass (Fig. 25–8).

The Zoobenthos and Macrophytes

As the macrophyte biomass increases, the zoobenthos becomes increasingly associated with the plants rather than with the sediments (Fig. 25–7). The epibenthic fauna is dominated by small species of chironomid lar-

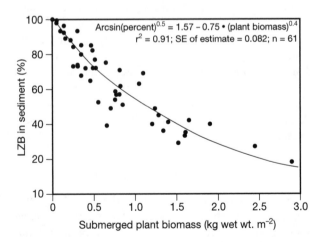

$$\text{Arcsin(percent)}^{0.5} = 1.57 - 0.75 \cdot \text{(plant biomass)}^{0.4}$$
$$r^2 = 0.91; \text{ SE of estimate} = 0.082; n = 61$$

Figure 25–7 Relationship between the proportion of the sediment littoral zoobenthos (LZB) collected with a 200 µm screen, and the biomass of submerged macrophytes in Lake Memphremagog (CA,US) along-shore–deepwater transects at the depth where the plant canopy height was maximal. *(After Rasmussen 1988.)*

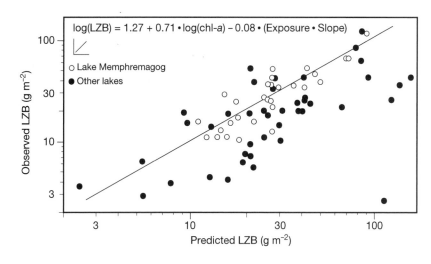

Figure 25–8 Observed littoral zoobenthic biomass (LZB) in Lake Memphremagog (CA,US) and other temperate zone lakes, and its prediction from a model developed for Lake Memphremagog. Note that empirical models normally show more scatter for systems in regions other than the one for which they were developed, because the additional systems differ in important environmental variables (e.g., aspects of morphometry, temperature, growing season, predators, etc.) that exhibited little or no variability (had little impact) at the spatial scale over which the model was developed. Empirical models such as these provide a basis for (1) experimental and observational work on the underlying processes and mechanisms; (2) better predictive models; and ultimately (3) better models based on the dynamic actuality. *(Modified after Rasmussen 1988.)*

vae and other insects, small amphipods (e.g., *Hyalella*), small gastropods (snails), and small oligochaetes (Cyr and Downing 1988, and Rasmussen 1993). The macrophytes serve as a habitat and their periphyton is an important source of food for the animals. Other epibenthic species supplement their diet by filtering suspended particles from the surrounding water column. Yet others feed largely or totally in the water column surrounding the plants or at the sediment–water interface and are more properly considered to be littoral zooplankton. The physical structure provided to the invertebrates by the macrophytes is important. Thus, a precipitous decline of macrophytes in a South African lake (Swartvlei) was followed by a rapid decrease in the macrobenthos, from a record high of 132 to a low of 35 g dry wt. m^{-2}, despite the development of extensive algal mats (Davies 1982). A similar response in a Polish lake is discussed in Section 24.10.

Filter-feeding epiphytic microcrustaceans living on the macrophytes, and pseudoplanktonic forms living among the macrophytes are part of a continuum of animal types that grade from strictly planktonic forms (true zooplankton) to the periphytic and epibenthic zoobenthos associated with firm substrates. The same zooplankton are also part of a second continuum, between strictly planktonic and strictly littoral forms. Finnish work shows 94 meiobenthic species of small

cladocerans and harpacticoid copepods living on or among the plants. The littoral zone species richness was much greater than for their counterparts in the more homogeneous pelagic zone of the same lakes. (See also Sec. 23.4, Kairesalo and Seppälä 1987, and Lehtovaara and Sarvola 1984.)

The Profundal Zone

From a sediment perspective, the zoobenthos of the deep water (**profundal zone**) inhabits the zone of sediment accumulation (Sec. 20.2). The sediments are fine (< 23 μm; silts and clays) and are in eutrophic lakes located in well-vegetated catchments dominated by organic matter. The resulting sediment-water content is typically high (~75–90% water), allowing easy burrowing. Decomposition at the sediment surface of organic particles derived from the epilimnion of eutrophic systems may at the same time allow sufficient microbial and zoobenthic respiration to yield surficial sediments low in dissolved oxygen (hypoxic), or lacking dissolved oxygen altogether (anoxic). Hypoxic conditions, depending on their duration, reduce or eliminate DO-sensitive species (Sec. 15.7), while sustained anoxic conditions not only reduce the abundance and seasonal and annual growth rates of individual species but also greatly lower species richness and community production (Jónasson 1972). In

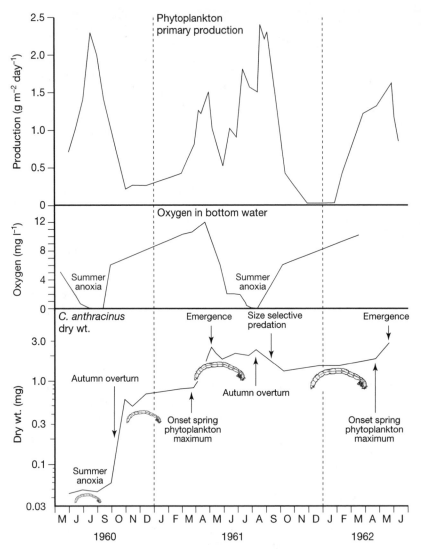

Figure 25–9 Seasonal changes in factors controlling growth of *Chironomus anthracinus* in the profundal zone of eutrophic Lake Esrom, DK. Depending on growth conditions, the midges pass through four larval stages in one or two years. They then pupate and emerge as flying adults, followed by reproduction. Growth is high during the autumn overturn of the first year, and during the subsequent spring phytoplankton maximum, characterized by high phytoplankton production and favorable dissolved oxygen conditions. A spring (April) and summer (August) growth peak is easily recognized. Growth declines rapidly during autumn and is low in winter. Oxygen at the bottom decreases to zero during summer stratification, with the period of anoxia considerably longer today as a result of eutrophication (see Fig. 15-9). The silhouettes show the relative size of the larvae. *(Modified after Jónasson 1978.)*

Lake Esrom (DK), characterized by an increasingly lengthy period of summer hypoxia/anoxia (Fig. 15–9), the community is composed of relatively few macrobenthic species (~20) able to cope.

▲ Apart from a few primarily carnivorous species (e.g., *Chaoborus*, Sec. 23.9), the vast majority of macrobenthic species and individuals in the profundal zone are omnivorous—feeding on suitable bacteria-coated detrital particles and sedimenting algae reaching the sediments (Jónasson 1972).[4]

Among-lake studies show that the profundal community biomass increases with increasing trophic status, followed by a decline at epilimnetic chl-*a* concentrations above about ~120 μg l⁻¹. The pattern probably reflects a positive effect of an increase in resources and a negative effect of low DO concentrations over the sediments during stratification in eutrophic lakes (Rasmussen and Kalff 1987). Other important among-system variables noted were the sampling depth (–), lake mean depth (–), slope (–),

---

[4]The usually little-studied benthic meiofauna (in the profundal zone of a Swedish lake) is dominated numerically by cyclopoid and harpacticoid copepods but includes a fair abundance of nematodes

and ostracod crustaceans (Goedkoop and Johnson 1996), whereas the well-studied fauna of Mirror Lake (US) is numerically dominated by nematodes (~680,000 m⁻²) in terms of biomass, and in production by chironomids (Strayer 1985.)

mean depth: maximum depth ratio (+), color (–), and temperature (+) (Rasmussen and Kalff 1987).

Chironomous anthracinus

It has long been recognized that the sedimentation of phytoplankton cells or detrital particles has important effects on the abundance and biomass of sediment-living invertebrates. During overturn periods, Langmuir and other currents (Chapter 12) carry phytoplankton produced in the euphotic zone toward the well-oxygenated sediment surface. In a now classic study, Jónasson (1972) demonstrated the positive effect of algal sedimentation and increased DO on the growth of *Chironomus anthracinus* (Fig. 25–9). Chironomids frequently dominate not only the macrobenthic invertebrate community biomass of the profundal zone of lakes at all latitudes but also the well-oxygenated littoral sediment of oligotrophic and of shallow unstratified eutrophic lakes (Johnson et al. 1989, Lindegaard 1994). As long as the sediments remain well oxygenated eutrophication results in increased larval survival and increased adult emergence (Welch et al. 1988).

The life cycle of *C. anthracinus* is normally two years in Lake Esrom (DK), but under favorable conditions (DO and food) at shallower depths they emerge after one year. There are four larval stages, with the last one followed by pupation. When they are ready to emerge the pupae become buoyant, float to the surface, metamorphose into adults (using the pupal case as a boat), and fly away. After swarming and mating, females release packets of eggs upon dipping their abdomen in the water. The packets sink slowly and in the process are distributed over the sediments by currents.

The abundance and population biomass of *C. anthracinus* in Lake Esrom follows a regular cycle (Fig. 25–10). The red-colored hemoglobin-rich chironomids consume detrital and algal organic particles plus the attached bacteria on the sediment surface and make a burrow to feed in the immediate anoxic layer below. Oscillating body movements set up a current that carries dissolved oxygen into the burrow tubes, permitting the organisms to feed, respire, and defecate below the oxic–anoxic interface during periods of high water column DO, in the process altering the redox potential and other sediment properties in the burrow walls (Sec. 16.2). Alternatively, *C. anthracinus* filter-feeds selectively on sedimenting or resuspended algal and detrital particles that are swept into the feeding

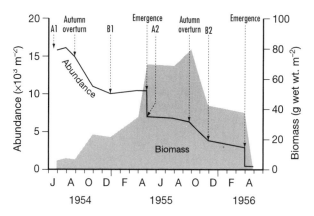

Figure 25–10 Survivorship curve showing the seasonal fluctuations in average abundance and population biomass of *Chironomus anthracinus* (Insecta) in the hypolimnion of Lake Esrom, DK (1954–1956) in relation to external factors shown by arrows. A1 and A2 show the beginning of the summer low-oxygen period (anoxia) in the first and second year of the two-year life cycle. B1 and B2 indicate the cessation of fish predation, which is linked to a seasonal decrease in water temperature. *(Modified after Jónasson 1972.)*

tubes by the currents created by body movements. During periods of lower DO the same chironomid species acts as surface-deposit feeders by extending their upper bodies out of the tubes to collect particles—including recently sedimented algae—from the sediment surface. They do this by first spreading a salivary secretion over the sediment like a net and then dragging it with the attached particles into the tubes (Jónasson 1972). Chironomid feeding is selective rather than indiscriminate. Gut analyses have shown that they contain some five times the quantity of bacteria found in the filtered suspended material (Johnson et al. 1989). Yet other chironomids are primarily predatory rather than omnivorous (e.g., *Procladius spp.*) and roam the sediment surface.

The zoobenthic community suffers high mortality rates that—in terms of community biomass—are usually more than offset by growth of the survivors (Fig. 25–10). Survivorship curves, together with information on increase in weight per individual, are used to compute production of cohorts of individual invertebrate species (Sec. 23.13, and Benke 1984). A large year-to-year variation in the population biomass of invertebrates is common, and is the outcome of large interyear differences in initial biomass (number of young produced), production, mortality, and emergence in the case of flying insects (Table 25–1 and Fig. 25–11).

Table 25–1 **Production, mortality (% of production in parentheses) and larval emergence (g m⁻² dry wt.) during three life cycles of the profundal *Chironomus anthracinus* (Insecta) in Lake Esrom, DK. Note the large among-year variation in production, mortality, and emergence.**

Parameter	Years		
	1954–1956	1956–1958	1958–1960
Initial biomass	0.6	0.4	0.7
Production	21	14	41
Mortality	7 (33%)	8 (57%)	22 (54%)
Emergence	15	6	20

Source: Modified from Jónasson 1972.

The hemoglobin-carrying *C. anthracinus* is able to maintain a high aerobic metabolism down to 2–3 mg O_2 l^{-1}, below this level it becomes increasingly dependent on the stored glycogen utilized in anaerobic metabolism (Sec. 15.7). Survival is greatly reduced during lengthy periods of hypoxia/anoxia, and only 50 percent of experimental animals survived two to five weeks under totally anoxic conditions (Hamburger et al. 2000). In Lake Esrom the time required for the onset of sediment anoxia has decreased by about a month since the 1950s (Fig. 15–9), but whether the observed decline of *C. anthracinus* is attributable to the longer period of anoxia cannot be clearly resolved as the result of an order of magnitude interyear variation in abundance (Fig. 25–11). However, hypolimnetic aeration (Sec. 17.6) of one basin of a two-basin eutrophic Canadian lake reduced the period of hypolimnetic anoxia from about four months to between three and six weeks and totally eliminated winter anoxia, thereby greatly raising the whole-lake abundance and

community biomass of *C. anthracinus* (Dinsmore and Prepas 1997). An aerobic sediment surface and the resulting lower mortality, plus a longer growing season, made a larger fraction of the lake bottom available for feeding, but also allows for a greater fish predation.

Growth and loss rates differ among years, among sites, and among species, and the community composition can be expected to show much seasonal and interannual variation.

25.5   Resource and Predation Control

Both resource availability (bottom-up) and predation (top-down) control help determine directly or indirectly the zoobenthic community biomass and composition under otherwise suitable environmental conditions. The impact of resource availability received a lot of attention during the so-called "eutrophication era" (1970–1985). Influential experimental

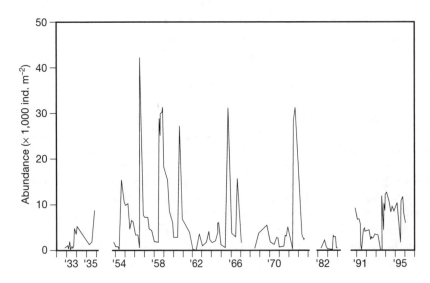

Figure 25–11 Changes in abundance of the midge *Chironomus anthracinus* (Insecta) in the deep profundal zone of Lake Esrom (LA = 17.3 km², $\bar{z}$ = 13.5 m), a Danish kettle lake, between 1932 and 1995. Note the large variation among years and decades. The life cycle is annual (univoltine) or may have two generations per year (bivoltine) in the upper hypolimnion, but the larvae have a two-year life cycle (hemivoltine) in the deep profundal zone with its long periods of hypoxia/anoxia. (*After Lindegaard et al. 1997.*)

work by Hall et al. (1970) on a series of replicated ponds (morphometry and climate held constant, no anoxia) elegantly demonstrated the impact of varying fertilization rates on the zoobenthic biomass and community structure. Similar conclusions were drawn from other observational and experimental studies (Kajak and Rybak 1966, Hanson and Peters 1984). But trophic indicators alone predict only a modest fraction (14–57 percent) of the among-system variation in zoobenthic biomass (Rasmussen and Kalff 1987) in observational studies, where the effect of important nontrophic variables cannot be held constant. It is therefore not surprising that empirical models of zoobenthic biomass, based on trophic status data from a variety of systems in a single climatic zone, are improved when exposure and underwater slope (morphometry) are considered as well (Fig. 25–8 and Rasmussen 1988). Even so, the evidence for the importance of food (resources) in determining community biomass and composition is not in dispute. Nor is the conclusion that predation and other sources of mortality affect the zoobenthos, but these losses tend to vary greatly over space and time and are not as well documented.

The macrozoobenthos is an important component of the diet of most fish species of commercial and recreational interest. An analysis of the literature shows that the stomach content of adult piscivorous fish such as lake trout and northern pike contain, on average, > 75 percent fish. But this underestimates the importance of zoobenthos to the extent that their fish prey was in turn dependent on the zoobenthos (Fig. 25–12). Adults of important prey fish such as perch and whitefish are typically largely benthivorous.

▲ Size-selective predation by fish helps structure the composition of the zoobenthos (Table 25–2) and affects the community abundance and biomass. Large

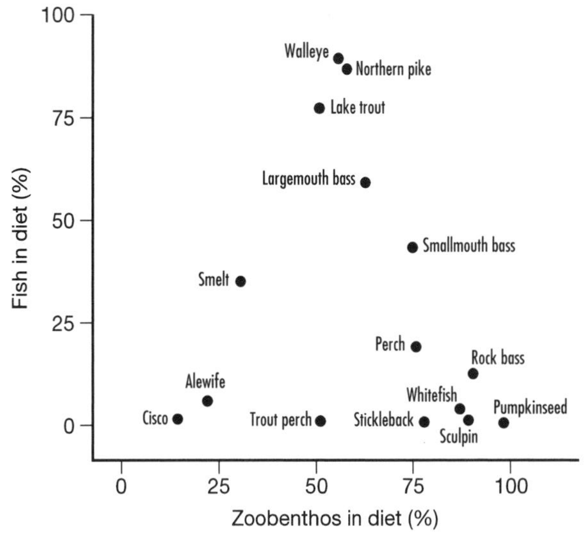

Figure 25–12 The relative importance of direct plus indirect use of the zoobenthos (by volume) in the average diet of fifteen North American fish, based on a literature analysis. Fish plus zoobenthos in the diet, totaling more than 100 percent, represents an indirect contribution of the zoobenthos via prey fish to the diet of piscivorous fish. Diet composition may vary greatly from the average, depending on differences in food web structure among lakes. (*Y. Vadeboncoeur and J. Vander Zanden, unpublished.*)

invertebrate species are disproportionately common in both fishless ponds and wetlands (Mittelbach 1988 and Mallory et al. 1994), while the predominance of small epiphytic invertebrates in most lakes and wetlands appears to be the result of size-selective fish predation on the larger forms (Strayer 1991). Furthermore, field manipulations have shown dramatic effects of benthivorous fish predation on the density of the benthos

Table 25–2 **Effects of fish on the mean body size of the macrozoobenthos. The "ratio" represents the mean body mass of the zoobenthos when fish were present divided by the mean body mass of the zoobenthos when fish were absent.**

Lake	Fish Species	Ratio
Pond, Michigan, US	bluegill	23%
Lawrence, Michigan, US	centrarchids	49%
Little Minnow, Ontario, CA	yellow perch	58%
Warniak, PL	carp, bream	79%
Ponds, New York, US	bluegill	"slightly smaller"
Third Sister, Michigan, US	mixed warmwater species	165%

Source: After Strayer 1991.

Table 25–3 **The estimated fate of macrozoobenthic production in several lakes. The data are approximate at best and the percentages do not sum to 100% (1 g C = 10 kcal). ND = no data.**

Lake	Zoobenthic Production g C m^{-2} yr^{-1}	fate %			
		Invertebrate Predation	Fish Predation	Bird Predation	Emergence
Myastro, BY	0.4	72	27	2	ND
Ovre Heimdalsvatn, NO	1.2	25–28	70	2	3
Naroch, BY	1.3	22	42	—	ND
Batorin, BY	1.3	46	19	4	ND
Pääjärvi, FI	1.9	50	40	—	6
Mirror, US	7.0	80	15	—	25
Mývatn, IS	21.0	5–52	43	9	48

Source: After Strayer 1991.

(Andersson et al. 1978). Yet, the effect of fish on the size structure of the benthos is generally modest compared to the effect of fish introductions on the size structure of zooplankton, suggesting that the sediments and macrophytic vegetation affords more protection from predation than is possible for plankton (Strayer 1991).

Zoobenthic production is not only lost to fish predation but also to predation by other invertebrates, as well as by birds such as grebes and ducks whose role as a "sink" for production has received little attention from limnologists and is likely underestimated (Table 25–3 and Sec. 26.9). The principal bird predator in Lake Mývatn, (IS, $\bar{z}$ = 2 m), is a large population of diving ducks feeding largely on two species of chironomids that together contribute about 87 percent of the macrozoobenthic net production (Lindegaard 1994). Finally, the emergence of insects and the dissipation of their production over land is a loss to aquatic systems (Fig. 23–20 and Strayer and Likens 1986).

▲   ## 25.6   The Zebra Mussel: A Keystone Species

Zebra mussels (*Dreissena* spp.) are small filter-feeding bivalves living in fresh or moderately saline (hyposaline) waters and native to the Caspian and Black Sea area of Eurasia. They spread into central and western Europe during the last 200 years after the construction of canals between river basins. The mussels have become an important component of the European zoobenthos and have been established for so long in most areas that very little is known about their effect on the native benthos.

During the 1980s dreissenid mussels (*D. polymorpha*) and the closely related quagga mussel (*D. bugensis*) were introduced into the Great Lakes of North America, presumably as hitchhikers on an ocean-crossing vessel that voided ballast water taken on in Europe. The zebra mussel has spread to all the Great Lakes and from there has colonized many connected waters, including the Hudson River and the Mississippi River basin. The mussels easily attach themselves to the hulls of boats by means of strong elastic threads (*byssal threads*) and thus spread quickly along transportation routes. The larvae (known as *veligers;* ~80–200 μm long) are also unique among freshwater bivalves in that they remain planktonic for several weeks and are widely distributed by water currents during this period. In the larval form they can be transported long distances in the bilge water of motorboats and in bait buckets. They complete their development in one to nine weeks and then temporarily settle on macrophytes and sediments or other hard surfaces—including floating objects—where they transform into the adult form. Using their byssal threads and an epoxy-like adhesive, the mussels can build large colonies several layers thick even on relatively soft sediments. Since the larvae prefer solid surfaces and moving water, they readily settle on submerged macrophytes, boat hulls, and water-intake pipes. They filter-feed, mature in one year, and live for several years (Strayer 1999).

▲ In the Laurentian Great Lakes the animals start reproducing in the second of their three-to-five year life span, producing as many as a million eggs per large female per year (Sprung 1990). There is now considerable concern among North American scientists about the effect of the dreissenids on the native mussels and ecosystem functioning. Water managers are concerned about the zebra mussel's ability to rapidly and efficiently block water-intake pipes and increase pipe erosion as the result of a buildup of bacteria that are nourished and physically protected by the mussels.

Data on abundance from areas of western and central Europe, where the mussel invaded around a century ago, leave unresolved whether abundances in North America can be expected to ultimately decline as the result of resource limitation, predation, and disease. However, maximum abundances (> 100,000 m^2 of recently settled young) do not greatly differ on the two continents. Furthermore, European values are of the same magnitude as those predicted by an empirical model developed for North America (Mellina and Rasmussen 1994).

The well-oxygenated erosional portion of the littoral and sublittoral sediments of lakes and equivalent sites in rivers are favored mussel habitat. *Dreissena* abundance among systems is, apart from resource availability, limited by the availability of suitable substrate and a sufficient dissolved calcium concentration (10–20 mg l^{-1}) for shell formation. (Mellina and Rasmussen 1994). A side-scan sonar survey in Lake Erie shows the deep water abundance there to range between 1,500 and 32,500 individuals per square meter on sand and sandy mud ($\bar{x}$ ~500–100 μm diameter) substrates in the extensive zone of sediment erosion (ZSE) between 10 m and 20 m (Berkman et al. 1998).

Ecosystem Impacts

The two mussel species are reorganizing community structure and energy flow in heavily colonized shallow lakes and rivers, in the process shifting the principal pathway of energy flow away from the plankton towards the benthos. The dreissenids affect not only the phytoplankton through their predation but also the zooplankton, directly by competition for algal food and bacteria, and indirectly by predation on rotifers, ciliates, small naupli (see Johannsson et al. 2000) and, presumably, veliger larvae. In the fresh water portion of the Hudson River (US) estuary, filtering by zebra mussels was responsible for an 85 percent decline in the phytoplankton biomass (chl-*a*), from a summer average of 30 μg l^{-1} prior to the invasion to < 5 μg l^{-1} when the mussels had become well-established in 1993 and 1994 (Caraco et al. 1997).

The reduction in phytoplankton biomass and community primary production in shallow, and most affected, Lake Erie has changed the planktonic food web structure and greatly increased not only the relative importance but also the absolute importance of benthic primary production—the result of increased transparency and the extension of the bottom of the littoral/euphotic zone into deeper water. An equivalent decrease in the turbidity of Saginaw Bay in Lake Huron (US, CA) increased the density, depth distribution, and area covered by submerged macrophytes and macroalgae after the mussels invaded (Skubinna et al. 1995).

Estimates of the relative importance of zooplankton production (including veliger larvae) and benthic production in Lake Erie indicates that in 1993, when the zebra mussels had become well established, the zooplankton contributed an average of only 15 percent of the pelagic plus benthic secondary production per square meter, with the balance contributed by the benthos. Although no good preinvasion data are available, a comparison of estimates of benthic biomass in the western basin in 1993 (351 g wet wt. m^{-2}, shell-free) with equivalent preinvasion data from 1979 (7 g wet wt. m^{-2}, but without the native unionid mussels) indicates a huge increase in benthic biomass and its importance to energy flow, with the dreissenids alone contributing > 90 percent of the benthic production (Johannsson et al. 2000).

It now appears that the invasion has not been, as was feared, at the expense of macrobenthic species diversity, except for the native unionid mussels whose shells are smothered by the invaders and who also compete with them for food, thereby threatening the survival of those species unable to find refuge in sites (e.g., wetlands) unsuited to the dreissenids. Experimental and observational research has shown that the dreissenids provide a favorable physical habitat and contribute increased particulate organic matter from mussel feces and pseudofeces for benthic invertebrates—other than native mussels—(Stewart et al. 1998) and previously abundant amphipods.

Predators on zebra mussels include a variety of fish, crayfish, crabs, turtles, coots, and diving ducks. Although some are efficient predators, none is known to exert much short-term or even local control over the mussels (Strayer 1999). The long-term effect of

Table 25–4 **Estimated production of macrozooplankton and macrozoobenthos in selected lakes, expressed as a percentage of the net organic carbon input to the lakes. Note (1) the generally modest mean depth of lakes for which good zoobenthic (plus zooplankton) data are available; (2) the shallowest and deepest waters are dominated by benthic and planktonic production, respectively; but (3) the relative significance of zoobenthic vs. zooplankton production is probably further modified by differences in predation suffered, water temperature, and periods of sediment anoxia, (probably of greatest importance). For sources of organic carbon see Table 22-7.**

Lake	$\bar{z}$(m)	Net Organic Carbon Inputs[1] (g m^{-2} yr^1)	Production (% of net organic carbon inputs)		Zoobenthos (% of zooplankton + zoobenthos)
			Macro-zooplankton	Macro-zoobenthos	
Tundra Pond (US)	<< 1	26[3]	0.8	7	90
Marion (CA)	2	110	0.9	3[2]	77
Mývatn (IS)	2	330[3]	1.3	6	82
Hjarbaek (DK)	2	—	—	—	86
Kiev Res. (UA)	4	280	7	6[4]	46
Mirror (US)	6	49	5	12[5]	71
Red (RU)	7	140	7	1.4	17
Findley (US)	8	12	4	6[2]	60
Mikolajskie (PL)	11	260	21	2	9
Esrom (DK)	12	160[3]	7	6	46
Pääjärvi (FI)	14	60	12	3[4]	20
Dalnee (RU)	32	260	22	1	4
Thingvallavatn (IS)	34	—	—	—	32

[1]Primary production and allochthonous inputs minus losses by outflow.
[2]Aquatic insects only.
[3]Excluding allochthonous inputs.
[4]Macrobenthos and some meiobenthos.
[5]All benthic metazoans.

Source: Modified after Strayer and Likens 1986, and Lindegaard 1994.

zebra mussels is a shifting of invertebrate production away from zooplankton in the water column, and thereby away from zooplanktivorous fish and their predators, to the benthos. The effect of all this on pelagic fish stocks in the Laurentian Great Lakes and elsewhere is an issue of great economic importance and scientific interest.

▲ 25.7 The Zoobenthos and Energy Flow in Lakes

The relative importance of the macrozoobenthos in invertebrate production is large compared to the macrozooplankton in the shallowest lakes examined (Table 25–4) but tends to decline with increasing lake depth. Shallow systems have, as mentioned earlier, a large sediment surface to overlying volume ratio. A thin water column allows little time for the decomposition of sedimenting organic particles, permitting a higher fraction of them to reach the sediments. Furthermore, the relative importance of macrozooplankton in shallow eutrophic lakes appears negatively affected by a disproportionate abundance of visually feeding zooplanktivorous fish (Fig. 23–16).

The considerable variation in the relative importance of the zoobenthos in the production of eutrophic lakes deep enough to stratify (Table 25–4) suggests that the importance of the macrozoobenthos in whole-lake invertebrate production is obscured by

Table 25–5 **The approximate energy budget of oligotrophic and moderately humic Lake Pääjärvi (FI, $\bar{z}$ = 14 m), and percent of total inputs and export (40 g C = 1 kJ) contributed by components. Note (1) that all whole-system budgets are based on a summation of components that require scaling up from inevitably small areas or volumes sampled, with imperfect techniques on a limited number of days, to an annual estimate of the whole system; (2) the budget does not balance; and (3) the output is much larger than the in situ inputs, pointing to an important role for the abundant allochthonous organic matter.**

	Inputs g C m^2 yr^{-1}	Total Inputs %		Outputs (losses) g C m^{-2} yr^{-1}	Total Outputs %
Allochthonous organic matter	62	63	Bacterioplankon respiration	47	37
Phytoplankton net production	27	28	Zooplankton respiration	17	14
Littoral primary production	9	9	Sediment bacterial respiration	14	11
			Macrobenthic and meiozoobenthic respiration	5	4
			Fish respiration	2	1
			Yield to fisheries	<< 1	<< 1
			Insect emergence	<< 1	<< 1
			Sediment storage	8	6
			Outflow	35	27
Total	98	100		128	100

Source: Modified after Sarvala et al. 1981.

differences in trophic status and lake morphometry that help determine the extent and duration of hypoxic or anoxic conditions over the sediments. A large variation among (formerly) USSR lakes in the ratio of macrozoobenthic respiration to phytoplankton production (Alimov 1982) may, similarly, have been attributable to much among-lake and among-year variation in the duration of the low DO period.

The macrozoobenthos is a crucially important direct and indirect source of food to fish (Fig. 25–12) and aquatic birds, who then help structure the size distribution of the benthos and the relative abundance of the different species (Sec. 26.9). However, based on energy flow considerations (Sec. 22.9) and supported by measurements in a Finnish lake, the contribution of the macrozoobenthos (and macrozooplankton) to whole-system metabolism is modest. In Lake Pääjärvi (FI), nearly three-quarters (72%) of the whole-system respiration was attributable to the bacteria, compared to about 20 percent for the macrozooplankton and six

percent for the macrobenthos plus meiozoobenthos (Table 25–5). The contribution of the sediment bacteria to the *sediment* energy-flow budget of Mirror Lake (US) was smaller (59%), and the contributions of the meiobenthos (27%) plus macrobenthos (14%) relatively more important. However, conclusions are, as always, dependent on the spatial and temporal scales of the research. They are further affected by the necessary assumptions made in scaling up measurements obtained at generally small spatial scales over relatively short periods to the system as a whole on an annual basis.

Highlights

- The zoobenthos is divided on the basis of size between the megabenthos (> 1,000 μm), the relatively well-studied macrobenthos (<1000–400 μm), the meiobenthos (400–100 μm), and the little investigated microbenthos (< 100 μm).

- The zoobenthos is numerically dominated by small species and is species-rich compared to the invertebrates of the open water (zooplankton).
- Species richness, abundance, community biomass, and production is highest in the littoral zone—the richest in resources, best-oxygenated, and warmest portion of aquatic systems—and is lowest in the profundal sediments of deep lakes.
- The structurally diverse macrophyte beds of wetlands, lakes, and rivers are important sites of organic matter accumulation and provide a large surface area for colonization by the zoobenthos and their prey; they also provide some protection from fish predators, and as a result they are characterized by the highest zoobenthic biomass per unit area and greatest species richness.
- Invertebrate and fish production (Ch. 26) is roughly predictable from abundance data, weight per individual, and temperature; biomass serves as a relatively easily obtained indicator of population and community production.
- The among-system zoobenthic biomass and production increases with resource supply (trophic status),

but is reduced in waters experiencing lengthy periods of hypoxia/anoxia.
- ▲ Macrozoobenthic net production is lost through mortality, insect emergence, and predation by invertebrates, fish, and birds.
- ▲ Zebra mussels, when abundant in rivers and shallow lakes, shift whole-system invertebrate production toward the sediments.
- ▲ There is a tendency for the macrozoobenthos to dominate the macrozoobenthos plus macrozooplankton biomass and production in shallow systems, characterized by a small water volume to sediment surface area, and for macrozooplankton to dominate deep systems. The pattern is weakened by variations in predation, sediment anoxia, and insect emergence.
- ▲ The macrozoobenthos (and macrozooplankton) play a crucially important direct and indirect role in the production of fish and aquatic birds, which in turn help structure the macrozoobenthic size distribution and relative species abundance.

26

Fish and Water Birds

26.1 Introduction

Fish play a major role in aquatic systems and are also of great economic importance. They have received much attention, albeit largely at the species level and therefore independent of other fish species and the ecosystem of which they are a component. There are two important reasons for the traditional isolation of research on fishes from the rest of limnology. First, the greater economic importance of some fish has encouraged more research on the life history and population dynamics of commercially important species in isolation from their noncommercial counterparts and the rest of limnology. The second reason is the vastly larger size of most fishes relative to the planktonic and benthic animals of interest to other aquatic scientists. The greater size of most fishes is associated with much greater longevity, slower biomass doubling time (Fig. 2–4), and much higher mobility and requires not only different sampling techniques but also study over different temporal scales (years to decades) and larger spatial scales (whole lakes plus their wetlands and inflowing rivers) than the equivalent scales of concern to most biological limnologists.

The lack of attention given to water birds is more understandable because birds are not exclusively aquatic even if most or all of their nutrition is obtained from aquatic systems; they exert an important effect on the abundance and community structure of macroinvertebrates and fish.

Fish Research and Limnology

The earliest work on the taxonomy of fishes and their distribution evolved in the first half of the 19th century into a broader-based study of fishes and their biology. This form of study came to be known as **ichthyology**. It began with descriptive studies by zoologists of the morphology and life history of individual species, including their early development, growth, and feeding habits. This focus gradually expanded to encompass research on fish physiology, diseases, culture, behavior, and population biology. The early fundamental research by ichthyologists formed the basis for more applied research after evidence of overfishing of commercially important stocks in both inland and marine waters became apparent in the late 19th and early 20th centuries. The applied research led to the development of a **fisheries science** that is concerned with the management of commercially important **stocks**, defined as groups of randomly mating individuals occupying the same spatial habitat. More specifically, it is concerned with the impact of fishing on exploited populations, the steps required to achieve and maintain optimum yield, and the replenishment of depleted stocks through fish culture. Fisheries science is practiced primarily in government-sponsored agencies. In North America, it was long taught in Schools or Departments of Fish and Game and Natural Resources programs that were physically and often conceptually isolated from the fundamental science departments (zoology and botany) that housed the limnologists (Magnuson 1991). With the development of ecology, including **fish ecology**, during the first 30 or 40 years of the 20th century, zoology departments sometimes supported a modest number of fish ecologists.

Communication and collaboration between these three groups of scientists has often been limited because the groups adhere to different scientific paradigms and typically work at different temporal, spatial, and hierarchical scales (Sec. 2.6). Fish ecologists tend to work at the hierarchical scale of populations, focusing on population structure and interactions between species, primarily from an evolutionary perspective. Fisheries scientists emphasize organization at the population level from an applied rather than fundamental perspective. Finally, biological limnologists study small organisms with doubling times ranging from hours to weeks; these organisms have typically been examined from a fundamental perspective over small spatial (ml, l, m^2) and temporal (hours, days) scales (Figs. 1–1 and 1–2). Nevertheless, most limnologists have—conceptually at least—placed their small-scale studies into a larger scale (ecosystem) perspective. All three groups have paid much more attention to the more easily sampled pelagic zone (open water) than the littoral zone of lakes, rivers, or wetlands. The traditional separation between fish ecology and fisheries science is becoming blurred now that an increasing number of ecologists are working on applied problems. Yet some of the long-established isolation between fundamental and applied aquatic science remains.

▲ The International Association for Theoretical and Applied Limnology (SIL) established in 1922 (Sec. 2.2) purposefully included the word "applied" in its title, but relatively few applied scientists found a home in an organization dominated by those with fundamental interests. Most hydrologists, water pollution biologists, fish ecologists, and virtually all fisheries scientists joined organizations that better represented their interests. As a result, only 10 percent to 15 percent of the papers presented at the SIL congresses in the half-century following World War II have been on fish (Northcote 1988). An even lower fraction of articles published in the pre-eminent international journal *Limnology and Oceanography* have been devoted to fish, and virtually all of those have addressed only one aspect—the top-down effect of fish on lower trophic levels.

Fish production and community structure are affected by the nutrient supply, something appreciated for centuries by those involved with aquaculture. During the last 50 years or so this has also been recognized by fisheries scientists and those limnologists interested in the effects of plant nutrients (resources) on fish yields (catches) and community production. Fish in turn directly affect the structure and energy flow at other trophic levels through their selective feeding on plants, invertebrates, and other fish. Fish indirectly influence the abundance and composition of algal and bacterial communities through their predation on macrozooplankton and benthic invertebrates, as well as through nutrient recycling and transformation of organic matter (Secs. 23.7 and 25.5).

▲ 25.6 Fish Species and Species Richness

Fish Species

The inland waters of the world contain nearly 11,000 fish species. These can be divided into two groups or classes: those with and those without jaws (a tiny fraction of the species). The jawed fishes are further divided into species with a cartilaginous skeleton (the sharks and rays) and those with a bony skeleton (the vast majority). Among the bony fishes two subclasses are relevant to limnologists: the lungfishes (*Dipnoi*) comprising only a few species, and the ray-finned fishes (*Actinopterygii*). The ray-finned fishes in turn are dominated by the evolutionarily advanced bony fishes, the (*Teleostei*) (Figs. 26–1 and 26–2). The teleost fishes, which have a symmetrical tail and a gas bladder used primarily for buoyancy control, make up ~94 percent of all inland water species. Among them the carps (Order Cypriniformes) dominate with ~3,000 species, followed by the catfishes (Order Siluriformes) with ~1,950 species, and the perch-like species (Order Perciformes) with ~950 species. The latter include temperate zone perch as well as the many species of low latitude cichlids.

The air-breathing lung fishes live in high temperature, low latitude wetlands that are likely to become periodically hypoxic or anoxic (Sec. 15.7). Some can even survive out of water in damp habitats while others survive droughts by burrowing in damp sediments (Lowe-McConnell 1982). In contrast, many of the teleost fish respond behaviorally to low dissolved oxygen concentrations (< ~4–7 mg l^{-1}) by migration or local movements and die when DO declines below species-specific minima (Sec. 15.7). However, a wide variety of low altitude fishes have developed anatomical and physiological adaptations that allow them to utilize (gulp) the air above the surface or use the thin (often only a few millimeters thick) well-oxygenated

Figure 26–1 Examples of genera of teleost fish from various river and lake systems of the world. (1) *Hypostomus*, S. America; (2) *Cyprinus*, now worldwide; (3) *Potamotrygon*, S. America; (4) *Perca*, originally N. America, Eurasia; (5) *Lates*, Africa; (6) *Labeo*, Africa, Asia; (7) *Osteoglossum*, S. America; (8) *Lepomis*, N. America; (9) *Salmo*, now worldwide; (10) *Clarias*, Africa, Asia; (11) *Esox*, Eurasia, N. America; (12) *Pseudoplatystoma*, S. America. *(After Welcomme 1985, Trautman 1957, and Lowe-McConnell, 1975.)*

surface layer (Kramer and McClure 1982, and Welcomme 1985). A relatively small number of low latitude cyprinid species are annual; the eggs survive the drying up of ponds and wetlands and hatch following inundation. Finally, the crucian carp (*Carassius carassius*) of Europe lives in shallow ponds that commonly experience winter anoxia. The fish survive in an inactive state by drawing upon previously deposited glycogen reserves and a capacity for anaerobic respiration (Sec. 15.7).

Species Richness

The determinants of species richness change, as does every other system attribute, with the scale examined. (Sec. 2.6) On a global scale, species numbers increase with increasing system surface area and decrease with increasing latitude and altitude (Barbour and Brown 1974). Similarly, the number of species increases with drainage-basin size on a global scale (Fig. 26–3) because at this spatial scale the impact of the nearly four

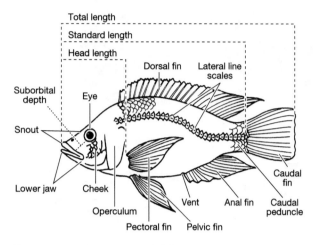

Figure 26–2 A spiny-rayed fish (*Oreochromis mossambicus*, Family Cichlidae) showing topographical features and how certain measurements are made. Some teleost fish species have two dorsal fins (see Fig. 26–1). The standard length is commonly assumed to be ~90 percent of total body length in spiny-rayed fish. *(After Ricker 1968.)*

orders of magnitude variations in drainage-basin size overwhelms the relatively modest effect of latitude on species richness—at least among the largely temperate zone and low latitude drainage basins examined to date. Larger basins, containing a number of smaller drainage basins, exhibit a wide range (variation) in catchment slopes (Sec. 8.5) and therefore contain

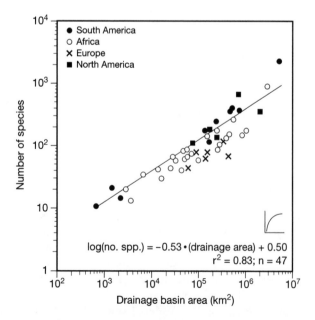

Figure 26–3 Relationship between drainage basin size and number of fish species. *(After Welcomme 1985.)*

rivers, lakes, and wetlands. Slowly flowing rivers and extensive wetlands in the low-sloped lower portions of large basins provide a wide range of habitats permitting ready colonization and recolonization after droughts, or following fish kills during periods of hypoxia. In contrast, smaller and more steeply sloped drainage basins typically contain relatively few but commonly deeper lakes and small rapidly flowing rivers that are much more difficult to ascend and colonize than lowland rivers. Drainage is normally good in steeper basins and, as a result, there is little or no wetland development. It is therefore not surprising that the number of fish species in a particular lake is related to the number of lakes and rivers in the drainage basin (Minns 1989).

A model of fish-species richness developed with data from 2,931 Ontario (CA) lakes is an example of how species richness changes with both spatial and range scales. It shows that not only lake size (habitat diversity) and latitude (climate, temperature, ice cover) but also altitude (climate, accessibility), dissolved organic carbon, and total aluminum concentration (Sec. 27.9) play a role in predicting species richness (Matuszek and Beggs 1988). Moreover, the Ontario lakes support fewer species per unit area than the rivers, a finding attributed to the greater habitat variability in river systems.

Habitat diversity is an important among system predictor of species richness in inland waters (Eadie and Keast 1984) and the frequently reported species number– lake area (Figs. 23–4, and 24–3) or species number– catchment area relationship (Fig. 26–3) is usually interpreted to represent a species number–habitat diversity model (Keller and Crisman 1990) in which larger systems provide greater habitat diversity. Although fish community biomass and production are related to the primary productivity of aquatic systems (Sec. 26.7), species richness is not. Richness is a function of the time available for the evolution of species, habitat variation, and the possibilities for colonization and subsequent survival (Watson and Balon 1984, Randall et al. 1995).

The total species pool available for colonization is small in the recently (~7–10 thousand years BP) deglaciated regions of Europe (200 native species) and North America (170 native species). The time available for speciation has been short on both continents with water temperatures too low in winter to allow year-round breeding. Feeding and growth are commonly reduced as well. Biogeographical barriers imposed by hills and mountains that divide catchments are another factor explaining the relative species poverty in the recently deglaciated portion of the Northern Hemisphere. **Endemic** species, those species evolved in and restricted to a particular waterbody, are absent. Consequently, the five relatively young Great Lakes of North America contain about 170 fish species while ancient Lake Malawi (east Africa) contains, depending on the systematist, nearly 1,000 species—including more than 600 species of cichlids (Order Perciformes, Family Cichlidae) of which 99 percent are endemic. However, time is not the only determinant of endemism. The endemic cichlids of shallow Lake Victoria (east Africa) may have evolved in a mere 12,000 years—based on evidence for the lake's desiccation at that time—unless they somehow survived the drought period in connected but less affected waterbodies.

26.3   Life-History Attributes and Population Dynamics of Age-0 Fish

The large variation among fish species in adult size, body shape, and food consumption is reflected in an equivalent diversity in life history. However, the life history of a species is the outcome not only of current conditions and constraints, but also of evolutionary (historical) constraints which confound a simple link between form, function, and environmental conditions.

Life history includes where a species lives (lotic vs lentic, littoral vs pelagic, pelagic vs profundal), how it feeds (visual vs filter-feeding), on what and where it feeds, food preferences at different ages, growth and mortality rates, number and size of eggs, age at maturity and longevity, breeding, brooding, and schooling behavior, as well as intraspecific interactions. These and other attributes are species-specific and are the subject of monographs and specialized journals on fish ecology, behavior, culture, and fisheries management. An understandable emphasis on selected species and population attributes[1] has resulted in much less attention paid to the functional grouping of species or community attributes than research on the plankton and benthos.

Changes in abundance are the result of changes in the **recruitment** (production) of young and mortality, with each affected by the biotic and abiotic environment. Today it is widely believed that, except where a stock is greatly depleted, the abundance of spawning fish and the **fecundity** (eggs individual^{-1} or population^{-1}) is less important than the survival of the larvae hatched. For example, the relationship between the population size of the walleye (*Stizostedion vitreum vitreum*) and the recruitment of young in Lake Erie (CA, US) varies dramatically between years (up to a hundredfold) over a 19-year period (Shuter and Koonce 1977).

Mortality Rates of Larval and Juvenile Fish

Fish larvae suffer enormous mortality rates of typically 5–25% d^{-1}. This is a result of both starvation, attributable to a poor match of the emergence of the larvae with that of their prey, and predation (Leggett 1986). Measured mortality rates from eggs to the larval stage for five species of age-0 (< 1 yr) fish (cyprinids) in the Thames River (GB) was > 99 percent (Mathews 1971), and 78 percent of the remainder between the **larval stage,** when the young lack a calcified skeleton and may not resemble the older fish, and the **juvenile stage** when they do resemble the adults. Juvenile mortality, is also enormous.[2] Long-term work on *young-of-the-year* (YOY, age-0, 0$^+$) yellow perch, *Perca flavescens*[3] in Oneida Lake (US) showed that the abundance of those age-0 fish that had survived from hatching in spring until August declined by an average of 84 percent, during the subsequent two months. However, as the surviving fish continued to grow, the population biomass declined less (60%) over that two-month period (Mills et al. 1989). At the same time, the interyear variation in weight gained by individual juveniles at the onset of fall ranged fivefold over an 11-year period (1.27–6.55 g), reflecting great interyear variations in food availability per individual, and a greater probability of winter starvation in years when the age-0 fish enter the winter as small individuals with low energy reserves. Furthermore, the smaller the size of the win-

---

[1]"With very notable exceptions, 'fish biologists' have tended to specialize at one particular level of organisation, most at the organism, some at the population, and fewer at the community level. Within a particular level one can arrange different workers along a continuum from the purely abiotic to the biotic . . . Sometimes the choice is made rationally, often it appears to be influenced in large measure by the biases within disciplines to which the worker happened to be subjected as a student." (Regier 1974.)

[2]"The majority of larval [and adult] fish estimates have low precision. Taking four replicated samples, the median number of samples taken in population studies of larval fish, one would have to count more than 30,000,000 larvae/sample on average to yield a $\bar{x}$CV [mean coefficient of variation] = 0.1 [or 10%] . . . Yet the median number of larval fish per sample in published studies is around 7." (H. Cyr et al. 1992.) See Peterman 1990 regarding the danger of drawing unwarranted conclusions about older fish as the result of small sample sizes or a large sampling variability.

ter survivors the greater the possibility that the now age-1 fish will continue to be subject to predation by adult perch and other fish (Mills and Forney 1983).[3]

Cohort Strength

Most fish populations show an enormous variation in **cohort strength**, the number of fish of any single age. Only small changes in the mortality of larvae and juveniles are required to produce large changes in cohort size. Where fish spawn during a single season the designation **year-class strength** is commonly used. Occasionally a very strong year-class may be responsible for the success of a commercial fishery over several years, or alternatively, the total or nearly total failure of one or more year-classes can result in years of poor catches. Long-term studies, mostly in Great Britain and the United States, have revealed strong perch cohorts in years when the abundance of reproductive adults was high but few potentially competing juveniles were present. Conversely there were few age-0 fish in the two to three years following the appearance of a strong year-class, apparently the result of cannibalism by juveniles and adult perch or competition for food (Kipling 1976, and Romare et al. 1999). Young fish commonly change habitat to reduce the risk of predation, cannibalism, or competition for food or habitat. They tend to avoid open water and select vegetated inshore areas that provide better protection from predation by larger fish (Tonn et al. 1992) and provide rich food resources.

Age-0 Fish and Zooplankton

The year-class (cohort) strength of age-0 fish, which are commonly zooplanktivorous, plus the cohort strength of species in which the adults too are zooplanktivorous, has a large and determining effect on the abundance of their zooplankton prey (Mills and Forney 1983) and thereby on zooplankton species composition and size distribution (Figs. 23–13, 23–15, and 23–16). As fish grow the **gape** size (mouth diameter) increases, enabling them to include or switch to progressively larger and energetically more efficient prey species, thereby imposing a temporally changing effect on the size structure and species composition of their prey. Densities of about two to three (age-0 perch m^{-3}) in perch-dominated lakes are sufficient to suppress the *Daphnia* biomass throughout the summer. The reduced zooplankton populations then exert a sufficiently reduced grazing effect to allow an elevated phytoplankton biomass, reduced water clarity (see Romare et al. 1999), and changes in algal species composition. Conversely, a reduction of fish stocks to < 100 kg ha^{-1} in some shallow eutrophic European reservoirs permitted an increase in the size of the macrozooplankton, while a further reduction to < 30 kg ha^{-1} somehow yielded a dense growth of filamentous algae in the littoral (Kubečka et al. 1998).

In the highly eutrophic shallow lakes where biomass manipulations are of greatest interest, a typically very high biomass of zooplanktivorous fish means that a high fraction (> 75%) of the abundant fish population must be removed to reduce their role in predation or nutrient recycling sufficiently to allow an increase in water column transparency (Meijer et al. 1999). Primarily benthivorous fish also exert a top-down effect by modifying the size distribution of the macrobenthos (Table 25–2) and they play a role in sediment resuspension and nutrient recycling. They excrete sediment-prey-derived nutrients into the pelagic zone, and compete with water birds for macrobenthos. The temperate zone gizzard shad (*Dorosoma cepedianum*) and a large number of low latitude fish species are **omnivorous**, feeding on a combination of detritus, plant material, and zooplankton or benthos, thereby helping to structure the biota. In the process, omnivorous fish may out-compete strictly zooplanktivorous age-0 fish of other species, including the larvae and juveniles of their piscivorous predators (Sec. 29.2).

▲ Predicting Larval Survival

Identifying the specific causes for a strong or weak year-class (cohort) is a subject of much research and speculation. The interannual variation is frequently attributed, in some species at least, to a temporal or spatial match or mismatch between the larvae and the abundance of their prey or predators during the difficult-to-define **crucial period**, when the larvae have absorbed most of their yolk sac and have to start feeding independently. But the still-small age-0 juveniles also remain highly susceptible to predation and starvation during periods when available resources per individual are poor. Unfortunately, resolving the reasons for a strong or weak year-class is not easy. Further-

---

[3]The frequent references here to perch and Section 26.6 are an attempt to provide the reader with examples for needed generalizations about fish biology as well as more detailed and species specific attributes of one well-studied and widespread temperate zone species.

more, an unusually strong age-0 cohort may suffer a high level of intracohort (intraspecific) competition for food, resulting in reduced growth and an elevated mortality. Furthermore, the a priori decision made by investigators as to when and where to sample may not match the crucial period or the location of the larvae well. Moreover, investigators must use sampling equipment appropriate not only for the larvae, and subsequently for the more motile older age-0 fish whose stomach content (fullness) is to be examined,[4] but also for the various potential predators of different sizes and behaviors.

Post-spawning water temperatures are a good indicator (predictor) of larval survival and year-class strength in the temperate zone (Craig 1987). Unfortunately, the various effects of temperature are impossible to interpret causally in nature. Temperature affects not only fish metabolism but also the match or mismatch between the zooplanktivorous larvae and juveniles on the one hand, and their zooplankton prey on the other. Important long-term work in northern England has shown that the particular match or mismatch is the outcome of yet another match or mismatch between the herbivorous macrozooplankton and the edible crop of nanoplankton and picoplankton on which they depend (Fig. 12–6). This has been linked to the timing of the onset and stability of lake stratification, which in northern England and western Europe is greatly influenced by changing ocean currents produced by the North Atlantic Oscillation (Sec. 5.4). Unfortunately, the number of large scale long-term studies that show the effect of abiotic factors is small compared to the more typical short-term studies, usually carried out over small spatial scales—and often short temporal scales as well—are geared to detect biotic but not abiotic interactions.[5]

Among-system differences in the growth rates of species (Fig. 26–4) in any single climatic zone and in aquaculture (Sec. 26.8), have been widely linked to ei-

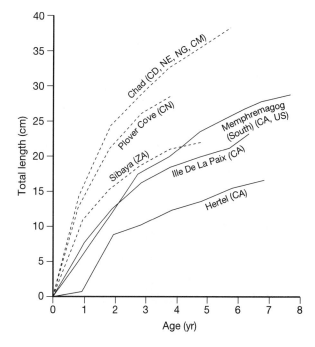

Figure 26–4   Growth of yellow perch (*Perca flavescens*) in three Canadian Lakes (solid lines), and tilapia (*Oreochromis mossambicus*) in Lake Sibaya (ZA) and Plover Cove, Hong Kong (CN) (dashed lines) and (*Oreochromis niloticus*) (broken dashed lines) in Lake Chad (CD, NE, NG, CM). Note that the perch in Lake Hertel and the tilapia in Lake Sibaya exhibit poor growth (stunting). *(CA data after Nakashima and Leggett 1975, other data after Lowe-McConnell, in Pullin and Lowe-McConnell 1982.)*

ther nutrient levels and primary production or to the resulting prey availability (Sec. 26.4). Growth enhancement is only possible where the individual growth rate had previously been food-limited and fish therefore exhibited *density-dependent growth*. Conversely, where populations are not food-limited and growth is *density independent*, the fish are feeding at the maximum rate possible. Any addition in food availability, as resulting from eutrophication, is typically expressed as a higher survival rate of larvae and juvenile fish (Mills and Chalanhuk 1987). Greater survival means a greater density (no. m^{-2}), implying less food per individual which results in reduced growth of the individuals and, once again, a density-dependent growth rate (Nakashima and Leggett 1975). Density-

---

[4]"The gear used in most studies (towed nets, pushed nets, high speed samplers) underestimates abundance of the larger age-0 fish, or are incapable of catching them, because the ability of the age-0 fish to avoid samplers increases exponentially with body length." (L. G. Rudstam et al. 1995.)

[5]"It is quite likely that aquatic ecologists have an inflated impression of biotic interactions. I say this because most small, temporal, taxonomic and environmental scale studies overemphasise the influence of specific interactions by ignoring abiotic variation, by accentuating the magnitude and duration of biotic interactions by studying few species in isolation." (Hinch 1991.) It is equally true that large

scale among-system studies, which are well-suited to reveal the impact of abiotic factors, underestimate the importance of biological interactions that are typically seen best over short temporal and spatial scales (Sec. 2.6).

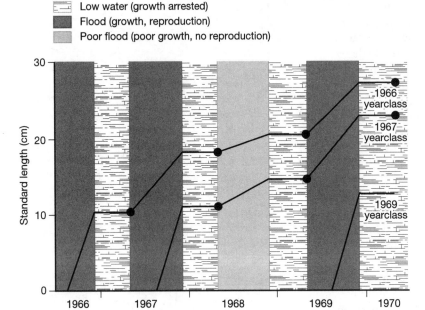

Low water (growth arrested)
Flood (growth, reproduction)
Poor flood (poor growth, no reproduction)

Figure 26–5 Growth of *Citharinus citharus* in the Senegal river floodplain system of subtropical west Africa (~16°N). *(Modified after C. Reizer 1974, in Welcomme 1985.)*

dependent growth usually reflects a shortage of suitable (high quality) prey but may also be the outcome of higher activity levels and a resulting higher metabolism in certain waters, leaving less food energy for growth. A higher activity level results in greater intraspecific or interspecific interactions among fishes, thereby affecting growth independent from food availability (Boisclair and Leggett 1989).

Predation and Survival

Survival is enhanced where food resources are abundant and where predation is low. But studies of food availability to a population must also consider the competition by juveniles and adults of other vertebrate and invertebrate species. For example, the phantom midge (*Chaoborus spp.*, Insecta) is another important predator of macrozooplankton as well as being prey for larger fish (Sec. 23.9). But most fish species are able, at all life stages, to shift to alternate types of prey and habitat when necessary.[6] Low predation rates and high survival rates are commonly seen in weedy waterways where young fish are able to hide from fish and

other vertebrate predators. A high survival rate may allow the development of very large populations of **stunted** (poorly growing) adult fish that mature at a small size. Fish are considered to be stunted when they are substantially smaller than equivalent adults from other populations in the same climatic region (Fig. 26–4).

26.4 Fish Growth: Determinants and Measurement

The strong seasonality in environmental conditions that is imposed at higher latitudes by a great seasonal variation in temperature, irradiance, and resource availability is reflected in seasonally changing growth rates, including a lengthy winter period of little or no growth and the possibility of experiencing under-ice oxygen depletion (Sec. 15.7). Seasonality at low latitudes is linked to seasonal floods and droughts affecting the habitat, resource availability, and deoxygenation of mid to low latitude river systems and their associated wetlands (Secs. 15.2, and 15.7), as well as the hypolimnia (Sec. 15.4). Both hydrology and temperature impose the observed seasonality at intermediate (subtropical) latitudes. Growth will be poor and reproduction nonexistent in tropical rivers and their associated wetlands if hydrological conditions are unfavorable during the breeding season (Fig. 26–5).

---

[6]"A survey of 33 lakes located on the Canadian Shield indicated that juvenile and adult brook charr *Salvelinus fontinales* (Mitchell) shift their food habits from zoobenthos to zooplankton and their spatial distribution from littoral to pelagic zone, when living sympatrically [sharing habitat] with white sucker *Catastomus commersoni* (Lácepède). (H. Venne and P. Magnan 1995.)

The average weight gain of 12 fish species in the Amazon flood plain was 60 percent higher during the rising water period than during the remainder of the year (Bayley 1988). The high primary and secondary productivity over the flood-inundation period produces the abundant fish stocks that sustain the floodplain populations and fishing during the dry season. Where seasonality is weak, as in low latitude lakes exhibiting relatively little variation in water level or resource availability, fish grow all year and breed repeatedly (*multiple spawners*). Multiple spawning involves the release of a fraction of the total egg stock. Parental care of the modest number of young produced each time is common. Conversely, *total spawners*, in which all the eggs ripen and are released within a relatively short period (weeks to several months) greatly dominate where seasonality is strong and only one breeding period is the rule. Total spawners rarely care for the often abundant number of young, and among the pelagic species parental care is usually restricted to scattering the eggs over suitable vegetation or gravel beds (Welcomme 1985, and Winemiller and Rose 1992).

Growth can be determined directly following the recapture of marked fish and measuring the increase in length, or the highly correlated increase in weight or energy content. However, growth rates are most commonly based on aging fish and describing yearly—sometimes monthly or daily—increments in length or weight (Fig. 26–4). The age is often obtained from changes in the pattern of calcium salt deposits laid down in the scales, bones of the inner ear (*otoliths*), gill cover (*operculum*), vertebrae, or fin spines as teleost fish grow. The pattern of ridges (*circuli*) denotes changes in growth rate resulting from changes in environmental conditions or spawning. Where the periods of relatively good and poor growth occur on an annual basis, the discontinuity in pattern between closely spaced circuli laid down in fall and winter or during periods of drought at low latitudes, and the widely spaced circuli laid down during the season of good growth, are called *annuli* and are used for aging (Casselman 1987) and back-calculating past growth rates.[7]

Where there is seasonality but clear annuli are lacking and where the scales are difficult to read, as in older or slow-growing fish, the bands of translucent and clear material in the otoliths provide a favorite alternative. However, where climatic conditions are relatively stable throughout the year and growth is continuous, aging is difficult or impossible. But if spawning periods are short, the different cohorts can be identified from the distinct size groupings that emerge when fish length vs frequency is plotted (Fig. 26–6). Indeed, by knowing the age of fish much can be learned about individuals and populations, including past growth, the age of maturity, and longevity.

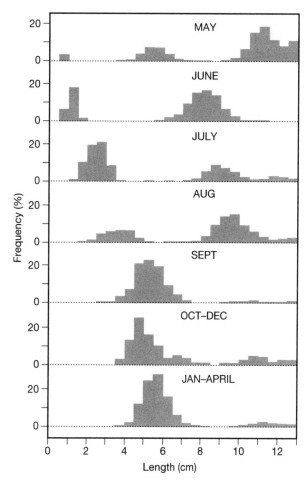

Figure 26–6 The frequency of newly hatched and one-year-old perch (*Perca fluviatilus*) in Lake Windermere (GB) as a function of their length distribution for each summer month, the combined autumn months and the combined spring months. (*After Le Cren 1947.*)

---

[7]"In the temperate zone [the annuli] are clearly associated with the winter cessation of growth. However, in the tropics more rings have often been recorded than would be expected if ring formation depended solely on a regular seasonal event. Care therefore has to be taken in interpreting rings in scales or other hard structures as indicators of age or time series." (Welcomme 1985.)

▲ Production: Biomass Ratio

Older (larger) individuals grow more slowly than young (smaller) fish with the result that the production (*P*) from the existing population biomass (*B*) declines with size, both within and among species (Fig. 26–7). The P:B ratio thus defines the growth rate (tissue growth plus reproductive products) of the identified biomass. The stability of the P:B ratio at a particular age or size is well-recognized and is commonly used to predict the growth of individuals from their biomass. The observed among-species differences in *biomass density* (kg ha^{-1}) may reflect real differences in the among-system productivity, biological (genetic) differences between species, or species–species and species–environment interactions. Sources of mortality, including fishing, may be high enough to allow enhanced growth of the survivors, thereby providing a possible explanation for unusually high P:B ratios reported in the literature. But reported P:B differences may, at least in part, be a scaling artifact resulting from differences in the capture efficiency of the sampling gear used or the area considered to be part of the habitat. The characteristic patterns seen in density or biomass density (Fig. 26–8) are not restricted to fish or other vertebrates but are equally evident among invertebrates (Fig.

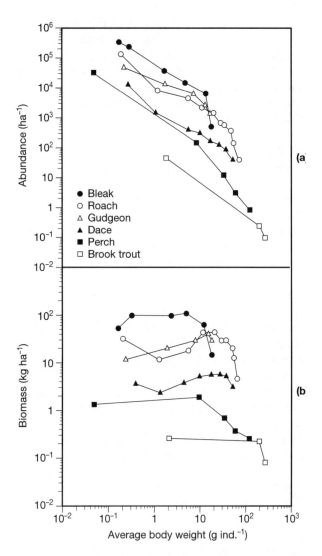

Figure 26–8 The relationship between average body weight and (a) the abundance of individuals per hectare, and (b) the biomass per hectare for six freshwater fishes. Note (1) the rapid decline in number with increasing weight but an almost constant population biomass until old age; and (2) the within and among species changes appear deceptively smaller than they are when plotted on a log scale (as here). *(Modified after Boudreau and Dickie 1989.)*

25–1, and Boudreau and Dickie 1989), and plants (Fig. 24–12).

Age, Density, and Biomass

Among populations, density declines systematically with increases in the weight (age) of the individuals with the result that the biomass density (kg ha^{-1}) of in-

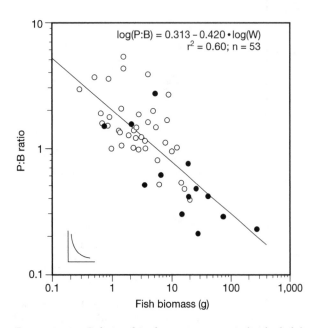

Figure 26–7 Relationship between mean individual fish biomass and the production:biomass ratios of fish communities in lakes (●) and rivers (○). *(Modified after Randall et al.*

dividual cohorts of age-1 and older fish, changes relatively little until late in life (Fig. 26–8) unless fishing or other mortality of the larger fish is unusually high. When information on abundance and age over time is combined with data on weight or length, it is possible to determine the production and structure of individual populations. The production of the different cohorts can be obtained by multiplying the P:B ratios of each by the relevant biomass density.

A detailed two-year study of the European roach (*Rutilus rutilus*, Family Cyprinidae, Order Cypriniformes), which is omnivorous as an adult, revealed a very rapid decline in abundance (no. m²) with age in the Thames River (GB), a decline that was most pronounced for the larvae and juvenile fish. However, the growth of the abundant age-0 roach was so large that the age-0 fish contributed more than half (66% and 73%) of the total lifetime production of the cohort. But nearly 100 percent of the biomass that was accumulated (the production) during the first year was lost, and it is evident that age-0 fish serve as a crucially important source of food for a variety of fish, bird, and other predators that may include humans.[8]

Young (small) fish have much higher specific (per g) feeding and respiration rates than older (larger) fish but a similar biomass density; they therefore have a disproportionate impact on lower trophic levels. Post (1990) estimates that for an equal biomass of perch larvae (0.01 g each) and adults (100 g each) at optimal temperature and maximum consumption rates, the larvae consume approximately 70 times the amount of food consumed by the adults. Fish, especially young fish, are at the same time major sources of regenerated nutrients through nutrient excretion. When abundant, fish may dominate the nutrient supply rate to the plankton during periods of low external nutrient loading, and when they feed in the littoral zone they subsidize the pelagic zone with nutrients derived from the littoral zone (Schindler et al. 1996). Fish (and invertebrates) therefore exert top-down (predation) and bottom-up (nutrient supply) effects that are difficult to distinguish.

Sampling Populations

Population estimates in inland waters are usually based on sampling with stationary *traps* or *gill nets*, *seines*, or *towed nets* (*trawls*) that are also used in modified form for commercial purposes (Fig. 26–9).[9] Experimental nets typically have sections with different mesh sizes (net openings) to catch species and individuals of different size. Government agencies set the minimum mesh size permitted for commercially important species to allow the juveniles and a fraction of the brood (breeding) stock to escape capture. Overfishing, unless strictly regulated, encourages the use of nets of progressively smaller mesh size in a natural, but in the long-term shortsighted, attempt to maintain catches.

Besides nets and traps, there are other techniques used to obtain population estimates. Among them is *poisoning* all the fish in small ponds or in an experimentally enclosed area of the littoral zone, followed by their collection and enumeration. Natural poisons have long been used by local fishermen at lower latitudes and are used in fisheries management for the removal of undesirable fish species prior to restocking with more desirable species. *Electroshocking*, which temporarily immobilizes the fish so they can be scooped up, is another commonly used sampling method (Fig. 26–10). It involves passing a sufficient current through the water between an electrode and an anode and works best for larger fish. *Echolocating*, an acoustic method, involves sending one or more beams of ultrasonic sound waves either downward at specific angles from a boat, or by a newer method horizontally from a transducer positioned a short distance below the water surface. When the beams are intercepted by fish (or large zooplankton, submerged macrophytes, and the bottom) the waves are bounced back to one or two transducers and shown on a screen or printout after they are processed (Misund 1997). The great advantages of this method are large sample sizes, the ease of obtaining them, plus the possibility of

---

[8]A disproportionately high human-imposed [fishing] mortality, rather than natural mortality, on the rapidly growing age-0 fish is the rule in many economically developing countries where the mesh size of nets and traps may be as small as 3.5 mm along the edge of each square and virtually all fish > 2–3 cm are retained. "While a fishery with small-meshed gear may be sustained by the smaller fish species, there is an associated danger that the larger fish [species] will become over-exploited and decline [or disappear]." (Hoggarth et al. 1999.)

---

[9]"All of the common problems of sampling error and bias must be faced by the fish biologist. They demand of him [or her] the fullest possible understanding of habitats and habits of the fishes to be sampled, and of the construction, operation and selectivity of sampling gear." (K. F. Lagler 1968.)

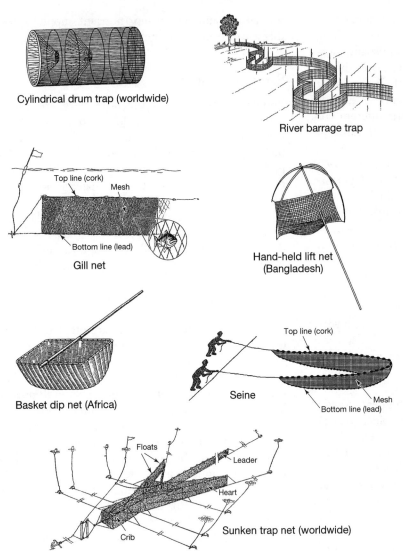

Cylindrical drum trap (worldwide)

River barrage trap

Figure 26–9   Examples of fish traps and nets from around the world.

Top line (cork)

Mesh

Bottom line (lead)

Gill net

Hand-held lift net (Bangladesh)

Top line (cork)

Mesh

Bottom line (lead)

Basket dip net (Africa)

Seine

Floats

Leader

Heart

Crib

Sunken trap net (worldwide)

tracking groups of pelagic individuals over time and space. The disadvantages include not knowing what species are recorded unless verified independently, and not knowing the fish size unless sophisticated dual-beam or split-beam instruments and data processing software are available. Moreover, vertical echosounding is not possible in the shallow disproportionately productive littoral zone of lakes and shallow rivers which are typically characterized by a much higher fish density and biomass per unit area than the open water (Keast and Harker 1977). Recent attempts to sample in the horizontal mode yield much higher densities (fish ha^{-1}) in shallow lakes than can be obtained from vertical beaming. Open water abundance in the horizontal mode in a number of mostly eu-

trophic lakes and shallow reservoirs is typically an order of magnitude larger than obtained by vertical beaming (Kubečka and Wittingerova 1998). Most of the fish were present in the upper few meters of the mixed layer but they apparently avoid the 4–5 m below the boat, providing a proposed explanation for the underestimation of near-surface assemblages using the vertical mode.

Population estimates of larger fish are most commonly obtained by using a *mark-recapture method*, first used in 1783 to estimate the human population of France. In fish research this involves catching and marking (with tags, clipping part of a fin, or absorbing fluorescent dye) n_1 fish at time 1. At time 2, n_2 fish are recovered of which m_2 are marked. Estimates of the

Figure 26–10   Electro-shocking a northern Borneo (Malaysia) rainforest stream. Note the net to collect the stunned fish. *(Photo courtesy of K. Martin-Smith)*

population size can be made using a number of formulations (Ricker 1975); one of the simplest is the *Peterson estimator*:

$$N = \frac{n_1 \times n_2}{m_2}.$$ EQ. 26.1

Better estimates are obtained by extending the experiment beyond the two-sample stage by also marking the unmarked fish in the second (third, etc.) sample along with the previously marked and recaptured fish. A larger number of marked fish provide a better estimate because the sample size of recaptured fish is typically quite small. Mark-recapture methods assume (1) a 100 percent random mixing of marked and unmarked fish, with all fish having the same probability of recapture; (2) the same mortality, if any, from natural causes or capture during the interval; and (3) no immigration or emigration. If these assumptions are reasonable—something normally assumed—the only errors that remain are the random errors associated with sampling. These can be estimated by calculating the 95 percent confidence interval (two times the standard error) around the estimate (Ricker 1975).

▲   26.5   Fisheries and Fisheries Management

The mark-recapture methods used primarily in research are impossible or impracticable in oceans or large lakes fished commercially with large modern boats that are equipped with sophisticated fishing gear, echolocators, and satellite navigation. The principal goal of fisheries biologists in very large waterbodies is to obtain reliable surrogate estimates (indicators) of population size or *maximum sustainable yield* (catch) without overfishing the stocks. This involves regular surveys of population abundance and structure of the relevant species, and sometimes of important prey species as well, to gauge food availability, using standard gear along standard transects. When standardized for time expended, the catch data yield the **catch per unit effort** (CPUE), another index of abundance (Figs. 26–11 and 23–16).

CPUE data, together with growth rate measurements for commercial species and an assumed estimate of the natural (nonfishing) mortality, are used in mathematical models to predict the maximum harvestable yield of particular cohorts in the ocean, and to a limited extent in some of the world's largest lakes. The traditional population-based models, in the absence of equivalent information on the many other species and environmental conditions, have had to assume that nonfishing mortality is a constant and fishing mortality is the only driving variable. The models project that population abundance can be controlled by simply increasing or decreasing fishing pressure (Regier 1982).

Predicting Sustainable Yields

The widespread failure of models to correctly predict the harvestable yield of overfished stocks is slowly encouraging the development of a variety of more comprehensive research and management models, including dynamic models that attempt to take into account environmental conditions and species interactions. Among many other modeling approaches in research are models based on bioenergetics which involve information on the food intake, digestion, excretion, and growth of particular species. Yet others use the relationship between organismal size and metabolism (see Eq. 26.2). There are also interesting attempts by behavioral ecologists to build simulation models based on the behavior and physiology of individuals in a population. Unfortunately, the impact of species–species and species–environment interactions on populations and communities remains poorly known.

The traditional fisheries management models, which view each population as a bit of reality that can be understood and modeled in isolation from competitors, predators, lower trophic levels, and the abiotic environment, have nevertheless been of some

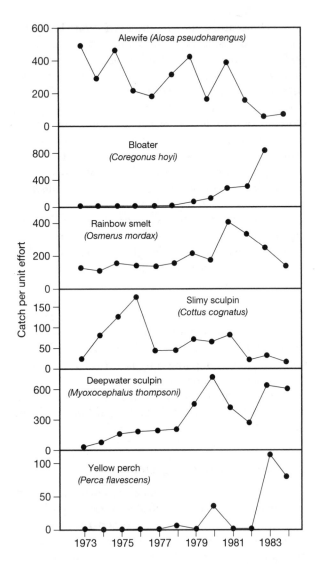

Figure 26–11 Mean number of six potential prey of the lake trout (*Salvelinus namaycush*) caught per 10-minute trawl tow in Lake Michigan. Catches represent > age-1 fish for all species except yellow perch where catches represent YOY fish. Note the scale differences. *(Modified after Eck and Wells 1987.)*

mortality) of the larger and most fecund age groups (cohorts) is large enough to bring about major changes in population structure. That kind of overfishing greatly changes not only the fecundity of the population or its interactions with other potential competitors for the resources but also changes in unknown ways the energy flow and food-web structure of the system.[10] The overfishing is sometimes referred to as **recruitment** or **biological overfishing**. It is increasingly common worldwide and leads to the collapse of stocks and economic hardship in the affected fishing communities, and has totally unpredictable consequences for the recovery of particular stocks and the possibility of managing them.

Managing Inland-Water Fisheries

Biological overfishing is a major problem in inland waters, especially in regions where management is lacking but demand for particular species, or simply for fish protein (food), exceeds the recruitment of some or all of the species involved. Management geared toward maximizing protein production is obviously not well served by the abundant literature on the modeling of individual stocks (species).

Fisheries management includes setting size and quantity (bag) limits for individuals in recreational fisheries, fishing gear restrictions, deciding the period particular species are allowed to be caught, setting fishing quotas in commercial fisheries, limiting the number of fishermen, and may also involve selectively stocking hatchery-reared fish. The great importance of maintaining good littoral and open water habitat conditions and high water quality—including restoration where it is required—is gradually becoming recognised in fisheries (and biodiversity) research. Additional management options include leasing fishing rights to a limited number of individuals with a stake in not overfishing, but this reduces employment in the fishing industry. However, attempts at management will fail if they are offset by increased fishing effort

utility but only where fish stocks are large and relatively stable and where overfishing was sufficiently modest to affect only the maximum harvest but not the production of future cohorts or the food-web structure. Modest overfishing under such conditions permits rapid recovery when the fishing pressure is reduced. This type of overfishing is sometimes known as **economic** or **growth overfishing** (Regier 1974). Unfortunately, traditional population models lack any predictive power when overfishing (human-imposed

[10]Severe overfishing is equivalent to a strong and sustained fish biomanipulation experiment involving a drastic reduction of fish in enclosures or small lakes, or equivalent to large-scale fish introductions. Strong manipulations produce a cascade of direct and indirect effects or *stressors* at other trophic levels, and typically have unpredictable and therefore surprising consequences (Sec. 23.7). Some effects of strong fish manipulations include diet or habitat shifts of the manipulated and unmanipulated species, and changes in the relative abundance, biomass, and growth rates of species (Persson and Hansson 1999, Christie et al. 1987.)

during the time allowed, often with more effective gear, or by habitat degradation.[11]

The catch statistics forming the basis for managing inland stocks primarily come from **creel census** (survey) data. Mathematical models akin to those used in the management of marine fisheries are not often used and are usually not cost-effective for managing the large number of small and greatly differing inland waters that may vary considerably in community structure, fishing pressure, and the fishing gear used. In inland waters predictive modeling is usually based on empirical relationships between regionally obtained estimates or informed guesses of the **yield** (catch), or the **maximum sustainable yield** (MSY, kg ha^{-1} yr^{-1}), and environmental variables (Sec. 26.7). The predictions made are for the average waterbody in the data set that was used to develop the regression models and are thus of only modest use in the management of individual waterbodies. These may not resemble the average waterbody and probably differ in the proportion of fish belonging to the desirable species. The management of individual systems requires estimates of stock sizes obtained by mark-recapture experiments, echolocating, creel census, or other methods, followed by modeling estimates of the MSY. But MSY estimates are typically based on limited data with considerable uncertainty attached to them and overestimates are common, allowing unsustainable harvesting and making subsequent recovery difficult. To reduce the risk, J. Caddy and H. Regier (2000) recommend using what is called the **optimal sustainable yield**, set at 10–20 percent below the MSY.

26.6   The Biology of a Temperate Zone Fish and a Tropical Fish: *Perch* and *Tilapia*

These two groups of generally shallow-water fishes share a number of functional attributes beyond the fact that they are both members of the Order Perciformes. The two species exhibit considerable plasticity in their feeding biology and preferred habitat. Both groups are able to cope relatively well with hypoxic conditions and are fecund, allowing them to increase their population size rapidly following their introduction to a new waterbody or after a period of high mortality. As a consequence, the two groups of species are widespread and common within their distribution range. Furthermore, they are of a size that makes them commercially attractive. But they differ in the emphasis accorded them in the literature. The large literature on perch largely concerns its biology in nature, whereas the tilapia literature is overwhelmingly based on research on their diet, growth, and genetics under culture conditions.

Perch

The Eurasian perch (now called *Perca fluviatilis*) invaded North America toward the end of the last ice age and evolved into *Perca flavescens*. But the species (Fig. 26–12) remain so similar in morphology, physiology, behavior, and ecology that they can be considered biologically equivalent (J. Thorpe 1977). Annual spawning occurs as early as February toward the southern limits of its distribution in Italy, Greece, and the southeastern United States, but as late as July in the northern boreal forest region of Canada, Scandinavia, and Siberia. Following courtship, the fish typically spawn at a depth of 0–3 m but as deep as 6 m. The females deposit their egg masses on substrates of sand, fine gravel, and submerged vegetation in lakes and large inflowing rivers where they are then fertil-

Figure 26–12   School of perch (*Perca flavescens*) over a bed of macroalgae (Characeae) in the St. Lawrence River (CA, US). *(Photo courtesy of C. Hudon and C. Carignan.)*

[11]"The management of fish stocks has a long history. It is also very political and the degree to which the recommendations of biologists are taken into consideration depends on the local jurisdiction. The managers of the resources have to consider the users who in many cases are the voters who put the lawmakers into office." (Craig 1987.)

"The major challenge in governance of fisheries is managing the fishers and not just the fish stocks." (J. Caddy and H. Regier 2000.)

ized. Egg production in fish is proportional to body weight and in perch this averages 15,000 eggs per 100 grams of body weight; large females can produce more than 100,000 eggs. In the central portion of its range it takes two to three years for males and two to five years for females to become part of the spawning population but ranges from two years in the southern United States to 10 years at its northern limits in Canada.

Egg mortality is high if eggs become coated with silt (siltation), exposed to anoxic conditions, subjected to fungal infection, or where predation by fish and crayfish is important. Egg mortality in acidified lakes increases when the pH declines below 5.5 during the spawning period, but it is only at pH 4 that hatching is reduced to about 60 percent. Mortality is complete at pH 3.5 in highly acidified waterways (Sec. 27.9). The ability of perch to reproduce successfully in acidified lakes is virtually unsurpassed and helps explain their abundance in such waters that, furthermore, now lack their traditional predators.

Age-0 Fish

Egg development is a function of water temperature, and in perch takes about 40 days at 8° but only ten days at 15° (Duarte and Alcaraz 1989). Newly hatched larvae (5–6 mm) are weak swimmers and currents carry them from the littoral to the limnetic (pelagic) zone where, at midlatitudes, they remain for four to eight weeks until fin development is sufficient to allow them to return to the littoral zone. In the limnetic zone they feed visually during the daytime and on moonlit nights on copepod nauplii, small copepodites, as well as on small cladocerans (Chapter 23) and even small fish larvae. With increasing size and gape the young are able to feed on progressively larger prey, including benthic invertebrates and fish (Mehner et al. 1998).

In the littoral zone, the young form shoals (schools) and feed and hide from predators within macrophyte beds. Alternatively, they become **demersal** (bottom living), in particular in shallow unstratified lakes. Cannibalism by age-1 perch commenced when the perch in a German reservoir reached 12–15 mm, a time when the gape size of the age-0 fish was large enough for them to be able to feed on large daphnids. Older perch contributed to the predation (cannibalism) when the age-0 cohort reached about 30 mm in mid June (Döerner et al. 1999).

Daily offshore migration of the shoals of larger larvae and juveniles toward dusk is followed by on-shore migration at dawn but ends when the juveniles reach about 30 mm, after which they remain in the littoral or demersal zone for the remainder of the growing season. There is ongoing debate about whether offshore migrations at dusk allow for pelagic feeding at a time when they are not as subject to visually feeding predators, and the return at dawn is a response to greater prey abundance in the littoral zone or warmer inshore waters during the daytime, or whether the migrations have a (genetic) evolutionary basis instead. Mortality of the larvae and age-0 juveniles is, as in other fish, enormous (Sec. 26.3), averaging ~ 6% d^{-1} (range 2–9%) and 2% d^{-1} in two shallow Scottish lakes. The age-0 perch, including the egg stage, contributed 88–96 percent of the total annual perch production in the lakes (Treasurer 1989). However, the contribution of age-0 (and often age-1) fish to annual fish production is not normally considered in work on production, (Sec. 26.7) which deals with older, larger animals caught in gill nets or traps with relatively coarse meshes.

The size at which surviving juveniles enter the winter season varies greatly among years and among lakes. The maximum range reported is 48–159 mm total length, but is more typically 50–100 mm, with the juveniles weighing 1–3 g wet wt. (Post and McQueen 1994). Interyear variation in cohort size and prey abundance appears to have a major impact on the resources available per individual and the resulting (density-dependent) growth of the young.

Winter mortality is typically high. In Oneida Lake (US) the mortality for cohorts of age-0 perch ranged from 72–99 percent of those fish that had survived the summer and autumn, with higher mortalities in years when the juveniles entered winter at a small size (Nielson 1980). More recent work has shown that winter mortality within a cohort is higher for small individuals and is further linked to the length of the ice-cover period (Post and Evans 1989). The northern limits of perch distribution are similarly believed to be set by a growing season too short to acquire the size and energy reserves needed to survive the long winters (Shuter and Post 1990).

Age-1 and Older Fish

The age-1 and older fish tend to be primarily littoral, feeding largely on benthic invertebrates supplemented by large zooplankton but there is considerable amongsystem variation in the extent that the limnetic zone is utilized; this is attributed to the relative food availability and predation pressure in the two zones. The density and structure of macrophyte beds (habitat complexity, plant growth form and density, degree of patchiness, dissolved oxygen concentration) has a big impact on the abundance and distribution of perch (and other species), their prey, and their predators (Eklöv 1977), with reduced fish abundances in the densest beds where mobility is much reduced (Fig. 24–18) and in unvegetated exposed areas.

Large perch (7–15 cm) have a sufficient gape size to become piscivorous on smaller fish as well as cannibalistic on age-0 and small age-1 juveniles in both the littoral and limnetic zones (Persson et al 1988). But the shift to a fish diet in adult perch is naturally contingent on a sufficient supply of invertebrates earlier in life, enabling them to grow to the size necessary to make the diet shift possible.

Perch feed at low water temperatures, including under ice, if the visibility and prey presence permits this. The species is also well-adapted to surviving hypoxic conditions (> 0.25–0.5 mg O_2 l^{-1}) in cold (< 4°) ice-covered waters. The lower survival limit in fish rises with temperature and the associated increased metabolism. In perch it rises to about 1.2 mg O_2 l^{-1} at 16° and 2 mg O_2 l^{-1} at 26°. Even so, growth and reproduction in fish continues to be negatively affected at dissolved oxygen (DO) levels well above survival limits (Sec. 15.7 and Craig 1987).

Tilapia

Tilapia is the common name applied to what are now recognized as a group of African fish (one species extends into the Middle East) and South American fish once classified in the genus *Tilapia* (Family Cichlidae). The tilapias are low latitude fish characterized not only by a similar morphology but also by a flexible omnivorous diet. Most of their food is obtained at the base of the food chain, making them energetically highly efficient. Most species share a common feeding and breeding biology based primarily on the littoral zone. Following the most recent reclassification, the several hundred species were split into three genera. The species most widely used in aquaculture are the mouth-brooding tilapias, now placed in the genus *Oreochromis* (Fig. 26–2). They include *O. mossambicus* (mozambique tilapia), originally restricted to the southeastern coastal region of Africa, and *O. aureus* (blue tilapia) originally limited to the Chad and Niger River basins of west Africa. *O. niloticus*, the nile tilapia, originated in the Nile and Rift Valleys of northeastern Africa but then naturally extended its range to the

Niger River and Lake Chad basins of western Africa. The three mouth-brooding species, now extensively used in low latitude aquaculture worldwide, have been widely introduced to waterways outside their natural range to increase fish yields, as well as for macrophyte or mosquito control. The species are so closely related that they can hybridize in culture and share a very similar life history.

Environmental Conditions

The desirability of tilapias in aquaculture stems from rapid year-round growth to a commercial size (Fig. 26–4) and the plasticity of their diet. In addition, their shoaling behavior permits high-density culture, while the ease of hybridization in captivity allows for an increased body size of hybrids and the production of unisex individuals to prevent crowding and stunting. Finally, their ability to thrive in fresh as well as saline waters is desirable in culture. *O. mossambicus* is the most *euryhaline* (salt tolerant; > 40‰), but the other species can be grown in only moderately less saline ponds. Their salinity tolerance is an adaptation for survival in the shallow waters they naturally occupy. These are subject to substantial seasonal and decade or longer periods of droughts and high evaporation resulting in a variable and periodically high salinity. Tilapia also copes well with high turbidity, a condition commonly encountered in their natural habitat where rivers carry high sediment loads to the lakes and wetlands during the rainy season. Tilapias are capable of coping with low dissolved oxygen levels, allowing them to survive periods of hypoxia in littoral zones and wetlands. But massive tilapia kills have been recorded in stratified lakes with anoxic hypolimnia when, during overturn, the whole water mass becomes anoxic. One such overturn in Lake George (Uganda) killed ~1.3 × 10^6 fish in a few hours (Pullin and Lowe-McConnell 1982).

Breeding Biology

The single most detailed life-history study has been on a stunted population of *O. mossambicus* in subtropical Lake Sibaya, a South African coastal lake at the southern and most temperate portion of its range (Bruton and Bolt 1975). Adult males move into the littoral zone in spring (September) to establish territories and dig nests, preferentially between sparsely spaced stems of emergent and submerged macrophytes at a depth of 0.5–5 m, but nests are found down to ~6 m among the stems of *Myriophyllum spicatum* and

Potamogeton pectinatus (Fig. 24–1). A few weeks after the females arrive, breeding starts. The breeding season extends over a three to four month period. During this period several spawnings take place. Breeding occurs year-round to some extent in the more tropical portions of the range, but there is a spawning peak during the period of rising water levels (Donnelly 1969).

After each batch of eggs is fertilized in the nest the female takes the 100–600 eggs (the number of eggs produced is a function of fish size) into her mouth and broods them in the densest, most secure portion of the macrophyte bed. The eggs hatch about two weeks later at a size of ~6 mm (standard length, SL, Fig. 26–2) and she mouthbroods (guards) the larvae (in Sibaya, typically about 200). Once the yolksacs are mostly absorbed, the larvae become free-swimming and start feeding. They are then allowed to leave the mouth, but the cloud of fry takes refuge there at the least sign of danger (Pullin and Lowe-McConnell 1982). When the larvae reach 9–10 mm they are released at the land–water interface at a depth of 1–3 cm where daytime water temperatures reach 32°, well within the preferred range of 27–33.5° (Donnelly 1969).

The larvae and juveniles usually form large shoals in water 1–15 cm deep, but go down to 2 m. The age-0 fish feed on detritus and the associated bacteria during the day, supplemented with small copepods, amphipods, and insects. Rising water levels during the rainy season provide a continuing rich supply of allochthonous detritus at the land–water interface while the emergent grasses and sedges provide protective cover against bird predation. As their size increases, the young (30–80 mm) also feed in somewhat deeper water on epiphytic algae, supplemented with epibenthic invertebrates, and bacteria-rich sediment detritus. Work on the same species elsewhere reveals great plasticity in diet between waterbodies (Pullin and Lowe-McConnell 1982). The same species introduced into eutrophic Sri Lanka reservoirs not only shows much among-reservoir variation in diet but an equal seasonal plasticity and ecological adaptibility within single reservoirs. However, a benthic diet dominated by detritus was most common in the rainy season and phytoplankton dominated the diet during the dry season (Maitipe and De Silva 1985). Other work shows that blue-green algae or a diet of both larval and adult insects, especially chironomids, are important.

At the end of the breeding season in Lake Sibaya (ZA), the stunted adults (> ~10–20 cm SL) congregate in the deeper portion of the macrophyte zone (4–7 m) before they and the juveniles move offshore into slightly deeper water during the cool season, but they make daytime forays into the littoral zone for feeding (Pullin and Lowe-McConnell 1982, and Fitzsimmons 1997).

26.7 Predicting Fish Biomass, Production, and Yield

Community fish biomass and production increases with trophic status and declines when the nutrient supply to the phytoplankton is reduced substantially following nutrient abatement (Fig. 26–13 and Table 26–1). However, large among-system and among-region differences in response to a particular nutrient supply indicate that factors other than the nutrient supply, or the linked primary production, play a major role in determining population and community biomass (Fig. 26–14). Differences in where fish are sampled (littoral vs pelagic) and in mesh sizes of nets or traps appear to have a major influence on the reported biomass, as must the assumptions made in scaling up from the small areas sampled to the system as a whole. Differences in the biomass per unit phosphorus ratio between studies (Fig. 26–14) must also reflect differences in system morphometry ($\bar{z}$, % littoral area), hydrology (flushing time, nutrient loading, Fig. 9–3), and the harvesting history of stocks. In addition, biomass estimates always exclude small species and the young of larger species, therefore underestimating the total biomass and thereby the importance of fish in energy flow and nutrient cycling. Fish abundance is typically higher in the less frequently studied littoral zone than in the less pro-

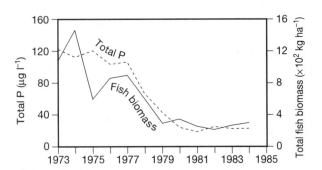

Figure 26–13 Change in total phosphorus and total fish biomass in Smith Mountain reservoir (Virginia, US) between 1973 and 1984. Note the rapid response of the fish community to the decline in the resource base for which total phosphorus serves as a surrogate. (*After Yurk and Ney 1989.*)

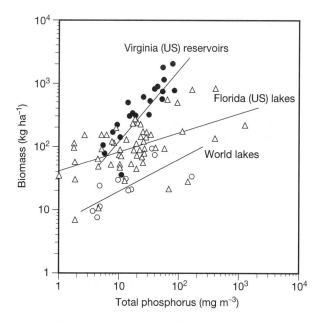

Figure 26–14 Relationship between total phosphorus concentrations and the pelagic fish biomass in Virginia (US) reservoirs, world lakes and subtropical Florida (US) lakes. Note that both the fish biomass and the rate of biomass increase per unit phosporus vary greatly among studies (see also Table 26–1). *(Virginia data, Yurk and Ney 1989; world data, Randall et al. 1995, Florida data, Bachmann et al. 1996.)*

Production and Yield

Estimates of **production**, defined as the elaboration of new tissue biomass during a time interval (kg ha^{-1} t^{-1}), including what is formed by individuals who do not survive the interval, is time-consuming and involves a variety of assumptions. The simplest and most reliable graphical method involves the construction of an *Allen curve*, a plot or computer-integration of the mean number of fish (or invertebrates) against the mean body weight of a cohort throughout its life. The area under the curve provides a measure of production (Ricker

1975). Reliable production estimates naturally depend on accurate estimates of abundance (no. ha^{-1}) and individual weight (kg) distribution of a particular species. If an estimate of community production is required instead, the same information must also be obtained for the other important species and then summed.

Fisheries management in the economically developed world typically involves the yield (the catch) of a few selected species and is furthermore concerned only with the biomass produced by those older fish of desirable size that survive the sampling interval. Even determining this is no simple matter. Moreover, extrapolating estimates made on inevitably few waterbodies to other systems requires accepting the normally untested assumption that the yield of a species from a given system represents the same fraction of the production as other waterbodies with which it is compared (Leach et al 1987). The time-consuming nature of production studies, as well as uncertainties about the reliability of yield statistics, has led to a search for easily measured correlates of production and yield that can serve as surrogates (proxies) at population and community scales.

Predicting Production and Yield

A recent analysis of 100 studies of largely temperate zone fish populations that exhibited a wide variety of feeding habits (Downing and Plante 1993) found that production (P, kg ha^1 yr^{-1}) was most strongly correlated with the population biomass (B, kg ha^{-1}) and secondly with the weight of the largest size-class of fish in a population (W, kg). The rate was further affected by water temperature, for which the mean air temperature of the region served as a surrogate:

$$\log(P) = 0.20 + 0.93 \cdot \log(B) - 0.19 \cdot \log(W) + 0.02(T)$$

$$R^2 = 0.88; n = 100; RMS = 0.084; p < 0.0201$$

EQ. 26.2

The slope coefficient average of log B (0.93) is close to one and this equation demonstrates that the average population P:B ratio does not change significantly with the population biomass.[12] This population model (Eq. 26.2) is similar to one reported for aquatic

ductive but better investigated pelagic zone of larger lakes (Werner et al. 1977). It is furthermore increasingly better recognized that many so-called pelagic species only survive because they receive an important energy subsidy by making feeding forays into the littoral zone (Schindler et al. 1996 and Vanni 1996). Rivers appear to be more productive than lakes on a per unit area basis (Randall et al. 1995) and because they are shallow, resemble the littoral zone of lakes and wetlands in terms of biomass and productivity.

---

[12]As production estimates (kg ha^{-1} yr^{-1}) are largely based on estimates of biomass change over time (Eq. 26.2) there is much covariation between the two measures. Such covariation detracts from the utility of P:B ratios (and all other ratios exhibiting considerable covariation) and the interpretability of predictions based on biomass (Fig. 26–7).

Table 26–1 **Equations predicting the fish biomass (B, kg ha^{-1}), fish production (P, kg ha^{-1} yr^{-1}), fish catch (C, kg ha^{-1} yr^{-1}) and fish catch per unit effort (CPUE, kg^{-1} night^{-1} standard gillnet^{-1}) as a function of environmental and morphoedaphic factors: total phosphorus (TP, µg l^{-1}), primary production of the water column (PP, g C m^{-2} yr^{-1}), gross primary production (GPP, g O$_2$ m^{-2} d^{-1}), temperature (T, °C), lake bottom dissolved oxygen (DOb, mg l^{-1}) total dissolved solids per meter depth (MEI, mg l^{-1} m^{-1}), flood plain area (FPA, km^2), or mean depth ($\bar{z}$, m); as a function of characteristics of the population: fish density (D, no. ha^{-1}), fish biomass (B, kg ha^{-1}) or mean weight of the fish community (W, g); and as a function of the fishing effort (E, man-day ha^{-1} yr^{-1}). Note the importance of the spatial scale examined (regional, global) on the type of predictive variable(s), and the often small sample sizes.**

Equation	r^2	n	Location	References
Biomass (kg ha^{-1})				
log (B)= 0.84 + 0.48 · log (TP)	0.51	11	Largely n. temp. lakes	Randall et al. 1995
log (B)= 1.07 + 1.14 · log (TP)	0.75	22	Virginia (US) reservoirs	Yurk & Ney 1989
log (B)= 1.55 + 0.32 · log (TP)	0.24	65	Florida (US) lakes	Bachmann et al. 1996
Production (kg ha^{-1} yr^{-1})				
log (P)= −0.86 + 0.65 · log (D)	0.80	53	Largely n. temp. lakes and rivers	Randall et al. 1995
log (P)= −0.98 + 0.67 · log (D)	0.74	11	Largely n. temp. lakes	Randall et al. 1995
log (P)= −0.57 + 0.59 · log (D)	0.64	42	Largely n. temp. rivers	Randall et al. 1995
log (P)= −0.37 + 1.09 · log (B)	0.80	22	Largely n. temp. lakes	Randall et al. 1995
log (P)= 0.38 + 0.89 · log (B)	0.74	51	Largely n. temp. rivers	Randall et al. 1995
log (P)= 0.30 − 0.38 · log (W) + 0.91 log (B)	0.83	11	Largely n. temp. lakes	Randall et al. 1995
log (P)= 0.51 − 0.33 · log (W) + 0.89 log (B)	0.80	42	Largely n. temp. rivers	Randall et al. 1995
log (P)= 0.60 + 0.58 · log (PP)	0.79	19	Largely n. temp. lakes	Downing et al. 1990
log (P)= −0.42 + 1.08 · log (B)	0.67	23	Largely n. temp. lakes	Downing et al. 1990
Catch (kg ha^{-1} yr^{-1})				
log (C)= 2.08 + 0.121 (GPP)	0.76	7	Chinese lakes and ponds	Liang et al. 1981
log (C)= 0.24 + 0.05(T) + 0.28 · log (MEI)	0.81	43	Global lakes	Schlessinger & Regier 1982
C*= 0.98 + 2.65 (FPA)	0.91	10	African rivers	Welcomme 1976
log (C**)= −1.80 + 2.70 (chl-*a*)	0.91	25	US lakes and reservoirs	Jones & Hoyer 1982
log (C)= −1.16 + 0.07 (TP) + 0.16 ($\bar{z}$)	0.96	21	Largely n. temp. lakes	Hanson & Leggett 1982
log (C)= 1.81 + 0.93 · log (E)	0.75	66	Cuban reservoirs	Quiros 1998
log (CPUE†)= 1.08 + 0.60 · log (TP) + 0.14 (DOb)	0.52	99	Argentinian lakes and reservoirs	Quiros 1990

*(kg ha^{-1} yr^{-1} km^{-1})
**Sport fish

invertebrates (Eq. 23.5). The model shows that fish production (P) increases systematically with population biomass (B), declines among species with increases in adult size, yielding a declining P:B ratio with increasing size or age (Fig. 26–7). Temperature has a positive effect on production, apparently by raising the P:B ratio. The positive effect of temperature suggests a higher annual production under suitable growing conditions at lower latitudes where the growing season is longer. Finally, the temperate-zone-based model implies that sustainable fish production will be higher for smaller species. The last conclusion is supported by continued high production of small species in low latitude floodplains subject to severe overfishing of large species.

Morphoedaphic Index of Yield

The most widely explored index of fish yield is the **morphoedaphic index** (MEI) developed by R. A. Ryder (see Ryder 1982), which uses the ratio of total dissolved solids (TDS, mg l^{-1}, or its correlate conductivity, Sec. 13.2) to mean depth ($\bar{z}$, m) ratio as a surrogate for the expected population yield of selected species (Table 26–1). The index was developed for a homogeneous set of medium to large forest lakes at low altitude in one climatic and geologic zone (Ontario, CA), with the lakes typically oligotrophic and subject to moderate fishing pressure. The MEI has also been found to be useful as a quick approximation of the yield of lakes, reservoirs, and river–wetland systems in other relatively homogeneous regions at low and mid latitudes (Toews and Griffith 1979, and Jenkins 1982). In these regions, the MEI is not confounded either by much variation in anthropogenic eutrophication, temperature, geology (morphometry), or greatly varying fishing practices.

▲ Mean depth has long been known to be an indicator of fish yield in relatively homogeneous sets of oligotrophic forest lakes (Fig. 7–3) that vary considerably in mean depth but only moderately in trophic status. Mean depth affects the distribution of organisms, lake stratification, light climate, relative size of the littoral zone, water temperature, the flushing rate (Fig. 7–2), nutrient supply (Fig. 9–3), and ultimately the catch. Ryder added the TDS to create the TDS: $\bar{z}$ ratio in an attempt to capture the effect of relatively modest differences in essential nutrients among lakes on fish yield. The easily measured TDS had previously been shown to be an indicator of plant nutrient concentration in fresh waters of low to moderate salinity that are

little affected by eutrophication (Chow-Fraser 1991). However, it is poorly resolved whether or where TDS contributes significantly to the prediction of fish yield, beyond that already contributed by mean depth and its own correlation with nutrient richness (Sec. 7.4).

Simple empirical regression models—including the MEI—make useful management predictions only for waterbodies similar to those used to build the model. It is not surprising that the MEI, based on lakes varying relatively little in trophic status but much in mean depth, is a poor predictor of fish yields among lowland systems varying much more in nutrient richness than mean depth. Large among-system variation in nutrient input or primary production in these systems allows measures of trophic status to serve as better surrogate predictors of fish production (Fig. 26–15) or yield than the depth-dominated MEI. Other primary production surrogates include the easy-to-measure chlorophyll-*a* concentration, as well as the more cumbersome measures of macrozooplankton or macrobenthic biomass (Table 26–1).

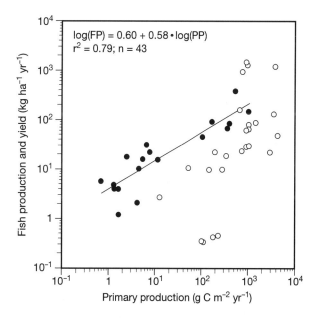

Figure 26–15 Relationship between community fish production (●) and yield (○), and phytoplankton production. The solid line is described by the equation. The four lowest points are fisheries-yield data from Lakes Superior, Huron, Michigan, and Ontario (CA, US). The four highest points are the average of three consecutive years of commercial yield from three Indian ponds and the annual yield from an unnamed carp pond in Israel. Note that the yield (catch) is typically lower than the fish production. (*After Downing et al. 1990.*)

Trophic Status, Yield, and Production

Modest eutrophication (increase in resource availability) allows a higher standing stock and higher fish production (Table 26–1), but normally without changes in the species composition (Nakashima and Leggett 1975). This does not apply when the nutrient loading is sufficient to produce hypoxic or anoxic hypolimnia that are no longer available to fish and, following overturn, cause fish kills in the mixed layer. Nor does it apply where an elevated phytoplankton biomass im-

poses sufficient light extinction to cause the disappearance of the submerged macrophytes (Fig. 24–8) and associated fish and invertebrates.

Work on shallow and productive Danish lakes reveals not only a systematic shift towards a greater zooplanktivorous fish biomass with increasing eutrophication but also a decline in the fraction composed of piscivorous fish (Fig. 26–16). Conversely, reductions in nutrient loading and primary production are quickly followed by a reduction in fish biomass (Fig. 26–13) and fish production (Fig. 26–15). Unfortunately, an aesthetically appealing high transparency is incompatible with high community fish production. These two incompatible goals are an increasingly divisive issue in the multi-use management of lakes and reservoirs in economically advanced countries.

There is no single relationship between primary production or other surrogates for trophic status and fish yield (Fig. 26–15 and Table 26-1), in part because fish catches in the Western world are largely restricted to specific species whose contribution to the total fish production is quite variable. Moreover, yields are greatly affected by the capture efficiency of the equipment that is used and the fishing effort. But the wide among-system variation in the relationship between trophic status and fish biomass (Fig. 26–14) indicates that factors other than those reflecting trophic status have an important effect on fish production and biomass. Even so, fish yields approach fish production in highly managed Asian fish ponds where the mortality is low and the available food is used optimally. This is made possible by stocking a variety of species with complementary feeding habits, thereby maximizing the use of the resources available.

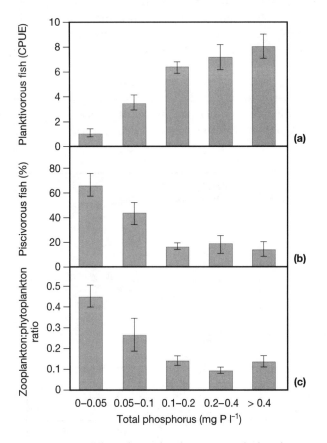

Figure 26–16 The relationship between total phosphorus in 65 shallow and mostly highly eutrophic Danish lakes and the mean ± SD of (a) planktivorous fish biomass (CPUE) collected in August using multiple mesh-sized gill nets (mesh sizes = 6.25–75 mm); (b) the fraction of total catch composed of piscivorous fish; and (c) the macrozooplankton to phytoplankton biomass ratio. Note (1) the typically very high TP concentrations and; (2) that the importance of piscivorous fish declines in more highly eutrophic waters; and (3) that systems dominated by zooplanktivorous fish have low macrozooplankton to phytoplankton biomass ratios. *(Modified after Jeppesen et al. 1999.)*

26.8   Aquaculture and Water Quality

Aquaculture contributed between 50 percent and 70 percent of the total global inland fish yield in 1997 (Fig. 26–17); the value is closer to 50 percent if the official catch statistics underestimate the total wild catch by a factor of roughly two as believed. Inland water **aquaculture**, defined as the farming of aquatic organisms, is growing rapidly while the global yield of wild caught fish is leveling off (Fig. 26–17). Asia contributed 84 percent of the global inland fish production in 1992 (FAO 1995).

Among the fish grown—most often at lower latitudes—85 percent are noncarnivorous species. Although their food conversion (growth) efficiency is low

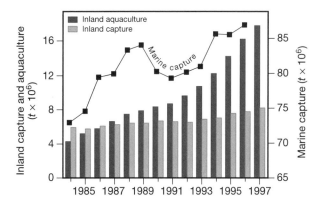

Figure 26–17 Inland capture and aquaculture of fish from 1984 to 1997, with marine capture for comparison. *(After FAO Review 1995.)*

($\leq$ 5%), herbivorous, detritivorous, and omnivorous fish benefit from an only one-step (trophic level) transfer of energy from the relatively abundant plant material directly to fish protoplasm. Carp species dominate fish production worldwide (Fig. 26–18), distantly followed by salmonids, catfishes, and tilapias. However, the culture of atlantic salmon (*Salmo salar*) and tilapia species and their hybrids is growing very rapidly.

▲ Types of Aquaculture

Aquaculture can be divided into four forms that are easy to separate conceptually but overlap in practice. **Extensive culture** involves raising organisms under relatively natural conditions, traditionally based on stocking selected species from the wild into ponds,

Figure 26–18 Indian farmers harvesting catla carp (*Catla catla*) from their village pond. *(Photo courtesy of F.A.O. Food and Agriculture Organization of the United Nations.)*

reservoirs, sections of rivers, or flooded rice fields. But hatchery-raised fish are increasingly used for stocking. Supplementary feed is normally not provided, but habitat improvement may be undertaken. Extensive culture is principally undertaken as a source of supplementary income in productive tropical areas. Fish yields of ~100–1,000 kg ha^{-1} yr^{-1} (Fig. 26–19) usually considerably exceed the highest wild catches of ~100–200 kg ha^{-1} yr^{-1} in South Asian floodplains and highly eutrophic Sri Lanka reservoirs.

Semi-intensive culture entails rearing fish under more controlled conditions, usually in ponds. Their diet is supplemented, and stocking densities are well above natural densities; the eggs or young are usually obtained from hatcheries. It is a favored culture approach in developing countries, often in conjunction with other aquatic species (invertebrates, ducks), combining relatively high yields (Fig. 26–19) with small capital costs and low economic risks.

Semi-intensive systems grade into **intensive culture** systems characterized by more intensive management and additional feeding and/or fertilizer application. Many of the intensive culture systems are **monocultures** (one species), but **polycultures** in which several fish species are raised together are increasingly common. The use of two or more species utilizing different food resources (plankton vs bottom feeders) make it possible to increase stocking densities without exceeding the carrying capacity of the system as a whole.[13] Pond culture in South and East Asia may involve adding livestock (chicken and pig) manure and crop wastes along with ducks in what is known as *integrated culture*.

In Chinese polyculture, as many as four or five carp species are stocked together, thereby optimally filling the various trophic levels and different habitats. Silver carp (*Hypophthalmichthys molitrix*) live in mid water and feed primarily on phytoplankton. Bighead carp (*H. nobilis*) also live in mid water but feed mostly on zooplankton. The grass carp (*Ctenopharyngodon idella*) feeds on macrophytes, supplemented by grass

---

[13]"The largest single waste-fed aquaculture system in the world is the Calcutta [India] sewage system, where water and sewage are fed into two lakes covering an estimated 2,500 hectares. After an initial bloom of algae, fish—principally [bottom feeding] carp and [plankton feeding] tilapia—are introduced, and additional sewage is fed into the lakes once each month. The system supplies about 7,000 metric tons of fish annually to the Calcutta market, or 2.8 metric tons [2,800 kg] per hectare per year." (World Resources 1992.)

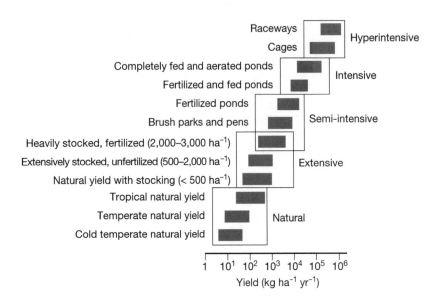

Figure 26–19 Yield from world capture and culture systems. Note that even cold temperate zone catches are high compared to those from the oligotrophic open oceans (0.03–0.7 kg ha^{-1} yr^{-1}). *(Modified after Welcomme and Bartley 1998.)*

and vegetable cuttings, while the black carp (*Mylopharyngodon piceus*) preys on molluscs in the sediments (Wootton 2000).

Selective breeding has long been used to increase growth rates. Hybridization of closely related tilapia species is widely practiced in order to gain hybrid vigor, more rapid growth, and a more desirable meat color, as well as to produce unisex offspring that prevent the production of a large number of poorly growing (stunted) individuals. South and East Asian yields from intensive culture reach well above 10,000 kg ha^{-1} yr^{-1} (Fig. 26–19).

Hyperintensive culture involves raising high-value species with high stocking densities in cages,[14] raceways (artificial stream channels) in which the water is replaced within hours, or enclosed tanks with water circulating systems. Diet, habitat, and water quality are controlled, and active disease prevention and control is necessary. Hyperintensive culture has much more in common with raising cattle in feedlots or industrial-style chicken production than with limnology. Except for prawn culture in coastal areas of the tropics, hyperintensive culture was practiced primarily in the economically advanced regions of the temperate zone using piscivorous fish (largely

salmonids). But a rapidly growing hyperintensive cage-culture industry using primarily nonpiscivorous fish has sprung up during the last decade, above all in East Asia, utilizing leased public waterways. Fish yields from hyperintensive cage-culture farming are exceptionally high (Fig. 26–19), but so is the initial investment and continuing operating cost. There are major financial risks from bad weather, floods, vandalism and theft, bird and mammal predation, disease, parasites, and market conditions that are of no professional concern to limnologists. However, what is of concern is the effluent that is produced, especially when it is released into oligotrophic waters.

Waste Production and Its Management

Waste production has been investigated best in hyperintensive culture systems used primarily for the production of juvenile salmonids in hatcheries and their subsequent cage-culture for the marketplace. Roughly 500 kg of solid wastes (uneaten feed and feces) are produced for each tonne (1,000 kg) of fish, as well as some 100 kg of total nitrogen and about 20 kg of phosphorus. Outputs of organic matter and nutrients are greatest in spring and summer when temperature and feeding rates are highest (Costa-Pierce 1996) but these vary considerably between operations. For example, about 30 percent of the carbon, 20 percent of the nitrogen, and 60 percent of the phosphorus inputs as food were lost in solid form from Scottish trout cages (Merican and Philips 1985). Liquid effluent concentrations have been reported as: 125 µg TP l^{-1} and

---

[14]"Cages can be broadly categorised into two types: fixed and floating. In the former, the cage bag is attached to posts driven into the lake or reservoir bottom while with the floating cages, the bag is secured to a floating collar. . . . As in other aquaculture systems, cage culture can be by extensive, semi-intensive or [hyper] intensive means." (M. C. M. Beveridge and J. A. Stewart, 1998 in Petr 1998.)

1,400 μg TN l^{-1} for a TN:TP ratio of about 11 (as mass), and 8 mg l^{-1} BOD (Cripps and Kelly 1996). Other compilations report TN:TP ratios averaging ~5 (as mass) (Costa-Pierce 1996). The different ratios probably reflect differences in food composition. The nutrient ratios show that effluents range from optimal for algal growth to slightly enriched in phosphorus—the element in shortest supply in most temperate zone oligotrophic systems (Table 8–1).

The export of phosphorus from large-scale salmon culture is of the same magnitude as the runoff from intensive temperate zone agriculture (Costa-Pierce 1996), but in contrast to agriculture, 100 percent of the wastes enters waterways. Although the area under agriculture is vastly larger, the relative contributions from fish farming are currently small but rapidly growing in both economically developed and developing countries. In some Western jurisdictions, waste collection has become obligatory, usually consisting of a sludge trap under cages to prevent a high input of particulate organic matter to the sediments. In a few other jurisdictions, there are either limits on the number of farms or their size to prevent algal blooms in the water or a BOD high enough to produce hypoxic or anoxic sediments. The particulate wastes that are released may increase the abundance and growth of wild fish in the area as well as their contact with antibiotics and disease.

Nutrient pollution is much less an issue in naturally eutrophic waters at low altitudes where the additional nutrients have little or no effect on already high algal growth rates and turbidity. There the waterbodies are typically too shallow to stratify and anoxic hypolimnia are not commonly an issue. Lastly, food production is generally considered a higher priority than the conservation of natural ecosystems or native species. In China, the fraction of waterbodies utilized for—and affected by—aquaculture already approaches 30 percent (Hu and Liu 1998) and is growing rapidly. But further growth in Asia is constrained by increasingly serious pollution of the waterways by sewage, deoxygenation, industrial toxins, and agriculturally derived nutrients and pesticides.

Fish Farming and Limnology

Fish farming impinges on limnology through its effect on (1) water quality; (2) the usurping of good natural fish habitat and destruction of wetlands for culture that impact on biodiversity and ecosystem functioning; (3) the removal of young fish of selected species

for culturing; and (4) the introduction of exotic fish species. There is further risk from (5) genetically altered stocks of farmed fish escaping and interbreeding with wild stocks (e.g., salmonids, tilapias) that are adapted to the local environment, or hybridizing with closely related species. The interbreeding problem, resulting in the loss of genetically well-adapted stocks, will be exacerbated when genetically manipulated (transgenic) species, specially adapted for efficient farming, become widely available. (6) Even where related species are absent, escaped or purposely released fish may colonize regions where they are nonnative, with unforeseen and sometimes disastrous consequences (see Sec. 23.8).[15] Moreover, escaped fish can—and do—introduce disease and parasites into wild populations (Arthington and Blühdorn 1996); and finally, (7) raising piscivorous fish typically requires a considerably greater input of wild fish caught for feed (in the form of fish meal and fish oil) than the fish biomass that is produced. Fish farmers of piscivorous fish may thus increase rather than decrease the pressure on the ocean fisheries providing the feed (Naylor et al. 2000).

▲   26.9   Water Birds

System morphometry, littoral zone vegetation structure, and trophic status have a major impact not only on organisms living below the waterline but also on the species richness and composition, abundance, and biomass of waterfowl (e.g., ducks, coots, loons, commorants), wading birds (e.g., herons), and those bird species otherwise associated with the littoral zone and the open water (e.g., kingfishers, swallows, fish eagles). As for other taxonomic groupings, species richness increases with the size of waterbodies and is attributed to greater habitat diversity in larger systems (Suter 1994).

System Morphometry and Trophic Status

Bird densities are typically highest in the shallow and highly productive littoral zones of lakes and rivers, and their associated wetlands. The land–water interface (riparian zone) is of great importance to birds

---

[15]"Most notable in recent years has been the [heated] exchange of views on the advantages and disadvantages of the introduction of nile perch in Lake Victoria [east Africa] which has raised production from Lake Victoria from 100,000t in 1980 to 450,000t in 1990 at the probable cost of the loss of several hundred [largely endemic] native species." (R. I. Welcomme and D. M. Bartley 1998.)

associated with water—for feeding, resting, hiding, and breeding. However, the riparian zone is increasingly threatened by physical alterations for agricultural, industrial, and recreational purposes. High nutrient or sediment inputs yielding a high light extinction may exert indirect negative effects through loss of the submerged macrophytes (Chapter 24) and a decreased ability to see prey.

Undisturbed lowland lakes and wetlands with disproportionally long (irregular) shorelines relative to the area of open water have extensive littoral zones and a high abundance and biomass of water birds (Fig. 7–10). However, the effect of shore length is scale dependent and most evident among oligotrophic systems that vary a great deal in shore length (morphometry) but relatively little in nutrients (resource availability). In contrast, trophic status emerges as a determinant of bird abundance and biomass in, for example, low topographical relief Florida (US) where trophic status varies more than system morphometry (Fig. 26–20). The large data scatter in the plots indicates that other, unmeasured variables play an even larger role in determining aquatic bird abundance and biomass. The unmeasured variables might include (1) human disturbance; (2) the size of adjoining or nearby wetlands; (3) underwater slope as an indicator of the area of sediment accessible or suitable for feeding; (4) the area of anoxic sediments incapable of harboring invertebrate or fish prey; (5) variation in the dominant predators; and (6) bird–bird and bird–fish competition for food. Moreover, bird abundance is imprecisely enumerated.

Bird–Fish Interactions

While fish-rich lakes attract more piscivorous birds, insectivorous birds and fish compete for benthic invertebrates. This competition is sufficient to raise questions about the advisability of introducing fish into wetlands that are important waterfowl breeding grounds (Bouffard and Hanson 1997). For example, Eriksson (1979) observed a greater use of fishless lakes in Sweden by a common duck (goldeneye, *Bucephala clangula*). Eadie and Keast (1982) reported a 71 percent and 80 percent overlap in prey type and prey size between goldeneye and perch. Experimental fish removal was followed by increased duck use, which was attributed to reduced competition for invertebrate prey (Eriksson 1979). A second example of an apparent fish–fowl interactions followed the introduction of roach (*Rutilus rutilus*, Family Cyprinidae) in a large eutrophic Irish lake. The subsequent progressive in-

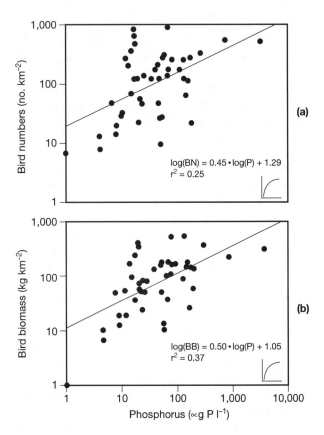

Figure 26–20   Relationship between water plus macrophyte phosphorus concentration in Florida lakes and (a) the average annual bird numbers, and (b) the average annual bird biomass. *(Modified from Hoyer and Canfield 1994.)*

crease in the abundance of the cyprinid and of the piscivorous great crested grebe (*Podiceps cristatus*) during the following decade was accompanied by the decline of the most abundant overwintering waterfowl species, the tufted duck (*Aythya fuligula*). The duck decline was attributed to increased competition with the fish for macrozoobenthos. Duck abundances increased again after the eventual decline of the roach population (Winfield et al. 1992).

Not all bird–fish interactions are the result of direct competition for resources. The aerial application of a fish poison (rotenone) to what had become a highly eutrophic lake with few submerged macrophytes and dominated by zooplanktivorous fish led to a marked change in the macrozooplankton and benthic macrovertebrate populations. Following the loss of the fish predator, the zooplankton shifted from an assemblage dominated by small species to one dominated by large daphnids. The benthos saw a significant increase in

Hyalella, a benthic amphipod resembling *Gammarus* (Fig. 25–3) and a favorite fish and duck prey (Hanson and Butler 1994). The associated large increase in migrating water birds in the Minnesota (US) lake was attributed not only to a more abundant benthos resulting from greatly reduced fish predation but also to zooplankton-induced greater water clarity, (z_{SD} 0.3–0.4 m to 1.2–1.5 m). The greater transparency allowed the macrophytes to recolonize, especially *Potanogeton pectinatus*, (Fig. 24–1), and *Vallisneria americana*, whose desirable seeds, tubers, and leaves helped draw large numbers of migrating birds in early autumn.

The role water birds play in structuring the aquatic biota has received little attention—apart from some largely qualitative data such as those described above—on relatively few waterbodies. An exception is some work on whole-system energy flow in limnologically well-characterized Lake Esrom (DK) (Chapter 25). Water birds there consumed 20 percent of the annual aboveground macrophyte production, but only ~4 percent of the annual zoobenthic production in the relatively steep-sloped lake (see also Table 25–3). The birds were estimated to be significant competitors of piscivorous fish, consuming roughly a quarter of the quantity of fish taken by the piscivorous fish (Woollhead, 1994).

Land–Water Interactions

It has become increasingly recognized that aquatic systems cannot be understood in isolation from their increasingly human-impacted drainage basins and the input of materials from the atmosphere. Water-associated birds (and other vertebrates such as frogs, turtles, and muskrats) straddle the terrestrial and aquatic domains, cycling organic matter and nutrients between them. By feeding on aquatic organisms, both fish and birds influence the abundance and community structure of the biota at the land–water interface and below the waterline.[16] Limnologists can make an important contribution to research on the management of water birds.

---

[16]"When planktivorous fish densities were low [in the eutrophic ponds, and macrozooplankton densities and transparency were high] mallards and muskrats made heavy use of the resulting abundant macrophytes and periphyton. Replacement of the macrophytes and periphyton by dense algal blooms associated with control of zooplankton by an abundance of planktivorous fish resulted in decreased use by muskrats and ducks and increased use by fish-eating birds including kingfishers, herons and egrets." (C. N. Spencer and D. L. King 1984.)

Highlights

- Fish species richness declines with increasing latitude and altitude but increases with drainage basin size and the size of waterbodies. As for other groups of organisms, the determinants of species richness and other fish attributes change with the temporal and spatial scale examined.

- Fish populations are characterized by orders of magnitude interannual variation in the survival of young, linked primarily to variation in food availability and predation. Small changes in mortality of age-0 fish lead to large changes in the abundance of cohorts.

- Age-0 fish typically yield more than half the lifetime production of a cohort. But nearly 100 percent of the biomass accumulated in the first year is lost to predation or starvation.

- Fish have an important and temporally changing effect on the size structure, abundance, and species composition of their prey. Age-0 fish, which commonly feed on zooplankton, thereby exert an indirect effect on the phytoplankton community structure.

- ▲ The age of age-1 and older fish is commonly determined through examination of the pattern of calcium deposits laid down in scales, bones, and fins which, together with size measurements, are used to obtain growth rates.

- Eutrophication results in an enhanced community fish production, but if severe, results in changes in fish species composition and fish kills.

- Fish density (n^0 ha^{-1}) and biomass density (kg ha^{-1}) tends to be greatest near-shore in systems with a well-developed littoral zone and in the near-surface layer, and tends to decline with depth in lakes.

- Overfishing of commercially important stocks is widespread and globally increasing, with unpredictable effects on the energy flow and food-web structure of the affected systems.

- The fraction of the global inland fish yield (catch) contributed by aquaculture is around 50 percent of the total and is rising.

- ▲ In culture, the highest fish yields are obtained from noncarnivorous species that benefit from a one-step (trophic level) transfer of energy from the relatively abundant plant material to fish protoplasm.

- ▲ Aquaculture can have important negative effects on aquatic systems.

- ▲ The biomass of water birds is greater in eutrophic than oligotrophic waters, but as for most freshwater fish, it is also greatly dependent on the presence of well-vegetated littoral zones and adjacent areas for breeding, hiding, and often for feeding.

27

Acidification of Waterways

27.1 Introduction

Acidification of the environment by the deposition of strong acids or compounds that subsequently form strong acids (acid precursors), has a major effect on the chemistry and biota of tens of thousands of lakes, rivers, wetlands, and drainage basins in northern Europe, northeastern North America (Fig. 27–1), parts of east Asia, and elsewhere on a smaller spatial scale. Aquatic acidification is defined by a decrease of the acid neutralizing capacity (ANC) of waterways (Sec. 14–4). Acidified waters undergo chemical and biological changes, and experience species losses or changes in the relative abundance of species. These changes are the combined outcome of hydrogen ions (H^+) deposited directly on water surfaces and the indirect effects of high H^+ concentrations on catchment soils and vegetation. High H^+ concentrations lead to the release of metals from soils and their subsequent stream transport to the receiving lakes and wetlands. High H^+ levels in waterways also release metals, including toxic metals, from the sediments directly into the overlying water. Humans and wildlife can be affected by drinking water high in toxic metals that is at the same time low in carbonates. People are indirectly affected by the demise of natural resources—through the loss of desirable fish or invertebrate species, forests, or irreparable damage to our architectural heritage when H^+ in precipitation dissolves the limestone and sandstone of ancient buildings and public monuments.

Not all lakes, rivers, and wetlands receiving acidic precipitation are affected. Some have sufficiently buffered drainage basins to resist acidification. Others are naturally acidic as the result of organic acids and are thus little changed by the addition of highly acidic precipitation.

Many waterbodies worldwide are acidified by industrial or mining effluents. A quarter of all acidic streams in acid-sensitive areas of the United States are the result of acidic mine drainage resulting from the oxidation of sulfur in the waste rock or "tailings" (Baker et al. 1991). However, the geology of drainage basins largely determines the acid neutralizing capacity (ANC) of inland water and how well waterways— and the soils in their drainage basins—will be protected against atmospheric acidification.

27.2 Sources and Distributions

The principal sources of acid precipitation are sulfur dioxide (SO_2) and nitrogen oxides (NO and NO_2), collectively denoted as NO_x, released during the combustion of coal, oil, gasoline, and the smelting of ores containing sulfur. Globally, between two and three times more SO_2 and NO_x are released into the atmosphere as anthropogenic emissions than naturally as inorganic elements (for example, sulfur from volcanoes) or organic compounds from soils, wetlands, and marine systems (Galloway 1995).

The emitted SO_2, NO_x, and their oxidation products, SO_4^{2-} and NO_3^-, have typical mean atmospheric residence times of one to three days. With a median transport velocity of about 400 km per day, the mean

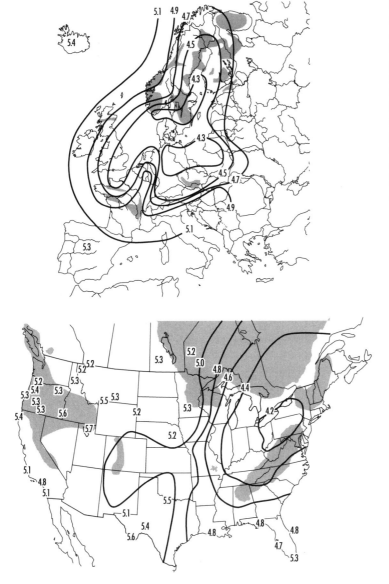

Figure 27–1 Volume-integrated average acidity (pH) of European and North American precipitation in 1985, which has seen fractional (0.2–0.3) increases in pH since that time. Episodically, a much lower precipitation pH is recorded. Areas of lowest pH are typically characterized by the highest average concentration of SO_4, NO_3, and trace metals in precipitation. Shaded areas indicate zones with a surficial geology that makes them particularly susceptible to acidification. *(World Resources 1988–1989, and Reuss et al. 1986.)*

transport distance can be 400 to 1,200 km. Consequently, these oxides are transported hundreds or even thousands of kilometers from the source of emission before falling on the Earth's surface (Schwartz 1989). The oxides are deposited as *wet deposition*—rain, or snow containing sulfuric acid (H_2SO_4), nitric acid (HNO_3), or hydrochloric acid (HCl)—or as *dry deposition* in the form of aerosols[1] (oxide particles) or gases

that impact vegetation, soils, and water surfaces. A major problem in acidification and contaminant research is the difficulty of measuring the dry deposition accurately.

Rainwater in north temperate regions little affected by human activity and the oceans typically has a pH between 5.5 and 6. The moderate acidity of the rainwater far from volcanic sources is primarily the result of the production and dissociation of H_2CO_3 in

---

[1]Aerosol-sized particles commonly have a diameter of 0.1—1.0 μm, but range in size from a cluster of a few molecules to particles > 20 μm in radius. The larger particles (> 5 μm) and their adsorbed contaminants (Ch. 28) do not remain airborne for long. Small particles

between ~0.2 and 2 μm are not readily dry-deposited and travel long distances before being deposited, probably by incorporation into precipitation.

the atmosphere (Sec. 14.2). The present acidity of precipitation in heavily industrialized regions of the Northern Hemisphere (Fig. 27–1), including central Japan and southern Korea, has increased ten to fiftyfold (decreased by 1–1.5 pH unit) from an estimated preindustrial background pH of 5.5–6.

▲ The loading of atmospheric acids is not a good measure of the acidifying power of the precipitation in portions of western and central Europe or the northeast and midwest United States where there is an appreciable release of NH_3 into the atmosphere from animal manure, sewage treatment plants, fertilizer, industrial processes and vehicle emissions. The NH_3 released produces OH^-, which neutralizes atmospheric H^+.

$$NH_3(gas) + H_2O \rightleftarrows NH_4^+ + OH^-.   \text{EQ. 27.1}$$

Ammonia also reacts with sulfuric and nitric acids and forms aerosols. The reaction consumes H^+ and thus serves as a source of alkalinity, yielding a more nearly neutral precipitation than predicted from atmospheric SO_4 and NO_3 levels alone.

$$HNO_3 + H_2SO_4 + 3NH_3 \rightarrow NH_4NO_3 + (NH_4)_2SO_4.$$

$$\text{EQ. 27.2}$$

But where NH_3 levels are disproportionately high, the pH of the precipitation appreciably underestimates the acidifying potential of NH_4^+. This is because the chemically and biologically mediated oxidation (nitrification, Sec. 18.3) of NH_4^+ in soils and waters releases H^+ and ultimately contributes to acidification, with one mole of NH_4^+ yielding two moles of hydrogen ions and one mole of nitrate:

$$NH_4^+ + 2O_2 \rightarrow 2H^+ + NO_3^- + H_2O   \text{EQ. 27.3}$$

(Eq. 11 in Table 27–1), unless the NO_3^- that is produced is taken up by the biota or denitrified by the microbes (Secs. 18.4 and Eq. 7, Table 27–1) in catchments, lake sediments, or at the land–water interface of rivers and wetlands (Sec. 8.4).

Sulfur and Nitrogen Oxide Emissions

The rapid increase in fossil-fuel consumption after World War II not only greatly raised the levels of SO_2 and NO_x in the atmosphere of industrialized countries but also the toxic trace-metal levels (Chapter 28). In both eastern North America and in Europe sulfur emissions more than doubled between 1900 and 1980 (Husar 1986, Dovland and Semb 1980), primarily as

the result of power plants burning coal and fuel oil to generate electricity (Fig. 27–2). Emission controls are responsible for the subsequent decline (averaging > 40% by the year 2000) in SO_2 emissions in the original group of OECD countries (United States, Canada, western Europe, and Japan).

Anthropogenic emissions of NO_x are derived primarily from the oxidation (fixation) of atmospheric N_2 gas in internal combustion engines rather than from the fuel itself. As a result, hydrogen ions derived from HNO_3 are particularly prevalent near urban areas. The magnitude of nitrogen emissions is much less certain than that of sulfur because it is much more difficult to quantify the many individual sources of NO_x emission (including cars and agricultural activities) than the relatively few major SO_2 emitters. Even so, a fairly recent estimate suggests an increase in NO_x emissions of 12–20 times in the eastern United States since 1900 (Husar 1986). While SO_2 emissions have declined greatly, NO_x emissions have risen slightly between 1980 and 1996 in the OECD countries as a whole. Increases in NO_x are reflected in measured increases in the absolute and relative importance of the NO_3 concentrations of the precipitation in highly populated regions. At present, SO_2 and NO_x emissions each contribute roughly half of the anthropogenically produced hydrogen ions in economically developed countries. Chloride production resulting from industrial activity and particularly from garbage incinerators can be a third locally important source of H^+.

In both eastern North America and northwestern Europe prevailing winds tend to be from the southwest. As a result, winds carrying high concentrations of H^+ blow from the heavily industrialized regions of both continents toward poorly buffered igneous drainage basins—the Canadian Shield in northeastern Canada and northern New England, and the Shield Region of Finno-Scandinavia—dotted with an enormous number of low salinity lakes that are readily acidified. These same winds carry heavy metals, organic compounds liberated during incomplete combustion processes, and particulates. Winds from the south, and in Europe from the southeast as well, are sufficiently common on both continents to allow the eastern seaboard of the United States and eastern Europe to serve as important additional source regions of H^+.

Deposition levels decline with distance from major sources, but a precipitation pH averaging less than 4.6–4.7 is sufficient to allow the acidification of sensitive catchments over large areas of igneous outcroppings on both continents (Sec. 27.10).

Table 27–1   **Selected processes affecting the acid neutralizing capacity (ANC) of aquatic systems and drainage basins, expressed per mole of CH_2O (reduction) or O_2 (oxidation) consumed (Δ ANC, organic) and per mole of inorganic substrate reduced (Δ ANC, inorganic).**

Processes	Reaction	Δ ANC (organic)	Δ ANC (inorganic)	In-lake and wetland Mechanism for Permanent Change
Reduction Processes				
Weathering	(1) $CaCO_3 + 2H^+ \rightleftarrows Ca^{2+} + CO_2 + H_2O$	—	+2	
	(2) $CaAl_2Si_2O_8 + 2H^+ \rightleftarrows Ca^{2+} + H_2O + Al_2Si_2O_5 (OH)_4$	—	+2	
	(3) $Al_2O_3 + 3H_2O + 6H^+ \rightleftarrows 2Al_3^+ + 6H_2O$	—	+6	
Ion exchange	(4) $2ROH + SO_4^{2-} \rightleftarrows R_2SO_4 + 2OH^-$	—	+2	
	5) $NaR + H^+ \rightleftarrows HR + Na^+$	—	+1	
Denitrification	6) $2CH_2O + NO_3^- + 2H^+ \rightleftarrows 2CO_2 + NH_4^+ + H_2O$	+1	+2	export of NH_4^+ via outflow
Denitrification	(7) $5CH_2O + 4NO_3^- + 4H^+ \rightleftarrows 5CO_2 + 2N_2 + 7H_2O$	+0.8	+1	release of N_2 to atmosphere
Manganese reduction	(8) $CH_2O + 2MnO_2 + 4H^+ \rightleftarrows CO_2 + 2Mn^{2+} + 3H_2O$	+4	+2	Mn^{2+} export via outflow
Iron reduction	(9) $CH_2O + 4FeO(OH) + 8H^+ \rightleftarrows CO_2 + 4Fe^{2+} + 7H_2O$	+8	+2	sediment burial as FeS_2
Sulfate reduction	(10) $2CH_2O + SO_4^{2-} + 2H^+ \rightleftarrows CO_2 + H_2S + 2H_2O$	+1	+2	sediment burial as FeS, FeS_2, organic sulfur, or H_2S gas release to atmosphere
Oxidation Processes				
Nitrification	(11) $NH_4^+ + 2O_2 \rightleftarrows NO_3^- + 2H^+ + H_2O$	–1	–2	
Manganese oxidation	(12) $2Mn^{2+} + O_2 + 3H_2O \rightleftarrows 2MnO_2 + 4H^+ + H_2O$	–4	–2	
Iron oxidation	(13) $4Fe^{2+} + O_2 + 6H_2O \rightleftarrows 4FeO(OH) + 8H^+$	–8	–2	
Sulfide oxidation	(14) $H_2S + 2O_2 \rightleftarrows SO_4^{2-} + 2H^+$	–1	–2	
Pyrite oxidation	(15) $FeS_2 + 3\frac{3}{4}O_2 + 3\frac{1}{2}H_2O \rightleftarrows Fe(OH_3) + 2SO_4^{2-} + 4H^+$	–1.1	–4	

Source: After Schnoor and Stumm 1985, and Davison 1987.

27.3   Acid-Sensitive Waters

The inland waters that are most sensitive to acidification are transparent with a very low salinity (conductivity < ~50 μS cm^{-1}), a correspondingly low HCO_3^- concentration, and a small acid neutralizing capacity (ANC) or alkalinity of < 50 μeq/l. Humic lakes in the same area commonly have a low alkalinity but are naturally acidic from the release of organic acids. Highly stained (rich in organic acid) lotic systems in the Amazon and Congo River basins (central Africa) have a pH of ~3.5–4.5 although far from sources of

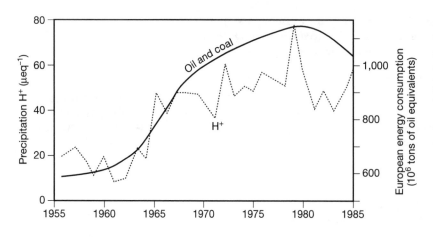

Figure 27–2 Weighted yearly mean H⁺ concentrations in precipitation at Lista, in southernmost Norway, and yearly energy consumption in Europe for the period 1955–1985. Emission of SO_2 has continued to decline, resulting in a ~ forty percent reduction in the mean H⁺ concentration of the precipitation. *(After Henriksen 1987.)*

anthropogenic H⁺ inputs. The most acid-sensitive transparent waters are characterized by a background pH between 5 and 7. Chemical sensitivity to acidification implies that small increments of added acid produce relatively large declines in pH. From an ecological rather than a chemical perspective, a sensitive waterbody is one in which additional inputs of acid are likely to result in biological changes prior to rapid lowering of the pH. Although the first changes to be noted are usually biological, these are subtle and often difficult to detect—therefore a chemical definition is usually required.

Kelso et al. (1986) estimated that there are more than 350,000 lakes in eastern Canada with an ANC < 50 μeq/l, of which more than 14,000 are acidified (pH < 4.7) with an ANC of ≤ 0 μeq/l. There are also a much more modest number of acidified lakes in the northeastern United States, primarily in the Adirondack region of New York. Sweden, with a much smaller landmass than eastern Canada, has about 85,000 lakes larger than 1 ha of which about 20,000 were acidified, plus around 90,000 km of acidified streams (Dickson 1985). In Norway, lakes and streams in an area of more than 33,000 km² have lost, or were losing, their fish stocks and seven major rivers have lost their salmon populations.

▲ In response to lack of information on the preacidification pH of lakes, paleolimnologists have succeeded in developing quantitative relationships (models) between the species composition of algal communities in surface sediments and the pH of overlying waters. The calibration data sets are then used to infer the lakewater pH at different times in the past on the basis of the species composition found in slices of dated cores (Sec. 20.5). Measurements show that most of the lakes in the Adirondack Mountains (US) had a pH close to 6

around 1900. Today the pH has declined between 1 (tenfold) and 2 (a hundredfold) units in the majority of lakes, with the greatest acidification occurring between 1920 and 1950 (Cumming et al. 1994). A similar decline is seen in Swedish west coast (Fig. 27–3) and south coast lakes (Wright and Gjessing 1976).

The extent of pH decline in affected regions is a function of both local geology (soils that determine the available buffering capacity) and the sulfur plus nitrogen oxide deposition rates, which decrease with distance from major source regions (Fig. 27–4). As pointed out above, not all acidic lakes in acid-sensitive areas are acidic because of acid precipitation. For example, about one-quarter of 1,180 lakes sampled in acid-sensitive areas of the United States were acidified by organic acids derived from their drainage basins

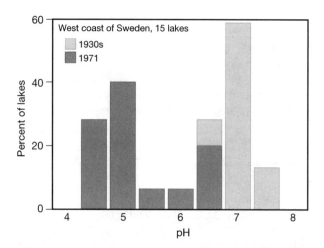

Figure 27–3 Changes in the fraction of lakes of different pH on the west coast of Sweden between the 1930s and 1971. *(After Wright and Gjessing 1976.)*

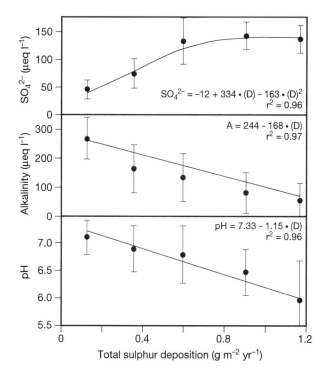

Figure 27–4 Relation between sulfur deposition (D) and the mean SO_4^{2-} concentration in lake waters, the alkalinity (A), and the pH of Ontario lakes on the Canadian Shield with increasing distance from the major source region. Error bars correspond to one standard deviation. *(After Neary and Dillon 1988.)*

(Baker et al. 1991). Canada and Finland also have large numbers of highly colored acidic lakes (pH < 5.3) containing very low levels of SO_4^{2-} that are naturally acidic as the result of a high input of organic acids (Kämäri et al. 1991). It is evident that in regions dominated by brown-water systems (humic, dystrophic), the role of organic acids must be evaluated before the effect of industrial acids can be determined.

27.4 Characteristics of Acid-Sensitive Waters and Catchments

Susceptibility to acidification is affected by:

1. the ability of catchment soils and rock to neutralize incoming acids;
2. lake morphometry and catchment attributes;
3. organic acids in runoff; and
4. neutralizing agents and processes within aquatic systems.

The ability of the catchment soils and rock to neutralize the incoming acids is the single most important factor distinguishing acidified from nonacidified lake districts in regions with low pH precipitation. The larger the fraction of a catchment that is covered by easily weathered carbonate-rich rocks and soils, the higher the ANC and salinity (conductivity) (Tables 13–1 and 13–2). Susceptible areas, denoted by a low ANC, are found in glaciated areas on igneous (e.g., quartzite, granite, basalt, or gneiss) or other highly insoluble bedrock rich in silicate minerals (e.g., sandstone), where the surface material is derived from material of similar geology. Rivers, lakes, and wetlands are readily protected where the insoluble bedrock is within the catchment overlain by a modest amount of calcareous drift (Sec. 13.2) brought in from other areas by glaciers.

Geologically susceptible regions impacted by acidification include much of northern Canada and Scandinavia, mountainous areas of central and western Europe, and the Adirondack Mountains of the northeastern United States (Fig. 27–1). Other major regions covered with soils containing little buffer include the Rocky Mountains of western North America, large parts of central and eastern China, all of Korea, and portions of Japan, New Zealand, northeastern Brazil, and South Africa.

In lake districts where surficial soils are of igneous origin, the extent of susceptibility is determined by lake morphometry, catchment attributes, and the absence of atmospheric dust rich in $CaCO_3$.[2] Streams and lakes positioned at the head of drainage basins (*headwater lakes*) are most at risk because the catchment areas feeding the inflowing streams or groundwater are small, the soils are thin, and most of the precipitation reaches the lake surfaces directly rather than as runoff carrying alkalinity (ANC) released from the soil and rock. Similarly, small, high latitude lake basins are frequently lined with peat in locations where catchment slopes are very low and bogs develop (Sec. 8.8), preventing the underlying soil and rock

---

[2]The near-neutral precipitation pH recorded in heavily industrialized regions of south-central China is attributed to atmospheric neutralization with $CaCO_3$-rich dust carried from upwind regions (H. B. Xue and J. L. Schnoor 1994). The Sahara desert is also a rich source of wind-blown alkaline material and of phosphorus. In North America, significant levels of acid-neutralizing dust are seen in precipitation in the semiarid plains east of the Rocky Mountains and in the midwest, which are characterized by exposed calcareous soils subject to wind erosion.

from coming into contact with the water. Finally, catchments dominated by glacially deposited sands or quartzite rock yield exceptionally low ANCs (Rapp et al. 1987). Pätilä (1986) demonstrated the importance of elevation, a surrogate for soil thickness, lake morphometry, and catchment attributes in determining lake pH in southern Finland:

$$pH = 7.51 - 0.029 \cdot [\text{elevation (m)}]$$
$$+ 1.012 \cdot [\text{lake surface area (km}^2)]$$
$$- 0.021 \cdot (\% \text{ of catchment}$$
$$\text{covered by sphagnum peat})$$

$$r^2 = 0.52; \; n = 52 \qquad \text{EQ. 27.4}$$

If the geology, position in the catchment, altitude, and lake morphometry were all that determined the sensitivity of waters to acidification, they would be easy to identify and the pH readily predictable. Unfortunately, the sensitivity is additionally affected by organic acids derived from soils and upstream wetlands. Sphagnum bogs (wetlands) in particular release organic acids that help lower the pH of the receiving streams and lakes. How much these lower the pH is determined by their supply and the buffering provided by the receiving water. The neutralizing agents are produced by microbially mediated oxidation–reduction reactions that are qualitatively well recognised (Table 27–1 and Fig. 16–2), but are not easily predicted in terms of effect without detailed study (Sec. 27.10).

27.5 ▲ Catchments and Lake Acidification: Wet and Dry Deposition

Limnologists have used input–output (mass balance) budgets of acids and acid precursors to quantify the role of catchments in modifying acid inputs to waterways. The principal uncertainty in such budgets is due to the inability of the widely used *bulk deposition collectors* to trap more than a fraction of the dry deposition. The collectors trap all the material that falls into the continuously open collector, including all wet deposition plus an unknown and variable fraction of dry deposition. As the dry deposition is difficult to measure, Wright and Johannessen (1980), among others, computed the dry deposition in southern Norway by assuming that atmospheric chloride was overwhelmingly derived from the oceans and not significantly retained by catchments. They found an average of 37 percent larger output of Cl in the outflowing streams than could be accounted for by the

Cl supply determined in wet deposition collectors alone. The 37 percent excess was assumed to reflect the chloride dry deposition and, by implication, to be indicative of the significance of the dry deposition of all the other major elements of marine origin.

In climatically drier eastern North America, the dry deposition appears in excess of the wet deposition in high emission areas, declining to about 20 percent of the total deposition in remote regions (Summers et al. 1986). Whatever the precise proportions at any one location, it is evident that impacted Scandinavian and North American catchments generally release much less H[+], and NO$_3^-$ (Fig. 27–5) to outflowing streams than they receive via the atmosphere, thereby protecting the receiving streams, wetlands, and lakes from the acidification potential of these ions.[3] It was originally believed that sulfur inputs and outputs balanced (Fig. 27–5) with the output of SO$_4^{2-}$ serving as the principal strong acid anion to the receiving waters. Unfortunately, the assumption that input equals output is incorrect. Outputs may be much lower than inputs during dry years and greatly exceed inputs during wet ones, confounding predictions based on changes in atmospheric deposition of SO$_4^{2-}$ (Sec. 27.12).

27.6 ▲ Neutralization and Buffering Processes in Catchments

Different processes dominate the assimilation (retention) of incoming H[+] at different pH values in both soils and water. In calcareous drainage basins, the incoming H[+], or the H[+] produced during soil respiration is immediately neutralized following the dissolution of additional calcium (or magnesium) carbonate (Eq. 1 in Table 27–1, and Eq. 14.3). In the process, liberated HCO$_3^-$ ions pass via the soil water to the waterways, while the CO$_2$ that is produced is lost to the atmosphere. The high HCO$_3$ content of water from calcareous drainage basins or those containing some CaCO$_3$-rich material neutralizes direct inputs of low pH precipitation on the water surface without markedly lowering the pH (Sec. 14.2). In such catchments ANC is almost totally attributable to the

---

[3]Deep oligotrophic lakes in southern Sweden show 50–100 percent increases in NO$_3^-$ between 1970 and 1986, as do streams in Norway. In some forested areas of the Czech Republic and Germany, the NO$_3^-$: NO$_3^-$ + SO$_4^{2-}$ output ratio was, prior to recent improvements, 0.4–0.5. This indicates that the catchments might have become NO$_3^-$-saturated following sustained high N loading and are now releasing NO$_3^-$, thereby contributing to lake and stream acidification. (Henriksen and Brakke 1988 and Sec. 18.1.)

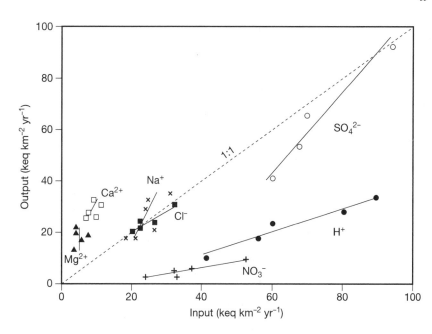

Figure 27–5 Specific inputs versus specific outputs of ions for each of five years (1973–1978) in an acidified Norwegian catchment, with the dry deposition fraction determined by the chloride correction method (see Sec. 27.5). The 1:1 line represents an equal input and output. Note (1) the retention of hydrogen ions and nitrate; (2) the net export of calcium and magnesium; and (3) an input–output ratio close to one for sulfate, chloride, and sodium. *(After Wright and Johannessen 1980.)*

CO_3/HCO_3 system. If the H^+ inputs are large and the accessible $CaCO_3$ deposits in the catchment are limited, a progressively increasing fraction of the HCO_3^- released is lost as CO_2 to the atmosphere and the HCO_3 of the outflowing streams and the pH of receiving lakes declines over time.

Sensitive Catchments

Sensitive igneous catchments that lack calcareous deposits and also lack bogs releasing organic acid typically have water with a pH of 5–6 when nonacidified. The CO_2 produced in respiration by the soil biota and plant roots combines with water to form H_2CO_3 (Sec. 14.2). Following the dissociation of the H_2CO_3, the H^+ produced is neutralized by the weathering of Na or K feldspar minerals—the principal component of igneous rock—and the HCO_3^- released. Slow weathering yields clay minerals composed primarily of insoluble aluminium silicates, but also releases moderate amounts of HCO_3^-, cations, and dissolved silica to waterways (Eq. 13.2, and Eq. 2 in Table 27–1).

The HCO_3^- released during weathering of hard igneous rock is sufficiently small to make the receiving lakes and rivers extremely sensitive to acidification. The H^+ in nonacidifying precipitation is primarily consumed (and retained) in the catchment soils by a cation exchange neutralization (Eqs. 4 and 5 in Table 27–1,) and little is released to the runoff (see Fig. 27–5). More specifically, some of the H^+ is taken up and removed in the process of dissolving the aluminium hydroxides, oxides, and SiO_2 contained in the

clay minerals produced during the breakdown of weathered silicates (Eq. 3 in Table 27–1). Other H^+ may be exchanged for cations on the soil particles. The Al ions that are released replace Ca, Mg, and other base cations on the negatively charged colloidal clay and humus particles of the soil matrix, with the released cations entering waterways.

The extent to which soils produced from igneous rock can serve as a buffer and protect the outflowing rivers and the receiving lakes and wetlands is a function of (1) the abundance of silicate and clay minerals to be weathered; (2) the extent to which negatively charged clay particles in soil and sediment are associated with base cations such as Ca^{2+}, Mg^{2+}, or NH_4^+ and Al^{3+} available to exchange places with the incoming H^+; and (3) the thickness of the soil or, more correctly, the *overburden* and flowpath contact time that the water has with the soil. The leaching (release) of base cations is consequently a function of the supply rate of the H^+ and water movement.[4]

When the supply of strong acid anions (primarily SO_4 and NO_3) is increased during acidification, there must be an equivalent increase in cation release to maintain the ionic charge balance. Initially the required cations come from exchangeable bases in the

---

[4]In acid-rain research, the major cations are, for historical reasons, referred to as *base cations*. They form the cation of strong bases [e.g., Ca $(OH)_2$, K (OH)] whereas those anions associated with strong acids (SO_4^{2-}, Cl^-, and NO_3^-) are referred to as the (strong) *acid anions*.

soil or sediments, thereby protecting the receiving waters from acidification. But when the base cation supply declines over time, they are replaced by the cations Al^{3+} and H^+ to maintain the ionic charge balance on the soil particles. A further H^+ input allows excess Al (plus other trace metals) and excess H^+ to be exported, raising their levels in streams, wetlands, and lakes. As the water pH declines to about 4.5–4.7, organic acids that had previously acted as natural sources of H^+ start buffering against further acidification. Similarly, Al oxides released to the water start to dissolve or dissociate and begin to serve as additional buffers against acidification.

$$Al(OH)_3 + 3H^+ \rightleftarrows Al^{3+} + 3H_2O \quad EQ.\,27.5$$

▲ 27.7 Buffering Capacity of Lakes, Rivers and Wetlands

The acid neutralizing capacity of water is of the greatest importance in determining the pH and assessing the buffering capacity of fresh water. It can be expressed most simply as:

$$ANC = [HCO_3^-] + 2[CO_3^{2-}] \quad EQ.\,27.6$$
$$+ [OH^-] \pm [Al^-] - [H^+] - \Sigma[Al^+]$$

Where $\Sigma[Al^+] = 3[Al^{3+}] + 2[AlOH^{2+}] + [AL(OH)_2^+] \dots$

CO_3^{2-} and OH^- are essentially absent from the generally low ANC oligotrophic waters of acid-sensitive areas (Fig. 14–1) and organic anion concentrations are negligible in clear water lakes, therefore Equation 27.6 can be simplified to:

$$ANC = [HCO_3^-] - H^+ - [Al^+] \quad EQ.\,27.7$$

and in aquatic systems with a pH > ~5.5 to:

$$ANC = [HCO_3] - H^+ \quad EQ.\,27.8$$

The sum of the charge of all strong base cations must balance the sum of their strong acid anion counterparts; the ANC can then be expressed as the difference between the sum of the base cation and the strong acid anions (unit: $\mu eq\,l^{-1}$)

$$ANC = [Ca^{2+}] + [Mg^{2+}] + [Na^+] \quad EQ.\,27.9$$
$$+ [K^+] + [NH_4^+] - [SO_4^{2}]^- + [Cl^-]^- + [NO_3^-]$$

This definition of ANC overlooks ionic aluminium and organic anions (A^-), whose role remains to be completely defined. When the supply of acid anions exceeds the base cations available, the ANC is < 0

$\mu eq\,l^{-1}$ and the waterways will be strongly acidic with a pH of < 5.5.

Microbes and ANC

Complex chemical processes are far from the only determinants of ANC in drainage basins and aquatic systems. Microbes also play a central role in increasing or reducing the ANC. For example, a microbe-mediated oxidation of reduced inorganic sulfur and nitrogen compounds results in the production of H^+, SO_4^{2-}, NO_3^-, Fe^{3+}, Mn^{4+} and a reduction in ANC (Eqs. 11–15 in Table 27–1). Conversely, acid deposition on bogs (peatlands) is normally largely neutralized by a NO_3^- uptake by plants and through the microbial reduction of SO_4^{2-}, NO_3^-, Fe^{3+}, and Mn^{4+} at oxyclines (Eqs. 6 - 10 in Table 27–1). The role of microbes as catalysts for the redox reactions that bring about a reduction in H^+ (increase in pH) is reversed when waterlogged soils are aerated, wetlands are drained, or when a series of dry years produce a lower water table. The aeration of previously anoxic soil and sediment permits microbially mediated oxidation of the reduced sulfur minerals and the release of H^+ (acidification) following rewetting (Eqs. 14 and 15 in Table 27–1).

Changes in Alkalinity

A seasonal reduction in base cation inputs to waterways from still-frozen soils during the spring melt of acidic snow has a short-term negative effect on pH and alkalinity (Molot et al. 1989). Conversely, a periodic pattern of summer anoxia in lakes or wetlands has a short-term positive effect on the ANC (in Fig. 19–1), but the alkalinity gained is temporary and is lost when oxic conditions return (Eqs. 11–15 in Table 27–1). Much more important is the permanent gain in alkalinity (ANC) obtained through burial of reduced redox elements or their loss to the outflow or atmosphere, or when the SO_4^{2-} or NO_3^- are taken up and stored in the forest vegetation; for example:

$$106CO_2 + 138H_2O + 16NO_3^- \quad EQ.\,27.10$$
$$\rightleftarrows (CH_2O)_{106}(NH_3)_{16} + 16OH^- + 138O_2$$

The in situ production of permanent alkalinity is of great importance in buffering (protecting) acidified waters from further acidification and aids in their recovery following a reduction in the deposition of acid precursors. With the reduction (removal of H^+) occurring just below the oxycline in the surficial sediments, it is not surprising that the sediment pH in acidified waterways is well above that recorded in the water column. The possibility for reduction of the oxidized form of redox elements is greater in slowly flushed lakes than streams or wetlands that are rapidly flushed, where a high fraction of oxidized elements that are entering leave before reduction is possible. Conversely, accumulating terrestrial vegetation removes significant amounts of base cations in the production of aboveground biomass, thereby adding to the acidification potential. Ammonium uptake by plants also results in soil acidification unless offset by denitrification.

27.8   Aluminum and Other Toxic Metals

Aluminum is present in great abundance in the Earth's crust where it accounts for about 8 percent of igneous rock (Table 13–2). Aluminum silicates (clays) are produced upon weathering. Further weathering yields aluminum oxides and hydroxides (Eqs. 2 and 3 in Table 27–1,); gibbsite [Al (OH)$_3$] is a fairly common mineral produced. Under the nearly neutral pH conditions that characterize most sediments, the low solubility of the Al oxides and hydroxide polymers produced is further reduced by organic complexing agents, as well as by the aluminum fluoride and sulfate complexes formed. Although the speciation of Al is very complex and highly pH dependent, it is evident that the solubility of Al is very low in nonacidified sediments and soils. Consequently, no more than a few µg l^{-1} are found dissolved in soil water and receiving fresh waters. The remainder is retained in the soil profile or deposited as insoluble precipitates in the sediments. Therefore, soils and sediments in igneous catchments contain large concentrations of precipitated Al oxides and hydroxides which, upon acidification, can be released to the water. In reality, even the small amount of the (so called) dissolved Al reported in near-neutral water is largely complexed with dissolved organic and inorganic compounds (e.g., Al fluoride and Al sulfate), with the result that virtually none is in the Al^{3+} form. However, Al solubility increases when the pH of sediments and soils declines below ~5.5 (Fig. 27–6). Then the Al that is released includes a labile (inorganic) *monomeric fraction* (Al^{3+}, Al (OH)$_3$, and Al (OH)$^{2+}$) plus simple inorganic complexes that are the most toxic to the biota (Fig. 27–7). However, there is much variation at any pH in the amount of soluble Al measured among rivers and lakes

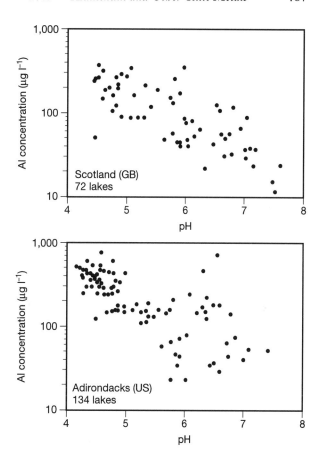

Figure 27–6   Total dissolved Al vs pH in lakes in acidified areas of Europe and North America. (*After Wright et al. 1980.*)

(see Fig. 27–6), indicating that more than pH alone determines Al levels.

▲ Aluminum and Organic Matter

Other than pH, an important determinant of Al solubility is the amount of dissolved organic matter (or its imperfect surrogate, water color). Work in Finland (Pätilä 1986) shows that the amount of total Al is disproportionately high (in humic lakes) at a pH > 6 because colloidal organic matter (DOM) complexes with colloidal Al, retarding its precipitation. Organically complexed Al is nonlabile and nontoxic (Driscoll 1980) and it is evident that the toxicity of a particular concentration of dissolved Al increases with declining pH.

The Norwegian survey of 1,000 lakes quantified the relationship between *chemically reactive Al* (RAl, mg l^{-1}) and environmental factors (Hongve 1993).

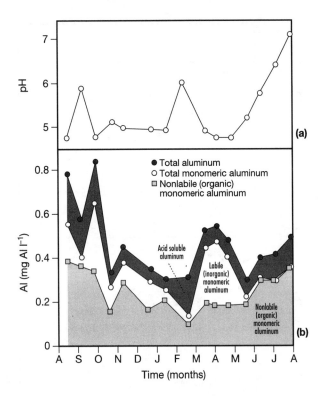

Figure 27–7   Observed temporal variation in the (a) pH and (b) aluminum fractions of a North Lake tributary, Canachagala Creek (US), from August 1977 to August 1978. (*After Driscoll 1980.*)

$$\log RAl = 1.82 - 0.53 \cdot pH$$
$$- 0.30 \cdot \log(DOC) \qquad\qquad EQ.\ 27.11$$
$$- 0.11 \cdot \log(\text{runoff})\ R^2 = 0.79;\ n = 1,005$$

Equation 27.11 states that as the pH and DOC decline, the reactive Al increases. The negative effect of runoff may reflect the associated reduction in organic acid loading and thus a reduced sorbing of Al to DOM, and increased photooxidation of DOM in more slowly flushed lakes.

The increase in total Al with declining pH is mirrored in the behavior of other metals (Fig. 27–8). In humic lakes, metal concentrations are higher than expected on the basis of pH (Pätilä 1986) as the result of sorbing to DOM.

Acidification and Transparency

Even though cause and effect cannot be resolved in observational studies, it is evident that the transparency of lakes is distinctly increased in part by in-

creased coagulation of the colored DOM upon acidification. It is also partly the result of increased coagulation (flocculation) of the DOM with sedimentary aluminum released upon acidification and partly of enhanced photo bleaching and oxidation of colored organic matter. Schofield (1972) was one of the first to note an increase in mean transparency from 5.6 m to 8.3 m in 14 New York (US) lakes after a 20-year pH decline from 5.9 to 5.0. An experimental example is the acidification of Lake 302S (ELA,CA) with H_2SO_4, which over a 10-year period lowered the pH of the lake from 6.0–6.7 to 4.5 and transformed the humic lake (DOC ~7.2 mg l^{-1}) into a highly transparent clear-water system (DOC ~1.5 mg l^{-1}) (Schindler et al. 1996). Conversely, a decrease in H^+ loading (increase in pH) will decrease DOM coagulation and decrease transparency. That is unless, as in Nellie Lake (Fig. 27–9), the effect of a long-term reduction in runoff and catchment export of the colored DOM exceeds the effect of the within lake increase in color (decrease in coagulation) (Sec. 11.12).

Increasing transparency has tremendous ramifications. It allows for increased penetration of light energy which leads to reduced near-surface heating (Sec. 10.8), thereby increasing the depth of the thermocline, as well as increasing the depth to which phytoplankton and benthic algal photosynthesis is possible (Sec. 11.12). However, the same increase in transparency also enhances the ultraviolet radiation (UVR) penetration (Fig. 10–7). The depth at which 1 percent of entering UVB remained increased ninefold from ~0.3 m to 2.8 m in experimentally acidified Lake 302S (Schindler et al. 1996). It remains unclear to what extent, if any, increases in UVR penetration and changes in stratification have contributed to the biotic changes observed during lake acidification.

27.9   Effects of Acidification on the Aquatic Biota

There is abundant literature on the effect of acidification on the biota—much of it obtained from surveys—which make the identification of underlying causes and mechanisms impossible. There is also substantial literature (largely on fish) examining the physiological responses to acidification under controlled laboratory conditions. Although the laboratory results are usually clear-cut and permit identification of the responsible physiological mechanisms, it remains impossible to as-

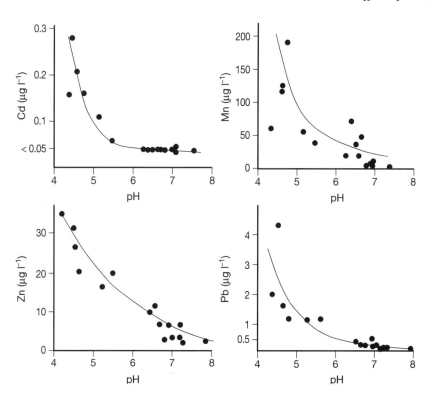

Figure 27–8 Metals in 16 lakes on the west coast of Sweden experiencing a similar atmospheric metal deposition but differing in pH. *(After Dickson 1985.)*

cribe observed changes in species composition and growth in nature to a particular mechanism. This is because the abundance, or even the presence, of particular organisms in nature may be affected by not only the acidity but also other stressors, including reactive Al and other trace metals, as well as by seasonal changes in DOC, food supply, predation, competition for resources, and other interactions between the species (Fig. 27–10). Furthermore, species of fish and their prey respond differently to acidification and, in addition, change their responses during the various life stages, making the importance of physiological mechanisms observed in the laboratory difficult to resolve in nature.[5] Yet, it is evident that aquatic food webs in acid-stressed systems usually become simpler due to the loss of acid-sensitive species even though the community biomass is often unaffected. Some of the effects are the result of the loss of sensitive prey or predator species.

The loss of species richness (biodiversity) commences when acidic deposition causes the pH of waterbodies to decline below pH 6.0, although not all taxonomic groups of organisms are equally affected. Despite the above qualifications, some broad patterns of response to acidification have become evident, showing the greatest losses in species richness in lakes with an original low ANC (< 50 μeq l[-1]). The synthesis below of the biotic responses to acidification is based on many sources, but Schindler et al. 1991, and Brezonik et al. 1993 are particularly relevant.

Figure 27–9 Nellie Lake an upland Canadian Shield lake in Ontario (LA = 2.6 km², z_{max} = 55 m, CA:LA = 5) with a transparency of ~29 m, pH ~4.6, and estimated preacidification pH ~6.8. *(Photo courtesy of E. Snucins.)*

---

[5]"Observed phenomena can be explained from many mechanisms, and different mechanisms can be responsible for a similar phenomenon as that produced by a single mechanism." (L. Kamp-Nielsen 1997)

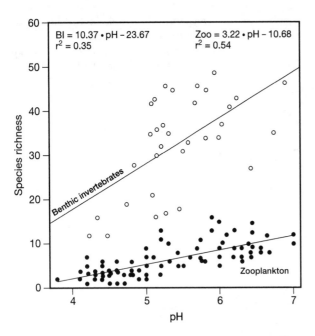

Figure 27–10   The relationship between pH and the number of crustacean macrozooplankton species (Zoo, ●) and the number of benthic macroinvertebrate species or genera in fishless lakes (BI, O) on the Canadian Shield. Note that no pattern was observed between the number of benthic macroinvertebrate species and pH in lakes with fish, indicating an important but highly variable effect of fish predation on species richness. *(Data from Sprules 1975, Confer et al. 1983, Yan and Miller 1984, and Jeffries 1997.)*

Microbial Activity and Composition

Although virtually nothing is known about the species composition, acidification appears to have little or no affect on the abundance of heterotrophic bacteria in the plankton. Work on sediment-oxygen uptake, glucose turnover, and bacterial abundance also shows no decline with increased acidification. There is, however, some evidence that leaves and litter from the drainage basins are less rapidly decomposed in acidified waters and thus tend to accumulate. Changes are evident in the fungal species composition of acidified streams.

Macrophytes

Many of the macrophyte species inhabiting susceptible lakes are acid resistant, thus they show little change and effects tend to be restricted to small lakes Fig. 24–4. There may be shifts in the relative impor-

tance of individual species. However, in some lakes a shift has been observed in which the *Lobelia* and *Isoetes* species (Fig. 24–3)—typically dominant in nonacidified soft-water lakes at higher latitudes—are replaced by *Sphagnum* mosses. This shift has been seen at a pH between 4.4 and 5.4 in both Sweden and North America.

Benthic algae

The development of extensive algae mats, particularly of filamentous green algae on the littoral sediment surface and the occurrence of loose cloudlike masses of filamentous green algae (metaphyton) in the littoral zone and among macrophytes is commonly observed upon acidification in the shallow shoreline waters of acidified forest lakes. The cloudlike masses become abundant when the pH declines to 5.6 and disappear when the pH rises to ≥ 5.8 (Jeffries 1997). An increase in benthic algae has been noted in artificially acidified lakes and streams in Europe and North America (Schindler et al. 1991). The causes for these increases are not well resolved, but have variously been attributed to decreases in invertebrate grazers, to an ability of the benthic algae to cope with the very low levels of inorganic carbon availability for photosynthesis, or simply to higher rates of benthic photosynthesis in the now more transparent lakes (Planas 1996).

Phytoplankton

Both surveys and experimental work have shown that the number of planktonic species become reduced, but neither the community biomass nor the primary productivity is much affected. The chrysophyte and other small flagellates and diatoms that usually dominate north temperate oligotrophic low alkalinity lakes (Fig. 21–5) tend to be replaced by larger dinoflagellates (Schindler et al. 1991, Almer et al. 1978). However, phytoplankton respond not only to the addition of H^+ but are also sensitive to Al, the specific anions that dominate, and selective grazing by zooplankton (Findlay and Kasian 1990). Paleolimnology uses the differential species sensitivity to acidification to infer pH conditions in dated sediment cores (Secs. 27.12 and 20.5).

Zooplankton

While the community biomass typically changes little, there is both an appreciable species loss and a change to less sensitive species. Doka et al. (1997) report an average 50 percent loss of macrozooplankton and rotifer species in eastern Canada when the pH declined to 4.8. Crustacean zooplankton and rotifers richness show a distinct impoverishment upon acidification (Fig. 27–10), and small species tend to become more dominant (Dillon et al. 1984). This is somewhat surprising because the loss of zooplanktivorous fish species, which feed preferentially on larger zooplankton, is expected to favor the larger zooplankton. But some zooplanktivorous fish species (e.g., perch) are more acid-tolerant than others and become highly abundant following the loss of their more acid-sensitive predators (Gunn and Mills 1998). The phantom midge (*Chaoborus*; Sec. 23.9), which is little affected by acidic conditions and feeds on midsize zooplankton and the small forms of larger species, typically becomes very abundant in the absence of fish predators. Paleolimnologists use the inverse relationship between fish and *Chaoborus* abunbance to infer the loss of fish after acidification, by using *Chaoborus* mouthparts preserved in dated sediment cores (Uutala and Smol 1996).

Benthic Macroinvertebrates

This group, which includes the large bottom-living crustaceans and molluscs, has many species that are sensitive to a lowered pH (Fig. 27–11). Among the crustaceans, the freshwater amphipod, *Gammarus lacustris* is rare at pH < 6.6 and absent at pH < 6.0 in Norwegian lakes (Økland and Økland 1980). Amphipods, crayfish, and other macroinvertebrates appear to experience problems with exoskeleton hardening under low pH conditions in both lotic and lentic system. Snails (gastropods) disappear when the pH declines to 5.2–5.0 and almost no Ca(HCO$_3$)$_2$ is available for shell formation. In contrast, the sowbug or waterlouse (*Asellus aquaticus*), which is abundant in Norwegian waters at pH 4.7–5.2, only disappears at pH 4.2.

A Norwegian survey of 1,500 waterbodies showed that molluscs (snails and mussels) are highly acid sensitive and some species disappears at pH > 6, but many more common species are lost at pH < 6.0. Very few of the mollusc species are able to maintain themselves at pH 4.4–4.6 (Fig. 27–11). Similar findings were made among the molluscs, leeches, and insects in the most affected areas of New England, US (Schindler et al. 1989).

While there are pronounced shifts in the relative abundance of the sediment-living immature stages of aquatic insects, there appears to be little change in the total invertebrate community biomass upon acidification. Among the insects, the midges (*Chironomus*

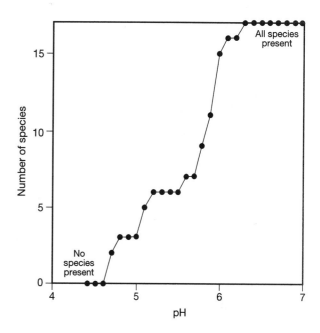

Figure 27–11   Tolerance limit of pH for 17 widespread benthic invertebrate species (important as fish food organisms) in acidified Norwegian lakes. *(After Økland and Økland 1980.)*

spp.)—which make up a large fraction of the benthic biomass in high and low latitude inland waters (Sec. 25.4)—are as a group little affected by acidification. There are even insects that do well under acidic conditions. Among them are the filter-feeding blackfly larvae (Family Simuliidae) of lake outlet streams, whirligig beetles (Family Gyrinidae), water boatmen (Family Corixidae), and damselflies (Order Zygoptera), probably the result of reduced predation from fish. Conversely, some species of mayflies (Order Ephemeroptera), caddisflies (Order Trichoptera), and stoneflies (Order Plecoptera) are particularly acid-sensitive and quickly decline following the onset of river and lake acidification and the linked increase in aluminium.[6] A loss of biodiversity and in particular of key prey species carries consequences for higher levels in the food chain (amphibians, fish, and birds).

---

[6]"The results indicate that many of the zoobenthic taxa frequently cited as being negatively affected by acidification instead reflect changes along a water-hardness gradient . . . While [particular] taxa may be physiologically susceptible to increasing H$^+$ levels, their natural distributions appear limited by water hardness, precluding them from soft-water lakes in danger from acidification." (S. P. Lonergan and J. B. Rasmussen 1996.)

Fish and Birds

In southern Norway, hit hard by lake acidification, 50 percent of the 2,823 lakes that once contained brown trout (*Salmo trutta*) had lost their stock by the mid-1970s (Sevaldrud et al. 1980). It is not surprising that the losses are most pronounced in lakes and streams with the lowest alkalinity. In the Adirondack Mountains of New York (US) the effect of acidification is most evident at higher altitudes, where the lakes and streams are located in small igneous catchments unable to provide appreciable buffering (Sec. 27.6). Some 50 percent of 217 New York lakes above 610 m have a pH < 5 and 90 percent of them lack fish altogether. Similarly, hundreds if not thousands of lakes in Canada are believed to have lost their fish stocks while many Norwegian streams and lakes now lack a fish biota altogether or have an impoverished one (Jensen and Snekvik 1972).

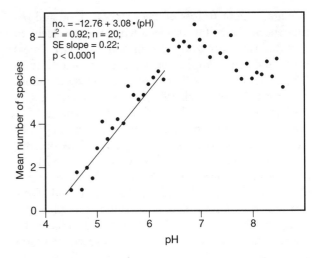

no. = −12.76 + 3.08 • (pH)
r^2 = 0.92; n = 20;
SE slope = 0.22;
p < 0.0001

Figure 27–12   Relationship between the mean number of fish species and pH for Ontario (CA) lakes. (*After Matuszek and Beggs 1988.*)

Community Structure and Acidification

The loss of fish populations is gradual because different species have different lower pH limits, but the decline is evident below about pH 6.5 (Figs. 27–11 and 27–12) and about a 10-20 percent species loss is apparent at pH 5.5 (Jeffries 1997). However, pH is usually coupled with other important environmental variables, such as Al and Ca concentration, and it is not clear from observational (nonexperimental) studies to what extent pH alone is responsible for the observed declines in fish and invertebrates.[7]

▲ The abundance and breeding success of piscivorous waterbirds, such as common loons (*Gavia immer*) and the common merganser (*Mergus merganser*), is indeed greater on higher pH waterbodies with low aluminium concentrations, but the birds also breed successfully in lakes with a pH as low as 4.5. The principal effect on the birds appears not to be pH-related but seems to be indirect, resulting from the reduction

or elimination of prey fish populations (resource availability) in low pH lakes (Doka et al. 1997, Jeffries 1997). However, low pH boreal forest lakes are also typically highly oligotrophic and the effect of acidity, trophic status (Kerekes et al. 1994), water color (on underwater vision), and high mercury levels in birds feeding on contaminated fish (Chapter 28) means that the reasons for low breeding success are difficult to separate mechanistically.

The lowest pH at which a fish species is recorded overestimates its tolerance limit because adults survive even though the tolerance limit for successful reproduction has been exceeded. Reproductive (recruitment) failure due to failed spawning, egg maturation, or death of recently hatched larvae is considered to be the most common cause of fish population losses. Unfortunately, recruitment failure is not readily evident to fishermen or biologists who study the adults of long-lived game fish. Surveys based on the presence of adults of a particular game fish are therefore not a good measure of whether a waterway or the particular species is affected by acidification. Studies on the age distribution of species are more useful and these typically show a shift to older individuals and species more resistant to low pH (e.g., perch) as the pH and associated environmental conditions exceed the environmental conditions tolerated for successful reproduction. Even though recruitment has become impaired, the growth rate of the older fish is sometimes even enhanced, presumably the result of reduced competition for food (Ryan and Harvey 1980).

---

[7]The problem of factors other than pH (e.g., lake size, DOC, reactive Al, invading species) having an effect on species richness has naturally been seized upon by industries emitting acids or acid precursors to dispute their responsibility and argue for further studies. Equivalent arguments were produced by the detergent industry in the 1970s, disputing the importance of phosphorus in lake eutrophication (Sec. 21.12). In all cases controls were imposed after the correlations were convincingly demonstrated even though the details of the cause and effect relationships remained unresolved. At the time controls are imposed, the political focus has shifted from arguments about cause to the empirical correlations.

Interpretations of species losses are further confounded by summer pH levels high enough to suggest that factors other than acidification are responsible for the observed changes, whereas the possibility of episodic (periodic) acidification during the spring melt of highly acidic snow, or much lower pH during periodic heavy summer rains is overlooked (Fig. 27–7). Short-term exposure to a sharply lowered pH during the snow-melt period may expose eggs and larvae to acutely toxic concentrations of H^+ or metals at a time when little or no sampling is being done.

Low pH waters normally have high Al concentrations (Fig. 27–6), of which the reactive monomeric fractions are particularly toxic. Much of the aluminum is derived from the catchments, but additional Al is released by sediments. Thus, the experimental acidification of streams (but not their catchments) with H_2SO_4 to a pH of about 4 is followed by a substantial release of Al to the water (Henriksen et al. 1988). Unfortunately, the toxic effect of Al (and other metals) observed in the laboratory (Parkhurst et al. 1990) is confounded in nature by its correlation with H^+, the availability and uptake of calcium, and the normal lack of information about the extent that the measured elevated levels of dissolved Al were in a labile and toxic monomeric form, rather than sorbed to DOC and nonlabile.

▲   27.10   Modeling the Acidification Process

A variety of models have been developed, ranging from simple empirical ones to complex dynamic models that consider catchment weathering processes and the chemical equilibria determining the ANC of receiving waters. Elaborate process-oriented models are appropriate for well-studied catchments and aquatic systems for which most of the necessary information is available (Reuss et al. 1986).

Mechanistic and Empirical Models

Probably the most influential of the process-based analytical models used today is MAGIC (Model of Acidification of Groundwater In Catchments). It is a simple mechanistic simulation model based on a mathematical representation of a small number of what are considered to be key physical and chemical processes controlling the equilibrium between major dissolved and adsorbed ions in soils and water quality. The moderate data requirements make it easy to apply

in comparison with more complex (comprehensive) models, but at the inevitable risk of overlooking important processes. Simulation modeling is used to predict changes in the surface-water quality response time to changes in atmosphere precipitation. Impressively, predictions made by Cosby et al. (1985) concerning the time required for acidification and the recovery of different soil types in well-characterized catchments are qualitatively close to what is being observed 20 years later.

Unfortunately, the processes dominating particular catchments are generally unknown and the most useful predictive model is a semiempirical model developed by Henriksen (1980). It focuses on measured alkalinity changes below the waterline rather than, as in MAGIC and other process models, on the catchment and sediment processes that produce the observed alkalinity. The Henriksen model has been refined (Wright 1983a, Henriksen and Brakke 1988), but as the model's underlying ideas and its utility are equally evident from the simpler original version it will be briefly discussed.

Henriksen (1980) visualized the acidification process as one in which the alkalinity of a waterbody is titrated by strong acid (H_2SO_4) deposited on the surface. He used alkalinity (ANC) as an integrator and indicator of the unexamined catchment processes that are responsible for most of the observed buffer capacity of inland waters. The titration with acid of a whole aquatic system in nature resembles the titration of a $Ca(HCO_3)_2$ solution in a beaker, a procedure used to determine the alkalinity of inland waters (Fig. 27–13 and Sec. 14.4).

The Henriksen model uses the term acidification in a special sense by defining it by the change in alkalinity (ANC).

acidification = preacidification alkalinity (ANC_0)
− present alkalinity (ANC).

Very few preacidification alkalinities are available and it is assumed that there is an electron neutrality (ion balance). Simplifying the most critical components of the cation-anion balance, it is possible to conclude that:

$$2[Ca^{2+}]* + 2[Mg^2]* = [HCO_3^-] - [H^+] \quad \text{EQ. 27.12}$$

the units are in $\mu eq\ l^{-1}$ and the asterisk (*) indicates the nonmarine derived concentrations. Therefore:

$$ANC_0 = [Ca^{2+} + Mg^{2+}]* \quad \text{EQ. 27.13}$$

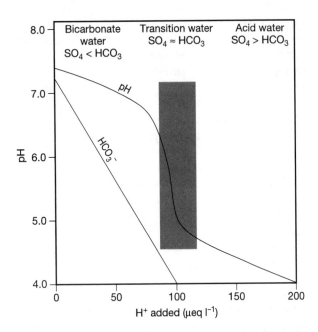

Figure 27–13 This titration curve in a beaker for bicarbonate solution at a given concentration (100 μeq/l) upon addition of a strong acid illustrates the acidification process and indicates the extreme sensitivity of poorly buffered (but nonacidified) waters (pH ~5–6) to acidification. Acid-sensitive clear-water lakes lie along the pH spectrum represented by the curve. With increasing acidification they shift from bicarbonate waters that serve as a buffer against acidification to transitional waters in which small H⁺ additions result in large pH changes and frequently fish kills, to acidic waters of relatively stable pH and lacking fish. *(After Henriksen 1980.)*

But because Ca^{2+} is normally by far the most important cation and is assumed to be in a fixed relationship with Mg^{2+} and other cations, the equation can be further simplified to:

$$ANC_0 = [Ca^{2+}]^* \qquad \text{EQ. 27.14}$$

While ANC declines during acidification the Ca is considered unaffected. If the assumptions underlying Equations 27.12–27.14 are correct, then water being acidified should have more $[Ca^{2+}]^*$ at any particular pH than nonacidified systems, which has been observed (Overrein et al. 1980). Although the pH: $[Ca]^*$ ratio or the $[Ca + Mg]^*$: HCO_3 ratio provide a useful indication as to whether acid sensitive waters have become acidified, they do not allow predictions about the expected effect of a change in H⁺ or $[SO_4]^*$ loading. To make such predictions possible Henriksen first

characterized the acidification of lakes (equally applicable to lotic and wetland systems) in three stages:

- *Stage 1:* The first stage in the titration of acid-sensitive waters is characterized by a definite decline in alkalinity but accompanied by a more modest decline in pH to a value no lower than 5.5–6.0 (Fig. 27–13). The HCO_3^- buffer system remains largely intact. Systems in this category are designated as *bicarbonate waters*.

- *Stage 2:* The HCO_3^- buffer system has been largely destroyed at this stage, and the pH shows large seasonal fluctuations. Such systems, known as *transition waters* (Fig. 27–13) are subject to periods of low pH while poorly buffered snow melts, or during large rainstorms in summer or fall that also prevent much water contact with the soil. Waters with a low preacidification HCO_3 concentration (< 50 μeq l⁻¹ or < ~50 μS cm⁻¹) and associated pH of 5.0–6.0 (Fig. 27–13) are particularly susceptible to acidification, periodically elevated aluminium concentrations, as well as to changes in the biota. Some of the present transition lakes and streams may have been bicarbonate waters whose alkalinity became reduced upon acidification. Other such systems, located in unaffected areas, never had a higher buffering capacity than currently exists; it is this latter category that is most vulnerable to acidification once the region starts receiving strongly acidic precipitation.

- *Stage 3:* This stage of acidification results in *acid waters* with a chronically depressed pH of < 5, elevated aluminium levels, a reduced fish species richness (Fig. 27–12), or the absence of fish.

To develop the predictive equilibrium model Henriksen (1980) first produced empirical relationships between concentrations in the water of $[SO_4]^*$ and $[Ca^* + Mg^*]$ for systems with a pH between 5.2 and 5.4 ($\bar{x}$ = 5.3) and between 4.6 and 4.8 ($\bar{x}$ = 4.7). The pH 4.7–5.3 range separates bicarbonate lakes (stage 1) and acid lakes (stage 3) from transition lakes (Fig. 27–14). Waters that plot on the pH 5.3 line have a $[Ca^* + Mg^*]$ almost totally balanced by $[SO_4]^*$. Below this line HCO_3 is lacking and the pH is a function of strong acids and aluminum. The last two steps needed to predict the equilibrium effect of acidifying precipitation on inland waters of different ANCs were to correlate the $[SO_4]^*$ in the precipitation with the $[SO_4]^*$ concentration in

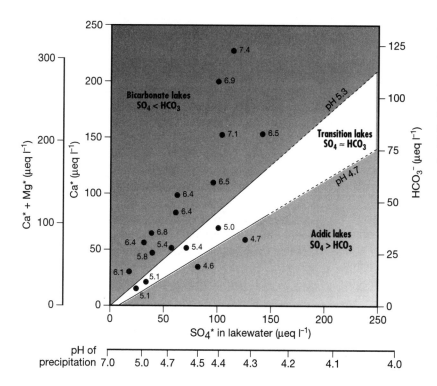

Figure 27–14 A nomograph to predict the pH of lakes uses the sum of nonmarine calcium and magnesium concentration (or calcium alone) and bicarbonate content as well as the nonmarine sulfate concentration of lake water or weighted average pH of the precipitation. Observed values presented are for large lakes in south Norway. *(Modified after Henriksen 1980.)*

the water and to link the observed mean H^+ concentration of the precipitation to the $[SO_4]^*$ in the precipitation. These two steps complete the scale on the nomograph (Fig. 27–14).

Predicting the Water pH

The nomograph, based on small lakes, is equally appropriate for categorizing much larger lakes (Fig. 27–14). It is apparent from the nomograph that more inland waters would be turning acidic if the pH of the precipitation were to decline further. It is also evident that a lake with, say, a $[Ca^{2+}]^*$ concentration of 50 μeq l^{-1} located in an area with an average precipitation pH of 4.7 will remain a bicarbonate lake, but that a precipitation decline to pH 4.5 yields a transition lake, and a precipitation pH decline to 4.3 ultimately yields an acidic lake. Conversely, the model predicts that a lake with a pH < 4.3 requires an increase in precipitation pH to 4.6–4.7 before the normal HCO_3^- buffering will reappear ultimately.

Work elsewhere has confirmed that a yearly average volume-integrated precipitation pH of 4.7 is a crit-

ical limit below which adverse effects are expected in geologically highly sensitive areas of North America and Europe (Wright 1983b). In slightly less sensitive areas, acidification problems occur where the yearly weighted average precipitation pH is 4.0–4.5, but where the precipitation may episodically reach pH 3.0.

Critical Loading

The utility of the Henriksen model, as tested by its predictive power, has been confirmed elsewhere in Europe and North America (Wright et al. 1980). All models are simplifications and have inherent limitations; the (steady-state) Henriksen model makes no statement about the length of time needed for water to acidify or recover following changes in the precipitation acidity. Nor does the original Henriksen model consider the effect of aluminum and nitrate deposition on buffering and pH, or the effect of high levels of dissolved organic matter (DOM) on the pH, or the measured cations.

The concentration of H^+ (pH) in precipitation and its $[SO_4]^*$ surrogate in water, are used in Scandinavia for modeling the expected responses of lakes to future changes in SO_2 emissions (Fig. 27–14). In North America this is usually accomplished by considering changes in the effect of the precipitation loading of $[SO_4]$ on the water pH. Thus, a United States-Canada memorandum of intent sets a target loading of 40 keq SO_4 km^{-2} yr^{-1} (20 kg ha^{-1} yr^{-1}) for eastern North America by 2010. But Gorham et al. (1984) showed that this target produces an average precipitation pH of only 4.4. They conclude that the $[SO_4]$ would have to be reduced to about 15 kg SO_4 ha^{-1} yr^{-1}, representing an integrated average precipitation pH of 4.6–4.7, to allow the most sensitive lakes to reach the projected target loading. Recent paleolimnological work on the most sensitive of Adirondack (US) lakes shows that these started to acidify around 1900 when estimated loads ranged between five and 10 kg ha^{-1} yr^{-1} (Cumming et al. 1994), indicating that the $[SO_4]$ reduction will have to be much greater (> 50% of the 20 kg ha^{-1} yr^{-1} target) to protect lakes in the most threatened regions. Their conclusion agrees with a critical load limit of < 8 kg ha^{-1} yr^{-1} for SO_4 that was proposed for the most sensitive Swedish lakes (Dickson 1985).

Critical Load Concept

Recognition that the effect of a particular wet-sulfate deposition rate changes with the sensitivity of the receiving waterbodies, and that a water pH of at least 6.0 is required to protect the aquatic biota gave rise to the *critical load concept* during the 1990s. This concept has become important for establishing policy and guiding management. The critical load is defined as the amount of wet-sulfate deposition that must not be exceeded in order to protect at least 95 percent of the lakes in a region with a historical pH > 6.0 from acidifying to a pH < 6.0. It is estimated that somewhere between 12,000 and 23,000 eastern Canadian lakes and their associated streams will continue to exceed the critical load when the present SO_4^{2-} target reductions are achieved by 2010 (Jeffries 1997).

The critical load concept has not been applied widely to NO_3 depositions which have changed little over the last 20 years. In boreal forest regions, the deposited NO_3 has been overwhelmingly retained by the soils and taken up by the vegetation in the typically nutrient-deficient drainage basins; but when the supplied NO_3 exceeds the demand of terrestrial and aquatic systems these become N-saturated. This has happened in some heavily impacted regions of central Europe, Scandinavia, and in a modest fraction of eastern Canadian catchments (Jeffries et al. 1998). The excess leaches into waterways and contributes to their acidification, providing a possible explanation for the rapid NO_3 increase that has been noted in some Norwegian lakes (Sec. 18.1).

27.11 Aquatic Management: Recovering from Acidification
▲

A temporary recovery of waterways can be produced by liming and is usually accomplished by adding fine-ground limestone ($CaCO_3$) to either the lake's surface, river edge, to an upstream wetland, or the catchment (Fig. 27–15). The finer the particle size, the more rapidly the limestone dissolves, with less of it lost from the water by sedimentation. Roughly 5 g l^{-1} of the most soluble agent and 10 g l^{-1} of coarser, less-soluble limestone is needed to raise the pH from 4.5 to 6.5. When added directly to a lake or wetland, the liming initially neutralizes the acids and permits aluminum precipitation (Fig. 27–6); the balance serves as a buffer against acids arriving via the atmosphere or inflowing streams. The duration of the recovery is largely determined from the easily computed flushing time for liming compounds, which are conservative substances (Eq. 9.3 and 9.4).

Apart from providing only temporary protection, liming is not a cheap solution either. Between 1976 and 1982, approximately 6,500 lakes and 6,000 km of river water were limed in Sweden. In recent decades

Figure 27–15 Neutralizing an acidified Canadian lake by adding fine limestone from the air. *(Photo courtesy of W. Keller.)*

about 200,000 tonnes have been applied each year at a cost of $25 million (US). Although the cost is too great and the results too temporary—particularly in areas of high runoff (flushing)—to encourage widespread use, it is evident that liming yields dramatic effects through a rapid rise in pH and decline of aluminum, mercury, and other metals following their precipitation from the water column (SMA 1982).

Liming and the Biota

The effect of liming on the biota is less well-understood. The reasons for this include the generation time of the organisms. Algae have a generation time of days and should be able to respond more rapidly to the changed conditions than some macrozooplankton which have generation times of weeks to a year or more, or the benthic insects with typically one generation per year at higher latitudes. Indeed, the number of algal species increases relatively quickly upon liming and shifts from acid-tolerant dinoflagellate species (Fig. 21–10) to a chrysophyte flagellate community similar to that found in oligotrophic but nonacidified lakes (Fig. 21–10, Eriksson et al. 1983). Yet the rate of algal change is less rapid than might be expected, which may be the result of an overriding nutrient limitation, similar to that observed in nonacidified oligotrophic lakes. Furthermore, liming raises the pH and the ANC beyond the levels encountered in other sensitive but nonacidified lakes and therefore does not represent a return to the original conditions.

The absence of a rapid and clear-cut community response to liming is not only the result of generation time and nutrient considerations but is also attributable to other environmental changes that follow both liming and a reduction in acidifying precipitation. These changes typically include reduced water clarity (more color) (Sec. 27.8) resulting in a reduced mixing depth (Sec. 11.7), therefore a reduction in the size of the euphotic zone, and a resulting reduction in benthic algae production. Another factor influencing the postneutralization community composition is the change in food-web structure following the renewed successful reproduction of surviving piscivorous fish species. Their offspring directly and indirectly affect the zooplankton and benthos (Gunn and Mills 1998).

The rate of recovery after liming or a major reduction in acidifying precipitation is a function of the dispersal ability and reproductive strategy of species. The recovery is most rapid for algae, followed by zoo-

plankton and most benthic invertebrates characterized by a short (≤ 1 yr) generation time. Yan et al. (1996) noted a full zooplankton recovery within 10 years in limed lakes that had been acidified to pH 5.7, but lakes that were strongly acidified (pH ≤ 4.5) and metal-contaminated remained impoverished even after 15 years. Finally, the recovery of large aquatic species (e.g., molluscs, fish) that were extirpated is impossible unless a species innoculum survived elsewhere in the drainage basin or they are stocked. A full biological recovery will therefore greatly lag a chemical recovery. A final factor influencing community recovery is the time required for species that were lost during acidification to invade a particular lake or stream from outside the drainage basin (Henrikson and Brodin 1995).

A few experimental studies have explored the use of inorganic fertilizer additions as an alternative to liming. The resulting increased removal of CO_2 (H_2CO_3) in photosynthesis raises the pH (Sec. 14.2) and the ANC. Davison et al. (1995) propose modest fertilization instead of liming for producing needed alkalinity. Fertilization, they argue, has minimal effects on the community plus the advantage of moderately increasing the productivity of the typically highly oligotrophic lakes.

27.12   The Future

The long-term solution to acidification is a major reduction in emissions of SO_2 and NO_x; the magnitude of this reduction in SO_2 can be predicted using the Henriksen model (Fig. 27–14).

While much remains to be learned about the response of inland waters and their catchments to acidification, it has become abundantly clear that a reduction in acid precursor emissions reduces the acid deposition on drainage basins in much the same way that a reduction in external nutrient loading ultimately reduces lake eutrophication (Sec. 17.6). Direct evidence for the effect of emission controls on lakes has been obtained by monitoring at Sudbury, Canada. The mining-smelting complex located there was once the single-largest point source of SO_2 emissions in the world; it reduced its emissions in recent decades by about 80 percent to about 5×10^5 tons SO_2 yr^{-1}. The local lakes have recovered in step with this reduction (Keller and Gunn 1995). For example, Swan Lake had a measured pH of 4.0 in 1977 which rose to 5.6 within 10 years, while the water transparency declined as the lake became more colored. The pH inferred from the community of chrysophyte algae seen in dated sediment cores nicely tracked the measured pH (Fig. 27–16), showing the lake to be well on its way to the preacidification pH of about 6.0. Lake trout (*Salvelinus namaycush*), a long-lived and moderately acid-resistant species (pH > ~5.1), reproduced successfully when the pH rose above 5.4–5.6 (Gunn and Mills 1998).

A second whole-ecosystem "experiment" involved the ~30 percent reduction in nitrogen and 40 percent reduction in sulfur emissions that occurred in central Europe following the political and economic changes initiated in 1989. The parallel decrease in deposition of SO_4^{2-}, NO_3^-, and NH_4^+ allowed Czech mountain lakes to show signs of chemical recovery in less than 10 years (Kopáček et al. 1998).

A third line of evidence for recovery from acidification is based on the recovery of a tiny Norwegian catchment following an experimental reduction in H^+ loading. The catchment was covered with a plastic roof and received nonacidic precipitation by means of a sprinkler system for four years (Wright et al. 1988). Compared to atmosphere-exposed reference systems, the catchment responded with a rapid decline in output of the strong acid anions SO_4 and NO_3, and a small rise in pH.

Evidence for Recovery

The unambiguous results seen after strong manipulations are not as yet widely observed at the scale of whole-lake districts in northeastern North America and Europe (Stoddard et al. 1999). It appears that the large reduction in SO_2 emissions (30–50%) following the institution of stricter emission controls on power plants (Hedin et al. 1987, and Rodhe and Rood 1986) is obscured by differences in the total historical sulfur and nitrogen deposited. Differences in soil buffering, as well as interannual and among-system differences in hydrology (Webster et al. 1990), flushing rates, and sediment-alkalinity generation impose enough variation to obscure a single response pattern. Superimposed on this are local increases in NO_x deposition, the effect of long-term changes in air temperature on soil weathering and the release of relevant ions (Sommaruga-Wögrath et al. 1997), and current regional differences in SO_4 emission and deposition. Variation in the depletion of cations in sensitive soils also results in variable neutralization of acids and recovery of the alkalinity, while no recovery is expected where sensitive catchments continue to receive loading too high to permit recovery. Finally, lack of response to reduced H^+ emissions has been attributed in part to concomitant reductions in the deposition of airborne base cations to catchments over parts of Europe and North America. Such reductions reduce the ANC of the atmosphere and catchment. Emission controls on fly-ash from coal combustion, limestone quarries, and cement-producing plants, plus reductions in calcareous road dust and wind-swept bare soils in eastern North America and western Europe may be responsible for a significant reduction in atmospheric cation loading.[2]

An evaluation of acidification trends in 111 eastern Canadian lakes (1983–1991) found no change in 60, continued acidification of 17, and only 34 that were recovering despite major reductions in sulfate loading (Clair et al. 1995). The recovery in Europe has been less ambiguous during the 1990s (Stoddard et al. 1999).

While many aspects of the acidification and recovery of acidified systems are poorly resolved and remain unpredictable, the information available is sufficient to show that recovery over the long-term is propor-

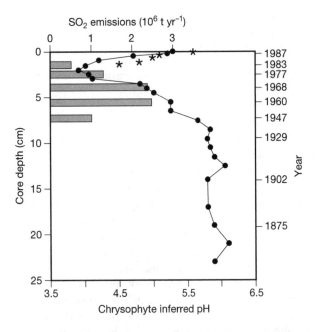

Figure 27–16 Changes in sulfur dioxide emission (bar graph) by the mining-smelting complex at Sudbury, CA and its effect on Swan Lake pH changes that were inferred from changes in the phytoplankton (chrysophyte) composition seen in dated cores. (*) represents the measured pH in the water. (*After Dixit et al. 1989.*)

tional to the reduction in emissions of acid precursors. Unfortunately, there has been insufficient long-term follow-up and acidification research funding has been severely reduced in most jurisdictions as priorities have shifted to concern about the effects of projected climate change and toxic chemicals (Chapter 28) on humans and the biota.[8]

Highlights

- Acidification of the environment by the deposition of strong acids or their precursors (SO_2, NO_3, NO_2) is defined by a decrease of the acid neutralizing capacity (ANC) which greatly affects the chemistry and biota of aquatic systems over large areas of northeastern North America, northern and central Europe, and elsewhere on a smaller spatial scale.
- The precipitation pH in industrialized nations of Europe, North America and Japan has declined tenfold to thirtyfold during the last century from a volume-integrated precipitation pH of 5–6 to 4.1–4.7, but is now slowly increasing in response to reduced SO_2 emissions.
- Acid precursors emitted into the atmosphere are transported long distances before deposition.
- Inland waters most sensitive to anthropogenic acidification have a low ANC (< 50 μeq l^{-1}) and are located in small drainage basins underlain by (1) high insoluble bedrock and surface material derived from the same geology; (2) siliceous sand; or (3) a peat layer separating the water from contact with the underlying soils.
- Aquatic systems located in calcareous drainage basins and those containing some calcareous deposits are usually well-protected from acidifying precipitation.
- A fraction of acidified waters in affected areas is the result of organic acids released from sphagnum bogs and organic-rich soils rather than anthropogenic emissions.
- The solubility of aluminum and trace metals in soils and sediments increases systematically as the pH declines below ~5.5, with certain labile forms of aluminum most toxic to the biota.
- The biota varies in its sensitivity to acidification and linked increases in reactive and toxic aluminum, but species richness declines predictably with a declining pH below 6.0. Among the most sensitive groups are the large bottom-living crustaceans and molluscs, as well as certain fish species.
- While a temporary recovery from acidification can be produced by liming ($CaCO_3$ addition)—thereby increasing the ANC and pH—a long-term recovery is contingent upon a reduction in emissions of acid precursors, a reduction now underway in Europe and North America.
- Recovery over the long-term will be proportional to reductions in emission of acid precursors, but the rate is a complex function of many factors, including catchment geology, soil buffering, the extent and duration of acidification, hydrology, temperature, microbial alkalinity-generation in the sediments, the generation time of organisms, and the dispersal ability of species.
- ▲ Present loading targets in North America and Europe would have to be reduced further to permit the recovery of thousands of the most sensitive lakes and associated streams.

---

[8]"Research on the effect of acid rain on water quality is beset by two severe problems: first is the paucity of data and the time and money required to obtain additional data; the second is the long, [years to decade] time scale of the responses that interest us." (B. J. Cosby et al. 1985)

C H A P T E R
28

Contaminants

28.1 Introduction

Limnology first became concerned about contaminants in the 1960s and 1970s when phosphorus and nitrogen pollution of waterways, a two-contaminant problem that manifests itself at the drainage basin scale, came to the fore (Chapters 16 and 17). The principal environmental issue of the 1980s was the acid-rain problem, evident over large portions of eastern North America and Europe. It has subcontinental to continental dimensions and is largely a four-contaminant issue (H^+, SO_x, NO_x, and AL, Chapter 27). Toxic chemicals are currently the principal environmental issue in limnology and their effects are evident locally, regionally, and globally (Fig. 28–1).

The solutions to the toxic chemical problem are much more complex and intractable than those associated with eutrophication and acidifying precipitation. Better wastewater treatment and agricultural practices can reverse eutrophication. Emission controls on power plants and vehicles are resulting in major reductions in H^+ or acid precursors emitted into the atmosphere of Western countries. In contrast, the number and variety of toxic chemicals that are subject to release in the environment is vast—nearly 80,000 synthetic organic chemicals plus a small number of toxic metals are now in daily use (Stumm et al. 1983).

Many contaminants are inadvertent by-products from industrial processes, including substances that affect the endocrine (hormone) system of aquatic invertebrates and vertebrates. Yet others are produced in nature during the microbial degradation of previously innocuous compounds, or during the chlorination of otherwise harmless natural organic compounds.

Many chemicals of greatest concern are fungicides, insecticides, herbicides, and other persistent biocides whose use is widespread. All of them are, or once were, of great benefit in the production and protection of food produced for a rapidly growing human population. However, their use has serious effects on species diversity and community structure as well as human health. Some 500–1,000 newly synthesized organic compounds are added annually to the vast pool already in commercial use, with no information available on the toxic effects, if any, on about 79 percent of the chemicals used. (Postel 1987). Even if nearly all of these chemicals were harmless, the remainder would still be large. The International Joint Commission on the Great Lakes of North America (IJC) has identified 362 chemicals in the Laurentian Great Lakes that are potentially toxic to plants and animals—including humans (IJC 1987). Eleven chemicals or groups of chemicals, termed "critical pollutants," were noted in this list. An almost identical list is covered by a 2001 United Nations treaty banning or greatly restricting the use of 12 *persistent organic pollutants* (POPs) (Table 28–1). Even this modest number of the most critical pollutants with widely different structures is far too large to be dealt with individually in a limnology textbook; entire books and symposiums are devoted to a single chemical. Nor is it possible to say anything definitive about the environmental impact of compounds that serve as endocrine disrupters, popularly known as "gender benders," that are present in wastewater effluents and industrial chemicals (e.g., plastics, cleaning

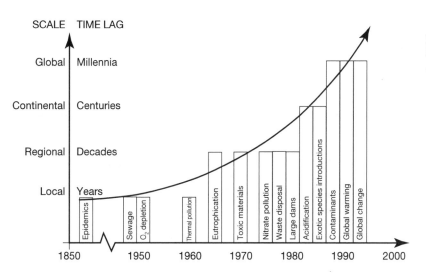

Table 28–1 **The 12 persistent organic pollutants to be banned or strictly controlled worldwide by a United Nations treaty to be signed in 2001(*) and the 11 "critical pollutants" on the International Joint Commission for the Laurentian Great Lakes "primary track," (Δ) selected on the basis of their chemical stability and resulting persistence, the quantity and type of use, the tendency to accumulate in organisms, and their toxicity.**

Chemical	Production and Release	Source
Dioxins (* Δ)	Unintentional	Polychlorinated dibenzo-dioxins (PCDDs; 75 congeners) and polychlorinated dibenzo-furans (PCDFs; 135 congeners) are created in the manufacture of herbicides used in agriculture
Furans (* Δ)	Unintentional	and forest management. Also produced as by-products during the combustion of chlorinated additives in fossil fuels and chlorine-containing medical and municipal wastes such as plastics, through production of pentachlorophenol (PCP) wood preservative, and in pulp and paper production processes that use chlorine to bleach wood and pulp. Some congeners are highly toxic (see Fig. 28–2); most have low acute toxicity.
Benzo (a) pyrene (Δ)	Unintentional	Product of the incomplete combustion of fossil fuels and wood, including forest fires, auto exhaust, and waste incineration. One of a large family of polycyclic aromatic hydrocarbons (PAHs).
DDT and its breakdown products, including DDE (* Δ)	Intentional	Insecticide used primarily for mosquito control in tropical areas. Sediment half-life of DDT is ~10–15 years at temperate latitudes.
Aldrin/Dieldrin (* Δ)	Intentional	Insecticide used extensively, especially on fruit and for termite control. Half-life in soil is about five years. Aldrin rapidly degrades to dieldrin.
Hexachlorobenzene (* Δ)	Unintentional	HCB is a by-product of the combustion of fuels that contain chlorinated additives, incineration of wastes that contain chlorinated substances, and in manufacturing processes using chlorine. Found as a contaminant in chlorinated pesticides (half-life in nature ~3–6 yrs).

(continued)

Table 28–1   (*continued*)

Chemical	Production and Release	Source
Alkylated lead (Δ)	Intentional	Used as a fuel additive and in solder, pipes, and paint.
	Unintentional	Released through combustion of leaded fuel, wastes, cigarettes, as well as from pipes, paint cans, and paint chips.
Mirex (* Δ)	Intentional	Fire retardant and pesticide used to control fire ants. Breaks down to a more toxic form, photomirex, in the presence of sunlight. Present sources are residuals from manufacturing sites, spills, and disposal in landfills. Highly persistent with a half-life up to 10 years.
Mercury (Δ)	Intentional	Used in metallurgy.
	Unintentional	By-product locally and regionally of chlor-alkali, paint, electrical equipment manufacturing processes, refuse incineration, and coal combustion.
Polychlorinated biphenyls (* Δ)	Intentional	PCBs used as insulating fluids in electrical capacitors and transformers, and in the production of hydraulic fluids and lubricants. Additives in paint, carbon-copy paper, and plastics. Once used as a vehicle for pesticide dispersal; 209 congeners varying greatly in toxicity. Half-life is from weeks to years.
	Unintentional	Primarily released to the environment through leakage, spills, and waste storage and disposal.
Toxaphene (* Δ)	Intentional	Insecticide used mostly on cotton, but also on cereal grain, fruits, nuts, and vegetables. Soil half-life from a year to 14 years. The most important congeners have 6–10 Cls. Substitute for DDT. Relatively volatile with global distribution.
Heptachlor (*)	Intentional	Used to control soil insects, termites, mosquitoes, and crop pests.
Chlordane (*)	Intentional	Insecticide used primarily for termite control. Soil half-life ~1 year.
Endrin (*)	Intentional	Insecticide and rodenticide used on cotton, rice, and corn. Soil half-life up to 12 years.

Source: After T.E. Colborn et al. 1990 in Anon 1991 and various sources.

products, spermicides, and pesticides are a few). These alter or block the hormone system of individual vertebrates (fish, crocodiles) and their offspring and can produce developmental (sexual) immune system changes, and possibly carcinogenic effects. The present chapter focuses on aspects of behavior in a small number of groups of organic chemicals and a few selected metals. Fortunately, the toxicity of these and other potentially toxic compounds is generally related to their physiochemical structure. Structure, therefore, gives a first estimate of the probable toxicity of existing chemicals, as well as those being considered for commercial use. The relationship between the structure and functioning of contaminants provides an important basis for this chapter. The physiochemical structure of chemicals provide the limnologist with some badly needed generalities and a foundation for a further exploration of the literature.

Most of the chemicals of concern are individually found in lake water in extremely low concentrations, often at the picogram per liter (10^{-12} g l^{-1}) or parts per trillion (ppt) level. The technology to make measurements at such low concentrations is both recent and expensive, and requires extreme precautions against contamination during field collection, storage, extraction, digestion, and analysis of the samples. An even greater problem is the variation over time and space of these chemicals in both water and the atmosphere.

In contrast to previous chapters, the present one draws most heavily on research carried out in a single region, the Laurentian Great Lakes basin of North America. This is not because there are few good studies elsewhere but because work in the Great Lakes region is unsurpassed in the variety and duration of investigations and has been carried out with largely standardized methods on both sides of the United States- Canadian border. The two federal governments have provided most of the large financial resources necessary, and the adjoining states and provinces have also collaborated to an unprecedented degree.

The Laurentian Great Lakes vary considerably in the amount of pollution that is experienced—remote Lake Superior is generally most pristine and downstream Lake Ontario is most polluted. The latter receives water from the Niagara River, which connects it to upstream Lake Erie. The Niagara River basin has been home to a large chemical industry dependent on its hydropower and as a result now holds some 1.2 million tonnes of contaminated materials stored in the 66 largest of a much largr number of chemical dump sites. A considerable number of these sites either leak or have the potential to leak chemicals into Lake Ontario.

Contaminants of Greatest Concern

Many of the chemicals of greatest concern worldwide are highly chlorinated organic molecules, known as organochlorines (OCs). These, together with some other groups of persistent organic chemicals, are known as **persistent organic pollutants** (POPs), which are defined as those with a half-life in water or sediments of greater than eight weeks. In other words, their activity decays to less than half the original concentration during this time period. While they have diverse chemical structures, most have a low solubility in water, high accumulation potential in fat, and resistance to degradation (high recalcitrance). The recalcitrant chemicals are either purposely synthesized as biocides (e.g., insecticides and herbicides) or inadvertently created in industrial processes and released. One example are the PCBs, a group of *polychlorinated hydrocarbons* (PCHs) produced commercially by the chlorination of biphenyl primarily for industrial use (Table 28–1). The mixtures of *polychlorinated biphenyls* (PCBs) that are produced vary greatly in composition and physical properties, depending importantly on the ratio of chlorine to biphenyl used and the production temperature (Fig. 28–2). The number of chlorine molecules and their positioning on the rings influences their properties and biological activity. The commercial mixtures also contain some inadvertently produced and exceptionally toxic polychlorinated dibenzo-p-dioxins and polychlorinated dibenzo-p-furans (PCDDs, PCDFs).

Some of the POPs of greatest concern are not themselves highly toxic at the concentrations normally encountered in nature. They are, however, highly persistent (resistant to microbial and chemical breakdown) and tend to accumulate in the biota (Sec. 28.8). Some of the POPs have become widely distributed

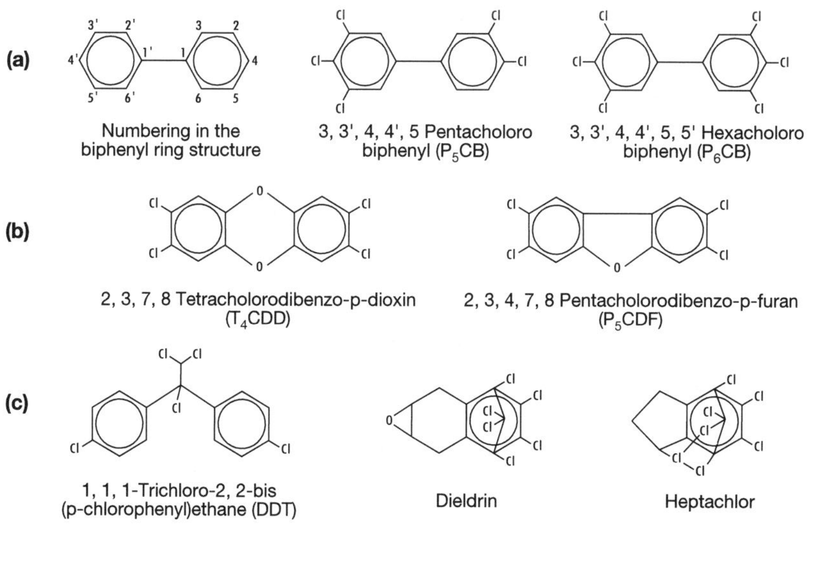

Figure 28–2 (a) The biphenyl ring structure chemical makeup of selected PCBs; (b) two closely related and the most toxic polychlorinated dioxins and furans; and (c) DDT, dieldrin, and heptachlor (see Table 28–1)

from sites of industrial and urban emissions via the atmosphere and waterways. Burning plastic and other natural or manufactured materials in municipal and medical waste facilities or landfills releases chlorine, polyvinyl chloride (PVC), acid gas (HCl), carbon dioxide, lead (e.g., in paints), and cadmium (used in the production of plastic) into the atmosphere. The HCl combined with other organic substances in the environment produces polychlorinated and carcinogenic dioxins and furans, among others (Piasecki et al. 1998). Other POPs such as biocides are, or have been, applied to agricultural fields, forests, and rice paddies—often from the air—with a great probability of atmospheric distribution. Many POPs are somewhat volatile (have a vapor phase) and are vented from soil and water into the atmosphere as a gas, or become attached to windblown particles. Other types of persistent contaminants, including dioxins and furans, are produced during the incomplete combustion of organic matter in forest fires and industrial processes, as well as during incomplete incineration of municipal and medical waste at temperatures < 1,200°C. Internal combustion engines also produce toxic organic compounds that are readily distributed to rivers, lakes, and wetlands via the atmosphere.

28.2   Toxic Substances

Toxic substances may have two or three other properties besides their **toxicity**: **bioaccumulation potential**, **persistence**, and **vapor phase** (volatility) (Fig. 28–3). The toxicity can be **acute**, resulting in a quick death, or be **chronic** (**sublethal**).

Chronic Toxicity

Chronic toxicity can result in abnormal hormone levels or unusual behavior that, if severe enough, results in the reduction or disappearance of a particular species over time. Alternatively, damage to the genetic material may affect fecundity, the incidence of malformations, or cancers, but may not necessarily cause

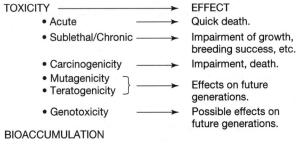

Figure 28–3 Working definitions for toxic substances, toxicity, bioaccumulation, persistence, and volatility. *(Modified after Kaiser 1984.)*

rapid death,[1] The division between the two types of toxicity is primarily conceptual because, if death is the endpoint, they differ only in their time scales.

A particular chronic toxicity need not affect reproductive success (fecundity) greatly or at all, making it particularly difficult to see its effect in nature. Thus, chronic toxicity revealed by an increased incidence of highly visible benign or carcinogenetic tumors in older specimens of, say, sediment-feeding fishes in a polluted bay need not affect their fecundity or the fecundity of younger fish without tumors. The abundance of a species is therefore not necessarily reduced by chronic toxicity. Nevertheless, an unusual abundance of tumors in fish or malformations in the mouth parts of sediment-living invertebrates for example, provide an important warning of environmental contamination.[2] However, chronic toxicity is particularly problematic because the effects may express them-

---

[1]Bacterial cultures are now widely used as test organisms to detect genotoxicants and mutagens. The now well-established relationship between DNA damage, mutagenicity, and carcinogenicity has been most extensively demonstrated with the *Salmonella*/microsome assay or "Ames test." Another more rapid colorimetric test ("SOS chromotest") measures damage to the DNA using a genetically modified strain of *E. coli*. It is increasingly used to examine the genotoxicity of industrial and wastewater effluents (White and Rasmussen 1996) and is well correlated with the mutagenicity and carcinogenicity measured with the Ames test.

[2]Toxicology as a science has its roots in concern about the effect of toxins on humans, using warmblooded animals as surrogates in testing. Aquatic toxicology evolved when a minority of toxicologists started using fish and invertebrates for, normally, acute toxicity testing of usually single chemicals in the laboratory. The last decade has seen major developments in chronic toxicity testing in the laboratory, followed by attempts to use the bioassays developed in nature. Those scientists, working primarily in nature, consider themselves "environmental toxicologists" or "ecotoxicologists," and are usually concerned with the effect of chronic toxins on wildlife populations and with "ecosystem health."

selves in terms of fecundity only after a lag period, making it extremely difficult to attribute a causal effect to a particular source, or to a particular concentration, or to any one ingredient of the cocktail of chemicals measured at the time of sampling. Furthermore, lag periods make it exceptionally difficult to distinguish the impact of one or more chronic toxins from those imposed by another of the many environmental factors affecting population and communities in nature. Finally, many of the chemicals of concern are normally present in water at concentrations at or below the detection limits, and usually in concentrations much smaller than those for which effects are observed in the laboratory.

Contaminant Attributes

Chronic pollutants that accumulate in the biota (*bioaccumulate*) are the result of direct uptake from the water or accumulation via food (Sec. 28.8), plus a rate of accumulation (storage) that is greater than the metabolism (loss rate). The concentrations of bioaccumulating toxins increase over time, particularly evident in longer-lived organisms. Resistance to degradation (*persistence*) is a second significant attribute of many toxic substances (Fig. 28–3). Highly persistent organic toxins such as DDT or some PCBs may decay only moderately over periods of decades. If toxins also have a *vapor phase* (volatility), they are transported via the atmosphere from areas of production to remote regions (Fig. 5–13). Pollutants volatile at lower temperatures, including the less-chlorinated PCB congeners, are most mobile. Rivers contribute importantly to the long-range transport of both volatile and nonvolatile compounds.

Cyanide in the past and most organic phosphorus or nitrogen-based biocides today are used as biocides precisely because they are acutely toxic but do not persist (Fig. 28–4). Others used resemble persistent OCs but with oxygen and sulfur substitutes for some of the chlorines. Their effect may be drastic, but it is short-lived relative to the lifespan of the target organisms, therefore their impact is localized. Compounds with a somewhat longer half-life may bioaccumulate to some extent but degrade quickly enough to limit or avoid accumulation in predators at the top of the food chain. Much more disturbing are substances that are not only chronically toxic but are also persistent and bioaccumulate (Fig. 28–3) because they can attain high concentrations in long-lived organisms. This particular group of compounds tends to have a low solubility in water and thus are **hydrophobic** (water-fearing) (Sec.

Substance	Toxicity	Bioaccumulation	Persistence
Cyanide-ion	X	–	–
Freon	–	–	X
Organo-phosphates	X	?	–
Organo-tins	X	?	?
Methyl-mercury-ion	X	X	?
PCBs	X	X	X
Mirex	X	X	X
Chloro-dioxins	X	X	X
Chloro-styrenes	?	?	?
Hexachlorobutadiene	X	X	X
Chloro-phenols	X	?	X

Figure 28–4 Examples of environmental contaminants with various degrees of toxicity, persistence, and bioaccumulation potential. Note: the vapor pressure, determining the extent that organic contaminants can be transported over long distances to previously pristine regions, is not considered here. (*After Kaiser 1984.*)

28.6). Hydrophobic contaminants sorb to gills, guts, and the external surfaces of living organisms and detritus, and are disproportionately retained when ingested. Their preferential association with particles containing lipids affects their distribution in nature and their bioavailability. Compounds that are **hydrophilic** (water-loving) usually remain dissolved in water. The links between the physiochemical structure of contaminants and their behavior in aquatic systems is discussed in Section 28.6.

▲ Many of the groups of persistent chemicals of greatest environmental concern, such as DDT and PCBs, are no longer produced or used in the Western world and are to be phased out or greatly restricted globally, but are so highly persistent (e.g., DDT and its breakdown products) that the aquatic biota remains contaminated, albeit at a slowly declining level. Virtually all of the PCBs ever produced are still in use or in storage and continue to be released through leakage, spills, or escape from waste storage sites. Restrictions on their use and disposal have often been weak or lacking altogether in the economically less-developed portions of the world.

Little is known about the in situ synthesis of particular toxic compounds from nontoxic or less toxic precursors, or about those toxins produced as a degradation product of another chemical product. Nor is much known about the additive effect of toxins that individually may not exhibit any effect at the typically very low concentrations observed in nature. Finally, little is known about the release (supply) rate of a large number of toxic chemicals inadvertently produced in low concentrations as a by-product of a wide variety of industrial processes during the production of generally low-purity industrial chemicals or biocides. Not only do the industrial processes used differ but the chemical stocks that are used in the production differ somewhat as well.

28.3   Sources of Contaminants

The highest concentrations of anthropogenic contaminants in lakes, rivers, and wetlands are normally found below industrial and urban *point sources*. Much more difficult to identify and control is the diffuse *nonpoint source* pollution, such as biocides used in agriculture, that enters waterways along with runoff from catchments or arrive via the atmosphere. Aquatic systems lacking either a point source or nonpoint source of pollution in their drainage basins (Sec. 4.1) are known as *remote* but are nevertheless contaminated with anthropogenically produced contaminants via the atmosphere.

Trace Metal Production and Distribution

Today more than half the quantities of most trace metals—including the most toxic ones (As, Cd, Hg, and Pb)—emitted to the atmosphere worldwide are anthropogenically produced (Table 28–2). Atmospheric dominance is greatest in the Northern Hemisphere where most industries are located. Remote lakes and wetlands there contain elevated levels of trace metals in their water and sediments that can only have arrived via the atmosphere (Nriagu 1990). In the case of mercury, a contaminant of great importance in a very large number of remote lakes and wetlands (Sec. 28.9), the two principal anthropogenic sources worldwide are coal combustion and refuse incineration, but other important sources include the chlorine alkali industrial process, waste incineration, landfills, copper and lead smelters, and cement manufacturers. For cadmium, the largest atmospheric source results from smelting metals other than iron. However, coal com-

Table 28–2   **Anthropogenic versus natural emission of trace metals (10^3kg yr^{-1}) to the global atmosphere in 1983. Median values with ranges in estimated emissions given in brackets.**

Element	Anthropogenic (Mean/Range)	Natural	Total	Median Anthropogenic %[2]
As	19 (12–26)	12 (0.9–23)	31 (13–49)	61
Cd	7.6 (3.1–12)	1.3 (0.2–2.6)	8.9 (3.2–15)	85
Cr	30 (7.3–54)	44 (4.5–83)	74 (12–134)	41
Cu	35 (20–51)	28 (2.3–54)	63 (22–105)	56
Hg	3.6 (0.9–6.2)	2.5 (0.1–4.9)	6.1 (1.0–11)	59
Mn	38 (11–66)	317 (52–582)	355 (63–648)	11
Mo	3.3 (0.8–5.4)	3.0 (0.1–5.8)	6.3 (0.93–11)	52
Ni	56 (24–87)	30 (3.0–57)	86 (27–144)	65
Pb	332[1] (289–376)	12 (1.0–23)	344 (290–399)	96
Sb	3.5 (1.5–5.5)	2.4 (0.1–4.7)	5.9 (1.6–10)	59
Se	6.3 (3.0–9.7)	9.3 (0.7–18)	16 (2.5–24)	42
V	86 (30–142)	28 (1.6–54)	114 (32–220)	75
Zn	132 (70–194)	45 (4.0–86)	177 (74–280)	66

[1]Declining rapidly with the widespread removal of lead from gasoline.
[2]Relative anthropogenic contribution varies greatly locally and regionally as a function of proximity to major sources and as a function of prevailing winds.

Source: After Nriagu 1989.

bustion and waste incineration in urban areas are also important sources of cadmium and other trace metals (see Nriagu and Pacyna 1988).

Production and Distribution of POPs

Among the most important sources of POPs are those used to control agricultural and forest pests, plus those produced inadvertently as either a by-product in the combustion of fuel or the low temperature incineration of wastes containing chlorine that yield chlorinated hydrocarbons. Among primarily industrial chemicals are the 209 different forms of polychlorinated biphenyls (PCBs) that were produced as commercial mixtures containing as many as 70–100 *congeners* (forms of different configuration, e.g., P_5CB and P_6CB, Fig. 28–2). They differ considerably in their physical and chemical properties with 13 exhibiting a dioxin-like toxicity. A last major and widespread group of persistent toxic chemicals discussed here are the *polycyclic hydrocarbons* (PAHs), among which the highly toxic benzo (a) pyrene stands out. PAHs are produced during the incomplete combustion of organic carbon in fossil fuels and wood (including forest fires), coal gasification, petroleum cracking, and in the production of furnace coke, carbon black, tarpitch, asphalt, as well as in waste incineration (Table 28–1).

Atmospheric Distribution

The classic evidence for the impact of humans on the atmospheric distribution of contaminants is the concentration of lead (Pb) accumulated in the Greenland ice cap (Fig. 28–5) and more recently measured in Antarctica. Lead levels in Greenland had already become slightly elevated as the result of human activity in Europe and Asia before the onset of the Industrial Revolution, but its meteoric rise is associated with the great increase in the burning of leaded gasoline after 1945. The more recent rapid decline is the result of the widespread removal of lead from fuel (Fig. 28–5). A correlated decline in cadmium and zinc further attests to the effectiveness of industrial emission controls in limiting environmental contamination (Boutron et al. 1991).

The role of the atmosphere in supplying organic contaminants and lead has probably been best assessed for the Laurentian Great Lakes (US, CA). Among these lakes, the data for remote Lake Superior are easiest to interpret because its drainage basin and shoreline remain largely pristine in terms of point and nonpoint (diffuse) sources of contaminants. Loading rates of important contaminants are relativity low, but 60 percent of the water received, ~66 percent of the nitrogen (Sec. 18.1), and 90 percent or more of certain toxins are attributable to a direct atmospheric deposition on the Lake Superior surface (Table 28–3). The effective retention of hydrophobic contaminants de-

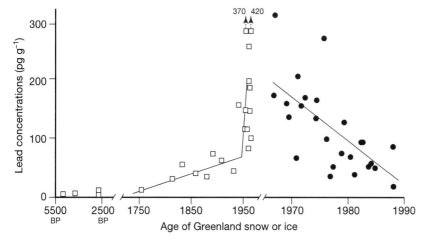

Figure 28–5 Changes in Pb concentrations in Greenland ice and snow from 5500 BP to the present from four sites. Note a great increase (~200 fold) from several thousand years ago to the mid-1960s, with increases evident even before the onset of the industrial revolution. Pb concentrations have decreased rapidly (by a factor of ~7.5) during the past twenty years, mainly as a consequence of the fall in the use of lead additives in gasoline. Note that modern values for Antarctica are smaller by a factor of more than 10. Elevated copper levels (also not shown), resulting from extensive mining and crude smelting are evident much earlier and date back to the time of the Greek republics (2,500 BP) and the Roman Empire (~2,000 BP) (Hong et al. 1996). *(Figure after Boutorn et al. 1991.]*

Table 28–3 **Approximate chemical loads to the Great Lakes of North America, and the percent obtained via direct atmospheric deposition on the water surface.**

Contaminant	Superior $\mu g\ m^{-2}\ yr^{-1}$	%	Michigan $\mu g\ m^{-2}\ yr^{-1}$	%	Huron $\mu g\ m^{-2}\ yr^{-1}$	%	Erie $\mu g\ m^{-2}\ yr^{-1}$	%[1]	Ontario $\mu g\ m^{-2}\ yr^{-1}$	%[1]
PCBs	7.4	90	11.8	58	10.6	63	96.9	7	127	6
t-DDT	1.1	97	1.1	98	1.5	71	12.3	10	5.6	23
benzo(a)pyrene	0.9	96	3.6	86	4.8	63	4.7	66	7.8	40
lead	2,939	97	9,362	99.5	7,167	94	21,808	40	21,300	51

[1]Upstream inputs and local sources greatly lower the relative atmospheric contribution of the chemicals in the two lower lakes.

Source: Modified after Strachan and Eisenreich 1988.

posited on catchments (<10 percent released) does mean that they will continue to be a source to waterways long after sources of atmospheric pollution have been controlled. It is evident that good databases and models of catchment–contaminant–water interactions are needed to predict the temporal course of atmospheric pollutants in aquatic systems.

The great importance of the atmosphere as a direct source of particular persistent organic contaminants and toxic trace metals in remote aquatic systems is much affected by the catchment area:water area ratio. Lakes with a disproportionately low ratio (e.g., Lake Superior CA:LA = 1.6) receive proportionately more of their water, nutrients, and contaminants from the atmosphere than lakes and rivers with a large catchment, assuming an identical catchment export coefficient (mg exported $m^{-2}\ yr^{-1}$). Contaminants that are particularly strongly adsorbed to particles in catchments have particularly low catchment export coefficients and the receiving waters obtain most of their contaminants via direct deposition on water surfaces from the atmosphere.

It is evident from Table 28–3 that at best a few percent of strongly sorbed POPs reach Lake Superior via rivers from the land. Most trace metals are also readily sorbed and very largely retained on nonacidified well-vegetated catchments everywhere, which implies that little particulate matter is exported to rivers and lakes (Table 28–4). However, enhanced toxic trace-metal solubilization and export from acidified catchments results in elevated trace-metal levels in the receiving waters (Sec. 27.8, and Tessier et al. 1985).

▲ 28.4   The Fate of Contaminants

Some hydrophobic chemical pollutants entering rivers and lakes are removed via the outflow, but the fraction retained by the receiving lake or wetland increases as the flushing rate declines or water residence time in-

Table 28–4 **Estimated catchment retention of cadmium, copper, and lead.**

	Percent Retention		
	Cd	Cu	Pb
New Hampshire, US	—	—	95–98
Illinois, US	—	—	96
Tennessee, US	67–100	72–100	97–99
New Jersey, US	—	51	88
Germany	69	61	95
Central Ontario, CA	> 51–> 91	85–90	95–99

Source: Schut and Evans 1986, and Blais and Kalff 1993.

creases (Chapter 8). Those chemicals most subject to adsorption are most readily removed from the water column by sedimentation and thus are least readily exported via the outflow.

Long-range Transport

Most organochlorines and organic mercury (methyl mercury) have a sufficiently high vapor pressure that a fraction of the persistent contaminants are lost from both the land and water to the atmosphere by volatilization; this route is particularly important in summer when water and soil temperatures are highest. But on an annual basis volatilization is expected to be most important from warm tropical lakes, wetlands, and their catchments. *Volatile organic compounds* (VOCs) are carried aloft by rising air masses and may be transported in the gas phase or adsorbed to particles for hundreds if not thousands of kms at the latitude of emission. They are also carried to higher latitudes and altitudes (see Fig. 5–13), either directly or following a temperature-driven transport–deposition–partial revolatization–transport sequence (Sec. 5.4). Deposition is enhanced when airborne POPs—in the gas phase and as aerosols—reach their condensation temperature at high latitudes, or in high mountain regions at a lower latitude. Deposition then prevails over revolatilization (degassing), whereas it is the reverse at low latitudes. Low temperatures further reduce rates of VOC degradation and increase the probability of a long-term storage. POPs accumulated in the snowpack or on plant and soil surfaces may become buried, revolatilized, or are transported to aquatic systems during periods of snow melt where they become subject to uptake and bioaccumulation (Sec. 28.8).

Sediments, lake ice, and the biota of arctic waters contain elevated levels of a wide variety of POPs, mercury, and radionuclides that resulted from atomic bomb testing or nuclear accidents (Lockhart et al. 1992). In winter, the North American arctic air often contains a haze of small aerosol particles (primarily "soot" or carbon black and sulfates) and gases derived from Eurasia and midlatitude North America. Desert dust particles (< 1 μm) from central Asia have been noted in Alaskan snow (Lockhart et al. 1992). "Brown snow" in the Canadian Arctic includes clay and soot particles plus adsorbed POPs that have been traced to Siberian and western China regions of Asia (Welch et al. 1991). However, with increasing distance from the North Pole, the contaminants in

boreal forest lakes are increasingly derived from lower latitude industrial or agricultural regions (see Fig. 5–12).

The extent that POPs in waterways are primarily lost by biodegradation by microbes, chemically or photochemically, is highly dependent on environmental conditions and the specific physiochemical characteristics of the particular chemical (Sec. 28.7).

Contaminant Concentrations over Time

Rough mass-balance calculations for the sum total of the most common PCB congeners and for lead show that the two contaminant types differ greatly in their principal loss processes in Lake Superior. The atmosphere is ultimately the principal sink (92%) of the more volatile PCBs.[3] As a consequence, only 13 percent (170 kg yr^{-1}) of total PCBs entering Lake Superior is permanently buried in the sediments or lost via the outflow, with the balance (87%, ~1,900 kg yr^{-1}) lost by volatilization (Strachan and Eisenreich 1988). The roughly 20 percent decline per year in the whole-water concentration of PCBs between 1980 and 1992 is attributed to a systematic decline in atmospheric inputs to Lake Superior. The lake is now serving as a source rather than as a sink of PCBs (Jeremiason et al. 1994). In contrast, the sediments rather than the atmosphere are the principal sink for lead, which is both strongly particle sorbing and not very volatile (Strachan and Eisenreich 1988).

The encouraging decline in concentration of the most abundant PCB congeners in Lake Superior is also evident for other POPs, measured in fish from North American lakes not subject to point-source contamination (Schmitt et al. 1985) as well as in sediments (Fig. 28–6). Consequently, there is encouraging evidence that contaminant levels will decline where regional or continent wide controls on emissions are instituted. The declines are most evident from analyses of the biota over time, and from concentration profiles in dated sediments. The declines are not as obvious from the much more variable and typically

---

[3]Low chlorinated (1–3 chlorine atoms) biphenyls are not only more volatile but also more readily degraded by chemical, photochemical, and metabolic processes. They consequently have a much shorter half-life (months) than the highly chlorinated (4–10 chlorine atoms) biphenyl congeners, with the most highly chlorinated congeners almost totally persistent and dominant in long-lived organisms and sediments close to sources.

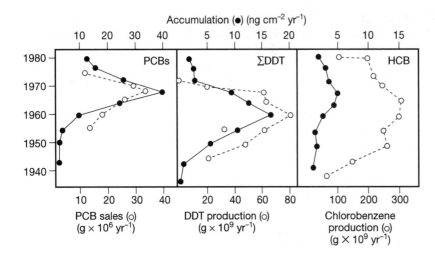

Figure 28–6 Relationship of chlorinated hydrocarbon accumulation with United States production and sales data in a Lake Ontario sediment core. Not only does the total PCB concentration change with sediment depth, but the composition changes as well, with the concentration of the less chlorinated and more biodegradable isomers decreasing with depth and the more highly chlorinated and persistent isomers increasing with depth (Oliver et al. 1989). As a result of a ban on North American sales in 1972 the fraction of ΣDDT in the surface sediments that consists of DDT, rather than one of its aerobic or anaerobic breakdown products, is modest. Possibly half of the most recently deposited PCBs appear to be attributable to an upward mixing by oligochaetes of deeper and more contaminated sediments. (*After Eisenreich et al. 1989.*)

low concentration measurements in the water or the atmosphere. Concentrations measured in precipitation vary considerably over time and among years at single stations. This, combined with a frequent but not universal decline with distance from source regions, and problems with measuring concentrations of contaminants in precipitation, at levels that are often close to the detection limits, has drawn attention to lake sediments and long-lived piscivorous fish or their predators (e.g., loons) as indicators of changes in contaminant loading. Both sediments and piscivorous fish contain substantial levels of contaminants accumulated over several to many years. The levels are sufficient to have permitted accurate measurements of change over the last several decades, something not equally realized for measurements made on water or precipitation samples.

Measurement of Contaminants

Major changes (improvements) over the years in analytical techniques have led to much lower detection limits and greater awareness of the possibilities of sample contamination during collection, storage, and analysis. These changes, together with temporal and spatial variation in contaminant concentrations in the water and air, have often made it difficult to conclude

unambiguously from atmospheric and water samples alone whether contamination levels have declined in the Great Lakes between the 1970s and 1990s. Two examples will suffice: Coale and Flegal (1989), using the best "clean" sampling, storing, and analytical techniques available, reported concentrations of Cu, Zn, Cd, and Pb two orders of magnitude smaller than those reported recently by good laboratories using modern techniques but less aware of the possibility of sample contamination. Second, progressive improvement in technique mistakenly left the impression that lakewater concentrations of dissolved lead in Lake Erie had declined rapidly from an apparent high of < 15,000 ng l^{-1} in 1965 to < 30 ng l^{-1} in 1988, whereas the actual concentration appears to have changed little. Reliable water measurements date only from the late 1980s.

There are few reliable quantitative data prior to about 1988 for toxic POPs in the water of the Laurentian Great Lakes and elsewhere because nearly all earlier results were below the detection limits in the small sample volumes (1–2 l) routinely collected at the time. Better historical data are available for the (higher) concentrations of hydrophobic contaminants sorbed to suspended particles and collected following filtration of water samples, and those stored in sediments.

28.5   The Sediment Record

The best long-term trends for persistent hydrophobic organic chemical and toxic trace metals come from dated sediment cores (Sec. 20.6). The sediment concentrations of most particle sorbing contaminants are usually high enough to be well above detection limits. Sediment trace-metal concentrations are furthermore typically high enough to have been affected little by contamination problems encountered during collection, storage, and analysis (Sec. 28.4). Cores also provide a time-integrated record that is usually not obscured, as in the water column, by a considerable short-term variation over time and space (Fig. 28–6).

Dated core profiles show a clear temporal trend in contaminant concentrations that are linked to production (Fig. 28–6). Trace-metal concentrations recorded in recent sediments are usually much higher than in preindustrial sediments when the concentrations were the result of natural fluxes from soil, volcanoes, fires, and sometimes modest early human activity. The ratio of the present sediment (or atmospheric) concentration to the preindustrial background concentration is known as the **enrichment factor** (EF).

▲ Contaminant Accumulation over Time and Space

Sediment EFs between 2.5 and 9 for a number of metals characterize a south German lake located downwind from an industrialized region (Table 28–5). But the metal enrichment factors are small compared with those measured for the sum total of the different polycyclic hydrocarbons (PAHs) and other POPs in both western Europe (Table 28–6) and eastern North America. The reason is that the background levels of biogenically produced POPs plus those produced by preindustrial humans, are very small compared to the levels encountered following the onset of extensive coal burning in Europe around 1800 (Fig. 28–7) and in upstate New York some 50 years later. The recent decline in PAHs on both continents appears to be related to a shift to high-temperature coal burning for electricity generation, which releases fewer PAHs (Charles and Norton 1986). The decline is abetted by increased use of oil, gas, and in some jurisdictions, increased use of nuclear power. However, contaminant concentrations recorded per unit dry weight of sediment show considerable interyear variations (Fig. 28–7). The variation does not so much represent interannual differences in emissions but rather differ-

Table 28–5 **Enrichment factors, background, and maximum values for heavy metals and polycyclic aromatic hydrocarbons in sediment cores from dystrophic Lake Huzenbach in the Black Forest region, Germany.**

	Enrichment Factor	Dry Substance (mg/kg)	
		Background value	Maximum value
Zn	7.3	22.8	167.6
Cd	9.0	0.17	1.5
Pb	7.5	13.4	100.2
Cu	3.5	8.0	28.0
Cr	2.5	21.1	53.6
Fl	> 520	< 0.005	2.60
BbF	> 676	< 0.005	3.38
BkF	> 1,320	< 0.005	6.53
BaP	> 1,396	< 0.005	6.98
BghiPer	> 1,900	< 0.005	9.50
IndP	> 334	< 0.005	1.67

Fl: fluoranthene; BbF: benzo (b) fluoranthene; BkF: benzo (k) fluoranthene; BaP: benzo (a) pyrene; BghiPer: benzo (ghi) perylene; IndP: Indeno (1, 2, 3-cd) pyrene.

Source: After Hilgers et al. 1993.

Table 28–6 **Mean percentage of genotoxins and fluorescent substances (expressed as benzo (a) pyrene equivalents) adsorbed to suspended particulate matter in the effluent from 40 industries along the St. Lawrence and Saguenay Rivers (Quebec, CA) and from two industries along the Fraser River (British Columbia, CA).**

Industry Type	Percent of Genotoxins Adsorbed to Suspended Particulates	Percent of Fluorescent Aromatics Adsorbed to Suspended Particulates
Surface finishing	86	23
Inorganic chemical production	85	43
Aluminum refining	84	92
Organic chemical production	72	55
Petroleum refining	58	92
Metal refining (not Al)	54	77
Sewage treatment	21	22
Pulp and paper	29	50

Source: Modified from White et al. 1996.

ences in runoff and associated sediment loading, with years of elevated sediment loading diluting the primarily atmospherically supplied contaminants.

The sediment enrichment factors for nonpoint source lakes close to industrialized regions may be high (Table 28–5) and may reflect both long-distance transport from distant source regions and transport from nearby sources. Much higher concentrations are naturally found in *point-source* polluted lakes, which are not explicitly dealt with here because their contamination is point-source specific. Contamination resulting from atmospheric wetfall and dryfall declines with distance from source regions within the temperate zone. Thus, Windsor and Hites (1979) noted a decline of three orders of magnitude in total PAH levels in sediments within the first 100 km from Boston (Massachusetts, US). Beyond 100 km the concentrations were similar to the 100 ppb that is observed in surficial sediments of lakes far from major sources, a concentration level that represents continent-wide levels of contamination. However, research on individual PCB congeners shows that the most chlorinated (most hydrophobic and persistent) congeners in the atmosphere are preferentially adsorbed to aerosol particles which are deposited closer to the source regions than the less chlorinated (less hydrophobic and less adsorbed) congeners. The less chlorinated PCBs are proportionately more in vapor phase, allowing a much longer atmospheric residence time and much wider distribution.[3]

28.6 Physical and Chemical Characteristics of Contaminants and Their Distribution in Nature

As mentioned earlier, the solubility of contaminants in water varies greatly. Those that dissolve readily in water are known as *hydrophilic* (water-liking), whereas those with a low solubility are known as *hydrophobic* (water-fearing) sorb readily onto detrital particles and living organisms. The solubility of contaminants in water affects not only their behavior in waterways but is also central to the effectiveness with which they are retained on drainage basins, and whether toxins emitted into the atmosphere are more likely to be sorbed to aerosol particles and deposited as dry deposition rather than be dissolved in wet deposition (Sec. 27.2).

Solubility in Water

The extent to which a contaminant is bioavailable upon its arrival in aquatic systems is mainly determined by its solubility in water. Bioavailability determines not only the extent that a toxin is available for uptake from the water but also the extent to which it is able to react with the metabolic machinery once on or in an organism. Hydrophobic contaminants that tightly sorb to colloids, or larger particles such as inorganic or organic aggregates (flocs), are much less available for direct uptake from the water than hydro-

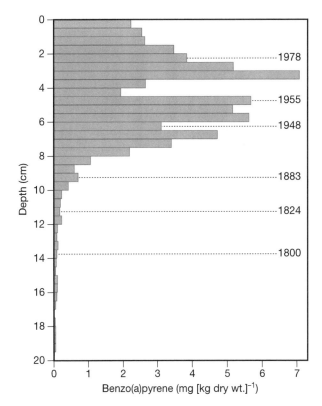

Figure 28–7 Vertical distribution of benzo(a)pyrene in the sediment of Lake Huzenbach, south Germany. Interyear variation in contaminant concentration, invariably expressed per unit sediment dry weight, is here and elsewhere usually attributable to a variable dilution by incoming sediments. Dilution that is linked to variation in runoff or sometimes to construction activities in the drainage basins, with runoff and sediment loading varying much more over a series of years than emission and deposition of particular contaminants. *(After Hilgers et al. 1993.)*

philic contaminants. Conversely, hydrophobic contaminants are more likely to be removed from the water by sedimentation and by filter-feeding organisms.

The extent that a given concentration of a chemical at equilibrium is sorbed to particles rather than dissolved in the water is described by a **distribution** or **partition coefficient** (K_d or K_p).

$$K_d \text{ or } K_p = \frac{\text{concentration on particles}}{\text{concentration in water}} \quad \text{EQ. 28.1}$$

Hydrophobic organic chemicals and, among others, the widespread radioisotope cesium 137 (^{137}Cs), have a high affinity for solids ($K_d \sim 10^3 - 10^7$). K_d is more commonly expressed in log units (log K_d 3–log K_d 7) with the partitioning affected by (1) the concentration of

the chemical in solution; (2) the concentration and size (available surface area) of sorbing particles; (3) the extent that the suspended particles are organic or have acquired an organic coating facilitating sorbtion; and (4) the temperature.

▲ Partition Coefficients in Nature

Since the original development of the K_d concept for soils, a wide variety of partition coefficients have been determined, describing the uptake and toxicity of contaminants in aquatic organisms at equilibrium. Toxic metals also partition between dissolved and particulate phases. In the Niagara River between Lake Erie and Lake Ontario, about 5 percent of the cadmium (log K_d ~3) but nearly 100 percent of the lead and zinc (log K_d ~7) is associated with suspended particles (Allan 1986). It should be understood that the proportion of the total contaminant load associated with particles increases with the number of particles. Highly transparent lakes and rivers consequently have a much higher fraction (often nearly total) of hydrophobic or otherwise sorbing chemicals in solution than the same chemicals in turbid waters.

Among the physiochemical properties of toxic metals determining partitioning (sorption) are the *charge radius*, *ionic radius*, and *electronegativity*. Electronegativity represents the power of an atom to attract electrons to itself in a covalent bond (Förstner and Wittmann 1979, and Stumm 1987). The greater it is, the lower the solubility and reactivity of the particular elements will be. Physical properties help determine how readily a particular metal (Fig. 28–8) or POP is scavenged from the water column by sinking particles, how long its water residence time will be, how tightly it is held by sediments, and how readily contaminants are released from particles in the gut of predators. Even so, sediment characteristics, temperature, and in particular, pH modify the solubility (K_d) of trace metals. Not only do the most toxic metals and the most worrisome persistent organochlorines have a high K_d and sorb strongly onto particles but so do sometimes genotoxic chemicals released in urban and industrial effluents (Table 28–6).

▲ 28.7 Toxicity and Its Prediction

Recognition that the structure of persistent organic contaminants and toxic metals bears a systematic relationship to their activity (behavior) permits a first a

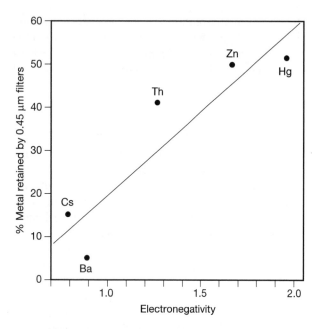

Figure 28–8   Relationship between trace metal properties and the average fraction of labeled trace metal scavenged by > 0.45 µm suspended particles over a 20-day period in a limnocorral experiment ($r^2 = 0.85$; $0.02 < P < 0.05$). *(After Jackson et al. 1980.)*

priori prediction about which of the very large number of different classes of anthropogenically produced compounds potentially toxic to the biota should receive particular attention. Similarly, within a class of

compounds the different congeners usually differ predictably in their activity.

Structure–activity Relationships

The relationship between physiochemical and structural properties of organic compounds and their activity—including toxicity and distribution—has been formalized in **quantitative structure activity relationships (QSARs)**. The most widely used physical property for predicting the sorption characteristics of particles and living organisms is the **1-octanol-water partition coefficient** (log K_{ow}) describing the equilibrium concentration between water and a solvent (octanol), with octanol serving as a proxy for the partitioning between the aqueous phase and the more lipidlike biophase (Fig. 28–9). The K_{ow} therefore provides an indication of (1) the uptake potential by organisms; (2) the potential for bioaccumulation from the water (Fig. 28–10); (3) the biomagnification potential in food webs; and (4) the residence time of persistent organic pollutants in the environment.

▲   Contaminant concentrations in large and long-lived organisms are frequently much higher than predicted from the K_{ow} (solubility in water), suggesting that food rather than uptake from the water is the principal pathway of contaminant accumulation in nature (see Sec. 28.8). Empirically determined K_{ow}s show, over a wide range of both toxicity and K_{ow}, a linear relationship between the acute toxicity measured and the log K_{ow}, for organisms as diverse as bacteria,

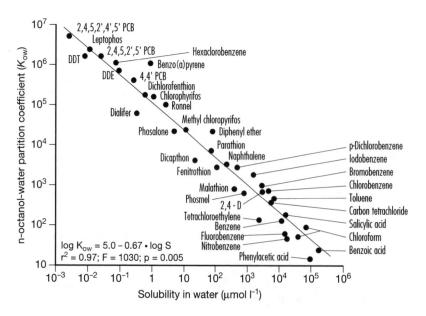

Figure 28–9   Relationship between the solubility of selected organic chemicals in water(s) and their partitioning between water and n-octanol, with the partition coefficient (K_{ow}) serving as a surrogate for the partitioning of the chemicals between the aqueous phase and the more lipid-like biophase. Data scatter is largely attributable to differences in temperature and analytical procedures. *(Modified after Chiou et al. 1977.)*

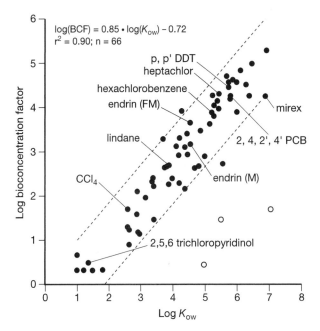

$\log(BCF) = 0.85 \cdot \log(K_{ow}) - 0.72$
$r^2 = 0.90$; $n = 66$

Figure 28–10 The relationship between bioconcentration factors (BCFs) for fathead minnows (FM), bluegills, rainbow trout, and mosquitofish (M) exposed to nonlethal chemical doses in water and the 1-octanol-water partition coefficient (K_{ow}) of the chemical, based on laboratory experiments of about one month duration. The 95 percent confidence interval for BCF predictions is shown. The regression equation excludes three outliers (○) which show little bioconcentration despite high K_{ow} values. Note, the BCFs are relative only in that they are based on a short-term uptake from water and not, as in nature, a relatively long-term uptake primarily from food. Also note variation imparted by different studies of the same chemical (e.g., endrin). *(Modified after Veith et al. 1979, and Stumm et al. 1983.)*

protozoa, algae, daphnids, shrimp, and fish (Nendza and Klein 1990). Chronic toxicity is related to the K_{ow} as well, but at contaminant concentrations about an order of magnitude below acute toxicity concentrations (McCarty and MacKay 1993). Chronic toxicity is commonly manifested as increased birth defects and embryo deaths.

The linear relationship between toxicity and K_{ow} breaks down for highly *lipophilic* toxins (log K_{ow} > ~5) which have a low solubility in water or suffer from reduced passage through membranes. As a result, high K_{ow} compounds are somewhat less toxic than suggested by their K_{ow}. The **lipophylicity** (fat-loving nature) of particular organic compounds determines the extent that an organic contaminant is preferentially stored rather than excreted. Among and within classes

of toxic hydrocarbon compounds of a particular hydrophobicity and associated lipophylicity, congeners that are most chlorinated are generally not only least volatile but also most toxic. These same congeners are furthermore least readily excreted by fish (Niimi 1986). Therefore, the highly chlorinated congeners are most readily stored by organisms and least subject to metabolic (enzymatic) and microbial degradation. The chlorine content and thus the toxicity of the hydrocarbon congeners that are retained consequently rises with trophic level in any region, but most of the responsible fractionation seems to occur at the bottom of food webs (Oliver et al. 1989).

QSAR Limitations

The principal limitation of QSAR modeling is shared with all other regression models (including the well-known total phosphorus–chl-*a* relationship, Fig. 8–16).[4] While QSAR models reveal patterns (regularity), the utility of the K_{ow} as a predictor becomes increasingly reduced as the degree of resolution (interval scale) over which the compounds are measured narrows. When the K_{ow} range that is examined declines to less than an order of magnitude, the K_{ow} toxicity is no longer well-predicted from the K_{ow}. Over a narrow interval scale other—often unmeasured—variables become more important (Sec. 2.6). A second limitation of QSARs is that they are species specific, the different types of test organisms vary in sensitivity to a particular toxin by as much as three orders of magnitude. This variation makes it difficult to generalize about the impact of a particular toxin on the community as a whole when it is based on tests made on one or a few species under controlled conditions. These differences further reflect experimental conditions, the age and physiological condition of the test organisms, the mode of toxic action, as well as the kinetics of toxin uptake and elimination.

Despite its limitations, the K_{ow} and other available QSARs play a crucially important role as a first indicator of the expected toxicity of any newly synthesized

---

[4]"One criticism often lodged against correlations such as QSARs is that they are not based on a fundamental understanding of the mechanisms of toxicity. They provide quick answers but not basic understanding. It is the greater understanding of toxicity mechanisms that can ultimately answer many of our toxicity questions. However, in the short run, we still need some guidance. And in fact QSARs used in conjunction with more fundamental research may serve both needs." (Blum and Speece 1990)

organic compound to be produced in quantity. Therefore, the K_{ow} is routinely required by regulatory agencies before new compounds are approved. Other information that is required includes the physiochemical characteristics of the compound (melting and boiling points, water solubility, and resistance to hydrolysis and other forms of chemical degradation) plus information on toxicity to a few specified organisms. The K_{ow} serves as a surrogate for the solubility of toxic compounds in water (Fig. 28–9). Thus they reflect the extent and strength of sorption to sedimenting particles (K_d) and the extent that persistent contaminants are expected to remain in the water rather than incorporate into the sediments, following sedimentation of contaminated particles.

Measurement of Ecosystem Toxicity

Major problems still beset our ability to assess the ecosystem toxicity of compounds (Kaiser 1984). Reasons for this include the use of only a few test species under laboratory conditions whose relevance and species-specific response to toxins cannot closely mimic those of the many species in natural communities operating under highly variable conditions. Nor are all of the life stages usually examined in toxicity testing. An assessment of ecosystem toxicity is further hampered by an emphasis on short-term *acute* toxicity tests, even though the concentrations of individual toxic substances in inland waters only occasionally reach acutely toxic levels. There are, furthermore, two types of acute toxicity: **nonspecific (nonreactive or narcotic) toxicity** and **specific (reactive or nonnarcotic) toxicity**. Most QSARs apply to the former, in which toxicity is not associated with a specific mechanism but instead is a function of the quantity of toxicant accumulated. Conversely, specific toxins affect a specific mechanism (e.g., inhibition of a specific metabolic pathway or enzyme production). Specific toxins appear to act as nonspecific toxins below their threshold for specific toxicity (McCarty and MacKay 1993). Two additional assessment problems remain: The first is the emphasis on the acute toxicity of individual compounds or elements even though most contaminants are normally released as complex mixtures. Contaminated waters usually contain a wide variety of organic and inorganic toxic contaminants with virtually nothing known about how they might interact. The second problem is a lack of well-developed tools to measure chronic toxicity or its effects in nature.

The development of rapid screening tests to detect the chronic toxicity (genotoxicity or mutagenicity) of environmental samples using bacterial cultures[1] are but one of a number of promising approaches in environmental toxicology complementing the use of QSARs. In another new approach, environmental toxicologists are starting to correlate toxic effects seen in nature to tissue concentrations of contaminants present in the biota, based on long-term toxicity tests on the same or surrogate species in the laboratory. This second way is a major advance over the traditional toxicological approach of simply exposing organisms to an elevated environmental concentration of a particular contaminant dissolved in water during short-term (acute) toxicity tests in the laboratory, employing **dose–response relationships**. Unfortunately, the relevance of mortality rate obtained from dose–response relationships is difficult to extrapolate to nature (Sec. 28.8).

Toxicity Equivalency Factors

Fish and fish-eating predators in nature are exposed to a variety of contaminants creating a great need for a measure of the sum total effect of the different contaminants on the biota. The recent development of **toxicity equivalency factors** (TEFs) is an important response to this need. The TEFs express the toxicity of individual or mixtures of, for example, polychlorinated hydrocarbons (PCHs), as a toxic equivalent to the extremely toxic tetrachlorodibenzo-p-dioxin (TCDD) standard, using a bioassay system as a chemical detector. The method is based on the fact that PCHs cause the induction of certain liver enzymes that in turn metabolize (detoxify) toxic congeners. The degree of decline of the enzymes, known as the **mixed function oxidase** (MFO) **system**, is a function of the experienced toxicity (Ludwig et al. 1993). TEFs are starting to permit a much better prediction of PCH toxicity in nature than is obtained from a measure of only the total concentration of PCHs in the biota (Fig. 28–11). TEFs measured in chicks of double-crested cormorants and caspian terns in the North American Great Lakes accurately predict the occurrence of deformations in the chicks. Yet other ecotoxicologists are measuring particular enzyme or hormone levels in fish as diagnostic tests for the "stress" imposed by low levels of contaminants in nature (Hontela et al. 1992).

Finally, levels of **metallothioneins**, low molecular-weight proteins with an affinity toward trace metals, are being used as a bioassay to estimate stress from

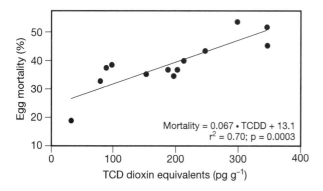

Figure 28–11 Correlation between concentrations of bioassay-derived TCD dioxin equivalents as determined in eggs in the laboratory, and the egg mortality rates in colonies of double-crested cormorants in the North American Great Lakes from 1986 to 1988. *(After Giesy et al. 1994.)*

toxic metal contamination in fish and invertebrates (Hamilton and Mehrle 1986).

▲ 28.8 Bioaccumulation and Biomagnification

Bioaccumulation is the net result of more rapid accumulation than release rate of persistent contaminants. The uptake involves the adsorption to the body surface or gills of lipophilic (high K_{ow}) or otherwise sorbing contaminants (e.g., toxic metals), or their uptake in food. The adsorption accumulation route is most important for small organisms because of their larger surface to volume ratio and elevated metabolic rates. However, the metabolic rate of organisms typically rises more slowly than their gain in weight (metabolism = weight$^{-0.75}$, Peters 1983b) and direct contaminant uptake from the water is less important for large organisms at the top of food webs.[5] The latter can be expected to obtain most of their often high levels of persistent contaminants as the result of consuming contaminated food. Reduced metabolic rates—including reduced excretion rates—of larger organisms allow for strong partitioning of persistent organic con-

taminants to tissue organic matter and above all to either lipids or proteins.

Adult lake trout, a game fish at the apex of the aquatic food web, appears to accumulate somewhere between 80 and 100 percent of their persistent organic contaminant burden from food and the balance, if any, from direct uptake from the water (see Borgmann and Whittle 1992). Fish size or age, often used as surrogates for the length of exposure to contaminants in the water, is therefore generally of lesser importance than food in determining the contamination of adult game fish. Within a climatic zone, their predators (e.g., gulls, loons, eagles, mink, otter) then have tissue levels of persistent organic contaminants (including organic mercury) much higher (by a factor of five) than expected from either environmental concentrations or the K_{ow} values of the compounds (Thomann 1989, Oliver and Niimi 1988).

Biomagnification and Bioaccumulation

The cumulative increase observed in the concentration of persistent contaminants at successively higher levels in food webs is known as **biomagnification** and the **biomagnification factor** (BMF) is obtained from the predator:prey concentration ratio, whereas the **bioaccumulation factor** (BAF) represents the organism concentration:environmental concentration ratio (Fig. 28–12). The bioaccumulation factor (BAF) from water to piscivorous fish for PCBs or dioxins is typically around 10^6, but is tenfold greater (10^7) for fish-eating gulls, and is close to 10^8 for eagles (Ludwig et al. 1993). For piscivorous fish (salmonids) in Lake Ontario the BAF for PCBs is roughly predictable from the K_{ow} (Oliver and Niimi 1988).

$$\log (\text{BAF}) = 1.07 \log (K_{ow}) - 0.21$$

$$r^2 = 0.86 \qquad n = 18 \qquad \text{EQ. 28.2}$$

Among other strongly bioaccumulating substances are organomercury (binds primarily to lipids), cesium-137 (^{137}Cs, binds to proteins), and strontium (^{90}Sr, substitutes for calcium in bone) with the two isotopes released during atomic bomb testing and nuclear accidents.

Food-web Structure and Biomagnification

While biomagnification explains the high concentration of persistent contaminants at the top of food webs, it cannot also explain the two orders of

---

[5]Laboratory work based on short-term dose–response experiments using contaminants dissolved in water rather than measuring the uptake from food had, until recently, led most toxicologists to consider the direct uptake from the water as the only, or principal, route of uptake.

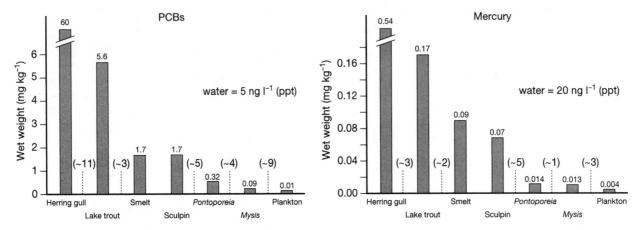

Figure 28–12   Bioaccumulation factors (BAF) of 10 million for PCBs and 27,000 for mercury represent the contaminant increase from water (including colloids) to herring gulls (eggs) at the top of the food web. Vertical dashed lines separate the presumed trophic levels. The bioaccumulation factor (BAF) represents the concentration ratio between predator and prey. Approximate biomagnification factors (BMF) between presumed trophic levels are presented in brackets. *(Modified from Anon. 1991.)*

magnitude *variation* in the average lake trout PCB concentration (0.015 – 3 μg g^{-1}) and mercury concentration (0.03 – 3.96 μg g^{-1}) in remote Ontario lakes (Rasmussen et al. 1990, Cabana et al. 1994), or the similarly wide variation in mercury levels seen in pike in Scandinavian lakes. The reasons for these wide variations is the subject of fundamental research that has strong management implications for human fish-consumption advisories. The often wide variation in trout or pike contamination in adjacent lakes, located equidistant from a distant source region, requires a different explanation. The single most important factor is among-system differences in trophic position (length of food chain).

Lake trout are least contaminated in lakes with the shortest food chains where, in the absence of other prey, the fish feed preferentially on certain macrozooplankton (e.g., the larger *Daphnia* species) and benthic invertebrates—termed class 1 lakes (see Fig 28–13). The fish are more contaminated where pelagic forage fish (e.g., smelt, whitefish) are also present and serving as an intermediate trophic level—called class 2 lakes. The fish are typically most contaminated in class 3 lakes that, in addition to the macrozooplankton and forage fish, also contain *Mysis*, a large and predacious crustacean zooplankton (Fig. 28–13 and Sec. 23.8). By having the longest food chain, class 3 lakes experience

the greatest biomagnification. The trophic position of prey and predator is more accurately predicted using a stable isotope ratio to obtain a measure of trophic position than it is from the inevitably limited number of examinations of the gut content (prey) of predators. Stable isotope analyses[6] are showing that fish and other vertebrates are more omnivorous (opportunistic) in terms of diet then was once thought (Vander Zanden and Rasmussen 1996).

Not only are PCB and Hg concentrations correlated with size in lake trout (Cabana et al. 1994) but so are PCB and DDT concentrations in a variety of pis-

---

[6]Stomach contents have long been used as a measure of diet, but this is time-consuming and only provides a momentary measure (unless repeated regularly). Limnologists increasingly use stable isotopes to obtain a measure of food categories, trophic structure, and energy flow in food webs. Ratios of the widely used heavy to light nitrogen (^{14}N and ^{13}N) and carbon (^{13}C and ^{12}C), isotopes express the per mil (‰) change with reference to standards (the atmosphere for N and a particular limestone for C) that are arbitrarily assigned a value of 0 ‰, with the deviation reported in delta units (δ). The δ^{15}N is typically assumed enriched by 1.7 to 3.4 (± 1 ‰) relative to its diet and is used to determine the trophic position of organisms in food webs. In contrast, δ^{13}C changes little (~1 ‰) as carbon flows through food webs and is used, often together with δ^{15}N, to differentiate between sources of energy having distinct signatures (e.g., phytoplankton, macrophyte, allochthonous matter).

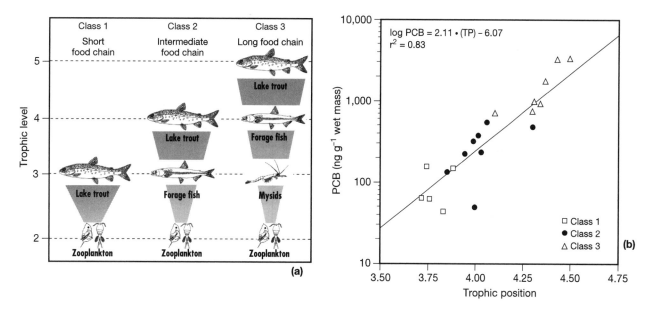

Figure 28-13 (a) Diagramatic representation of how lake trout may belong to one of three trophic classes depending on their choice of diet. (b) PCB levels in organisms of three trophic classes plotted against dietary estimates of lake trout trophic position, based on the stable isotope (^{15}N:^{14}N) ratio. *(After Vander Zanden and Rasmussen 1996.)*

civorous fish (Rowan and Rasmussen 1992), indicating a consistent pattern of bioaccumulation and biomagnification of different persistent contaminants.

28.9   Mercury and the Mercury Cycle

The mercury contamination of lakes is a widespread problem over large areas of the north temperate zone. Some 40,000 out of 83,000 Swedish lakes (> 0.01 km^2) have elevated Hg levels (> 0.5 mg kg^{-1}) in pike. About 10,000 of the lakes have pike mercury levels sufficiently high (> 1 mg kg^{-1}) to be a potential health hazard to frequent human consumers (Lindqvist et al. 1991), and even more so to wildlife species that are largely or totally dependent on fish, but unlike humans they cannot change their diet. In the United States, fish Hg levels high enough to elicit warnings from government agencies exist for portions of 26 states. For example, in Wisconsin, over 30 percent of over 300 mostly remote lakes contain Hg concentrations exceeding the 0.5 mg kg^{-1} state health advisory limit (Fitzgerald and Watras 1989). Consumption warnings have furthermore been issued for approximately 90 percent of Ontario lakes, 90 percent of lakes in northeastern Minnesota, and for about 10,000

Michigan lakes. Elevated Hg levels apparently are not restricted to lakes on the Precambrian Shield or boreal forest regions of North America and Europe because Hg-contaminated fish are also reported from poorly buffered (low pH) Florida waters (Lange et al. 1993).

Mercury Poisoning

It has long been recognized that Hg poisoning was an occupational hazard, for example to mercury miners. However, two major poisonings during the last 50 years of people eating Hg-contaminated food (Japan and Iraq) has increased concerns about Hg as a health hazard. In the 1960s and 1970s, much of the work on Hg as an environmental rather than human contaminant was done in Sweden where organomercury compounds were widely used as biocides. An alarming decline in the abundance of some seed-eating birds after World War II, and in birds of prey feeding on seed-eating rodents was linked to the use of Hg as a biocide. Later, elevated Hg levels were noted in the early 1960s in freshwater fish in poorly buffered lakes. Today it is evident that mercury levels in the larger piscivorous fish in remote North American and European lakes can be sufficiently elevated (> 1.5 mg kg^{-1}) to expose humans, and particularly fish-eating wildlife to worrisomely high levels of Hg. In Ontario, Scheuhammer and Blancher (1994) estimated that 30 percent of the acidified lakes contained prey-sized fish with concentrations of methylmercury high enough to affect the reproduction of the common loon.

Contamination and Distance from Sources

Pike mercury levels in Swedish lakes have increased from a probable natural background level of about 0.05–0.3 mg kg^{-1} in nonacidified lakes to commonly about 0.5–1 mg kg^{-1} today. The increases are of the same magnitude (~3–5 times) as the increases seen in North American lake sediments (enrichment factor, EF, Sec. 28.5) between sediments recently deposited and background levels (Fig. 28–14, and Lindqvist et al. 1991). The increased contamination of both fish and sediments is attributed to (1) increased atmospheric transport of Hg from industrialized regions to remote areas more than 1,000–2,000 km from the source region: (2) a long atmospheric residence time of between six months and two years (Iverfeldt 1991); and (3) increases in acidification of poorly buffered lakes downwind from industrial and urban sources (Chapter 27).

A decline in contamination with distance from sources means that contaminant deposition rates also decline with distance from sources. Indeed, southern Scandinavia's sediment levels have increased much more (~7 times) than in northern Sweden (Johansson 1985). Even so, recently constructed reservoirs in boreal forest regions far from sources of Hg tend to have exceptionally highly contaminated fish, at least during the first decades following construction. The high Hg levels are at least in part attributable to the release of Hg, largely derived from elsewhere over the last several centuries, from the recently flooded and now anoxic soils (Bodaly et al. 1984), although rock sources may be important as well. Northern pike (*Esox lucius*), a top predator in one such shallow subarctic Quebec reservoir has Hg concentrations averaging ~3 mg kg^{-1} compared to ~0.4 mg kg^{-1} for natural unimpounded lakes in the area (Chevalier et al. 1997).

▲ Mercury Determination

The relatively abundant data on total Hg (t-Hg) in the biota and sediments stands in sharp contrast to the paucity of data in waters and the atmosphere. Sample

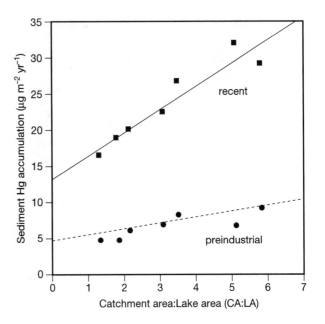

Figure 28–14 Whole-basin sediment Hg accumulation rates as a function of the catchment area to lake area ratio (CA:LA) for seven lakes in Wisconsin and Minnesota (US). Recent rates represent the last decade (≈ 1982–1992) and preindustrial rates are those before about 1850. The intercept of each regression line predicts the average rate of atmospheric Hg deposition on lakes in the absence of a drainage basin while the ratio of the slope to the intercept estimates the proportion of the atmospheric flux transported to the lakes from their catchments. Note not only the nearly threefold higher mercury sedimentation in recent sediments, but also the disproportionately higher accumulation rate in larger catchments that are probably most impacted by changes in land use. *(After Swain et al. 1992.)*

contamination during collecting and processing (Sec. 28.4) has retarded the understanding of mercury bioaccumulation rates and pathways as well as the mechanisms regulating the fluxes. Recent research, using ultra-clean trace-metal free-collecting protocols, storage protocols, and analysis protocols for total Hg (t-Hg) yields estimates that are between 20 and 100 times lower than those obtained without such extreme precautions (Fitzgerald and Watras 1989). Even fewer interpretable data are available for methyl mercury (MeHg) concentrations, the most toxic bioaccumulating form of organic (alkyd) mercury for which virtually no reliable data exist prior to 1989 (Bloom 1989). Finally, the quantitative techniques required to measure the in situ methylation and demethylation rates of inorganic mercury have not yet been developed.

Both t-Hg and MeHg are highly particle reactive with log K_ds of 5–7 in aquatic systems (Watras et al. 1994). MeHg has the higher K_d and, being particularly lipophilic, is less excreted and therefore bioaccumulates preferentially. For game fish feeding at the top of the food web, mercury is therefore almost totally (95%) in the MeHg form (Westöö 1966, Grieb et al. 1990).

Environmental Factors and Contamination

Attempts to link the relatively abundant fish Hg data to environmental factors have repeatedly shown that in the north temperate zone fish Hg levels in remote waters rise with increasing water color (a surrogate for DOC) and with increases in acidity below a pH of about 6 (Håkanson et al. 1990, and Grieb et al. 1990). Acidity and water color are highly correlated in most temperate zone systems, making it impossible to assess the relative importance of each. But as water color is primarily determined by the size and slope of the drainage basins (Eq. 8.2), and as both total Hg and MeHg are highly sorbed to the dissolved organic matter (DOM), it is believed that humic systems receive not only more t-Hg from their drainage basins but also proportionately more MeHg than do more transparent lakes, rivers, and wetlands.

Laboratory experiments on bacterially mediated methylation and demethylation, supplemented by a few measurements of the mercury species in the atmosphere, water, and sediments have yielded the conceptual view of the mercury cycle presented in Fig. 28-15. Ionic mercury (Hg^{2+}) is a highly electronegative element (Fig. 28-8) which forms stable and durable complexes with organic matter. Methylation is primarily mediated by microbes and is key in the biogeochemical cycling of mercury because, as mentioned above, MeHg is the principal form accumulated and transported by the biota. Microbial methylation appears to have evolved to detoxify the Hg^{2+} and facilitate its escape by increasing Hg mobility. This is accomplished by the transfer of a methyl group to Hg^{2+}, yielding the more soluble (mobile) monomethyl mercury (CH_3Hg^+) that is capable of diffusing slowly from sediments and soils, and by the transfer of CH_3 to CH_3Hg^+ to yield a more insoluble but also more volatile dimethyl mercury [$(CH_3)_2Hg^+$]. The microbially induced reduction of Hg^{2+} to the much more volatile Hg^0 (metal) form is a third mechanism to permit a detoxification of (loss from) the microbial environment (Wood 1987). Volatilized mercury has a sufficiently long atmospheric residence time to allow for Hemispheric-wide dispersion before redeposition. Finally, when H_2S is available a chemical reaction with Hg yields relatively volatile $(CH_3)_2Hg$, with the precipitation of the balance as insoluble and nontoxic HgS (Fig. 28-15).

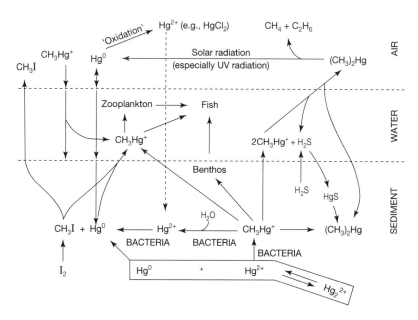

Figure 28–15 A schematic view of the mercury cycle in aquatic systems showing those reactions catalyzed by bacteria, the chemical reduction by H_2S, and the photochemistry of organomercury compounds. Not shown are the adsorbed (complexed) inorganic mercury and CH_3Hg^+ entering the water from catchments. (*Modified after Wood 1987.*)

28.10　Toxic Chemicals, Environmental Health, and Aquatic Management

The only effective way to reduce the supply of toxic chemicals to the environment is to prevent their release to the atmosphere and waterways. Toxic chemicals that are persistent and also bioaccumulate are of particular concern because of their potential for biomagnification in food webs. The human consumption of contaminated fish and wildlife is an important issue, particularly for subsistence fishermen and isolated communities of native people with a traditional diet high in piscivorous fish who have to balance contamination risks against a nutritious diet. However, piscivorous fish and fish-eating wildlife (e.g., loons, kingfishers, gulls, eagles, mink, otter), cannot change their fish diet and are far more at risk.

There is considerable debate as to the seriousness of the contamination of piscivorous fish and their predators in terms of health but there is enough evidence to show that certain wildlife species are at least subtly affected in the more polluted portions of the Laurentian Great Lakes. The increases in the frequency of deformities and reproductive failures in aquatic birds, including embryo survival (Giesy et al. 1994), that have been observed are similar to those seen in laboratory studies of the same or a surrogate species exposed to PCBs and other POPs (Ludwig et al. 1993).[7]

Highlights

- Among the chemicals of greatest concern are highly chlorinated organic molecules, known as organochlorines, that are persistent (degrade slowly) and accumulate in the biota. Those *persistent organic pollutants* (POPs) that are somewhat volatile or sorbed to airborne particles are carried for hundreds if not thousands of kilometers from areas of production to remote regions.
- Toxicity can be acute, resulting in quick death. More common is chronic (sublethal) toxicity, whose effects are much more difficult to quantify and, because it expresses its effects with lags, makes it difficult to link chronic toxicity to particular sources.
- More than half of the quantities of most trace metals (including toxic As, Cd, Hg, and Pb) globally emitted to the atmosphere are anthropogenically produced.
- The relative importance of the atmosphere as a direct source of particular POPs and toxic trace metals in remote aquatic systems is greatly affected by the catchment area:water area ratio.
- Contaminants that dissolve readily in water are known as hydrophilic, while those with a low solubility are known as hydrophobic. The latter sorb readily onto detrital particles and living organisms, are less bioavailable for direct uptake from the water than hydrophilic contaminants, and are more likely to be removed from the water by sedimentation and filter-feeding organisms.
- A systematic relationship between the structure of persistent organic contaminants and toxic metals and their activity—including their toxicity, uptake, and bioaccumulation potential—and distribution has been formalized in *quantitative structure and activity relationships* (QSARs) that have predictive power.
- ▲ The recent development of *toxicity equivalency factors* (TEFs) allows the sum total effect of different contaminants in nature to be compared.
- ▲ Important differences between adjacent remote waterways in the contamination of particular piscivorous fish species with, for example, PCBs or mercury, are linked in part to differences in the number of steps in the respective food chains (food-web structure), allowing for different degrees of biomagnification.
- The only effective way to reduce the supply of toxins to aquatic systems is to prevent their release to waterways and the atmosphere.

---

[7]"This study indicates that the protection of wildlife species will require much more stringent regulations than those currently recommended to protect the human population from cancer. The protection of wildlife populations and humans from subtle effects should be given equal priority to that of protecting human populations from cancer." (Ludwig et al. 1993)

C H A P T E R

29

Reservoirs

29.1 Introduction

The vast majority of reservoirs in existence today are relatively small (volume $< \sim 10^6$ m^3)—they are called *impoundments*, or *dams* in southern Africa and Australia (see Sec. 6.8)—and are used primarily for irrigation, water supply, and also for fish farming. The smallest ones are usually constructed for domestic use, watering livestock, irrigation, or fish production. However, about 948 large-volume reservoirs, those with dams at least 15 m high, were under construction in 1994 (Anon. 1997) contributing to the about 50,000 such dams worldwide in 2000.

Most reservoirs are constructed by damming rivers in regions where evaporation approaches or exceeds precipitation (Fig 5–9), with the water typically derived from local rivers during rainy periods. Permanent lakes are uncommon as the result of low local runoff and normally few basins deep enough to hold water yearround. River discharge into semiarid and arid zone (dryland) reservoirs shows large seasonal and interyear variation. Large reservoirs in the low runoff portions of the world are built principally for irrigation, flood control, and water supply to urban areas, but at times also for improved navigation (as on the Volga River, Sec. 5.8), fish culture, or recreation. Large-volume reservoirs, built primarily for hydroelectric power generation, are constructed in regions of low as well as high precipitation, where landscape topography and river discharge make damming possible and economically attractive (Fig. 29–1).

The fraction of the world's runoff temporarily retained behind dams has been growing rapidly, with large-volume reservoirs (> 0.5 km^3) holding some 20 percent ($\sim 8,400$ km^3) of the global mean runoff. Reservoir water storage has risen to about seven times the volume of water in rivers (Vörösmarty et al. 1997).

Water retention by reservoirs has major ramifications for river discharge and its timing, for sediment and nutrient retention behind dams (Sec. 9.6), for the biota of the extensively modified rivers and wetlands upstream and downstream from the reservoirs, and for the affected people. The negative environmental consequences and social impacts of large reservoirs are becoming gradually better appreciated and a priori *Environmental Impact Assessments* (EIAs) are required and are more rigorous than in the past.

Western limnologists have traditionally paid little attention to reservoirs; the first limnology textbook on them in English was only published in 1990 (Thornton et al. 1990). Until recently, possibly the single most historically important set of publications in a west European language are the three slim volumes reporting the long-term careful study of a number of Czech hydroelectric/drinking water reservoirs (Hrbáček 1966, Hrbáček and Straškraba 1973a and b). These (dated) volumes remain useful for introducing Western readers to a vast Russian and east European literature that has been largely inaccessible to the Western world. As early as 1963, 613 papers had been written on the shallow eutrophic Rybinsk Reservoir on the Volga River (RU) (Straškraba and Straškrabová 1969).

Figure 29–1 The Itaipu hydroelectric dam forming a 200 km-long reservoir (LA: 1347 km^2; $\bar{z}$ = 22 m, 12max = 115m, $\bar{x}$ WRT = 35d)) on the Paraná River, on the border of Brazil and Paraguay. The spillway (in operation) prevents flooding of the dam during high water. *(Photo courtesy of Itaipu Binacional.)*

29.2   Natural Lakes and Reservoirs

Natural and man-made lakes everywhere have much in common.[1] They largely share the same species and have identical habitats (e.g., the pelagic zone). At the scale of redox reactions, predator–prey interactions, and convective mixing, along with the techniques used to study them, they are fundamentally the same as lakes. Yet, there are major differences between, for example, temperate zone reservoirs and natural lakes on the same continent (Table 29–1). Some of the differences are linked to location, with most United States reservoirs located at lower latitudes (between 33°N and 42°N) than the natural glaciated lakes that dominate in better-watered regions with which they are compared. Most of the natural lakes studied are dimictic, whereas most of the reservoirs are polymictic, or monomictic when deep enough to stably stratify. Furthermore, reservoirs (and any natural lakes) at a lower latitude usually experience a longer growing season, longer stratification period, higher water tempera-

tures, lack an icecover, and usually receive a much greater input of particulate inorganic matter from their less well-vegetated catchments than dimictic lakes. Other differences are given in Tables 29–1 and 29–2.

Most of the best-studied dimictic lakes are headwater lakes with a typically ellipsoid sinusoid shape (Chapter 7) and their maximum depth away from the outflow. A relatively even distribution of inflowing streams around natural lakes results in relatively rapid mixing of their incoming water with the receiving water (Fig. 29–2). In contrast, man-made lakes are often elongated as the result of their location in drowned river valleys. The inundation of smaller tributary valleys during filling frequently makes them *dendritic* in outline (Fig. 29–3 and Sec. 7.5).

Reservoirs are constructed at the bottom end of preferably large drainage basins to maximize their water-collecting potential. In the United States, their drainage ratios (CA:LA) average nearly three times that of natural lakes (Table 29–1 and Sec. 9.5) and also differ in many other aspects (Table 29–2). Drainage ratios of semiarid zone storage reservoirs worldwide are typically much larger. Larger catchments guarantee higher inputs of particulate matter, nutrients (Tables 29–1 and 29–2), and contaminants.

Most of a reservoir's water enters at the upstream end at a principal inflowing river. The presence of one principal inflow, together with the elongated shape of many rapidly flushed hydroelectric reservoirs, implies a unidirectional through-flow of a coherent water mass, known as *plug flow*. In contrast, the typical lake is much more quickly mixed and is often considered to resemble an *instantaneously mixed reactor*.[2] The dendritic or otherwise irregularly shaped outline of reservoirs built in river valleys, broad valleys, or low relief plains (Fig. 29–3) further implies that the many lateral subbasins will also differ greatly from the central basin of the same reservoir in mean depth, flushing rate, sediment loading, turbidity, and biota.

The attributes of natural and man-made lakes overlap extensively, but the differences are important as well (Table 29–2). Reservoirs invariably have the greatest depth and possibility for stable stratification near the dam, whereas most natural lakes have one or more "deep holes" elsewhere. A more important difference

---

[1]Reservoirs and man-made lakes, terms used interchangeably here, are not really synonymous because a large number of natural lakes have their outflow modified by a dam and, technically speaking, can be called reservoirs.

[2]The *assumed* relatively rapid mixing of most natural lakes is probably an important reason for the traditionally small amount of attention devoted by limnologists to water movements and horizontal variation.

Table 29–1 **Comparison of the geometric means of selected variables in natural United States lakes and reservoirs, and median values in primarily semiarid zone reservoirs worldwide. ND = no data**

Variable	Natural Lakes n = 309	Reservoirs n = 107	World Reservoirs n = 113
Drainage area (km^2)	222	3,228	1,281
Lake area (km^2)	6	34	13
Catchment Area: Lake Area	33	93	166
Maximum depth (m)	11	20	30
Mean depth (m)	4	7	10
Water residence time (yr)	0.7	4	1.1
Areal water load (m yr^{-1})	6.5	19	ND
P-loading (g m^{-2} yr^{-1})	0.9	1.7	1.3
N-loading (g m^{-2} yr^{-1})	18	28	ND

Source: From Cooke et al. 1986, and Thornton and Rast 1993.

is the subsurface outflow of most hydroelectric reservoirs, which affects their temperature and dissolved oxygen profiles and increases their nutrient retention (Figs. 29–2 and 9–7).

The release of hypolimnetic water results in reduced hypolimnetic WRT, leaving less time for dissolved oxygen levels to decline during stratification than in natural lakes. The profundal sediments of temperate zone hydroelectric reservoirs are consequently more oxygenated than in natural lakes or surface-release reservoirs. The release of usually colder (at higher latitudes) and nutrient-poorer hypolimnetic

Table 29–2 **Comparison of the characteristics of mainstream reservoirs and stratifying lakes.**

Characteristic	Lakes	Reservoirs
Quantitative (absolute) differences		
Origin	natural	anthropogenic
Geological age	old (≥ Pleistocene)	mostly young (< 60 years)
Aging	slow	rapid (first few years)
Formed by filling	depressions	river valleys
Shape	regular	dendritic
Shore development ratio	low	high
Maximum depth	nearly central	extreme (at the dam)
Bottom sediments	more autochthonous	more allochthonous
Longitudinal gradients	wind-driven	flow-driven
Outlet depth	surface	deep
Qualitative (relative) differences		
Catchment: Lake area	lower	higher
Water retention time	longer	shorter
Coupling with catchment	lesser	greater
Morphometry	U-shaped	V-shaped
Level fluctuations	smaller	larger
Hydrodynamics	more regular	highly variable
Causes of pulses	natural	man-made operation
Water management system	rare	common

Source: Modified after Straškraba 1996.

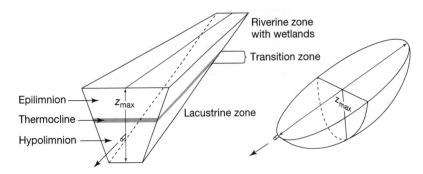

Figure 29–2 Simplified representation of the morphometry, zonation, and stratification of a mainstream reservoir in a drowned valley; and the morphometry of a typical glacial lake. Whether such reservoirs exhibit a temperature stratification depends on the climate and flushing rate; very rapidly flushed reservoirs (WRT < ~10 days) are too turbulent to allow stratification. *(Modified after Straškraba et al. 1993.)*

water affects the temperature and nutrient levels in both the outflowing river and the inflow temperature of a second, not too distant, downstream reservoir. There, the colder denser water entering it plunges downward in summer until it reaches water of equal density at the thermocline or the sediments, whichever it encounters first. It then continues to flow horizontally as an *intrusion current* (Sec. 12.7). As a result, the nutrients it carries will be largely unavailable to the phytoplankton in the euphotic zone of the lower reservoir, lowering its productivity.

Regulated Rivers

Rivers influenced by upstream dams, known as **regulated rivers**, may manifest the influence of the damming for many, even hundreds of kilometers downstream (Hart and Allanson 1984). The rivers are also much affected by whether the dam has a surface outlet or a deep release outlet. A surface outlet permits the export of large quantities of plankton, nutrients, and (at low latitudes) large floating macrophytes to the receiving river. The exports are commonly much reduced in dams with a deep release outlet. When the hypolimnia are anoxic and also contain toxic H_2S, reservoirs have additional negative effects on the downstream biota.

In semiarid zones in particular, trapping large quantities of sediments derived from poorly vegetated drainage basins results in periodically or permanently turbid reservoir water. High sedimentation within a reservoir has negative effects on the benthos, reduces visual feeding efficiencies, and may result in juvenile fish mortality (Ward and Davies 1984). Sediment loads of the outflowing rivers are usually greatly reduced (Table 9–4), thereby increasing their erosive capacity as the result of the reduced density of the outflowing water. The enhanced erosion, together with a modified flow regime, results in excavation (degradation) of the original river channel, affecting

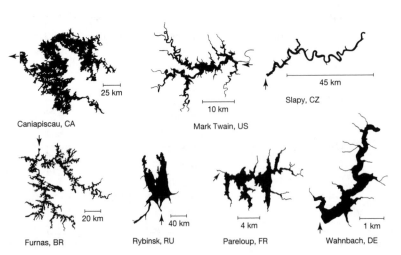

Figure 29–3 Surface shape of selected reservoirs. Note the highly dendritic nature of the reservoirs in flooded river valleys in low relief landscapes. Arrows indicate the locations of dams, except in the Caniapiscau (Quebec, CA) reservoir which lacks a dam on its immediate outflow. It was created by damming a large river system on an extremely low slope (Precambrian) landscape, creating a large, shallow and dendritic reservoir with many islands. The arrow in this reservoir indicates where water flows into the downstream portion of a large reservoir system.

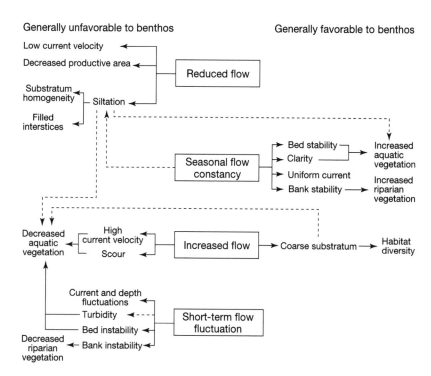

Figure 29–4 Potential effects of various flow patterns below dams on ecological factors that have an important influence on the stream benthos. Dashed lines indicate less definite relationships. *(After Ward and Davies 1984.)*

sediment characteristics, lowering the groundwater table, and impacting the feeding and spawning habitats of fishes and the biota of the rivers in general (Fig. 29–4).

Trapping sediments and their adsorbed nutrients behind dams removes the river's role as a fertilizing agent to previously undisturbed and seasonally flooded wetlands, crop-growing areas downstream, and the receiving estuaries. Dams are furthermore often insurmountable barriers to the upstream migration of fish (e.g., salmon and sturgeon) unless suitable fish passageways are available. For example, declines in salmon stocks in the North American northwest, or sturgeon in the Caspian Sea (Sec. 5.8) are linked directly and indirectly to dam construction. Much less is known about the effect of reservoirs on the migration of invertebrate and vertebrate species moving along the sediment surface.

▲ Discharge Regulation and Water Diversion

The magnitude of the effects of river regulation by dams are both river-specific and general (Fig. 29–5). The management of hydroelectric and irrigation dams create short-term flow fluctuations but reduced seasonal amplitudes. For example, the maximum and

minimum discharge ratio for the Nile River declined from 12:1 to 2:1 following construction of the Aswan High dam (EG, SD). Flood control dams also reduce seasonal variations in discharge. Reductions in seasonal discharge variation reduce the size of seasonally flooded wetlands and may allow the downstream development of dense within-channel growths of benthic algae and macrophytes that interfere with river flow. Reductions in discharge favor invertebrate species (including mosquitoes that in the tropics carry malaria) that thrive under a seasonal flow constancy (Fig. 29–5).[3] Where the landscape relief is modest, wetlands develop at the upstream end of reservoirs.

An even larger threat than discharge regulation to rivers and their biota stems from the diversion of water, primarily for agriculture, captured in reservoirs

---

[3]"Seasonal constancy of water flow [at lower latitudes] favors multivoltine [several generations per year] species, such as Simuliidae [blackflies], more than their natural invertebrate predators and this has led to explosive population increases of pest species. Benthic algal and macrophyte growth is enhanced by reduced flow variation. Dense growths of such aquatic plants can cause severe reductions in stream flow and this in turn favors the development of large populations of mosquitoes. Both these examples support the idea of intermediate disturbance of the river system by using periodic flow disruptions to prevent any one species from dominating." (Ward and Davies 1984)

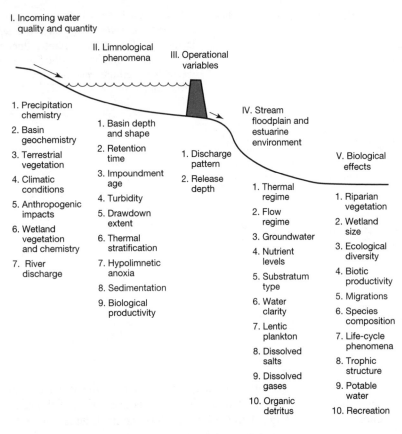

Figure 29–5 Major factors and phenomena influencing the environment below dams and the resultant effects on the biota. The effect of a reservoir on the downstream river increases with increasing water residence time and with increasing productivity, but is modest in rapidly flushed reservoirs. *(Modified after Ward and Davies 1984.)*

built upon dryland (arid or semiarid zone) rivers. Thirty-seven of the world's major rivers flow through drylands, characterized by < 500 mm precipitation annually. Such rivers exhibit enormous interyear differences in discharge, with a mean coefficient of variation of 99 percent compared to 20–30 percent in other parts of the world (Kingsford 2000). One example of the effect of diversion is when the two major rivers entering the Aral Sea dried up (Sec. 5.7). Other examples include the lower reaches of the Yellow River (Huang-Ho, CN) drying up for an average of 70 days yr^{-1} during the ten years before 1996, and as long as 122 days in 1995. Diversions from the Ganges River in India mean that water no longer reaches the sea for all or parts of the year, while in California (US) diversions are responsible for the destruction of about 90 percent of the wetlands in California's Central Valley. Eighty percent of the river flow in the huge Murray-Darling river basin (AU) has been diverted, resulting in the widespread disappearance of wetlands and floodplains. Water diversion and the resulting shrinking of downstream wetlands certainly have dramatic negative effects not only on the aquatic biota and water birds but also on human communities dependent on rivers and wetlands for agriculture, livestock grazing and fishing (Kingsford 2000).

Reservoirs: Their Age and the Biota

Reservoirs differ fundamentally from lakes in that they are mostly very young (< 60 years) whereas natural higher latitude lakes—on which limnological research has concentrated—were created by glacial activity some 10,000 years ago (Table 29–1). Reservoirs are often populated with unusual fish communities derived from inflowing rivers and upstream lakes. This encourages much stocking of selected lentic species because riverine species are usually not well-adapted to lentic environments. In contrast, the biotic interactions within natural lakes have been shaped by evolution over millenia (MacLean and Magnuson 1977).

Among the most commonly introduced sport-fish species in United States reservoirs are the largemouth bass (*Micropterus salmoides*), crappie (*Pomoxis spp.*), and channel catfish (*Ictalurus punctatus*). However, the most commonly introduced prey species in the reservoirs is the gizzard shad (*Dorosoma cepedianum*), which often accounts for 30 percent or more of the total fish biomass (Stein et al. 1995). The adult shad is herbivorous as well as zooplanktivorous and feeds by pumping water through its gills while swimming. The young-of-the-year (YOY or age 0+) shad influence the structure of fish communities by reducing the growth, survival, and abundance of coexisting species. The shad larvae feed on protozoans and rotifers, shifting to crustacean zooplankton when somewhat bigger, and feeding on phytoplankton, protozoa, rotifers, and sediment detritus (including bacteria and algae) when larger than about 3 cm. Where abundant, shad and other such omnivorous fish compete successfully for macrozooplankton with the larvae of piscivorous game fish. Omnivorous species therefore indirectly affect the piscivorous species by influencing the abundance of their zooplanktivorous prey who, as visual feeders, are handicapped under turbid conditions (O'Brian 1990). Shad and other sediment-feeding fish can play a significant role in recycling nutrients to the phytoplankton following excretion (Vanni 1996).

North temperate zone food-chain models are based on top-down control of planktivorous fish by piscivorous species and are unlikely to be useful in predicting the impact of gizzard shad or other omnivorous fish species that are typically dominant in lower and low latitude lakes and reservoirs. The ability of omnivorous fish, such as tilapia (Sec. 26.6), to feed on phytoplankton, zooplankton, and detritus allows them to exert a still not well understood influence on the structure of food webs and energy flow, differing fundamentally from the widely discussed patterns seen in higher latitude waters (Vanni 1996).

29.3   The River–Lake–Reservoir Continuum

Both artificial and natural lakes lie along gradients of climate, geology, morphometry, flushing, chemistry, and biology that preclude a clear separation between lakes and reservoirs. The single most useful continuum along which to conceptualize and model the functioning of reservoirs is to envision them as intermediates between lakes and rivers (Fig. 29–6) in terms of their water residence time (yr) or flushing rate (yr^{-1}).

Large reservoirs are of three basic types (Fig. 29–6). **Mainstream** or **run-of-the-river reservoirs** are most riverlike (WRT = days to weeks) and are most often constructed to generate electricity. Mainstream reservoirs experience little or no temperature

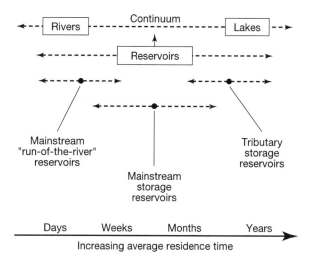

Figure 29–6 Reservoirs occupy an intermediate position between rivers and natural lakes on a continuum of aquatic ecosystems. The extent of riverine influence and the hydraulic retention time determine the relative positions of impoundment types (e.g., mainstream run-of-the-river, mainstream storage, tributary storage) along the river–lake continuum. *(Modified after Kimmel and Groeger 1984.)*

stratification if the water residence time is < ~10 days (Straškraba et al. 1993), no seasonal lowering of the water level, and the least modification of the riverine biota within and below the reservoir. As a group, mainstream reservoirs have received the least attention from limnologists. Exceptions include certain reservoir cascades or chains of reservoirs (e.g., Volga, Colorado, and Vltava Rivers). **Tributary storage reservoirs** built on small rivers have the longest water residence time (WRT, typically from a month to a year or more) and the greatest possibility of temperature stratification where morphometry and climate allow this. These reservoirs have a surface outflow and are usually built for flood control and irrigation. They periodically experience a *drawdown* (water-level lowering, Sec. 29.6) when much more water is removed than enters. However, drawdowns are not restricted to reservoirs. Natural drawdowns, a result of evaporation during the dry season, are experienced most severely by natural lakes in semiarid and arid regions. The tributary storage reservoir is the most common type worldwide and has received the most limnological attention among reservoir types, precisely because they most resemble natural lakes in flushing (OECD 1982) and other attributes. **Mainstream storage reservoirs** hold an intermediate position (WRT = typically a few months), but become more riverine (lotic) during high runoff periods and more lacustrine (lentic) at other times.

The WRT of both reservoirs and rapidly flushed lakes is highly sensitive to seasonal variation in runoff and evaporation. The effect of this is once again most evident in low runoff regions. An example is the Lake McIlwaine reservoir (ZW). It has an average WRT of about one year, but varies from a *calculated* theoretical maximum of 12 years during months of severe droughts to an *observed* four months during exceptionally high rainfall years (Ballinger and Thornton 1982).

▲ 29.4 Water Residence Time and Plankton Growth Rates

The plankton diversity of rapidly flushed reservoirs and lakes declines when the flushing (dilution) rate exceeds the growth rate (doubling time) of the most rapidly growing species at a particular temperature and they experience *washout*. First to be affected are the freshwater copepods whose development time (egg to egg) under optimal conditions is about 30 days at 10°, 14 days at 20°, and 7.5 days at 25°. Their development time is about 25 percent slower than cladocerans at the same temperature. Development time of the typically much smaller rotifers is around 1.5 days at 25° under otherwise optimal conditions (Allan 1976), explaining the dominance of rotifers (Basu and Pick 1996) and fast-growing protozoa in rapidly flushed lotic systems and reservoirs. The sequential washout

of the different zooplankton with increased flushing is expected to have an effect on the community structure of their phytoplankton and protozoan prey before these more rapidly growing prey organisms themselves become directly affected by flushing. Macrozooplankton encountered in rapidly flushed systems must be derived from embayments characterized by a longer WRT, or from adjacent wetlands (see Fig. 29–7). The first macrozooplankton to appear as the flushing rate declines are small rapidly reproducing, parthenogenic cladocerans (Sec. 23.3)—a large *Daphnia* species made its appearance in a cool Scottish lake when the WRT consistently exceeded about 18 days. The number of individuals became unrelated to flushing when the WRT exceeded about four weeks (Brook and Woodward 1956), reflecting the importance of factors other than WRT (e.g., predation and food) over longer time scales.

Washout of phytoplankton commences at lower WRT in turbid, deeply mixed, or cold reservoirs where conditions for photosynthesis are much less favorable than in shallow systems of low turbidity at summer temperatures. As the WRT declines within and among systems, the flushing rate progressively exceeds the growth rates of more of the species until the community becomes restricted to a few small species with very high growth potential (Fig. 21–12). Under favorable conditions of light, temperature, and nutrients, the minimum WRT required to allow the most rapidly growing species to reach a biomass maximum is around 5–7 days in the relatively few temperate

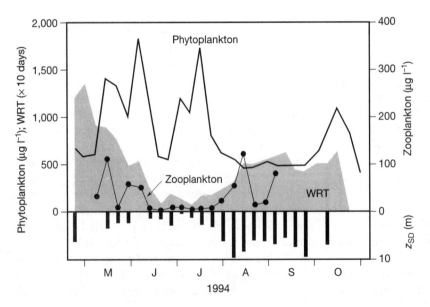

Figure 29–7 Seasonal changes in water residence time (WRT), total phytoplankton, and total macrozooplankton biomass and transparency in polymictic Glenmore Reservoir (Alberta, CA; LA = 4.6 km², $\bar{z}$ = 6.1 m). Note that the macrozooplankton abundance tends to be low during periods of most rapid flushing, whereas the phytoplankton biomass in the particular reservoir tends to be greatest during the periods of highest irradiance and lowest turbidity ($z_{eu} > z_{mix}$) and most rapid (~ 10 days) flushing (greatest nutrient loading, and wash-out of slower growing plankton). (*After Watson et al. 1996.*)

zone lakes and reservoirs that have been investigated (Fig. 29–7, Straškraba and Javornický 1973, and Søballe and Threlkeld 1985).

Field observations show that the effect of washout on the summer composition of phytoplankton communities becomes evident first when the WRT declines to below 60–100 days in temperate zone waterbodies (Kimmel et al. 1990). The wide range is attributable to differences in irradiance received at the water surface, the transparency, the thickness of the mixed layer subject to flushing, the z_{eu}:z_{mix} ratio (Sec. 10.11), and the temperature, which together determine the doubling time of phytoplankton (Sec. 21.11) and directly or indirectly affect the growth of their predators.

▲ 29.5   Reservoir Zonation: A Conceptual View

Mainstream reservoirs exhibit pronounced longitudinal zonation that is largely absent from the typically shallow and windswept storage reservoirs that do not occupy a well-defined river valley. The same zonation is also lacking in well-mixed tributary storage reservoirs exhibiting much longer WRTs and therefore enhanced possibilities for water mixing. Mainstream reservoirs without many arms (bays) exhibit pronounced longitudinal gradients in flow velocity, water residence time, suspended solids, mixed-layer light climate, nutrient concentration, and the composition and productivity of the biota. The three zones present in an idealized mainstream storage reservoir are: a *riverine zone*, a *transition zone*, and a *lacustrine zone* (Fig. 29–8).

Riverine Zone

In semiarid regions and crop-growing areas characterized by much bare soil, the shallow riverine zone is particularly turbid during periods of high river discharge when much inorganic matter is washed in from the land. Inorganic nutrient levels are also typically maximal in the riverine zone, but the light climate is unfavorable during high discharge and production is low. The zoobenthos and fish community of the riverine zone resemble that of the inflowing rivers.

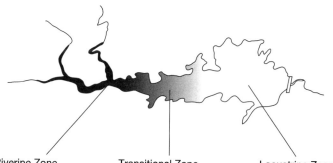

Riverine Zone	Transitional Zone	Lacustrine Zone
Narrow, channelized basin	Broader, deeper basin	Broad, deep, lakelike basin
Relatively high flow and rapid water flushing	Reduced flow and flushing	Little flow and slowest water flushing
High suspended solids; turbid; low light availability; $z_{eu} < z_{mix}$	Reduced suspended solids; less turbid; light availability increased	Relatively clear; more light at depth; $z_{eu} > z_{mix}$
Nutrient supply by advection; relatively high nutrients	Advective nutrient supply reduced	Nutrient supply by internal recycling; relatively low nutrients
Light limited planktonic primary production; relatively high benthic primary production	Planktonic primary production (m^{-3}) relatively high; relatively low benthic primary production	Nutrient-limited planktonic primary production; relatively high planktonic primary production (m^{-2}); relatively low benthic primary production
Algal cell losses primarily by sedimentation and advection	Algal cell losses by sedimentation and grazing	Algal cell losses often primarily by grazing
Very low macrozooplankton; growth rate < flushing rate	Variable macrozooplankton; growth rate < to > flushing rate	Intermediate macrozooplankton; growth rate > flushing rate
Organic matter supply primarily allochthonous; P < R	Organic matter supply allochthonous to autochthonous	Organic matter supply most autochthonous; P > to < R
Coarse sediments (low water content) and associated benthos	Fine inorganic and organic sediments and associated biota	Finest sediments (high water content) and associated biota
Dissolved oxygen high over sediments	Dissolved oxygen high when unstratified, low when stratified (thin hypolimnion)	Dissolved oxygen variable; thick hypolimnion but anoxic in tropics

Figure 29–8 Longitudinal zonation in environmental factors controlling light and nutrient availability for phytoplankton production, algal productivity, standing crop, organic matter supply, and trophic status in an idealized temperate reservoir. During low water phases the transitional zone may be lacustrine and the riverine zone a wetland, while during periods of rapid flushing the transitional zone may extend to the dam. (*Modified after Kimmel and Groeger 1984.*)

Transition Zone

In the deeper transition zone (Fig. 29–8), the sediment advection (transport) rate declines and transparency increases as the water residence time increases, resulting in enhanced particle sedimentation. Phytoplankton primary production may, as the result of favorable light and nutrient climates, be high in near-surface waters, and the depth-integrated primary production (mg C m^2 d^{-1}) may be maximal during periods of reduced turbidity. Temperature stratification develops if the water depth, the WRT, and temperature allow this. If stratified, the hypolimnion of the transition zone will be thin and vulnerable to deoxygenation because organic-matter sedimentation rates in organic-rich catchments tend to be substantial in this zone (Cole and Hannan 1990). High primary production may allow large populations of filter-feeding macrozooplankton during the growing season if the flushing rate does not exceed their growth rate (P. L. Pirozhnikov 1961, in Brandl 1973), and when fish predation is modest.

In turbid reservoirs, the relative and absolute ability of the different macrozooplankton species to cope with the negative effects of abundant inorganic particles on feeding help determine the zooplankton species composition (Fig. 29–9). But macrozooplankton loss rates imposed by visually feeding predators are reduced as well (Marzolf 1990). South African scientists have done most of the best work on links between inorganic turbidity and the biota, including the negative effect of suspended solids on macroinvertebrates and fish production (Allanson et al. 1990).

Lacustrine Zone

The lacustrine zone of mainstream reservoirs is stereotypically more transparent and lower in nutrient levels than the transition zone. Even so, depth-integrated rates of phytoplankton primary production (mg C m^{-2} t^{-1}) may be as high or higher than in the typically nutrient-richer but more turbid transition zone as the result of a more favorable *effective light climate* and z_{eu}:z_{mix} ratio (Eq. 10.6 and Fig. 10–12). Higher transparency further permits greater development of submerged macrophytes (Fig. 24–5). As in natural lakes, macrophytes are most abundant where the underwater slope is low (Fig. 24–6) and particularly where the lacustrine portions of reservoirs experience little drawdown (Sec. 29.6).

The conceptual picture presented above does not fit most tributary reservoirs at mid and low latitudes. In these, a relatively short period of rapid flushing or

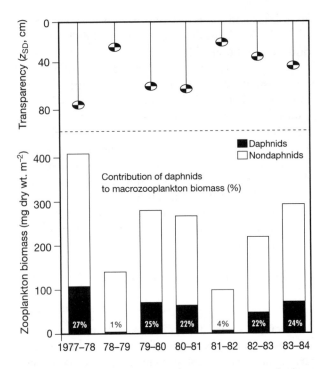

Figure 29–9 Weighted annual mean macrozooplankton standing stock in relation to the weighted mean transparency values in the le Roux Reservoir (ZA) between 1977 and 1984. Note the severe depression of the filter-feeding daphnids during years of high inorganic turbidity. (*After Hart 1988.*)

high turbidity in the riverine zone during periods of high runoff is followed by long periods of low discharge by inflowing rivers and much surface evaporation. The shallow riverine zone is then transformed into a wetland dominated by macrophytes and their associated biota (Chapters 24 and 25), and the reservoirs lack a clear transition zone. Slowly flushed tributary reservoirs are limnologically virtually indistinguishable from natural lakes of the same depth in the same region. Nor does the conceptual picture fit tributary reservoirs that are dendritic (see Fig. 29–3), with arms (basins) varying a great deal in turbidity, stratification, water residence time, and biota.

▲ 29.6   Drawdowns

Storage reservoirs experience an annual drawdown. The drawdown is primarily a response to seasonally changing needs for hydroelectric power or irrigation water. Drawdowns of 2–4 m or more are common in shallow and low-sloped reservoirs, leaving large areas of sediment exposed to desiccation.

Fluctuations in water level, whether natural or artificial, have dramatic impacts on emergent and submerged macrophytes, their benthos (Palomäki 1994, and Chapter 25), and fish (Chapter 26). In fact, drawdowns are used in relatively shallow United States reservoirs that have a large littoral zone for submerged macrophyte control (Chapter 24, and Cooke et al. 1993). In the temperate zone the duration and timing of the previous year's drawdown determines the macrophyte biomass and species composition the following year. Those macrophyte species capable of maintaining living roots during a drawdown and those growing from stored seeds (*seedbanks*) are able to cope and thrive (Nilsson and Keddy 1988).

The abundance and age distribution of fish is much affected by postspawning water level changes because most reservoir fish use the vegetated littoral zone for spawning, feeding, and hiding by the larvae and YOY. A drawdown of as little as 50 cm immediately following spawning will desiccate a large fraction of the eggs in higher latitude reservoirs, thereby eliminating whole year-classes of fish having only one annual spawning period (Duncan and Kubečka, 1995). Summer drawdowns are used in some United States reservoirs to increase the growth of largemouth bass by forcing YOY prey fish out of dense macrophyte beds where the large predators do not readily venture (O'Brian 1990).

Anthropogenic drawdowns in shallow reservoirs and natural drawdowns of shallow lakes and wetlands at low latitudes, resulting from high evapotranspiration losses (Fig. 5–9), expose large areas to desiccation, organic-matter oxidation and denitrification, freezing at higher latitudes, and growth of terrestrial vegetation. These impact sediment chemistry (E_h, pH, nutrients) and the sediment biota composition and activity. Considerable quantities of inorganic nutrients, organic matter, and soil bacteria are released upon subsequent reflooding, at a time when inputs of catchment-derived inorganic nutrients and organic matter is also maximal. The result is a typically positive relationship between bacterial abundance and water levels in shallow (former) USSR reservoirs (S. I. Kuznetsov et al. 1966, in Procházková et al. 1973).

▲   ## 29.7   Reservoir Aging and the Trophic Upsurge

The limnological characteristics of newly created reservoirs usually changes significantly for a number of years after the land is first flooded, with the extent of these changes a function of the land area flooded, water level changes, soil and vegetation characteristics, and climate. *Aging* is linked to oxidation of the terrestrial vegetation and flooded soil organic matter, as well as to shifts in the composition and abundance of the biota during its transition from a primarily riverine (lotic) to a lake (lentic) assemblage.

During the early years after it is filled, nutrient concentrations and productivity at all trophic levels are usually elevated. This is known as the **trophic upsurge**. The subsequent decline, **trophic depression**, may similarly last for a period of years to decades but will ultimately yield a more stable community composition and productivity, reflecting both the lower nutrient-supply rate and the new competition and predation relationships established between the species.

The trophic upsurge in the Klíčava reservoir (CZ) has been particularly well documented since it was filled between 1952 and 1955 (Fig. 29–10). Straškraba et al. (1993) recently reinterpreted the patterns observed based on more recent insights about the importance of top-down effects. It is now thought that the abundant macrozooplankton noted shortly after filling was attributable to low predation pressure at that time by zooplanktivorous fish, which in turn enabled a large macrozooplankton predation pressure to retard the phytoplankton upsurge. Where the reservoir fish fauna is composed only of poorly adapted lotic species, fish production and predation on macrozooplankton will remain low until lentic species invade or are introduced. Thus fish production in Sri Lankan reservoirs remained low until after the introduction of the well-adapted *Tilapia* (Sec. 26.6).

Tropical Reservoirs

Modest hypolimnetic dissolved oxygen (DO) declines during the trophic upsurge in the Klíčava reservoir to a minimum of 5 mgO_2 l^{-1} are attributed to preconstruction clearing of trees and shrubs and generally cool water temperatures. The observed relatively high DO concentrations contrast with a complete hypolimnetic, and often epilimnetic, anoxia in newly established reservoirs in the humid and well-vegetated tropics where the forest vegetation is left standing (Fig. 29–11, and Tundisi et al. 1993).

Anoxia in tropical reservoirs is abetted by the dendritic nature of most of the reservoirs (Fig. 29–3) and the resulting reduced fetch (Secs. 11.3 and 11.6), low DO solubility at high temperatures (Sec. 15.2), and relatively low wind speeds characterizing the conti-

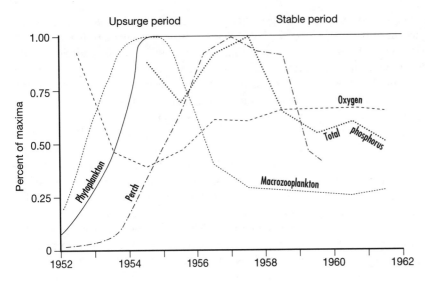

Figure 29–10 The aging process in Klíčava Reservoir (CZ) during and following filling in 1958. The maxima reached were: hypolimnetic, oxygen (13 mg l^{-1}), total phosphorus (25 μg l^{-1}), phytoplankton (10^6 org. l^{-1}), zooplankton (7 kg N ha^{-1}) and perch (1,500 ind. ha^{-1}). *(Modified after Straškraba et al. 1993.)*

nental portions of the tropics (Fig. 5–11). An near-surface oxycline and high concentrations of reduced sulfur just below may, in the early years, allow the development of a plate of photosynthetic sulfur-oxidizing bacteria utilizing the abundant reduced sulfur compounds released following bacterial degradation of sulfur-rich amino acids in the vegetation (Dumestre et al. 1999). Even after decay of the original flooded vegetation anoxic conditions continue to develop, at least in the hypolimnia (Sec. 15.2).

Most of the reservoirs created in ancient well-eroded tropical landscapes are of necessity relatively shallow and therefore polymictic (Sec. 11.2). Periodic destratifications (overturns) lead to frequent fish kills as the result of mixing highly anoxic hypolimnetic water, which often contains large quantities of toxic H_2S gas and reduced (oxygen-utilizing) compounds (CH_4, FeS, Mn, NH_4^+, and Sec. 15.4) into the epilim-

nia. The outflowing rivers are anoxic after the periodic overturns with a long-term negative effects on their species composition and system functioning. Many newly established tropical and subtropical reservoirs are not only largely anoxic but become partially or totally covered with large free-floating macrophyte species able to benefit from the nutrients released during the trophic upsurge (e.g., water hyacinth *Eichhornia crassipes*, and the false water fern *Salvinia molesta*; Fig. 24–1).

Volga River Reservoirs

Some of the best-studied man-made lakes are the large but shallow ($\bar{z}$ = ~7 m) series of *cascade reservoirs* in the Volga River (RU). The widespread anoxic conditions commonly observed in tropical hypolimnia following filling did not develop because a modest amount of

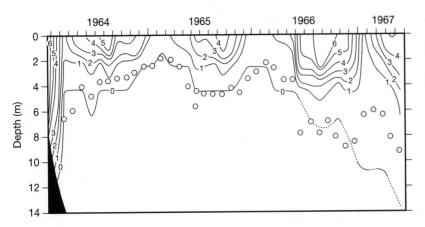

Figure 29–11 Brokopondo Reservoir, SR (LA = 362 km^2, z_{max} = 19 m). Dissolved oxygen distribution (mg l^{-1}) over depth and time during the first four years after filling. Open circles show the depth of the euphotic zone, which largely coincides with the 0 mg l^{-1} contour. *(After van der Heide 1982.)*

and due to generally rapid flushing and cooler water temperatures. The sediments remained generally well oxygenated, making the development of an abundant macrobenthic community possible during the trophic upsurge. The macrobenthos permitted a particularly productive fishery at the time. The shallow reservoirs developed large submerged macrophyte beds in the extensive littoral and wetland zones created (Straškraba et al. 1993). The reservoirs did not develop the extensive floating rafts of water hyacinth or false water fern so common at lower latitudes. However, rich soils and agricultural activity have guaranteed eutrophic conditions even after the end of the upsurge period (summer z_{SD} = ~1.5 m, TN = ~1 mg l^{-1}, TP = ~110 μg l^{-1}; Litvinov and Mineeva 1994).

Boreal Reservoirs

The last group of reservoirs to be mentioned are the large but generally shallow hydroelectric reservoirs constructed in recent decades in the low relief boreal forest regions of North America and Eurasia. The trophic upsurge is small in these regions of thin nutrient-poor soils overlying hard igneous rock. Windswept cool waters have maintained high DO concentrations in summer, but permit localized deoxygenation under the winter ice cover in shallow areas where bogs with a thick layer of organic matter are flooded.

An unpleasant and unexpected side effect of construction has been greatly elevated levels of mercury in long-lived piscivorous fish, such as lake trout and pike, during the first several decades (Sec. 28.9). Work on some Canadian reservoirs showed the onset of a decline in the mercury content of their principal prey, the shorter-lived zooplanktivorous lake whitefish, seven years after flooding; a decline that is attributed to gradual depletion of the available soil mercury but not yet seen in the piscivorous fish (Chevalier et al. 1997). Even so, elevated sediment mercury levels in natural lakes nearby indicate that mercury levels are likely to remain high, especially in piscivorous fish at the top of the food chain.

29.8   Large Reservoirs and Their Impacts

The construction of big dams has been a mixed benefit, particularly at low latitudes in the economically less-developed portions of the world. The benefits of hydroelectric power are large but go primarily to urban areas far downstream from the dams. Between 30 and 40 percent of the world's 271 million irrigated hectares—which play a major role in global food production—rely on dams for water, but their benefits come at a large environmental cost that has received disproportionately little attention. The cost includes the disruption of river flow, the loss of vast areas of wetlands and their biota following the termination of the large seasonal floods, and the desiccation of dryland rivers downstream as a result of water diversion for irrigation. Furthermore, the dams interfere with the migration of fish and invertebrates, causing reduced biodiversity, while reduced discharge to the oceans results in salinization of coastal wetlands. Other losers have been politically weak indigenous people—dependent on the rivers, wetlands, and land flooded by reservoirs, for hunting, fishing, farming, livestock raising and for maintaining their culture. Limnologists working for consulting firms contribute their technical expertise to the *Environmental Impact Assessments* that are required before international lenders provide the necessary funds for construction. Two principal difficulties bedevil reservoir construction: The benefits and costs tend to go to different groups of people, and the typically unanticipated but major environmental costs have received far too little attention (Straškraba et al. 1993). Whether a growing international pressure for the development of a greater number of smaller, individually less intrusive dams is on balance less environmentally (and socially) destructive must be region- and site-specific. Whatever the case, limnologists with a broad environmental perspective, and an ability to collaborate and argue the impacts with economists, sociologists, and engineers, will be in demand.

Highlights

- Reservoirs are variously constructed for irrigation, electricity generation, flood control, water supply to urban areas, improved navigation, fish culture, recreation, or some combination of the above.
- Most reservoirs are small and built for irrigation by damming rivers in low rainfall (low runoff) regions.
- Reservoirs and natural lakes have much in common, but reservoirs tend to have larger drainage basins than natural lakes of similar size in the same region. They further differ in that reservoirs are usually recently constructed and have their greatest depth at the dam.

- Large reservoirs are of three basic types. Mainstream or run-of-the-river reservoirs are constructed to generate electricity and are characterized by rapid flushing, little or no stratification, no drawdown, water release via a deep outlet, and are most riverlike. Tributary storage reservoirs are built on smaller rivers and are used principally for irrigation and flood control. They experience slower flushing, may be stratified if deep enough, experience periodic drawdowns, have a surface outlet, and most resemble natural lakes in the area. Mainstream storage reservoirs have an intermediate position, becoming more lotic during periods of high runoff and more lentic at other times.
- ▲ Mainstream storage reservoirs, lying in well-defined river valleys, exhibit pronounced longitudinal zonation with gradients in flow velocity, water residence time, suspended solids, mixed layer light climate, and the biota.
- ▲ Flushing rates greater than the doubling time of planktonic organisms results in washout.
- Rivers and their biota are greatly affected by the insertion of a lentic system which follows dam construction, and dams also interfere with migrations. Modifications in the timing and volume of the discharge regime, plus trapping materials behind dams affects the biota downstream. However, the single largest threat to rivers, their biota, and downstream wetlands stem from the large-scale water diversion for agriculture, resulting in a greatly reduced discharge or zero discharge in dryland rivers and the loss of downstream wetlands, and their biota.
- ▲ Exposure and desiccation of the littoral zone following drawdowns has pronounced effects on the biota.
- The construction of big dams has been a mixed blessing, particularly in the economically less-developed portions of the world.

Appendix 1 **International Organization for Standardization (ISO, Geneva) Country Codes mentioned in the text.**

Code	Country	Code	Country
AF	Afghanistan	KZ	Kazakhstan
Al	Albania	LA	Laos
AM	Armenia	LK	Sri Lanka
AO	Angola	LT	Lithuania
AQ	Antarctica	LU	Luxembourg
AR	Argentina	MD	Moldova
AT	Austria	MG	Madagascar
AU	Australia	ML	Mali
AZ	Azerbaijan	MM	Myanmar (Burma)
BD	Bangladesh	MN	Mongolia
BE	Belgium	MW	Malawi
BF	Burkina Faso	MX	Mexico
BG	Bulgaria	MY	Malaysia
BI	Burundi	MZ	Mozambique
BJ	Benin	NA	Namibia
BO	Bolivia	NE	Niger
BR	Brazil	NG	Nigeria
BT	Bhutan	NI	Nicaragua
BW	Botswana	NL	Netherlands
BY	Belarus	NO	Norway
CA	Canada	NP	Nepal
CH	Switzerland	NZ	New Zealand
CI	Cote d'Ivoire	PA	Panama
CL	Chile	PE	Peru
CM	Cameroon	PH	Philippines
CN	China	PK	Pakistan
CO	Colombia	PL	Poland
CU	Cuba	PY	Paraguay
CZ	Czech Republic	RO	Romania
DE	Germany	RU	Russian Federation
DK	Denmark	RW	Rwanda
EG	Egypt	SD	Sudan
ES	Spain	SK	Slovakia
ET	Ethiopia	SL	Sierra Leone
FI	Finland	SR	Suriname
FR	France	SE	Sweden
GB	United Kingdom	SY	Syrian Arab Republic
GH	Ghana	TD	Chad
GN	Guinea	Th	Thailand
GT	Guatemala	TJ	Tajikistan
HK (CN)	Hong Kong	TM	Turkmenistan
HR	Croatia	TR	Turkey
HU	Hungary	TZ	Tanzania
ID	Indonesia	UA	Ukraine
IE	Ireland	UG	Uganda
IL	Israel	US	United States
IN	India	UY	Uruguay
IQ	Iraq	UZ	Uzbekistan
IR	Iran	VE	Venezuela
IS	Iceland	VN	Vietnam
IT	Italy	YU	Yugoslavia
JO	Jordan	ZA	South Africa
JP	Japan	ZM	Zambia
KE	Kenya	ZR	Zaire
KG	Kyrgyzstan	ZW	Zimbabwe
KH	Cambodia		

Source: IATA Airline Coding Directory.

Appendix 2   **Conversion factors: milligrams per liter × F₁ = milliequivalents per liter, milligrams per liter × F₂ = millimoles per liter; for selected elements and reported species.**

Element and Reported Species	F_1	F_2
Aluminium (Al^{3+})	0.11119	0.03715
Ammonium (NH_4^+)	0.05544	0.05544
Arsenic (As)	—	0.01334
Barium (Ba^{2+})	0.01456	0.00728
Bicarbonate (HCO_3^-)	0.01639	0.01639
Boron (B)	—	0.09250
Cadmium (Cd^{2+})	0.01779	0.00890
Calcium (Ca^{2+})	0.04990	0.02495
Carbonate (CO_3^{2-})	0.03333	0.01666
Cesium (Cs^+)	0.00752	0.00752
Chloride (Cl^-)	0.02821	0.02821
Chromium (Cr)	—	0.01923
Cobalt (Co^{2+})	0.03394	0.01697
Copper (Cu^{2+})	0.03147	0.01574
Fluoride (F^-)	0.05264	0.05264
Hydrogen (H^+)	0.99216	0.99216
Hydroxide (OH^-)	0.05880	0.05880
Iron (Fe^{2+})	0.03581	0.01791
Iron (Fe^{3+})	0.05372	0.01791
Lead (Pb^{2+})	0.00965	0.00483
Magnesium (Mg^{2+})	0.08229	0.04114
Manganese (Mn^{2+})	0.03640	0.01820
Mercury (Hg)	—	0.00499
Molybdenum (Mo)	—	0.01042
Nickel (Ni)	—	0.01704
Nitrate (NO_3^-)	0.01613	0.01613
Nitrite (NO_2^-)	0.02174	0.02174
Phosphate (PO_4^{3-})	0.03159	0.01053
Potassium (K^+)	0.02558	0.02558
Selenium (Se)	—	0.01266
Silica (SiO_2)	—	0.01664
Sodium (Na^+)	0.04350	0.04350
Strontium (Sr^{2+})	0.02283	0.01141
Sulfate (SO_4^{2-})	0.02082	0.01041
Sulfide (S^{2-})	0.06238	0.03119
Zinc (Zn^{2+})	0.03059	0.01530

Source: After Hem 1985.

Bibliography

Agustí, S., 1991. Allometric Scaling of Light Absorption and Scattering by Phytoplankton Cells. *Can. J. Fish. Aquat. Sci.* 48: 763–767.

Agustí, S., and J. Kalff, 1989. The Influence of Growth Conditions on the Size Dependence of Maximal Algal Density and Biomass. *Limnol. Oceanogr.* 34: 1104–1108.

Agustí, S., C.M. Duarte, and D.E. Canfield, Jr., 1990. Phytoplankton Abundance in Florida Lakes: Evidence for the Lack of Nutrient Limitation. *Limnol. Oceanogr.* 35:181–188.

Ahl, T., 1980. Variability in Ionic Composition of Swedish Lakes and Rivers. *Arch. Hydrobiologia* 89: 5–16.

Ahlgren, G., 1970. Limnological Studies of Lake Norrviken, a Eutrophicated Swedish Lake. Phytoplankton and its Production. *Schweiz. Z. Hydrol.* 32: 353–396.

Ahlgren, G., 1983. Comparison of Methods for Estimation of Phytoplankton Carbon. *Arch. Hydrobiologia* 98: 489–508.

Ahlgren, G., 1988. Comparison of Algal C^{14} Uptake and Growth Rate In Situ and In Vitro. *Verh. Int. Ver. Limnol.* 23: 1898–1907.

Ahlgren, I., 1980. A Dilution Model Applied to a System of Shallow Eutrophic Lakes After Diversion of Sewage Effluents. *Arch. Hydrobiologia* 89: 17–32.

Ahlgren, I., 1988. *Nutrient Dynamics and Trophic State Response of Two Eutrophicated Lakes After Reduced Nutrient Loading.* Pp. 79–97. In G. Balvay (ed.) Eutrophication and Lake Restoration. Water Quality and Biological Impacts. Limnological Symposium, Thonon-les-Bains.

Ahlgren, I., T. Frisk and L. Kamp-Nielsen, 1988. Empirical and Theoretical Models of Phosphorus Loading, Retention and Concentration vs. Lake Trophic State. *Hydrobiologia* 170: 285–303.

Aizaki, M., A. Otsuki, T. Fukushima, M. Hosomi, and K. Muraoka, 1981. Application of Carlson's Trophic State Index to Japanese Lakes and Relationships Between the Index and Other Parameters. *Verh. Int. Ver. Limnol.* 21: 675–681.

Alabaster, J.S., and R. Lloyd (ed), 1982. *Water Quality Criteria for Freshwater Fish, 2nd ed.* London: Butterworths.

Aladin, N.V., I.S. Plotnikov, and A.A. Filippov, 1993. Alteration of the Aral Sea Ecosystem by Human Impact. *Hydrobiologia J.* 29: 22–31.

Alimov, A.F., 1982. Productivity of Invertebrate Macrobenthos Communities in Continental Waters of the USSR (A Review). *Hydrobiologia J.* 18: 5–15.

Allan, J.D., 1976. Life History Patterns in Zooplankton. *Amer. Naturalist.* 110: 165–180.

Allan, J.D., 1995. *Stream Ecology: Structure and Function of Running Waters.* London: Chapman and Hall.

Allan, R.J., 1986. *The Role of Particulate Matter in the Fate of Contaminants in Aquatic Ecosystems.* Inland Waters Dir., Nat. Res. Inst. Canada Centre for Inland Waters. *Env. Can. Sci. Ser. #142:* Ontario.

Allanson, B.R., R.C. Hart, and J.H. O'Keeffe, 1990. *Inland Waters of Southern Africa: An Ecological Perspective.* Dordrecht: Kluwer Academic.

Allen, S.E., A. Carlisle, E.J. White, and C.C. Evans, 1967. The Plant Nutrient Content of Rainwater. *J. Ecology* 56: 497–504.

Almer B., W. Dickson, C. Ekström, and E. Hörnström, 1978. Sulfur Pollution and the Aquatic Ecosystem. Pp. 271–311. In J.O. Nriagu (ed.), Sulfur in the Environment. Part II: Ecological Impacts. New York: John Wiley and Sons.

Andersen, J.M., 1982. Effect of Nitrate Concentration in Lake Water on Phosphate Release from the Sediment. *Water Res.* 16: 1119–1126.

Anderson, M.R., and J. Kalff, 1986. Nutrient Limitation of *Myriophyllum spicatum* Growth In Situ. *Freshwater Biol.* 16: 735–743.

Anderson, N.J., 1993. Natural vs. Anthropogenic Change in Lakes: The Role of the Sediment Record. *Trends in Ecol. and Evol.* 8: 356–361.

Andersson, G., H. Berggren, G. Cronberg, and C. Gelin, 1978. Effects of Planktivorous and Benthivorous Fish on Organisms and Water Chemistry in Eutrophic Lakes. *Hydrobiologia* 59: 9–15.

Andradóttir, H., and Nepf, H.M. 2000. Thermal Mediation by Littoral Wetlands and Impact on Lake Instrusion Depth. *Water Resour. Res.,* 36:725–735

Andrew, T.E., 1983. The Estimation of Secondary Production in a Natural Population of *Daphnia hyalina* (Leydig) Using Alternative Methods of Computation. *Hydrobiologia* 107: 3–18.

Anon. 1991 *Synopsis: Toxic Chemicals in the Great Lakes and Associated Effects* Min. of Supply and Services Canada, Ottawa.

Anon. 1998 *Water Power and Dam Construction, 1997 Yearbook.* Wilmington: Wilmington Business Publ.

Appenzeller, A.R., and W.C. Leggett, 1995. An Evaluation of Light-mediated Vertical Migration of Fish Based on Hydroacoustic Analysis of the Diel Vertical Movements of Rainbow Smelt (*Osmerus mordax*). *Can. J. Fish. Aquat. Sci.* 52: 504–511.

Arai, T., 1981. Climatic and Geomorphological Influences on Lake Temperature. *Verh. Int. Ver. Limnol.* 21: 130–134.

Archibald, C.G.M., and M.S. Muller, 1987. Physical and Chemical Characteristics of Bulk Precipitation in the Richards Bay Area, with Special Reference to Lake Mzingazi. *S. African J. of Sci.* 83: 700–704.

Arndt, H., 1993. Rotifers as Predators on Components of the Microbial Web (Bacteria, Heterotrophic Flagellates, Ciliates)—A Review. *Hydrobiologia* 255/256: 231–246.

Arthington, A.H., and D.R. Blühdorn, 1996. *The Effects of Species Interactions Resulting from Aquaculture Operations.* Pp. 114–139. In D.J. Baird, M.C.M. Beveridge, L.A. Kelly, and J.F. Muir (eds.), Aquaculture and Water Resource Management. Oxford: Blackwell Science.

Arts, M.T., R.D. Robarts, F. Kasai, M.J. Waiser, V.P. Tumbler, A.J. Plante, H. Rai, and H.J. de Lange, 2000. The Attenuation of Ultraviolet Radiation in High Dissolved Organic Carbon Waters of Wetlands and Lakes on the Northern Great Plains. *Limnol. Oceangr.* 45: 290–299.

Ashton, P.J., 1979. Nitrogen Fixation in a Nitrogen-limited Impoundment. *J. Water Poll. Contr. Fed.* 51: 570–579.

Assaf, G., R. Gerard, and A.L. Gordon, 1971. Some Mechanisms of Oceanic Mixing Revealed in Aerial Photographs, *J. Geophys. Res.* 76:6550–6572.

Azam, F., T. Fenchel, J.G. Field, J.S. Gray, L.A. Meyer-Reil, and F. Thingstad, 1983. The Ecological Role of Water-column Microbes in the Sea. *Mar. Ecol. Prog. Ser.* 10: 257–263.

Babin, J., and E.E. Prepas, 1985. Modelling Winter Oxygen Depletion Rates in Ice-covered Temperate Zone Lakes in Canada. *Can. J. Fish. Aquat. Sci.* 42: 239–249.

Bachmann R.W., B.L. Jones, D.D. Fox, M. Hoyer, L.A. Bull, and D.E. Canfield, Jr., 1996. Relations Between Trophic State Indicators and Fish in Florida (U.S.A.) Lakes. *Can. J. Fish. Aquat. Sci.* 53: 842–855.

Baigún, C., and M.C. Marinone, 1995. Cold-temperate Lakes of South America: Do They Fit Northern Hemisphere Models? *Arch. Hydrobiologia* 135: 23–51.

Baines, S.B., and M.L. Pace, 1991. The Production of Dissolved Organic Matter by Phytoplankton and its Importance to Bacteria: Patterns Across Marine and Freshwater Systems. *Limnol. Oceanogr.* 36: 1078–1090.

Baines, S.B., and M.L. Pace, 1994. Relationships Between Suspended Particulate Matter and Sinking Flux Along a Trophic Gradient and Implications for the Fate of Planktonic Primary Production. *Can. J. Fish. Aquat. Sci.* 51: 25–36.

Baker, L.A., A.T. Herlihy, P.R. Kaufmann, and J.M. Eilers, 1991. Acidic Lakes and Streams in the United States: The Role of Acidic Deposition. *Science* 252: 1151–1154.

Balistrieri, L.S., J.W. Murray, and B. Paul, 1992. The Cycling of Iron and Manganese in the Water Column of Lake Sammamish, Washington. *Limnol. Oceanogr.* 37: 510–528.

Ballinger, B.R., and J.A. Thornton., 1982. *The Hydrology of the Lake McIlwaine Catchment.* Pp. 34–41. In J.A. Thornton (ed.), Lake McIlwaine: The Eutrophication and Recovery of a Tropical African Man-made Lake. Monogr. Biol. 49. The Hague: W. Junk Publ.

Balvay, G., M. Gawler, J.P. Pelletier, 1990. *Lake Trophic Status and the Development of the Clear-water Phase in Lake Geneva.* Pp. 580–591. In M.M. Tilzer, and C. Serruya, Large Lakes: Ecological Structure and Function. Berlin: Springer Verlag.

Banens, R.J., 1990. Occurrence of Hypolimnetic Blooms of the Purple Sulfur Bacterium, *Thiopedia rosea* and the Green Sulfur Bacterium, *Chlorobium limicola* in an Australian Reservoir. *Aust. J. Mar. Freshwater. Res.* 41: 223–235.

Barbiero, R.P., and E.B. Welch, 1992. Contribution of Benthic Blue-green Algal Recruitment to Lake Populations and Phosphorus Translocation. *Freshw. Biol.* 27: 249–260.

Barbour, C.D., and J.H. Brown, 1974. Fish Species Diversity in Lakes. *Amer. Nat.* 108: 473–489.

Barica, J., 1975. Summerkill Risk in Prairie Ponds and Possibilities of its Prediction. *J. Fish. Res. Board Can.* 32: 1283–1288.

Barko, J.W., and R.M. Smart, 1981. Comparative Influences of Light and Temperature on the Growth and Metabolism of Selected Submersed Freshwater Macrophytes. *Ecol. Monogr.* 51: 219–213.

Barko, J.W., M.S. Adams, and N.L. Clesceri, 1986. Environmental Factors and Their Consideration in the Management of Submersed Aquatic Vegetation: A Review. *J. Aquat. Plant Monogr.* 24: 1–10.

Bärlocher, F. (ed.), 1992. *The Ecology of Aquatic Hyphomycetes. Ecological Studies #94.* Heidelberg: Springer-Verlag.

Barton, D.R., W.D. Taylor, and R.M. Biette, 1985. Dimensions of Riparian Buffer Strips Required to Maintain Trout Habitat in Southern Ontario Streams. *N. Amer. J. Fish. Man.* 5: 364–378.

Basu, B.K., and F.R. Pick, 1995. Longitudinal and Seasonal Development of Planktonic Chlorophyll-*a* in the Rideau River, Ontario. *Can. J. Fish. Aquat. Sci.* 52: 804–815.

Basu, B.K., and F.R. Pick, 1996. Factors Regulating Phytoplankton and Zooplankton Biomass in Temperate Rivers. *Limnol. Oceanogr.* 41: 1572–1577.

Basu, B.K., and F.R. Pick, 1997. Factors Related to Heterotrophic Bacterial and Flagellate Abundance in Temperate Rivers. *Aquat. Microb. Ecol.* 12: 123–129.

Battarbee, R.W., 1991. *Recent paleolimnology and Diatom-based Environmental Reconstruction.* Pp. 129–174. In L.C. Shane, K. and E. J. Cushing (eds.), Quaternary Landscapes. Minneapolis: University of Minnesota Press.

Bayley, P.B., 1988. Factors Affecting Growth Rate of Young Tropical Floodplain Fishes: Seasonality and Density-dependence. *Env. Biol. Fishes* 21: 127–142.

Beadle, L.C., 1981. *The Inland Waters of Tropical Africa: An Introduction to Tropical Limnology,* 2nd ed. New York: Longman Inc.

Beaumont, P., 1975. *Hydrology.* Pp. 1–38. In B.A. Whitton (ed.), 1975. River Ecology: Studies in Ecology, vol. 2. Los Angeles: Univ. of Cal. Berkley Press.

Beckel, A.L., 1987. Breaking New Waters: A Century of Limnology at the University of Wisconsin. *Trans. Wisc. Acad. of Sci., Arts, and Lett.* Special Edition.

Bedard, C., and R. Knowles, 1991. Hypolimnetic O_2 Consumption, Denitrification, and Methanogenesis in a Thermally Stratified Lake. *Can. J. Fish. Aquat. Sci.* 48: 1048–1054.

Bedford, B.L., 1996. The Need to Define Hydrology Equivalence at the Landscape Scale for Freshwater Wetland Mitigation. *Ecology Appl.* 6: 57–68.

Bell, S.G., and G.A. Codd, 1994. Cyanobacterial Toxins and Human Health. *Rev. Med. Microbiol.* 5: 256–264.

Benke, A.C., 1984. *Secondary Production of Aquatic Insects.* Pp. 289–322. In V.H. Resh, and D.M. Rosenberg (eds.), The Ecology of Aquatic Insects. New York: Praeger Publ.

Benke, A.C., 1992. Concepts and Patterns of Invertebrate Production in Running Waters. *Verh. Int. Ver. Limnol.* 25: 15–38.

Benndorf, J., 1990. Conditions for Effective Biomanipulation; Conclusions Derived from Whole-lake Experiments in Europe. *Hydrobiologia* 200/201: 187–203.

Bennett, E.B., 1986. The Nitrifying of Lake Superior. *Ambio* 15: 272–275.

Benoy, G.A., and J. Kalff, 1999. Sediment Accumulation and Pb Burdens in Submerged Macrophyte Beds. *Limnol. Oceanogr.* 44: 1081–1090.

Benson, B.B., and D. Krause, 1980. The Concentration and Isotopic Fractionation of Gases Dissolved in Freshwater in Equilibrium with the Atmosphere, 1. Oxygen. *Limnol. and Oceanogr.* 25: 662–671.

Berg, K., 1951. The Content of Limnology Demonstrated by F.A. Forel and August Thienemann on the Shore of Lake Geneva. *Verh. Int. Ver. Limnol.* 11: 41–57.

Bergh, Ø., K.Y. Børsheim, G. Bratbak, and M. Heldal, 1989. High Abundance of Viruses Found in Aquatic Environments. *Nature* 340: 467–468.

Bergmann, M.A., and H.E. Welch, 1990. Nitrogen Fixation by Epilithic Periphyton in Small Arctic Lakes in Response to Experimental Nitrogen and Phosphorus Fertilization. *Can. J. Fish. Aquat. Sci.* 47: 1545–1550.

Bergström, A.K., and M. Jansson, 2000. Bacterioplankton in Humic Lake Örträsket in Relation to Input of Bacterial Cells and Input of Allochthonous Organic Carbon. *Microb. Ecology* 39:101–115.

Berkman, P.A., M.A. Haltuch, and E. Tichich, 1998. Zebra Mussels Invade Lake Erie Muds. *Nature* 393: 27–28.

Berner, E.K., and R.A. Berner, 1987. *The Global Water Cycle: Geochemistry and Environment.* Englewood Cliffs: Prentice Hall.

Berninger, U.G., D.A. Caron, and R.W. Sanders, 1992. Mixotrophic Algae in Three Ice-covered Lakes of the Pocono Mountains, USA. *Freshw. Biol.* 28: 263–272.

Best, M.D., and K.E. Mantai, 1978. Growth of *Myriophyllum*: Sediment or Lake Water as the Source of Nitrogen and Phosphorus. *Ecology* 59: 1075–1080.

Beutel, M.W., and A.J. Horne, 1999. A Review of the Effects of Hypolimnetic Oxygenation on Lake and Reservoir Water Quality. *Lake and Reserv. Man.* 15: 285–279.

Bierhuizen, J.F.H., and E.E. Prepas, 1985. Relationship Between Nutrients, Dominant Ions, and Phytoplankton Standing Crop in Prairie Saline Lakes. *Can. J. Fish. Aquat. Sci.* 42: 1588–1594.

Bierman, Jr., V.J., D.M. Dolan, R. Kasprzyk, and J.L. Clark, 1983. Retrospective Analysis of the Response of Saginaw Bay, Lake Huron, to Reductions in Phosphorus Loadings. *Environ. Sci. Technol.* 18: 23–31.

Biggs, W., 1986. *Radiation Measurement*. Pp. 3–20. In W.G. Genser (ed.), Advanced Agriculture Instrumentation, Design and Use. Dordrecht: M. Nijhoff Publ.

Bilby, R.E., 1981. Role of Organic Debris Dams in Regulating the Export of Dissolved and Particulate Matter from a Forested Watershed. *Ecology* 62: 1234–1243.

Bird, D.F., and J. Kalff, 1984. Empirical Relationships Between Bacterial Abundance and Chlorophyll Concentration in Fresh and Marine Waters. *Can. J. Fish. Aquat. Sci.* 41: 1015–1023.

Bird, D.F., and J. Kalff, 1986. Bacterial Grazing by Planktonic Lake Algae. *Science* 231: 493–495.

Bird, D.F., and J. Kalff, 1987. Algal Phagotrophy: Regulating Factors and Importance Relative to Photosynthesis in *Dinobryon* (Chrysophyceae). *Limnol. Oceanogr.* 32: 277–284.

Bird, D.F., and J. Kalff, 1993. Protozoan Grazing and the Size–Activity Structure in Limnetic Bacterial Communities. *Can. J. Fish. Aquat. Sci.* 50: 370–380.

Birge, E.A., and C. Juday, 1929. Transmission of Solar Radiation by Waters of the Inland Lakes. *Trans. Wis. Acad. Sci. Arts Lett.* 24: 509–580.

Birge, E.A., and C. Juday, 1932. Solar Radiation and Inland Lakes, Fourth Report. Observations on 1931. *Trans. Wis. Acad. Sci. Arts Lett.* 27: 523–562.

Bisson, P.A., T.P. Quinn, G.H. Reeves, and S.V. Gregory, 1992. *Best Management Practices, Cumulative Effects, and Long-term Trends in Fish Abundance in Pacific Northwest River Systems*. Pp. 189–233. In R.J. Naiman (ed.), Watershed Management: Balancing Sustainability and Environmental Change. New York: Springer-Verlag.

Björk, S., S. Fleischer, W. Granéli, L. Bengtsson, L. Leonardson, G. Andersson, and W. Ripl, 1979. Lake Management: Studies and Results at the Institute of Limnology in Lund. *Arch. Hydrobiol. Beih. Ergebn. Limnol.* 13: 31–55.

Black, R.W., 1980. The Nature and Causes of Cyclomorphosis in a Species of the *Bosmina longirostris* Complex. *Ecology* 61: 1122–1132.

Blais, J.M., and J. Kalff, 1993. Atmospheric Loading of Zn, Cu, Ni, Cr and Pb to Lake Sediments: The Role of Catchment, Lake Morphometry, and Physio-chemical Properties of the Elements. *Biogeochem.* 23: 1–22.

Blais, J.M., and J. Kalff, 1995. The Influence of Lake Morphometry on Sediment Focusing. *Limnol. Oceanogr.* 40: 582–588.

Blais, J.M., D.W. Schindler, D.C.G. Muir, L.E. Kimpe, D.B. Donald, and B. Rosenberg, 1998. Accumulation of Persistent Organochlorine Compounds in Mountains of Western Canada. *Nature* 395: 685–688.

Blais, J.M., J. Kalff, R.J. Cornett, and R.D. Evans, 1995. Evaluation of ^{210}Pb Dating in Lake Sediments Using Stable Pb, *Ambrosia* pollen and ^{137}Cs. *J. Paleolimnol.* 13: 169–178.

Blaxter, J.H.S., 1970. Fishes. Pp. 213–320. In O. Kinne (ed.), Marine Ecology: A Comprehensive, Integrated Treatise on Life in Oceans and Coastal Waters, Vol.1. London: John Wiley and Sons.

Blindow, I., 1992. Long- and Short-term Dynamics of Submerged Macrophytes in Two Shallow Eutrophic Lakes. *Freshw. Biol.* 28: 15–27.

Bloesch, J., 1995. Mechanisms, Measurement and Importance of Sediment Resuspension in Lakes. *Mar. Freshwat. Res.* 46: 295–304.

Bloesch, J., 1996. Towards a New Generation of Sediment Traps and a Better Measurement/Understanding of Settling Particle Flux in Lakes and Oceans: A Hydrodynamical Protocol. *Aquat. Sci.* 58: 283–296.

Bloesch, J., P. Stadelmann, and H. Bührer, 1977. Primary Production, Mineralization, and Sedimentation in the Euphotic Zone of Two Swiss Lakes. *Limnol. Oceanogr.* 22: 511–526.

Bloesch, J., and N.M. Burns, 1980. A Critical Review of Sedimentation Trap Techniques. *Schweiz. Z. Hydrol.* 42: 15–55.

Bloesch, J., and M. Sturm, 1986. *Settling Flux and Sinking Velocities of Particulate Phosphorus (PP) and Particulate Organic Carbon (POC) in Lake Zug, Switzerland*. Pp. 481–490. In P.G. Sly (ed), Sediments and Water Interactions. New York: Springer-Verlag.

Bloesch, J., and U. Uehlinger, 1986. Horizontal Sedimentation Differences in a Eutrophic Swiss Lake. *Limnol. Oceanogr.*, 31: 1094–1109.

Bloesch, J., P. Bossard, H. Bührer, H.R. Bürgi, and U. Uehlinger, 1988. Can Results from Limnocorral Experiments be Transferred to in Situ Conditions? *Hydrobiologia*, 159: 297–308.

Bloesch, J., and H.R. Bürgi, 1989. Changes in Phytoplankton and Zooplankton Biomass and Composition Reflected by Sedimentation. *Limnol. Oceanogr.* 34: 1048–1061.

Bloesch, J., and U. Uehlinger, 1990. Epilimnetic Carbon Flux and Turnover of Different Particle Size Classes in Oligo-mesotrophic Lake Lucerne, Switzerland. *Arch. Hydrobiologia* 118: 403–419.

Blom, G., E.H.S. Van Duin, and J.E. Vermaat, 1994. *Factors Contributing to Light Attenuation in Lake Veluwe*. Pp. 158–174. In W. Van Vierssen, M. Hootsmans, and J. Vermaat (eds.), Lake Veluwe, a Macrophyte-dominated System Under Eutrophication Stress. Dordrecht: Kluwer Acad. Publ.

Bloom, N., 1989. Determination of Picogram Levels of Methylmercury by Aqueous Phase Ethylation, Followed by Cryogenic Gas Chromatography with Cold Vapour Atomic Fluorescence Detection. *Can. J. Fish Aquat. Sci.* 46: 1131–1140.

Blum, D.J.W., and R.E. Speece, 1990. Determining Chemical Toxicity to Aquatic Species. *Environ. Sci. Technol.* 24: 284–293.

Blumenshine, S.C., Y. Vadeboncoeur, D.M. Lodge, K.L. Cottingham, and S.E. Knight, 1997. Benthic–Pelagic Links: Responses of Benthos to Water-column Nutrient Enrichment. *J. North Amer. Benthol. Soc.* 16: 466–479.

Bocking, S., 1990. Stephen Forbes, Jacob Reighard, and the Emergence of Aquatic Ecology in the Great Lakes Region. *J. Hist. Biol.* 23: 461–498.

Bodaly, R.A., R.E. Hecky, and R.J.P. Fudge, 1984. Increases in Fish Mercury Levels in Lakes Flooded by the Churchill River Diversion, Northern Manitoba. *Can. J. Fish. Aquat. Sci.* 41: 682–691.

Boersma, M., 2000. The Nutritional Quantity of P-limited Algae for *Daphnia*. *Limnol. Oceanogr.* 45: 1157–1161.

Boisclair, D., and W.C. Leggett, 1989. Among-population Variability of Fish Growth: III. Influence of Fish Community. *Can. J. Fish. Aquat. Sci.* 46: 1539–1550.

Boland, K.T., and D.J. Griffiths, 1995. *Water Column Stability as a Major Determinant of Shifts in Phytoplankton Composition: Evidence from Two Tropical lakes in Northern Australia*. pp. 89–99. In Schiemer, F., and K.T. Boland (eds). Perspectives in tropical limnology, SPB Acad. Publ., Amsterdam.

Borg, H., P. Andersson, and K. Johansson, 1989. Influence of Acidification on Metal Fluxes in Swedish Forest Lakes. *Sci. Total Env.* 87/88: 241–253.

Borgmann, U., and D.M. Whittle, 1992. Bioenergetics and PCB, DDE and Mercury Dynamics in Lake Ontario Lake Trout (*Salvelinus namaycush*): A Model Based on Surveillance Data. *Can. J. Fish. Aquat. Sci.* 49: 1086–1096.

Boring, L.R., W.T. Swank, J.B. Waide, and G.S. Henderson, 1988. Sources, Fates, and Impacts of Nitrogen Inputs to Terrestrial Ecosystems: Review and Synthesis. *Biogeochem.* 6: 119–159.

Boston, H.L., and W.R. Hill, 1991. Photosynthesis–Light Relations of Stream Periphyton Communities. *Limnol. Oceanogr.* 36: 644–656.

Boström, B., M. Jansson, and C. Forsberg, 1982. Phosphorus Release from Lake Sediments. *Arch. Hydrobiologia Beih. Ergebn. Limnol.* 18: 5–59.

Boström, B., J.M. Andersen, S. Fleischer, and M. Jansson, 1988. Exchange of Phosphorus Across the Sediment–Water Interface. *Hydrobiologia* 170: 229–244.

Boström, B., A.K. Pettersson, and I. Ahlgren, 1989. Seasonal Dynamics of a Cyanobacteria-dominated Microbial Community in Surface Sediments of a Shallow Eutrophic Lake. *Aquat. Sci.* 51: 153–178.

Bothwell, M.L., D.M.J. Sherbot, and C.M. Pollock, 1994. Ecosystem Response to Solar Ultraviolet-B Radiation: Influence of Trophic-level Interactions. *Science* 265: 97–100.

Bottrell, H.H., A. Duncan, Z.M. Gliwicz, E. Grygierek, A. Herzig, A. Hillbricht-Ilkowska, H. Kurasawa, P. Larssen, and T. Weglenska, 1976. A Review of Some Problems in Zooplankton Production Studies. *Norw. J. Zool.* 24: 419–456.

Boudreau, P.R., and L.M. Dickie, 1989. Biological Model of Fisheries Production Based on Physiological and Ecological Scalings of Body Size. *Can. J. Fish. Aquat. Sci.* 46: 614–623.

Bouffard, S.H., and M.A. Hanson, 1997. Fish in Waterfowl Marshes: Waterfowl Managers' Perspective. *Wildlife Soc. Bull.* 25: 146–157.

Boulton, A.J., 1993. Stream Ecology and Surface-hyporheic Hydrologic Exchange: Implication, Techniques and Limitations. *Aust. J. Mar. Freshwat. Res.* 44: 553–564.

Boutorn, C.F., U. Görlach, J.P. Candelone, M.A. Bolshov, and R.J. Delmas, 1991. Decrease in Anthropogenic Lead, Cadmium and Zinc in Greenland Snows Since the Late 1960s. *Nature* 353: 153–156.

Bouvy, M., R. Arfi, P. Cecchi, D. Corbin, M. Pagano, L. Saint-Jean, and S. Thomas, 1998. Trophic Coupling Between Bacterial and Phytoplanktonic Compartments in Shallow Tropical Reservoirs (Ivory Coast, West Africa). *Aquat. Microb. Ecology* 15: 25–37.

Bowling, L.C., M.S. Steane, and P.A. Tyler, 1986. The Spectral Distribution and Attenuation of Underwater Irradiance in Tasmanian Inland Waters. *Freshwater Biol.* 16: 313–335.

Boyce, F.M., 1974. Some Aspects of Great Lakes Physics of Importance to Biological and Chemical Processes. *J. Fish. Res. Board Canada*, 31: 689–730.

Boyce, F.M., 1977. Response of the Coastal Boundary Layer on the North Shore of Lake Ontario to a Fall Storm. *J. Phys. Oceanogr.*, 7: 719–732.

Boyce, F.M., W.M. Schertzer., P.F. Hamblin, and C.R. Murthy, 1991. Physical Behavior of Lake Ontario with Reference to Contaminant Pathways and Climate Change. *Can. J. Fish Aquat. Sci.* 48: 1517–1528.

Bradbury, J.P., B. Leyden, M. Salgado-Labouriau, W.M. Lewis Jr., C. Schubert, M.W. Binford, D.G. Frey, D.R. Whitehead, and F. Weibezahn, 1981. Late Quaternary Environmental History of Lake Valencia, Venezuela. *Science* 214: 1299–1305.

Bradley, R.S., H.F. Diaz, J.K. Eischeid, P.D. Jones, P.M. Kelly and C.M. Goodess, 1987. Precipitation Fluctuations over Northern Hemisphere Land Areas Since the Mid-19th Century. *Science* 237: 171–175.

Brandl, Z., 1973. *Relation Between the Amount of Net Zooplankton and the Depth of Station in the Shallow Lipno Reservoir*. Pp. 7–51. In J. Hrbáček, and M. Straškraba (eds.), Hydrobiological Studies, 3. Prague: Acad. Publ. Hs. Czechoslovak. Acad. of Sci.

Breen, C.M., K.H. Rogers, and P.J. Ashton, 1988. *Vegetation Processes in Swamps and Flooded Plains*. Pp. 223–247. In J.J. Symoens, (ed.), Vegetation of Inland Waters. Vol. 15/1, Handbook of Vegetation Science. Dordrecht: Kluwer Acad. Publ.

Bretschko, G., 1995. River/Land Ecotones: Scales and Patterns. *Hydrobiologia* 303: 83–91.

Brezonik, P., 1996. *Organizing Paradigms for the Study of Inland Aquatic Ecosystems*. Pp. 181–202. In NRC, Freshwater Ecosystems: Revitalizing Educational Programs in Limnology. Washington: Nat. Acad. Press.

Brezonik, P.L., and C.L. Harper, 1969. Nitrogen Fixation in Some Anoxic Lacustrine Environments. *Science* 164: 1277–1279.

Brezonik, P.L., J.G. Eaton, T.M. Frost, P.J. Garrison, T.K. Kratz, C.E. Mach, J.H. McCormick, J.A. Perry, W.A. Rose, C.J. Sampson, B.C.L. Schelley, W.A. Swenson, and K.E. Webster, 1993. Experimental Acidification of Little Rock Lake, Wisconsin: Chemical and Biological Changes over the pH Range 6.1 to 4.7. *Can. J. Fish Aquat. Sci.* 50: 1101–1121.

Brinkhurst, R.O. (ed.), 1974. *The Benthos of Lakes*. London: MacMillan Press.

Broberg, O., and G. Persson, 1984. External Budgets for Phosphorus, Nitrogen and Dissolved Organic Carbon for the Acidified Lake Gårdsjön. *Arch. Hydrobiol.* 99: 160–175.

Brönmark, C., 1985. Freshwater Snail Diversity: Effects of Pond Area, Habitat Heterogeneity and Isolation. *Oecologia* 67: 127–131.

Brook, A.J., and W.B. Woodward, 1956. Some Observations on the Effects of Water Inflow and Outflow on the Plankton of Small Lakes. *J. Animal Ecology* 25: 22–35.

Brooks, A.S., and D.N. Edgington, 1994. Biogeochemical Control of Phosphorus Cycling and Primary Production in Lake Michigan. *Limnol. Oceanogr.* 39: 961–968.

Brooks, J.L., and S.I. Dodson, 1965. Predation, Body Size, and Composition of Plankton. *Science* 150: 28–35.

Bruning, K., 1991. Effects of Temperature and Light on the Population Dynamics of the *Asterionella-Rhizophydium* Association. *J. Plankton Res.* 13: 707–719.

Brunskill, G.J., and D.W. Schindler, 1971. Geography and Bathymetry of Selected Lake Basins, Experimental Lakes Area, Northwestern Ontario. *Can. J. Fish. Res. Bd.* 28: 139–155.

Brunskill, G.J., and S.D. Ludlam, 1969. Fayetteville Green Lake, New York, I. Physical and Chemical Limnology. *Limnol. Oceanogr.* 14: 817–829.

Bruton, M.N., and R.E. Bolt, 1975. Aspects of *Tilapia mossambicus* Peters (Pisces: Cichlidae) in a Natural Freshwater Lake (Lake Sibaya, South Africa). *J. Fish. Biol.* 7: 423–445.

Bryson, R.A., and F.K. Hare (eds.), 1974. *World Survey of Climatology. Vol. 11: Climates of North America.* Amsterdam: Elsevier.

Burke, C.M., and H.R. Burton, 1988. The Ecology of Photosynthetic Bacteria in Burton Lake, Vestfold Hills, Antartica. *Hydrobiologia* 165: 1–11.

Burnison, B.K., and G.G. Leppard, 1983. Isolation of Colloidal Fibrils from Lake Water by Physical Separation Techniques. *Can. J. Fish. Aquat. Sci.* 40: 373–381.

Butturini, A., and F. Sabater, 1998. Ammonium and Phosphate Retention in a Mediterranean Stream: Hydrological Versus Temperature Control. *Can. J. Fish. Aquat. Sci.* 55: 1938–1945.

Cabana, G., A. Tremblay, J. Kalff, and J.B. Rasmussen, 1994. Pelagic Food Chain Structure in Ontario Lakes: A Determinant of Mercury Levels in Lake Trout (*Salvelinus namaycush*). *Can. J. Fish. Aquat. Sci.* 51: 381–389.

Cabrera, S., and V. Montecino, 1984. The Meaning of the Euphotic Chlorophyll-*a* Measurement. *Verh. Int. Ver. Limnol.* 22: 1328–1331.

Calheiros, D.F., and S.K. Hamilton, 1998. Limnological Conditions Associated with Natural Fish Kills in the Pantanal Wetland of Brazil. *Verh. Int. Ver. Limnol.* 26: 2189–2193.

Canfield, Jr., D.E., 1983. Prediction of Chlorophyll-*a* Concentrations in Florida Lakes: The Importance of Phosphorus and Nitrogen. *Water Res. Bull.* 19: 255–262.

Canfield, Jr., D.E., and R.W. Bachmann, 1981. Prediction of Total Phosphorus Concentrations, Chlorophyll-*a* and Secchi Depths in Natural and Artificial Lakes. *Can. J. Fish. Aquat. Sci.* 38: 414–423.

Canfield, Jr., D.E., S.B. Linda, and L.M. Hodgson, 1985. Chlorophyll–Biomass–Nutrient Relationships for Natural Assemblages of Florida Phytoplankton. *Water Res. Bull.* 21: 381–391.

Canfield, Jr., D.E., E. Phlips, and C.M. Duarte, 1989. Factors Influencing the Abundance of Blue-green Algae in Florida Lakes. *Can. J. Fish. Aquat. Sci.* 46: 1232–1237.

Capblancq, J., 1982. *Phytoplancton et Production Primaire.* Pp. 1–48. In R. Pourriot, J. Capblancq, P. Champ, and J.A. Meyer (eds.), Ecologie du Plancton des Eaux Continentales. Paris: Masson.

Capone, D.G., and R.P. Kiene, 1988. Comparison of Microbial Dynamics in Marine and Freshwater Sediments: Contrasts in Anaerobic Carbon Catabolism. *Limnol. Oceanogr.* 33: 725–749.

Caraco, N.F., J.J. Cole, and G.E. Likens, 1991. Phosphorus Release from Anoxic Sediments: Lakes that Break the Rules. *Verh. Int. Ver. Limnol.* 24: 2985–2988.

Caraco, N.F., J.J. Cole, and G.E. Likens, 1992. New and Recycled Primary Production in an Oligotrophic Lake: Insights for Summer Phosphorus Dynamics. *Limnol. Oceanogr.* 37: 590–602.

Caraco, N.F., J.J. Cole, and G.E. Likens, 1993. Sulfate Control of Phosphorus Availability in Lakes. A test and Re-evaluation of Hasler and Einsele's Model. *Hydrobiologia* 253: 275–280.

Caraco, N.F., J.J. Cole, P.A. Raymond, D.L. Strayer, M.L. Pace, S.E.G. Findlay, and D.T. Fischer, 1997. Zebra Mussel Invasion in a Large, Turbid River: Phytoplankton Response to Increased Grazing. *Ecology* 78: 588–602.

Carignan, R., 1982. An Empirical Model to Estimate the Relative Importance of Roots in Phosphorus Uptake by Aquatic Macrophytes. *Can. J. Fish. Aquat. Sci.* 39: 243–247.

Carignan, R., and J. Kalff, 1980. Phosphorus Sources for Aquatic Weeds: Water or Sediments? *Science* 207: 987–989.

Carignan, R., and D. Planas, 1994. Recognition of Nutrient and Light Limitation in Turbid Mixed Layers: Three Approaches Compared in the Paraná Floodplain (Argentina). *Limnol. Oceanogr.* 39: 580–596.

Caron, D.A., F.R. Pick, D.R.S. Lean, 1985. Chroococcoid Cyanobacteria in Lake Ontario: Vertical and Seasonal Distribution During 1982. *J. Phycol.*, 21: 171–175.

Carpenter, S.R., 1983. Lake Geometry: Implications for Production and Sediment Accretion Rates. *J. Theor. Biol.* 105: 273–286.

Carpenter, S.R., and J.F. Kitchell, 1987. Plankton Community Structure and Limnetic Primary Production. *Am. Nat.* 124: 159–172.

Carpenter, S.R., and J.F. Kitchell, 1992. Trophic Cascade and Biomanipulation: Interface of Research and Management—A Reply to the Comment by DeMelo et al. *Limnol. Oceanogr.* 37: 208–213.

Carpenter, S.R., J.F. Kitchell, and J.K. Hodgson, 1985. Cascading Trophic Interactions and Lake Productivity. *Bioscience* 35: 634–639.

Carr, N.G., and B.A. Whitton (eds.), 1982. *The Biology of Cyanobacteria. Botanical Monographs, Vol. 19.* Oxford: Blackwell Sci. Publ.

Casselman, J.M., 1987. *Determination of Age and Growth.* Pp. 209–242. In A.H. Weatherly, and H.S. Gill (eds.), The Biology of Fish Growth. London: Acad. Press.

Cattaneo, A., 1983. Grazing on Epiphytes. *Limnol. Oceanogr.* 28: 124–132.

Cattaneo, A., 1987. Periphyton in Lakes of Different Trophy. *Can. J. Fish. Aquat. Sci.* 44: 296–303.

Cattaneo, A ., 1993. Size Spectra of Benthic Communities in Laurentian Streams. *Can. J. Fish. Aquat. Sci.* 50: 2659–2666.

Cebrián, J., and C.M. Duarte, 1994. The Dependence of Herbivory on Growth Rate in Natural Plant Communities. *Funct. Ecology* 8: 518–525.

Chambers, P.A., 1987. Light and Nutrients in the Control of Aquatic Plant Community Structure, Vol. II, In situ Observations. *J. Ecology* 75: 621–628.

Chambers, P.A., 1987. Nearshore Occurrence of Submersed Aquatic Macrophytes in Relation to Wave Action. *Can. J. Fish. Aquat. Sci.* 44: 1666–1669.

Chambers, P.A., and J. Kalff, 1985. Depth Distribution of Biomass of Submersed Aquatic Macrophyte Communities in Relation to Secchi Depth. *Can. J. Fish. Aquat. Sci.* 42: 701–709.

Chambers, P.A., and E.E. Prepas, 1988. Underwater Spectral Attenuation and its Effect on the Maximum Depth of Angiosperm Colonization. *Can. J. Fish. Aquat. Sci.* 45: 1010–1017.

Chambers, P.A., E.E. Prepas, H.R. Hamilton, and M.L. Bothwell, 1991. Current Velocity and its Effect on Aquatic Macrophytes in Flowing Water. *Ecology Appl.* 1: 249–257.

Charles, D.F., and J.P. Smol, 1994. *Long-term Chemical Changes in Lakes: Quantitative Inference Using Biotic Remains in the Sediment Record.* Pp. 3–31. In L.A. Baker (ed.), Environmental Chemistry of Lakes and Reservoirs. Washington: Amer. Chem. Soc.

Charles, D.F., and S.A. Norton, 1986. *Paleolimnological Evidence for Trends in Atmospheric Deposition of Acids and Metals.* Pp. 335–431. In NRC, Acid Deposition: Long-term Trends. Washington, DC: Nat. Acad. Press.

Charlton, M.N., 1980. Hypolimnion Oxygen Consumption in Lakes: Discussion of Productivity and Morphometry Effects. *Can. J. Fish. Aquat. Sci.* 37: 1531–1539.

Chételat, J., F.R. Pick, A. Morin, and P.B. Hamilton, 1999. Periphyton Biomass and Community Composition in Rivers of Different Nutrient Status. *Can. J. Fish. Aquat. Sci.* 56: 560–569.

Chevalier, G., C. Dumont, C. Langlois, and A. Penn, 1997. Mercury in Northern Quebec: Role of the Mercury Agreement and Status of Research and Monitoring. *Water, Air, and Soil Poll.* 97: 75–84.

Chiou, C.T., V.H. Freed, D.W. Schmedding, and R.L. Kohnert, 1977. Partition Coefficient and Bioaccumulation of Selected Organic Chemicals. *Environ. Sci. Tech.* 11: 475–478.

Chorus, I., and J. Bertram (eds.), 1999. *Toxic Cyanobacteria in Water: A Guide to Their Public Health Consequences, Monitoring, and Management.* London: E & FN Spon. Publ.

Chow-Fraser, P., 1991. Use of the Morphoedaphic Index to Predict Nutrient Status and Algal Biomass in Some Canadian Lakes. *Can. J. Fish. Aquat. Sci.* 48: 1909–1918.

Christensen, P.B., and J. Sørensen, 1986. Temporal Variation of Denitrification Activity in Plant-covered, Littoral Sediment from Lake Hampen, Denmark. *Appl. Environ. Microbiol.* 51: 1174–1179.

Christie, W.J., K.A. Scott, P.G. Sly, and R.H. Strus, 1987. Recent Changes in the Aquatic Food Web of Eastern Lake Ontario. *Can. J. Fish. Aquat. Sci.* 44(Suppl.2): 37–52.

Christoffersen, K., B. Riemann, A. Klysner, and M. Søndergaard, 1993. Potential Role of Fish Predation and Natural Populations of Zooplankton in Structuring a Plankton Community in Eutrophic Lake Water. *Limnol. Oceanogr.* 38: 561–573.

Cimbleris, A.C.P., and J. Kalff, 1998. Planktonic Bacterial Respiration as a Function of C:N:P Ratios Across Temperate Lakes. *Hydrobiologia* 384: 89–100.

Clair, T.A., P.J. Dillon, J. Ion, D.S. Jeffries, M. Papineau, and R.F. Vet, 1995. Regional Precipitation and Surface Water Chemistry Trends in Southeastern Canada (1983–1991). *Can. J. Fish. Aquat. Sci.* 52: 197–212.

Cloern, J.E., 1987. Turbidity as a Control on Phytoplankton Biomass and Productivity in Esturaries. *Cont. Shelf. Res.* 7: 1367–1381.

Coale, K.H., and A.R. Flegal, 1989. Copper, Zinc, Cadmium and Lead in Surface Waters of Lakes Erie and Ontario. *Sci. Total Env.* 87/88: 297–304.

Coordinating Committee on Great Lakes Basic Hydraulic and Hydraulic Data. *Lake Erie Outflows, 1900–1998.* Env. Canada, Ontario/ US Army Corps of Eng., Detroit.

Colborn, T.E., A. Davidson, S.N. Green, R.A. Hodge, C.I. Jackson, R.A. Liroff, 1990. *Great Lakes, Great Legacy?* In Min. of Supply and Services Canada, 1991, The State of Canada's Environment. Ottawa.

Cole, G.A., 1968. *Desert Limnology.* Pp. 423–486. In G.W. Brown, Jr. (ed.), Special Topics on the Physical and Biological Aspects of Arid Regions. Desert Limnology, Vol. 1. New York: Acad. Press.

Cole, G.A., 1994. *Textbook of Limnology, 4th ed.* Illinois: Waveland Press.

Cole, J.J., S. Findlay, and M.L. Pace, 1988. Bacterial Production in Fresh and Saltwater Ecosystems: A Cross-system Overview. *Mar. Ecology Progr. Sci.* 43: 1–10.

Cole, J.J., M.L. Pace, N.F. Caraco, and G.S. Steinhart, 1993. Bacterial Biomass and Cell Size Distributions in Lakes: More and Larger Cells in Anoxic Waters. *Limnol. Oceanogr.* 38: 1627–1632.

Cole, J.J., B.L. Peierls, N.F. Caraco, and M.L. Pace, 1993. *Nitrogen Loading of Rivers as a Human-driven Process.* Pp. 141–157. In M.J. McDonnell, and S.T.A. Pickett (eds.), Humans as Components of Ecosystems: The Ecology of Subtle Human Effects and Populated Areas. New York: Springer-Verlag.

Cole, J.J., N.F. Caraco, G.W. Kling, and T.K. Kratz, 1994. Carbon Dioxide Supersaturation in the Surface Waters of Lakes. *Science* 265: 1568–1570.

Cole, T.M., and H.H. Hannan, 1990. *Dissolved Oxygen Dynamics.* Pp. 71–107. In K.W. Thornton, B.L. Kimmel, and F.E. Payne (eds.), Reservoir Limnology: Ecological Perspectives. New York: John Wiley and Sons.

Confer, J.L., G.L. Howick, M.H. Corzette, S.L. Kramer, S. Fizgibbon, and R. Landesberg, 1978. Visual Predation by Planktivores. *Oikos* 31: 27–37.

Confer, J.L., T. Kaaret, and G.E. Likens, 1983. Zoolplankton Diversity and Biomass in Recently Acidified Lakes. *Can. J. Fish. Aquat. Sci.* 40: 36–42.

Cook, C.D.K., 1996. *Aquatic Plant Book.* New York: SPB Acad. Publ.

Cook, R.E., 1977. Raymond Lindeman and the Trophic-dynamic Concept in Ecology. *Science* 198: 22–26.

Cook, R.B., 1984. Distributions of Ferrous Iron and Sulfide in an Anoxic Hypolimnion. *Can. J. Fish. Aquat. Sci.* 41: 286–293.

Cooke, G.D., E.B. Welch, S.A. Peterson, and P.R. Newroth (eds.), 1986. *Lake and Reservoir Restoration, 1st ed.* Boston: Butterworth Publ.

Cooke, G.D., E.B. Welch, S.A. Peterson, and P.R. Newroth, 1993. *Restoration and Management of Lakes and Reservoirs, 2nd ed.* Boca Raton: Lewis Publ.

Cooke, G.W., and R.J.B. Williams, 1973. Significance of Man-made Sources of Phosphorus: Fertilizers and Farming. *Water Res.* 7: 19–33.

Cornett, R.J., and F.H. Rigler, 1980. The Areal Hypolimnetic Oxygen Deficit: An Empirical Test of the Model. *Limnol. Oceanogr.* 25: 672–679.

Cornett, R.J., and F.H. Rigler, 1987. Decomposition of Seston in the Hypolimnion. *Can. J. Fish. Aquat. Sci.* 44: 146–151.

Cosby, B.J., G.M. Hornberger, J.N. Galloway, and R.F. Wright, 1985. Time Scales of Catchment Acidification. *Environ. Sci. Technol.* 19: 1144–1149.

Costanza, R., R. d'Arge, R. de Groot, S. Farber, M. Grasso, B. Hannon, K. Limburg, S. Naeem, R. O'Neill, J. Faruelo, J. R. Raskin, P. Sutton, and M. van den Belt, 1997. The Value of the World's Ecosystem Services and Natural Capital. *Nature* 387: 253–260.

Costa-Pierce, B.A., 1996. *Environmental Impacts of Nutrients from Aquaculture: Towards the Evolution of Sustainable Aquaculture Systems.* Pp. 81–113. In D.J. Baird, M.C.M. Beveridge, L.A. Kelly, and J.F. Muir (eds.), Aquaculture and Water Resources Management. Oxford: Blackwell Sci. Publ.

Cotner, J.B., 2000. Intense Winter Heterotrophic Production Stimulated by Benthic Resuspension. *Limnol. Oceanogr.* 45: 1672–1676.

Cottrell, M.T., and C.A. Suttle, 1995. Dynamics of a Lytic Virus Infecting the Photosynthetic Marine Picoflagellate *Micromonas pusilla. Limnol. Oceanogr.* 40: 730–739.

Coveney, M.F., and R.G. Wetzel, 1995. Biomass, Production, and Specific Growth Rate of Bacterioplankton and Coupling to Phytoplankton in an Oligotrophic Lake. *Limnol. Oceanogr.* 40: 1187–1200.

Cowling, E.B., 1982. Acid Precipitation in Historical Perspective. *Env. Sci. and Technol.* 16: 110–123.

Craig, J., 1987. *The Biology of Perch and Related Fish.* Kent: Croon Helm.

Crill, P.M., 1996. Latitudinal Differences in Methane Fluxes from Natural Wetlands. *Mitt. Int. Ver. Limnol.* 25: 163–171.

Cripps, S.J., and L.A. Kelly, 1996. *Reduction in Wastes from Aquaculture.* Pp. 166–201. In D.J. Baird, M.C.M. Beveridge, L.A. Kelly, and J.F. Muir (eds.), Aquaculture and Water Resources Management. Oxford: Blackwell Sci. Publ.

Cuffney, T.F., and J.B. Wallace, 1988. Particulate Organic Matter Export from Three Headwater Streams: Discrete Versus Continuous Measurements. *Can. J. Fish. Aquat. Sci.* 45: 2010–2016.

Cullen, P., and C. Forsberg, 1988. Experiences with Reducing Point Sources of Phosphorous to Lakes. *Hydrobiologia* 170: 321–336.

Cullen, J.J., and M.R. Lewis, 1988. The Kinetics of Algal Photoadaptation in the Context of Vertical Mixing. *J. Plankton Res.* 10: 1039–1063.

Cumming, B.F., K.A. Davey, J.P. Smol, and H.J.B. Birks, 1994. When Did Acid-sensitive Adirondack Lakes (New York, USA) Begin to Acidify and Are They Still Acidifying? *Can. J. Fish. Aquat. Sci.* 51: 1550–1568.

Currie, D.J., 1990. Large-scale Variability and Interactions Among Phytoplankton, Bacterioplankton and Phosphorus. *Limnol. Oceanogr.* 35: 1437–1455.

Curtis, P.J., and H.E. Adams, 1995. Dissolved Organic Matter Quantity and Quality from Freshwater and Saltwater Lakes in East-central Alberta. *Biogeochem.* 30: 59–76.

Curtis, P.J., and D.W. Schindler, 1997. Hydrologic Control of Dissolved Organic Matter in Low-order Precambrian Shield Lakes. *Biogeochem.* 36: 125–138.

Cushing, C.E., K.W. Cummins, and G.W. Minshall (eds.), 1995. *Ecosystems of the World, Vol. 22: River and Stream Ecosystems.* Amsterdam: Elsevier.

Cuthbert, I.D., and P. del Giorgio, 1992. Toward a Standard Method of Measuring Color in Freshwater. *Limnol. Oceanogr.* 37: 1319–1326.

Cyr, H., and J.A. Downing, 1988. The Abundance of Phytophilous Invertebrates on Different Species of Submerged Macrophytes. *Freshwater Biol.* 20: 365–374.

Cyr, H., and M.L. Pace, 1993. Magnitude and Patterns of Herbivory in Aquatic and Terrestrial Ecosystems. *Nature* 361: 148–150.

Davies, B.R., 1982. Studies on the Zoobenthos of Some Southern Cape Coastal Lakes. Spatial and Temporal Changes in the Benthos of Swartvlei, South Africa, in Relation to Changes in the Submerged Littoral Macrophyte Community. *J. Limnol. Soc. Sth. Afr.* 8: 33–45.

Davies-Colley, R.J., 1988. Mixing Depths in New Zealand Lakes. *New Zealand J. Mar. Freshw. Res.* 22: 517–527.

Davison, I.R., 1991. Environmental Effects on Algal Photosynthesis: Temperature. *J. Phycol.* 27: 2–8.

Davison, W., 1987. Internal Elemental Cycles Affecting the Long-term Alkalinity Status of Lakes: Implications for Lake Restoration. *Schweiz. Z. Hydrol.* 49: 186–201.

Davison, W., and G. Seed, 1983. The Kinetics of the Oxidation of Ferrous Iron in Synthetic and Natural Waters. *Geochim. Cosmochim. Acta* 47: 67–79.

Davison, W., and E. Tipping, 1984. Treading in Mortimer's Footsteps: The Geochemical Cycling of Iron and Manganese in Esthwaite Water. *Freshw. Biol. Assoc. Annu. Rep.* 52: 91–101.

Davison, W., D.G. George, and N.J.A. Edwards, 1995. Controlled Reversal of Lake Acidification by Treatment with Phosphate Fertilizer. *Nature* 377: 504–507.

Dawidowicz, P., J. Pijanowska, and K. Ciechomski, 1990. Vertical Migration of *Chaoborus* Larvae is Induced by the Presence of Fish. *Limnol. Oceanogr.* 35: 1631–1637.

Day, J.A., and B.R. Davies, 1986. *The Amazon River System.* Pp. 289–318. In B.R. Davies, and K.F. Walker (eds.), The Ecology of River Systems. Monographiae Biologicae, 60. Dordrecht: Dr. W. Junk Publ.

de Bernardi, R., A. Calderone, and R. Mosello, 1996. Environmental Problems in Italian Lakes, and Lakes Maggiore and Orta as Succesful Examples of Correct Management Leading to Restoration. *Verh. Int. Ver. Limnol.* 26: 123–138.

De Mora, S.J., S. Demers, and M. Vernet (eds.), 2000. *Effects of UV Radiation on Marine Ecosystems.* New York: Cambridge Univ. Press.

de Oliveira, M.D., and D.F. Calheiros, 2000. Flood Pulse Influence on Phytoplankton Communities of the South Pantanal Floodplain, Brazil. *Hydrobiologia* 427: 101–112.

Dearing, J.A., and I.D.L. Foster, 1993. *Lake Sediments and Geomorphological Processes: Some Thoughts.* Pp. 5–14. In J. McManus, and R.W. Duck (eds.), Geomorphology and Sedimentology of Lake Reservoirs. New York: John Wiley and Sons.

del Giorgio, P.A., and R.H. Peters, 1993. Balance Between Phytoplankton Production and Plankton Respiration in Lakes. *Can. J. Fish. Aquat. Sci.* 50: 282–289.

del Giorgio, P.A., and R.H. Peters, 1994. Patterns in Planktonic P:R Ratios in Lakes: Influence of Lake Trophy and Dissolved Organic Carbon. *Limnol. Oceanogr.* 39: 772–787.

del Giorgio, P.A., and J.M. Gasol, 1995. Biomass Distribution in Freshwater Plankton Communities. *Am. Nat.* 146: 135–152.

del Giorgio, P.A., and G. Scarborough, 1995. Increase in the Proportion of Metabolically Active Bacteria Along Gradients of Enrichment in Freshwater and Marine Plankton: Implications for Estimates of Bacterial Growth and Production Rates. *J. Plankton Res.* 17: 1905–1924.

del Giorgio, P.A., and J.J. Cole, 1998. Bacterial Growth Efficiency in Natural Aquatic Systems. *Ann. Rev. Ecol. Syst.* 29: 503–541.

DeMelo, R., R. France, and D.J. McQueen, 1992. Biomanipulation: Hit or Myth? *Limnol. Oceanogr.* 37: 192–207.

Demers, E., and J. Kalff, 1993. A Simple Model for Predicting the Date of Spring Stratification in Temperate and Subtropical Lakes. *Limnol. Oceanogr.* 38: 1077–1081.

DeMott, W.R., and R.D. Gulati, 1999. Phosphorus Limitation in *Daphnia*: Evidence from a Long-term Study of Three Hypereutrophic Dutch Lakes. *Limnol. Oceanogr.* 44: 1557–1564.

den Heyer, C., and J. Kalff, 1998. Organic Matter Mineralization Rates in Sediments: A Within- and Among-lake Study. *Limnol. Oceanogr.* 43: 695–705.

den Oude, P.J., and R.D. Gulati, 1988. Phosphorus and Nitrogen Excretion Rates of Zooplankton from the Eutrophic Loosdrecht Lakes, with Notes on Other P Sources for Phytoplankton Requirements. *Hydrobiologia* 169: 379–390.

Denny, P., 1980. Solute Movement in Submerged Angiosperms. *Biol. Rev.* 55: 65–92.

Desortová, B., 1983. Relationship Between Chlorophyll-*a* Concentration and Phytoplankton Biomass in Several Reservoirs in Czecholslovakia. *Int. Revue ges. Hydrobiologia* 66: 153–169.

Devito, K.J., P.J. Dillon, and B.D. Lazerte, 1989. Phosphorus and Nitrogen Retention in Five Precambrian Shield Wetlands. *Biogeochem.* 8: 185–204.

Diamond, J., 1986. *Overview: Laboratory Experiments, Field Experiments, and Natural Experiments.* Pp. 3–22. In J. Diamond, and T.J. Case (eds.), Community Ecology. New York: Harper and Row Publ.

Dickson, W., 1985 *Acid Rain in Sweden: Effects on Lake Systems.* Pp. 341–346. In Proc. Europ. Water Poll. Control Assoc., Lakes Pollution and Recovery. Rome.

Dillon, P.J., and F.H. Rigler, 1974. The Phosphorus–Chlorophyll Relationship in Lakes. *Limnol. Oceanogr.* 19: 767–773.

Dillon, P.J., and L.A. Molot, 1997. Dissolved Organic and Inorganic Carbon Mass Balances in Central Ontario Lakes. *Biogeochem.* 36: 29–42.

Dillon, P.J., and L.A. Molot, 1997. Effect of Landscape Form on Export of Dissolved Organic Carbon, Iron, and Phosphorus from Forested Stream Catchments. *Water Resources Res.* 33: 2591–2600.

Dillon, P.J., N.D. Yan, and H.H. Harvey, 1984. Acidic Deposition: Effects on Aquatic Ecosystems. *CRC Critical Rev. in Env. Control.* 13: 167–194.

Dillon, P.J., R.D. Evans, and L.A. Molot, 1990. Retention and Resuspension of Phosphorus, Nitrogen, and Iron in a Central Ontario Lake. *Can. J. Fish. Aquat. Sci.* 47: 1269–1274.

Dini, M.L., and S.R. Carpenter, 1992. Fish Predators, Food Availability and Diel Vertical Migration in *Daphnia. J. Plankton Res.* 14: 359–377.

Dinsmore, W.P., and E.E. Prepas, 1997. Impact of Hypolimnetic Oxygenation on Profundal Macro-invertebrates in a Eutrophic Lake in Central Alberta, II. Changes in *Chironomus* spp. Abundance and Biomass. *Can. J. Fish. Aquat. Sci.* 54: 2170–2181.

Dixit, S.S., A.S. Dixit, and J.P. Smol, 1989. Lake Acidification Recovery Can be Monitored Using Chrysophycean Microfossils. *Can. J. Fish. Aquat. Sci.* 46: 1309–1312.

Dobson, H.F.H., 1985. *Lake Ontario Water Chemistry Atlas. Scientific Series #139.* Burlington: Inland Waters Directorate, Nat. Water Res. Inst., and Canada Centre for Inland Waters.

Dodds, W.K., V.H. Smith, and B. Zander, 1997. Developing Nutrient Targets to Control Benthic Chlorophyll Levels in Streams: A Case Study of the Clark Fork River. *Water Res.* 31: 1738–1750.

Dodson, S., 1990. Predicting Diel Vertical Migration of Zooplankton. *Limnol. Oceanogr.* 35: 1195–1200.

Dodson, S., 1992. Predicting Crustacean Zooplankton Species Richness. *Limnol. Oceanogr.* 37: 848–856.

Dodson, S.I., S.E. Arnott, and K.L. Cottingham, 2000. The Relationship in Lake Commnunities Between Primary Productivity and Species Richness. *Ecology* 81: 2662–2679.

Döerner, H., A. Wagner, and J. Benndorf, 1999. Predation by Piscivorous Fish on Age-0 Fish: Spatial and Temporal Variability in a Biomanipulated Lake (Bautzen Reservoir, Germany). *Hydrobiologia* 408/409: 39–46.

Doka, S.E., M.L. Mallory, D.K. McNicol, C.K. Minns, 1997. Species Richness and Species Occurence of Five Taxonomic Groups in Relation to pH and Other Lake Characteristics in Southeastern Canada. *Can. Tech. Rep. Fish Aquat. Sci.* 2179: 62pp.

Dokulil, M., 1979. *Optical Properties, Colour and Turbidity.* Pp. 151–167. In H. Löffler (ed.), Neusiedlersee: The Limnology of a Shallow Lake in Central Europe. The Hague: W. Junk Publ.

Donnelly, B.G., 1969. A Preliminary Survey of Tilapia Nurseries on Lake Kariba During 1967/68. *Hydrobiologia* 34: 195–206.

Dovland, H., and A. Semb, 1980. *Atmospheric Transport of Pollutants.* Pp. 14–21. In D. Drabløs, and A. Tollan (eds.), Ecological Impact of Acid Precipitation. Oslo: SNSF-Project #1432.

Downes, M.T., 1988. Aquatic Nitrogen Transformations at Low Oxygen Concentrations. *Appl. Environ. Microbiol.* 54: 172–175.

Downing, J.A., and F.H. Rigler (eds.), 1984. *A Manual on Methods for the Assessment of Secondary Productivity in Fresh Waters,* 2nd ed. Oxford: Blackwell Sci. Publ.

Downing, J.A., and M.R. Anderson, 1985. Estimating the Standing Biomass of Aquatic Macrophytes. *Can. J. Fish. Aquat. Sci.* 42: 1860–1869.

Downing, J.A., C. Plante, and S. Lalonde, 1990. Fish Production Correlelated with Primary Productivity, Not the Morphoedaphic Index. *Can. J. Fish. Aquat. Sci.* 47: 1929–1936.

Downing, J.A., and E. McCauley, 1992. The Nitrogen: Phosphorus Relationship in Lakes. *Limnol. Oceanogr.* 37: 936–945.

Downing, J.A., M. McClain, R. Twilley, J.M. Melack, J. Elser, N.N. Rabalais, W.M. Lewis, Jr., R.E. Turner, J. Corredor, D. Soto, A. Yanez-Arancibia, J.A. Kopaska, R.W. Howarth, 1999. The Impact of Accelerating Land-use Change on the N-cycle of Tropical Aquatic Ecosystems: Current Conditions and Projected Changes. *Biogeochem.* 46: 109–148.

Driscoll, C.T., 1980. *Aqueous Speciation of Aluminum in the Adirondack Region of New York State, USA.* Pp. 214–215. In D. Drabløs, and A. Tollan (eds.), Ecological Impact of Acid Precipitation. Oslo: SNSF-Project #1432.

Duarte, C.M., 1992. Nutrient Concentration of Aquatic Plants: Patterns Across Species. *Limnol. Oceanogr.* 37: 882–889.

Duarte, C.M., and J. Kalff, 1986. Littoral Slope as a Predictor of the Maximum Biomass of Submerged Macrophyte Communities. *Limnol. Oceanogr.* 31: 1072–1080.

Duarte, C.M., J. Kalff, and R.H. Peters, 1986. Patterns in Biomass and Cover of Aquatic Macrophytes in Lakes. *Can. J. Fish. Aquat. Sci.* 43: 1900–1908.

Duarte, C.M., and J. Kalff, 1987. Latitudinal Influences on the Depths of Maximum Colonization and Maximum Biomass of Submerged Angiosperms in Lakes. *Can. J. Fish. Aquat. Sci.* 44: 1759–1764.

Duarte, C.M., and J. Kalff, 1987. Weight–Density Relationships in Submerged Macrophytes. The Importance of Light and Plant Geometry. *Oecologia* 72: 612–617.

Duarte, C.M., and M. Alcaraz, 1989. To Produce Many Small or Few Large Eggs: A Size-independent Reproductive Tactic of Fish. *Oecologia* 80: 401–404.

Duarte, C.M., and J. Kalff, 1989. The Influence of Catchment Geology and Lake Depth on Phytoplankton Biomass. *Arch. Hydrobiol.* 115: 27–40.

Duarte, C.M., and J. Kalff, 1990. Biomass Density and the Relationship Between Submerged Macrophyte Biomass and Growth Form. *Hydrobiologia* 196: 17–23.

Duarte, C.M., and J. Kalff, 1990. Patterns in the Submerged Macrophyte Biomass of Lakes and the Importance of the Scale of Analysis in the Interpretation. *Can. J. Fish. Aquat. Sci.* 47: 357–363.

Duarte, C.M., S. Agusti, and D.E. Canfield, Jr., 1992. Patterns in Phytoplankton Community Structure in Florida Lakes. *Limnol. Oceanogr.* 37: 155–161.

Duarte, C.M., D. Planas, and J. Peñuelas, 1994. *Macrophytes Taking Control of an Ancestral Home.* Pp. 59–79. In R. Margalef, (ed.), Limnology Now: A Paradigm of Planetary Problems. Amsterdam: Elsevier.

Duarte, C.M., J.M. Gasol, and D. Vaqué, 1997. Role of Experimental Approaches in Marine Microbial Ecology. *Aquat. Microb. Ecol.* 13: 101–111.

Dubay, C.I., and G.M. Simmons, Jr., 1979. The Contribution of Macrophytes to the Metalimnetic Oxygen Maximum in a Montane, Oligotrophic Lake. *Amer. Midland Nat.* 101: 108–117.

Dudgeon, D., 1992. Endangered Ecosystems: A Review of the Conservation Status of Tropical Asian Rivers. *Hydrobiologia* 248: 167–191.

Dumestre, J.F., A. Vaquer, P. Gosse, S. Richard, and L. Labroue, 1999. Bacterial Ecology of a Young Equatorial Hydroelectric Reservoir (Petit Saut, French Guiana). *Hydrobiologia* 400: 75–83.

Dumont, H., 1995. Ecocide in the Caspian Sea. *Nature* 377: 673–674.

Duncan, A., and J. Kubečka, 1995. Land/Water Ecotone Effects in Reservoirs on the Fish Fauna. *Hydrobiologia* 303: 11–30.

Eadie, B.J., R.L. Chambers, W.S. Gardner, and G.L. Bell, 1984. Sediment Trap Studies in Lake Michigan: Resuspension and Chemical Fluxes in the Southern Basin. *J. Great Lakes Res.,* 10: 307–321.

Eadie, J.M., and A. Keast, 1982. Do Goldeneye and Perch Compete for Food? *Oecologia* 5: 225–230.

Eadie, J.M., and A. Keast, 1984. Resource Heterogeneity and Fish Species Diversity in Lakes. *Can. J. Zool.* 62: 1689–1695.

Eberly, W.R., 1964. Further Studies on the Metalimnetic Oxygen Maximum, With Special Reference to its Occurrence Throughout the World. *Invest. Indiana Lakes and Streams* 6: 103–139.

Eck, G.W., and L. Wells, 1987. Recent Changes in Lake Michigan's Fish Community and Their Probable Causes, With Emphasis on the Role of the Alewife (*Alosa pseudoharengus*). *Can. J. Fish. Aquat. Sci.* 44: 53–60.

Edmondson, W.T., 1966. Changes in the Oxygen Deficit of Lake Washington. *Verh. Int. Ver. Limnol.* 16: 153–158.

Edmondson, W.T., 1977. *Lake Washington.* Pp. 288–300. In L. Seyb, and K. Randolph (eds.), North American Project—A Study of U.S. Water Bodies. A Report for the Organization for Economic Cooperation and Development (EPA-600/3-77-086). Environm. Res. Lab. Corvallis.

Edmondson, W.T., 1998. *Perspectives in Plankton Studies.* Pp. 331–361. In R. deBernardi, G. Giussani, and L. Barbanti (eds.), Scientific Perspectives in Theoretical and Applied Limnology. Mem.1st Ital. Idrobiol. 47: 1–378.

Edmondson, W.T., and J.T. Lehman, 1981. The Effect of Changes in the Nutrient Income on the Condition of Lake Washington. *Limnol. Oceangr.* 26: 1–29.

Edwards, R.T., 1987. Sestonic Bacteria as a Food Source for Filtering Invertebrates in Two Southeastern Blackwater Rivers. *Limnol. Oceanogr.* 32: 221–234.

Eilers, J.M., T.J. Sullivan, and K.C. Hurley, 1990. The Most Dilute Lake in the World? *Hydrobiologia* 199: 1–6.

Eisenreich, S.J., P.D. Capel, J.A. Robbins, and R. Bourbonniere, 1989. Accumulation and Diagenesis of Chlorinated Hydrocarbons in Lacustrine Sediments. *Environ. Sci. Technol.* 23: 1116–1126.

Eklöv, P., 1997. Effect of Habitat Complexity and Prey Abundance on the Spatial and Temporal Distributions of Perch (*Perca fluviatilis*) and Pike (*Esox lucius*). *Can J. Fish. Aquat. Sci.* 54: 1520–1531.

Elser, J., and J. Urabe, 1999. The Stoichiometry of Consumer-driven Nutrient Recycling: Theory, Observations, and Consequences. *Ecology* 80: 735–751.

Elser, J.J., E.R. Marzolf, and C.R. Goldman, 1990. Phosphorous and Nitrogen Limitation of Phytoplankton Growth in the Freshwaters of North America: A Review and Critique of Experimental Enrichments. *Can. J. Fish. Aquat. Sci.* 47: 1468–1477.

Elser, J.J., W.F. Fagan, R.F. Denno, D.R. Dobberfuhl, A. Folarin, A. Huberty, S. Interlandi, S.S. Kilham, E. McCauley, K. L. Schulz, E.H. Siemann, R.W. Sterner, 2000. Nutritional Constraints in Terrestrial and Freshwater Food Webs. Nature 408:578–580.

Elwood, J.W., J.D. Newbold, R.V. O'Neill, R.W. Stark, and P.T. Singley, 1981. The Role of Microbes Associated with Organic and Inorganic Substrates in Phosphorus Spiralling in a Woodland Stream. *Verh. Int. Ver. Limnol.* 21: 850–856.

Emery, K.O., and G.T. Csanady, 1973. Surface Circulation of Lakes and Nearly Land-locked Seas. *Proc. Nat. Acad. Sci. USA*, 70: 93–97.

Emmenegger, L., D.W. King, L. Sigg, and B. Sulzberger, 1998. Oxidation Kinetics of Fe(II) in a Eutrophic Swiss Lake. *Env. Sci. Technol.* 32: 2990–2996.

Enell, M., and S. Löfgren, 1988. Phosphorus in Interstitial Water: Methods and Dynamics. *Hydrobiologia* 170: 103–132.

Engel, S., 1987. The Restructuring of Littoral Zones. *Lake Res. Man.* 3: 235–242.

Enríquez, S., C.M. Duarte, and K. Sand-Jensen, 1993. Patterns in Decomposition Rates Among Photosynthetic Organisms: The Importance of Detritus C:N:P Content. *Oecologia* 94: 457–471.

Eriksen, C.H., 1963. Respiratory Regulation in *Ephemera simulans* Walker and *Hexagenia limbata* (Serville) (Ephemeroptera). *J. Exp. Biol.* 40: 455–467.

Erikson, R., K. Vammen, A. Zelaya, and R.T. Bell, 1998. Distribution and Dynamics of Bacterioplankton Production in a Polymictic Tropical Lake (Lago Xolotlan, Nicaragua). *Hydrobiologia* 382: 27–39.

Eriksson, F., E. Hörnström, P. Mossberg, and P. Nyberg, 1983. Ecological Effects of Lime Treatment of Acidified Lakes and Rivers in Sweden. *Hydrobiologia* 101: 145–163.

Eriksson, M.O.G., 1979. Competition Between Freshwater Fish and Goldeneyes *Bucephala clangula* (L.) for Common Prey. *Oecologia* 41: 99–107.

Eugster, H.P., and L.A. Hardie, 1978. *Saline Lakes*. Pp. 237–293. In A. Lerman (ed.), Lakes: Chemistry, Geology, Physics. New York: Springer-Verlag.

Evans, M.S., and D.W. Sell, 1983. Zooplankton Sampling Strategies for Environmental Studies. *Hydrobiologia* 99: 215–223.

Fagerbakke K.M., M. Heldal, and S. Norland, 1996. Content of Carbon, Nitrogen, Oxygen, Sulfur and Phosphorus in Native Aquatic and Cultured Bacteria. *Aquat. Microb. Ecology* 10: 15–27.

Fahnenstiel, G.L., L. Sicko-Goad, D. Scavia, and E. F. Stoermer, 1986. Importance of Picoplankton in Lake Superior. *Can. J. Fish. Aquat. Sci.* 43: 235–240.

FAO, 1995. *Review of the State of World Fishery Resources: Inland Capture Fisheries*. FAO Fisheries Circular #885. Rome.

FAO, 1995. *Review of the State of World Fishery Resources: Aquaculture*. FAO Fisheries Circular #886. Rome.

FAO, 1997. *FAO Aquaculture Newsletter:16*. Rome.

Fast, A.W., and P.A. Tyler, 1981. The Re-establishment of Meromixis in Hemlock Lake, Michigan, After Artificial Destratification. *Int. Rev. Ges. Hydrobiologia* 66: 665–674.

Fee, E.J., 1979. A Relation Between Lake Morphometry and Primary Productivity and its Use in Interpreting Whole-lake Eutrophication Experiments. *Limnol. Oceanogr.* 24: 401–416.

Fee, E.J., R.E. Hecky, and H.A. Welch, 1987. Phytoplankton Photosynthesis Parameters in Central Canadian Lakes. *J. Plankton Res.* 9: 305–316.

Feminella, J.W., M.E. Power, and V.H. Resh, 1989. Periphyton Responses to Invertebrate Grazing and Riparian Canopy in Three Northern California Coastal Streams. *Freshw. Biol.* 22: 445–457.

Fenchel, T., and T.H. Blackburn, 1979. *Bacterial and Mineral Cycling*. New York: Acad. Press.

Ferguson, A.J.D., and D.M. Harper, 1982. Rutland Water Phytoplankton: The Development of an Asset or a Nuisance. *Hydrobiologia* 88: 117–133.

Fernando, C.H., and J. Holcík, 1982. The Nature of Fish Communities: A Factor Influencing the Fishery Potential and Yield of Tropical Lakes and Reservoirs. *Hydrobiologia* 97: 127–140.

Ferris, J.M., and H.R. Burton, 1988. The Annual Cycle of Heat Content and Mechanical Stability of Hypersaline Deep Lake, Vestfold Hills, Antarctica. *Hydrobiologia* 165: 115–128.

Findenegg, I., 1964. Types of Planktic Primary Production in the Lakes of the Eastern Alps as Found by the Radioactive Carbon Method. *Verh. Int. Ver. Limnol.* 15: 352–359.

Findenegg, I., 1967. Die Verschmutzung Österreichischer Alpenseen aus Biologisch-chemischer Sicht. *Berichte zur Raumforschung und Raumplanung* 11: 3–12.

Findlay, D.L., and S.E.M. Kasian, 1990. Phytoplankton Communities of Lakes Experimentally Acidified with Sulfuric and Nitric Acids. *Can. J. Fish. Aquat. Sci.* 47: 1378–1386.

Finlay, B.J., K.J. Clarke, A.J. Cowling, R.M. Hindle, A. Rogerson, and U.G. Berninger, 1988. On the Abundance and Distribution of Protozoa and Their Food in a Productive Freshwater Pond. *Europ. J. Protistol.* 23: 205–217.

Finney, B.P., I. Gregory-Eaves, J. Sweetman, M.S.V. Douglas, and J.P. Smol, 2000. Impact of Climatic Change and Fishing on Pacific Salmon Abundance Over the Past 300 Years. Science 290: 795–799.

Fischer, J.M., and T.M. Frost, 1997. Indirect Effects of Lake Acidification on *Chaoborus* Population Dynamics: The Role of Food Limitation and Predators. *Can. J. Fish. Aquat. Sci.* 54: 637–646.

Fisher, S.G., 1994. *Pattern, Process and Scale in Freshwater Systems: Some Unifying Thoughts*. Pp. 575–591. In P.S. Giller, A.G. Hildrew, and D.G. Raffaelli (eds.), Aquatic Ecology: Scale, Pattern and Process. Oxford: Blackwell Sci. Publ.

Fitzgerald, W.F., and C.J. Watras, 1989. Mercury in the Surficial Waters of Northern Wisconsin Lakes. *Sci. Tot. Env.* 87/88: 223–232.

Fitzsimmons, K. (ed.), 1997. *Tilapia aquaculture. Proc. 4th Int. Symp. Tilapia Aquaculture*. Ithaca: Northeast Reg. Agr. Eng. Serv.

Fleischer, S., and L. Stibe, 1991. Drainage Basin Management—Reducing River Transported Nitrogen. *Verh. Int. Ver. Limnol.* 24: 1753–1755.

Flett, R.J., D.W. Schindler, R.D. Hamilton, and N.E.R. Campbell, 1980. Nitrogen Fixation in Canadian Precambrian Shield Lakes. *Can. J. Fish. Aquat. Sci.* 37: 494–505.

Forsberg, B.R., A.H. Devol, J.E. Richey, L.A. Martinelli, and H. dos Santos, 1988. Factors Controlling Nutrient Concentrations in Amazon Floodplain Lakes. *Limnol. Oceanogr.* 33: 41–56.

Forsberg, C., 1985. *Lake Recovery in Sweden*. Pp. 272–281. In European Water Poll. Control Assoc. Int. Congress Proc., Lakes: Pollution and Recovery. Rome.

Forsberg, C., S.O. Ryding, A. Claesson, and A. Forsberg, 1978. Water Chemical Analyses and/or Algal Assay?—Sewage Effluent and Polluted Lake Water Studies. *Mitt. Int. Ver. Limnol.* 21: 352–363.

Forsberg, C., and S.O. Ryding, 1979. Correlations Between Discharge and Transport of Elements. Studies From Six Catchment Areas. *Vatten* 35: 291–300.

Forsberg, C., and S.O. Ryding, 1980. Eutrophication Parameters and Trophic State Indices in 30 Swedish Waste-receiving Lakes. *Arch. Hydrobiol.* 89: 189–207.

Förstner, U., and G.T.W. Wittmann, 1979. *Metal Pollution in the Aquatic Environment*. Berlin: Springer-Verlag.

France, R.L., 1991. Empirical Methodology for Predicting Changes in Species Range Extension and Richness Associated With Climate Warming. *Int. J. Biometeorol.* 34: 211–216.

France, R.L., 1995. Carbon-13 Enrichment in Benthic Compared to Planktonic Algae: Foodweb Implications. *Mar. Ecology Prog. Ser.* 124: 307–312.

Francisco, D.E., R.A. Mah, and A.C. Rabin, 1973. Acridine Orange Epifluorescence Technique for Counting Bacteria in Natural Waters. *Trans. Amer. Micros. Soc.* 92: 416–421.

Frantz, T.C., and A.J. Cordone, 1967. Observations on Deepwater Plants in Lake Tahoe, California and Nevada. *Ecology* 48: 709–714.

Franzmann, P.D., P.P. Deprez, H.R. Burton, and J. van den Hoff, 1987. Limnology of Organic Lake, Antarctica, A Meromictic

Lake that Contains High Concentrations of Dimethyl Sulfide. *Aust. J. Mar. Freshw. Res.* 38: 409–417.

Freeze, R.A., and J.A. Cherry, 1979. *Groundwater.* Englewood Cliffs: Prentice Hall.

Frey, D.G., 1963. *Limnology in North America.* Madison: Univ. of Wisconsin Press.

Frey, D.G., 1966. *Limnology.* Pp. 297–320. In Ind. Acad Sci., The Indiana Sesquicentennial Volume: Natural Features of Indiana. Indianapolis.

Frey, D.G., (ed.), 1955. Längsee: A History of Meromixis. *Mem. Ist. Ital. Idrobiol.* 8(suppl.): 141–164.

Froelich, P.N., G.P. Klinkhammer, M.L. Bender, N.A. Luedtke, G.R. Heath, and D. Cullen, P. Dauphin, B. Hartman, and V Maynard, 1979. Early Oxidation of Organic Matter in Pelagic Sediments of the Eastern Equatorial Atlantic: Suboxic Diagenesis. *Geochim. Cosmochim. Acta.* 43: 1075–1090.

Frost, T.M., D.L. DeAngelis, S.M. Bartell, D.J. Hall, and S.H. Hurlbert, 1988. *Scale in the Design and Interpretation of Aquatic Community Research.* Pp. 229–258. In S.R. Carpenter (ed.), Complex Interactions in Lake Communities. New York: Springer-Verlag.

Fuhrman, J.A., 1999. Marine Viruses and Their Biogeochemical and Ecological Effects. *Nature* 399: 541–548.

Fukami, K., B. Meier, and J. Overbeck, 1991. Vertical and Temporal Changes in Bacterial Production and Its Consumption by Heterotrophic Nanoflagellates in a North German Eutrophic Lake. *Arch. Hydrobiol.* 122: 129–145.

Fulton, R.J., and J.T. Andrews (eds.), 1987. The Laurentide Ice Sheet. *Geographie Physique et Quaternaire* 41: 179–318.

Gächter, R., and J. Bloesch, 1985. Seasonal and Vertical Variation in the C:P Ratio of Suspended and Settling Seston of Lakes. *Hydrobiologia* 128: 193–200.

Gächter, R., J.S. Meyer, and A. Mares, 1988. Contribution of Bacteria to Release and Fixation of Phosphorus in Lake Sediments. *Limnol. Oceanogr.* 33: 1542–1558.

Gächter, R., and J.S. Meyer, 1993. The Role of Microorganisms in Mobilization and Fixation of Phosphorus in Sediments. *Hydrobiologia* 253: 103–121.

Gächter, R., and A. Wüest, 1993. Effects of Artificial Aeration on Trophic Status and Hypolimnetic Oxygen Concentration in Lakes. *EAWAG-News* 34: 25–30.

Gächter, R., and B. Wehrli, 1998. Ten Years of Artificial Mixing and Oxygenation: No Effect on the Internal Phosphorous Loading of Two Eutrophic Lakes. *Env. Sci. Technol.* 32: 3659–3665.

Gal, G., E.R. Loew, L.G. Rudstam, and A.M. Mohammadian, 1999. Light Diel Vertical Migration: Spectral Sensitivity and Light Avoidance by *Mysis relicta. Can. J. Fish. Auqat. Sci.* 56: 311–322.

Gala, W.R., and J.P. Giesy, 1991. Effects of Ultraviolet Radiation on the Primary Production of Natural Phytoplankton Assemblages in Lake Michigan. *Ecotoxicol. and Env. Safety* 22: 345–361.

Galat, D.L., 1986. Organic Carbon Flux to a Large Salt Lake: Pyramid Lake, Nevada, USA. *Int. Rev. Ges. Hydrobiol.* 71: 621–654.

Galat, D.L., and R.L. Jacobsen, 1985. Recurrent Aragonite Precipitation in Saline-Alkaline Pyramid Lake, Nevada. *Arch. Hydrobiologia* 105: 137–159.

Galat, D.L., and J.P. Verdin, 1988. Magnitude of Blue-green Algal Blooms in a Saline Desert Lake Evaluated by Remote Sensing: Evidence for Nitrogen Control. *Can. J. Fish. Aquat. Sci.* 45: 1959–1967.

Galloway, J.N., 1995. Acid Deposition: Perspectives in Time and Space. *Water, Air and Soil Poll.* 85: 15–24.

Ganf, G.G., and A.J. Horne, 1975. Diurnal Stratification, Photosynthesis and Nitrogen-fixation in a Shallow, Equatorial Lake (Lake George, Uganda). *Freshw. Biol.* 5: 13–39.

Gardner, W.S., B.J. Eadie, J.F. Chandler, C.C. Parrish, and J.M. Malczyk, 1989. Mass Flux and "Nutritional Composition" of Settling Epilimnetic Particles in Lake Michigan. *Can. J. Fish. Aquat. Sci.* 46: 1118–1124.

Gasol, J.M., and D. Vaqué, 1993. Lack of Coupling Between Heterotrophic Nanoflagellates and Bacteria: A General Phenomenon Across Aquatic Systems? *Limnol. Oceanogr.* 38: 657–665.

Gasol, J.M., A.M. Simons, and J. Kalff, 1995. Patterns in the Top-down Versus Bottom-up Regulation of Heterotrophic Nanoflagellates in Temperate Lakes. *J. Plankton Res.* 17: 1879–1903.

Gebre-Mariam, Z., and W.D. Taylor, 1989. Seasonality and Spatial Variation in Abundance, Biomass and Activity of Heterotrophic Bacterioplankton in Relation to Some Biotic and Abiotic Variables in an Ethiopian Rift-valley Lake (Awassa). *Freshw. Biol.* 22: 355–368.

Geller, W., 1986. Diurnal Vertical Migration of Zooplankton in a Temperate Great Lake (L. Constance): A Starvation Avoidance Mechanism? *Arch. Hydrobiologia* 74(Suppl.): 1–60.

Geller, W., 1992. The Temperature Stratification and Related Characteristics of Chilean Lakes in Midsummer. *Aquat. Sci.* 54: 37–57.

Geller, W., and H. Müller, 1981. The Filtration Apparatus of Cladocera: Filter Mesh-sizes and Their Implications of Food Selectivity. *Oecologia* 49: 316–321.

George, D.G., 1981. Wind-induced Water Movements in the South Basin of Windermere. *Freshw. Biol.,* 11:37–60.

George, D.G., and R.W. Edwards, 1973. Daphnia Distribution within Langmuir Circulations. *Limnol. Oceanogr.,* 18: 798–800.

George, D.G., and G.P. Harris, 1985. The Effect of Climate on Long-term Changes in the Crustacean Zooplankton Biomass of Lake Windermere, U.K. *Nature* 316: 536–539.

George, D.G., and D.H. Jones, 1987. Catchment Effects on the Horizontal Distribution of Phytoplankton in Five of Scotland's Largest Freshwater Lochs. *J. Ecol.* 75: 43–59.

George, D.G., D.P. Hewitt, J.W.G Lund, and W.J.P. Smyly, 1990. The Relative Effects of Enrichment and Climate Change on the Long-term Dynamics of *Daphnia* in Esthwaite Water, Cumbria. *Freshw. Biol.* 23: 55–70.

Gerten, D., and R. Adrian, 2000. Climate-driven Changes in Spring Plankton Dynamics and the Sensitivity of Shallow Polymictic Lakes to the North Atlantic Oscillation. *Limnol. Oceangr.* 45: 1058–1066.

Gibbs, M.M., 1987. *Groundwater Contributions to Water and Nutrient Budgets.* Pp. 167–171. In W.N. Vant (ed.), Lake Managers Handbook. Water and Soil Misc. Publ. #103. Wellington: Nat. Water Soil Cons. Auth.

Gibbs, R.J., 1970. Mechanisms Controlling World Water Chemistry. *Science* 170: 1088–1090.

Gibbs, R.J., 1992. A Reply to the Comment of Eilers et al. *Limnol. Oceanogr.* 37: 1338–1339.

Gibson, C.E., Y. Wu, S.J. Smith, and S.A. Wolfe-Murphy, 1995. Synoptic Limnology of a Diverse Geological Region: Catchment and Water Chemistry. *Hydrobiologia* 306: 213–227.

Giesy, J.P., J.P. Ludwig, and D.E. Tillitt, 1994. Deformities in Birds of the Great Lakes Region: Assigning Causality. *Env. Sci. Technol.,* 28: 128–135.

Gleick, P.H., 1993. *Water in Crisis: A Guide to the World's Freshwater Resources.* New York: Oxford Univ. Press.

Gliwicz, Z.M., 1986a. Predation and the Evolution of Vertical Migration in Zooplankton. *Nature* 320: 746–748.

Gliwicz, Z.M., 1986b. A Lunar Cycle in Zooplankton. *Ecology* 67: 883–897.

Gliwicz, Z.M., A. Hillbricht-Ilkowska, and T. Weglenska, 1978. Contribution of Fish and Invertebrate Predation to the Elimination of Zooplankton Biomass in Two Polish Lakes. *Verh. Int. Ver. Limnol.* 20: 1007–1011.

Gliwicz, Z.M., and A. Rykowska, 1992. "Shore Avoidance" in Zooplankton: A Predator-induced Behavior or Predator-induced Mortality? *J. Plankton Res.* 14: 1331–1342.

Gloor, M., A. Wüest, and M. Münnich, 1994. Benthic Boundary Mixing and Resuspension Induced by Internal Seiches. *Hydrobiologia*, 284: 59–68.

Goedkoop, W.,and R.K. Johnson, 1996. Pelagic–Benthic Coupling: Profundal Benthic Community Response to Spring Diatom Deposition in the Mesotrophic Lake Erken. *Limnol. Oceanogr.* 41: 636–647.

Goedkoop, W., K.R. Gullberg, R.K. Johnson, and I. Ahlgren, 1997. Microbial Response of a Freshwater Benthic Community to a Simulated Diatom Sedimentation Event: Interactive Effects of Benthic Fauna. *Microb. Ecol.* 34: 131–143.

Goldman, C.R., 1981. Lake Tahoe: Two Decades of Change in a Nitrogen-deficient Oligotrophic Lake. *Verh. Int. Ver. Limnol.* 21: 45–70.

Goldman, C.R., 1993. The Conservation of Two Large Lakes: Tahoe and Baikal. *Verh. Int. Ver. Limnol.* 25: 388–391.

Goldman, C.R., 2000. Four Decades of Change in Two Subalpine Lakes. Verh. Int. Ver. Limnol. 27: 7–26.

Golterman, H.L., 1975. *Chemistry.* Pp. 39–80. In B.A. Whitton (ed.), River Ecology. Berkeley: Univ. of Cal. Press.

Golterman, H.L., and F.A. Kouwe, 1980. *Chemical Budgets and Nutrient Pathways.* Pp. 85–140. In E.D. Le Cren, and R.H. Lowe-McConnell (eds.), The Functioning of Freshwater Ecosystems. Cambridge: Cambridge Univ. Press.

Golterman, H.L., R.S. Clymo, and M.A.M. Ohnstad, 1978. *Methods for Physical and Chemical Analysis of Fresh Waters, 2nd ed.* IBP Handbook #8. Oxford: Blackwell Sci. Publ.

Golterman, H.L., R.S. Clymo, E.P.H. Best, and J. Lauga, 1988. *Methods of Exploration and Analysis of the Environment of Aquatic Vegetation.* Pp. 31–61. In J.J. Symoens (ed.), Handbook of Vegetation Science: Vegetation of Inland Waters, 15/1. Dordrecht: Kluwer Acad. Publ.

Golubev, G.N., 1996. Caspian and Aral Seas: Two Different Paths of Environmental Degradation. *Verh. Int. Ver. Limnol.* 26: 159–166.

Gorham, E., 1953. Some Early Ideas Concerning the Nature, Origin and Development of Peat Lands. *J. Ecology* 41: 257–274.

Gorham, E., 1964. Morphometric Control of Annual Heat Budgets in Temperate Lakes. *Limnol. Oceanogr.* 9: 525–529.

Gorham, E., 1996. Wetlands: An Essential Component of Curricula in Limnology. Pp. 234–246. In NRC, Freshwater Ecosystems: Revitalizing Educational Programs in Limnology. Washington, DC: Nat. Acad. Press.

Gorham, E., W.E. Dean, and J.E. Sanger, 1983. The Chemical Composition of Lakes in the North-central United States. *Limnol. Oceanogr.* 28: 287–301.

Gorham, E., F.B. Martin, and J.T. Litzau, 1984. Acid Rain: Ionic Correlations in the Eastern United States, 1980–1981. *Science* 225: 407–409.

Gorham, E., and F.M. Boyce, 1989. Influence of Lake Surface Area and Depth Upon Thermal Stratification and the Depth of the Summer Thermocline. *J. Great Lakes Res.* 15: 233–245.

Gorlenko, V.M., G.A. Dubinina, and S.I. Kuznetsov, 1983. *The Ecology of Aquatic Micro-organisms.* Stuttgart: E. Schweizerbant'sche Verlagsbuchhandlung.

Graham, L.E., and L.W. Wilcox, 2000. *Algae.* Upper Saddle River: Prentice Hall.

Green, J., 1993. Diversity and Dominance in Planktonic Rotifers. *Hydrobiologia* 255/256: 345–352.

Green, J., and H. Kramadibrata, 1988. A Note on Lake Goang, An Unusual Acid Lake in Flores, Indonesia. *Freshw. Biol.* 20: 195–198.

Gregory, K.J., and D.E. Walling, 1973. *Drainage Basin Form and Process: A Geomorphological Approach.* London: Edward Arnold Publ.

Grieb, T.M., C.T. Driscoll, S.P. Gloss, C.L. Schofield, G.L. Bowie, and D.B. Porcella, 1990. Factors Affecting Mercury Accumulation in Fish in the Upper Michigan Peninsula. *Environ. Toxicol. Chem.* 9: 919–930.

Grigorovich, I.A., O.V. Pashkova, Y.F. Gromova, and C.D.A. van Overdyk, 1998. *Bythotrephes longimanus* in the Commonwealth of Independent States: Variability, Distribution and Ecology. *Hydrobiologia* 379: 183–198.

Grobbelaar, J.U., 1985. Phytoplankton Productivity in Turbid Waters. *J. Plankton Res.* 7: 653–663.

Groeger, A.W., and B.L. Kimmel, 1984. *Organic Matter Supply and Processing in Lakes and Reservoirs.* Pp.282–285. In EPA Report #440/5/84-001, Lake and Reservoir Management. Washington, DC.

Grossant, H.P., and M. Simon, 1998. Bacterial Colonization and Microbial Decomposition of Limnetic Organic Aggregates (Lake Snow). *Aquat. Microb. Ecol.* 15: 127–140.

Güde, H., 1990. *Bacterial Production and the Flow of Organic Matter in Lake Constance.* Pp. 489–502. In M.M.Tilzer, and C. Serruya (eds.), Large Lakes: Ecological Structure and Function. New York: Springer Verlag.

Gunkel, V.G., and A. Sztraka, 1986. Untersuchungen zum Verhalten von Schwermetallen in Gewässern. II. Die Bedeutung der Eisen- und Mangan- Remobilisierung für die Hypolimnishe Anreicherung von Schwermetallen. *Arch. Hydrobiologia* 106: 91–117.

Gunn, J.M., and K.H. Mills, 1998. The Potential for Restoration of Acid Damaged Lake Trout Lakes. *Restoration Ecology* 6: 390–397.

Guy, M., W.D. Taylor, and J.C.H. Carter, 1994. Decline in Total Phosphorus in the Surface Water of Lakes During Summer Stratification, and its Relationship to Size Distribution of Particles and Sedimentation. *Can. J. Fish. Aquat. Sci.* 51: 1330–1337.

Haga, H., T. Nagata, and M. Sakamoto, 1995. Size-fractionated NH_4^+ Regeneration in the Pelagic Environments of Two Mesotrophic Lakes. *Limnol. Oceanogr.* 40: 1091–1099.

Hairston, N.G., 1996. Zooplankton Egg Banks as Biotic Reservoirs in Changing Environments. *Limnol. Oceanogr.* 41: 1087–1092.

Håkanson, L., 1977. On Lake Form, Lake Volume and Lake Hypsographic Survey. *Geografiska Annaler* 59A: 1–29.

Håkanson, L., 1981. *A Manual of Lake Morphometry.* Berlin: Springer-Verlag.

Håkanson, L., and T. Ahl, 1976. *Vättern—Recenta Sediment och Sedimentkemi, [Lake Vättern—Recent Sedimentary Deposits and Sediment Chemistry].* NLU Rapport 88. Uppsala: Statens Naturvårdverk.

Håkanson, L., and M. Jansson, 1983. *Principles of Lake Sedimentology.* Berlin: Springer-Verlag.

Håkanson, L., and B. Karlsson, 1984. On the Relationship Between Regional Geomorphology and Lake Morphometry—A Swedish Example. *Geografiska Annaler* 66: 103–119.

Håkanson, L., T. Andersson, and Å. Nilsson, 1990. Mercury in Fish in Swedish Lakes: Linkages to Domestic and European Sources of Emission. *Water, Air and Soil Poll.* 50: 171–191.

Hall, D.J., W.E. Cooper, and E.E. Werner, 1970. An Experimental Approach to the Production Dynamics and Structure of Freshwater Animal Communities. *Limnol. Oceanogr.* 15: 839–928.

Hall, G.H., 1986. *Nitrification in Lakes.* Pp. 127–156. In J.I. Prosser (ed.), Nitrification. Oxford: IRL Press.

Hall, K.J., and T.G. Northcote, 1990. Production and Decomposition Processes in a Saline Meromictic Lake. *Hydrobiologia* 197: 115–128.

Hall, R.I., and J.P. Smol, 1999. *Diatoms as Indicators of Lake Eutrophication.* Pp. 128–168. In E.F. Stoermer, and J.P. Smol (eds.), The Diatoms: Applications for the Environmental and Earth Sciences. Cambridge: Cambridge Univ. Press.

Hamburger, K., P.C. Dall, C. Lindegaard, and I.B. Nilson, 2000. Survival and Energy Metabolism in an Oxygen-deficient Environment. Field and Laboratory Studies on the Bottom Fauna from the Profundal Zone of Lake Esrom, Denmark. *Hydrobiologia* 432: 173–188.

Hamilton, D.P., and S.F. Mitchell, 1988. Effects of Wind on Nitrogen, Phosphorous and Chlorophyll in a Shallow New Zealand Lake. *Verh. Int. Ver. Limnol.* 23: 624–628.

Hamilton, S.J., and P.M. Mehrle, 1986. Metallothionein in Fish: Review of its Importance in Assessing Stress from Metal Contaminants. *Trans. Amer. Fish. Soc.* 115: 596–609.

Hamilton, S.K., S.J. Sippel, and J.M. Melack, 1995. Oxygen Depletion and Carbon Dioxide and Methane Production in Waters of the Pantanal Wetland of Brazil. *Biogeochem.* 30: 115–141.

Hamilton, S.K., S.J. Sippel, D.F. Calheiros, and J.M. Melack, 1997. An Anoxic Event and Other Biogeochemical Effects of the Pantanal Wetland on the Paraguay River. *Limnol. Oceanogr.* 42: 257–272.

Hamilton-Taylor, J., and W. Davison, 1995. *Redox-driven Cycling of Trace Elements in Lakes.* Pp. 217–263. In A. Lerman, D.M. Imboden, and J.R. Gat (eds.), Physics and Chemistry of Lakes. Berlin: Springer-Verlag.

Hammer, U.T., 1986. *Saline Lake Ecosystems of the World. Monographiae Biologicae, vol. 59.* Dordrecht: W. Junk Publ.

Hammer, U.T., J.S. Sheard, and J. Kranabetter, 1990. Distribution and Abundance of Littoral Benthic Fauna in Canadian Prairie Saline Lakes. *Hydrobiologia* 197: 173–192.

Hanarato, T., and M. Yasumo, 1989. Zooplankton Community Structure Driven by Vertebrate and Invertebrate Predators. *Oecologia* 81: 450–458.

Haney, J.F., 1973. An In Situ Examination of the Grazing Activities of Natural Zooplankton Comunities. *Arch. Hydrobiol* 72: 87–132.

Haney, J.F., A. Craggy, K. Kimball, and F. Weeks, 1990. Light Control of Evening Migration by *Chaoborus punctipennis* Larvae. *Limnol. Oceanogr.* 35: 1068–1078.

Hanna, M., 1990. Evaluation of Models Predicting Mixing Depth. *Can. J. Fish. Aquat. Sci.* 47: 940–947.

Hanson, J.M., and W.C. Leggett, 1982. Empirical Prediction of Fish Biomass and Yield. *Can. J. Fish. Aquat. Sci.* 39: 257–263.

Hanson, J.M., and R.H. Peters, 1984. Empirical Prediction of Crustacean Zooplankton Biomass and Profundal Macrobenthos Biomass in Lakes. *Can. J. Fish. Aquat. Sci.* 41: 439–445.

Hanson, M.A., M.G. Butler, J.L. Richardson, and J.L. Arndt, 1990. Indirect Effects of Fish Predation on Calcite Supersaturation, Precipitation and Turbidity in a Shallow Prairie Lake. *Freshw. Biol.* 24: 547–556.

Hanson, M.A., and M.G. Butler, 1994. Responses to Food Web Manipulation in a Shallow Waterfowl Lake. *Hydrobiologia* 279/280: 457–466.

Hansson, L.A., 1996. Algal Recruitment from Lake Sediments in Relation to Grazing, Sinking, and Dominance Patterns in the Phytoplankton Community. *Limnol. Oceanogr.* 41: 1312–1323.

Happey-Wood, C., 1988. *Ecology of Freshwater Planktonic Green Algae.* Pp.175–226. In C.D. Sandgren (ed.), Growth and Reproductive Strategies in Freshwater Phytoplankton. Cambridge: Cambridge Univ. Press.

Happey-Wood, C.M., 1991. Temporal and Spatial Patterns in the Distribution and Abundance of Pico-, Nano- and Microphytoplankton in an Upland Lake. *Freshw. Biol.* 26: 453–480.

Harding, W.R., 1997. Phytoplankton Primary Production in a Shallow, Well-mixed, Hypertrophic South African Lake. *Hydrobiologia* 344: 87–102.

Hardy, F.J., K.S. Shortreed, and J.G. Stockner, 1986. Bacterioplankton, Phytoplankton, and Zooplankton Communities in a British Columbia Coastal Lake Before and After Nutrient Reduction. *Can. J. Fish. Aquat. Sci.* 43: 1504–1514.

Harremoës, P., E. Arvin, and M. Henze, 1985. *Wastewater Nutrient Removal—A State-of-the-Art Review.* Pp. 120–127. In Proc. Int.

Congress of the European Water Poll. Control Assoc., Lakes: Pollution and Recovery. Rome.

Harris, G.P., 1978. Photosynthesis, Productivity, and Growth: The Physiological Ecology of Phytoplankton. *Arch. Hydrobiologia Beih. Ergeb. Limnol.* 10: 1–171.

Harris, G.P., 1980. *The Measurement of Photosynthesis in Natural Populations of Phytoplankton.* Pp. 129–187. In I. Morris (ed.), The Physiological Ecology of Phytoplankton. Oxford: Blackwell Sci. Publ.

Harris, G.P., 1986. *Phytoplankton Ecology: Structure, Function and Fluctuation.* London: Chapman and Hall.

Harris, G.P., 1994. Pattern, Process and Prediction in Aquatic Ecology. A Limnological View of Some General Ecological Problems. *Freshw. Biol.* 32: 143–160.

Harris, G.P., 1999. This is Not the End of Limnology (or of Science): The World May Well Be a Lot Simpler Than We Think. *Freshw. Biol.* 42: 689–706.

Harris, G.P., and J.N.A. Lott, 1973. Observations of Langmuir circulations in Lake Ontario. *Limnol. Oceanogr.* 18: 584–589.

Hart, B.T., P. Bailey, R. Edwards, K. Hortle, K. James, A. McMahon, C. Meredith, and K. Swadling, 1991. A Review of the Salt Sensitivity of the Australian Freshwater Biota. *Hydrobiologia* 210: 105–144.

Hart, R.C., 1988. Zooplankton Feeding Rates in Relation to Suspended Sediment Content: Potential Influences on Community Structure in a Turbid Reservoir. *Freshw. Biol.* 19: 123–139.

Hart, R.C., and B.R. Allanson, 1984. *Limnological Criteria for Management of Water Quality in the Southern Hemisphere. South African Nat. Sci. Prog. Report #93.* Pretoria: CSIR.

Hasler, A.D., and W.G. Einsele, 1948. Fertilization for Increasing Productivity of Natural Inland Waters. *Trans. N. Amer. Wildl. Conf.* 13: 527–555.

Havel, J.E., 1987. *Predator-induced Defenses: A Review.* Pp. 263–278. In W.C. Kerfoot, and A. Sih (eds.), Predation: Direct and Indirect Impacts on Aquatic Communities. Hanover: New England Press.

Hawkes, H.A., 1975. *River Zonation and Classification.* Pp. 312–374. In B.A. Whitton (ed.), River Ecology. Berkeley: Univ. Cal. Press.

Hayes, F.R., B.L. Reid, and M.L. Cameron, 1958. Lake Water and Sediment. II. Oxidation-reduction Relations at the Mud–Water Interface. *Limnol. Oceanog.* 3: 308–317.

Healey, F.P., and L.L. Hendzel, 1979. Indicators of Phosphorous and Nitrogen Deficiency in Five Algae in Culture. *Can. J. Fish. Res. Bd.* 36: 1364–1369.

Healey, F.P., and L.L. Hendzel, 1980. Physiological Indicators of Nutrient Deficiency in Lake Phytoplankton. *Can. J. Fish. Aquat. Sci.* 37: 442–453.

Heath, C.W., 1988. Annual Primary Productivity of an Antarctic Continental Lake: Phytoplankton and Benthic Algal Mat Production Strategies. *Hydrobiologia* 165:77–87.

Hecky, R.E., 1978. The Kiru-Tanganyika Basin: The Last 14,000 Years. *Pol. Arch. Hydrobiologia* 25: 159–165.

Hecky, R.E., and H.J. Kling, 1987. Phytoplankton Ecology of the Great Lakes in the Rift Valleys of Central Africa. *Arch. Hydrobiologia Beih.* 25: 197–228.

Hecky, R.E., and P. Kilham, 1988. Nutrient Limitation of Phytoplankton in Freshwater and Marine Environments: A Review of Recent Evidence on the Effects of Enrichment. *Limnol. Oceanogr.* 33: 796–822.

Hedin, L.O., G.E. Likens, and F.H. Bormann, 1987. Decrease in Precipitation Acidity Resulting from Decreased SO_4^{2-} Concentration. *Nature* 325: 244–246.

Hedin, L.O., J.C. von Fischer, N.E. Ostrom, B.P. Kennedy, M.G. Brown, and G.P. Robertson, 1998. Thermodynamic Constraints on Nitrogen Transformations and Other Biogeochemical Processes at Soil–Stream Interfaces. *Ecology* 79: 684–703.

Hem, J.D., 1985. *Study and Interpretation of the Chemical Characteristics of Natural Water, 3rd ed.* U.S. Geological Survey Water-Supply Paper #2254. Washington, DC.

Henriksen, A., 1980. *Acidification of Freshwaters—A Large-scale Titration.* Pp. 68–74. In D. Drabløs, and A. Tollan (eds.), Ecological Impact of Acid Precipitation. SNSF-Project #1432, Oslo.

Henriksen, A., 1987. *Freshwater Acidification in Norway—A "Direct Response" Process?* Pp. 187–197. In Acidification and Water Pathways, Vol. II. Norway Nat. Comm. Hydrol., Oslo.

Henriksen, A., and D.F. Brakke, 1988. Increasing Contributions of Nitrogen to the Acidity of Surface Waters in Norway. *Water, Air, and Soil Poll.* 42: 183–271.

Henriksen, A., and D.F. Brakke, 1988. Sulfate Deposition to Surface Waters. Estimating Critical Loads for Norway and the Eastern United States. *Environ. Sci. Technol.* 22: 8–14.

Henriksen, A., B.M. Wathne, E.J.S. Røgenberg, S.A. Norton, and D.F. Brakke, 1988. The Role of Stream Substrates in Aluminum Mobility and Acid Neutralization. *Wat. Res.* 22: 1069–1073.

Henrikson, L., and Y.W. Brodin, 1995. *Liming of Acidified Surface Waters—A Swedish Synthesis.* Berlin: Springer-Verlag.

Henson, E.B., A.S. Bradshaw, and D.C. Chandler, 1961. *The Physical Limnology of Cayuga Lake, New York.* NY State Coll.of Ag., Memoir 378.

Herdendorf, C.E., 1982. Large Lakes of the World. *J. Great Lakes Res.* 8: 379–412.

Hessen, D.O., and T. Andersen, 1992. The Algae–Grazer Interface: Feedback Mechanisms Linked to Elemental Ratios and Nutrient Cycling. *Arch. Hydrobiologia Beih.* 35: 111–120.

Hessen, D.O., K. Nygaard, K. Salonen, and A. Vähätalo, 1994. The Effect of Substrate Stoichiometry on Microbial Activity and Carbon Degradation in Humic Lakes. *Env. Int.* 20: 67–76.

Hilgers, E., H. Thies, and W. Kalbfus, 1993. A Lead-210 Dated Sediment Record on Heavy Metals, Polycyclic Aromatic Hydrocarbons and Soot Spherules for a Dystrophic Mountain Lake. *Verh. Int. Ver. Limnol.* 25: 1091–1094.

Hillbricht-Ilkowska, A., Z. Kajak, J. Ejsmont-Karabin, A. Karabin, and J. Rybak, 1975. Ecosystem of the Mikolajskie Lake. The Utilization of the Consumers Production by Invertebrate Predators in Pelagic and Profundal Zones. *Pol. Arch. Hydrobiol.* 22: 53–64.

Hinch, S.G., 1991. Small- and Large-scale Studies in Fisheries Ecology: The Need for Cooperation Among Researchers. *Fisheries* 16: 22–27.

Hobbie, J.E., 1961. Summer Temperatures in Lake Schrader, Alaska. *Limnol. Oceanogr.* 6: 326–329.

Hobbie, J.E., 1984. *Polar Limnology.* Pp. 63–105. In F.B. Taub (ed.), Ecosystems of the World, Vol. 23: Lakes and Reservoirs. Amsterdam: Elsevier.

Hobbie, J.E., 1988. A Comparison of the Ecology of Planktonic Bacteria in Fresh and Salt Water. *Limnol. Oceanogr.* 33: 750–764.

Hobbie, J.E., R.J. Daley, and S. Jasper, 1977. Use of Nucleopore Filters for Counting Bacteria by Fluorescence Microscopy. *Appl. Environ. Microbiol.* 33: 1225–1228.

Hodell, D.L., and C.L. Schelske, 1998. Production, Sedimentation and Isotopic Composition of Organic Matter in Lake Ontario. *Limnol. Oceanogr.* 43: 200–214.

Hodgman, C.D., R.C. Weast, and S.M. Selby (eds.), 1959. *Handbook of Physics and Chemistry, 42nd ed.* Cleveland: Chem. Rubber Publ.

Hofer, M., W. Aeschbach-Hertig, U. Beyerle, S. B. Haderlein, E. Hoehn, T.B. Hofstetter, A. Johnson, R. Kiffer, A. Ulrich, and D.M. Imboden, 1997. Tracers as Essential Tools for the Investigation of Physical and Chemical Processes in Groundwater Systems. *Chimia* 51: 941–946.

Hoggarth, D.D., V.J. Cowan, A.S. Halls, M. Aeron-Thomas, J.A. McGregor, C.A. Garaway, A.I. Payne, R.L. Welcomme, 1999.

Management Guidelines for Asian Floodplain River Fisheries. FAO Fish. Tech. Pap. #384/1. Rome.

Höhener, P., and R. Gächter, 1994. Nitrogen Cycling Across the Sediment–Water Interface in a Eutrophic, Artificially Oxygenated Lake. *Aquat. Sci.* 56: 115–132.

Hong, S., J.P. Candelone, C.C. Patterson, and C.F. Boutron, 1996. History of Ancient Copper Smelting Pollution During Roman and Medieval Times Recorded in Greenland Ice. *Science* 272: 246–249.

Hongve, D., 1993. Total and Reactive Aluminum Concentrations in Non-turbid Norwegian Surface Waters. *Verh. Int. Ver. Limnol.* 25: 133–136.

Hontela, A., J.B. Rasmussen, C. Audet, and G. Chevalier, 1992. Impaired Cortisol Stress Response in Fish from Environments Polluted by PAHs, PCBs and Mercury. *Arch. Environ. Contam. Toxicol.* 22: 278–283.

Horie, S. (ed.), 1984. *Lake Biwa. Monographiae Biologicae, Vol. 54.* Dordrecht: Dr. W. Junk Publ.

Horie, S. (ed.), 1987. *History of Lake Biwa.* Kyoto: Inst. of Paleolimn. and Paleoenv. on Lake Biwa.

Horn, H., and W. Horn, 1993. Sedimentary Losses in the Reservoir Saidenbach: Flux and Sinking Velocities of Dominant Phytoplankton Species. *Int. Rev. Ges. Hydrobiologia* 78: 39–57.

Horn, W., 1981. Phytoplankton Losses Due to Zooplankton Grazing in a Drinking Water Reservoir. *Int. Rev. Ges. Hydrobiol.* 66: 787–810.

Horn, W., and H. Horn, 1995. Interrelationships Between Crustacean Zooplankton and Phytoplankton: Results from 15 Years of Field Observations at the Mesotrophic Saidenbach Reservoir (Germany). *Hydrobiologia* 307: 231–238.

Hough, J.L., 1958. *Geology of the Great Lakes.* Urbana: Univ. Illinois Press.

Howard-Williams, C., 1983. Wetlands and Watershed Management: The Role of Aquatic Vegetation. *J. Limnol. Soc. sth. Afr.* 9: 54–62.

Howard-Williams, C., 1985. Cycling and Retention of Nitrogen and Phosphorus in Wetlands: A Theoretical and Applied Perspective. *Freshw. Biol.* 15: 391–431.

Howard-Williams, C., and G.M. Lenton, 1975. The Role of the Littoral Zone in the Functioning of a Shallow Tropical Lake Ecosystem. *Freshw. Biol.* 5: 445–459.

Howard-Williams, C., and B.R. Allanson, 1981. An Integrated Study on Littoral and Pelagic Primary Production in a Southern African Coastal Lake. *Arch. Hydrobiologia* 92: 507–534.

Howard-Williams, C., and W.F. Vincent, 1983. *Plants of the Littoral Zone.* Pp. 73–83. In D.J. Forsyth, and C. Howard-Williams (eds.), Lake Taupo: Ecology of a New Zealand Lake. Wellington: DSIR.

Howard-Williams, C., K. Law, C.L. Vincent, J. Davies, and W.F. Vincent, 1986. Limnology of Lake Waikaremoana with Special Reference to Littoral and Pelagic Primary Producers. *New Zealand J. Mar. and Freshw. Res.* 20: 583–597.

Howarth, R.W., R. Marino, J. Lane, and J.J. Cole, 1988. Nitrogen Fixation in Freshwater, Estuarine, and Marine Ecosystems. Rates and Importance. *Limnol. Oceanogr.* 33: 669–687.

Howarth, R.W., G. Billen, D. Swaney, A. Townsend, A. Jaworski, N. Lajtha J.A. Downing, R. Elmgren, N. Caraco, T. Jordan, F. Berendse, J. Freney, V. Kudeyarov, P. Murdoch, and Z. Zhao-Liang, 1996. Regional Nitrogen Budgets and Riverine N & P Fluxes for the Drainages to the North Atlantic Ocean: Natural and Human Influences. *Biogeochemistry* 35:75–139.

Hoyer, M.V., and D.E. Canfield, Jr., 1994. Bird Abundance and Species Richness on Florida Lakes: Influence of Trophic Status, Lake Morphology, and Aquatic Macrophytes. *Hydrobiologia* 279/280: 107–119.

Hrbáček, J., 1958. Typologie und Produktivität der Teichartigen Gewässer. *Verh. Int. Ver. Limnol.* 13: 394–399.

Hrbáček, J. (ed.), 1966. *Hydrobiological Studies 1*. Prague: Acad. Publ. House of the Czechoslovak Acad. of Sci.

Hrbáček, J., and M. Straškraba (eds.), 1973a. *Hydrobiological Studies 2*. Prague: Acad. Publ. House of the Czechoslovak Acad. of Sci.

Hrbáček, J., and M. Straškraba (eds.), 1973b. *Hydrobiological Studies 3*. Prague: Acad. Publ. House of the Czechoslovak Acad. of Sci.

Hu, B., and Y. Liu, 1998. *The Development of Cage Culture and its Role in Fishery Enhancement in China*. Pp. 255–262. In T. Petr (ed.), Inland Fishery Enhancements. FAO Fish. Tech. Pap. #374. Rome.

Hughes, R.M., and J.M. Omernik, 1981. *Use and Misuse of the Terms Watershed and Stream Order*. Pp. 320–326. In L.A. Krumholz (ed.), Warmwater Streams Symposium: A National Symposium on Fisheries Aspects of Warmwater Streams. Lawrence: Allen Press.

Hunt, R.L., 1975. *Food Relations and Behaviour of Salmonid Fishes*. Pp. 137–151. In A.D. Hasler (ed.), Coupling of Land and Water Systems. New York: Springer-Verlag.

Huntley, M., 1985. Experimental Approaches to the Study of Vertical Migration of Zooplankton. *Contrib. Marine Sci.* 27(Suppl.): 71–90.

Hurrell, J.W., 1995. Decadal Trends in the North Atlantic Oscillation: Regional Temperatures and Precipitation. *Science* 269: 676–679.

Huryn, A.D., 1996. An Appraisal of the Allen Paradox in a New Zealand Trout Stream. *Limnol. Oceanogr.* 41: 243–252.

Husar, R.B., 1986. *Emission of Sulfur Dioxide and Nitrogen Oxides and Trends for Eastern North America*. Pp. 48–92. In NRC, Acid Deposition: Long-term Trends. Washington: Nat. Acad. Press.

Huszar, V.L. de M., and C.S. Reynolds, 1997. Phytoplankton Periodicity and Sequence of Dominance in an Amazonian Floodplain Lake (Lago Batata, Pará, Brazil): Responses to Gradual Environmental Change. *Hydrobiologia* 346: 169–181.

Hutchinson, G.E., 1941. Limnological Studies in Connecticut: IV. The Mechanisms of Intermediary Metabolism in Stratified Lakes. *Ecology Monogr.* 11: 21–60.

Hutchinson, G.E., 1957. *A Treatise on Limnology, Vol I: Geography, Physics and Chemistry*. New York: John Wiley and Sons.

Hutchinson, G.E., 1961 The Paradox of the Plankton. *Amer. Nat.* 95: 137–145.

Hutchinson, G.E., 1964. The Lacustrine Microcosm Reconsidered. *Amer. Sci.* 52: 334–341.

Hutchinson, G.E., 1967. *A Treatise on Limnology, Vol II: Introduction to Lake Biology and the Limnoplankton*. New York: John Wiley and Sons.

Hutchinson, G.E., 1975. *A Treatise on Limnology, Vol. III: Limnological Botany*. New York: John Wiley and Sons.

Hutchinson, G.E., 1979. *The Kindly Fruits of the Earth: Recollections of an Embryo Ecologist*. New Haven: Yale Univ. Press.

Hynes, H.B.N., 1970. *The Ecology of Running Waters*. Liverpool: Liverpool Univ. Press.

Hynes, H.B.N., 1975. The Stream and Its Valley. *Verh. Int. Ver. Limnol.* 19: 1–15.

Idso, S.B., and R.G. Gilbert, 1974. On the Universality of the Poole and Atkins Secchi Disk-Light Extinction Equation. *J. Appl. Ecology* 11: 399–401.

Imberger, J., 1985. Thermal Characteristics of Standing Waters: An Illustration of Dynamic Processes. *Hydrobiologia*, 125: 7–29.

Imberger, J., and J.C. Patterson, 1990. Physical Limnology. *Adv. Appl. Mech.* 27: 303–475.

Imboden, D.M., and S. Emerson, 1978. Natural Radon and Phosphorus as Limnologic Tracers: Horizontal and Vertical Eddy Diffusion in Greifensee. *Limnol. Oceanogr.*, 23: 77–90.

Imboden, D.M., U. Lemmin, T. Joller, and M. Schurter, 1983. Mixing Processes in Lakes: Mechanisms and Ecological Relevance. *Schweiz. Z. Hydrol.*, 45:11–44.

Imboden, D.M., and A. Wüest, 1995. *Mixing Mechanisms in Lakes*. pp 83–138. In Lerman, A., D.M. Imboden, and J.R. Gat (eds). Physics and Chemistry of Lakes 2nd ed. Springer-Verlag, Berlin.

Infante, A., O. Infante, M. Márquez, W.M. Lewis, and F.H. Weibezahn, 1979. Conditions Leading to Mass Mortality of Fish and Zooplankton in Lake Valencia, Venezuela. *Acta Cient. Venezolana* 30: 67–73.

Int. Lake Env. Comm., *Data Book of World Lake Environments, 1987–1989* (Vols. 1–3). Otsu.

International Joint Commission, 1987. *Report on Great Lakes Water Quality*. Windsor: Great Lakes Water Qual. Bd.

Irvine, K., 1997. Food Selectivity and Diel Vertical Distribution of *Chaoborus edulis* (Diptera, Chaoboridae) in Lake Malawi. *Freshw. Biol.* 37: 605–620.

Istvánovics, V., J. Padisák, K. Pettersson, and D.C. Pierson, 1994. Growth and Phosphorus Uptake of Summer Phytoplankton in Lake Erken (Sweden). *J. Plankton Res.* 16: 1167–1196.

Iverfeldt, Å., 1991. Occurrence and Turnover of Atmospheric Mercury over the Nordic Countries. *Water, Air and Soil Poll.* 56: 251–265.

Iversen, T.M., B. Kronvang, C.C. Hoffmann, M. Søndergaard, and H.O. Hansen, 1995. *Restoration of Aquatic Ecosystems and Water Quality*. Pp. 63–69. In H. Skotte Møller (ed.), Nature Restoration in the European Union: Proceedings of a Seminar, Denmark, 1995. Min. of Env. and Energy, The Nat. Forest and Nature Agency. Copenhagen.

Jackson, J.K., and V.H. Resh, 1989. Distribution and Abundance of Adult Aquatic Insects in the Forest Adjacent to a Northern California Stream. *Environ. Entomol.* 18: 278–283.

Jackson, T.A., G. Kipphut, R.H. Hesslein, and D.W. Schindler, 1980. Experimental Study of Trace Metal Chemistry in Softwater Lakes at Different pH Levels. *Can. J. Fish. Aquat. Sci.* 37: 387–402.

Jacobs, J., 1987. Cyclomorphosis in *Daphnia*. *Mem. Ist. Ital. Idrobiol.* 45: 325–352.

Jäger, P., and J. Röhrs, 1990. Coprecipitation of Phosphorus with Calcite in the Eutrophic Wallersee (Alpine Forestland of Salzburg, Austria). *Int. Rev. Ges. Hydrobiol.* 75: 153–173.

James, M.R., and D.J. Forsyth, 1990. Zooplankton–Phytoplankton Interactions in a Eutrophic Lake. *J. Plankton Res.* 12: 455–472.

James, W.F., and J.W. Barko, 1990. Macrophyte Influences on the Zonation of Sediment Accretion and Composition in a North-temperate Reservoir. *Arch. Hydrobiol.* 120: 129–142.

James, W.F., and J.W. Barko, 1991. Estimation of Phsophorus Exchange between Littoral and Pelagic Zones During Nighttime Convective Circulation. *Limnol. Oceanogr.*, 36: 179–187.

Janse, J.H., T. Aldenberg, and P.R.G. Kramer, 1992. A Mathematical Model of the Phosphorus Cycle in Lake Loosdrecht and Simulation of Additional Measures. *Hydrobiologia* 233: 119–136.

Jansson, M., 1979. Nutrient Budgets and the Regulation of Nutrient Concentrations in a Small Sub-arctic Lake in Northern Sweden. *Freshwat. Biol.* 9: 213–231.

Jaworski, N.A., R.W. Howarth, and L.J. Hetling, 1997. Atmospheric Deposition of Nitrogen Oxides onto the Landscape Contributes to Coastal Eutrophication in the Northeast United States. *Env. Sci. Technol.* 31: 1995–2004.

Jeffries, D.S. (ed), 1997. *Canadian Acid Rain Assessment, vol. 3: The Effects on Canada's Lakes, Rivers and Wetlands*. Burlington: Env. Canada.

Jeffries, D.S., S.E. Doka, M.L. Mallory, F. Norouzian, A. Storey, and I. Wong, 1998. Aquatic Effects of Acidic Deposition in Canada: Present and Predicted Future Situation. *J. Water Sci.* (Special): 129–143.

Jellison, R., J. Romero, and J.M Melack, 1998. The Onset of Meromixis During Restoration of Mono Lake, California: Unintended Consequences of Reducing Water Diversions. *Limnol. Oceanogr.* 43: 706–711.

Jenkins, R.M., 1982. The Morphoedaphic Index and Reservoir Fish Production. *Trans. Amer. Fish. Soc.* 111: 133–140.

Jensen, H.S., P. Kristensen, E. Jeppesen, and A. Skytthe, 1992. Iron: Phosphorus Ratio in Surface Sediment as an Indicator of Phosphate Release from Aerobic Sediments in Shallow Lakes. *Hydrobiologia* 235/236: 731–743.

Jensen, J.P., E. Jeppesen, K. Olrik, and P. Kristensen, 1994. Impact of Nutrients and Physical Factors on the Shift from Cyanobacterial to Chlorophyte Dominance in Shallow Danish Lakes. *Can. J. Fish. Aquat. Sci.* 51: 1692–1699.

Jensen, K.W., and E. Snekvik, 1972. Low pH Levels Wipe Out Salmon and Trout Population in Southernmost Norway. *Ambio* 1: 223–225.

Jeppesen, E., J.P. Jensen, P. Kristensen, M. Søndegaard, E. Mortensen, O. Sortkjær, and K. Olrik, 1990. Fish Manipulation as a Lake Restoration Tool in Shallow, Eutrophic, Temperate Lakes 2: Threshold Levels, Long-term Stability and Conclusions. *Hydrobiologia* 200/201: 219–227.

Jeppesen, E., M. Søndergaard, E. Kanstrup, B. Petersen, R.B. Erikren, M. Hammershøj, J. Mortensen, J.P. Jensen, and A. Have, 1994. Does the Impact of Nutrients on the Biological Structure and Function of Brackish and Freshwater Lakes Differ? *Hydrobiologia* 275/276: 15–30.

Jeppesen, E., J.P. Jensen, M. Søndergaard, T. Lauridsen, L.J. Pedersen, and L. Jensen, 1997a. Top-down Control in Freshwater Lakes: The Role of Nutrient State, Submerged Macrophytes and Water Depth. *Hydrobiologia* 342/343: 151–164.

Jeppesen, E., M. Søndergaard, Mo. Søndergaard, and K. Christofferson (eds.), 1997b. *The Structuring Role of Submerged Macrophytes in Lakes.* New York: Springer-Verlag.

Jeppesen, E., M. Søndergaard, B. Kronvang, J.P. Jensen, L.M. Svendsen, and T.L. Lauridsen, 1999. Lake and Catchment Management in Denmark. *Hydrobiologia* 395/396: 419–432.

Jeppesen, E., T. Lauridsen, S.F. Mitchell, and C.W. Burns, 1997. Do Planktivorous Fish Structure the Zooplankton Communities in New Zealand Lakes? *New Zealand J. of Mar. and Freshw. Res.* 31: 163–173.

Jeremiason, J.D., K.C. Hornbuckle, and S.J. Eisenreich, 1994. PCBs in Lake Superior, 1978–1992: Decreases in Water Concentrations Reflect Loss by Volatilization. *Env. Sci. and Technol.* 28: 903–914.

Jewson, D.H., 1976. The Interaction of Components Controlling Net Phytoplankton Photosynthesis in a Well-mixed Lake (Lough Neagh, Northern Ireland). *Freshw. Biol.* 6: 551–576.

Jin, X.C., 1994. An Analysis of Lake Eutrophication in China. *Mitt. Int. Ver. Limnol.* 24: 207–211.

Johannsson, O.E., R. Dermott, D.M. Graham, J.A. Dahl, E.S. Millard, D.D. Myles, and J. LeBlanc, 2000. Benthic and Pelagic Secondary Production in Lake Erie After the Invasion of *Dreisena spp.* with Implications for Fish Production. *J. Great Lakes Res.* 26: 31–54.

Johansson, J.A., 1983. Seasonal Development of Bacterioplankton in Two Forest Lakes in Central Sweden. *Hydrobiologia* 101: 71–88.

Johansson, K., 1985. Mercury in Sediment in Swedish Forest Lakes. *Verb. Int. Ver. Limnol.* 22: 2359–2363.

Johnson, P.L., 1984. *Thoughts on Selection and Design of Reservoir Aeration Devices.* Pp. 537–541. In Lake and Reservoir Management: Proceedings of the Third Annual Conference of the North American Lake Management Society. US EPA Report #440/5-84-001. Washington, DC.

Johnson, R.K., and T. Wiederholm, 1992. Pelagic–Benthic Coupling—The Importance of Diatom Interannual Variability for Population Oscillations of *Monoporeia affinis*. *Limnol. Oceanogr.* 37: 1596–1607.

Johnson, R.K., B. Boström, and W. van de Bund, 1989. Interactions Between *Chironomus plumosus* (L.) and the Microbial Community in Surficial Sediments of a Shallow, Eutrophic Lake. *Limnol. Oceanogr.* 34: 992–1003.

Johnson, T.C., 1984. Sedimentation in Large Lakes. *Ann. Rev. Earth Planet Sci.* 12: 179–204.

Johnston, N.T., C.J. Perrin, P.A. Slaney, and B.R. Ward, 1990. Increased Juvenile Salmonid Growth by Whole-river Fertilization. *Can. J. Fish. Aquat. Sci.* 47: 862–872.

Johnstone, I.M., and P.W. Robinson, 1987. Light Level Variation in Lake Tutira After Transient Sediment Inflow and its Effect on the Submersed Macrophytes. *New Zealand J. of Mar. and Freshw. Res.* 21: 47–53.

Jónasson, P.M., 1972. Ecology and Production of the Profundal Benthos in Relation to Phytoplankton in Lake Esrom. *Oikos* 14(suppl.): 1–148.

Jónasson, P.M., 1978. Zoobenthos of Lakes. *Verb. Int. Ver. Limnol.* 20: 13–37.

Jónasson, P.M., 1993. Lakes as a Basic Resource for Development: The Role of Limnology. *Mem. Ist. Ital. Idrobl.* 52: 9–26.

Jones, J.G., and B.M. Simon, 1981. Differences in Microbial Decomposition Processes in Profundal and Littoral Lake Sediments, with Particular Reference to the Nitrogen Cycle. *J. Gen. Microbiol.* 123: 297–312.

Jones, J.R., and R.W. Bachmann, 1978. Trophic Status of Iowa Lakes in Relation to Origin and Glacial Geology. *Hydrobiologia* 57: 267–273.

Jones, J.R., B.P. Borofka, and R.W. Bachmann, 1976. Factors Affecting Nutrient Loads in Some Iowa Streams. *Water Res.* 10: 117–122.

Jones, J.R., and R.W. Bachmann, 1976. Prediction of Phosphorus and Chlorophyll Levels in Lakes. *J. Water Poll. Control Fed.* 48: 2176–2182.

Jones, J.R., and M.V. Hoyer, 1982. Sportfish Harvest Predicted by Summer Chlorophyll-*a* Concentration in Midwestern Lakes and Reservoirs. *Trans. Amer. Fish. Soc.* 111: 176–179.

Jones, R.A., W. Rast, and G.F. Lee, 1979. Relationship Between Summer Mean and Maximum Chlorophyll-*a* Concentration in Lakes. *Env. Sci. Technol.* 13: 869–870.

Jones, R.I., 1977. Factors Controlling Phytoplankton Production and Succession in a Highly Eutrophic Lake (Kinnego Bay, Lough Neagh). III. Interspecific Competition in Relation to Irradiance and Temperature. *J. Ecol.* 65: 579–586.

Jones, R.I., 1977. Factors Controlling Phytoplankton Production and Succession in a Highly Eutrophic Lake (Kinnego Bay, Lough Neagh). II. Phytoplankton Production and its Chief Determinants. *J. Ecol.* 65: 561–577.

Jones, R.I., 1992. The Influence of Humic Substances on Lacustrine Planktonic Food Chains. *Hydrobiologia* 229: 73–91.

Jones, R.I., J.M. Young, A.M. Hartley, and A.E. Bailey-Watts, 1996. Light Limitation of Phytoplankton Development in an Oligotrophic Lake—Loch Ness, Scotland. *Freshw. Biol.* 35: 533–543.

Jordan, T.E., D.L. Correll, and D.E. Weller, 1997. Relating Nutrient Discharge from Watershed to Land Use and Streamflow Variability. *Water Resources Res.* 33: 2579–2590.

Jørgensen, S.E., L. Kamp-Nielsen, and L.A. Jørgensen, 1986. Examination of the Generality of Eutrophication Models. *Ecology Modelling* 32: 251–266.

Jungwirth, M., S. Muhar, and S. Schmutz, 1995. The Effects of Recreated Instream and Ecotone Structures on the Fish Fauna of an Epipotamal River. *Hydrobiologia* 303: 195–206.

Junk, W.J., and G.E. Weber, 1996. Amazon Floodplains: A Limnological Perspective. *Verb. Int. Ver. Limnol.* 26: 149–157.

Jupp, B.P., and D.H.N. Spence, 1977. Limitations of Macrophytes in a Eutrophic Lake, Loch Leven. *J. Ecology* 65: 431–446.

Jürgens, K., H. Arndt, and K.O. Rothhaupt, 1994. Zooplankton-mediated Changes of Bacterial Community Structure. *Microb. Ecol.* 27: 27–42.

Jürgens, K., S.A. Wickham, K.O. Rothhaupt, and B. Santer, 1996. Feeding Rates of Macro- and Microzooplankton on Heterotrophic Nanoflagellates. *Limnol. Oceanogr.* 41: 1833–1839.

Jürgens, K., J. Pernthaler, S. Schalla, and R. Amann, 1999. Morphological and Compositional Changes in a Planktonic Bacterial Community in Response to Enhanced Protozoan Grazing. *Appl. Env. Microbiol.* 65: 1241–1250.

Jürgens, K., and K. Šimek, 2000. Functional Response and Particle Selection of *Halterra* cf. *grandinella*, a Common Freshwater Oligotrichous Ciliate. *Aquat. Microb. Ecol.* 22: 57–68.

Kairesalo, T., and T. Seppälä, 1987. Phosphorus Flux Through a Littoral Ecosytem: The Importance of Cladoceran Zooplankton and Young Fish. *Int. Rev. Ges. Hydrobiol.* 72: 385–403.

Kaiser, K.L.E., 1984. Organic Contaminants in the Environment: Research Progress and Needs. *Env. Int.* 10: 241–250.

Kajak, Z., and J.I. Rybak, 1966. Production and Some Trophic Dependences in Benthos Against Primary Production and Zooplankton Production of Several Masurian Lakes. *Verh. Int. Ver. Limnol.* 16: 441–451.

Kajak, Z., and A. Hillbricht-Ilkowska (eds.), 1972. *Productivity Problems of Freshwaters. IBP-UNESCO 1970 Symposium on Productivity Problems of Freshwaters, Kazimierz, Poland.* Kraków: PWN-Polish Sci. Publ.

Kalff, J., 1967. Phytoplankton Abundance and Primary Production Rates in Two Arctic Ponds. *Ecology* 48: 558–565.

Kalff, J., 1983. Phosphorous Limitation in Some Tropical African Lakes. *Hydrobiologia* 100: 101–112.

Kalff, J., H.E. Welch, and S.K. Holmgren, 1972. Pigment Cycles in Two High-arctic Canadian Lakes. *Verh. Int. Ver. Limnol.* 18: 250–256.

Kalff, J., H.J. Kling, S.H. Holmgren, and H.E. Welch, 1975. Phytoplankton, Phytoplankton Growth and Biomass Cycles in an Unpolluted and in a Polluted Polar Lake. *Verh. Int. Ver. Limnol.* 19: 487–495.

Kalff, J., and R. Knoechel, 1978. Phytoplankton and Their Dynamics in Oligotrophic Lakes. *Ann. Rev. Ecology Syst.* 9: 475–495.

Kalff, J., and S. Watson, 1986 Phytoplankton and Its Dynamics in Two Tropical Lakes: A Tropical and Temperate Zone Comparison. *Hydrobiologia* 138: 161–176.

Kämäri, J., M. Forsius, P. Kortelainen, J. Mannio, and M. Verta, 1991. Finnish Lake Survey: Present Status of Acidification. *Ambio* 27: 23–27.

Kamp-Nielson, L., and J.M. Anderson, 1977. A Review of the Literature on Sediment: Water Exchange of Nitrogen Compounds. *Prog. Water Technol.* 8: 393–418.

Kawai, T., A. Otsuki, M. Aizaki, and I. Nishikawa, 1985. Phosphate Release from Sediment into Aerobic Water in a Eutrophic Shallow Lake, L. Kasumigaura. *Verh. Int. Ver. Limnol.* 22: 3316–3322.

Keast, A., and J. Harker, 1977. Fish Distribution and Benthic Invertebrate Biomass Relative to Depth in an Ontario Lake. *Env. Biol. Fish.* 2: 235–240.

Keeney, D.R., 1973. The Nitrogen Cycle in Sediment–Water Systems. *J. Env. Qual.* 2: 15–29.

Keeney, D.R., and T.H. DeLuca, 1993. Des Moines River Nitrate in Relation to Watershed Agricultural Practices: 1945 Versus 1980s. *J. Env. Qual.* 22: 267–272.

Keller, A.E., and T.L. Crisman, 1990. Factors Influencing Fish Assemblages and Species Richness in Subtropical Lakes and a Comparison with Temperate Lakes. *Can. J. Fish. Aquat. Sci.* 47: 2137–2146.

Keller, W., and M. Conlon, 1994. Crustacean Zooplankton Communities and Lake Morphometry in Precambrian Shield Lakes. *Can. J. Fish. Aquat. Sci.* 51: 2424–2434.

Keller, W., and J.M. Gunn, 1995. *Lake Water Quality Improvements and Recovering Aquatic Communities.* Pp. 87–80. In J.M. Gunn, Restoration and Recovery of an Industrial Region: Progress in Restoring the Smelter-damaged Landscape near Sudbury, Canada. New York: Springer-Verlag.

Kelly, C.A., and D.P. Chynoweth, 1981. The Contributions of Temperature and of the Iinput of Organic Matter in Controlling Rates of Sediment Methanogenesis. *Limnol. Oceanogr.* 26: 891–897.

Kelly, C.A., J.W.M. Rudd, and D.W. Schindler, 1988. Carbon and Electron Flow Via Methanogenesis, SO_4^{2-}, NO_3^-, and Mn^{4+} Reduction in the Anoxic Hypolimnia of Three Lakes. *Arch. Hydrobiol. Beih.* 31: 333–344.

Kelso, J.R.M., C.K. Minns, J.E. Gray, and M.L. Jones, 1986. Acidification of Surface Waters in Eastern Canada and Its Relationship to Aquatic Biota. *Can. Spec. Publ. Fish . Aquat. Sci.* 87.

Kelts, K., and K.J. Hsü, 1978. *Freshwater Carbonate Sedimentation.* Pp. 295–323. In A. Lerman (ed.), Lakes: Chemistry, Geology, Physics. New York: Springer-Verlag.

Kerekes, J., R. Tordor, A. Nieuwburg, and L. Risk, 1994. Fish-eating Bird Abundance in Oligotrophic Lakes in Kejimkujik National Park, Nova Scotia, Canada. *Hydrobiologia* 279/280: 57–61.

Ketelaars, H.A.M., F.E. Lambregts-van de Clundert, C.J. Carpentier, A.J. Wagenvoort, and W. Hoogenboezem, 1999. Ecological Effects of the Mass Occurrence of the Ponto-Caspian Invader, *Hemimysis anomala* G. O. Sars, 1907 (Crustacea: Mysidacea), in a Freshwater Storage Reservoir in the Netherlands, with Notes on Its Autecology and New Records. *Hydrobiologia* 394: 233–248.

Kilham, P., 1982. Acid Precipitation: Its Role in the alkalization of a Lake in Michigan. *Limnol. Oceanogr.* 27: 856–867.

Kimmel, B.L., and A.W. Groeger, 1984. *Factors Controlling Primary Production in Lakes and Reservoirs: A Perspective.* Pp. 277–281. In EPA Report #440/5/84-001, Lake and Reservoir Management. Washington, DC.

Kimmel, B.L., R.M. Gersberg, L.J. Paulson, R.P. Axler, and C.R. Goldman, 1978. Recent Changes in the Meromictic Status of Big Soda Lake, Nevada. *Limnol. Oceanogr.* 23: 1021–1025.

Kimmel, B.L., O.T. Lind, and L. J. Paulson, 1990. *Reservoir Primary Production.* Pp. 133–193. In K.W. Thornton, B.L. Kimmel, and F.E. Payne (eds.), Reservoir Limnology: Ecological Perspectives. New York: John Wiley and Sons.

Kindler, J., 1998. Linking Ecological and Developmental Objectives: Trade-offs and Imperatives. *Ecology Appl.* 8: 591–600.

Kingsford, R.T., 2000. Protecting Rivers in Arid Regions or Pumping them Dry? *Hydrobiologia* 427: 1–11.

Kingsland, S.E., 1985. *Modelling Nature: Episodes in the History of Population Ecology.* Chicago: Univ. of Chicago Press.

Kipling, C., 1976. Year-class Strengths of Perch and Pike in Windermere. *Rep. Freshwat. Biol. Assoc.* 44: 68–75.

Kirk, J.T.O., 1994. *Light and Photosynthesis in Aquatic Ecosystems, 2nd ed.* Cambridge: Cambridge Univ. Press.

Kirschner, A.K.T., and B. Velimirov, 1999. Benthic Bacterial Secondary Production Measured Via Simultaneous ^{3}H-thymidine and ^{14}C-leucine Incorporation, and Its Implications for the Carbon Cycle of a Shallow Macrophyte-dominated Backwater System. *Limnol. Oceanogr.* 44: 1871–1881.

Kitchell, J.A., and J.F. Kitchell, 1980. Size-selective Predation, Light Transmission, and Oxygen Stratification: Evidence from the Recent Sediments of Manipulated Lakes. *Limnol. Oceanogr.* 25: 389–402.

Kitchell, J.F. (ed.), 1992. Food Web Management: A Case Study of Lake Mendota. New York: Springer-Verlag.

Kjeldsen, K., T.M. Iversen, J. Thorup, and T. Winding, 1998. Benthic Algal Biomass in an Unshaded First-order Lowland Stream: Distribution and Regulation. *Hydrobiologia* 377: 107–122.

Klemer, A.R., and A.E. Konopka, 1989. Cases and Consequences of Blue-green Algal (Cyanobacterial) Blooms. *Lake and Res. Man.* 5: 9–19.

Kling, G.W., 1987. Seasonal Mixing and Catastrophic Degassing in Tropical Lakes, Cameroon, West Africa. *Science* 237: 1022–1024.

Kling, G.W., 1988. Comparative Transparency, Depth of Mixing, and Stability of Stratification in Lakes of Cameroon, West Africa. *Limnol. Oceanogr.* 33: 27–40.

Kling, G.W., M.A. Clark, H.R. Compton, J.D. Devine, W.C. Evans, A.M. Humphrey, J.F Lockwood, and M.L. Tuttle, 1987. The 1986 Lake Nyos Gas Disaster in Cameroon, West Africa. *Science* 236: 169–175.

Knoechel, R., and J. Kalff, 1975. Algal Sedimentation: The Cause of a Diatom–Blue-green Succession. *Verh. Int. Ver. Limnol.* 19: 745–754.

Knoechel, R., and J. Kalff, 1978. An In Situ Study of the Productivity and Population Dynamics of Five Freshwater Planktonic Diatom Species. *Limnol. Oceanogr.* 23: 195–218.

Knoechel, R., and L.B. Holtby, 1986. Cladoceran Filtering Rate: Body Length Relationships for Bacterial and Large Algal Particles. *Limnol. Oceanogr.* 31: 195–200.

Knoechel, R., and C.E. Campbell, 1988. Physical, Chemical, Watershed and Plankton Characteristics of Lakes on the Avalon Peninsula, Newfoundland, Canada: A Multivariate Analysis of Interrelationships. *Verh. Int. Ver. Limnol.* 23: 282–296.

Knowles R., 1982. Dentrification. *Microbiol. Rev.* 46: 43–70.

Knowles, R., and D.R.S. Lean, 1987. Nitrification: A Significant Cause of Oxygen Depletion Under Winter Ice. *Can. J. Fish. Aquat. Sci.* 44: 743–749.

Kobayashi, T., R.J. Shiel, P. Gibbs, and P.I. Dixon, 1998. Freshwater Zooplankton in the Hawkesbury-Nepean River: Comparison of Community Structure with Other Rivers. *Hydrobiologia* 377: 133–145.

Koenings, J.P., and J.A. Edmundson, 1991. Secchi Disk and Photometer Estimates of Light Regimes in Alaskan Lakes: Effects of Yellow Color and Turbidity. *Limnol. Oceanogr.* 36: 91–105.

Kopáček, J., L. Procházková, and J. Hejzlar, 1997. Trends and Seasonal Patterns of Bulk Deposition of Nutrients in the Czech Republic. *Atmos. Env.* 31: 797–808.

Kopáček, J., J. Hejzlar, E. Stuchlík, J. Fott, and J. Veselý, 1998. Reversibility of Acidification of Mountain Lakes After Reduction in Nitrogen and Sulphur Emissions in Central Europe. *Limnol. Oceanogr.* 43: 357–361.

Kopylov, A.P., P.S. Kuzin, and E.P. Senkov. 1978. *General Information on Hydrology*. Pp. 23–41. In V.I. Korzun (ed.), World Water Balance and Water Resources of the Earth. Paris: Unesco Press.

Kortmann, R.W., and P.H. Rich, 1994. Lake Ecosystem Energetics: The Missing Management Link. *Lake and Res. Man.* 8: 77–97.

Kortmann, R.W., G.W. Knoecklein, and C.H. Bonnell, 1994. Aeration of Stratified Lakes: Theory and Practice. *Lake Res. Man.* 8: 99–120.

Korzun, V.I. (ed.), 1978. *World Water Balance and Water Resources of the Earth*. Paris: Unesco Press.

Kosarev, A.N., and E.A. Yablonskaya, 1994. *The Caspian Sea*. The Hague: SPB Acad. Publ.

Koschel, R., J. Benndorf, G. Proft, and F. Recknagel, 1983. Calcite Precipitation as a Natural Control Mechanism of Eutrophication. *Arch. Hydrobiol.* 98: 380–408.

Koste, W., and E.D. Holloway, 1993. A Short History of Western European Rotifer Research. *Hydrobiologia* 255/256: 557–572.

Kowalczewski, A., and T. Ozimek, 1993. Further Long-term Changes in the Submerged Macrophyte Vegetation of the Eutrophic Lake Mikolajskie (North Poland). *Aquat. Bot.* 46: 341–345.

Kramer, D.L., and M. McClure, 1982. Aquatic Surface Respiration, a Widespread Adaptation to Hypoxia in Tropical Freshwater Fishes. *Env. Biol. Fish.* 7: 47–55.

Kratz, T.K., K.E. Webster, C.J. Bower, J.J. Magnuson, and B.J. Benson, 1997. The Influence of Landscape Position on Lakes in Northern Wisconsin. *Freshw. Biol.* 37: 209–217.

Krause-Jensen, D., and Sand-Jensen, K., 1998. Light Attenuation and Photosynthesis of Aquatic Plant Communities. *Limnol. Oceanogr.* 43: 396–407.

Krishnaswami, S., and D. Lal, 1978. *Radionuclide Limnochronology*. Pp. 153–177. In A. Lerman (ed.), Lakes: Chemistry, Geology, Physics. New York: Springer-Verlag.

Kubečka, J., and M. Wittingerova, 1998. Horizontal Beaming as a Crucial Component of Acoustic Fish Stock Assessment in Freshwater Reservoirs. *Fish. Res.* 35: 99–106.

Kubečka, J., J. Seda, A. Duncan, J. Matěna, H.A.M. Ketelaars, and P. Visser, 1998. Composition and Biomass of the Fish Stocks in Various European Reservoirs and Ecological Consequences. *Int. Rev. Hydrobiologia* 83(special iss.): 559–568.

Kudoh, S., and M. Takahashi, 1990. Fungal Control of Population Changes of the Planktonic Diatom *Asterionella formosa* in a Shallow Eutrophic Lake. *J. Phycol.* 26: 239–244.

Kufel, L., A. Prejs, J.I. Rybak (eds.), 1997. Shallow Lakes '95: Trophic Cascades in Shallow Freshwater and Brackish Lakes. *Hydrobiologia* 342/343:

Kuhn, A., C.A. Johnson, and L. Sigg, 1994. *Cycles of Trace Elements in a Lake with a Seasonally Anoxic Hypolimnion*. Pp. 473–497. In L.A. Baker (ed.), Environmental Chemistry of Lakes and Reservoirs. Washington: Amer. Chem. Soc.

Kullenberg, G., C.R. Murthy, and H. Westerberg. 1974. *Vertical Mixing Characteristics in the Thermocline and Hypolimnion Regions of Lake Ontario (IFYGL)*. pp. 425–434. In Proc. 17th Conf. Great Lakes Res. Intl. Assoc. Great Lakes Res., Ann Arbor.

Kuznetsov, S.I., 1970. *The Microflora of Lakes and its Geochemical Activity*. Austin: Univ. of Texas Press.

Lachance, M., B. Bobée, and Y. Grimard, 1985. Sensitivity of Southern Quebec Lakes to Acidic Precipitation. *Water, Air, and Soil Poll.* 25: 115–132.

Laird, K.R., S.C. Fritz, K.A. Maasch, and B.F. Cumming, 1996. Greater Drought Intensity and Frequency Before A.D. 1200 in the Northern Great Plains, USA. *Nature* 384: 552–554.

Lamontagne, S., and D.W. Schindler, 1994. Historical Status of Fish Populations in Canadian Rocky Mountain Lakes Inferred From Subfossil *Chaoborus* (Diptera: Chaoboridae) Mandibles. *Can. J. Fish. Aquat. Sci.* 51: 1376–1383.

Lampert, W., 1985. *The Role of Zooplankton: An Attempt to Quantify Grazing*. Pp. 54–62. In Proc. Int. Congress of the European Water Poll. Control Assoc.; Lakes, Pollution and Recovery. Rome.

Lampert, W., 1989. The Adaptive Significance of Diel Vertical Migration of Zooplankton. *Funct. Ecol.* 3: 21–27.

Landers, D.H., R.H. Hughes, S.G. Paulsen, D.P. Larsen, and J.M. Omerick, 1998. How Can Regionalization and Survey Sampling Make Limnological Research More Relevant? *Verh. Int. Ver. Limnol.* 26: 2428–2436.

Lange, T.R., H.E. Royals, and L.L. Connor, 1993. Influence of Water Chemistry on Mercury Concentration in Largemouth Bass from Florida Lakes. *Trans. Amer. Fish. Soc.* 122: 74–84.

Langeland, A., 1982. Interactions Between Zooplankton and Fish in a Fertilized Lake. *Holarctic Ecol.* 5: 273–310.

Langeland, A., 1988. Decreased Zooplankton Density in a Mountain Lake Resulting from Predation by Recently Introduced *Mysis relicta*. *Verh. Int. Ver. Limnol.* 23: 419–429.

Larsen, D.P., and H.T. Mercier, 1976. Phosphorus Retention Capacity of Lakes. *J. Fish. Res. Bd. Can.* 33: 1742–1750.

Larsen, D.P., J. Van Sickle, K.W. Malueg, and P.D. Smith, 1979. The Effect of Wastewater Phosphorus Removal on Shagawa Lake, Minnesota: Phosphorus Supplies, Lake Phosphorus, and Chlorophyll-*a*. *Water Res.* 13: 1259–1272.

Larson, D.W., 1984. The Crater Lake Study: Detection of Possible Optical Deterioration of a Rare, Unusually Deep Caldera Lake in Oregon, USA. *Verh. Int. Ver. Limnol.* 22: 513–517.

Larson, D.W., 2000. Waldo Lake, Oregon: Eutrophication of a Rare, Ultraoligotrophic, High-mountain Lake. *Lake and Res. Man.* 16: 2–16.

Larson, D.W., C.N. Dahm, and N.S. Geiger, 1987. Vertical Partitioning of the Phytoplankton Assemblage in Ultraoligotrophic Crater Lake, Oregon, USA. *Freshw. Biol.* 18: 429–442.

Larsson, P., 1978. The Life Cycle Dynamics and Production of Zooplankton in Øvre Heimdalsvatn. *Holarct. Ecol.* 1: 162–218.

Lasenby, D.C., T.C. Northcote, and M. Furst, 1986. Theory, Practise, and Effects of *Myses relicta* Introductions to North American and Scandinavian Lakes. *Can. J. Fish. Aquat. Sci.* 43: 1277–1284.

Lauridsen, T., and D. Lodge, 1996. Avoidance of *Daphinia magna* of Fish and Macrophytes: Chemical Cues and Predator-mediated Use of Macrophyte Habitiat. *Limnol. Oceanogr.* 41: 794–798.

Lauridsen, T.L., L.J. Pedersen, E. Jeppesen, and M. Søndergaard, 1996. The Importance of Macrophyte Bed Size for Cladoceran Composistion and Horizontal Migration in a Shallow Lake. *J. Plankton Res.* 18: 2283–2294.

Lauscher, F., 1951. Über die Verteilung der Windgeschwindigkeit auf der Erde. *Arch. F. Meteorologie, Geophysik und Biolimatologie* B2: 427–435.

Laybourn-Parry, J., S.J. Perriss, G.G.R. Seaton, and J. Rohozinski, 1997. A Mixotrophic Ciliate as a Major Contributor to Plankton Photosynthesis in Australian Lakes. *Limnol. Oceanogr.* 42: 1463–1467.

Le Cren, E.D., 1947. The Determination of the Age and Growth of the Perch (*Perca fluviatilis*) from the Opercular Bone. *J. Anim. Ecol.* 16: 188–204.

Leach, J.H., L.M. Dickie, B.J. Shuter, U. Borgmann, J. Hyman, and W. Lysack, 1987. A Review of Methods for Predicting Potential Fish Production with Application to the Great Lakes and Lake Winnipeg. *Can. J. Fish. Aquat. Sci.* 44(Suppl. 2): 471–485.

Lebo, M.E., J.E. Reuter, C.R. Goldman, and C.L. Rhodes, 1994. Interannual Variability of Nitrogen Limitation in a Desert Lake: Influence of Regional Climate. *Can. J. Fish. Aquat. Sci.* 51: 862–872.

Lee, G.F., and R.A. Jones, 1984. Summary of US OECD Eutrophication Study. Results and Their Application to Water Quality Management. *Verh. Int. Ver. Limnol.* 22: 261–267.

Leggett, W.C., 1986. *The Dependence of Fish Larval Survival on Food and Predator Densities*. Pp. 117–137. In S. Skreslet (ed.), The Role of Freshwater Outflow in Coastal Marine Ecosystems. Berlin: Springer-Verlag.

Lehman, J.T., and C.E. Cáceres, 1993. Food-web Responses to Species Invasion by a Predatory Invertebrate: *Bythotrephes* in Lake Michigan. *Limnol. Oceanogr.* 38: 879–891.

Lehtovaara, A., and J. Sarvola, 1984. Seasonal Dynamics of Total Biomass and Species Composition of Zooplankton in the Littoral Zone of an Oligotrophic Lake. *Verh. Int. Ver. Limnol.* 22: 805–810.

Leibovich, S., 1983. The Form and Dynamics of Langmuir Circulations. *Annu. Rev. Fluid Mech.* 15: 391–427.

Lellák, J., 1966. Influence of the Removal of the Fish Population on the Bottom Animals of the Five Elbe Backwaters. *Hydrobiol. Studies* 1: 323–380.

Lemly, A.D., R.T. Kingsford, and J.R. Thompson, 2000. Irrigated Agriculture and Wildlife Conservation: Conflict on a Global Scale. *Environ. Man.* 25: 485–512.

Leopold, L.B., M.G. Wolman, and J.P. Miller, 1992. *Fluvial Processes in Geomorphology*. New York: Dover Publ.

Lerman, A., and A.B. Hull, 1987. Background Aspects of Lake Restoration: Water Balance, Heavy Metal Content, Phosphorus Homeostasis. *Schweiz. Z. Hydrol.* 49: 148–169.

Lesack, L.F.W., R.E. Hecky, and J.M. Melack, 1984. Transport of Carbon, Nitrogen, Phosphorus, and Major Solutes in the Gambia River, West Africa. *Limnol. Oceanogr* 29: 816–830.

Lewis, W.M., Jr., 1983. A Revised Classification of Lakes Based on Mixing. *Can. J. Fish. Aquat. Sci.* 40: 1779–1787.

Lewis, W.M., Jr., 1986. Nitrogen and Phosphorus Runoff Losses from a Nutrient-poor Tropical Moist Forest. *Ecology* 67: 1275–1282.

Lewis, W.M., Jr., 1987. Tropical Limnology. *Ann. Rev. Ecol. Syst.* 18: 159–184.

Lewis, W.M., Jr., 1996. *Tropical Lakes: How Latitude Makes a Difference*. Pp. 43–64. In F. Schiemer, and K.T. Boland (eds.), Perspectives in Tropical Limnology. Amsterdam: SPB Acad. Publ.

Lewis, W.M., Jr., 2000. Basis for the Protection and Management of Tropical Lakes. *Lakes and Reservoirs: Res. Manage.* 5: 35–48.

Lewis, W.M., Jr., and F.H. Weibezahn, 1983. Phosphorus and Nitrogen Loading of Lake Valencia. *Acta Cient. Venezolana* 34: 345–349.

Lewis, W.M., Jr., S.W. Chisholm, C.F. D'Elia, E.J. Fee, N.G. Hairston, J.E. Hobbie, G. Likens, S. Threlkeld, and R. Wetzel, 1995. Challenges for Limnology in the United States and Canada: An Assessment of the Discipline in the 1990s. *Amer. Soc. of Limnol. and Oceanogr. Bull.* 4: 1–20.

Liang, Y., J.M. Melack, J. Wang, 1981. Primary Production and Fish Yields in Chinese Ponds and Lakes. *Trans. Amer. Fish. Soc.* 110: 346–350.

Light, J.J., J.C. Ellis-Evans, and J. Priddle, 1981. Phytoplankton Ecology in an Antarctic Lake. *Freshw. Biol.* 11: 11–26.

Likens, G.E., 1984. Beyond the Shoreline: A Watershed-ecosystem Approach. *Verh. Int. Ver. Limnol.* 22: 1–22.

Likens, G.E., 1985. An Experimental Approach for the Study of Ecosystems. *J. of Ecology* 73: 381–396.

Likens, G.E., F.H. Bormann, R.S. Pierce, J.S. Eaton, and N.M. Johnson, 1977. *Biogeochemistry of a Forested Ecosystem*. New York: Springer-Verlag.

Lind, O.T., and L. Dávalos-Lind, 1991. Association of Turbidity and Organic Carbon with Bacterial Abundance and Cell Size in a Large, Turbid, Tropical Lake. *Limnol. Oceanogr.* 36: 1200–1208.

Lind, O.T., T.H. Chrzanowski, and L. Dávalos-Lind, 1997. Clay Turbidity and the Relative Production of Bacterioplankton and Phytoplankton. *Hydrobiologia* 353: 1–18.

Lindegaard, C., 1994. The Role of the Zoobenthos in Energy Flow in Two Shallow Lakes. *Hydrobiologia* 275/276: 313–322.

Lindegaard, C., P.C. Dall, and P.M. Jónasson, 1997. *Long-term Patterns of the Profundal Fauna in Lake Esrom*. Pp. 39–53. In K. Sand-Jensen, and O. Pedersen (eds.), Freshwater Biology: Priorities and Development in Danish Research. Copenhagen: G.E.C. Gad.

Lindeman, R.L., 1942. The Trophic-dynamic Aspect of Ecology. *Ecology* 23: 399–418.

Lindholm, T., and J.A.O. Meriluoto, 1991. Recurrent Depth Maxima of the Hepatotoxic Cyanobacterium *Oscillatoria agardhii*. *Can. J. Fish. Aquat. Sci.* 48: 1629–1634.

Lindqvist, O., K. Johansson, M. Aastrup, A. Andersson, L. Bringmark, G. Hovsenius, L. Håkanson, Å. Iverfeldt. M. Meili, and B. Timm, 1991. Mercury in the Swedish Environment: Recent Research on Causes, Consequences and Corrective Methods. *Water, Air, Soil. Poll.* 55: 1–261.

Litvinov, A.S., and N.M. Mineeva, 1994. Characteristics of the Summer Hydrological Regimen and Chlorophyll Distribution in the Volga River Reservoirs. *Int. Rev. Ges. Hydrobiol.* 79: 229–234.

Livingstone, D.A., 1963. *Alaska, Yukon, Northwest Territories, and Greenland*. Pp. 559–574. In D.G. Frey (ed.), Limnology in North America. Illinois: Univ. of Wisconsin Press.

Livingstone, D.A., and J.M. Melack, 1984. *Some Lakes of Subsaharan Africa*. Pp. 467–497 In F.B. Taub, Ecosystems of the World, vol. 23: Lakes and Reservoirs. Amsterdam: Elsevier.

Livingstone, D.M., 1999. Ice Break-up on Southern Lake Baikal and Its Relationship to Local and Regional Air Temperatures in

Siberia and the North Atlantic Oscillation. *Limnol. Oceangr.* 44: 1486–1497.

Locke, A., and W.G. Sprules, 1993. Effects of Acidification on Zooplankton Populations and Community Dynamics. *Can. J. Fish. Aquat. Sci.* 50: 1238–1247.

Lockhart, W.L., R. Wagemann, B. Tracey, D. Sutherland, and D.J. Thomas, 1992. Presence and Implications of Chemical Contaminants in the Freshwaters of the Canadian Arctic. *Sci. Total Environ.* 122: 165–243.

Loczy, S., R. Carignan, and D. Planas, 1983. The Role of Roots in Carbon Uptake by the Submerged Macrophytes *Myriophyllum spicatum, Vallisneria americana* and *Heteranthera dubia*. *Hydrobiologia* 98: 3–7.

Lodge, D.M., 1991. Herbivory on Freshwater Macrophytes. *Aquat. Bot.* 41: 195–224.

Lodge, D.M., and J.G. Lorman, 1987. Reductions in Submersed Macrophyte Biomass and Species Richness by the Crayfish *Orconectes rusticus. Can. J. Fish. Aquat. Sci.* 44: 591–597.

Lodge, D.M., S.C. Blumenshine, and Y. Vadeboncoeur, 1998. *Insights and Application of Large-scale, Long-term Ecological Observations and Experiments*. Pp. 202–235. In W.J. Resetarits, Jr., and J. Bernardo (eds.), Experimental Ecology: Issues and Perspectives. New York: Oxford Univ. Press.

Loeb, S.L., 1981. An In Situ Method for Measuring the Primary Productivity and Standing Crop of the Epilithic Periphyton Community in Lentic Systems. *Limnol. Oceanogr.* 26: 394–399.

Löffler, H., 1997. Längsee: A History of Meromixis; 40 Years Later: Homage to Dr. D.G. Frey. *Verh. Int. Ver. Limnol.* 26: 829–832.

Lotter, A.F., and H.J.B. Birks, 1997. The Separation of the Influence of Nutrients and Climate on the Varve Time-series of Baldeggersee, Switzerland. *Aquat. Sci.* 59: 362–375.

Lovley, D.R., 1991. Dissimilatory Iron-III and Manganese-IV Reduction. *Microbiol. Rev.* 55: 259–287.

Lowe, R.L., 1996. *Periphyton Patterns in Lakes*. Pp. 57–76. In R.J. Stevenson, M.L. Bothwell, and R.L. Lowe (eds.), Algal Ecology, Freshwater Benthic Ecosystems. San Diego: Acad. Press.

Lowe-McConnell, R.H., 1975. *Fish Communities in Tropical Freshwaters: Their Distribution, Ecology, and Evolution*. London: Longman.

Lowe-McConnell, R.H., 1982. *Tilapias in Fish Communities*. Pp. 83–113. In R.S.V. Pullin, and R.H. Lowe-McConnell (eds.), The Biology and Culture of Tilapias. Manila: ICLARM.

Ludwig, J.P., J.P. Giesy, C.L. Summer, W. Bowerman, R. Aulerich, S. Bursian, H.J. Auman, P.D. Jones, L.L. Williams, D.E. Tillitt, and M. Gilbertson, 1993. A Comparison of Water Quality Criteria for the Great Lakes Based on Human and Wildlife Health. *J. Great Lakes Res.* 19: 789–807.

Luecke, C., 1986. A Change in the Pattern of Vertical Migration of *Chaoborus flavicans* After the Introduction of Trout. *J. Plankton Res.* 8: 649–657.

Luecke, C., M.J. Vanni, J.J. Magnuson, J.F. Kitchell, and P.T. Jacobson, 1990. Seasonal Regulation of *Daphnia* Populations by Planktivorous Fish: Implications for the Spring Clear-water Phase. *Limnol. Oceanogr.* 35: 1718–1733.

Lund, J.W.G., 1954. The Seasonal Cycle of the Plankton Diatom, *Melosira italica* (Ehr.) Kütz. Subsp. *Subartica* O. Müll. *J. Ecology* 42: 151–179.

Maberly, S.C., 1996. Diel, Episodic and Seasonal Changes in pH and Concentrations of Inorganic Carbon in a Productive Lake. *Freshw. Biol.* 35: 579–598.

MacArthur, R.H., and J.H. Connell, 1966. *The Biology of Populations*. New York: John Wiley and Sons.

MacDonald, W.W., 1956. Observations on the Biology of *Chaoborids* and *Chironomids* in Lake Victoria and on the Feeding Habits of the 'Elephant-snout Fish' (*Mormyrus kannume* Forsk). *J. Anim. Ecology* 25: 36–53.

MacIntyre, S., K.M. Flynn, R. Jellison, and J.R. Romero, 1999. Boundary Mixing and Nutrient Fluxes in Mono Lake, California. *Limnol. Oceanogr.* 44: 512–529.

MacIntyre, S., and J.M. Melack, 1988. Frequency and Depth of Vertical Mixing in an Amazon Floodplain Lake (Lake Calado, Brazil). *Verh. Int. Ver. Limnol.* 23: 80–85.

Mackay, R., and J. Kalff, 1969. Seasonal Variation in Standing Crop and Species Diversity of Insect Communities in a Small Québec Stream. *Ecology* 50: 101–109.

MacLean, J., and J.J. Magnuson, 1977. Species Interactions in Percid Communities. *J. Fish. Res. Bd. Canada* 34: 1941–1951.

Madigan, M.T., 1988. *Microbiology, Physiology, and Ecology of Phototrophic Bacteria*. Pp. 39–111. In A.J.B. Zehnder, Biology of Anaerobic Microorganisms. New York: John Wiley and Sons.

Madsen, T.V., and K. Sand-Jensen, 1991. Photosynthetic Carbon Assimilation in Aquatic Macrophytes. *Aquat. Bot.* 41: 5–40.

Magnuson, J.J., 1991. Fish and Fisheries Ecology. *Ecology Appl.* 1: 13–26.

Mahon, R., 1984. Divergent Structure in Fish Taxocenes of North Temperate Streams. *Can. J. Fish. Aquat. Sci.* 41: 330–350.

Maiss, M., J. Ilmberger, and K.O. Münnich, 1994. Vertical Mixing in Überlingersee (Lake Constance) Traced by SF_6 and Heat. *Aquat. Sci.,* 56: 329–347.

Maitipe, P., and S.S. De Silva, 1985. Switches Between Zoophagy, Phytophagy and Detritivory of *Sarotherodon mossambicus* (Peters) Populations in Twelve Man-made Sri Lankan Lakes. *J. Fish. Biol.* 26: 49–61.

Mallory, M.L., P.J. Blancher, P.J. Weatherhead, and D.K. McNicol, 1994. Presence or Absence of Fish as a Cue to Macroinvertebrate Abundance in Boreal Wetlands. *Hydrobiologia,* 279/280: 345–351.

Mandych, A.F., 1995. *Enclosed Seas and Large Lakes of Eastern Europe and Middle Asia*. Amsterdam: SPB Acad. Publ.

Manny, B.A., W.C. Johnson, and R.G. Wetzel, 1994. Nutrient Addition by Waterfowl to Lakes and Reservoirs: Predicting Their Effects on Productivity and Water Quality. *Hydrobiologia* 279/280: 121–132.

Maranger, R., and D.F. Bird, 1995. Viral Abundance in Aquatic Systems: A Comparison Between Marine and Fresh Waters. *Mar. Ecology Prog. Ser.* 121: 217–226.

Markager, S., W.F. Vincent, and E.P.Y. Tang, 1999. Carbon Fixation by Phytoplankton in High Arctic Lakes: Implications of Low Temperature for Photosynthesis. *Limnol. Oceanogr.* 44: 597–607.

Marsden, M.W., 1989. Lake Restoration by Reducing External Phosphorus Loading: The Influence of Sediment Phosphorus Release. *Freshw. Biol.* 21: 139–162.

Marshall, C.T., and R.H. Peters, 1989. General Patterns in the Seasonal Development of Chlorophyll-*a* for Temperate Lakes. *Limnol. Oceanogr.* 34: 856–867.

Martens, K., 1997. Speciation in Ancient Lakes. *Trends Ecol. Evol.* 12: 177–182.

Martí, E., and F. Sabater, 1996. High Variability in Temporal and Spatial Nutrient Retention in Mediterranean Streams. *Ecology* 77: 854–869.

Marzolf, E.R., P.J. Mulholland, and A.D. Steinman, 1998. Reply: Improvements to the Diurnal Upstream–Downstream Dissolved Oxygen Change Technique for Determining Wholestream Metabolism in Small Streams. *Can. J. Fish Aquat. Sci.* 55: 1786–1787.

Marzolf, G.R., 1990. *Reservoirs as Environments for Zooplankton*. Pp. 195–208. In K.W. Thornton, B.L. Kimmel, and F.E. Payne (eds.), Reservoir Limnology: Ecological Perspectives. New York: John Wiley and Sons.

Mathews, C.P., 1971. Contribution of Young Fish to Total Production of Fish in the River Thames near Reading. *J. Fish Biol.* 3: 157–180.

Matuszek, J.E., and G.L. Beggs, 1988. Fish Species Richness in Relation to Lake Area, pH, and Other Abiotic Factors in Ontario Lakes. *Can. J. Fish. Aquat. Sci.* 45: 1931–1941.

Mazumder, A., 1994. Phosphorus–Chlorophyll Relationships Under Contrasting Herbivory and Thermal Stratification: Predictions and Patterns. *Can. J. Fish. Aquat. Sci.* 51: 390–400.

Mazumder, A., W.D. Taylor, D.J. McQueen, D.R.S. Lean, 1990. Effects of Fish and Plankton on Lake Temperature and Mixing Depth. *Science* 247: 312–315.

McCarty, L.S., and D. Mackay, 1993. Enhancing Ecotoxicological Modeling and Assessment. *Env. Sci. and Technol.* 27: 1719–1728.

McCauley, E., and J. Kalff, 1981. Empirical Relationships Between Phytoplanton and Zooplankton Biomass. *Can. J. Fish. Aquat. Sci.* 38: 458–463.

McCauley, E., J.A. Downing, and S. Watson, 1989. Sigmoid Relationships Between Nutrients and Chlorophyll Among Lakes. *Can. J. Fish. Aquat. Sci.* 46: 1171–1175.

McGowan, J.A., and P.W. Walker, 1979. Structure in the Copepod Community of the North Pacific Central Gyre. *Ecology Monogr.* 49: 195–226.

McIntosh, R.P., 1985. *The Background of Ecology: Concept and Theory.* Cambridge: Cambridge Univ. Press.

McKnight, D., D.F. Brakke, and P.J. Mulholland (eds.), 1996. Freshwater Ecosystems and Climate Change in North America. *Limnol. Oceanogr.* 41: 815–1149.

McLaren, I.A., 1974. Demographic Strategy of Vertical Migration by a Marine Copepod. *Am. Nat.* 108: 91–102.

McQueen, D.J., and N.D. Yan, 1993. Metering Filtration Efficiency of Freshwater Zooplankton Hauls: Reminders from the Past. *J. Plankton. Res.* 15: 57–65.

McQueen, D.J., J.R. Post., and E.L. Mills, 1986. Trophic Relationships in Freshwater Pelagic Ecosystems. *Can. J. Fish. Aquat. Sci.* 43: 1571–1581.

Mehner, T., D. Bauer, and H. Schultz, 1998. Early Omnivory in Age-0 Perch (*Perca fluviatilis*)—A Key for Understanding Long-term Manipulated Food Webs? *Verh. Int. Ver. Limnol.* 26: 2287–2289.

Meijer, M.L., I. de Boois, M. Scheffer, R. Portielje, and H. Hosper, 1999. Biomanipulation in Shallow Lakes in the Netherlands: An Evaluation of 18 Case Studies. *Hydrobiologia* 408/409: 13–30.

Meijer, M.L., M.W. de Haan, A.W. Breukelaar, and H. Buiteveld, 1990. Is Reduction of the Benthivorous Fish an Important Cause of High Transparency Following Biomanipulation in Shallow Lakes? *Hydrobiologia* 200/201: 303–315.

Melack, J.M., 1978. Morphometric, Physical and Chemical Features of the Volcanic Crater Lakes of Western Uganda. *Arch. Hydrobiologia* 84: 430–453.

Melack, J.M., and P. Kilham, 1974. Photosynthetic Rates of Phytoplankton in East African Alkaline, Saline Lakes. *Limnol. Oceanogr.* 19: 743–755.

Mellina, E., and J.B. Rasmussen, 1994. Patterns in the Distribution and Abundance of Zebra Mussel (*Dreissena polymorpha*) in Rivers and Lakes in Relation to Substrate and Other Physiochemical Factors. *Can. J. Fish. Aquat. Sci.* 51: 1024–1036.

Merican, Z.O., and M.J. Phillips, 1985. Solid Waste Production from Rainbow Trout, *Salmo gairdneri* (Richardson), Cage Culture. *Aquacult. Fish. Man.* 16: 55–69.

Meybeck, M., 1979. Concentrations des eaux Fluviales en Éléments Majeurs et Apports en Solution aux Océans. *Rev. Géol. Dyn. Géogr. Phys.* 21: 215–246.

Meybeck, M., 1982. Carbon, Nitrogen, and Phosphorus Transport by World Rivers. *Amer. J. Sci.* 282: 401–450.

Meybeck, M., 1995. *Global Distribution of Lakes.* Pp. 1–35. In A. Lerman, D.M. Imboden, and J.R. Gat (eds.), Physics and Chemistry of Lakes, 2nd ed. Berlin: Springer-Verlag.

Meybeck, M., and R. Helmer, 1989. The Quality of Rivers: From Pristine Stage to Global Pollution. *Palaeogeog., Palaeoclim., PalaeoEcol.* 75: 283–309.

Meybeck, M., D.V. Chapman, and R. Helmer, 1990. *Global Freshwater Quality: A First Assessment.* Oxford: Blackwell Scientific.

Meyer, J.L., 1990. A Blackwater Perspective on Riverine Ecosystems. *Biosci.* 40: 643–651.

Micklin, P.P., 1988. Desiccation of the Aral Sea: A Water Management Disaster in the Soviet Union. *Science* 241: 1170–1176.

Middelboe, A.L., and S. Markager, 1997. Depth Limits and Minimum Light Requirements of Freshwater Macrophytes. *Freshw. Biol.* 37: 553–568.

Mill, A.J.B., 1980. Colloidal and Macromolecular Forms of Iron in Natural Waters. A review. *Env. Technol. Lett.* 1: 97–108.

Milliman, J.D., and R.H. Meade, 1983. World-wide Delivery of River Sediment to the Oceans. *J. of Geol.* 91: 1–21.

Mills, E.L., and J.L. Forney, 1983. Impact on *Daphnia pulex* of Predation by Young Yellow Perch in Oneida Lake, New York. *Trans. Amer. Fish. Soc.* 112: 154–161.

Mills, E.L., D.M. Green, and A. Schiavone, Jr., 1987. Use of Zooplankton Size to Assess the Community Structure of Fish Populations in Freshwater Lakes. *North Amer. J. Fish. Man.* 7: 369–378.

Mills, E.L., R. Sherman, and D.S. Robson, 1989. Effect of Zooplankton Abundance and Body Size on the Growth of Age-0 Yellow Perch (*Perca flavescens*) in Oneida Lake, New York. *Can. J. Fish. Aquat. Sci.* 46: 880–886.

Mills, E.L., J.H. Leach, J.T. Carlton, and C.L. Secor, 1993. Exotic Species in the Great Lakes: A History of Biotic Crises and Anthropogenic Introductions. *J. Great Lakes Res.* 19: 1–54.

Mills, K.H., and S.M. Chalanhuk, 1987. Population Dynamics of Lake Whitefish (*Coregonus clupeaformis*) During and After the Fertilization of Lake 226, the Experimental Lakes Area. *Can. J. Fish. Aquat. Sci.* 44: 55–63.

Minns, C.K., 1989. Factors Affecting Fish Species Richness in Ontario Lakes. *Trans. Amer. Fish. Soc.* 118: 533–548.

Miranda, L.E., and K.B. Hodges, 2000. Role of Aquatic Vegetation Coverage on Hypoxia and Sunfish Abundance in Bays of a Eutrophic Reservoir. *Hydrobiologia* 427: 51–57.

Misund, O.A., 1997. Underwater Acoustics in Marine Fisheries and Fisheries Research. *Rev. Fish. Biol.* 7: 1–34.

Mitchell, P., and E. Prepas (eds.), 1990. *Atlas of Alberta Lakes.* Edmonton: Univ. of Alberta Press.

Mitsch, W.J., and J.G. Gosselink, 1986 and 1993. *Wetlands.* New York: Van Nostrand Reinhold.

Mittelbach, G.G., 1988. Competition Among Refuging Sunfishes and Effects of Fish Density on Littoral Zone Invertebrates. *Ecology* 69: 614–623.

Moeller, R.E., J.M. Burkholder, and R.G. Wetzel, 1988. Significance of Sedimentary Phosphorous to a Rooted Submersed Macrophyte [*Najas flexilis* (Wild.) Rostk. and Schmidt] and its Algal Epiphytes. *Aquat. Bot.* 32: 261–281.

Molot, L.A., and P.J. Dillon, 1993. Nitrogen Mass Balances and Denitrification Rates in Central Ontario Lakes. *Biogeochem.* 20: 195–212.

Molot, L.A., and P.J. Dillon, 1996. Storage of Terrestrial Carbon in Boreal Lake Sediments and Evasion to the Atmosphere. *Global Biogeochem. Cycles* 10: 483–492.

Molot, L.A., P.J. Dillon, and B.D. Lazerte, 1989. Factors Affecting Alkalinity Concentrations of Streamwater During Snowmelt in Central Ontario. *Can. J. Fish. Aquat. Sci.* 46: 1658–1666.

Molot, L.A., P.J. Dillon, B.J. Clark, and B.P. Neary, 1992. Predicting End-of-Summer Oxygen Profiles in Stratified Lakes. *Can. J. Fish. Aquat. Sci.* 49: 2363–2372.

Monismith, S.G., J. Imberger, and M.L. Morison, 1990. Convective Motions in the Sidearm of a Small Reservoir. *Limnol. Oceanogr.* 35: 1676–1702.

Montesinos, E., and H. van Gemerden, 1986. *The Distribution and Metabolism of Planktonic Phototrophic Bacteria.* Pp. 349–359. In F. Megusar, and M. Gantar (eds.), Perspectives in Microbial Ecology. Ljubljana: Slovene Soc. for Microbiol.

Moore, M.V., 1988. Differential Use of Food Resources by Instars of *Chaoborus punctipennes. Freshw. Biol.* 19: 249–268.

Moran, M.A., and R.E. Hodson, 1990. Bacterial Production on Humic and Nonhumic Components of Dissolved Organic Carbon. *Limnol. Oceanogr.* 35: 1744–1756.

Moran, M.A., and R.G. Zepp, 1997. Role of Photoreactions in the Formation of Biologically Labile Compounds from Dissolved Organic Matter. *Limnol. Oceanogr.* 42: 1307–1316.

Morant, P.D., 1983. Wetland Classification: Towards an Approach for Southern Africa. *J. Limnol. Soc. sth. Afr.* 9: 76–84.

Morgan, N.C., 1980. *Secondary Production.* Pp. 247–340. In E.D. Le Cren, and R.H. Lowe-McConnell (eds.), The Functioning of Freshwater Ecosystems. Cambridge: Cambridge Univ. Press.

Mori, S., Y. Saijo, and T. Mizuno, 1984. *Limnology of Japanese Lakes and Ponds.* Pp. 303–329. In F.B. Taub (ed.), Ecosystems of the World, Vol. 23: Lakes and Reservoirs. Amsterdam: Elsevier.

Moring, J.R., 1975. *Oregon Dept. Fish Wildlife, Fishery Research Report #9, Part 2.* In J.R. Karr, and I.J. Schlosser (1978), Water Resources and the Land–Water Interface: Water Resources in Agricultural Watersheds Can be Improved by Effective Multidisciplinary Planning. Science 201: 229–234.

Morris, D.P., H. Zagarese, C.E. Williamson, E.G. Balseira, B.R. Hargreaves, and B. Modunutti, 1995. The Attenuation of Solar UV Radiation in Lakes and the Role of Dissolved Organic Carbon. *Limnol. Oceanogr.* 40: 1381–1391.

Mortimer, C.H., 1941. The Exchange of Dissolved Substances Between Mud and Water in Lakes (Parts I and II). *J. Ecology* 29: 280–329.

Mortimer, C.H., 1942. The Exchange of Dissolved Substances Between Mud and Water in Lakes (Parts III and IV, summary, and references). *J. Ecology* 30: 147–201.

Mortimer, C.H., 1956. *An Explorer of Lakes.* Pp. 165–206. In G.C. Sellery, E.A. Birge: A Memoir. Madison: Univ. Wisconsin Press.

Mortimer, C.H., 1971. Chemical Exchanges Between Sediments and Water in the Great Lakes—Speculations on Probable Regulatory Mechanisms. *Limnol. Oceanogr.* 16: 387–404.

Mortimer, C.H., 1974. Lake Hydrodynamics. *Mitt. Int. Ver. Limnol.*, 20: 124–197.

Mortimer, C.H., 1981. The Oxygen Content of Air-saturated Fresh Waters over Ranges of Temperature and Atmospheric Pressure of Limnological Interest. *Mitt. Int. Ver. Limnol.* 22: 1–23.

Mosello, R., C. Bonacina, A. Carollo, V. Libera, and G.A. Tartari, 1986. Acidification Due to In-lake Ammonia Oxidation: An Attempt to Quantify the Proton Production in a Highly Polluted Subalpine Italian Lake (Lake Orta). *Mem. Ist. Ital. Idrobiol.* 44: 47–71.

Moss, B., 1994. Brackish and Freshwater Shallow Lakes—Different Systems or Variations on the Same Theme? *Hydrobiologia* 275/276: 1–14.

Mulholland, P.J., E.R. Marzolf, J.R. Webster, D.R. Hart, and S.P Hendricks, 1997. Evidence that Hyporheic Zones Increase Heterotrophic Metabolism and Phosphorus Uptake in Forest Streams. *Limnol. Oceanogr.* 42: 443–451.

Munger, J.W., and S.J. Eisenreich, 1983. Continental-scale Variations in Precipitation Chemistry. *Environ. Sci. Technol.* 17: 33A–43A.

Murthy, C.R., 1976. Horizontal Diffusion Characteristics in Lake Ontario. *J. of Phys. Oceanogr.*, 6: 76–84.

Muttkowski, R.A., 1918. The Fauna of Lake Mendota: A Qualitative and Quantitative Survey with Special Reference to the Insects. *Trans. Wisc. Acad. Sci. Lett.* 19: 374–482.

Naegeli, M.W., and U. Uehlinger, 1997. Contribution of the Hyporheic Zone to Ecosystem Metabolism in a Prealpine Gravel-bed River. *J. N. Amer. Benthol. Soc.* 16: 794–804.

Nagata, T., 1988. The Microflagellate–Picoplankton Food Linkage in the Water Column of Lake Biwa. *Limnol. Oceanogr.* 33: 504–517.

Naiman, R.J., and H. Décamps (eds.), 1990. *The Ecology and Management of Aquatic-terrestrial Ecotones.* Paris: Parthenon Publ.

Nakano, S., 1994. Estimation of Phosphorus Release Rate by Bacterivorous Flagellates in Lake Biwa. *Jap. J. Limnol.* 55: 201–211.

Nakashima, B.S., and W.C. Leggett, 1975. Yellow Perch (*Perca flavescens*) Biomass Responses to Different Levels of Phytoplankton and Benthic Biomass in Lake Memphremagog, Quebec-Vermont. *Can. J. Fish. Res. Bd.* 32: 1785–1797.

Nalewajko, C., 1966. Photosynthesis and Excretion in Various Planktonic Algae. *Limnol. Oceanogr.* 11: 1–10.

NAS, 1969. *Eutrophication: Causes, Consequences, Correctives.* Washington, DC: Nat. Acad. of Sci.

Naumann, E., 1919. Några Synpunkter Angående Limnoplanktons Ökologi med Särskild Hänsyn till Fytoplankton. *Svensk Bot. Tidskr.* 13: 129–163. (English transl. by the Freshw. Biol. Assoc., #49.)

Nauwerck, A., 1963. Die Beziehungen Zwishen Zooplankton und Phytoplankton im See Erken. *Symb. Bot. Ups.* XVII: 165. Uppsala.

Nauwerck, A., 1978. *Bosmina obtusirostris* SARS im Latnjajaure. *Arch. Hydrobiologia* 82: 387–418.

Naylor, R.L., R.J. Goldburg, J.H. Primavera, N. Kautsky, M.C.M. Beveridge, J. Clay, C. Folke, J. Lubchenco, H. Mooney, and M. Troell, 2000. Effect of Aquaculture on World Fish Supplies. *Nature* 405: 1017–1024.

Neary, B.P., and P.J. Dillon, 1988. Effects of Sulfur Deposition on Lake-water Chemistry in Ontario, Canada. *Nature* 333: 340–343.

Needham, J.G., and P.R. Needham, 1962. *A Guide to the Study of Freshwater Biology*, 5th ed. San Fancisco: Holden-Day.

Neill, W.E., 1984. Regulation of Rotifer Densities by Crustacean Zooplankton in an Oligotrophic Montane Lake in British Columbia. *Oecologia* (Berlin) 61: 175–181.

Neill, W.E., 1985. The Effect of Herbivore Competition Upon the Dynamics of *Chaoborus* Predation. *Arch. Hydrobiologia Beih.* 21: 483–491.

Neill, W.E., 1988. *Complex Interactions in Oligotrophic Lake Food Webs: Responses to Nutrient Enrichment.* Pp. 31–44. In S.R. Carpenter (ed.), Complex Interactions in Lake Communities. New York: Springer-Verlag.

Neill, W.E., 1994. *Spatial and Temporal Scaling and the Organization of Limnetic Communities.* Pp. 189–231. In P.S. Giller, A.G. Hildrew, and D.G. Raffaelli (eds.), Aquatic Ecology—Scale, Pattern and Process. Oxford: Blackwell Sci. Publ.

Nelson, J.S., 1994. *Fishes of the World.* New York: John Wiley and Sons.

Nendza, M., and W. Klein, 1990. Comparative QSAR Study on Freshwater and Estuarine Toxicity. *Aquat. Toxicol.* 17: 63–74.

Nepf, H.M., and Oldham, C.E., 1997. Exchange Dynamics of a Shallow Contaminated Wetland. *Aquat. Sci.* 59: 193–213.

Nesbitt, L.M., H.P. Riessen, and C.W. Ramcharan, 1996. Opposing Predation Pressures and Induced Vertical Migration Responses in Daphnia. *Limnol. Oceanogr.* 41: 1306–1311.

Neumann, J., 1959. Maximum Depth and Average Depth of Lakes. *Can. J. Fish. Res. Bd.* 16: 923–927.

Newbold, J.D., J.W. Elwood, R.V. O'Neill, and A.L. Sheldon, 1983. Phosphorus Dynamics in a Woodland Stream Ecosystem: A Study of Nutrient Spiralling. *Ecology* 64: 1249–1265.

Newman, R.M., 1991. Herbivory and Detritivory on Freshwater Macrophytes by Invertebrates: A Review. *J. N. Am. Benthol. Soc.* 10: 89–114.

Nicholls, K.H., 1976. Nutrient–Phytoplankton Relationships in the Holland Marsh, Ontario. *Ecology Monogr.* 46: 179–199.

Nicholls, K.H., 1998. El Niño, Ice Cover, and Great Lakes Phosphorus: Implications for Climate Warming. *Limnol. Oceanogr.* 43: 715–719.

Nicholls, K.H., and P J. Dillon, 1978. An Evaluation of Phosphorous–Chlorophyll–Phytoplankton Relationships for Lakes. *Int. Rev. Ges. Hydrobiol.* 63: 141–154.

Nichols, S.A., 1991. The Interaction Between Biology and the Management of Aquatic Macrophytes. *Aquat. Bot.* 41: 225–252.

Nichols, S.A., 1992. Depth, Substrate, and Turbidity Relationships of Some Wisconsin Lake Plants. *Trans. Wisc. Acad. Sci. Art. Lett.* 80: 97–118.

Nielsen, S.L., 1993. A Comparison of Aerial and Submerged Photosynthesis in Some Danish Amphibious Plants. *Aquat. Bot.* 45: 27–40.

Nielsen, S.L., and K. Sand-Jensen, 1990. Allometric Scaling of Maximal Photosynthetic Growth Rate to Surface/Volume Ratio. *Limnol. Oceanogr.* 35: 177–181.

Nielsen, S.L., and K. Sand-Jensen, 1991. Variation in Growth Rates of Submerged Rooted Macrophytes. *Aquat. Bot.* 39: 109–120.

Nielson, L.A., 1980. Effect of Walleye (*Stizostedion vitreum vitreum*) Predation on Juvenile Mortality and Recruitment of Yellow Perch (*Perca flavescens*) in Oneida Lake, New York. *Can. J. Fish. Aquat. Sci.* 37: 11–19.

Niemi, Å., 1982. Dynamics of Phytoplankton in the Brackish-water Inlet Pojoviken, Southern Coast of Finland. *Hydrobiologia* 86: 33–39.

Niimi, A.J., 1986. Biological Half-lives of Chlorinated Diphenyl Ethers in Rainbow Trout (*Salmo gairdneri*). *Aquat. Toxicol.* 9: 105–116.

Nilsson, C., and P.A. Keddy, 1988. Predictability of Change in Shoreline Vegetation in a Hydroelectric Reservoir, Northern Sweden. *Can. J. Fish. Aquat. Sci.* 45: 1896–1904.

Nilsson, L., 1978. Breeding Waterfowl in Eutrophicated Lakes in South Sweden. *Wildfowl* 29: 101–110.

Nishri, A., and S. Ben-Yaakov, 1990. The Solubility of Oxygen in the Dead Sea. *Hydrobiologia* 197: 99–104.

Nixdorf, B., 1994. Polymixis of a Shallow Lake (Großer Müggelsee, Berlin) and Its Influence on Seasonal Phytoplankton Dynamics. *Hydrobiologia* 275/276: 173–186.

Nontji, A., 1994. The Status of Limnology in Indonesia. *Mitt. Int. Ver. Limnol.* 24: 95–113.

Northcote T.G., 1988. Fish in the Structure and Function of Freshwater Ecosystems: A "Top-down" View. *Can. J. Fish. Aquat. Sci.* 45: 361–379.

Northcote, T.G., and K.J. Hall, 1983. Limnological Contrasts and Anomalies in Two Adjacent Saline Lakes. *Hydrobiologia* 105: 179–194.

Nriagu, J.O., 1989. A Global Assessment of Natural Sources of Atmospheric Trace Metals. *Nature* 338: 47–49.

Nriagu, J.O., 1990. Global Sources, Pathways and Sinks of Metals. *Env. Canada's Nat. Water Res. Inst. Digest* 1990(spring/summer): 6–7.

Nriagu, J.O., and J.M. Pacyna, 1988. Quantitative Assessment of Worldwide Contamination of Air, Water and Soils by Trace Metals. *Nature* 333: 134–139.

Nürnberg, G.K., 1984. The Prediction of Internal Phosphorus Load in Lakes with Anoxic Hypolimnia. *Limnol. Oceanogr.* 29: 111–124.

Nürnberg, G.K., 1985. Availability of Phosphorus Upwelling from Iron-rich Anoxic Hypolimnia. *Arch. Hydrobiol.* 104: 459–476.

Nürnberg, G.K., 1988. A Simple Model for Predicting the Date of Fall Turnover in Thermally Stratified Lakes. *Limnol. Oceanogr.* 33: 1190–1195.

Nürnberg, G., and R.H. Peters, 1984. The Importance of Internal Phosphorus Load to the Eutrophication of Lakes with Anoxic Hypolimnia. *Verh. Internat. Ver. Limnol.* 22: 190–194.

Nürnberg, G.K., and P.J. Dillon, 1993. Iron Budgets in Temperate Lakes. *Can. J. Fish. Aquat. Sci.* 50: 1728–1737.

O'Brian, W.J., 1990. *Perspectives of Fish in Reservoir Limnology.* Pp. 209–225. In K.W. Thornton, B.L. Kimmel, and F.E. Payne (eds.), Reservoir Limnology: Ecological Perspectives. New York: John Wiley and Sons.

OECD, 1982. *Eutrophication of Waters—Monitoring, Assessment and Control.* Paris: Org. for Econ. Cooperation and Dev.

Økland, J., and K.A. Økland, 1980. *pH Level and Food Organisms for Fish: Studies of 1,000 Lakes in Norway.* Pp. 326–327. In D. Drabløs, and A. Tollan (eds.), Ecological Impact of Acid Precipitation. Oslo: SNSF-Project #1432.

Oliver, B.G., and A.J. Niimi, 1988. Trophodynamic Analysis of Polychlorinated Biphenyl Congeners and Other Chlorinated Hydrocarbons in the Lake Ontario Ecosystem. *Env. Sci. Technol.* 22: 388–397.

Oliver, B.G., M.N. Charlton, and R.W. Durham, 1989. Distribution, Redistribution, and Geochronology of Polychlorinated Biphenyl Congeners and Other Chlorinated Hydrocarbons in Lake Ontario Sediments. *Env. Sci. Technol.* 23: 200–208.

Olrik, K., 1998. Ecology of Mesotrophic Flagellates with Special Reference to Chrysophyceae in Danish Lakes. *Hydrobiologia* 369/370: 329–338.

Opuszynksi, K., and J.V. Shireman,1995. *Herbivorous Fishes: Culture and Use for Weed Management.* Boca Raton: CRC Press.

Osborne, P.L., 1991. Seasonality in Nutrients and Phytoplankton Production in Two Shallow Lakes: Waigani Lake, Papua New Guinea, and Barton Broad, Norfolk, England. *Int. Rev. Ges Hydrobiol.* 76: 105–120.

Ostrovsky,m I., Y.Z. Yacobi, P. Walline, and I. Kalikhman, 1996. Seiche-induced Mixing: Its Impact on Lake Productivity. *Limnol. Oceanogr.,* 41: 323–332.

Otsuki, A., and R.G. Wetzel, 1972. Coprecipitation of Phosphate with Carbonates in a Marl Lake. *Limnol. Oceanogr.* 17: 763–767.

Overmann, J., J.T. Beatty, K.J. Hall, N. Pfenning, and T.G. Northcote, 1991. Characterization of a Dense, Purple Sulfur Bacterial Layer in a Meromictic Salt Lake. *Limnol. Oceanogr.* 36: 846–859.

Overrein, L.N., H.M. Seip, and A. Tollan, 1980. *Acid Precipitation— Effects on Forest and Fish.* Oslo-Ås: SNSF-Project Final Report.

Ozimek, T., and A. Kowalczewski, 1984. Long-term Changes of the Submerged Macrophytes in Eutrophic Lake Mikolajskie (North Poland). *Aquat. Bot.* 19: 1–11.

Pace, M.L., and J.D. Orcutt, Jr., 1981. The Relative Importance of Protozoans, Rotifers and Crustaceans in a Freshwater Zooplankton Community. *Limnol. Oceanogr.* 26: 822–830.

Pachur, H.J., and S. Kröpelin, 1987. Wadi Howar: Paleoclimatic Evidence from an Extinct River System in the Southeastern Sahara. *Science* 237: 298–300.

Padisák, J., 1992. Seasonal Succession of Phytoplankton in a Large Shallow Lake (Balaton, Hungary)—A Dynamic Approach to Ecological Memory, Its Possible Role and Mechanisms. *J. Ecology* 80: 217–230.

Paerl, H.W., 1988. Nuisance Phytoplankton Blooms in Coastal, Estuarine, and Inland Waters. *Limnol. Oceanogr.* 33: 823–847.

Paerl, H.W., 1990. Physiological Ecology and Regulation of N_2 Fixation in Natural Waters. *Advances Microb. Ecol.* 11: 305–344.

Palmstrom, N.S., R.E. Carlson, and G.D. Cooke, 1988. Potential Links Between Eutrophication and the Formation of Carcinogens in Drinking Water. *Lake Res. Man.* 4: 1–15.

Palomäki, R., 1994. Response by Macrozoobenthos Biomass to Water Level Regulation in some Finnish Lake Littoral Zones. *Hydrobiologia* 286: 17–26.

Parkhurst, B.R., H.L. Bergman, J. Fernandez, D.D. Gulley, J.R. Hockett, and D.A. Sanchez, 1990. Inorganic Monomeric Aluminum and pH as Predictors of Acidic Water Toxicity to Brook Trout (*Salvelinus fontinalis*). *Can. J. Fish. Aquat. Sci.* 47: 1631–1640.

Pastorak, R.A., 1981. Prey Vulnerability and Size Selection by *Chaoborus* Larvae. *Ecology* 62: 1311–1324.

Patalas, K., 1984. Mid-summer Mixing Depths of Lakes of Different Latitudes. *Verh. Int. Ver. Limnol.* 22: 97–102.

Pätilä, A., 1986. Survey of Acidification by Airborne Pollutants in 52 Lakes in Southern Finland. *Aqua Fennica* 16: 203–210.

Patterson, J.H., 1976. The Role of Environmental Heterogenity in the Regulation of Duck Populations. *J. Wild. Man.* 40: 22–32.

Pauli, H.R., 1990. *Seasonal Succession of Rotifers in Large Lakes.* Pp. 459–474. In M.M. Tilzer, and C. Serruya (eds.), Large Lakes: Ecological Structure and Function. Berlin: Springer-Verlag.

Pedrós-Alió, C., and T.D. Brock, 1985. Zooplankton Dynamics in Lake Mendota: Short-term Versus Long-term Changes. *Freshw. Biol.* 15: 89–94.

Pedrós-Alió, C., and R. Guerrero, 1991. Abundance and Activity of Bacterioplankton in Warm Lakes. *Verh. Int. Ver. Limnol.* 24: 1212–1219.

Pedrós-Alió, C., and R. Guerrero, 1993. Microbial Ecology in Lake Cisó. *Advances Microb. Ecology* 13: 155–209.

Pedrós-Alió, C., J.I. Calderón-Paz, and J.M. Gasol, 2000. Comparative Analysis Shows that Bacterivory, Not Viral Lysis, Controls the Abundance of Heterotrophic Prokaryotic Plankton. *FEMS Microb. Ecology* 32: 157–165.

Pennington, W., 1978. Responses of Some British Lakes to Past Changes in Land Use on Their Catchments. *Verh. Int. Ver. Limnol.* 20: 636–641.

Pennington, W., R.S. Cambray, and E.M. Fisher, 1973. Observations on Lake Sediments Using Fallout ^{137}Cs as a Tracer. *Nature* 242: 324–326.

Pennycuick, C.J., 1988. *Conversion Factors: SI Units and Many Others.* Chicago: Univ. of Chicago Press.

Pernthaler, J., B. Sattler, K. Šimek, A. Schwarzenbacher, and R. Psenner, 1996. Top-down Effects on the Size–Biomass Distribution of a Freshwater Bacterioplankton Community. *Aquat. Microb. Ecol.* 10: 255–263.

Pernthaler, J., T. Posch, K. Šimek, J. Vrba, R. Amann, and R. Psenner, 1997. Contrasting Bacterial Strategies to Coexist with a Flagellate Predator in an Experimental Microbial Assemblage. *Appl. Env. Microbiol.* 63: 596–601.

Pernthaler, J., F.O. Glöckner, S. Unterholzner, A. Alfreider, R. Psenner, and R. Amann, 1998. Seasonal Community and Population Dynamics of Pelagic Bacteria and Archaea in a High Mountain Lake. *Appl. Env. Microbiol.* 64: 4299–4306.

Persson, A., and L.A. Hansson, 1999. Diet Shift in Fish Following Competitive Release. *Can. J. Fish. Aquat. Sci.* 56: 70–78.

Persson, G., and O. Broberg, 1985. Nutrient Concentrations in the Acidified Lake Gårdsjön: The Role of Transport and Retention of Phosphorus, Nitrogen and DOC in Watershed and Lake. *Ecology Bull.* (Stockholm) 37: 158–175.

Persson, L., G. Andersson, S.F. Hamrin, and L. Johansson, 1988. *Predator Regulation and Primary Production Along the Productivity Gradient of Temperate Lake Ecosystems.* Pp: 45–65. In S.R. Carpenter (ed.), Complex Interactions in Lake Communities. New York: Springer-Verlag.

Peterman, R.M., 1990. Statistical Power Analysis Can Improve Fisheries Research and Management. *Can. J. Fish. Aquat. Sci.* 47: 2–15.

Peters, R.H., and M. Bergmann, 1982. A Comparison of Different Phosphorus Fractions as Predictors of Particulate Pigment Levels in Lake Memphremagog and its Tributaries. *Can. J. Fish. Aquat. Sci.* 39: 785–790.

Peters, R.H., 1983. Ecological Implications of Body Size. New York: Cambridge Univ. Press.

Peters, R.H., 1990. *Pathologies in Limnology.* In R. de Bernardi, G. Giussani, and L. Barbanti (eds.), Scientific Perspectives in Theoretical and Applied Limnology. *Mem. Ist. Ital. Idrobiol.* 47: 181–217.

Peters, R.H., and J.A. Downing, 1984. Empirical Analysis of Zooplankton Filtering and Feeding Rates. *Limnol. Oceanogr.* 29: 763–784.

Peters, R.H., E. Demers, M. Koelle, and B.R. MacKenzie, 1994. The Allometry of Swimming Speed and Predation. *Verh. Int. Ver. Limnol.* 25: 2316–2323.

Petersen, Jr., R.C., G.M. Gíslason, and L.B.M. Vought, 1995. *Rivers of the Nordic Countries.* Pp. 295–341. In C.E. Cushing, K.W. Cummins, and G.W. Minshall (eds.), Ecosystems of the World #22: River and Stream Ecosystems. Amsterdam: Elsevier.

Peterson, B.J., J.E. Hobbie, A.E. Hershey, M.A. Lock, T.E. Ford, J.R. Vestal, M. Hullar, R. Ventullo, and G. Volk, 1985. Transformation of a Tundra River from Heterotrophy to Autotrophy by Addition of Phosphorus. *Science* 229: 1383–1386.

Petr, T., 1998. *Inland Fishery Enhancements.* FAO Fish. Tech. Pap. #374. Rome.

Petrova, N.A., 1986. Seasonality of *Melosira*-plankton of the Great Northern Lakes. *Hydrobiologia* 138: 65–73.

Pfennig, N., 1989. *Ecology of Phototrophic Purple and Green Sulfur Bacteria.* Pp. 97–116. In H.G. Schlegel, and B. Bowien (eds.), Autotrophic Bacteria. New York: Springer-Verlag.

Piasecki, B., D. Rainey, and K. Fletcher, 1998. Is Combustion of Plastics Desirable? *Amer. Sci.* 86: 364–373.

Pickett, R.L., and D.A. Dossett, 1979. Mirex and the Circulation of Lake Ontario, *J. Phys. Oceanogr.*, 9: 441–445.

Pieczyńska, E., T. Ozimek, and J.I. Rybak, 1988. Long-term Changes in Littoral Habitats and Communities in Lake Mikolajskie (Poland). *Int. Rev. Ges. Hydrobiologia* 73: 361–378.

Pierson, D.C., K. Pettersson, and V. Istvanovics, 1992. Temporal Changes in Biomass Specific Photosynthesis During the Summer: Regulation by Environmental Factors and the Importance of Phytoplankton Succession. *Hydrobiologia* 243/244: 119–135.

Pilskaln, C.H., and T.C. Johnson, 1991. Seasonal Signals in Lake Malawi Sediments. *Limnol. Oceanogr.* 36: 544–557.

Pinay, G., H. Décamps, E. Chauvet, and E. Fustec, 1990. *Function of Ecotones in Fluvial Systems.* Pp. 141–169. In R.J. Naiman, and H. Décamps (eds.), The Ecology and Management of Aquaticterrestrial Ecotones. Paris: Parthenon Publ.

Pinel-Alloul, B., and Y. Letarte, 1993. Relationships Between Small and Large Bacterioplankton and Primary Producers in Québec Lakes. *Verh. Int. Ver. Limnol.* 25: 321–324.

Planas, D., 1996. *Acidification Effects.* Pp. 497–530. In R.J. Stevenson, M.L. Bothwell, and R.L. Rowe, Algal Ecology, Freshwater Benthic Ecosystems. San Diego: Acad. Press.

Plante, C., and J.A. Downing, 1989. Production of Freshwater Invertebrate Populations in Lakes. *Can. J. Fish. Aquat. Sci.* 46: 1489–1498.

Pomeroy, L.R., and W.J. Wiebe, 1988. Energetics of Microbial Food Webs. *Hydrobiologia* 159: 7–18.

Porter, K.G., Y.S. Feig, and E.F. Vetter, 1983. Morphology, Flow Regimes, and Filtering Rates of *Daphnia, Ceriodaphnia* and *Bosmina* Fed Natural Bacteria. *Oecologia* 58: 156–163.

Porter, K.G., E.B. Sherr, B.F. Sherr, M. Pace, and R.W. Sanders, 1985. Protozoa in Planktonic Food Webs. *J. Protozool.* 32: 409–415.

Post, J.R., 1990. Metabolic Allometery of Larval and Juvenile Yellow Perch (*Perca flavescens*): In Situ Estimates and a Bioenergetic Model. *Can. J. Fish. Aquat. Sci.* 47: 554–560.

Post, J.R., and D.O. Evans, 1989. Size-dependent Overwinter Mortality of Young-of-the-Year Yellow Perch (*Perca flavescens*): Laboratory, In Situ Enclosure, and Field Experiments. *Can. J. Fish. Aquat. Sci.* 46: 1958–1968.

Post, J.R., and D.J. McQueen, 1994. Variability in First-year Growth of Yellow Perch (*Perca flavescens*): Predictions From a Simple Model, Observations, and an Experiment. *Can. J. Fish. Aquat. Sci.* 51: 2501–2512.

Postel, S., 1987. *Stabilizing Chemical Cycles*. Pp. 157–176. In State of the World, A Worldwatch Institute Rep. on Progr. Toward a Sustainable Soc. New York: W.W. Norton.

Poulin, M., P.B. Hamilton, M. Proulx, 1995. Catalogue des Algues d'eau Douce du Québec, Canada. *Can. Field Nat.* 109: 27–110.

Prairie, Y.T., 1989. Statistical Models for the Estimation of Net Phosphorus Sedimentation in Lakes. *Aquat. Sci.* 51: 192–210.

Prairie, Y.T., and J. Kalff, 1986. Effect of Catchment Size on Phosphorus Export. *Water Res. Bull.* 22: 465–470.

Prairie, Y.T., and J. Kalff, 1988. Particulate Phosphorus Dynamics in Headwater Streams. *Can. J. Fish. Aquat. Sci.* 45: 210–215.

Prairie, Y.T., C.M. Duarte, and J. Kalff, 1989. Unifying Nutrient–Chlorophyll Relationships in Lakes. *Can. J. Fish. Aquat. Sci.* 46: 1176–1182.

Prairie Y.T., C. de Montigny, and P.A. Del Giorgio, 2001. Anaerobic Phosphorus Release from Sediments: A Paradigm Revisited. *Verh. Int. Ver. Limnol.* 27 (in press).

Prézelin, B.B., M.M. Tilzer, O. Schofield, and C. Haese, 1991. The Control of the Production Process of Phytoplankton by the Physical Structure of the Aquatic Environment with Special Reference to Its Optical Properties. *Aquat. Sci.* 53: 136–186.

Pridmore, R.D., 1987. *Phytoplankton Response to Changed Nutrient Concentrations*. Pp. 183–194. In W.N. Vant (ed.), Lake Managers Handbook. Misc. Publ. #103. Wellington: New Zealand Min. Works and Dev.

Pridmore, R.D., and G.B. McBride, 1984. Prediction of Chlorophyll-a Concentrations in Impoundments of Short Hydraulic Retention Time. *J. Env. Manage.* 19: 343–350.

Pridmore, R.D., W.N. Vant, and J.C. Rutherford, 1985. Chlorophyll–Nutrient Relationships in North Island Lakes (New Zealand). *Hydrobiologia* 121: 181–189.

Procházková, L., V. Straškrabová, and J. Popovský, 1973. *Changes of Some Chemical Constituents and Bacterial Numbers in Slapy Reservoir During Eight Years*. Pp. 83–154. In J. Hrbáček, and M. Straškraba (eds.), Hydrobiological Studies 2. Prague: Acad. Publ. House of the Czechoslovak Acad. of Sci.

Procházková, L., and P. Blažka, 1989. Ionic Composition of Reservoir Water in Bohemia: Long-term Trends and Relationships. *Arch. Hydrobiologia Beih.* 33: 323–330.

Prosser, J.I. (ed.), 1986. *Nitrification*. Oxford: IRL Press.

Psenner, R., and R. Sommaruga, 1992. Are Rapid Changes in Bacterial Biomass Caused by Shifts from Top-down to Bottom-up Control? *Limnol. Oceanogr.* 37: 1092–1100.

Puckridge, J.T., F. Sheldon, K.F. Walker, and A.J. Boulton, 1998. Flow Variability and the Ecology of Large Rivers. *Mar. Freshw. Res.* 49: 55–72.

Pullin, R.S.V., and R.H. Lowe-McConnell (eds.), 1982. *The Biology and Culture of Tilapias*. Manila: ICLARM.

Quay, P.D., W.S. Broecker, R.H. Hesslein, and D.W. Schindler, 1980. Vertical Diffusion Rates Determined by Tritium Tracer Experiments in the Thermocline and Hypolimnon of Two Lakes. *Limnol. Oceanogr.* 25: 201–218.

Quinn, J.M., and C.W. Hickey, 1990. Magnitude of Effects of Substrate Particle Size, Recent Flooding, and Catchment Development on Benthic Invertebrates in 88 New Zealand Rivers. *New Zealand J. Mar. and Freshw. Res.* 24: 411–427.

Quirós, R., 1990. Factors Relating to the Variance of Residuals in Chlorophyll–Total Phosphorus Regressions in Lakes and Reservoirs of Argentina. *Hydrobiologia* 200/201: 343–355.

Quirós, R., 1990. Prediction of Relative Fish Biomass in Lakes and Reservoirs of Argentina. *Can. J. Fish Aquat. Sci.* 47: 928–939.

Quirós, R., 1998. *Reservoir Stocking in Latin America, An Evaluation*. Pp. 91–117. In T. Petr (ed.), Inland Fishery Enhancements. FAO Fish. Tech. Pap. #374. Rome.

Ragotzkie, R.A., 1978. *Heat Budgets of Lakes*. Pp. 1–19. In A. Lerman (ed.), Lakes: Chemistry, Geology, and Physics. New York: Springer-Verlag.

Rahel, F.J., C.J. Keleher, and J.L. Anderson, 1996. Potential Habitat Loss and Population Fragmentation for Cold Water Fish in the North Platte River Drainage of the Rocky Mountains: Response to Climate Warming. *Limnol. Oceanogr.* 41: 1116–1123.

Randall, R.G., J.R.M. Kelso, and C.K. Minns, 1995. Fish Production in Freshwaters: Are Rivers More Productive than Lakes? *Can. J. Fish. Aquat. Sci.* 52: 631–643.

Rapp, G., Jr., B.W. Liukkonen, J.D. Allert, and J.A. Sorensen, 1987. Geologic and Atmospheric Input Factors Affecting Watershed Chemistry in Upper Michigan. *Env. Geol. Water Sci.* 9: 155–171.

Rasmussen, J.B., 1988. Habitat Requirements of Burrowing Mayflies (*Ephemeridae: Hexagenia*) in Lakes, with Special Reference to the Effects of Eutropication. *J. N. Amer. Benthol. Soc.* 7: 51–64.

Rasmussen, J.B., 1988. Littoral Zoobenthic Biomass in Lakes and Its Relationship to Physical, Chemical, and Trophic Factors. *Can. J. Fish. Aquat. Sci.* 45: 1436–1447.

Rasmussen, J.B., 1993. Patterns in the Size of Littoral Zone Macroinvertebrate Communities. *Can. J. Fish. Aquat. Sci.* 50: 2192–2207.

Rasmussen, J.B., and J. Kalff, 1987. Empirical Models for Zoobenthic Biomass in Lakes. *Can. J. Fish. Aquat. Sci.* 44: 990–1001.

Rasmussen, J.B., L. Godbout, and M. Schallenberg, 1989. The Humic Content of Lake Water and Its Relationship to Watershed and Lake Morphology. *Limnol. Oceanogr.* 34: 1336–1343.

Rasmussen, J.B., D.J. Rowan, D.R.S. Lean, and J.H. Carey, 1990. Food Chain Structure in Ontario Lakes Determines PCB Levels in Lake Trout (*Salvelinus namaycush*) and Other Pelagic Fish. *Can. J. Fish. Aquat. Sci.* 47: 2030–2038.

Rasmussen, J.B., and D.J. Rowan, 1997. Wave Velocity Thresholds for Fine Sediment Accumulation in Lakes, and Their Effect on Zoobenthic Biomass and Composition. *J. N. Am. Benthol. Soc.* 16: 449–465.

Raspopov, I.M,. I.N. Andronikova, O.N. Dotsenko, E.A. Kurashov, G.I. Letanskaya, V.E. Panov, M.A. Rychkova, I.V. Telesh, O.A. Tchernykh and F.F. Vorontsov, 1996. Littoral Zone of Lake Ladoga: Ecological State Evaluation. *Hydrobiologia* 322: 39–47.

Raven, J.A., and J. Beardall, 1981. Respiration and Photorespiration. *Can. Bull. Fish. Aquat. Sci.* 210: 55–82.

Rawson, D.S., 1939. Some Physical and Chemical Factors in the Metabolism of Lakes. *Pub. Amer. Assoc. Adv. of Sci.* 10: 9–26.

Reckhow, K.H., and S.C. Chapra, 2001. *Engineering Approaches for Lake Management, Vol I: Data Analysis and Empirical Modeling*. Ann Arbor: Ann Arbor Sci. Publ. (in press).

Reddy, K.R., W.H. Patrick, Jr., and C.W. Lindau, 1989. Nitrification–Denitrification at the Plant–Root–Sediment Interface in Wetlands. *Limnol. Oceanogr.* 34: 1004–1013.

Redfield, A.C., B.H. Ketchum, and F.A. Richards, 1963. *The Influence of Organisms on the Chemical Composition of Seawater*. Pp. 26–77. In M.N. Hill (ed.), The Sea: Ideas and Observations on Progress in the Study of the Seas, Vol. 2: The Composition of Seawater, Comparative and Descriptive Oceanography. New York: Interscience.

Regier, H.A., 1974. Fish Ecology and Its Application. *Mitt. Int. Ver. Limnol.* 20: 273–286.

Regier, H.A., 1982. *Training Course on the Management of Small-scale Fisheries in the Inland Waters of Africa : Conceptual Framework and*

Approaches for the Acquisition of Key Resource Information. FAO Fish. Circ. #752. Rome.

Regier, H.A., J.A. Holmes, and D. Pauly, 1990. Influence of Temperature Changes on Aquatic Ecosystems: An Interpretation of Empirical Data. *Trans. Am. Fish. Soc.* 119: 373–389.

Reice, S.R., 1994. Nonequilibrium Determinants of Biological Community Structure. *Amer. Sci.* 82: 424–435.

Reifsnyder, W.E., and H.W. Lull, 1965. *Radiant Energy in Relation to Forests.* US Dept. of Ag. Tech. Bull. #1344. Washington DC:

Resh, V.H., and D.M. Rosenberg, 1984. *Introduction.* Pp. 1–8. In V.H. Resh, and D.M. Rosenberg (eds.), The Ecology of Aquatic Insects. New York: Praeger Publ.

Reuss, J.O., N. Christophersen, and H.M. Seip, 1986. A Critique of Models for Freshwater and Soil Acidification. *Water, Air and Soil Poll.* 30: 909–930.

Reynolds, C.S., 1984a. Phytoplankton Periodicity: The Interaction of Form, Function and Environmental Variability. *Freshw. Biol.* 14: 111–142.

Reynolds, C.S., 1984b. *The Ecology of Freshwater Phytoplankton.* Cambridge: Cambridge Univ. Press.

Reynolds, C.S., 1989. *Physical Determinants of Phytoplankton Succession.* Pp. 9–56. In U. Sommer (ed.), Plankton Ecology: Succession in Plankton Communities. New York: Springer-Verlag.

Reynolds, C.S., 1990. Temporal Scales of Variability in Pelagic Environments and the Response of Phytoplankton. *Freshw. Biol.* 23: 25–53.

Reynolds, C.S., 1993. Scales of Disturbance and Their Role in Plankton Ecology. *Hydrobiologia* 249: 157–171.

Reynolds, C.S., 1997. *Vegetation Processes in the Pelagic: A Model for Ecosystem Theory.* Oldendorf: Ecology Inst.

Reynolds, C.S., and A.E. Walsby, 1975. Water-blooms. *Biol. Rev.* 50: 437–481.

Richman, S., and S.I. Dodson, 1983. The Effect of Food Quality on Feeding and Respiration by *Daphnia* and *Diaptomus*. *Limnol. Oceanogr.* 28: 948–956.

Ricker, W.E., 1968. *Methods for Assessment of Fish Production in Fresh Waters. IBP Handbook #3.* Oxford: Blackwell Sci. Publ.

Ricker, W.E., 1975. *Computation and Interpretation of Biological Statistics of Fish Populations.* Can. Bull. Fish. Res. Bd. #191. Ottawa: Department of Environment, Fisheries and Marine Services.

Rigler, F.H., 1973. *A Dynamic View of the Phosphorus Cycle in Lakes.* Pp. 539–72. In E.J. Griffith, A. Beeton, J.M. Spencer, and D.T. Mitchell (eds.), Environmental Phosphorus Handbook. New York: John Wiley and Sons.

Rigler, F.H., 1982. The selection between fisheries management and limnology. *Trans. Amer. Fish. Soc.* 111: 121–132.

Riley, E.T., and E.E. Prepas, 1985. Comparison of the Phosphorus–Chlorophyll Relationships in Mixed and Stratified Lakes. *Can. J. Fish. Aquat. Sci.* 42: 831–835.

Ringelberg, J., B.J.G. Flik, D. Lindenaar, and K. Royackers, 1991. Diel Vertical Migration of *Daphnia hyalina* (Sensu Latiori) in Lake Maarsseveen, Part 1. Aspects of Seasonal and Daily Timing. *Arch. Hydrobiol.* 121: 129–145.

Robarts, R.D., and T. Zohary, 1984. *Microcystis aeruginosa* and Underwater Light Attenuation in a Hypertrophic Lake (Hartbeespoort Dam, South Africa). *J. Ecology* 72: 1001–1017.

Robarts, R.D., P.J. Ashton, J.A. Thornton, H.J. Taussig, and L.M. Sephton, 1982. Overturn in a Hypertrophic, Warm, Monomictic Impoundment (Hartbeespoort Dam, South Africa). *Hydrobiologia* 97: 209–224.

Rocha, O., and A. Duncan, 1985. The Relationship Between Cell Carbon and Cell Volume in Freshwater Algal Species Used in Zooplanktonic Studies. *J. Plankton Res.* 7: 279–294.

Rocha, O., S. Sendacz, and T. Matsumura-Tundisi, 1995. *Composition, Biomass and Productivity of Zooplankton in Natural Lakes and Reservoirs of Brazil.* Pp. 151–165. In J.G. Tundisi, C.E.M.

Bicudo, and T. Matsumura-Tundisi (eds.), Limnology in Brazil. Rio de Janeiro: Braz. Acad. Sci., Braz. Limnol. Soc.

Roden, E.E., and R.G. Wetzel, 1996. Organic Carbon Oxidation and Suppression of Methane Production by Microbial Fe(III) Oxide Reduction in Vegetated and Unvegetated Freshwater Wetland Sediments. *Limnol. Oceanogr.* 41: 1733–1748.

Rodhe, H., and M.J. Rood, 1986. Temporal Evolution of Nitrogen Compounds in Swedish Precipitation Since 1955. *Nature* 321: 762–764.

Rodhe, W., 1949. The Ionic Composition of Lake Waters. *Verh. Int. Ver. Limnol.* 10: 377–386.

Rodhe, W., 1974. Limnology Turns to Warm Lakes. *Arch. Hydrobiol.* 73: 537–546.

Rodhe, W., 1975. The SIL Founders and our Fundament. *Verh. Int. Ver. Limnol.* 19: 16–25.

Rodhe, W., 1979. The Life of Lakes. *Arch. Hydrobiologia Beih.* 13: 5–9.

Rojo, C., M. Alvarez-Cobelas, and M. Arauzo, 1994. An Elementary Structural Analysis of River Phytoplankton. *Hydrobiologia* 289: 43–55.

Romare, P., E. Bergman, and L.A. Hansson, 1999. The Impact of Larval and Juvenile Fish on Zooplankton and Algal Dynamics. *Limnol. Oceanogr.* 44: 1655–1666.

Rørslett, B., 1991. Principal Determinants of Aquatic Macrophyte Richness in Northern European Lakes. *Aquat. Bot.* 39: 173–193.

Roth, N.E., 1994. *Land Use, Riparian Vegetation, and Stream Ecosystem Integrity in an Agricultural Watershed* (Univ. Mich. Thesis). In J.D. Allan, 1995. Stream Ecology: Structure and Function of Running Waters. London: Chapman and Hall.

Rowan, D.J., and J. Kalff, 1991. The Limnological Implications of Catchment Sediment Load. *Verh. Int. Ver. Limnol.* 24: 2980–2984.

Rowan, D.J., and J.B. Rasmussen, 1992. Why Don't Great Lakes Fish Reflect Environmental Concentrations of Organic Contaminants?—An Analysis of Between-lake Variability in the Ecological Partitioning of PCBs and DDT. *J. Great Lakes Res.* 18: 724–741.

Rowan, D.J., J. Kalff, and J.B. Rasmussen, 1992a. Estimating the Mud Deposition Boundary Depth in Lakes from Wave Theory. *Can. J. Fish. Aquat. Sci.* 49: 2490–2497.

Rowan, D.J., J. Kalff, and J.B. Rasmussen, 1992b. Profundal Sediment Organic Content and Physical Character Do Not Reflect Lake Trophic Status, but Rather Reflect Inorganic Sedimentation and Exposure. *Can. J. Fish. Aquat. Sci.* 49: 1431–1438.

Rozengurt, M.A., and J.W. Hedgpeth, 1989. The Impact of Altered River Flow on the Ecosystem of the Caspian Sea. *CRC Crit. Rev. in Aquat. Sci.* 1: 337–362.

Rudd, J.W.M., C.A. Kelly, and A. Furutani, 1986. The Role of Sulfate Reduction in Long-term Accumulation of Organic and Inorganic Sulfur in Lake Sediments. *Limnol. Oceanogr.* 31: 1281–1291.

Rudd, J.W.M., C.A. Kelly, D.W. Schindler, and M.A. Turner, 1988. Disruption of the Nitrogen Cycle in Acidified Lakes. *Science* 240: 1515–1517.

Rudstam, L.G., R.C. Lathrop, and S.R. Carpenter, 1993. The Rise and Fall of a Dominant Planktivore: Direct and Indirect Effects on Zooplankton. *Ecology* 74: 303–319.

Russell-Hunter, W.D., 1969. *A Biology of Higher Invertebrates. Current Concepts in Biology Series.* London: MacMillan.

Rutherford, J.C., R.B. Williamson, and A.B. Cooper, 1987. *Nitrogen, Phosphorus, and Oxygen Dynamics in Rivers.* Pp. 139–165. In A.B. Viner, (ed.), Inland Waters of New Zealand. Wellington: DSIR Bull. #241.

Ruttner, F., 1963. *Fundamentals of Limnology (3rd ed.).* Translated by D.G. Frey, and F.E.J. Frey. Toronto: Univ. of Toronto Press.

Ryan, P.M., and H.H. Harvey, 1980. Growth Responses of Yellow Perch, *Perca flavescens* (Mitchill), to Lake Acidification in the La Cloche Mountain Lakes of Ontario. *Env. Biol. Fish.* 5: 97–108.

Rybak, J.I., 1969. Bottom Sediments of the Lakes of Various Trophic Type. *Ekol. Pol. Ser.* A17: 611–662.

Ryder, R.A., 1982. The Morphoedaphic Index—Use, Abuse, and Fundamental Concepts. *Trans. Am. Fish. Soc.* 111: 154–164.

Ryder, R.A., and J. Pesendorfer, 1989. *Large Rivers are More than Flowing Lakes: A Comparative Review.* Pp. 65–85. In D.P. Dodge (ed.), Proceedings of the International Large River Symposium. Ottawa: Can. Spec. Publ. Fish. Aquat. Sci. #106.

Ryding, S.O., 1980. Monitoring of Inland Waters. OECD Eutrophication Programme: The Nordic Project. Report from the Working Group for Eutrophication Research: NORDFORSK.

Ryding, S.O., and C. Forsberg, 1976. *Sediments as a Nutrient Source in Shallow Polluted Lakes.* Pp. 227–234. In H.L. Golterman (ed.), Interactions Between Sediments and Fresh Water. The Hague: Dr. W. Junk Publ.

Rzóska, J., 1980. *History and Development of the Freshwater Production Section of IBP.* Pp. 7–12. In E.D. Le Cren, and R.H. Lowe-McConnell, The Functioning of Freshwater Ecosystems. Cambridge: Cambridge Univ. Press.

Sabater, F., J.L. Meyer, and R.T. Edwards, 1993. Longitudinal Pattern of Dissolved Organic Carbon Concentration and Suspended Bacterial Density Along a Blackwater River. *Biogeochem.* 21: 73–93.

Saggio, A., and J. Imberger, 1998. Internal Wave Weather in a Stratified Lake. *Limnol. Oceanogr.* 43: 1780–1795.

Sakamoto, M., 1966. Primary Production by Phytoplankton Community in Some Japanese Lakes and Its Dependence on Lake Depth. *Arch. Hydrobiologia* 62: 1–28.

Salonen, K., L. Arvola, and M. Rask, 1984. Autumnal and Vernal Circulation of Small Forest Lakes in Southern Finland. *Verh. Int. Ver. Limnol.* 22: 103–107.

Sander, B.C., and J. Kalff, 1993. Factors Controlling Bacterial Production in Marine and Freshwater Sediments. *Microb.Ecology* 26: 79–99.

Sanders, R.W., 1991. Mixotrophic Protists in Marine and Freshwater Ecosystems. *J. Protozool.* 38: 76–81.

Sanders, R.W., K.G. Porter, S.J. Bennett, and A.E. DeBiase, 1989. Seasonal Patterns of Bacterivory by Flagellates, Ciliates, Rotifers, and Cladocerans in a Freshwater Planktonic Community. *Limnol. Oceanogr.* 34: 673–687.

Sanders, R.W., D.A. Caron, and U.G. Berninger, 1992. Relationships Between Bacteria and Heterotrophic Nanoplankton in Marine and Fresh Waters: An Inter-ecosystem Comparison. *Mar. Ecology Prog. Ser.* 86: 1–14.

Sand-Jensen, K., 1997. *Macrophytes as Biological Engineers in the Ecology of Danish Streams.* Pp. 74–101. In K. Sand-Jensen, and O. Pedersen (eds.), Freshwater Biology: Priorities and Development in Danish Research. Copenhagen: G.E.C. Gad Publ.

Sand-Jensen, K., 1989. Environmental Variables and Their Effect on Photosynthesis of Aquatic Plant Communities. *Aquat. Bot.* 34: 5–24.

Sand-Jensen, K., and M. Søndergaard, 1979. Distribution and Quantitative Development of Aquatic Macrophytes in Relation to Sediment Characteristics in Lake Kalgaard, Denmark. *Freshw. Biol.* 9: 1–11.

Sand-Jensen, K., and J. Borum, 1991. Interactions Among Phytoplankton, Periphyton, and Macrophytes in Temperate Freshwaters and Estuaries. *Aquat. Bot.* 41: 137–175.

Sand-Jensen, K., and T.V. Madsen, 1991. Minimum Light Requirements of Submerged Freshwater Macrophytes in Labratory Experiments. *J. Ecology* 79: 749–764.

Sand-Jensen, K., and J.R. Mebus, 1996. Fine-scale Patterns of Water Velocity Within Macrophyte Patches in Streams. *Oikos* 76: 169–180.

Sand-Jensen, K., and O. Pedersen, (eds.) 1997. *Freshwater Biology: Priorities and Development in Danish Research.* Copenhagen: G.E.C. Gad Publ.

Sarvala, J., V. Ilmavirta, L. Paasivirta, and K. Salonen, 1981. The Ecosystem of the Oligotrophic Lake Pääjärvi, 3. Secondary Production and an Ecological Energy Budget of the Lake. *Verh. Int. Ver. Limnol.* 21: 454–459.

Sas, H., 1989. *Lake Restoration by Reduction of Nutrient Loading: Expectations, Experiences, Extrapolations.* St. Augustin: Acad. Verlag Richarz.

Saunders, and J. Kalff, 2001a. Denitrification in the Sediments of Lake Memphremagog, Canada-USA. *Water Res.* (in press).

Saunders, D., and J. Kalff, 2001b. Nitrogen Retention in Wetlands, Lakes and Rivers. *Hydrobiologia* 443: 205–212.

Schallenberg, M., and J. Kalff, 1993. The Ecology of Sediment Bacteria in Lakes and Comparisons with Other Aquatic Ecosystems. *Ecology* 74: 919–934.

Schallenberg, M., J. Kalff, and J.B. Rasmussen, 1989. Solutions to Problems in Enumerating Sediment Bacteria by Direct Counts. *Appl. Env. Microbiol.* 55: 1214–1219.

Schaller, T., H.C. Moor, and B. Wehrli, 1997. Reconstructing the Iron Cycle from the Horizontal Distribution of Metals in the Sediment of Baldeggersee. *Aquat. Sci.* 59: 326–344.

Schanz, F., 1982. Light Conditions in Lake Zurich 1979–1981. Part I: Secchi Disk Transparency. *Vierteljahrs. Naturf. Gesell. Zürich* 127: 357–367.

Schanz, F., 1983. Light Conditions in Lake Zurich, 1980/81. Part II: Surface Effects. *Arch. Hydrobiologia* 97: 501–508.

Scheffer, M., 1998. *Ecology of Shallow Lakes.* Population and Community Biology Series 22: London: Chapman and Hall.

Scheffer, M., and J. Beets, 1994. Ecological Models and the Pitfalls of Causality. *Hydrobiologia* 275/276: 115–124.

Scheider W.A., W.R. Snyder, and B. Clark, 1979. Deposition of Nutrients and Major Ions by Precipitation in Southcentral Ontario. *Water, Air and Soil Poll.* 12: 171–185.

Schelske, C.L., A. Peplow, M. Brenner, and C.N. Spencer, 1994. Low-background Gamma Counting: Applications for ^{210}Pb Dating of Sediments. *J. Paleolimnol.* 10: 115–128.

Schelske, C.L., H.J. Carrick, and F.J. Aldridge, 1995. Can Wind-induced Resuspension of Meroplankton Affect Phytoplankton Dynamics? *J. N. Am. Benthol. Soc.* 14: 616–630.

Scheuhammer, A.M., and P.J. Blancher, 1994. Potential Risk to Common Loons (*Gavia immer*) from Methylmercury Exposure in Acidified Lakes. *Hydrobiologia* 279/280: 445–455.

Schiff, S.L., R. Aravena, S.E. Trumbore, M.J. Hinton, R. Elgood, and P.J. Dillon, 1997. Export of DOC from Forested Catchments on the Precambrian Shield of Central Ontario: Clues from ^{13}C and ^{14}C. *Biogeochem.* 36: 43–65.

Schindler, D.E., S.R. Carpenter, K.L. Cottingham, X. He, J.R. Hodgson, J.F. Kitchell, and P.A. Soranno, 1996. *Food Web Structure and Littoral Zone Coupling to Pelagic Trophic Cascades.* Pp. 96–105. In G.A. Polis, and K.O. Winemiller (eds.), Food Webs: Integration of Patterns and Dynamics. New York: Chapman and Hall.

Schindler, D.W., 1969. Two Useful Devices for Vertical Plankton and Water Sampling. *Can. J. Fish Res. Bd.* 26: 1948–1955.

Schindler, D.W., 1997. Widespread Effects of Climatic Warming on Freshwater Ecosystems in North America. *Hydrol. Proc.* 11: 1043–1067.

Schindler, D.W., H. Kling, R.V. Schmidt, J. Prokopowich, V.E. Frost, R.A. Reid, and M. Capel, 1973. Eutrophication of Lake 227 by Addition of Phosphate and Nitrate: The Second, Third and Fourth Years of Enrichment, 1970, 1971, and 1972. *Can. J. Fish. Res. Bd.* 30: 1415–1440.

Schindler, D.W., H.E. Welch, J. Kalff, G.J. Brunskill, and N. Kritsch, 1974. Physical and Chemical Limnology of Char Lake, Cornwallis Island (75°N Lat.). *Can. J. Fish. Res. Bd.* 31: 585–607.

Schindler, D.W., R.W. Newbury, K.G. Beatty, and P. Campbell, 1976. Natural and Chemical Budgets for a Small Precambrian

Lake Basin in Central Canada. *Can. J. Fish. Res. Bd.* 33: 2526–2543.

Schindler, D.W., S.E.M. Kaisan, and R.H. Hesslein, 1989. Biological Impoverishment in Lakes of the Midwestern and Northeastern United States from Acid Rain. *Env. Sci. Technol.* 23: 573–580.

Schindler, D.W., K.G. Beaty, E.J. Fee, D.R. Cruikshank, E.R. DeBruyn, D.L. Findley, G.A. Lindsey, J.A. Shearer, M.P. Stainton, and M.A. Turner, 1990. Effects of Climatic Warming on Lakes of the Central Boreal Forest. *Science* 250: 967–970.

Schindler, D.W., T.M. Frost, K.H. Mills, P.S.S. Chang, I.J. Davies, D.L. Findlay, D.F. Malley, J.A. Schearer, M.A. Turner, P.J. Garrison, C.J. Watras, K. Webster, J.M. Gunn, P.L. Brezonik, and W.A. Swenson, 1991. Comparison Between Experimentally- and Atmospherically-acidified Lakes During Stress and Recovery. *Proc. Roy. Soc. Edinburgh* 97B: 193–226.

Schindler, D.W., S.E. Bayley, B.R. Parker, K.G. Beatty, D.R. Cruikshank, E.J. Fee, E.U. Schindler, and M.P. Stainton, 1996. The Effects of Climatic Warming on the Properties of Boreal Lakes and Streams at the Experimental Lakes Area, Northwestern Ontario. *Limnol. Oceanogr.* 41: 1004–1017.

Schindler, D.W., P.J. Curtis, B.R. Parker, and M.P. Stainton, 1996. Consequences of Climate Warming and Lake Acidification for UV-B Penetration in North American Boreal Lakes. *Nature* 379: 705–708.

Schlesinger, D.A., and H.A. Regier, 1982. Climatic and Morphedaphic Iindices of Fish Yields from Natural Lakes. *Trans. Amer. Fish. Soc.* 111: 141–150.

Schlesinger, W.H., 1997. *Biogeochemistry: An Analysis of Global Change*, 2nd ed. San Diego: Acad. Press.

Schlesinger, W.H., and J.M. Melack, 1981. Transport of Organic Carbon in the World's Rivers. *Tellus* 33: 172–187.

Schmitt, C.J., J.L. Zajicek, and M.A. Ribick, 1985. National Pesticide Monitoring Program: Residues of Organochlorine Chemicals in Freshwater Fish, 1980–1981. *Arch. Environ. Contam. Toxicol.* 14: 225–260

Schnoor, J.L., and W. Stumm, 1985. *Acidification of Aquatic and Terrestrial Systems.* Pp. 311–338. In W. Stumm (ed.), Chemical Processes in Lakes. New York: John Wiley and Sons.

Schofield, C.O., 1972. The Ecological Significance of Air-pollution-induced Changes in Water Quality of Dilute-lake Districts in the Northeast. *Trans. N. E. Fish. Wildlife Conf.* pp. 98–112.

Scholtz, C.A., and B.R. Rosendahl, 1988. Low Lake Stands in Lakes Malawi and Tanganyika, East Africa, Delineated with Multifold Seismic Data. *Science* 240: 1645–1648.

Schuiling, R.D., 1976. *Source and Composition of Lake Sediments.* Pp. 12–18. In H.L. Golterman (ed.), Interactions Between Sediments and Fresh Water. The Hague: Dr. W. Junk Publ.

Schut, P.H., and R.D. Evans, 1986. Variation in Trace Metal Exports from Small Canadian Shield Watersheds. *Water, Air, Soil Poll.* 28: 225–237.

Schwartz, S.E., 1989. Acid Deposition: Unraveling a Regional Phenomenon. *Science* 243: 753–763.

Schweizer, A., 1997. From Littoral to Pelagial: Comparing the Distribution of Phytoplankton and Ciliated Protozoa Along a Transect. *J. Plankton. Res.* 19: 829–848.

Scott, D.F., and R.E. Smith, 1997. Preliminary Empirical Models to Predict Reductions in Total and Low Flows Resulting from Aforestation. *Water S. A.* 23: 135–140.

Seda, J., and J. Kubečka, 1997. Long-term Biomanipulation of Rimov Reservoir (Czech Republic). *Hydrobiologia* 345: 95–108.

Seitzinger, S.P., 1988. Denitrification in Freshwater and Coastal Marine Ecosystems: Ecological and Geochemical Significance. *Limnol. Oceanogr.* 33: 702–724.

Sellers, P., C.A. Kelly, J.W.M. Rudd, and A.R. MacHutchon, 1996. Photodegradation of Methylmercury in Lakes. *Nature* 380: 694–697.

Sephton, D.H., and G.P. Harris, 1984. Physical Variability and Phytoplankton Communities, VI. Day-to-Day Changes in Primary Productivity and Species Abundance. *Arch. Hydrobiol* 102: 155–175.

Serruya, C., and H. Leventer, 1984. *Lakes and Reservoirs of Israel.* Pp. 357–384. In F.B. Taub (ed.), Ecosystems of the World, Vol. 23: Lakes and Reservoirs. Amsterdam: Elsevier.

Setaro, F.V., and J.M. Melack, 1984. Responses of Phytoplankton to Experimental Nutrient Enrichment in an Amazon Floodplain Lake. *Limnol. Oceanogr.* 29: 972–984.

Sevaldrud, I.H., I.P. Muniz, S. Kalvenes, 1980. *Loss of Fish Populations in Southern Norway. Dynamics and Magnitude of the Problem.* Pp. 350–351. In D. Drabløs, and A. Tollan (eds.), Ecological Impact of Acid Precipitation. Oslo: SNSF-Project #1432.

Shapiro, J., 1990. Biomanipulation: The Next Phase—Making It Stable. *Hydrobiologia* 200/201: 13–27.

Shaw, R.D., and E.E. Prepas, 1990. Groundwater–Lake Interactions, I. Accuracy of Seepage Meter Estimates of Lake Seepage. *J. of Hydrol.* 119: 105–120.

Shaw, R.D., A.M. Trimbee, A. Minty, H. Fricky, and E.E. Prepas, 1989. Atmospheric Deposition of Phosphorus and Nitrogen in Central Alberta with Emphasis on Narrow Lake. *Water, Air, and Soil Poll.* 43: 119–134.

Sheldon, R.W., A. Prakash, and W.H. Sutcliffe, Jr., 1972. The Size Distribution of Particles in the Ocean. *Limnol. Oceanogr.* 17: 327–340.

Sherwood, J.E., F. Stagnitti, M.J. Kokkinn, and W.D. Williams, 1992. A Standard Table for Predicting Equilibrium Dissolved Oxygen Concentrations in Salt Lakes Dominated by Sodium Chloride. *Int. J. Salt Lake Res.* 1: 1–6.

Shiklomanov, I.A., 1993. *World Fresh Water Resources.* Pp. 13–24. In P.H. Gleick, Water in Crisis: A Guide to the World's Fresh Water Resources. New York: Oxford Univ. Press.

Shuter, B.J., and J.F. Koonce, 1977. A Dynamic Model of the Western Lake Erie Walleye (*Stizostedion vitreum vitreum*) Population. *Can. J. Fish. Res. Bd.* 34: 1972–1982.

Shuter, B.J., D.A. Schlesinger, and A.P. Zimmerman. 1983. Empirical Predictors of Annual Surface Water Temperature Cycles in North American Lakes. *Can. J. Fish. Aquat. Sci.* 40: 1838–1845.

Shuter, B.J., and H.A. Regier, 1989. *The Ecology of Fish and Populations: Dealing with Interactions Between Levels.* Pp. 33–49. In C.D. Levings, L.B. Holtby, and M.A. Henderson (eds.), Proceedings of the National Workshop on Effects of Habitat Alteration on Salmonid Stocks. Ottawa: Can. Spec. Publ. Fish. Aquat. Sci. #105.

Shuter, B.J., and J.R. Post, 1990. Climate, Population Viability, and the Zoogeography of Temperate Fishes. *Trans. Am. Fish. Soc.* 119: 314–336.

Sicko-Goad, L., E.F. Stoermer, and G. Fahnenstiel, 1986. Rejuvenation of *Melosira granulata* (Bacillariophyceae) Resting Cells from the Anoxic Sediments of Douglas Lake, Michigan, 1. Light Microscopy and ^{14}C Uptake. *J. Phycol.* 22: 22–28.

Sieburth, J.M., V. Smetacek, and J. Lenz, 1978. Pelagic Ecosystem Structure: Heterotrophic Compartments of the Plankton and Their Relationship to Plankton Size Fractions. *Limnol. Oceanogr.* 23: 1256–1263.

Sigg, L., C.A. Johnson, and A. Kuhn, 1991. Redox Conditions and Alkalinity Generation in a Seasonally Anoxic Lake (Lake Greifen). *Mar. Chem.* 36: 9–26.

Šimek, K., and V. Straškrabová, 1992. Bacterioplankton Production and Protozoan Bacterivory in a Mesotrophic Reservoir. *J. Plankt. Res.* 14: 773–787.

Šimek, K., J. Bobková, M. Macek, J. Nedoma and R. Psenner, 1995. Ciliate Grazing on Picoplankton in a Eutrophic Reservoir During the Summer Phytoplankton Maximum: A Study at the Species and Community Level. *Limnol. Oceanogr.* 40: 1077–1090.

Šimek, K., P. Kojecká, J. Nedoma, P. Hartman, J. Vrba, and J.R. Dolan, 1999. Shifts in Bacterial Composition Associated with Different Microzooplankton Size Fractions in a Eutrophic Reservoir. *Limnol. Oceanogr.* 44: 1634–1644.

Simola, H., 1981. Sedimentation in a Eutrophic Stratified Lake in S. Finland. *Ann. Bot. Fennici* 18: 23–36.

Simola, H., I. Hanski, and M. Liukkonen, 1990. Stratigraphy, Species Richness and Seasonal Dynamics of Plankton Diatoms During 418 Years in Lake Lovojärvi, South Finland. *Ann. Bot. Fennici* 27: 241–259.

Simon, M., 1987. Biomass and Production of Small and Large Free-living and Attached Bacteria in Lake Constance. *Limnol. Oceanogr.* 32: 591–607.

Simon, M., B.C. Cho, and F. Azam, 1992. Significance of Bacterial Biomass in Lakes and the Ocean: Comparison to Phytoplankton Biomass and Biogeochemical Implications. *Mar. Ecology Progr. Ser.* 86: 103–110.

Simonich, S.L., and R.A. Hites, 1995. Global Distribution of Persistent Organochlorine Compounds. *Science* 269: 1851–1854.

Siole, H. (ed.), 1984. *The Amazon: Limnology and Landscape Ecology of a Mighty Tropical River and Its Basin.* Monogr. Biol., Vol. 56. Dordrecht: Junk Publ.

Sissenwine, M., 1984. *Why Do Fish Populations Vary?* Pp. 59–94. In R.M. May (ed.), Exploitation of Marine Communities. Berlin: Springer-Verlag.

Sivonen, K., S.I. Niemelä, R.M. Niemi, L. Lepistö, T.H. Luoma, and L.A. Räsänen, 1990. Toxic Cyanobacteria (Blue-green Algae) in Finnish Fresh and Coastal Waters. *Hydrobiologia* 190: 267–275.

Skubinna, J.P., T.G. Coon, and T.R. Batterson, 1995. Increased Abundance and Depth of Submerged Macrophytes in Response to Decreased Turbidity in Saginaw Bay, Lake Huron. *J. Great Lakes Res.* 221: 476–487.

SMA, 1982. *Acidification Today and Tomorrow.* Stockholm, Swedish Ministry of Agriculture.

Smayda, T.J., 1970. The Suspension and Sinking of Phytoplankton in the Sea. *Oceanogr. Mar. Biol. Ann. Rev.* 8: 353–414.

Smith, I.R., 1975. Turbulence in Lakes and Rivers. *Scientific Publication No. 29.* Freshw. Biol. Assoc.

Smith, I.R., 1979. Hydraulic Conditions in Isothermal Lakes. *Freshw. Biol.,* 9: 119–145.

Smith, I.R., and I.J. Sinclair, 1972. Deep-water Waves in Lakes. *Freshw. Biol.* 5: 387–399.

Smith, R.C., J.E. Tyler, and C.R. Goldman, 1973. Optical Properties and Color of Lake Tahoe and Crater Lake. *Limnol. Oceanogr.* 18: 189–199.

Smith, V.H., 1982. The Nitrogen Phosphorus Dependence of Algal Biomass in Lakes: An Empirical and Theoretical Analysis. *Limnol. Oceanogr.* 27: 1101–1112.

Smock, L.A., J.E. Gladden, J.L. Riekenberg, L.C. Smith, and C.R. Black, 1992. Lotic Macroinvertebrate Production in Three Dimensions: Channel Surface, Hyporheic, and Floodplain Environments. *Ecology* 73: 876–886.

Smol, J.P., 1990. Paleolimnology: Recent Advances and Future Challenges. *Mem. Ist. Ital. Idrobiol.* 47: 253–284.

Smol, J.P., S.R. Brown, and R.N. McNeely, 1983. Cultural Disturbances and Trophic History of a Small Meromictic Lake from Central Canada. *Hydrobiologia* 103: 125–130.

Smol, J.P., and J.R. Glew, 1992. *Paleolimnology.* Pp 551–564. In Encyclopedia of Earth System Science, Vol. 3. San Diego: Acad. Press.

Snead, R.E., 1980. *World Atlas of Geomorphic Features.* Huntington: Robert E. Krieger Publ.

Søballe, D.M., and S.T. Threlkeld, 1985. Advection, Phytoplankton Biomass, and Nutrient Transformations in a Rapidly Flushed Impoundment. *Arch. Hydrobiol.* 105: 187–203.

Søballe, D.M., and B.L. Kimmel, 1987. A Large-scale Comparison of Factors Influencing Phytoplankton Abundance in Rivers, Lakes and Impoundments. *Ecology* 68: 1943–1954.

Sollins, P., G. Spycher, and C. Topik, 1983. Processes of Soil Organic-matter Accretion at a Mudflow Chronosequence, Mt. Shasta, California. *Ecology* 49: 227–254.

Somlyódy, L., 1995. Water Quality Management: Can We Improve Integration to Face Future Problems? *Wat. Sci. Tech.* 31: 249–259.

Sommaruga, R., and R.D. Robarts, 1997. The Significance of Autotrophic and Heterotrophic Picoplankton in Hypertrophic Ecosystems. *FEMS Microbiol. Ecology* 24: 187–200.

Sommaruga, R., and D. Conde, 1997. Seasonal Variability of Metabolically Active Bacterioplankton in the Euphotic Zone of a Hypertrophic Lake. *Aquat. Microb. Ecology* 13: 241–248.

Sommaruga, R., I. Obernosterer, G.J. Herndl, and R. Psenner, 1997. Inhibitory Effect of Solar Radiation on Thymidine and Leucine Incorporation by Freshwater and Marine Bacterioplankton. *Appl. Env. Microbiol.* 63: 4178–4184.

Sommaruga-Wögrath, S., K.A. Koinig, R. Schmidt, R. Sommaruga, R. Tessadri, and R. Psenner, 1997. Temperature Effects on the Acidity of Remote Alpine Lakes. *Nature* 387: 64–67.

Sommer, U., 1984. Sedimentation of Principal Phytoplankton Species in Lake Constance. *J. Plankton Res.* 6: 1–14.

Sommer, U., 1988. Some Size Relationships in Phytoflagellate Motility. *Hydrobiologia* 161: 125–131.

Sommer, U., Z.M. Gliwicz, W. Lampert, and A. Duncan, 1986. The PEG-model of Seasonal Succession of Planktonic Events in Fresh Waters. *Arch. Hydrobiologia* 106: 433–471.

Søndergaard, M., 1989. Phosphorus Release from a Hypertrophic Lake Sediment: Experiments with Intact Sediment Cores in a Continuous Flow System. *Arch. Hydrobiologia* 116: 45–59.

Søndergaard, M., P. Kristensen, and E. Jeppesen, 1993. Eight Years of Internal Phosphorous Loading and Changes in the Sediment Phosphorous Profile of Lake Søbygaard, Denmark. *Hydrobiologia* 253: 345–356.

Søndergaard, M., L. Bruun, T. Lauridsen, E. Jeppesen, and T.V. Madsen, 1996. The Impact of Grazing Waterfowl on Submerged Macrophytes: In Situ Experiments in a Shallow Eutrophic Lake. *Aquat. Bot.* 53: 73–84.

Søndergaard, M., and B. Moss, 1999. *Impact of Submerged Macrophytes on Phytoplankton in Shallow Freshwater Lakes.* Pp. 115–132. In E. Jeppesen, M. Søndergaard, and K. Christoffersen (eds.), The Structuring Role of Submerged Macrophytes in Lakes. New York: Springer-Verlag.

Soranno, P.A., S.R. Carpenter, and R.C. Lathrop, 1997. Internal Phosphorus Loading in Lake Mendota: Response to External Loads and Weather. *Can. J. Fish. Aquat. Sci.* 54: 1883–1893.

Sorokin, Y.I., 1970. Interrelation Between Sulfur and Carbon Turnover in Meromictic Lakes. *Arch. Hydrobiologia* 66: 391–446.

Spence, D.H.N., 1967. Factors Controlling the Distribution of Freshwater Macrophytes with Particular Reference to the Lochs of Scotland. *J. Ecology* 55: 147–170.

Spigel, R.H., and J. Imberger, 1980. The Classification of Mixed-Layer Dynamics in Lakes of Small to Medium Size. *J. Phys. Oceanogr.,* 10: 1104–1121.

Spigel, R.H., and J. Imberger, 1987. Mixing Processes Relevant to Phytoplankton Dynamics in Lakes. *New Zealand J. Marine Freshw. Res.,* 21: 361–377.

Spodniewska, I., 1983. Ecological Characteristics of Lakes in North-eastern Poland Versus Their Trophic Gradient, VI. The Phytoplankton of 43 Lakes. *Ekol. Pol.* 31: 353–381.

Sprules, W.G., 1975a. Factors Affecting the Structure of Limnetic Crustacean Zooplankton Communities in Central Ontario Lakes. *Verh. Int. Ver. Limnol.* 19: 635–643.

Sprules, W.G., 1975b. Midsummer Crustacean Zooplankton Communities in Acid-stressed Lakes. *Can. J. Fish. Res. Board.* 32: 389–395.

Sprung, M., 1990. Costs of Reproduction: A Study on Metabolic Requirements of the Gonads and Fecundity of the Bivalve *Dreissena polymorpha. Malacologia.* 32: 267–274.

St. Louis, V.L., J.W.M. Rudd, C.A. Kelly, K.G. Beatty, N.S. Bloom, and R.J. Flett, 1994. Importance of Wetlands as a Source of Methyl Mercury to Boreal Forest Ecosystems. *Can. J. Fish. Aquat. Sci.* 51: 1065–1076.

Stadelmann, P., 1971. Stickstoffkreislauf und Primärproduktion im Mesotrophen Vierwaldstättersee (Horwer Bucht) und im Eutrophen Rotsee, mit Besonderer Berücksichtigung des Nitrats als Limitierenden Faktors. *Schweiz. Z. Hydrol.* 33: 1–65.

Stanford, J.A., and J.V. Ward, 1993. An Ecosystem Perspective of Alluvial Rivers: Connectivity and the Hyporheic Corridor. *J. N. Amer. Benthol. Soc.* 12: 48–60.

Staubitz, W.W., and P.J. Zarriello, 1989. Hydrology of Two Headwater Lakes in the Adirondack Mountains of New York. *Can. J. Fish. Aquat. Sci.* 46: 268–276.

Stauffer, R.E., 1987. A Comparative Analysis of Iron, Manganese, Silica, Phosphorus, and Sulfur in the Hypolimnia of Calcareous Lakes. *Wat. Res.* 21: 1009–1022.

Stauffer, R.E., and G.F. Lee, 1973. *The Role of Thermocline Migration in Regulating Algal Blooms.* Pp. 73–82. In E.J. Middlebrooks, D.H. Falkenborg, T.E. Maloney (eds.), Modeling the Eutrophication Process. Ann Arbor: Ann Arbor Sci. Publ.

Steel, J.A., 1997. Scope and Limitation in Algal Modelling—An Example from the Thames Valley Reservoirs. *Hydrobiologia* 349: 27–37.

Stein, R.A., D.R. DeVries, and J.M. Dettmers, 1995. Food-web Regulation by a Planktivore: Exploring the Generality of the Trophic Cascade Hypothesis. *Can. J. Fish. Aquat. Sci.* 52: 2518–2526.

Steinberg, C.E., and H.M. Hartmann, 1987. Planktonic Bloomforming Cyanobacteria and the Eutrophication of Lakes and Rivers. *Freshw. Biol.* 20: 179–289.

Steinman, A.D., 1996. *Effects of Grazers on Freshwater Benthic Algae.* Pp. 341–373. In R.J. Stevenson, M.L. Bothwell, and R.L. Lowe (eds.), Algal Ecology: Freshwater Benthic Ecosystems. San Diego: Acad. Press.

Stemberger, R.S., and J.J. Gilbert, 1984. Spine Development in the Rotifer *Keratella cochlearis*: Induction by Cyclopoid Copepods and *Asplanchna*. *Freshw. Biol.* 14: 639–647.

Sterner, R.W., 1989. *The Role of Grazers in Phytoplankton Succession.* Pp. 107–170. In U. Sommer (ed.), Plankton Ecology: Succession in Plankton Communities. Berlin: Springer-Verlag.

Sterner, R.W., 1990. Lake Morphometry and Light in the Surface Layer. *Can. J. Fish. Aquat. Sci.* 47: 687–692.

Stevens, R.J., and C.E. Gibson, 1976. *Sediment Release of Phosphorus in Lough Neagh, Northern Ireland.* Pp. 343–347. In H.L. Golterman (ed.), Interactions Between Sediments and Fresh Water. The Hague: Dr. W. Junk Publ.

Stevens, R.J., and D.A. Stewart, 1981. The Effect of Sampling Interval and Method of Calculation on the Accuracy of Estimated Phosphorus and Nitrogen Loads in Drainage Water from Two Different-sized Catchment Areas. *Rec. of Agr. Res.* (Northern Ireland) 29: 29–38.

Stevenson, J.C., 1988. Comparative Ecology of Submersed Grass Beds in Freshwater, Estuarine, and Marine Environments. *Limnol. Oceanogr.* 33: 867–893.

Stevenson, R.J., M.L. Bothwell, and R.L. Lowe (eds.), 1996. *Algal Ecology: Freshwater Benthic Ecosystems.* San Diego: Acad. Press.

Stewart, R., 1964. On the Estimation of Lake Depth from the Period of the Seiche. *Limnol. Oceanogr.*, 9: 606–607.

Stewart, T.W., J.G. Miner, and R.L. Lowe, 1998. Macroinvertebrate Communities on Hard Substrates in Western Lake Erie: Structuring Effects of *Dreissena*. *J. Great Lakes Res.* 24: 868–879.

Stiller, M., and Y.C. Chung, 1984. Radium in the Dead Sea: A Possible Tracer for the Duration of Meromixis. *Limnol. Oceanogr.* 29: 574–586.

Stockner, J.G., and J.W.G. Lund, 1970. Live Algae in Postglacial Lake Deposits. *Limnol. Oceanogr.* 15: 41–58.

Stockner, J.G., and K.R.S. Shortreed, 1976. Autotrophic Production in Carnation Creek, a Coastal Rainforest Stream on Vancouver Island, British Columbia. *Can. J. Fish. Res. Bd.* 33: 1553–1563.

Stockner, J.G., and K.R.S. Shortreed, 1985. Whole-lake Fertilization Experiments in Coastal British Columbia Lakes: Empirical Relationships Between Nutrient Inputs and Phytoplankton Biomass and Production. *Can. J. Fish Aquat. Sci.* 42: 649–658.

Stockner, J.G., and K.G. Porter, 1988. *Microbial Food Webs in Freshwater Planktonic Ecosystems.* Pp. 69–83. In S.R. Carpenter (ed.), Complex Interactions in Lake Communities. New York: Springer-Verlag.

Stoddard, J.L., D.S. Jeffries, A. Lükewille, T.A. Clark, P.J. Dillon, C.T. Driscoll, M. Forsius, M. Johannessen, J.S. Kahl, J.H. Kellogg, A. Kemp, J. Mannio, D. Monteith, P.S. Murdoch, S. Patrick, A. Rebsdorf, B.L. Skjelkvåle, M. Stainton, T. Traaen, H. van Dam, K.E. Wester, J. Wieting, and A. Wilander, 1999. Regional Trends in Aquatic Recovery from Acidification in North America and Europe. *Nature* 401: 575–578.

Strachan, W.M.J., and S.J. Eisenreich, 1988. *Mass Balancing of Organic Contaminants in the Great Lakes: The Role of Atmospheric Deposition.* IJC, Workshop on the Estimation of Atmospheric Loadings of Toxic Chemicals to the Great Lakes Basin, Windsor, Ontario.

Straile, D., 1997. Gross Growth Efficiencies of Protozoan and Metazoan Zooplankton and Their Dependence on Food Concentration, Predator–Prey Weight Ratio and Taxonomic Group. *Limnol. Oceanogr.* 42: 1375–1385.

Straškraba, M., 1980. *The Effects of Physical Variables on Freshwater Production: Analyses Based on Models.* Pp. 13–84. In E.D. Le Cren, and R.H. Lowe-McConnell (eds.), The Functioning of Freshwater Ecosystems. Cambridge: Cambridge Univ. Press.

Straškraba, M., 1996. Lake and Reservoir Management. *Verh. Int. Ver. Limnol.* 26: 193–209.

Straškraba, M., and V. Straškrabová, 1969. *Eastern European Lakes.* Pp. 65–97. In Int. Symp. on Eutrophication, Eutrophication: Causes, Consequences, Correctives. Washington: Nat. Acad. Sci.

Straškraba, M., and P. Javornický, 1973. *Limnology of Two Re-regulation Reservoirs in Czechoslovakia.* Pp. 249–316. In J. Hrbáček, and M. Straškraba (eds.), Hydrobiological Studies, 2. Prague: Acad. Publ. House of the Czechoslovak Acad. of Sci.

Straškraba, M., J.G. Tundisi, and A. Duncan, 1993. a. *State-of-the-art of Reservoir Limnology and Water Quality Management.* Pp. 213–288. In M. Straškraba, J.G. Tundisi, and A. Duncan (eds.), Comparative Reservoir Limnology and Water Quality Management. Dordrecht: Kluwer Acad. Publ.

Straškraba, M., P. Blažka, Z. Brandl, P. Hejzlar, J. Komárková, J. Kubečka, I. Nesměrák, L. Procházková, V. Straškrabová, and V. Vyhnálek, 1993. b. *Framework for Investigation and Evaluation of Reservoir Water Quality in Czechoslovakia.* Pp 169–212. In M. Straškraba, J.G. Tundisi, and A. Duncan (eds.), Comparative Reservoir Limnology and Water Quality Management. Dordrecht: Kluwer Acad. Publ.

Strayer, D.L., 1985. The Benthic Micrometazoans of Mirror Lake, New Hampshire. *Arch. Hydrobiologia* 72(Suppl.): 287–426.

Strayer, D.L., 1991. Perspectives on the Size Structure of Lacustrine Zoobenthos, Its Causes and Its Consequences. *J. N. Amer. Benth. Soc.* 10: 210–221.

Strayer, D.L., 1999. Effects of Alien Species on Freshwater Molluscs in North America. *J. N. Amer. Benth. Soc.* 18: 74–98.

Strayer, D.L., and G.E. Likens, 1986. An Energy Budget for the Zoobenthos of Mirror Lake, New Hampshire. *Ecology* 67: 303–313.

Strayer, D.L., S.E. May, P. Nielsen, W. Wollheim, and S. Hausam, 1997. Oxygen, Organic Matter and Sediment Granulometry as Controls on Hyporheic Animal Communities. *Arch. Hydrobiol.* 140: 131–144.

Stumm, W., 1987. *Impact of Resource Use on the Hydrosphere and Aquatic Ecosystems*. Pp: 377–398. In D.J. McLaren, and B.J. Skinner (eds.), Resources and World Development. Chichester: John Wiley and Sons.

Stumm, W., and P. Baccini, 1978. *Man-made Chemical Perturbation of Lakes*. Pp. 91–126. In A. Lerman (ed.), Lakes: Chemistry, Geology, Physics. New York: Springer-Verlag.

Stumm, W., R. Schwarzenbach, and L. Sigg, 1983. From Environmental Analytical Chemistry to Ecotoxicology—A Plea for More Concepts and Less Monitoring and Testing. *Angew. Chem. Int. Ed. Engl.* 22: 380–389.

Stumm, W., and J.J. Morgan, 1996. *Aquatic Chemistry: Chemical Equilibria and Rates in Natural Waters, 3rd ed.* New York: John Wiley and Sons.

Summers, P.W., V.C. Bowersox, and G.J. Stensland, 1986. The Geographical Distribution and Temporal Variations of Acidic Deposition in Eastern North America. *Wat. Air Soil Pollut.* 31: 523–535.

Suter, W., 1994. Overwintering Waterfowl on Swiss Lakes: How are Abundance and Species Richness Influenced by Trophic Status and Lake Morphology? *Hydrobiologia* 279/280: 1–14.

Sverdrup, H.U., M.W. Johnson, and R.H. Fleming, 1942. *The Oceans: Their Physics, Chemistry, and General Biology*. New York: Prentice Hall.

Swain, E.B., D.R. Engstrom, M.E. Brigham, T.A. Henning, and P.L. Brezonik, 1992. Increasing Rates of Atmospheric Mercury Deposition in Midcontinental North America. *Science* 257: 784–787.

Talling, J.F., 1957. The Phytoplankton Population as a Compound Photosynthetic System. *New Phytol.* 56: 133–149.

Talling, J.F., 1969. The Incidence of Vertical Mixing, and Some Biological and Chemical Consequences, in Tropical African Lakes. *Verh. Int. Ver. Limnol.* 17: 998–1012.

Talling, J.F., 1993. Comparative Seasonal Changes, and Inter-annual Variability and Stability, in a 26-Year Record of Total Phytoplankton Biomass in Four English Lake Basins. *Hydrobiologia* 268: 65–98.

Talling, J.F., and I.B. Talling, 1965. The Chemical Composition of African Lake Waters. *Int. Rev. Ges. Hydrobiologia* 50: 421–463.

Talling, J.F., R.B. Wood, M.V. Prosser, and R.M. Baxter, 1973. The Upper Limit of Photosynthetic Productivity by Phytoplankton: Evidence from Ethiopian Soda Lakes. *Freshw. Biol.* 3: 53–76.

Talling, J.F., 1971. The Underwater Light Climate as a Controlling Factor in the Production Ecology of Freshwater Phytoplankton. *Mitt. Int. Verh. Limnol.* 19: 214–243.

Talling, J.F., and J. Lemoalle, 1998. *Ecological Dynamics of Tropical Inland Waters*. Cambridge Univ. Press, Cambridge.

Tátrai, I., J. Oláh, G. Paulovits, K. Mátyás, B.J. Kawiecka, V. Józsa, and F. Pekár, 1997. Biomass Dependent Interactions in Pond Ecosystems: Responses of Lower Trophic Levels to Fish Manipulations. *Hydrobiologia* 345: 117–129.

Tereshenkova, T.V., 1985. Environmental Effects on Specific Rate of Photosynthesis in Small Fertilized Lakes (Northwest of the USSR). *Int. Rev. Ges. Hydrobiologia* 70: 453–470.

Tessier, A., F. Rapin, and R. Carignan, 1985. Trace Metals in Oxic Lake Sediments: Possible Adsorption onto Iron Oxyhydroxides. *Geochim. Cosmochim. Acta* 49: 183–194.

Tezuka, Y., 1990. Bacterial Regeneration of Ammonium and Phosphate as Affected by the Carbon: Nitrogen:Phosphorus Ratio of Organic Substrates. *Microb. Ecology* 19: 227–238.

Theil-Nielsen, J., and M. Søndergaard, 1999. Production of Epiphytic Bacteria and Bacterioplankton in Three Shallow Lakes. *Oikos* 86: 283–292.

Thienemann, A., 1925. *Die Binnengewässer Mitteleuropas: Eine Limnologische Einführung*. Stuttgart: E. Schweizerbart'sche Verlagsbuchhandlung.

Thomann, R.V., 1989. Bioaccumulation Model of Organic Chemical Distribution in Aquatic Food Chains. *Env. Sci. Technol.* 23: 699–707.

Thomas, R.L., 1983. Lake Ontario Sediments as Indicators of the Niagara River as a Primary Source of Contaminants. *J. Great Lakes Res.*, 9: 118–124.

Thomas, P.A., and P.M. Room, 1986. Successful Control of the Floating Weed *Salvinia molesta* in Papua New Guinea: A Useful Biological Invasion Neutralizes a Disastrous One. *Env. Cons.* 13: 242–248.

Thompson, J.B., S. Schultz-Lam, T.J. Beveridge, and D.J. Des Marais, 1997. Whiting Events: Biogenic Origin Due to the Photosynthetic Activity of Cyanobacterial Picoplankton. *Limnol. Oceanogr.* 42: 133–141.

Thornton, J.A., 1987. A Review of Some Unique Aspects of the Limnology of Shallow Southern African Man-made Lakes. *GeoJournal* 14: 339–352.

Thornton, J.A., and W. Rast, 1993. *A Test of Hypotheses Relating to the Comparitive Limnology and Assessment of Eutrophication in Semiarid Man-made Lakes.* Pp. 1–24. In M. Straškraba, J.G. Tundisi, and A. Duncan (eds.), Comparative Reservoir Limnology and Water Quality Management. Dordrecht: Kluwer Acad. Publ.

Thornton, K.W., 1984. *Regional Comparisons of Lakes and Reservoirs: Geology, Climatology and Morphology.* Pp. 261–265. In Lake and Reservoir Management, EPA Report #440/5/84-001. Washington, DC.

Thornton, K.W., B.L. Kimmel, and F.E. Payne, 1990. *Reservoir Limnology: Ecological Perspectives*. New York: John Wiley and Sons.

Thorp, J.H., 1994. The Riverine Productivity Model: An Heuristic View of Carbon Sources and Organic Processing in Large River Ecosystems. *Oikos* 70: 305–308.

Thorpe, S. A., 1971. Experiments on the Instability of Stratified Shear Flows: Miscible Fluids. *J. Fluid Mech.*, 46: 299–319.

Thurman, E.M., 1985. *Organic Geochemistry of Natural Waters*. Dordrecht: Junk Publ.

Tilzer, M.M., and C. Serruya, 1990. *Large Lakes: Ecological Structure and Function*. Berlin: Springer-Verlag.

Timms, B.V., 1992. *Lake Geomorphology*. Adelaide: Gleneagles Publ.

Tipping, E., C. Woof, and D. Cooke, 1981. Iron Oxide from a Seasonally Anoxic Lake. *Geochim. Cosmochim. Acta* 45: 1411–1419.

Tipping, E., D.W. Thompson, and W. Davison, 1984. Oxidation Products of Mn(II) in Lake Waters. *Chem. Geol.* 44: 359–383.

Toews, D.R., and J.S. Griffith, 1979. Empirical Estimates of Potential Fish Yield for the Lake Bangweulu System, Zambia, Central Africa. *Trans. Amer. Fish. Soc.* 108: 241–252.

Toivonen, H., and T. Lappalainen, 1980. Ecology and Production of Aquatic Macrophytes in the Oligotrophic, Mesohumic Lake Suomunjärvi, Eastern Finland. *Ann. Bot. Fennici* 17: 69–85.

Tonn, W.M., C.A. Paszkowski, and I.J. Holopainen, 1992. Piscivory and Recruitment: Mechanisms Structuring Prey Populations in Small Lakes. *Ecology* 73: 951–958.

Townsend, C.R., 1991. *Community Organization in Marine and Freshwater Environments*. Pp. 125–144. In R.S.K. Barnes, and K.H. Mann (eds.), Fundamentals of Aquatic Ecology. Oxford: Blackwell Sci. Publ.

Tranvik, L.J., and M.G. Höfle, 1987. Bacterial Growth in Mixed Cultures on Dissolved Organic Carbon from Humic and Clear Waters. *Appl. Env. Microbiol.* 53: 482–488.

Trautman, M.B., 1957. *The Fishes of Ohio*. Columbus: Ohio State Univ. Press.

Treasurer, J.W., 1989. Mortality and Production of 0+ Perch, *Perca fluviatilis* L., in Two Scottish Lakes. *J. Fish Biol.* 34: 913–928.

Triska, F.J., J.H. Duff, and R.J. Avanzino, 1993. Patterns of Hydrological Exchange and Nutrient Transformation in the Hyporheic Zone of a Gravel-bottom Stream: Examining Terrestrial–Aquatic Linkages. *Freshw. Biol.* 29: 259–274.

Tundisi, J.G., T. Matsumura-Tundisi, and M.C. Calijuri, 1993. *Limnology and Management of Reservoirs in Brazil.* Pp. 25–55. In M. Straškraba, J.G. Tundisi, and A. Duncan (eds.), Comparative Reservoir Limnology and Water Quality Management. Dordrecht: Kluwer Acad. Publ.

Turner, R.R., E.A. Laws, and R.C. Harriss, 1983. Nutrient Retention and Transformation in Relation to Hydraulic Flushing Rate in a Small Impoundment. *Freshw. Biol.* 13: 113–127.

Twiss, M.R., JC. Auclair, and M.N. Charlton, 2000. An Investigation into Iron-stimulated Phytoplankton Productivity in Epipelagic Lake Erie During Thermal Stratification Using Trace Metal Clean Techniques. *Can. J. Fish. Aquat. Sci.* 57: 86–95.

Uehlinger, U., 1986. Bacteria and Phosphorus Regeneration in Lakes. An Experimental Study. *Hydrobiologia* 135: 197–206.

Uehlinger, U., and J. Bloesch, 1989. Primary Production of Different Phytoplankton Size Classes in an Oligo-mesotrophic Swiss Lake. *Arch. Hydrobiol.* 116: 1–21.

Uehlinger, U., and M.W. Naegeli, 1998. Ecosystem Metabolism, Disturbance, and Stability in a Prealpine Gravel Bed River. *J. N. Amer. Benthol. Soc.* 17: 165–178.

UNEP, 1991. *Freshwater Pollution.* Nairobi.

Ungar, I.A., 1974. *Inland Halophytes of the United States.* Pp. 235–306. In R.S. Reimold, and W.H. Queen (eds.), Ecology of Halophytes. New York: Acad. Press.

Urabe, J., 1990. Stable Horizontal Variation in the Zooplankton Community Structure of a Reservoir Maintained by Predation and Competition. *Limnol. Oceanogr.* 35: 1703–1717.

Urban, N.R., 1994. *Retention of Sulfur in Lake Sediments.* Pp. 323–369. In L.A. Baker (ed.), Environmental Chemistry of Lakes and Reservoirs. Washington: Amer. Chem. Soc.

Urban, N.R., S.E. Bayley, and S.J. Eisenreich, 1989. Export of Dissolved Organic Carbon and Acidity from Peatlands. *Water Resources Res.* 25: 1619–1928.

Urban, N.R., E. Gorham, J.K. Underwood, F.B. Martin, and J.G. Ogden, 1990. Geochemical Processes Controlling Concentrations of Al, Fe, and Mn in Nova Scotia Lakes. *Limnol. Oceanogr.* 35: 1516–1534.

Uutala, A.J., and J.P. Smol, 1996. Paleolimnological Reconstruction of Long-term Changes in Fisheries Status in Sudbury Lakes. *Can. J. Fish Aquat. Sci.* 53: 174–180.

Vallentyne, J.R., 1957. Principles of Modern Limnology. *Amer. Sci.* 45: 218–244.

Vallentyne, J.R., 1974. *The Algal Bowl: Lakes and Man.* Ottawa: Dept. of the Env., Fish. and Mar. Serv.

van der Heide, J., 1982. *Lake Brokopondo Filling Phase Limnology of a Man-made Lake in the Humid Tropics.* PhD thesis. Free University of Amsterdam.

Van der Vlugt, J.C., P.A. Walker, J. Van der Does, and A.J.P. Raat, 1992. Fisheries Management as an Additional Lake Restoration Measure: Biomanipulation Scaling-up Problems. *Hydrobiologia* 233: 213–224.

van der Weijden, C.H., and J.J. Middelburg, 1989. Hydrogeochemistry of the River Rhine: Long Term and Seasonal Variability, Elemental Budgets, Base Levels and Pollution. *Wat. Res.* 23: 1247–1266.

van Donk, E., 1989. *The Role of Fungal Parasites in Phytoplankton Succession.* Pp. 171–194. In U. Sommer (ed.), Plankton Ecology: Succession in Plankton Communities. Berlin: Springer-Verlag.

van Donk, E., M.P. Grimm, R.D. Gulati and J.P.G. Klein-Breteler, 1990. Whole-lake Food-web Manipulation as a Means to Study Community Interactions in a Small Ecosystem. *Hydrobiologia* 200/201: 275–289.

van Donk, E., R.D. Gulati, A. Iedema and J.T. Meulemans, 1993. Macrophyte-related Shifts in the Nitrogen and Phosphorus Contents of the Different Trophic Levels in a Biomanipulated Shallow Lake. *Hydrobiologia* 251: 19–26.

van Donk, E., E. De Deekere, J.G.P. Klein-Breteler, and J.T. Meulemans, 1994. Herbivory by Waterfowl and Fish in a Biomanipulated Lake: Effect on Long Term Recovery. *Verh. Int. Ver. Limnol.* 25: 2139–2143.

van Gemerden, H., E. Montesinos, J. Mas, and R. Guerrero, 1985. Diel Cycle of Metabolism of Phototrophic Purple Sulfur Bacteria in Lake Cisó (Spain). *Limnol. Oceanogr.* 30: 932–943.

van Gemerden, H., and J. Mas, 1995. *Ecology of Phototrophic Sulfur Bacteria.* Pp. 49–85. In R.E. Blankenship, M.T. Madigan, and C.E. Bauer (eds.), Anoxygenic Photosynthetic Bacteria. Dordrecht: Kluwer Acad. Publ.

Van Nieuwenhuyse, E.E., and J.R. Jones, 1996. Phosphorus–Chlorophyll Relationship in Temperate Streams and Its Variation with Stream Catchment Area. *Can. J. Fish. Aquat. Sci.* 53: 99–105.

Vander Zanden, M.J., and J.B. Rasmussen, 1996. A Trophic Position Model of Pelagic Food Webs: Impact on Contaminant Bioaccumulation in Lake Trout. *Ecol. Monogr.* 66: 451–477.

Vanderploeg, H.A., B.J. Eadie, J.R. Liebig, S.J. Tarapchak, and R.M. Glover, 1987. Contribution of Calcite to the Particle-size Spectrum of Lake Michigan Seston and Its Interactions with the Plankton. *Can. J. Fish. Aquat. Sci.* 44: 1898–1914.

Vanni, M.J., 1996. *Nutrient Transport and Recycling by Consumers in Lake Food Webs: Implications for Algal Communities.* Pp. 81–95. In G.A. Polis, and K.O. Winemiller, Food Webs, Integration of Patterns and Dynamics. New York: Chapman and Hall.

Vannote, R.L., G.W. Minshall, K.W. Cummins, J.R. Sedell, and C.E. Cushing, 1980. The River Continuum Concept. *Can. J. Fish. Aquat. Sci.* 37: 130–137.

Vaqué, D., and M.L. Pace, 1992. Grazing on Bacteria by Flagellates and Cladocerans in Lakes of Contrasting Food-web Structure. *J. Plankton Res.* 14: 307–321.

Vaqué, D., J.M. Gasol, and C. Marrasé, 1994. Grazing Rates on Bacteria: The Significance of Methodology and Ecological Factors. *Mar. Ecology Prog. Ser.* 109: 263–274.

Vareschi, E., 1982. The Ecology of Lake Nakuru (Kenya), III. Abiotic Factors and Primary Production. *Oecologia* 55: 81–101.

Vareschi, E., and J. Jacobs, 1985. The Ecology of Lake Nakuru, VI. Synopsis of Production and Energy Flow. *Oecologia* 65: 412–424.

Veith, G.D., D.L. DeFoe, and B.V. Bergstedt, 1979. *Measuring and Estimating the Bioconcentration Factor of Chemicals in Fish.* In W. Stumm, R. Schwarzenbach, and L. Sigg (1983), From Environmental Analytical Chemistry to Ecotoxicology—A Plea for More Concepts and Less Monitoring and Testing. Angew. Chem. Int. Ed. Engl. 22: 380–389.

Verbolov, V.I., V.N. Sinyukovich, and N.L. Karpysheva, 1989. Water and Mass Exchange in the Lake Baikal and Storage Reservoirs of the Angara Cascade. *Arch. Hydrobiologia Beih.* 33: 35–40.

Vincent, W.F., 1987. *Antarctic Limnology.* Pp. 379–412. In A.B.Viner (ed.), Inland Waters of New Zealand. Wellington: DSIR Bull. #241.

Vincent, W.F., and C.L. Vincent, 1982. Factors Controlling Phytoplankton Production in Lake Vanda (77° S). *Can. J. Fish. Aquat. Sci.* 39: 1602–1609.

Vincent, W.F., and D.J. Forsyth, 1987. Geothermally Influenced Waters. Pp. 349–377. In A.B. Viner (ed.), *Inland Waters of New Zealand.* Wellington: DSIR Bull. #241.

Vincent, W.F., R. Rae, I. Laurion, C. Howard-Williams, and J.C. Priscu, 1998. Transparency of Antarctic Ice-covered Lakes to Solar UV Radiation. *Limnol. Oceanogr.* 43: 618–624.

Vincent, W.F., and J.A. Hobbie, 2000. Ecology of Arctic Lakes and Rivers. Pp 197–231. In M. Nuttall, and T.V. Callaghan (eds.), *The Arctic: Environment, People, Policies.* London: Harwood Acad. Publ.

Viner, A.B., 1984. Resistance to Mixing in New Zealand Lakes. *New Zealand J. Mar. Freshw. Res.* 18: 73–82.

Viner, A.B., 1985. Conditions Stimulating Planktonic N_2-Fixation in Lake Rotongaio. *New Zealand J. Mar. and Freshw. Res.* 19: 139–150.

Viner, A.B., 1985. Thermal Stability and Phytoplankton Distribution. *Hydrobiologia* 125: 47–69.

Viner, A.B. (ed.), 1987. *Inland Waters of New Zealand, DSIR Bull. #241.* Wellington: DSIR Sci. Information Publ.

Vinyard, G.L., and W.J. O'Brien, 1976. Effects of Light and Turbidity on the Reactive Distance of Bluegill (*Lepomis macrochirus*). *Can. J. Fish. Res. Bd* 33: 2845–2849.

Vitousek, P.M., and R.W. Howarth, 1991. Nitrogen Limitation on Land and in the Sea: How Can it Occur? *Biogeochem.* 13: 87–115.

Vitousek, P.M., J.D. Aber, R.W. Howarth, G.E. Likens, P.A. Matson, D.W. Schindler, W.H. Schlesinger, and G.D. Tilman, 1997. Human Alteration of the Global Nitrogen Cycle: Sources and Consequences. *Ecol. Appl.* 7: 737–750.

Vogel, S., 1994. *Life in Moving Fluids: The Physical Biology of Flow, 2nd ed.* Princeton: Princeton Univ. Press.

Vollenweider, R.A., 1968. *Scientific Fundamentals of the Eutrophication of Lakes and Flowing Waters, with Particular Reference to Nitrogen and Phosphorus as Factors in Eutrophication, Technical Report DAS/CSI/68.27.* Paris: Org. for Econ. Cooperation and Dev.

Vollenweider, R.A., 1976. Advances in Defining Critical Loading Levels for Phosphorus in Lake Eutrophication. *Mem. Ist. Ital. Idrobiol.* 33: 53–83.

Vollenweider, R.A., 1979. The Nutrient Loading Concept as a Basis for the External Manipulation of the Process of Eutrophication of Lakes and Reservoirs. *J. Water and Wastewater Res.* 12: 46–56.

Vollenweider, R.A., and J. Kerekes, 1980. The Loading Concept as Basis for Controlling Eutrophication Philosophy and Preliminary Results of the OECD Programme on Eutrophication. *Prog. Wat. Tech.* 12: 5–38.

Vörösmarty, C.J., K.P. Sharma, B.M. Fekete, A.H. Copeland, J. Holden, J. Marble, and J.A. Lough, 1997. The Storage and Aging of Continental Runoff in Large Reservoir Systems of the World. *Ambio.* 26: 210–219.

Voskresensky, K.P., 1978. *Water of the Earth.* Pp. 42–56. In V.I. Korzun (ed.), World Water Balance and Water Resources of the Earth. Paris: Unesco Press.

Walker, K.F., and G.E. Likens, 1975. Meromixis and a Reconsidered Typology of Lake Circulation Patterns. *Verh. Int. Ver. Limnol.* 19: 442–458.

Walmsley, R.D., M. Butty, H. Van Der Piepen, and D. Grobler, 1980. Light Penetration and the Interrelationships Between Optical Parameters in a Turbid Subtropical Impoundment. *Hydrobiologia* 70: 145–157.

Ward, J.V., and B.R. Davies, 1984. *Stream Regulation.* Pp. 32–63. In R.C. Hart, and B.R. Allanson (eds.), Limnological Criteria for Management of Water Quality in the Southern Hemisphere. Pretoria: S. African Nat. Sci. Prog. Rep. #93.

Ward, J.V., G. Bretschko, M. Brunke, D. Danielopol, J. Gibert, T. Gonser, and A.G. Hildrew, 1998. The Boundaries of River Systems: The Metazoan Perspective. *Freshw. Biol.* 40: 531–569.

Watras, C.J., N.S. Bloom, R.J.M. Hudson, S. Gherini, R. Munson, S.A. Claas, K.A. Morrison, J. Hurley, J.G. Wiener, W.F. Fitzgerald, R. Mason, G. Vandal, D. Powell, R. Rada, R. Rislov, M. Winfrey, J. Elder, D. Krabbenhoff, A.W. Andren, C. Babiarz, D.B. Porcella and J.W. Huckabee, 1994. *Sources and Fates of Mercury and Methylmercury in Wisconsin Lakes.* Pp. 153–177. In C.J. Watras, and J.W. Huckabee (eds.), Mercury Pollution: Integration and Synthesis. Boca Raton: Lewis Publ.

Watson, D.J., and E.K. Balon, 1984. Structure and Production of Fish Communities in Tropical Rain Forest Streams of Northern Borneo. *Can. J. Zool.* 62: 927–940.

Watson, S., and E. McCauley, 1988. Contrasting Patterns of Net- and Nanoplankton Production and Biomass Among Lakes. *Can. J. Fish. Aquat. Sci.* 45: 915–920.

Watson, S., and E. McCauley, 1992. Sigmoid Relationships Between Phosphorus, Algal Biomass, and Algal Community Structure. *Can. J. Fish. Aquat. Sci.* 49: 2605–2610.

Watson, S., E. McCauley, E. Hardisty, E. Hargesheimer, and J. Dixon, 1996. Chrysophyte Blooms in Oligotrophic Glenmore Reservoir (Calgary, Canada). *Nova Hedwigia (Beih.)* 114: 193–217.

Watson, S.B., E. McCauley, and J.A. Downing, 1997. Patterns in Phytoplankton Taxonomic Composition Across Temperate Lakes of Different Nutrient Status. *Limnol. Oceanogr.* 42: 487–495.

Webster, J.R., J.B. Wallace, and E.F. Benfield, 1995. *Organic Processes in Streams of the Eastern United States.* Pp. 117–187. In C.E. Cushing, K.W. Cummins, and G.W. Minshall (eds.), Ecosystems of the World, 22: River and Stream Ecosystems. Amsterdam: Elsevier.

Webster, K.E., and A.D. Newell, L.A. Baker, and P.L. Brezonik, 1990. Climatically Reduced Rapid Acidification of a Softwater Seepage Lake. *Nature* 347: 374–376.

Weilenmann, U., C.R. O'Melia, and W. Stumm, 1989. Particle Transport in Lakes: Models and Measurements. *Limnol. Oceanogr.* 34: 1–18.

Weiler, R.R., 1981. Chemistry of the North American Great Lakes. *Verh. Int. Ver. Limnol.* 21: 1681–1694.

Weiler, R.R., and V.K. Chawla, 1969. Dissolved Mineral Quality of Great Lakes Waters. *Proc. 12th Conf. Great Lakes Res.* 801–818.

Weimbauer, M.G., and M.G. Höfle, 1998. Significance of Viral Lysis and Flagellate Grazing as Factors Controlling Bacterioplankton Production in a Eutrophic Lake. *Appl. Env. Microbial.* 64: 431–438.

Weiss, R.F., E.C. Carmack, and V.M. Koropalov, 1991. Deep-water Renewal and Biological Production in Lake Baikal. *Nature* 349: 665–669.

Weisse, T., and J.G. Stockner, 1993. Eutrophication: The Role of Microbial Food Webs. *Mem. Ist. Ital. Idrobiol.* 52: 133–150.

Weisse, T., and H. Müller, 1990. *Significance of Heterotrophic Nanoflagellates and Ciliates in Large Lakes: Evidence from Lake Constance.* Pp. 540–555. In M.M. Tilzer, and C. Serruya, Large Lakes: Ecological Structure and Function. Berlin: Springer-Verlag.

Welch, H.E., and J. Kalff, 1974a. Benthic Photosynthesis and Respiration in Char Lake. *Can. J. Fish. Res Bd.* 31: 609–620.

Welch, H.E., and J. Kalff, 1974b. Phytoplankton Production in Char lake, a Natural Polar Lake and in Meretta Lake, a Polluted Polar Lake, Cornwallis Island, Northwest Territories. *Can. J. Fish. Res. Bd.* 31: 621–636.

Welch, H.E., and M.A. Bergmann, 1985. Water Circulation in Small Arctic Lakes in Winter. *Can. J. Fish. Aquat. Sci.* 42: 506–520.

Welch, H.E., J.A. Legault, and M.A. Bergmann, 1987. Effects of Snow and Ice on the Annual Cycles of Heat and Light in Saqvaqjuac Lakes. *Can. J. Fish. Aquat. Sci.* 44: 1451–1461.

Welch, H.E., J.K. Jorgenson, and M.F. Curtis, 1988. Emergence of Chironomidae (Diptera) in Fertilized and Natural Lakes at Saqvaqjuac, N.W.T. *Can. J. Fish Aquat. Sci.* 45: 731–737.

Welch, H.E., D.C.G. Muir, B.N. Billeck, W.L. Lockhart, G.J. Brunskill, H.J. Kling, M.P. Olson, and R.M. Lemoine, 1991. Brown Snow: A Long Range Transport Event in the Canadian Arctic. *Env. Sci. Technol.* 25: 280–286.

Welcomme, R.L., 1976. Some General and Theoretical Considerations on the Fish Yield of African Rivers. *J. Fish Biol.* 8: 351–364.

Welcomme, R.L., 1985. *River Fisheries.* FAO Fish. Tech. Pap. #262. Rome:

Welcomme, R.L., and D.M. Bartley, 1998. *An Evaluation of Present Techniques for the Enhancement of Fisheries.* Pp. 1–36. In T. Petr (ed.), Inland Fishing Enhancements. FAO Fish. Tech. Pap. #374. Rome.

Werner, E.E, D.J. Hall, D.R. Laughlin, D.J. Wagner, L.A. Wilsmann, and F.C. Funk, 1977. Habitat Partitioning in a Freshwater Fish Community. *Can. J. Fish. Res. Bd.* 34: 360–370.

Westlake, D.F., 1980. *Primary Production.* Pp. 141–246. In E.D. LeCren, and R.H. Lowe-McConnell (eds.), The Functioning of Freshwater Ecosystems. Int. Biol. Prog. #22. Cambridge: Cambridge Univ. Press.

Westlake, D.F., 1982. *The Primary Productivity of Water Plants.* Pp. 165–180. In J.J. Symoens, S.S. Hooper, and P. Compère (eds.), Studies on Aquatic Vascular Plants. Brussels: Roy. Bot. Soc. Belgium.

Westöö, G., 1966. Determination of Methylmercury Compounds in Foodstuffs I: Methylmercury Compounds in Fish, Identification and Determination. *Acta Chem. Scand.* 20: 2131–2137.

Wetzel, R.G., 1983. *Limnology, 2nd ed.* Philadelphia: Saunders Coll. Publ.

Wetzel, R.G., 1990. Land–Water Interfaces: Metabolic and Limnological Regulators. *Verh. Int. Ver. Limnol.* 24: 6–24.

Wetzel, R.G., and G.E. Likens, 2000. *Limnological Analyses, 3rd ed.* New York: Springer-Verlag.

White, D.S., 1993. Perspectives on Defining and Delineating Hyporheic Zones. *J. N. Am. Benthol. Soc.* 12: 61–69.

White, P.A., J. Kalff, J.B. Rasmussen, and J.M. Gasol, 1991. The Effect of Temperature and Algal Biomass on Bacterial Production and Specific Growth Rate in Freshwater and Marine Habitats. *Microb. Ecology* 21: 99–118.

White, P.A., and J.B. Rasmussen, 1996. SOS Chromotest Results in a Broader Context: Empirical Relationships Between Genotoxic Potency, Mutagenic Potency and Carcinogenic Potency. *Env. Mol. Mutagen.* 27: 270–305.

White, P.A., J.B. Rasmussen, and C. Blaise, 1996. Sorption of Organic Genotoxins to Particulate Matter in Industrial Effluents. *Env. Mol. Mutagen.* 27: 140–151.

Whitton, B.A. (ed.), 1975. River Ecology: Studies in Ecology, vol 2. Los Angeles: Cal. Berkeley Press.

Whitton, B.A., A. Aziz, P. Francis, J.A. Rother, J.W. Simon, and Z.N. Tahmida, 1988. Ecology of Deepwater Rice-fields in Bangladesh, I. Physical and Chemical Environment. *Hydrobiologia* 169: 3–22.

Widdel, F., 1988. *Microbiology and Ecology of Sulfate- and Sulfur-reducing Bacteria.* Pp. 469–585. In A.J.B. Zehnder (ed.), Biology of Anaerobic Microorganisms. New York: John Wiley and Sons.

Wiederholm, T., 1978. Long-term Changes in the Profundal Benthos of Lake Malaren. *Verh. Int. Ver. Limnol.* 20: 818–824.

Willén, E., 1976. A Simplified Method of Phytoplankton Counting. *Br. Phycol. J.* 11: 265–278.

Williams, D.J., and K.W. Kuntz, 1999. *Spatial Distribution and Temporal Trends of Selected Parameters, with Emphasis on 1996–1997 Results.* Lake Superior Surv. Prog. Rep. #EHD/ECB-OR/99-01/I. Burlington: Env. Can.

Williams, W.D., 1967. *The Chemical Characteristics of Lentic Surface Waters in Australia: A Review.* Pp. 18–77. In A.H. Weatherley, Australian Inland Waters and Their Fauna. Canberra: Australia Nat. Univ. Press.

Williams, W.D., 1986. Conductivity and Salinity of Australian Salt Lakes. *Aust. J. Mar. Freshw. Res.* 37: 177–182.

Williams, W.D., 1988. Limnological Imbalances: An Antipodean Viewpoint. *Freshw. Biol.* 20: 407–420.

Williams, W.D., 1993. The Worldwide Occurrence and Limnological Significance of Falling Water-levels in Large, Permanent Saline Lakes. *Verh. Int. Ver. Limnol.* 25: 980–983.

Williams, W.D., 1995. Inland Lakes of Brine: Living Worlds Within Themselves. *Biol.* 42: 57–60.

Williams, W.D., 1996. The Largest, Highest and Lowest Lakes of the World: Saline Lakes. *Verh. Int. Ver. Limnol.* 26: 61–79.

Williams, W.D., 1998. *Management of Inland Saline Waters, Guidelines of Lake Management, Vol. 6.* Shiga: Int. Lake Environ. Com. Found. and UN Environ. Progr.

Williamson, C.E., R.S. Stemberger, D.P. Morris, T.M. Frost, and S.G. Paulsen, 1996. Ultraviolet Radiation in North American Lakes: Attenuation Estimates from DOC Measurements and Implications for Plankton Communities. *Limnol. Oceanogr.* 41: 1024–1034.

Winberg, G.G., 1972. *Some Interim Results of Soviet IBP Investigations on Lakes. IBP-UNESCO Symposium on Productivity Problems of Freshwaters, Kazimierz, Poland, 1970.* Pp. 363–381. In Z. Kajak, and A. Hillbricht-Ilkowska (eds.), Productivity Problems of Freshwaters. Kraków: PWN-Polish Sci. Publ.

Windolf, J., E. Jeppesen, J.P. Jensen, and P. Kristensen, 1996. Modelling of Seasonal Variation in Nitrogen Retention and In-lake Concentration: A Four-year Mass Balance Study in 16 Shallow Danish Lakes. *Biogeochem.* 33: 25–44.

Windsor, J.G., and R.A. Hites, 1979. Polycyclic Aromatic Hydrocarbons in Gulf of Maine Sediments and Nova Scotia Soil. *Geochim. Cosmochim. Acta.* 43: 27–33.

Winemiller, K.O., and K.A. Rose, 1992. Patterns of Life-history Diversification in North American Fishes: Implications for Population Regulation. *Can. J. Fish. Aquat. Sci.* 49: 2196–2218.

Winfield, I.J., D.K. Winfield, and C.M. Tobin, 1992. Interactions Between the Roach, *Rutilus rutilus*, and Waterfowl Populations of Lough Neagh, Northern Ireland. *Environ. Biol. Fish.* 33: 207–214.

Wissmar, R.C., J.E. Richey, and D.E. Spyridakis, 1977. The Importance of Allochthonous Particulate Carbon Pathways in a Subalpine Lake. *Can. J. Fish. Res. Bd.* 34: 1410–1418.

Wnorowski, A.U., 1992. Taste and Odours in the Aquatic Environment: A Review. *Water S. A.* 18: 203–214.

Wood, J.M., 1987. Biological Processes Involved in the Cycling of Elements Between Soil or Sediments and the Aqueous Environment. *Hydrobiologia* 149: 31–42.

Wood, R.B., R.M. Baxter, and M.V. Prosser, 1984. Seasonal and Comparative Aspects of Chemical Stratification in Some Tropical Crater Lakes, Ethiopia. *Freshw. Biol.* 14: 551–573.

Woollhead, J., 1994. Birds in the Trophic Web of Lake Esrom, Denmark. *Hydrobiologia* 279/280: 29–38.

Wootton, R.J., 2000. *Fish Ecology.* New York: Chapman and Hall.

World Resources Institute, 1988. World Resources 1988–1989: An Assessment of the Resource Base that Supports the Global Economy. New York: Basic Books.

World Resources Institute, 1990. World Resources 1990–1991: A Guide to the Global Environment. Oxford: Oxford Univ. Press.

World Resources Institute, 1992. World Resources 1992–1993: A Guide to the Global Enviroment. Oxford: Oxford Univ. Press.

Wright, R.F., 1983a. *Predicting Acidification of North American Lakes, Acid Rain Res. Rep. #4.* Oslo: Norway Inst. Water Res.

Wright, R.F., 1983b. Acidification of Freshwaters in Europe. *Water Qual. Bull.* 8: 137–142.

Wright, R.F., and E.T. Gjessing, 1976. Acid Precipitation: Changes in the Chemical Composition of Lakes. *Ambio* 5: 219–223.

Wright, R.F., N. Conroy, W.T. Dickson, R. Harriman, A. Henriksen, and C.L. Schofield, 1980. *Acidified Lake Districts of the World: A Comparison of Water Chemistry of Lakes in Southern Norway, Southern Sweden, Southwestern Scotland, the Adirondack Mountains of New York, and Southeastern Ontario.* Pp. 377–379. In D. Drabløs, and A. Tollan (eds.), Ecological Impact of Acid Precipitation. Oslo: SNSF-Project #1432.

Wright, R.F., and M. Johannessen, 1980. *Input–Output Budgets of Major Ions at Gauged Catchments in Norway.* Pp. 250–251. In D. Drabløs, and A. Tollan (eds.), Ecological Impact of Acid Precipitation. Oslo: SNSF-Project #1432.

Wright, R.F., E. Lotse, and A. Semb, 1988. Reversibility of Acidification Shown by Whole-catchment Experiments. *Nature* 334: 670–675.

Wylie, G.D., and J.R. Jones, 1987. Diel and Seasonal Changes of Dissolved Oxygen and pH in Relation to Community Metabolism of a Shallow Reservoir in Southeast Missouri. *J Freshw. Ecology* 4: 115–125.

Wylie, J.L., and D.J. Currie, 1991. The Relative Importance of Bacteria and Algae as Food Sources for Crustacean Zooplankton. *Limnol. Oceanogr.* 36: 708–728.

Xie, P., T. Iwakuma, and K. Fujii, 1998. Studies on the Biology of *Chaoborus flavicans* (Meigen) (Diptera: Chaoboridae) in a Fish-free Eutrophic Pond, Japan. *Hydrobiologia* 368: 83–90.

Yagi, A., 1988. Dissolved Organic Manganese in the Anoxic Hypolimnion of Lake Fukami-ike. *Jap. J. Limnol.* 49: 149–156.

Yan, N.D., 1986. Empirical Prediction of Crustacean Zooplankton Biomass in Nutrient-poor Canadian Shield Lakes. *Can. J. Fish. Aquat. Sci.* 43: 788–796.

Yan, N.D., and G.E. Miller, 1984. *Effects of Deposition of Acids and Metals on Chemistry and Biology of Lakes near Sudbury, Ontario.* Pp. 243–282. In J. Nriagu (ed.), Environmental Impacts of Smelters. New York: John Wiley and Sons.

Yan, N.D., W. Keller, K.M. Somers, T.W. Pawson, and R.E. Girard, 1996. The Recovery of Crustacean Zooplankton Communities from Acidification: Comparing Manipulated and Reference Lakes. *Can. J. Fish Aquat. Sci.* 53: 1301–1327.

Yu, N., and D.A. Culver, 2000. Can Zebra Mussels Change Stratification Patterns in a Small Reservoir? *Hydrobiologia*, 431: 175–184.

Yurk, J.J., and Ney, J.J., 1989. Phosphorus–Fish Community Biomass Relationships in Southern Appalachian Reservoirs: Can Lakes be too Clean for Fish? *Lake and Res. Man.* 5: 83–90.

Zehnder, A.J.B., and W. Stumm, 1988. *Geochemistry and Biogeochemistry of Anaerobic Habitats.* Pp. 1–38. In A.J.B. Zehnder (ed.), Biology of Anaerobic Microorganisms. New York: John Wiley and Sons.

Inland Waters Index

Subject Index

NOTE: An *n* following a page number indicates a footnote.